ENCYCLOPEDIA OF LIFE SCIENCE
VOLUME I

ENCYCLOPEDIA OF
LIFE SCIENCE
VOLUME I

KATHERINE CULLEN, Ph.D.

Facts On File
An imprint of Infobase Publishing

ENCYCLOPEDIA OF LIFE SCIENCE

Copyright © 2009 by Katherine Cullen, Ph.D.

Facts On File, Inc.
An imprint of Infobase Publishing
132 West 31st Street
New York NY 10001

Library of Congress Cataloging-in-Publication Data

Cullen, Katherine E.
Encyclopedia of life science / Katherine Cullen.
p. cm.
Includes bibliographical references and index.
ISBN-13: 978-0-8160-7008-4
ISBN-10: 0-8160-7008-3
1. Life sciences—Encyclopedias, Juvenile. I. Title.
QH309.2.C85 2009
570.3—dc22   2008016224

Text design by Annie O'Donnell
Illustrations by Richard Garratt, Chris and Elisa Scherer, Melissa Ericksen, and Sholto Ainslie
Photo research by Suzanne M. Tibor

Printed in China

CP Hermitage 10 9 8 7 6 5 4 3 2 1

This book is printed on acid-free paper.

CONTENTS

ACKNOWLEDGMENTS

I would like to express appreciation to Frank K. Darmstadt, executive editor, for his critical review of this manuscript, wise advice, patience, and professionalism. Thank you to Richard Garratt, Melissa Ericksen, Chris Scherer, and Elisa Scherer who created the illustrations that accompany the entries in this work, and to Suzie Tibor, who performed the photo research. I would like to express my sincere gratitude to Ann E. Hicks for her constructive comments and suggestions regarding the entries throughout this text. The guest essayists deserve recognition for generously donating time to share their expert knowledge in the numerous highlighted areas of interest. The staff of the main branch of the Medina County District Library was extremely helpful in obtaining hundreds of documents and print resources to aid me in my research, as were Alison Ricker and Jennifer Schreiner in showing me around the impressive science library at Oberlin College and introducing me to the available electronic resources. Appreciation is also extended to the production and copyediting departments and the many others who helped in the production of this project. Thank you all.

INTRODUCTION

Encyclopedia of Life Science is a two-volume reference intended to complement the material typically taught in high school biology and in introductory college biology courses. The substance reflects the fundamental concepts and principles that underlie the content standards for life science identified by the National Committee on Science Education Standards and Assessment of the National Research Council for grades 9–12. Within the category of life science, these include the cell; the molecular basis of heredity; biological evolution; interdependence of organisms; matter, energy, and organization in living systems; and the behavior of organisms. The National Science Education Standards (NSES) also place importance on student awareness of the nature of science and the process by which modern scientists gather information. To assist educators in achieving this goal, other subject matter discusses concepts that unify the life sciences with physical science and Earth and space science: science as inquiry, technology and other applications of scientific advances, science in personal and social perspectives including topics such as natural hazards and global challenges, and the history and nature of science. A listing of entry topics organized by the relevant NSES content standards and an extensive index will assist educators, students, and other readers in locating information or examples of topics that fulfill a particular aspect of their curriculum.

Encyclopedia of Life Science provides historical perspectives, portrays science as a human endeavor, and gives insight into the process of scientific inquiry by incorporating biographical profiles of people who have contributed significantly to the development of the sciences. Processes that shape the natural world and life within it are also discussed. Instruments and methodology-related entries focus on the tools and procedures used by scientists to gather information, conduct experiments, and perform analyses. Other entries summarize the major branches and subdisciplines of life science or describe selected applications of the information and technology gleaned from life science research. Pertinent topics in all categories collectively convey the relationship between science and individuals and science and society.

The majority of this encyclopedia comprises more than 200 entries covering NSES concepts and topics, theories, subdisciplines, biographies of people who have made significant contributions to the life sciences, common methods, and techniques relevant to modern science. Entries average more than 2,000 words each (some are shorter, some longer), and most include a cross-listing of related entries and a selection of recommended further readings. In addition, one dozen guest essayists contributed essays covering a variety of subjects—contemporary topics of particular interest and specific themes common to the life sciences. Approximately 150 photographs and 150 line art illustrations accompany the text, depicting difficult concepts, clarifying complex processes, and summarizing information for the reader. A chronology outlines important events in the history of the field, and a glossary defines relevant scientific terminology. The back matter of *Encyclopedia of Life Science* contains a list of additional print and Web resources for readers who would like to explore the discipline further. Readers can find a periodic table of the elements and common metric and temperature conversions in the appendixes.

I have been involved in research and teaching life sciences for almost two decades. After obtaining my doctorate in molecular biology from Vanderbilt University, I became a postdoctoral fellow in the department of biochemistry at Chandler Medical Center of the University of Kentucky. After spending several years on the biology faculty at Transylvania University, I moved to northern Ohio, where I currently teach as a visiting faculty member at Oberlin College. Throughout the years I have taught numerous life science subjects, including general biology, microbiology, genetics, cell and molecular biology, immunology, and human reproductive biology. My career has allowed me to continue to explore the exciting and constantly evolving life sciences. I hope that this encyclopedia serves you as a valuable reference, and that you will learn as much from referring to it as I have from writing it.

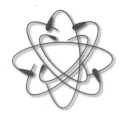

Entries Categorized by National Science Education Standards for Content (Grades 9–12)

When relevant, an entry may be listed under more than one category. For example, Sidney Altman, who studies RNA, is listed under both Life Science Content Standard C: The Molecular Basis of Heredity and Content Standard G: The History and Nature of Science. Biographical entries, topical entries, and entries that summarize a subdiscipline may all appear under Content Standard G: The History and Nature of Science when a significant portion of the entry describes a historical perspective of the subject. Subdisciplines are listed separately under the category Subdisciplines, which is not a NSES category, but are also listed under the related content standard category.

SCIENCE AS INQUIRY (CONTENT STANDARD A)
bioinformatics
cell culture
centrifugation
chromatography
cloning of DNA
data presentation and analysis
dissection
DNA sequencing
electrophoresis
metric system
microscopy
polymerase chain reaction
radioactivity
recombinant DNA technology
RNA interference
scientific investigation
scientific theory
spectrophotometry
X-ray crystallography

LIFE SCIENCE (CONTENT STANDARD C): THE CELL
aging
Altman, Sidney
biochemical reactions
biochemistry

biological membranes
biomolecules
Boveri, Theodor
Brock, Thomas
Calvin, Melvin
cancer, the biology of
cell biology
cell communication
cellular metabolism
cellular reproduction
chromosomes
Duve, Christian de
embryology and early animal development
enzymes
eukaryotic cells
gene expression
genetics
Golgi, Camillo
Ingenhousz, Jan
Just, Ernest
Leeuwenhoek, Antoni van
McClintock, Barbara
Mendel, Gregor
molecular biology
Morgan, Thomas Hunt
nutrition

organic chemistry, its relevance to life science
origin of life
Pasteur, Louis
Pauling, Linus
photosynthesis
plant form and function
Priestley, Joseph
prokaryotic cells
RNA interference
Schleiden, Matthias
Schwann, Theodor
Virchow, Rudolf
water, its biological importance
Wilmut, Sir Ian

LIFE SCIENCE (CONTENT STANDARD C): THE MOLECULAR BASIS OF HEREDITY
Altman, Sidney
Avery, Oswald
Boveri, Theodor
Cech, Thomas
Chargaff, Erwin
Chase, Martha
chromosomes
Crick, Francis
deoxyribonucleic acid (DNA)

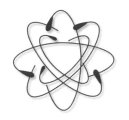

ENTRIES A–G

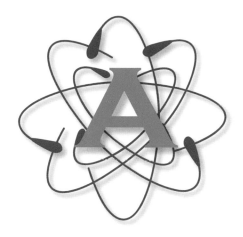

acquired immunodeficiency syndrome (AIDS) Since it was first reported in 1981, acquired immunodeficiency syndrome (AIDS) has become a worldwide epidemic. Caused by the human immunodeficiency virus (HIV), the life-threatening disease affects the specific immune system, destroying the host's ability to fight infections and developing cancers. According to the Joint United Nations Programme on HIV/AIDS (UNAIDS) 2007 AIDS epidemic update, in 2007 an estimated 33.2 million people were living with HIV, 2.5 million people became newly infected, and 2.1 million people died of AIDS. Though a person can live unaffected by HIV for many years, most do eventually develop AIDS. No cure for AIDS currently exists, though drugs that help fight HIV infection and its associated diseases are available. Because the viral particles are present in the semen, vaginal secretions, and blood of infected persons, HIV is transmitted through intimate sexual contact, contact with tainted blood, or sharing contaminated needles or syringes during intravenous drug use. The virus can also cross the placenta during pregnancy and infect an unborn fetus and is secreted in breast milk. Avoiding all contact with contaminated bodily fluids is the only sure way to prevent infection.

Stark discrepancies exist in global trends for the HIV/AIDS pandemic. The UNAIDS estimates that 22.5 million people infected with HIV (68 percent of the global total), including 90 percent of the children infected with HIV, live in sub-Saharan Africa. In comparison, 1.3 million (4 percent of the global total) live in North America. In sub-Saharan Africa, 61 percent of affected adults are women compared with 50 percent globally, and in numerous countries, more than 20 percent of pregnant women visiting prenatal clinics are infected. Without treatment, 35 percent of the children born to those women will become infected. Antiviral therapy effectively reduces mother-to-child transmission, but only 5 percent of women receive this therapy. Across Asia and Eastern Europe, infections due to injected drug use and commercial prostitution are increasing. Socioeconomic factors such as women's being forced to have sex or the refusal of men to wear a condom during sex, particularly among married women whose husbands have extramarital sex, also contribute.

MOLECULAR BIOLOGY OF HIV

Two HIV variants exist, HIV-1 and HIV-2: the first is prominent in the United States and Europe whereas the second occurs mostly in West Africa. Both are enveloped ribonucleic acid (RNA) viruses that encode a total of nine genes. The envelope contains surface glycoproteins (gp) called spikes. One that is 160 kilodaltons (gp 160) can be broken down into two smaller fragments, one that is 120 kilodaltons (gp 120) and another that is 41 kilodaltons (gp 41). The gp 120 spike is easily shed by the virus and interacts with the CD4 receptor on the T lymphocytes to gain entry. The smaller glycoprotein, gp 41, is embedded in the envelope. HIV is a retrovirus, one of a family of viruses that reproduce by synthesizing deoxyribonucleic acid (DNA) from the RNA genome, then inserting that DNA into a host chromosome. In addition to the RNA genome, the viral particles contain specific enzymes inside their capsids, including reverse transcriptase, which can read an RNA template and synthesize DNA from it. As soon as the virus penetrates the host cell, the genome is converted to DNA that is subsequently inserted into the host genome, where it can remain in the latent state for years. Stimulated by an unknown cause, the cell becomes activated and starts replicating the virus. Newly assembled virions are released and seek other cells to infect. This process destroys the T cells. At first, the body compensates by stepping up its synthesis of new T cells, but the virus can replicate faster than the body can replenish its T cells. The virus can also infect monocytes, macrophages, and B cells but does not kill these cells, so they serve as an additional and continual source of new virions.

SYMPTOMS

Within a few weeks of becoming infected, a person might develop flulike symptoms including a fever, headache, sore throat, and swollen lymph nodes. Since these symptoms are associated with numerous infections, the person might not be aware he or she is infected with HIV. Once inside the body, HIV attacks white blood cells, specifically T helper cells that have CD4 receptors. B and T lymphocytes mount a strong immune response and practically clear the virus from circulation, but the virus persists in the lymph nodes. During a period called clinical latency, years can pass without the infected person's experiencing any additional symptoms, but the number of healthy T lymphocytes decreases, until symptoms such as swollen lymph nodes, fever, diarrhea, weight loss, and fatigue develop and persist. HIV destroys the very cells whose job it is to fight the virus. The person loses the ability to fight infections and is said to have full-blown AIDS when he or she develops an opportunistic infection and has a CD4 T cell count of less than 200 per cubic millimeter of blood (normal range is between 500 and 1,800). An opportunistic infection is one that is caused by a microorganism that is pervasive in the environment and to which people are generally resistant, but that affects someone with an impaired immune system. Common opportunistic infections found in AIDS patients include toxoplasmosis, histoplasmosis, *Pneumocystis* pneumonia, herpes, hepatitis C, candidiasis, bacterial diarrheas, and tuberculosis. Because T lymphocytes are involved in fighting cancers, certain types of cancers including Kaposi's sarcoma, cervical cancer, and lymphoma are also frequently found in AIDS patients.

DIAGNOSIS AND TREATMENT

Infection with HIV is diagnosed by screening a sample of blood for the presence of antibodies against the virus. The enzyme-linked immunosorbent assay (ELISA) is one method that detects antibodies specific for viral proteins, but it takes a few weeks to get results. The production of a detectable level of antibodies by the specific immune system can take up to eight weeks after the initial exposure to the virus, so this test is not very informative if performed immediately after infection. Another procedure called a western blot analysis can detect the presence of HIV proteins in the blood and is used to confirm a positive antibody screen result. Two newer tests that can be performed in the doctor's office require a blood sample obtained from a finger prick or a swab sample of the fluids around the gum tissue. Results from these tests can be obtained in only 20 minutes.

Once infection is diagnosed, the physician will perform additional tests to assess the progression of the disease. One such test measures the viral load, the amount of viral particles present in the blood. A lower viral load correlates with a better prognosis for the patient.

When AIDS first appeared in the early 1980s, there were not any drugs to treat it and very few drugs to treat the numerous associated opportunistic infections. There is still no cure for AIDS and the drugs that are now available cause serious side effects and are enormously expensive, but the available treatments have increased the quality of life and the life expectancy or those infected with HIV. One category of drugs that targets HIV is the group that inhibits transcription of the viral genome. Reverse transcriptase inhibitor drugs work by inhibiting the enzyme and halting the life cycle of the virus. The nucleoside reverse transcriptase inhibitors are recognized by the enzyme, bind the active site, and are incorporated into DNA, but once incorporated, they do not allow additional nucleotides to become bonded to it. Azidothymidine (AZT), an example of a drug that blocks reverse transcriptase, was the first drug approved for treating HIV infection. Unfortunately, many strains of HIV have developed resistance to the drug. Nonnucleoside inhibitors bind the enzyme at a site other than the active site and inactivate it. Another group of drugs, the protease inhibitors, work by interfering with a viral protein called HIV protease. Without this protein the viral particles are not assembled properly and are noninfectious. One new type of drug are the fusion inhibitors. These work by inhibiting the viruses from fusing with the host cell membranes, stopping viral replication. In August 2007 the U.S. Food and Drug Administration approved a new drug called Selzentry that works by blocking a receptor, CCR_5, that the virus often uses to gain entry into the host's white blood cells. The long-term effects of this drug made by Pfizer are unknown. The recommended treatment is a cocktail, a combination of three or more drugs with different mechanisms of action. Called highly active antiretroviral therapy (HAART), this strict regimen is followed to overcome potential resistance. HIV mutates very rapidly, so it can quickly become resistant to a once-effective drug. The virus can only bear so many mutations at once without losing its infectiveness. Another difficulty in developing an effective chemotherapeutic approach is the fact that HIV persists in resting memory T cells. Since the current drugs all slow or stop viral replication rather than inactivate all viral particles, complete eradication is not possible.

To date, no vaccine to prevent AIDS has been approved. Besides the problem of the rapid mutation rate for the virus, antibodies do not seem effective against HIV. Consider that diagnosis of infection is based on the presence of antibodies against HIV proteins. Diagnosed individuals produce antibodies, but the presence of antibodies does not prevent AIDS. Many scientists around the world are involved in the development of a preventative vaccine and a therapeutic vaccine to boost the immune system of HIV-positive individuals. Clinical trials are under way for both, but success does not seem very near.

Most recently, in January 2008, researchers from Harvard Medical School published results from a study led by the geneticist Stephen J. Elledge that led to the identification of 273 proteins that HIV requires for survival and replication in human cells. Previous studies had only identified 36 human proteins necessary for the virus to enter host cells and replicate; thus these new findings open the door for potential new drug targets.

See also HOST DEFENSES; IMMUNE SYSTEM DISORDERS; INFECTIOUS DISEASES; VIRUSES AND OTHER INFECTIOUS PARTICLES.

FURTHER READING

Bardham-Quallen, Sudipta. *AIDS*. Farmington Hills, Mich.: Thomson Gale, 2005.
Centers for Disease Control and Prevention, National Center for HIV/AIDS, Viral Hepatitis, STD, and TB Prevention, Division of HIV/AIDS Prevention. "Living with HIV/AIDS." Available online. URL: http://www.cdc.gov/hiv/resources/brochures/livingwithhiv.htm. Updated June 21, 2007.
Libman, Howard, and Harvey J. Makadon, eds. *HIV*. 2nd ed. Philadelphia: American College of Physicians, 2003.
Marlink, Richard G., and Alison G. Kotin. *Global AIDS Crisis: A Reference Handbook*. Santa Barbara, Calif.: ABC-CLIO, 2004.
Matthews, Dawn D. *AIDS Sourcebook*. 3rd ed. Detroit: Omnigraphics, 2003.
MayoClinic.com. "HIV/AIDS." Available online. URL: http://www.mayoclinic.com/health/hiv-aids/DS00005. Updated June 20, 2008.
U.S. Department of Health and Human Services. "AIDS info." Available online. URL: http://aidsinfo.nih.gov/. Accessed January 14, 2008.
U.S. National Library of Medicine, National Institutes of Health. "HIV/AIDS Information." Available online. URL: http://sis.nlm.nih.gov/hiv.html. Updated July 16, 2008.
Watstein, Sarah Barbara, and Stephen E. Stratton. *The Encyclopedia of HIV and AIDS*. 2nd ed. New York: Facts On File, 2003.

addiction, the biology of   Although the definition of *addiction* is a compulsion to engage in any sort of behavior despite detrimental consequences (e.g., gambling, shopping, instant messaging), the term most commonly refers to physical dependence on a habit-forming chemical substance such as alcohol, heroin, or nicotine. Whereas an addict initially makes a choice whether or not to drink alcohol, use a drug, or smoke a cigarette, the phenomenon of addiction is a biological problem. Considered a disease, addiction stems from the attempt to avoid or overcome a negative consequence from not using the chemical substance. For example, smoking crack cocaine produces a pleasurable feeling called a high. After the drug wears off, the person experiences a low that another dose of the drug will prevent. Substance addiction not only changes a person's behavior, but also alters the cellular physiological characteristics of the brain in a manner that increases the body's tolerance to the substance and fosters continued use of it.

According to the National Institute on Drug Abuse (NIDA), the abuse of illegal drugs and alcohol and tobacco contribute to the deaths of more than 540,000 people in the United States every year. The psychiatric disorder of addiction impacts the lives of many more people, as children, spouses, siblings, friends, and parents of addicts suffer as well. Substance abuse negatively affects a person's health, but it can also damage a person's social life, family life, economic situation, and legal record. So why do people use potentially habit-forming substances? Reasons vary depending on the person and the situation, but common causes include desire to feel good, peer pressure, relief of anxiety or stress, for energy, for relaxation, out of curiosity, for pain relief, emotional problems, or mental illness. In the beginning users often feel they are in control and can stop using at any time, but the physiological changes that make quitting difficult can occur quickly.

Though no race, gender, ethnicity, socioeconomic status, religious affiliation, or other similar factor is impervious to addiction, chemical substances do not affect all individuals equally. Children do not inherit alcoholism or an addiction to pain pills, but evidence suggests that the tendency to become dependent on chemical substances runs in families. Certain genes have been found to play a role in addiction. According to the Genetic Science Learning Center at the

University of Utah, the following several studies have shown that certain genes play a role in addiction:

- A certain allele of the dopamine receptor is more common in people addicted to alcohol or cocaine.
- Expression of higher levels of the *Mpdz* gene in mice reduces withdrawal symptons from barbiturates.
- Absence of the serotonin receptor gene *Htr1b* is associated with increased attraction to cocaine and alcohol.
- Mice with low levels of neuropeptide Y or with a defective *Per2* gene drink more alcohol.
- Mice lacking the *Creb* gene develop morphine dependence less often.
- The reward response to morphine or cocaine is reduced in mice lacking the cannaboid receptor gene *Cnr1* or a gene encoding part of the nicotinic cholinergic receptors.

NIDA reports that between 40 and 60 percent of an individual's vulnerability to developing an addiction may be genetic. Thus two people who have taken the same drugs in the same doses the same number of times may not be equally affected. One may become addicted and the other not. Age does seem to play a role—people who start using a chemical substance as a child or young adult are more likely to become addicted to it. Scientists believe this partly relates to the fact that the brain continues to develop through adolescence into early adulthood. Besides biological risk factors, someone's upbringing, home situation, social factors, and other events that are going on in a person's life can also affect one's ability to control substance use.

DRUG TYPES

A drug is a chemical that affects the body's structure or function, and only some drugs are addictive. Addictive drugs fall into the following seven major classes:.

- Nicotine is a stimulant present in tobacco leaf and is found in cigarettes and chewing tobacco.
- Alcohol, barbiturates, benzodiazepines, and volatile substances used in sniffing can cause direct damage to the brain. Inhalants are particularly toxic and can damage the heart, kidneys, and lungs in addition to the brain. They are more harmful than addictive and can cause death within minutes.
- Opiates include opium, morphine, heroin (made by chemically treating morphine),

codeine, and synthetic opiates used as painkillers (such as oxycodone and Demerol). These drugs cause a feeling of pleasure followed by a sense of well-being or calmness and are highly addictive.
- Cocaine and amphetamines are stimulants. The effects of cocaine are short-lasting; thus abusers often binge, or take several doses within a single session. Amphetamines cause feelings of euphoria and alertness. The effects last longer.
- Cannabis, including any of the preparations made from the herb hemp, such as marijuana or hashish, is the most commonly abused illicit substance. The effects are impaired memory and learning, inability to focus, and lack of coordination.
- Caffeine, most commonly found in coffee, tea, and soft drinks, is the only addictive substance whose use is not prohibited in children.
- Hallucinogens include many naturally occurring products (such as psilocybin in magic mushrooms and mescaline in cactus) and synthetic compounds (such as lysergic acid diethylamide [LSD], MDMA or ecstasy, and phencyclidine [PCP] or angel dust). These drugs cause mind-altering effects, and the effects are unpredictable. In the short term, blood pressure, body temperature, and heart rate all increase, and sweating, appetite loss, dry mouth, and tremors can also result.

Because nicotine, alcohol, and caffeine are legal and therefore more socially acceptable, the term *addict* is rarely used to refer to a person who habitually uses these substances. Society may refer to such a person as a heavy smoker, an alcoholic, or a heavy coffee drinker rather than an addict, but the biological causes and effects of the addiction to these substances and to prescribed medications are the same as for illegal substances.

Different drugs have different routes for entering the body: injection directly into a vein, a muscle, or underneath the skin; ingestion though the mouth; inhalation into the lungs; or intranasal, in which the person snorts the substance, and it enters the bloodstream through the nasal mucosa. Drugs that are inhaled, injected, or snorted generally produce an effect more rapidly than ingested substances, and their effects wear off more quickly. The difference in how one feels before and after using the substance is very noticeable. Because of this, inhaled or injected drugs are typically more addictive than ingested substances.

EFFECTS ON THE BRAIN

The brain is the least understood organ in the human body. Made up of neurons and supporting cells, the brain serves as the control center of the nervous system. The brain stem, located at the base of the back of the head, regulates many involuntary physiological activities such as the heartbeat and respiration. The cerebral cortex is the convoluted surface layer of the cerebrum, which is the largest portion of the brain. Different regions of the cerebral cortex are responsible for specific functions, such as sensory processing, thinking, making decisions, and problem solving. The limbic system is a group of structures including the hippocampus, the hypothalamus, and the amygdala that are located underneath the cortex and control emotions and feelings of pleasure and motivation. Drugs that alter mood target this area of the brain.

The neurons that carry sensory input to the brain, within the brain, and from the brain to different parts of the body communicate with one another through chemical signals. Individual neurons connect to other neurons to form neural pathways and networks. The space between the end of one neuron and the start of another is called the synapse. When stimulated to do so, a neuron will release chemicals called neurotransmitters into the synapse. By diffusion, the neurotransmitters will reach the next neuron in the pathway and bind to specific receptors on its cell membrane. The binding triggers a chain of events in the postsynaptic neuron that will cause a particular response depending on the type of neurotransmitter. The action of the neurotransmitter may be stimulatory or inhibitory and might trigger muscular contraction or the release of certain hormones, additional neurotransmitters, or other chemical substances that have a specific physiological effect—for example, an increase in heart rate or blood sugar levels or slowed reflexes. Shortly after the release of the neurotransmitter, transporters bind and carry the neurotransmitter back into the neurons that released it, or, in some cases, enzymes degrade the neurotransmitters. Either way, the signal is terminated.

One neurotransmitter that plays a key role in addiction is dopamine, which belongs to the family of three neurotransmitters called catecholamines because they consist of a six-carbon catechol ring and an amino group. The other two catecholamines are epinephrine and norepinephrine, and all are synthesized from the amino acid tyrosine. Dopamine plays a role in motor coordination, but evidence suggests that it also functions in motivation, reward, and behavior reinforcement. The latter function relates to the importance of dopamine in substance addiction.

Prescribed and illicit drugs both interfere with the natural brain chemistry described. A physician may prescribe a drug for medical reasons. For example, an inadequate response by the brain to stimulation by the neurotransmitter serotonin may result in a psychiatric disorder such as anxiety or depression. Treatment may involve medications (called SSRIs, for selective serotonin reuptake inhibitors) that inhibit the reuptake of serotonin by binding to the transporters that carry serotonin back into the presynaptic cell, giving the serotonin a longer period to stimulate the receptors on the postsynaptic neurons. Another example of a medical use for drugs may be to prevent the perception of pain by the nervous system during surgery.

Addictive drugs work by either mimicking or blocking the effect of a natural neurotransmitter or by increasing the levels of neurotransmitter in a neural synapse. Drugs that mimic natural neurotransmitters resemble the neurotransmitter structurally such that the specific target receptors bind the drug and activate the postsynaptic neuron. By this mechanism, a drug can induce a response in the absence of the normal physiological stimulus. The effect of the drug, however, can differ from that of the natural neurotransmitter; for example, the duration may be longer or other neural pathways may be activated simultaneously, leading to an altered combined effect. Opiates are an example of a type of addictive drug that works by mimicking the structure of a natural neurotransmitter. Endogenous opioids (i.e., endorphins, enkephalins, dynorphins) are natural neural polypeptides that bind to pain receptors located along sensory pathways. At the point where the neural pathway reaches the spinal cord of the central nervous system, neurons release a peptide neurotransmitter called neurokinin that stimulates the pain pathway leading to the brain, where the stimulus is perceived as pain. The receptors that normally bind the endogenous opioids also recognize and bind morphine and other opiates. When these drugs bind, they block the pain pathway at the spinal cord. These drugs also work in the brain by causing an indifference to pain, so the patient may be aware of pain but not care about it. Another mechanism by which drugs act is by stimulating the inappropriate release of higher than normal levels of an endogenous neurotransmitter or preventing its reuptake. Either of these mechanisms results in an amplified signal to the postsynaptic neuron. Amphetamines and cocaine work in this manner.

In order for a person to seek and administer a drug without a medical reason, the drug must generate an enjoyable effect: in other words, the result must reinforce the habit of use. Neurobiologists have identified a specific area deep within the middle of the brain that produces a pleasurable and satisfying feeling when electrically stimulated. When electrodes were placed in different locations within rat brains, and the rats had access to a lever that stimulated the electrodes when pressed, the rats repeatedly pressed the lever when the electrodes penetrated this so-called reward pathway. This experimental method is called intracranial self-stimulation (ICSS), and similar experiments have been performed on humans during research on epilepsy. The tract of nerves believed to be stimulated is called the mesolimbic dopaminergic pathway. The neurons originate in an area called the ventral tegmental area (VTA) and extend into the nucleus accumbens (NAc) of the frontal cortex. When stimulated, the participating neurons release dopamine in the frontal cortex, where dopamine receptors are located. All addictive drugs, whether acting directly or indirectly, have been shown to increase synaptic dopamine levels in the NAc region of the brain; thus this phenomenon may explain the powerful reinforcement that leads to addiction, which can be viewed as self-initiated stimulation of this dopaminergic pathway. Scientists have also found evidence that blocking the dopamine receptors in the NAc prevents the reinforcement effect. This pathway normally functions in reinforcing behaviors that have positive survival and reproductive value, such as eating palatable food and engaging in behaviors that promote successful mating. Conditions such as hunger and sexual arousal increase the motivation and the likelihood of carrying out behaviors that will result in satiating this pathway. Addictive drugs are believed to act in the same way—when taken, they stimulate this reward pathway, and the early symptoms of drug withdrawal cue the behaviors that will result in the consequence of seeking and obtaining the drug.

The location of action of many addictive drugs has been mapped. Cocaine acts at the end of the neurons that release dopamine by blocking the transporters that normally return the neurotransmitter to the interior of the cell. Some evidence suggests that cocaine may also act elsewhere, but how and where are not known. Amphetamines enter dopamine neurons through the reuptake transporters and cause dopamine to be released into the synapse. One long-term effect of overstimulation by dopamine from prolonged drug use is a decreased production of dopamine. Opiates act earlier in the pathway on inhibitory receptors called mu opioid receptors, and they prevent the release of other neurotransmitters. The specific effect depends on the particular neurotransmitter. For example, this may result in the inhibition of the release of a neurotransmitter that controls or limits dopaminergic neuron stimulation.

Nicotine and alcohol may activate the reward pathway at the VTA by either stimulating the release of natural opioid peptides or directly stimulating dopaminergic neurons. Cannabis can act in the VTA or directly on the NAc. Hallucinogens, such as PCP, can also act directly on the NAc.

After repeated administration, an individual may develop homeostatic adaptations such as tolerance, or reduction in the response to the drug. The brain adjusts to the continued exposure of a drug to the point where the stimulus eventually is unnoticed, just as a person who buys a house near a railroad track adjusts to the whistle of oncoming trains. As one's tolerance for a drug increases, the individual must use higher dosages to achieve the same desired effect. This is due to physiological and biochemical changes that result from the flooding of dopamine in certain areas of the brain. These changes can affect areas that control judgment, decision making, learning and memory, and behavior control.

Neuroscientists have characterized the long-term physiological changes that chronic use of some drugs causes. For example, chronic morphine use increases the levels of certain cellular enzymes (adenylyl cyclases) that play a role in cellular signaling. These changes have a significant effect on cellular physiology and increase the intrinsic excitability of some neurons.

ENDING THE ADDICTION

In addition to struggling with the compulsion to continue using an addictive substance, an individual trying to stop taking a drug often suffers a range of negative physiological and emotional symptoms. In some cases, such as with opiates, the nerve cells in the brain become so used to the presence of the drug that they cannot function normally in its absence—the cells become overactive in an attempt to compensate. When the individual has developed a physical dependence, stopping the drug abruptly can cause withdrawal sickness. The urge to take another dose becomes very intense as the levels of the addictive chemical in the blood gradually decrease during the withdrawal period. The nature and intensity of the withdrawal symptoms vary depending on the specific drug and the synapses and neural circuits affected, but the range includes increased heart rate, increased blood pressure, sweating, tremors, depression, insomnia, fatigue, irritability, nausea, restlessness, general malaise, and muscle pain. In the case of alcohol, barbiturates, and benzodiazepines, abrupt withdrawal can be fatal. In most cases, though the withdrawal period may cause extreme discomfort, pain, or misery, the body does eventually recover. Images of brains of methamphetamine addicts show that

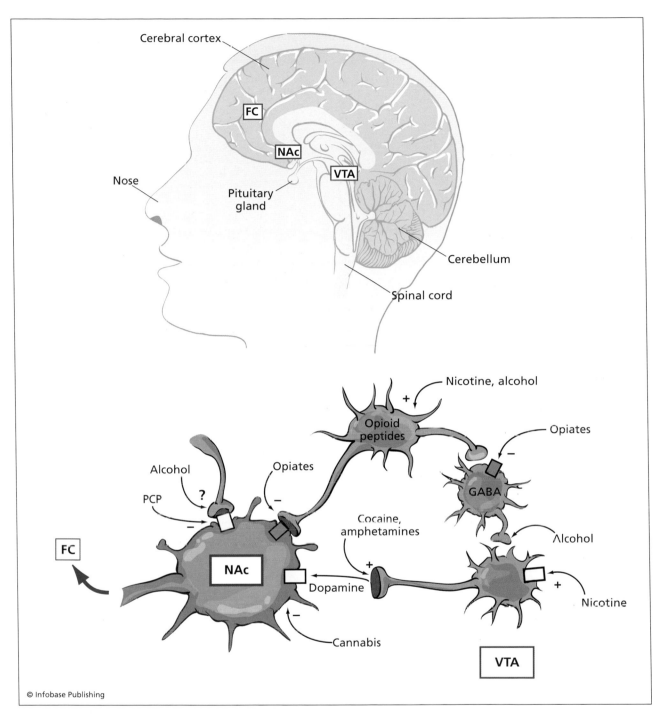

Many addictive drugs are known to act on the mesolimbic dopaminergic pathway, also called the reward pathway, that naturally plays a role in reinforcing certain behaviors that aid in survival.

although the amount of dopamine transporter was significantly lower than in a normal, healthy brain, the levels appeared near normal after 14 months of abstinence. Even after detoxification is complete—meaning all traces of the drug and its metabolites are absent from the body—a recovering addict must also deal with or change behavioral patterns and environmental conditions that contributed to

the development of the addiction. This process may take months, years, or a lifetime.

Addiction is a treatable but chronic disease—relapses occur among 40 to 60 percent of drug-addicted patients. The most effective treatment includes a combination of medication (when available) and behavioral therapy. The goal of some medications is to reduce the withdrawal symptoms so the individual is more likely to continue abstaining from use. Other medications help reduce the cravings for the drug while the brain gradually adjusts to its absence.

Recent research that has demonstrated a physiological basis for substance addiction has changed the way society views addicts and health care professionals approach treating addiction. Though much progress has been made in understanding the chemical and neural effects of addictive chemicals, the molecular, cellular, systems, and behavioral levels are still unknown. Medical researchers are also actively investigating novel pharmacological agents that have the potential to help addicts gain control of their compulsions, reduce the effects of withdrawals, and change the brain chemical processes to alter the effect of the substance to make it undesirable. Because prevention is the best strategy for dealing with addiction, increased educational efforts about the harmful nature of drugs and dangers of addiction must accompany the improved understanding of the biology of addiction if the number of lives that are affected by addiction is to be decreased.

See also CELL COMMUNICATION; NERVOUS SYSTEM.

FURTHER READING

Goldstein, Avram. *Addiction: From Biology to Drug Policy.* 2nd ed. New York: Oxford University Press, 2001.

Hyman, Steven E., Robert C. Malenka, and Eric J. Nestler. "Addiction: The Role of Reward-Related Learning and Memory." *Annual Review of Neuroscience* 29 (2006): 565–598.

National Institute on Drug Abuse home page. Available online. URL: http://www.nida.nih.gov/. Accessed January 22, 2008.

U.S. Department of Human Health and Human Services. Substance Abuse and Mental Health Services Administration home page. Available online. URL: http://www.samhsa.gov. Updated January 22, 2008.

aging Aging is the normal process of growing older. Numerous adverse effects often accompany the aging process. Some signs of aging, such as wrinkles, hair loss, or graying hair, may not be desirable, but they do not have any deleterious effects on someone's health. On the other hand, harmful conditions and diseases, such as arthritis, Alzheimer's disease, cataracts, cancer, bone fractures, hypothermia, forgetfulness, cardiovascular disease, high blood pressure, strokes, slowed reflexes, decreased strength, vision and hearing loss, decreased immune system function, and osteoporosis, occur more frequently as individuals age. Some of these develop as a consequence of the accumulation of years of damage from poor health habits including improper diet, drug or alcohol misuse, or lack of exercise, but many healthy people who have lived exemplary lifestyles and followed preventative health care recommendations still experience problems commonly associated with aging. Because deterioration in cell number and function manifests itself in symptoms experienced at the organism level, gerontology, the study of aging and age-related problems, involves research on molecules, cells, tissues, and organ systems. Research on the complex biological nature of aging and age-related diseases is active. By studying aging at several levels, gerontologists hope to achieve a better understanding of the causes of aging and perhaps learn how to delay or prevent the common adverse effects.

Senescence, the progressive deterioration caused by biochemical and physical changes associated with aging, increases the risk of mortality as one gets older. Effects of senescence accumulate, affecting body functions, and eventually result in death. According to death registration data collected by the Centers for Disease Control and Prevention and the National Center for Health Statistics, a person born in the United States in the year 1900 had a life expectancy of 47.3 years. That number rose to 77.8 years for people born in the year 2004. While expecting such a large jump over the course of the 21st century seems unrealistic, life expectancies continue to rise slowly as medical researchers learn more about the chronic diseases that ultimately lead to death.

PROGRAMMED THEORIES OF AGING

Gerontologists believe that, to some degree, biochemical, genetic, and physiological characteristics all contribute to the aging process. Many intrinsic factors seem to play a role in determining life span. Just as hormonal and other chemical signals control and coordinate other life cycle stages including embryogenesis, development, and puberty, they may also bring about senescence. Genes also clearly play a role, as different animal species have different average life spans. The gastrotrich, a type of aquatic animal, lives only three days, shrews live slightly more than one year, while giant tortoises live an average of 177 years. In addition, evidence suggests that longevity runs in families. Individuals with parents and grandparents who lived long lives have a greater probability of living long lives themselves. Environmental and ecological factors also contribute to aging and affect life span; extrinsic factors such as climate, food availability, and the niche an organism fills in an ecosystem affect an animal's survival. Gerontologists study all the factors that affect one's life span, but only recently have they begun to make significant advances toward understanding the intrinsic

NORMAL EFFECTS OF AGING

What Is Affected	General Change
arteries	lose elasticity, require more force to move blood through circulation
bladder	capacity declines
body fat	gradually increases until middle age, stabilizes until late in life, then decreases
bones	bone mineral is lost; bones weaken
brain	loss of some axons, diminished function
hearing	decline in hearing
heart	muscle thickening, diminished function
kidneys	less efficient at filtering wastes from blood
lungs	vital capacity decreases
muscles	muscle mass declines in absence of regular exercise
sight	difficulty focusing on near or fine objects, increased susceptibility to glare
skin	becomes thinner and loses elasticity

mechanisms of the physical decline associated with aging—in other words, the cellular events that lead to the symptoms characteristic of aging. Different theories proposed to explain aging generally fall into one of two categories: programmed theories or error theories. The programmed theories have in common the basic notion that biology controls the life span of a cell.

As a starting point, researchers have identified several genes that appear to affect longevity. One way to approach this avenue of investigation is to look for genes that are expressed either more or less as an organism grows older. One gene, *lag-1*, characterized in yeast, affects the number of cell divisions yeast can undergo before dying off. The average number of generations is about 21, but yeast that expressed higher than normal quantities of this gene averaged 28 generations. Current research aims to clone the related gene in humans and to study the function of the gene product. Overexpression of the *sir2* gene, which functions to stabilize DNA in yeast, increases the life span of fruit flies. Extrinsic factors purported to increase life span, such as caloric restriction and exposure to the antioxidant resveratrol, act by increasing levels of the Sir2 protein. The homologue for this gene in humans, *sirt1*, has been found to function similarly, though through more complex mechanisms. The protein helps cells withstand periods of stress that would normally lead to apoptosis, or programmed cell death. Another gene called *INDY* (for I'm Not Dead Yet), also studied in fruit flies, nearly doubles their normal life span from 37 to 70 days or longer when mutated. The mechanism by which the mutant gene extends life seems to be by decreasing the normal activity of the fly by restricting the calories absorbed by the cells. A gene from the roundworm *Caenorhabditis elegans*, *daf-2*, encodes a protein similar to the human insulin receptor. The hormone insulin plays a role in glucose utilization in metabolism, which is another important avenue of research for understanding senescence. Another longevity gene appears to regulate gene expression in *C. elegans*; the gene product is an enzyme that changes the structure of DNA and, as a result, turns off the synthesis of proteins encoded by other genes. Researchers have identified many other longevity genes in model organisms, and current research goals include finding more of those genes, characterizing the protein products of the genes, investigating the regulation of these genes, and examining the function of the proteins.

According to the programmed longevity theory of aging, cells have predetermined life spans. Cells do have finite life spans when grown in vitro; after a limited number of cell divisions, replication stops. The cells continue to be active metabolically, but they no longer proliferate. They change their gene expression patterns, and in some cases their presence becomes harmful. The cellular mechanism that limits the number of cell divisions before senescence sets in is currently unknown, but continued research on longevity genes may shed light on this. One explanation involves telomeres, looped structures located at the ends of linear chromosomes that contain numerous repeated copies of a sequence of DNA. In humans and other vertebrates this sequence is TTAGGG. Its

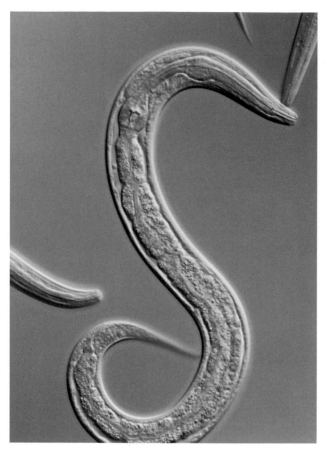

Researchers use animals such as the roundworm *Caenorhabditis elegans,* shown here, to study the aging process. *(Sinclair Stammers/Photo Researchers, Inc.)*

purpose is to permit replication of the DNA to proceed all the way to the end. DNA is a double-stranded molecule, resembling a ladder with the rails on the sides representing each strand. Unlike ladder rails, however, the two strands of DNA are antiparallel, meaning the two strands run in opposite directions, as if one rail on the ladder has a supportive base at the bottom and a cushioned tip at the top, but the other rail has the reverse—the supportive base at the top and a cushioned tip on the bottom. This is significant because the enzyme that replicates DNA only works in one direction and the replication system's mechanism that compensates for this requires extra length at one end of the double-stranded DNA molecule. Without telomeres, one of the strands would lose a bit of its sequence at the end during each round of replication. The telomeric repeats do not encode any protein; they just ensure that genes located at the ends of chromosomes do get replicated. With each round of DNA replication that precedes cell division, the telomeres shrink, eventually becoming too small to perform their job. At this point, the cell is said to have reached replicative senescence and will no longer divide. While these findings initially generated much excitement among gerontologists, no correlation between cellular senescence from decreasing telomere length and longevity of a whole organism has been established. In fact, some animals with short telomeres have longer life spans than other animals with longer telomeres. Many cell biologists continue to pursue research in this area, however, because of its initial promise and the as-of-yet unexplained connection of cellular senescence and aging.

The endocrine system may be an important regulator for mechanisms involved in programmed aging. Specialized glands and tissues produce and secrete chemicals called hormones that travel throughout the circulatory system and act on other target tissues or cells. For example, the pituitary gland located at the base of the brain secretes hormones that act on the ovaries and uterus in females to coordinate the complicated events of the menstrual cycle. Hormones regulate nearly all aspects of growth, metabolism, development, reproduction, and homeostasis, so it is logical that they would regulate aging as well. Some events associated with aging are definitely under hormonal control. For example, as a human female ages, she loses her ability to reproduce. This process is called menopause and results when the hypothalamus ceases to release the necessary stimulatory hormones. The concentration of many hormones decreases with age, but replacement therapy does not stop aging. Hormonal imbalance seems to explain many phenomena associated with aging, but medical researchers have yet to turn this information into practices that halt the aging process.

The immunological theory of aging states that self-destruction is programmed into the immune system. In addition to preventing and fighting infections, finding and removing foreign substances from the body, and repairing damaged tissues, the immune system functions to survey the body for cells that are abnormal, such as potential cancerous cells or cells that have been damaged by viruses, and then destroys them. In order to perform this task, the immune system must be able to differentiate between the body's normal cells and abnormal or foreign cells (such as pathogenic microorganisms). According to the immunological theory of aging, the immune system loses this ability over time and begins to attack its own healthy cells and tissues. As the immune system function declines, the body also becomes more vulnerable to infections and cancer.

ERROR THEORIES OF AGING
The error theories of aging purport that senescence results from damage to cells or loss of function over time due to continuous exposure to environmen-

tal assaults such as radiation, toxic chemicals, and free radicals. The chromosomes located inside the nucleus of the cells that make up a living organism contain the organism's genetic information. Genes composed of deoxyribonucleic acid (DNA) encode all the information necessary for the cells to make proteins and other biomolecules that compose cells and tissues and that perform all the functions necessary for life, such as growth and metabolism. DNA is copied every time the cell divides to create new cells as a normal part of growth, repair, and tissue maintenance. Regions that encode genes along a molecule of DNA contain specific sequences of four alternating nucleotide bases (abbreviated *A, C, G,* and *T*) that encode for specific sequences of amino acids, the subunits from which proteins are synthesized. Alterations in the DNA sequence called mutations can lead to changes in the amino acid sequence of the proteins. Cells have mechanisms for repairing the majority of mutations, but given the sheer enormity of the task of replicating the DNA during cellular reproduction—the sequences within the nucleus of each cell in humans are 3 billion nucleotides long—inevitably, a number of the mutations become permanent. Some mutations are minor and have no noticeable effect, but others cause significant problems, even resulting in cell death. Error theory states that over time, the number of mutations accumulates to dangerous levels, killing cells and, eventually, the entire organism.

Atoms are most stable when all the electrons are paired. Free radicals, also called reactive oxygen species, are atoms or molecules that have an unpaired electron, making them unstable. Though some cells use free radicals in a controlled manner to perform useful functions such as destroying pathogens, free radicals are often dangerous to living cells because they readily combine with other molecules or strip off electrons from other molecules including DNA or proteins in order to pair their lone electron. If a free radical steals an electron from one molecule, that molecule then may try to replace it and may take the electron off another molecule, initiating a chain reaction inside the cell. Damage from free radicals causes cells and eventually tissues and organs to lose their ability to function. Free radicals are created during the production of adenosine triphosphate (ATP), the form of energy used by cells, from oxygen and organic molecules during aerobic respiration. Sunlight, tobacco smoke, and other substances ingested with food or inhaled with air also cause the formation of free radicals. Mitochondria, the cellular organelles that carry out cellular respiration and ATP synthesis, are particularly vulnerable to damage, and damaged mitochondria are less efficient. These organelles contain their own DNA, which encodes for proteins that function in cellular respiration. The

constant exposure of the mitochondrial DNA to free radicals causes mutations in the mitochondrial genes, leading to an overall decrease in efficiency of ATP production over the years. Because cellular respiration is a necessary metabolic process, free radicals are unavoidable, but cells have some means for dealing with them. In particular, one enzyme called superoxide dismutase catalyzes the neutralization of oxygen free radicals produced during aerobic respiration. Fruit flies that have been genetically engineered to express higher than normal quantities of this enzyme live longer than normal fruit flies, suggesting there might be a link between free radicals and senescence. Unfortunately, the cellular defenses are not foolproof, and some free radicals escape them and cause damage. Evidence correlating the action of free radicals with senescence includes the association of genetic defects that accelerate the accumulation of free radicals with premature aging. *C. elegans* that had this genetic defect exhibited normal life spans when exposed to a synthetic antioxidant drug. Antioxidants are substances that can inhibit oxidation reactions (reactions involving the loss of electrons) by safely donating their own electrons to the radicals without incurring damage in the process. This makes the radical less reactive. Free radicals may also contribute to the development of cancer, neurodegenerative diseases, cataracts, and atherosclerosis. Many antiaging remedies include dietary supplements or lotions that contain antioxidants such as the vitamins C and E.

The rate of living theory closely relates to the free radical theory of aging. Caloric restriction, the only demonstrated means of extending life in mammals, prolongs life spans by slowing an individual's metabolic rate. The goal is to reduce total calories but not to let the body become dangerously emaciated, and maintaining a balanced diet is still important to prevent malnutrition. Basically, severely reducing the amount of calories an individual takes in decreases the activity level, and the individual lives longer. One idea behind this is that reducing the amount of food reduces the levels of exposure to toxic metabolic by-products, such as oxygen free radicals. Mice that consume a healthy diet containing 30 percent fewer calories than other mice live 40 percent longer. Researchers have found that all of the body's physiological systems exhibit delayed age-related degeneration as a result of caloric restriction. Studies on the effects of caloric restriction on aging in primates are currently in progress. Even if caloric restriction is proved to delay aging and improve health in humans, it is not a realistic antiaging therapy, as people are very resistant to drastic changes in their eating habits, as shown by the inability of many to maintain long-term restricted diets in order to improve cardiovascular health or to lose weight. However, understanding the mechanism by which caloric restriction works could lead to other therapies or treatments of age-related deterioration.

Cross-linking of proteins occurs when glucose molecules attach themselves to proteins, causing them to bind together. The accumulation of cross-linked proteins can stiffen tissues, causing them to lose elasticity and flexibility. Tissues that have lots of collagen, a common component of connective tissue, are particularly affected. Vision becomes blurred as collagen in the eye lens becomes more rigid. Arteries lose elasticity, leading to an increase in blood pressure. Lungs cannot expand as far, reducing their maximal capacity. Tendons lose their flexibility, limiting range of motion, stiffening joints, and increasing risk of injury. Thus cross-linking leads to the deterioration of several body functions that are associated with aging.

AGE-RELATED DISEASES

Some illnesses or diseases become more common as people age. The causes of some are known and others are unknown. Alzheimer's disease, cancer, arthritis, osteoporosis, diabetes, and cardiovascular disease are examples of conditions that result from impaired function due to deterioration of the body's tissues with age.

According to the Centers for Disease Control and Prevention (CDC) National Center for Health Statistics (NCHS), Alzheimer's disease rose to the fifth leading cause of death among people aged 65 and older in 2003. Alzheimer's is a degenerative brain disease of unknown origin that typically appears in people over the age of 60, and the risk increases with increasing

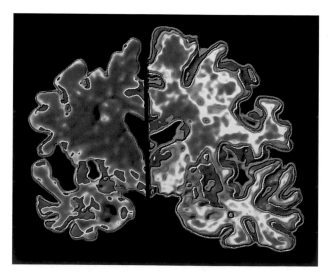

Alzheimer's disease causes degeneration of nerve tissue and nerve cell death, leading to shrinkage, as shown here on the left, compared to a normal brain on the right *(A. Pakieka/Photo Researchers, Inc.)*

age. The symptoms typically begin as mild then progressively worsen and include memory loss, impaired thinking, disorientation, mood disturbances, and personality changes. Abnormal clumps called amyloid plaques and tangled webs of nerve fibers in the cerebral cortex are characteristic signs of Alzheimer's. Several genetic mutations have been associated with the development of Alzheimer's. One factor appears to be a protein that carries cholesterol in blood circulation and is encoded by the apolipoprotein E (ApoE) gene. A team of researchers led by Huntington Potter from the University of South Florida found that ApoE speeds up the formation of amyloid plaques in the brain by converting harmless amyloid-beta protein into the toxic fibrous deposits and that this conversion is associated with the cognitive decline. No cure for Alzheimer's exists, but medications are available that slow the progression in some cases.

Cancer, when cells grow out of control to form tumors or neoplasms, can strike a person of any age, but as one grows older, its risk increases. According to the NCHS, malignant neoplasms were the second leading cause of death in 2003 for persons 65 years of age or older. The goal of many antiaging therapies is to increase the life spans of cells, but the body has evolved natural mechanisms for limiting cellular life span for a good reason. The greater number of times a cell line undergoes DNA replication and division, the greater the number of accumulated genetic mutations. Cancer results from the accumulation of mutations in genes that control the cell cycle. Proteins that stimulate cell division and those that inhibit it normally work together to ensure that cells replicate in a highly regulated and appropriate manner. Biochemists have identified and characterized numerous genes that play a role in cell cycle control. Tumor suppressor genes encode proteins that limit or inhibit cell proliferation under appropriate circumstances. For example, a cell should not reproduce when its DNA contains a lot of unrepaired mutations because cancer can result from the accumulation of numerous genetic mutations. Thus the normal function of tumor suppressor genes is to prevent the reproduction of cells that have a high potential for developing into tumors. The retinoblastoma gene is one tumor suppressor gene that acts by preventing mitosis; when mutated, the protein product of this gene (often abbreviated RB) prevents cellular senescence and leads to tumor formation. The protein p53, also encoded by a tumor suppressor gene, acts by binding to DNA and altering the expression patterns of numerous other proteins. More than half of all cancers are associated with a mutation in the gene encoding the p53 protein that leads to the protein's inactivation. Studying antiproliferative genes will help cell biologists to understand not only

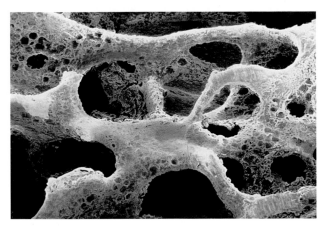

Scanning electron microscopy of bones affected by osteoporosis reveals numerous round dark areas void of bone tissue. *(Professor Pietro M. Motta/Photo Researchers, Inc.)*

cancer, but the process of cellular senescence. A connection also exists between cancer and telomeres. In cancerous cells, the telomeres behave abnormally and do not shrink with each round of replication. An enzyme called telomerase keeps replacing the telomeric sequences, preventing this mechanism from causing cellular senescence, allowing the cancer cells to divide uncontrollably.

Arthritis includes a number of diseases that affect the joints and are the leading cause of disability in elderly people. The most common type in older people is osteoarthritis, which results when the cartilage that cushions the bones at joints becomes worn as a result of extensive use and gravity. This degenerative disease causes joint pain, inflammation, and stiffness. Other common forms of arthritis include rheumatoid arthritis, which results from the immune system's attacking the lining of the joints, and gout, which results from the accumulation of uric acid crystals in the joints and connective tissue. Treatments for arthritis depend on the particular type of arthritis and include exercises and medications to reduce pain and swelling.

Osteoporosis is a disease that causes the bones to weaken and easily break. Throughout one's life, bone tissue breaks down and is replaced with new bone tissue, but as one grows older, more breaks down than is replaced, and strength decreases. Taking vitamin D and calcium supplements, getting plenty of exercise, and not smoking help prevent the breakdown of bone tissue and assist in building new bone tissue.

Diabetes results when the level of glucose (a simple sugar) in the blood is too high. The pancreas produce insulin, a hormone that helps cells to absorb glucose from the bloodstream. Type I diabetes is typically seen in children, young adults, or people below the age of 30, when their body stops producing insulin. In Type II diabetes, which is most common in people over the age of 40, the individual produces insulin, but the body stops responding to it properly. High blood glucose levels can lead to serious medical problems, such as heart disease and stroke. Diabetes is the seventh leading cause of death in people over the age of 65 according to the NCHS. Fat buildup, a normal effect of aging, also increases the levels of glucose in the blood; thus weight loss is one means to control Type II diabetes. Monitoring one's diet and increasing physical activity are also ways to manage this disease, and medications also help many people control their diabetes.

Cardiovascular disease includes numerous different conditions affecting the heart and blood vessels and can strike anyone, but the risk increases with age. Heart disease consistently ranks as the leading cause of death in persons of all ages, and stroke ranks third. The heart muscle weakens with age, and the blood vessels lose their elasticity, making it harder to propel the blood throughout the body. Atherosclerosis is a disease in which plaque deposits of dead cells and cholesterol build up in the arteries, narrowing the diameter of the vessel. Proper diet and exercise can prevent or reduce the risk of cardiovascular disease, which is a common, but not an inevitable consequence of aging. Many effective medications are also available.

See also CANCER, THE BIOLOGY OF; CHROMOSOMES; POINT MUTATIONS.

FURTHER READING

Clark, W. R. *A Means to an End: The Biological Basis of Aging and Death.* New York: Oxford University Press, 2002.

Dollemore, Doug. *Aging under the Microscope: A Biological Quest.* Bethesda, Md.: National Institute on Aging, National Institues of Health, June 2006.

Panno, Joseph. *Aging: Theories and Potential Therapies.* New York: Facts On File, 2005.

Scientific American Special Edition: The Science of Staying Young. Special edition. 14 (3) (June 2004).

Sinclair, David A., and Lenny Guarente. "Unlocking the Secrets of Longevity Genes." *Scientific American* 294 (March 2006): 48–57.

U.S. National Institutes of Health, National Institute on Aging home page. Available online. URL: http://www.nia.nih.gov/. Accessed January 17, 2008.

agriculture Agriculture is the science or process of producing food, feed, and fiber (cotton, wool, flax, and hemp) through the systematic cultivation of plants and raising of domesticated animals. In addition to food products for humans and animals, agriculture includes cut flowers, ornamentals, timber, fertilizers, leather, chemicals, fuels, and drugs such as tobacco, opium, and biopharmaceuticals. More than 40 percent of the world's population works in the field of agriculture.

Agriculture is an application of life science as it depends on and utilizes many principles from botany, zoology, genetics, microbiology, molecular biology, and more. Other sciences, such as chemistry and engineering, also contribute to agriculture. While the term *agriculture* refers to the activities performed to transform the environment for raising animals or growing crops for human use, agricultural science is a multidisciplinary field that encompasses the study of numerous subjects that are related to agriculture. Agronomy is the research and development related to studying and improving field-crop production and soil management. An agricultural scientist might research methods for improving crop yields, determine the best way to package dairy products, design and develop new pesticides, study waste management, or predict nitrogen needs of a field based on the basis of ecological models. Related disciplines include fields as varied as agricultural engineering, aquaculture, food science, environmental science, soil science, agrophysics, horticulture, agricultural economics, animal science, and irrigation management.

Archaeologists and anthropologists have traced the origins of agricultural practices to approximately 9500 B.C.E. in the area of the Middle East including Egypt, Mesopotamia, and the Levant. The first domesticated crops were cereal grains. Root crops and legumes were first cultivated 8,000 to 9,000 years ago, followed by vegetables, and oil, fiber, and fruit crops. Plants used for decoration and drugs were not domesticated until about 2,000 years ago. The gradual development of agriculture was a crucial step in the development of civilizations. Without having to wander in search of animals and plants for food, people could settle in a single location, an action that led to the formation of towns or cities. By 5000 B.C.E. the Sumerians, a civilization in the southern part of Mesopotamia, were cultivating land on a large scale, monocropping, and using labor and irrigation. The invention of the plow around 4000 B.C.E. was another major technological advance that led to crop planting in rows and increased efficiency, allowing populations to become denser. The earliest simple plows were simply forked sticks and timbers that later developed into heavy plows that had a coutler for cutting a thin strip in the earth, a share to slice the soil, and a mouldboard that turned the soil. Later wheels and seats were added. In the Middle Ages, the Persian Muslims made several advances that led to the methods used in more modern farming. These revolutionary techniques included the development of sophisticated irrigation, or watering systems; the use of the scientific method to improve farming techniques; giving incentives to and rewarding productive laborers with land and shares in the harvest; and the introduction of new crops, such as sugar cane, citrus fruits, and cotton, that were attractive as commercial products to people in other geographical locations. The later development of more specialized plows and the concept of crop rotation continued to improve farming efficiency. During the late 15th and 16th centuries, crops spread by exchange between the Old and New Worlds. Tomatoes, squash, maize, peanuts, potatoes, and tobacco were introduced to Europe. The Americas began to depend on wheat, rice, coffee, cattle, horses, and sheep. Agriculture played a significant role in the Industrial Revolution of the 18th and 19th centuries. New means of transportation facilitated the shipping of farm products and reduced the incidence of starvation. In turn, the increasing populations helped to meet the demand for new laborers in all the growing industries, including agriculture. Inventions such as the cotton gin, the mechanical reaper, threshing machines, mowing machines, and better plows further advanced farming. In the 20th century, tractors and trucks, refrigeration, the development of hybrid crops, and genetic engineering to improve varieties have all benefited agricultural industries.

Today crop improvement remains a major goal of agricultural science. Continued advancements in traditional farming methods include technology that has led to newer equipment. Better machines have improved yields while requiring less effort. Agronomists continue to search for new ways to control pests and weeds. Understanding the relationships among living organisms has led to natural mechanisms of pest control. The concept behind natural pest control is to identify the natural enemy of the crop-destroying pest. For example, ladybugs consume aphids, whiteflies, mealybugs, scales, mites, and many other soft-bodied insects in addition to bollworms, broccoli worms, cabbage moths, and tomato hornworms. The bacteria *Bacillus thuringiensis* produces a paralytic toxin (Bt toxin) that, when ingested by insects, paralyzes their digestive tract and eventually kills the pests. Though the term *natural* implies these methods cannot be harmful, ecologists must carefully consider the effect that introducing new species or altering the numbers of present species will have on the ecological community before implementing biological means of pest control. Animal husbandry, the breeding and raising of animals for their meat or products such as eggs or wool, uses selective breeding programs to create organisms that yield either larger quantities of products or more desirable products. Plant breeding programs accelerate the natural process of evolution

(continues on page 19)

THE COLONY COLLAPSE DISORDER CRISIS FOR HONEYBEES

by Zachary Huang, Ph.D.
Michigan State University

Honeybees (*Apis mellifera*) not only produce honey, the natural sweetener that is considered wholesome and healthy, but also beeswax, pollen, venom, royal jelly, and propolis. Beeswax is used for candle making, batik, and sculpturing; pollen, royal jelly, and propolis are used as nutraceuticals or health foods. Venom can be used for pharmacological studies or for treating autoimmune diseases such as rheumatoid arthritis and multiple sclerosis. While these products from honeybees are impressive, the most significant contribution of honeybees to human society is actually the pollination services they provide for agriculture. Approximately 130 crops grown in the United States depend on honeybees for pollination. Most fruits and vegetables need honeybees either to set fruit or to produce better fruit (more symmetric fruits, sweeter, or with a higher oil content) or to allow seed production (most vegetables). Examples of bee-pollinated fruits include almonds, apples, cherries, blueberries, pears, peaches, strawberries, raspberries, and the melons (squashes, cucumbers, pumpkins, and watermelons). Vegetables that require bees for seed productions include broccoli, onions, cabbages, and radishes. A Cornell study from 2001 estimated the total value of pollination to these crops in the United States to be about $15 billion per year. The total value of honeybee products (honey, pollen, beeswax, etc.) ranges from about 1/50 to 1/80 of the value of pollination. The almond crop alone in California requires about 1.2 million colonies each year, about 50 percent of the total number of colonies in the country. The almond growers paid about $150–$200 per colony in the year of 2007 for pollination, usually at two to three colonies per acre. Typical prices for pollinating apple and cherry orchards are about $50–$70 per colony.

THREATS TO HONEYBEES

Honeybees, as do most other organisms, face the threat of diseases and pests attacking them. The main threats to honeybees are pests, diseases, and pesticides, but other threats also exist. Global trade has complicated beekeeping by introducing new pests and diseases that have required new pesticides or antibiotics for control. The stagnant honey prices and increased cost for maintaining bees have caused the downward trend of colony numbers in the United States.

Pests

By far the worst threat to honeybee health, in the United States and worldwide, is the ectoparasite varroa mite (*Varroa destructor*). This mite, formerly referred to as *Varroa jacobsoni* in the literature, was introduced into the United States around 1987. Apiculturists initially thought the parasite was *V. jacobsoni* but later found that all mites in North America differed genetically from the real *V. jacobsoni* by about 8 percent. Denis Anderson named the newly identified North American species *V. destructor*. Within a few years almost all feral honeybees (i.e., unmanaged bees that nest inside tree holes and in the walls of people's houses) died off because of the mites. Managed honeybees rely upon pesticide use to reduce the mite population, or they too will die after one to two years. The mites suck blood from adult and immature bees but only reproduce on bee pupae, which are "capped" by a layer of wax by worker bees. Varroa mites have been known to transmit many types of bee viruses (e.g., deformed winged virus, cloudy wing virus, Kashmir virus, slow paralysis virus, and acute bee paralysis virus). One theory is that mites do not kill bees: the viruses they harbor and transmit do the killing.

Another mite pest is the tracheal mite (*Acarapis woodi*), an internal parasite that lives and reproduces inside the breathing tubes (tracheas) of the adult bees. First introduced to the United States around 1983, the tracheal mite also wiped out about 50 percent of managed bees during the first few years. The mites weaken the bees so that they live shorter lives and have trouble returning from their cleaning flights in early spring. Bees normally do not defecate inside hives; they wait for a 55°F (12.8°C) day in February or March to fly out and defecate. Typical symptoms of this mite are small clusters or no bees at all in the early spring when most or all bees die while away from the colony (flying out and not returning).

Diseases

Honeybees also suffer diseases that are caused by viruses, bacteria, parasites, or fungi. American foulbrood is a larval disease caused by *Paenibacillus larvae,* the endospores of which can survive decades on bee combs. Infected larvae die at late larvae stage, just before they pupate, or during the pupation process (under the sealed cells). The antibiotic tetracycline (marketed as Terramycin®) used to treat the infection effectively, but the bacterium is now mostly resistant to it. A new chemical, tylosin (marketed as Tylan®), was registered in 2006 for the control of *Paenibacillus. Nosema apis,* a protozoan parasite that infects the midgut cells of bees, causes nosema disease. This parasite forms spores that can survive many years inside honey and wax and only germinates inside the midgut of bees. Infection causes death of epithelial cells of the midgut, basically rendering bees unable to digest pollen to absorb proteins. Colonies often die during winter because sick bees defecate inside the hive and spread the disease. In the summer, infected workers forage earlier and live shorter lives, causing reduced honey yield. In 1996 a new species of *Nosema* was discovered in Asia and named *Nosema ceranae* because entomologists thought at the time that it only infected

(continues)

(continued)

the Asian hive bees, *Apis cerana*. In 2005 scientists in Taiwan found that it can also infect *Apis mellifera*. Soon European scientists found the same species (*Nosema ceranae*) infecting their bees and thought the new disease was the main culprit causing massive bees die-offs, especially in Spain.

Pesticides

There are two routes for pesticides to get into honeybees. One route is pesticides applied internally to the colonies by beekeepers to kill pests and disease organisms. Many types of pesticides are used against the varroa mites, partly because of resistance developed by the parasite. Apistan™, a pyrethroid (a type of insecticide), has been used since the late 1980s and currently is largely ineffective against the mites. Checkmite +, an organophosphate (which is more toxic to humans than pyrethroid), was introduced about 10 years ago, and mites are also becoming resistant to this. Other treatments include organic acids (acetic acid, formic acid, and oxalic acid) and essential oils (thymol in both ApiGard and Apilife Var). Other pesticides that are applied externally to the hive, those used on crops and on ground covers or water in an attempt to control other insects, also affect honeybees. Pesticides sprayed onto flowers directly can be picked up by bees on flowers; systemic insecticides are absorbed by plants and then transported to nectar and pollen and collected by bees. French beekeepers observed the mysterious "mad bee disease" around 1999, when foragers became disoriented and simply could not find their home. Beekeepers suspected a pesticide, Gaucho (imidacloprid), that was used for seed treatment for crops such as sunflowers. At high concentrations, the chemical can disrupt learning and memory of bees, but numerous studies between 2000 and 2003 concluded that there is not enough imidacloprid in the nectar and pollen from the seed-treated plants to have caused behavioral changes or toxic effects in bees.

Other Threats

In addition to pests, diseases, and insecticides, other stresses abound for the honeybees. There are concerns of decreased genetic variation among *A. mellifera* simply because of the well-developed queen rearing system in the United States. Queens are mass-produced from a few "queen-producers" (all closely related to one another), sometimes tens of thousands per year, and distributed around the country. As a result of this practice, if a particular line of bees is vulnerable to certain diseases or pests, they can represent a sizable proportion of the total bees in the country and be wiped out. Beekeepers also transport bees long distances for pollination purposes. For example, bees from Michigan or New York travel by truck all the way to California for almond pollination. Bee colonies are stacked together (as many as 400 of them) onto the same truck, covered with nets, and transported across four time zones. Moving the bees in this manner stresses them. Bees do not even have good tempers when moved two miles away—they dislike vibrations from cars or trucks and being confined. (Imagine being stuck in a black box elbow to elbow with 40,000 siblings!) Other unanswered questions include whether bees get jetlag (yes, foragers do sleep at night) and whether moving them makes them more prone to damage from diseases and pests.

THE CCD CRISIS

As if not enough assaults are directed at honeybees, the single most important pollinator nationally and globally, they are currently suffering from yet another problem. Starting around October 2006, some beekeepers observed the rapid loss of bees in many colonies. One beekeeper, for example, lost 90 percent of his 9,000 colonies. The symptoms were very different from those caused by known pests, diseases, or pesticide kills. Instead of a gradual weakening, in which case the colonies would have very few brood and adult bees left, the new ailment seems to deplete bees very quickly. A colony full of 40,000 workers on one day after four to eight days may have only one queen and a few hundred young bees left, with five to six frames of sealed brood (immature stages of bees, including eggs, larvae, and pupae). No dead adult bees are found inside the hive or in close proximity. Furthermore, opportunistic pests, such as wax moths or small hive beetles, which attack weak colonies or empty hives, do not invade these colonies. Nor are the very weakened colonies "robbed" by nearby honeybees. Robbing is a behavior by honeybees to remove honey from nearby, usually weaker, colonies when resources are scarce. Honeybee scientists termed this ailment CCD, shortened from *colony collapse disorder,* because it was not clear what caused these symptoms and whether it was a disease or simply a symptom of stresses or pesticide poisoning.

While to most people these symptoms are totally new, digging back into old papers one finds a report published in 1987 describing "disappearing disease," and yet another paper describes colonies with only brood left, but it was published in 1897! If what is happening today is the same thing as in 1987 and 1897, then it suggests the causes are not genetically modified organisms (GMOs), or pesticides, or varroa. Unfortunately, no samples were saved during the 1987 episode so researchers might never know whether the "disappearing disease" is the same as CCD.

NONCAUSES

In order to solve the problem of disappearing honeybees, scientists have eliminated several potential causes.

- *Cell phones.* Cell phones can be safely excluded from the list of potential causes of CCD. One German study placed a base set of a wireless phone under a honeybee colony and noticed some behavioral changes. This observation was blown up, leading people to claim that cell phones must cause disorientation in honeybees, preventing them from being able to return to their hives, resulting in CCD. The CCD working group did not find any association between cell phone towers (which have much stronger signals than cell phones) and the cases of CCD.

- *Tracheal mites.* When tracheal mite infestation is serious, colonies can dwindle in the fall, and sometimes no bees are left in the colony come springtime. This seems similar to CCD, but in tracheal mite–infested colonies, there is no brood and sometimes there is a small cluster of dead bees left in the colony (occasionally there are no bees). The disappearance of bees due to tracheal mites happens only in early spring, when mite-infested bees fly out and die in the field.

- *Nosema disease. Nosema*-infected colonies also show much higher winter mortality rates, but usually there is a small cluster of bees inside the hive and the symptoms do not occur in the fall. After learning that a new species of *Nosema* found in Europe and Taiwan was attacking the Western honeybees, scientists thought that *Nosema ceranae* might be the culprit. Usually when a new disease or parasite strikes a new host, it causes massive die-offs (as the varroa and tracheal mite did) because the host has not evolved to have adaptations for resisting the new disease or parasite and therefore the virulence is much higher. However, by using a highly specific method (polymerase chain reaction, PCR), most *Nosema* samples in the United States have been identified as *Nosema ceranae*. Not only that, but examination of *Nosema* samples collected seven or eight years ago revealed that they also were *Nosema ceranae*. Scientists do not know when *N. ceranae* replaced *N. apis* in the United States, but because *N. ceranae* has been here longer than five years, it is unlikely that it suddenly is causing new symptoms like those of CCD.

- *Varroa mites.* Colonies severely infested with varroa mites usually develop parasitic mite syndrome (PMS), characterized by spotty brood and gradual weakening over a longer period, not within four to six days, as CCD does. However, it still is possible a new type of virus is transmitted by varroa mites and is causing CCD.

- *Genetically modified organisms (GMOs).* Scientists found no obvious association between CCD occurrences and regions where a certain type of genetically modified plants are grown. Plants containing genes responsible for the production of the insecticidal protein known as Bt (named for *Bacillus thuringiensis*) are safe for honeybees, mainly because the Bt endotoxins are highly specific against certain groups of insects (e.g., moths and beetles), and no known Bt acts against hymenopterans (ants, bees, and wasps).

- *Food supplements.* Beekeepers feed honeybees honey substitutes such as cane or beet sugar, high-fructose corn syrup (either spring or fall), and pollen substitutes in the early spring. The CCD working group did not find any association of this practice with CCD. However, high-fructose corn syrup can shorten the life of caged bees (Zachary Huang, unpublished results) and may not be the optimal diet for honeybees.

WHAT IS LEFT?

After eliminating all of the possibilities discussed, a limited number of possibilities remain. Currently scientists who study CCD list four possible causes.

- *New or reemerging pathogens.* A paper published in the highly regarded journal *Science* in September 2007 (Cox-Foster et al.) showed that a new virus, IAPV (Israeli acute paralysis virus), was a "marker" for the occurrence of CCD. IAPV was first described in Israel in 2004 and was said to cause bees to have shivering wings, paralysis, and then death outside the hive. The Cox-Foster paper did not conclude that the virus was causing CCD because no healthy colonies were inoculated to show that the virus causes the exact symptoms. This step is needed to identify any cause of disease and is part of the so-called Koch's postulates. The *Science* paper caused a controversy because it implicated the new virus and because the route of the virus was suspected to be the importation of Australian package bees. A senator called for an embargo of Australian bees, and Australian beekeepers were not happy they were blamed. A more recent paper published in the December 2007 issue of *American Bee Journal* provided evidence that the Australian bees were not to blame, and importation was allowed for spring 2008. This paper found that IAPV was in the United States prior to the Australian bee importation. Scientists continue to debate the importance of this virus because the authors of the *Science* paper found that the IAPV that was present in United States was slightly different from those found in Australian bees.

- *New bee pests or parasites. Nosema ceranae* does not seem to be the cause since it has been in the country for at least eight to nine years. Again if the cause of CCD is a new insect pest, which would be relatively large, it should have been observed and identified by now.

- *Environmental and/or nutritional stresses.* Bees are trucked long distances across many time zones, climate is changing through global warming, bees are pushed very hard to pollinate multiple crops, and some of the crops perhaps do not provide adequate nutrition (especially protein) to bees. Beekeepers also feed bees high-fructose corn syrup, which in some instances has high amounts of hydroxymethylfurfuraldehyde (HMF), a chemical that is toxic to bees. The list goes on and on. Could these stresses cause bees simply not to return to their colonies? This is plausible but unlikely.

- *Pesticides.* Beekeepers have been putting all kinds of pesticides, including pyrethroids, organophosphates, organic acids, essential oils, and

(continues)

(continued)

who-knows-what else, into bee colonies. Some of these chemicals will remain stable inside beeswax for a long time and be slowly released into larval food, affecting bees. Studies have shown that insecticides inside beeswax can cause reduction in both the survival rate and the body weight of newly emerged queens and reduced sperm numbers in drones. It is not clear how these chemicals affect worker behavior, especially their learning and memory.

It is most likely that a combination of different factors causes the CCD problem. For example, bees may be stressed from an inadequate pollen supply, fed high-fructose corn syrup, be exposed to many kinds of pesticides reaching a high enough level inside the colonies, and face existing pathogens or parasites (such as *Nosema* or viruses transmitted by the varroa mite), and suddenly workers are affected such that they fly out of the hive and die en masse out in the field.

LESSON LEARNED?

As important as the honeybees are for agriculture, the United States has not really invested much on them. The U.S. Department of Agriculture has four honeybee laboratories (stationed at Beltsville, Maryland; Weslaco, Texas; Baton Rouge, Louisiana; and Tucson, Arizona), with a total funding of about $10 million per year. This is a small number compared to investment in other animals (such as poultry and cattle) and considering that bees are valued at about $15 billion per year. The CCD crisis is a wakeup call for the scientists and the general public that they cannot take the honeybees for granted. CCD generated many reports in the news media, and the general public now has a much better understanding of the importance of honeybees to our food supply. As a result of the CCD crisis, Congress is considering several bills to help fund research on honeybees and other pollinators.

Because the honeybee is an introduced species to North America, it is unlikely that any native plant species depends totally on honeybees for pollination. In other words, honeybees are not technically needed for the "natural" ecosystem in North America. However, people do need honeybees for agricultural purposes, precisely because agriculture is no longer "natural." This author has seen hundreds of other bees (mostly andrenids, some halictids, and megachilids), if not thousands of them, hovering above a single plum tree at the Michigan State University campus. The same thing was seen on flowering cherry trees in front of a botanic garden in South Carolina. Enough diversity and abundance of those solitary bees exist for a few trees in a backyard orchard. The problem is that with hundreds of acres of almonds, apples, cherries, and blueberries, with very few other flowers around (a result of herbicides, tilling, and mowing), there will never be enough of these alternative pollinators. The problem lies in the way of modern agriculture—monoculture on a grand scale. Increasing the diversity of pollinators for crops would be helpful, so that farmers do not rely so heavily on one species of insect that is not even native here. But the truth is, as long as this manner of large-scale monoculturing is maintained, fruit and vegetable growers will continue to rely heavily on honeybees for pollination. The leafcutter bees (*Megachile rodundata*) can do a great job of pollination, but they are specialized for seed production in alfalfa. Farmers have begun to use the blue orchard bees (*Osmia lignaria*) or horn-faced bees (*Osmia cornifrons*) for cherry and apple pollination in some states; however, it will take many years, if it ever happens, to culture them so as to reach the numbers needed for the 600,000 acres of almonds in California alone (valued about $2 billion in 2005). Honeybees are ideal for the modern way of agriculture: large in numbers (30,000 per hive), easily movable (one can transport 400 hives per truck), efficiently managed because

much of their biology is known, and relatively easy population to rebuild, even if 50 percent of colonies are lost.

That is, if only scientists can find out what causes CCD and prevent future bee losses.

FURTHER READING

Almond Board of California home page. Available online. URL: http://www.almondboard.com. Accessed February 13, 2008.

Chen, Yanping, and Jay D. Evans. "Historical Presence of Israeli Acute Paralysis Virus in the United States." *American Bee Journal* (December 2007). Available online. URL: http://www.dadant.com/documents/ChenandEvansarticlefromDec07ABJ.pdf. Accessed February 14, 2008.

Cox-Foster, D. L. et al. "A Metagenomic Survey of Microbes in Honey Bee Colony Collapse Disorder." *Science* 318, no. 5848 (2007): 283–287.

Huang, Zachary. Cyberbee.net home page. Available online. URL: http://www.cyberbee.net. Accessed February 13, 2008.

———. "Zach's Bee Photos." Available online. URL: http://photo.bees.net/gallery/Disease-and-Pests. Accessed February 13, 2008.

Maori, E., E. Tanne, and I. Sela. "Reciprocal Sequence Exchange between Non-Retro Viruses and Hosts Leading to the Appearance of New Host Phenotypes." *Virology* 362, no. 2 (June 5, 2007): 342–349.

Mid-Atlantic Apiculture, Research and Extension Consortium. "Colony Collapse Disorder." Available online. URL: http://www.ento.psu.edu/MAAREC/ColonyCollapseDisorder.html Updated November 20, 2007.

Official Web site of Barbara Boxer, United States Senator from California. "Senators Introduce Legislation to Maintain and Protect American's Bees and Native Pollinators." Press release dated June 26, 2007. Available online. URL: http://boxer.senate.gov/news/releases/record.cfm?id=277777.

Wikipedia. "Nosema ceranae." Available online. URL: http://en.wikipedia.org/wiki/Nosema_ceranae. Accessed February 13, 2008.

(continued from page 14)
by using artificial selection to create new plant varieties that are more resistant to certain pests and diseases, have improved yields, and have higher tolerances to environmental stresses.

Newer methods including biotechnology and genetic engineering have transformed agriculture. Laboratory researchers can genetically modify plants to suit their purposes, a much faster process than artificial selection. Transgenic plants are created by introducing one or more new genes into an organism, a process called transformation. In the past this was accomplished by hybridization, cross-pollinating different plant types. Recombinant deoxyribonucleic acid (DNA) technology now accomplishes the same task and even allows for the introduction to plants of genes from organisms belonging to completely different kingdoms. The new gene can be inserted either by gold particle bombardment or by insertion of the new gene into a plasmid vector and infecting the plant with an engineered *Agrobacterium tumefaciens*

strain that carries the vector. The use of recombinant DNA technology has led to varieties with longer shelf lives, disease resistance, herbicide resistance, pest resistance, drought resistance, resistance to nitrogen starvation, and nutritional improvement.

With all the benefits agricultural advancements have given to society, agricultural technology has also caused some problems. Widespread use of chemicals as fertilizers, pesticides, and herbicides has raised concern for the health of the environment and its natural resources. While hunger still threatens certain populations around the globe, other populations have overgrown, forcing the cultivation of lands that once supported lively diverse biological communities and leading to the deterioration of many ecosystems. Another concern voiced by society is the escape of transgenes into the environment. For example, could herbicide-resistant genes move into the natural weed populations, eliminating their effectiveness for treating crops? The gene that encodes Bt toxin can be inserted into plant

Agricultural researchers perform controlled experiments in laboratory settings, in greenhouses, and in the field. These investigators are examining the effects of herbicides and fertilizers on crop size. *(Maximilian Stock Ltd./ Photo Researchers, Inc.)*

genomes so farmers would not have to spray them, but would the increased exposure lead to the selection for insect populations that are resistant to it?

Governments have developed policies regarding the goals and methods used in agriculture in order to correct some problems and prevent other potential future problems. Many of the policies are concerned with food safety, the production of sufficient quantities for a given population, food quality, conservation, environmental impact of farming practices, and economic stability. President Abraham Lincoln founded the United States Department of Agriculture (USDA) in 1862 to assist America's farmers and ranchers. Today the role of the USDA has expanded to include many other responsibilities such as developing anti-hunger programs, ensuring the safety of the food supply and of drinking water, and maintaining the health of the land through sustainable management. The concept of sustainable agriculture has been accepted and is being integrated into traditional agricultural practices. Sustainable agriculture considers the health of the environment, economic profitability, and economic equity while balancing the current needs with the needs of the future generations.

See also BIOTECHNOLOGY; ENVIRONMENTAL CONCERNS, HUMAN-INDUCED; ENVIRONMENTAL SCIENCE; GENETIC ENGINEERING; PLANT DIVERSITY; PLANT FORM AND FUNCTION; POPULATION ECOLOGY.

algae Algae are a diverse group of aquatic photosynthetic eukaryotic organisms that make up one of two subkingdoms in the traditional kingdom Protista. (The other subkingdom is Protozoa.) Newer classification schemes place some algae in their own kingdoms and others in the plant kingdom. Algae are characteristically phototrophic, meaning they obtain their energy from sunlight, but vary widely in terms of cell structure and organization, size, and habitats. Algal cells contain the structures and organelles typical of eukaryotic cells, including a nucleus, mitochondria, an endomembrane system, and plastids (organelles that carry out photosynthesis). Some types of algae are unicellular, while others are filamentous or live in colonial arrangements. Algae are a main component of plankton, the mass of mostly microscopic organisms that float freely in aquatic environments, but they also live in soil, on rocks, or even in symbiotic relationships with other organisms.

Algae possess pigments that capture light of specific wavelengths, which the cells convert into chemical energy in the form of organic compounds. All algae contain chlorophyll a, the same pigment found in plants, but different types of algae contain different amounts of accessory pigments that absorb light of different wavelengths.

Algae are like photosynthetic plants but contain no true leaves, stems, roots, flowers, or veins and require moist or aquatic habitats. *(Simon Fraser/ Photo Researchers, Inc.)*

ALGAL CLASSIFICATION

The familiar term *algae* refers to all photosynthetic organisms that are not plants or bacteria. Unlike bacteria, algae are eukaryotic, and unlike plants, algae lack roots, leaves, and flowers. The more than 22,000 distantly related algal species have traditionally been categorized into seven main divisions as a matter of convenience rather than biological or evolutionary significance: Chrysophyta, Pyrrophyta, Euglenophyta, Chlorophyta, Rhodophyta, Phaeophyta, and Xanthophyta. Considerations for categorizing the algae include the combination of photosynthetic pigments contained in the plastids, the cellular structure and organization, composition of the cell wall, the presence or absence of flagella, reproductive means, and motility. The field of modern systematics depends on evolutionary relationships to classify organisms, but comparison of the deoxyribonucleic acid (DNA) sequences of the many diverse groups of photosynthetic protists shows that they have deeply divergent ancestries.

More modern classification schemes reflect the belief that eukaryotic algae developed by three different endosymbioses. The first evolutionary line includes the algae with chloroplasts that possess two membranes, Glaucophyta, Rhodophyta, and Chlorophyta. The second line includes algae with chloroplasts that are surrounded by an additional membrane of chloroplast endoplasmic reticulum, Euglenophyta and Dinophyta. The third line includes algae that possess two membranes of chloroplast endoplasmic reticulum, continuous with the outer membrane of the nuclear envelope: Cryptophyta, Bacillariophyta, Chrysophyta, Prymnesiophyta, Xanthophyta, Eustigmatophyta, Rhaphidophyta, and Phaeophyta. A brief overview of some of the most common algal divisions follows.

Members of the division Chrysophyta, commonly known as the golden algae, often contain the photosynthetic pigments chlorophyll c and carotenoids such as fucoxanthin that give them a yellowish brown color in addition to the chlorophyll a that is common to all algae. Most golden algae are unicellular and free-swimming, but colonial and filamentous forms also exist. The cell walls of these mostly marine organisms contain silica compounds and pectin in addition to cellulose. In the absence of light or in the abundance of food, chrysophytes can become facultatively heterotrophic and feed on bacteria and diatoms. Closely related to Chrysophyta are the diatoms, organisms with two protective overlapping shell halves, like a box with a lid, that form, what is called diatomaceous earth over millions of years. The glassy shells contain pores that allow material to flow between the internal and external environments of the cells and give the organisms intricate decorative patterns. Chrysophytes store their food reserves as oil droplets, giving buoyancy to the structures positioning them close to the surface of a body of water so they can capture sunlight to fulfill their energy needs. With sufficient energy and nutrients, diatoms can reproduce asexually daily; however, this leads to progressively smaller progeny. Occasional sexual reproduction allows a full-sized organism to develop.

The division Pyrrophyta (fire algae) consists of the dinoflagellates, unicellular algae that also possess chlorophyll c but that more closely resemble ciliated protists than other algae. As their name suggests, the dinoflagellates have a set of perpendicularly arranged flagella, one at the posterior end and another located within a groove and that spins as the cell swims. As diatoms do, these organisms produce oil droplets, but also store energy as starch. Some dinoflagellates are free-swimming, but some live in symbiotic relationships. For example, many corals have dinoflagellates living inside their tissues. The dinoflagellates undergo photosynthesis, releasing the organic products and oxygen to the corals, which, in turn, excrete waste products from which the dinoflagellates extract nutrients such as phosphates and nitrates in addition to carbon dioxide for their own metabolism. Dinoflagellates are known for their tendency to form algal blooms, a phenomenon that results from nutrient and light conditions that support overgrowth of the organisms. Dinoflagellates live in both marine water and freshwater and mostly reproduce by asexual reproduction.

Members of Euglenophyta, the third strictly unicellular division of algae, are mostly freshwater and contain chlorophylls a and b and carotenoids. This branch of life surprised early biologists, who believed that all life-forms belonged in one of two categories, plants or animals. These organisms have a flagella, a light-sensitive eyespot that helps the organism seek an environment conducive to photosynthesis, and numerous chloroplasts. They store their food as starch. Individual Euglenid cells have been known to survive after losing their chloroplasts. They absorb nutrients from their environments and produce nonphotosynthetic, colorless progeny.

The division Chlorophyta contains the most diverse members and is perhaps the most widely known of the algae. As do plants, these green algae have both chlorophylls a and b and β-carotene as their photosynthetic pigments. Most of the green algae are unicellular, freshwater organisms, but some are multicellular or marine, and others live in moist terrestrial environments. This form of algae is the type that coats the surfaces of ponds. Filamentous green algae form mats that float on top of water. One such organism, *Spirogyra,* undergoes sexual reproduction by conjugation in order to prevent death from dehydration or freezing temperatures. Strands of filaments line alongside one another, form connections, and produce zygotes that can withstand the unfavorable conditions until the next season.

Red algae, belonging to the phylum Rhodophyta, contain phycobilins, photosynthetic pigments that capture green and blue light that penetrates deep into the water, allowing the red algae to survive at greater depths than other algae. Some red algae have cell walls made of calcium carbonate and play a role in building coral reefs. Their life cycles are complex; mature sporophytes undergo meiosis to produce haploid spores that grow into haploid gametophytes, which produce either eggs or sperm. Fertilization results in a diploid sporophyte body, completing the cycle.

Phaeophyta, the brown algae, contain chlorophylls a and c and fucoxanthin and have structures similar to those of plants. Holdfasts anchor the brown algae to seabeds or rocks, stemlike stipes bend with the waves, and blades, also called fronds, resemble leaves and contain the photosynthetic organelles. Gas-filled floats provide buoyancy to keep the blades nearer to the sunlight. Brown and red algae together constitute seaweed. Some brown algae can reach up to 100 feet (30 m) in length.

The Xanthophyta, or yellow-green algae, primarily inhabit freshwater. They lack chlorophyll b and the brown pigment fucoxanthin but do have chlorophyll c. The composition of the cell wall in Xanthophytes is unclear, but the shape often consists of two overlapping cylindrical halves. Many yellow-green algae are unicellular and possess flagella, but colonial and filamentous forms also exist.

ALGAL REPRODUCTION

Algae exhibit a variety of reproductive strategies. Asexually, some employ mitosis, while others produce new organisms by fragmentation of cells from colonies or from multicellular aggregates or by producing spores that develop into mature organisms. Flagellated motile spores are common to algae that live in aquatic environments, and nonmotile spores are characteristic of terrestrial algae. Sexual reproduction involves meiosis, the production of haploid gametes that combine to form diploid zygotes, a process that results in greater genetic variation. Isogamous unions occur when identical gametes combine during fertilization. During heterogamous fertilization, distinct male and female gamete types combine. Some algae undergo a process called alternation of generations, during which haploid generations, gametophytes, alternate with diploid generations, sporophytes.

ECOLOGICAL SIGNIFICANCE

As primary producers, algae fill a vital niche in many ecosystems. Not only do they play an important role in aquatic food webs and serve to produce most of the planet's oxygen, they also provide a significant source of iodine and protein for many human societies. Given certain conditions, the beneficial role played by algae can turn threatening when a sudden increase in the availability of nutrients supports their overgrowth and upsets the delicate ecological balance of marine organisms in the community by disturbing the food webs. As the algae die, other aquatic microorganisms decompose the organic matter, absorbing much of the available oxygen in the process. The increased oxygen demand by the microbial decomposers decreases the availability of oxygen for the other inhabitants, resulting in the death of other community members.

The organism *Gymnodinium breve,* a dinoflagellate, causes outbreaks of a phenomenon known as red tide. During the spring and fall, the waters of the shores of the Pacific, Gulf Coast, and New England states churn, carrying an abundance of nutrients to the surface. These conditions allow *Gymnodinium breve* to thrive, giving the water a reddish appearance. The algae secrete products that are toxic to fish and other marine organisms and that concentrate in shellfish. Though the toxin is harmless to shellfish, it can be harmful to humans who ingest it, causing neuromuscular problems such as numbness.

ECONOMIC IMPORTANCE

Algae are economically important in a number of ways. Many Eastern cultures use red and brown algae as a food source. In Asia, nori, a red alga, is used in sushi and also as a component of soups.

Some algae are eaten directly as a vegetable or added to sweetened jellies. The brown alga commonly known as kelp is a good source of iodine, and other seaweeds provide necessary minerals and vitamins. Rhodophyta such as *Gelidium* and *Gracilaria* are an important source of agar, a substance that microbiologists use to thicken culture media and also to produce the gelatin capsules that contain drugs or vitamins. Extracts from a purple seaweed, *Carrageen,* are used in the food production, cosmetics, and pharmaceutical industries. Kelps also provide algin, a gelling agent used in many foods such as ice cream and in substances such as toothpaste.

See also BIOLOGICAL CLASSIFICATION; EUKARYA; EUKARYOTIC CELLS; PHOTOSYNTHESIS; PLANT DIVERSITY; PLANT FORM AND FUNCTION; PROTOZOA; REPRODUCTION.

FURTHER READING

Jackson, John, ed. *Encyclopedia of the Aquatic World.* Vol. 1. New York: Marshall Cavendish, 2004.

Phycological Society of America home page. Available online. URL: http://www.psaalgae.org/. Accessed January 13, 2008.

Sigee, David C. *Freshwater Microbiology: Biodiversity and Dynamic Interactions of Microorganisms in the Aquatic Environment.* Hoboken, N.J.: John Wiley & Sons, 2005.

Altman, Sidney (1939–) Canadian and American *Molecular Biologist* Sidney Altman received the 1989 Nobel Prize in chemistry, shared with the American molecular biophysicist Thomas Cech, for discovering that ribonucleic acid (RNA) had catalytic abilities. Biologists previously thought the role of RNA was simply to convey the information embedded in deoxyribonucleic acid (DNA) to the protein synthesis machinery. The revolutionary finding that RNA actively performs biochemical catalysis not only forced life scientists to redefine enzymes, but also led to a greater understanding of the evolution of the current roles of DNA, RNA, and proteins in living cells.

EDUCATION

Sidney Altman was born on May 7, 1939, in Montreal, Quebec, Canada, to parents who encouraged reading and supported education as a means of creating opportunity. The atomic bomb sparked an interest in science in six-year-old Sidney, and as a teenager he became enthralled by the predictive power of science after reading *Explaining the Atom* by Selig Hecht. At the age of 17, he moved to Cambridge to study science at the Massachusetts Institute of Technology. He completed his bachelor of science degree in physics in 1960, having written his senior thesis on nuclear physics. During his last semester he enrolled in an introductory course in molecular biology, not knowing that his own future contributions would help to shape this emerging field.

After graduation Altman began a graduate program in physics at Columbia University, but he left after one and one-half years and later ended up in a biophysics program at the University of Colorado Medical Center in Denver. He worked in the laboratory of Leonard Lerman, where he studied the effects of derivatives of the chemical mutagen acridine on DNA replication of bacteriophage T4. After completing his doctoral degree in 1967, Altman performed postdoctoral research in the laboratory of Matthew Meselson at Harvard University, studying the role of an endonuclease (an enzyme that cleaves within a segment of nucleic acid) in the replication and recombination of T4 DNA. He moved to Cambridge, England, in October 1969 to join the team of Sydney Brenner and Francis Crick at the Medical Research Council (MRC), a venture he later described as "scientific heaven." Altman thought his research project at MRC would involve the structure determination of transfer RNA (tRNA) using biophysical methods. During protein synthesis, tRNA molecules carry amino acids to the ribosome, where they become incorporated in the nascent polypeptide. Each tRNA has an anticodon, a sequence of three nucleotides that recognizes and specifically binds to a triplet codon on the messenger RNA (mRNA), ensuring that the correct amino acid is added to the growing chain.

RESEARCH LEADING TO NOBEL PRIZE

By the time Altman arrived at Cambridge, another group had solved the basic structure of a yeast transfer RNA (tRNA), one that was available in large enough quantities for crystallographic studies, and Altman had to pick another research topic. In order to gain supporting evidence for the proposed tRNA structure, Altman decided to take a genetic approach to study the structure and function of bacterial tRNA^{Tyr}, the tRNA that carries the amino acid tyrosine. By examining acridine-induced tRNA mutants and suppressors of those mutants, he started working down a pathway that would eventually lead to his receiving the Nobel Prize. Mutation studies provide useful information for structure-function studies. When a mutation causes the loss of a particular function, studies of the associated suppressor mutations provide clues to the mechanism by which the gene product functions. The tRNA molecules in the original mutants did not function normally during translation. Suppressor mutations are mutations that fully or partially restore the function lost as a

Sidney Altman shared the 1989 Nobel Prize in chemistry for his work on catalytic RNA. *(Michael Marsland/Yale University)*

result of the first mutation. The information gained can support or guide complementary structural studies. Altman hoped to learn whether or not altering the spatial relationships in the tRNA by adding or deleting nucleotides (a change that acridine induces) would alter the function of the molecule.

Molecular biologists knew that the synthesis of tRNA involved a processing step during which part of the 5' and 3' ends were removed, resulting in the release of a shorter mature tRNA molecule. One group of suppressor mutants that Altman created by treatment with the mutagenic chemical acridine never generated mature tRNA^{Tyr}; only the long precursor strands were present. Properties of the mutants suggested to Altman that the tRNA gene might be duplicated, since the mutants he used did not contain the usual suppressor tRNAs. He also thought that the bacteria transcribed the mutant gene and that isolating the mutant transcript would help elucidate the pathway for tRNA biosynthesis. In 1970 Altman successfully isolated and purified a homogeneous preparation of the transcript for the precursor tRNA^{Tyr} from mutant cells that had been grown in the presence of radioactive phosphate. Subsequent treatment of the precursor tRNA molecules with bacterial cell extracts supplied by coworker Hugh Robertson resulted in shorter radiolabeled tRNA molecules, suggesting an enzymatic activity that was present in the extracts removed a portion of the precursor molecule. They named the activity that removed a portion of the 5' end of the precursor tRNA ribonuclease P (RNase P). Ribonucleases are enzymes that cleave RNA, and the *P* stood for "precursor." This accomplishment helped Altman obtain a faculty position the following year at Yale University in New Haven, Connecticut, in the Department of Molecular, Cellular, and Developmental Biology, where he continued his research.

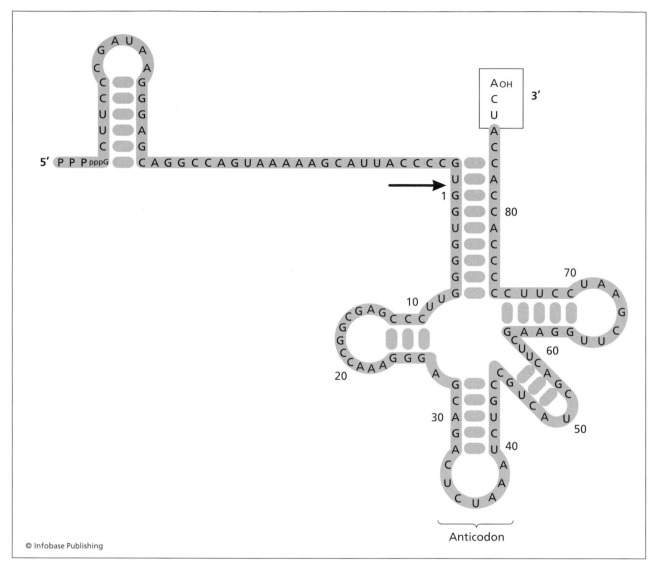

During tRNA biosynthesis, the enzyme ribonuclease P removes a portion of the 5' end by cleaving at the position indicated by the red arrow.

The RNase P isolated by Altman and Robertson would eventually lead Altman into his Nobel Prize–winning research. Characterization revealed that the enzymatic activity was specific, in contrast to other known ribonuclease activities that nonspecifically cleaved nucleotides one at a time from the end of an RNA molecule. Robertson, Altman, and another colleague from MRC named John Smith published their initial characterization of RNase P in 1972, having noted that even their purest fractions contained nucleic acid. Because all known enzymes were proteins, the purification procedure had been optimized for isolating a protein. When RNA appeared in the purified enzyme preparations, everyone assumed the RNA was a contaminant. A few years later, a graduate student in Altman's lab, Benjamin Stark, showed that a high-molecular-weight RNA molecule copurified with the enzyme, and that this RNA was essential for the enzymatic activity. When nucleases digested all the RNA, the enzyme no longer functioned, suggesting that RNA played a critical role in the catalysis. This RNA was named M1 RNA. Others showed that RNase P also participated in the metabolic pathways for the synthesis of other tRNA molecules in *Escherichia coli,* and that RNase P activity was present in other organisms including humans. In 1983 Altman's laboratory published research demonstrating that purified bacterial M1 RNA had catalytic activity under certain buffer conditions and that it exhibited other characteristics of typical enzymes, such as being unchanged during a reaction, and that it had a rate dependent on substrate concentration, was only necessary in small concentrations, and was stable. This was the first demonstration of biological catalysis performed by RNA. Not all enzymologists accepted this finding at first. In eukaryotic cells the RNA component of RNase P is not sufficient for catalysis; protein components are necessary for the enzymatic activity.

At about the same time, Thomas Cech, a molecular biophysicist at the University of Colorado in Boulder, discovered that a specific RNA molecule isolated from *Tetrahymena,* a ciliated protozoan, catalyzed its own cleavage in vitro in the absence of proteins. The RNA removed a segment of itself, then rejoined the remaining pieces. In 1989, the Karolinska Institute in Stockholm, Sweden, awarded Altman and Cech the Nobel Prize in chemistry for their work revealing the catalytic properties of RNA.

EXTERNAL GUIDE SEQUENCES

After performing extensive analysis of the mechanism by which M1 RNA catalyzed cleavage of the 5' end of precursor tRNA transcripts, Altman turned his expertise on RNA function beyond its serving as a passive messenger of genetic information to the problem of bacterial resistance to antibiotics. Physicians are having a difficult time treating many infections because the bacteria responsible for them have developed a resistance to antibiotics that once successfully destroyed the pathogenic bacteria. In some cases, the cause of the bacterial resistance to the antibiotics is the expression of a specific protein that has the ability to inactivate the antibiotic. In order to synthesize that protein, the bacteria must first transcribe the gene from the DNA to produce an mRNA transcript, which then is translated into a polypeptide. Altman and his colleagues made plasmids containing synthetic genes that encoded for small pieces of RNA that they called external guide sequences (EGSs) that were specifically designed to recognize the 5' untranslated region of the mRNAs from the antibiotic resistance genes. They inserted these plasmids into *E. coli* strains that exhibited resistance to antibiotics. These EGSs bound the target mRNA, forming a structure that the bacterial RNase P then hydrolyzed; thus the protein was not synthesized. Though the bacteria were originally resistant to antibiotic treatment, after transformation with the plasmids containing the EGS, they exhibited antibiotic sensitivity. This research was published in the prestigious journal *Proceedings of the National Academy of Sciences* in 1997.

The following year Altman's lab published similar results describing the use of EGSs directed against viral RNA. Viral infections result when a virus enters a host cell and directs that cell to synthesize viral proteins in order to replicate itself. In order to synthesize the viral proteins, the cell must first transcribe the viral genes to produce mRNA transcripts, which are then translated into polypeptides. The insertion

into cultured mouse cells of synthetic genes for EGSs specific for the polymerase and nucleocapsid genes of influenza virus led to cleavage of the mRNAs for those genes by endogenous RNase P and efficiently inhibited viral replication. The goal of this research, which is also being carried out in cultured human cells, is to apply this strategy to treat viral illnesses, such as influenza, the common cold, or other diseases caused by a genetic mutation or the inappropriate expression of a protein. Interest in RNase P has also risen because of the presence of antibodies against two of its subunits in humans with autoimmune diseases, in which the immune system begins to attack the patient's own tissues.

ALTMAN'S INFLUENCE

Altman's pioneering work on the unique properties of RNA molecules opened the door for others to make discoveries just as amazing as RNA catalysis. Molecular biologists are learning that posttranscriptional processing is an important means of regulating the expression of many genes. The American molecular biologists Andrew Fire and Craig Mello received the 2006 Nobel Prize in physiology, or medicine, for their discovery of RNA interference, a means by which double-stranded RNA silences genes in nematodes, plants, and animals. RNA elements, called riboswitches, in the 5' untranslated region of mRNA fold into structures that interact with metabolites to interfere with the completion of transcription or with translation. One such riboswitch mechanism involves self-cleavage of the mRNA in response to the metabolite binding.

Since joining the faculty at Yale in 1971, Altman served as chairperson of the Department of Molecular, Cellular, and Developmental Biology from 1983 to 1985, and then as dean of Yale College from 1985 to 1989. Today Altman continues his research at Yale, where he holds the chair of Sterling Professor of Biology, as well as being a professor of chemistry. The broad goal of research in his laboratory is gene regulation by posttranscriptional RNA processing. In addition to the Nobel Prize, Altman received the Rosenstiel Award for Basic Biomedical Research and the National Institutes of Health Merit Award in 1989 and the Yale Science and Engineering Association Award in 1990.

Altman married Ann Köner in 1972, and they have two grown children, a son named Daniel and a daughter named Leah.

See also BIOMOLECULES; CECH, THOMAS; GENE EXPRESSION; RNA INTERFERENCE.

FURTHER READING

Altman, Sidney. "The Road to RNase P." *Nature Structural Biology* 7, no. 10 (October 2000): 827–828.

Altman, Sidney, and Leif Kirsebom. "Ribonuclase P." In *The RNA World.* 2nd ed., edited by Raymond F. Gesteland, Thomas R. Cech, and John F. Atkins, 351–380. Cold Spring Harbor, N.Y.: Cold Spring Harbor Laboratory Press, 1999.

Guerrier-Takada, Cecilia, Kathleen Gardiner, Terry Marsh, Norman Pace, and Sidney Altman. "The RNA Moiety of Ribonuclease P is the Catalytic Subunit of the Enzyme." *Cell* 35, no. 3 (December 1983): 849–857.

Guerrier-Takada, Cecilia, Reza Salavati, and Sidney Altman. "Phenotypic Conversion of Drug-Resistant Bacteria to Drug Sensitivity." *Proceedings of the National Academy of Sciences USA* 94 (August 1997): 8,468–8,472.

Jarrous, Nayef, Paul S. Eder, Cecilia Guerrier-Takada, Christer Hoog, and Sidney Altman. "Autoantigenic Properties of Some Protein Subunits of Catalytically Active Complexes of Human Ribonuclease P." *RNA* 4, no. 4 (1998): 407–417.

Kresge, Nicole, Robert D. Simoni, and Robert L. Hill. "Ribonuclease P and the Discovery of Catalytic RNA: The Work of Sidney Altman." *Journal of Biological Chemistry* 282, no. 7 (February 16, 2007): e5–e7.

The Nobel Foundation. "The Nobel Prize in Chemistry 1989." Available online. URL: http://nobelprize.org/nobel_prizes/chemistry/laureates/1989/index.html. Accessed January 22, 2008.

Plehn-Dujowich, Debora, and Sidney Altman. "Effective Inhibition of Influenza Virus Production in Cultured Cells by External Guide Sequences and RNase P." *Proceedings of the National Academy of Sciences USA* 95 (June 1998): 7,327–7,332.

Robertson, Hugh D., Sidney Altman, and John D. Smith. "Purification and Properties of a Specific *Escherichia coli* Ribonuclease which Cleaves a Tyrosine Transfer Ribonucleic Acid Precursor." *Journal of Biological Chemistry* 247, no. 16 (1972): 5,243–5,251.

Stark, Benjamin C., Ryszard Kole, Emma J. Bowman, and Sidney Altman. "Ribonuclease P: An Enzyme with an Essential RNA Component." *Proceedings of the National Academy of Sciences USA* 75 (August 1, 1978): 3,717–3,721.

anatomy   Anatomy is the branch of life science concerned with the structure of organisms. Because structure is closely associated with function at all levels of hierarchy in the life sciences, knowledge of anatomy allows one to understand better how organisms carry out life's necessary functions such as respiration, metabolism, and reproduction. Physiology, the branch of life science that deals with the functions of organisms or life processes, complements anatomy, and the two fields are often taught in concert.

Anatomy can focus on the morphology of different types of organisms, such as plant anatomy, vertebrate anatomy, or more specifically human anatomy. Health care professionals in fields such as medicine, dentistry, and physical and occupational therapy, and those who work in veterinary medicine, must study human or animal anatomy in order to understand disease and to help their patients remain healthy. Anatomic anomalies are deviations from normal patterns for a species. For example, while the normal pattern for a human includes 10 fingers and 10 toes, approximately 0.2 percent of children are born with more, a condition called polydactyly. Individuals with an extra digit may experience different degrees of affected mobility, but the condition is not dangerous. Other anatomic anomalies can prevent normal function and cause harm to a patient or may indicate a serious medical condition. For example, Marfan syndrome results from a mutant gene for the protein fibrillin, a component necessary for the formation of elastic fibers in connective tissue. One dangerous complication caused by the improperly formed connective tissue is a leaky mitral or aortic valve in the heart, which can lead to an irregular pulse, shortness of breath, fatigue, and a potentially fatal aortic aneurism. Individuals with Marfan syndrome have atypically long, slender limbs and fingers, a symptom that in itself is not harmful in any way. But recognition of this characteristic anatomical feature could aid a physician in making a timely diagnosis of this disorder.

Even when studies are limited to a particular organism, such as humans, the field of anatomy can be divided into specialties focusing on a specific region or system or on different levels of organization. Gross anatomy is the study of anatomy on the macroscopic level, the organs and aspects of the tissues that can be examined without the aid of a microscope. Conversely, cytology, the study of cellular architecture, and histology, the study of tissues, depend heavily on microscopy. A developmental anatomist studies the successive changes in body structure as an organism progresses from a fertilized egg to an embryo (a subspecialty called embryology) and then into a fetus and even into adulthood. In regional anatomy, one focuses on a limited region such as the head or the abdomen. Systemic anatomy includes study of physiological systems such as the integumentary, skeletal, muscular, nervous, endocrine, cardiovascular, lymphatic, respiratory, digestive, excretory, and reproductive systems.

Comparative anatomy involves the study and comparison of the structures of animal species and often examines the relationships among different anatomical strategies animals employ for carrying out life processes and the difference in their environments in which they live or the niches they fill. For example, consider that both aquatic and terrestrial animals require oxygen to undergo cellular respiration, but the means by which they obtain oxygen gas from their external environment and transport it to their body tissues differ greatly. Different morphologies have evolved, depending on the needs of that particular species. The structure of a cell, tissue, organ, or organ system determines the function that structure can accomplish, but over long periods, different environments select for slightly modified structures that are better suited for those conditions. Thus structure determines the immediate potential function, but in the long term, the suitability of different functional adaptations will determine which structures prevail.

One means to study an organism's anatomy is by dissection, the systematic process of separating, taking apart, or exposing the parts for scientific examination. Teaching laboratories in high schools, universities, and medical colleges rely on dissection as a means to teach anatomy. Though preserved specimens are still a far cry from living organisms, dissections provide more realistic anatomical information than learning from observing two-dimensional sketches in a textbook or even three-dimensional images on a computer.

Anatomic imaging allows physicians to examine internal structures without surgery. X-rays are a form of electromagnetic radiation with short wave lengths. When an X-ray source is aimed at a region of the body, bones and other dense material in the path absorb some of the radiation, but it passes through softer tissues and reaches a piece of X-ray film positioned behind the body part of interest. The X-rays expose the film, and after developing the film, bones will appear white, and the exposed areas of the film will appear darker, creating a two-dimensional image for a radiologist to examine. Ultrasound imaging produces sonograms using high-frequency radio waves emitted from a device held next to the skin. The handheld device both transmits and receives the sound waves, which bounce off internal body structures and are analyzed by a computer. Ultrasound is frequently used to examine muscles, tendons, and other internal structures as well as to visualize a fetus during pregnancy. Computed tomography (CT), formerly known as computed axial tomography (CAT), generates a three-dimensional image by processing a series of subsequent or stacked two-dimensional X-ray images taken at repeated interval angles and reconstructing a digital image of the internal structures. Another method for obtaining internal structural information, magnetic resonance imaging (MRI), is better for examining softer tissues than CT scans, as it uses radio frequencies rather than ionizing radiation to collect data. The patient is placed in a large magnetic field and subject to radio waves,

causing the orientation of the numerous hydrogen protons (mostly from water molecules) to align, a process that occurs at different rates for different types of tissues. A computer analyzes this information to generate images of cross sections of the body. All of these technologies allow medical personnel to examine the inside of a patient's body to look for structural abnormalities or even tumors that may be associated with particular conditions or diagnoses.

See also ANIMAL FORM; DISSECTION; EMBRYOLOGY AND EARLY ANIMAL DEVELOPMENT; PHYSIOLOGY; PLANT FORM AND FUNCTION.

FURTHER READING

Bailey, Jill. *Animal Life: Form and Function in the Animal Kingdom.* New York: Oxford University Press, 1994.

Melloni, June L. *Melloni's Illustrated Review of Human Anatomy.* 3rd ed. Cambridge: Cambridge University Press, 2006.

Seeley, Rod R., Trent D. Stephens, and Philip Tate. *Anatomy and Physiology.* 7th ed. New York: McGraw Hill, 2006.

animal behavior Animal behaviors are the actions performed by an individual or species in response to its environment, including almost everything an animal does. Many behaviors involve a muscular activity, such as running away from a predator or building a nest. Other actions that are not as easily observed, such as the secretion of hormones to attract a mate or associating a certain color with danger, are also considered behaviors. Some behaviors are innate, or genetically determined, while others are learned. Ethology, the study of animal behavior, aims to understand the evolution, development, and control of the many behaviors exhibited by animals.

Understanding the mechanism by which an animal responds to a stimulus, in addition to the reason for it, helps form a more complete picture of the behavior. Ultimate questions address why an animal behaves in a specific way and examine the relationship of the behavior to fitness, or the reproductive success of an organism. Proximate questions address the physical adaptation that allows for the response, such as the responsible genes, biochemical pathways, anatomical structures, or physiological mechanisms, and the influence of development and maturation of the animal on the behavior. The answers to both ultimate and proximate questions make sense in light of the evolutionary process of natural selection. To summarize, animals exhibit different characteristics, including behaviors, determined by their genetic makeup. Individuals pass on some of their characteristics to their offspring via genes. Characteristics that improve one's success at transmitting his or her genes to the next generation relative to other members of the species will become more common in the population over time. Natural selection favors behaviors that increase an organism's chance of surviving to adulthood, finding and selecting a fit mate, producing offspring, and raising those offspring to adulthood.

The environment, both the physical environment and other members of the ecological community, profoundly influences the traits that persist in a population. Resources, such as nutrients, or hormones affect the development of a behavioral trait, often by altering the expression of certain genes at key times during development. For example, the level of the hormone testosterone present during embryogenesis of male mice has been shown to affect aggression levels exhibited later in life. Sensory experiences, such as olfaction, can also affect the development of a behavioral trait. Exposure to a certain nest smell soon after the emergence of a paper wasp from a brood cell allows one to recognize individuals that shared that nest. Later in life, the wasp treats differently individuals that do and do not exhibit that specific scent, acting more aggressively toward individuals from a different nest.

The execution of any behavior involves sophisticated neural and hormonal controls. The animal must initially obtain information from the environment, such as hearing a song or seeing a color, a task often involving sensory reception. After the brain processes the input, the animal then responds, perhaps by pursuing a mate or fleeing a possible predator. The control of such behavioral responses often involves the expression of specific gene products during a learning experience or the activity of certain neural pathways that generate a motor response.

FIXED ACTION PATTERNS AND IMPRINTING

Fixed action patterns (FAPs) are innate, stereotypical responses triggered by a well-defined simple stimulus, called a sign stimulus or a releaser. Once initiated, the pattern continues until reaching completion. Web building in spiders and egg retrieval by geese are examples of fixed action patterns. An example of a FAP in humans is yawning; seeing another person yawning triggers yawning in the observer. As instincts, these behavior patterns require no previous experience with the stimulus cues and are functional from the first time they are performed.

Imprinting, on the other hand, involves learning, usually early in life during a sensitive period. Exposure to a certain stimulus influences the behavioral development of an individual. While the tendency to respond is innate, the stimulus is environmental. In a classical example, baby geese formed an association with the pioneering ethologist Konrad Lorenz, whom they observed moving away and calling for them soon after hatching. This experience somehow affects the development of the geese's nervous system. Later in life, the male geese developed a preference for humans, like Lorenz, as sexual mates. One can surmise the ultimate cause to be that young who imprint on and follow their mother will receive parental care and learn species-specific survival behaviors and skills. Thus imprinting leads to a greater chance of survival and reproductive fitness.

MOVEMENTS

Directed movements include genetically controlled simple movements that cover very short distances, perhaps a fraction of an inch, to those covering thousands of miles. Kinesis is a simple movement that lacks direction. An organism that typically moves in a random direction, such as by turning, will continue turning until it reaches favorable conditions. Adverse conditions can increase the movement, which increases the chance of finding a more favorable environment. Kinesis differs from taxis such as chemotaxis (movement toward a chemical or nutrient) or phototaxis (movement directed toward or away from light), because in kinesis, the sensors do not distinguish the direction from which the stimulus originates. Sow bugs, a type of terrestrial crustacean, demonstrate kinesis by responding to dry conditions with increased random movement. When the sow bugs encounter moist conditions, their activity decreases, increasing their chance of staying in the moist environment, which their bodies prefer. Taxis, on the other hand, is directed movement. For example, maggots have light-sensitive eyespots on both sides of their head. When light intensity is greater on one side than the other, the maggot moves away from the source of illumination until both sides are equally stimulated.

Taxis and kinesis are behaviors exhibited by animals over a short distance. Migration, while still genetically determined, is more complex and involves the seasonal movement from one biome to another, covering distances of several hundred to thousands of miles. Researchers have shown that genetic differences help explain the migratory behavior in birds. The behavior improves the animal's reproductive success, since the migratory patterns are often associated with breeding and nesting in addition to finding food and provide optimal environmental conditions for the different stages of the animal's life cycles. Examples of migratory animals include birds, termites, fish, crabs, wildebeest, turtles, monarch butterflies, and dolphins.

COMMUNICATION

The development and exhibition of many animal behaviors involve communication, the transmission

Millions of western sandpipers stop over at Grays Harbor, Washington, during their spring migration from the lagoons and coastal estuaries from California to Peru to the tundra of western Alaska for nesting. *(F. Stuart Westmorland/Photo Researchers, Inc.)*

of information. Animals usually accomplish this via visual, auditory, chemical, electrical, or tactile signals. Successful communication requires sending, receiving, interpreting, and responding to a signal. Honeybees put on an elaborate visual display to communicate information to nest mates about the direction, distance, and odor of newly discovered floral patches. The bees that witness the dance receive the visual sensory input and translate it to locate the food source. Mating songs produced by male fruit flies vibrating their wings are an example of an auditory signal. The female fruit flies recognize the structure of the songs from members of their own species by distinguishing such details as the length of time between the pulses of the wing vibrations. Chemical signals often include pheromones, odor-emitting stimuli. Domestic cats use pheromones contained in their urine to mark their territory, sometimes causing a problem for the pet owners, but animals in the wild often leave behind pheromones to mark the boundaries of their territories as well. When other animals detect the smell, they know the territory is already claimed. Structures called ampullae of Lorenzini found on marine sharks and rays aid in navigation and can detect weak electrical fields generated by other creatures moving nearby. Tactile communication includes behaviors that involve touching, such as grooming in primates, a behavior that promotes bonding.

FEEDING/FORAGING BEHAVIOR

One would expect natural selection to favor successful foraging behaviors, those that aid in finding and obtaining or capturing food. Much of an animal's anatomy and physiology depends on how that species procures its nutrition. Any movement performed by the animal to obtain the food is considered feeding behavior. A variety of mechanisms for drawing in small particles for food include using cilia to sweep food particles toward the body opening, as in sponges; secreting mucus to trap small particles, as in some snails; and using tentacles to paralyze and pull in prey, as in jellyfish. Animals can obtain larger particles by a variety of methods. For example, earthworms burrow, snakes catch and swallow small organisms whole, mammals and crustaceans catch and chew, spiders and sea starts catch and begin digestion externally and then ingest much smaller particles, and snails use their radula to scrape algae off rocks and pull them into their mouth. Fluid

Primates, such as these bonobos, groom each other as a form of bonding. *(Connie Bransilver/Photo Researchers, Inc.)*

feeders exhibit unique behaviors, such as mosquitoes piercing and sucking or just sucking as butterflies do. Feeding behaviors can also be classified on the basis of the degree of discrimination of food sources. Filter feeders take in everything of the appropriate size, whereas selective feeders use sensory input to make food choices based on available selections.

No matter what mechanisms an animal employs to obtain food, it costs energy to seek and obtain nutrition. Moving greater distances to find prey or to capture prey, having to work harder to break open a shell, or breaking down a larger mass of food to extract the energy from it all will influence what food sources an animal will pursue. The animal has no choice whether or not to expend energy but can make decisions regarding the amount of energy spent to obtain certain types of food. Optimal foraging theory purports that organisms will consume the most energy while expending the least amount of energy, predicting that evolution will select for foraging behaviors that optimize this ratio. One factor that affects foraging behavior is the need to minimize the risk of predation. This risk may affect behavioral decisions such as whether an animal forages during the day or night, under shelter or out in the open; how far from safety an animal will venture; how often an animal eats; and whether foraging is performed individually or in a group. Other behaviors related to feeding behaviors include social behaviors such as feeding as part of a courtship ritual, communication about food sources, cooperating to cultivate or obtain a food source, or feeding as part of parental caring for young.

COURTSHIP AND MATING

Many behaviors are involved in attracting potential mates and successfully mating with them. Songs, complicated dances, postures, athletic displays, and other similar rituals are examples of courtship behaviors intended to convince a potential partner of the individual's worthiness for mating. Male bowerbirds build elaborate structures called bowers by clearing an area of ground; padding it with moss, leaves, and the like; or building a teepeelike bower from twigs and sticks. The bird collects colorful trinkets such as flowers, shells, and berries to adorn the bower and then waits for a female to approach. When she does, he dances in front of his bower and shows off his decorations. If she is impressed, they will mate inside the bower. Afterward, he cleans up by returning all his objects to their designated places and waits for another female to approach, mating with as many as he can attract. Feeding behaviors are often incorporated into a courtship ritual; for example, male European nursery spiders capture prey, wrap it in silk, hold it in their teeth for presentation to a female. If the food gift is acceptable, she will bite into it, and while she eats it, the male will mate with her. Domestic cocks call and peck at the ground, even when food is not present, in order to attract a female. Aggressive (or agonistic) displays warn others off, such as when a strange male approaches the territory of another. Just as animals often fight to gain access to a particular territory or food resource, males also often fight to compete for access to a female partner.

Different species employ different mating strategies. Whether a species is mostly monogamous, meaning an individual only mates with one other individual, or polygamous, meaning an individual mates with many partners, influences the types of behaviors that attract mates. Whether or not a male is sure that his sperm fertilized the female's eggs, something that is more difficult to know in the case of internal fertilization than external fertilization, also influences mating behavior. To ensure his own genes are passed on to the offspring, a male might guard or protect his female mating partner from other males or even physically remove preexisting sperm from a female's reproductive tract before mating. This also affects the degree to which a male participates in the parental care of offspring. Males will participate in parental care more often when they are certain the offspring are their own.

Sexual selection is a special type of natural selection that works on characteristics that improve an organism's ability to compete for mates against other members of the same species. Sexual selection results in dimorphic species, in which the males and females differ in their appearance beyond the necessary differences in their primary sexual reproductive organs. For example, males birds are often more colorful than females of the same species. Understanding why natural selection would favor a characteristic that would make an organism more visible to predators is difficult. Why would a female prefer such a mate? Some ethologists have suggested that colorful plumage might be an indicator of good health, something a female would desire in the father of her offspring in order for her offspring also to be healthy and to continue passing on her genes. By selecting such a mate, whatever the reason, the genes that influenced the female's decision to mate with males displaying that characteristic, as well as the genes for that characteristic itself, are passed on to the next generation.

PARENTAL CARE

Selection favors adaptations that improve the transmission of an individual's genes to future generations. An individual not only needs to reproduce, but needs the offspring to survive and also reproduce. In some species, both parents participate in caring for the offspring; in others only the mother or father does. The benefit to an individual in ensuring the survival of offspring is increased fitness, as the offspring pass on their genes to the grandchildren, and on to future generations, but this care has a great cost in terms of time and energy. Evolution has created a variety of different strategies for parental care. Most involve the mother more than the father. One reason for this is that fathers do not always know for sure that the offspring resulted from eggs fertilized by their sperm or sperm from other males. Also, males generally increase their fitness by reproducing with several mates, and the time spent caring for offspring is time they cannot spend seeking and obtaining additional mates.

Exceptions do exist, however, in which males of a species invest more in parental care than the females. Male bluegills build nests to attract females, then after mating, they stay there for a week to protect the fertilized eggs, occasionally fanning them to ensure they receive enough oxygen. While female emperor penguins go in search of food, the fathers guard their eggs between their feet under a fold of loose skin that blankets the egg from the Antarctic cold. The males all huddle together to help stay warm and forgo feeding for two months, guarding their eggs until their mates return.

Other examples of parental care include mother scorpions, who, after giving birth to their live young, carry their offspring around on their backs until they molt for the first time, up to two weeks after birth. Many frogs carry their broods in pouches on their backs or on the sides of their bodies. Male Darwin's frogs stay near the eggs they fertilize for three weeks. When the embryos begin to rotate in the translucent eggs, the male encloses them in his mouth, and they slide down into his vocal sac. The eggs soon hatch while inside the vocal sac, and the tadpoles continue to develop while the father provides nourishment and a safe, comfortable environment. After the young metamorphosize into frogs, they crawl into the father's mouth, causing him to gag, and the young frogs hop out. Australian social spider mothers perform the ultimate sacrifice for their young. After their offspring hatch, they give them food in the form of large insects, often weighing up to 10 times the mother's weight. When winter arrives and food becomes scarce, the mother allows her spiderlings to feed on her nutrient-rich blood, and then they inject her with venom and cannibalize her. While some examples are more extreme than others, parental care involves considerable investment, but one that yields a worthwhile evolutionary return, increased fitness.

See also ANIMAL COGNITION AND LEARNING; ECOLOGY; ETHOLOGY; EVOLUTION, THEORY OF; SOCIAL BEHAVIOR OF ANIMALS.

FURTHER READING
Alcock, John. *Animal Behavior: An Evolutionary Approach.* 8th ed. Sunderland, Mass.: Sinauer, 2005.
Crump, Marty. *Headless Males Make Great Lovers.* Chicago: University of Chicago Press, 2005.
Dugatkin, Lee Alan. *Principles of Animal Behavior.* New York: W. W. Norton, 2003.
Forsyth, Adrian. *A Natural History of Sex: The Ecology and Evolution of Mating Behavior.* Ontario, Canada: Firefly Books, 2001.
McFarland, David. *Animal Behavior: Psychobiology, Ethology, and Evolution.* 3rd ed. Harlow, England: Pearson, 1999.

animal cognition and learning Broadly defined, cognition is the ability of an animal's nervous system to perceive, store, process, and use information gathered through the senses. An animal's behavior depends on its cognitive experiences. Animal learning is the acquisition of a behavioral tendency based on specific experiences. When discussed in the context of human behavior, cognition often refers to a mental or intellectual awareness, and learning implies the gain of knowledge or a skill, but as a subfield of zoology, animal cognition and learning focus on actions and behaviors that can be experimentally examined and explained in terms of adaptive value within the context of an evolutionary framework. Most learned behaviors in animals are adaptations that help the animal find food, escape predators, or reproduce.

Habituation is a simple type of learning in which animals become accustomed to repeated exposure to a stimulus that does not provide any useful information. They become used to the stimulus and stop responding to it. This can occur at different levels of nervous system function. The sensory organs may stop sending stimulatory signals to the brain, or the animal may continue to perceive the stimuli, but the brain stops initiating a physiological response. Ethologists, scientists who study animal behavior, depend on habituation in order to observe animals in their natural environments. At first the animals may act cautious or afraid of the researcher, but after continued exposure, the animals begin to ignore the presence of the investigator. The ultimate causation for this behavior, or the evolutionary explanation habituation exists, may be to preserve the animal's energy for real dangers.

Spatial learning refers an animal's ability to recognize and distinguish differences in the spatial structure or details of arrangement within a particular environment. Nikolaas Tinbergen, one of the founders of the science of animal behavior, performed a well-known experiment examining the ability of a digger wasp to recognize her nest. To test his hypothesis that the female wasp used visual landmarks to locate her underground nests, he placed a ring of pinecones around a hole in the ground that served as an entrance to a nest. After the mother flew away, he moved the pinecones and set them up in the same ringed orientation but to the side of the nest. When the female wasp returned, she flew to the center of the ring of shifted pinecones, demonstrating that she used visual cues to remember the location of her nests. In this illustration, the adaptive value of spatial learning is to increase reproductive success. Cognitive mapping is another strategy animals use to navigate their environments. More complicated than spatial learning, cognitive mapping involves the creation of an internal representation, such as a mental image, of spatial relationships among objects in an environment. For example, some birds collect and store food in hundreds of different hiding places. Rather than remember visual cues about every single cache, the bird may follow a general rule, such as that the food is always stored at the base of a tree or under a rock. Though spatial learning and cognitive mapping are difficult to distinguish experimentally, the advantage of having the ability to form a cognitive map would be reducing the amount of detail to memorize while still being able to locate food stores more efficiently than by having to search for them randomly. This would be similar to recognizing a pattern to an arithmetic sequence and applying a formula to determine a number at a given position in the series rather than memorizing the entire sequence of numbers.

Associative learning occurs when an animal begins to recognize a connection between two events, such as a behavior and a reward. Classical conditioning is a type of associative learning in which an animal develops a response to a previously neutral condition through repeated combined presentation. Unconditioned or primary stimuli are stimuli such as food or pain to which an animal reacts without training. The response is completely inherent and physiological. Conditioned or secondary stimuli are usually arbitrary stimuli that an experimenter or animal trainer has associated with an unconditioned stimulus. If the unconditioned and conditioned stimuli are presented together repeatedly, then the animal will become conditioned to associate the two stimuli and over time will respond to the unconditioned stimulus when presented with the conditioned stimulus. For example, a cat owner may want to train the cat to stay off the kitchen counters when hearing fingers snapping. When the owner sees the cat on the counter, he squirts water at it and snaps his fingers at the same time. Being squirted with water naturally startles a cat (no training necessary for this response), so the water is an unconditioned stimulus. The sound of snapping fingers means nothing to the cat initially, but over time the cat becomes conditioned to associate the snapping with an oncoming squirt of water. After repeated simultaneous exposure to both the unconditioned and the conditioned stimulus, the sound of snapping fingers eventually produces the startled response even in the absence of water.

Operant conditioning is similar to classical conditioning in that both are associative learning processes, but whereas classical conditioning results in the association of two stimuli, operant conditioning results in an animal's associating a behavior with a consequence, either a reward or a punishment. In classical conditioning, the animal responds to a stimulus, but in operant conditioning the animal learns that a consequence follows a voluntary response and thus affects whether or not the behavior will happen again. Reinforcements are actions that increase a behavior by either presenting a reward or removing an unpleasant stimulus. Punishments decrease the frequency of a behavior by either presenting an unpleasant stimulus or taking away a favorable stimulus. If a bear turns over a rock and finds numerous ants to eat, the bear will learn that the behavior of looking under rocks will bring about the food reward of ants and will turn over rocks more frequently. On the other hand, if an animal eats a plant that tastes bitter or makes it sick, the animal will learn to recognize and avoid eating that plant in the future.

Research has shown that some animals have advanced cognitive abilities that allow them to perform tasks such as categorizing objects and solving problems. For example, after training baboons to match alphanumeric characters displayed in different fonts or typefaces, the animals could correctly recognize and place the same letter or number into the appropriate category when presented in a new font. Animal brains have adapted to deal with specific problems that their ancestors encountered; cognitive adaptations would have increased their ability to find food or survive challenges in order to reproduce successfully. Such abilities are demonstrated in pigeons, squirrel monkeys, baboons, gorillas, and chimpanzees, which have been shown to learn how to categorize through training as in the preceding example, but some studies show that other animals have this innate ability, as do humans. Researchers have observed young chimpanzees and infant macaques spontaneously sorting objects, such as models of animals and furniture or vehicles, according to perceptual properties. One can imagine that this adaptation would allow an animal to categorize an animal never seen before into a "dangerous" category and allow it to respond appropriately at the first encounter, or to recognize a potential new food source by its similarities to other familiar food sources, thus giving it an advantage if the familiar food source becomes scarce. The nervous systems of many animals allow them to solve complex problems, something that requires integrating prior experiences, observation, and insight. Different animals are capable of solving problems that require different degrees of problem solving skills. For example, if food is placed behind a barrier, can the animal figure out how to go around the barrier? Can an animal figure out how to get at food hanging from a string? The use of tools also demonstrates advanced cognitive abilities—creating a mental representation of an object to plan how to use it to achieve a goal or solve a problem. For example, a chimpanzee may stack boxes to climb and reach food placed high out of reach or a woodpecker finch may use a cactus spine to pry underneath the bark of a tree to access grubs hiding underneath.

Animals learn some of these complex behaviors by observing other animals performing similar tasks. Inherited or innate abilities and environmental influences both affect the ability of an animal to learn a complex behavior. The degree to which genetics or environment plays a role varies from behavior to behavior and from species to species. Genetics may exert control by setting a sensitive period during which an animal must learn a certain behavior, such as a species-specific song of a bird. For example, without being exposed to either real or recorded sparrow songs during the first 50 days of life, a white-crowned sparrow will not learn the adult song of its species. Other birds, however, such as New World flycatchers, develop the species-specific song even if raised in isolation; the behavior is completely innate.

Whether animals are capable of thinking is difficult to determine experimentally. Because they cannot communicate with humans through language, scientists have limited means to explore this aspect of animal behavior. Some people have taught gorillas and chimpanzees to use sign language to communicate, and the animals can convey basic needs and wants using signs. Some claim that the animals have expressed hopes and desires as well. Other investigators think that chimpanzees demonstrate self-recognition and awareness when they look in mirrors and perform actions on themselves, such as picking something out of their hair while looking in the mirror. Studies exploring concepts such as thought, self-awareness, and consciousness extend beyond connections between an animal's nervous system and its behavior and therefore beyond the realm of life science research and into the social science field of psychology. Though understanding human behavioral stimuli and responses can be useful in generating hypotheses concerning animal behaviors, life science researchers must be careful not to anthropomorphize, or attribute human behaviors, motivations, and feelings to animals.

See also ANIMAL BEHAVIOR; ETHOLOGY.

FURTHER READING

Dugatkin, Lee Alan. *Principles of Animal Behavior.* New York: W. W. Norton, 2003.

McFarland, David. *Animal Behavior: Psychobiology, Ethology, and Evolution.* 3rd ed. Harlow, England: Pearson, 1999.

Shettleworth, Sara J. *Cognition, Evolution, and Behavior.* New York: Oxford University Press, 1998.

Vauclair, Jacques. "Categorization and Conceptual Behavior in Nonhuman Primates." In *The Cognitive Animal: Empirical and Theoretical Perspectives on Animal Cognition*, edited by Marc Bekoff, Colin Allen, and Gordon M. Burghardt. Cambridge, Mass.: MIT Press, 2002.

animal form   The kingdom Animalia includes more than 35 recognized phyla of multicellular organisms as diverse as jellyfish, dragonflies, and elephants. Despite apparent differences in size and shape, animals share many characteristics, as they are derived from a common evolutionary ancestor. Most animals are diploid, meaning they contain two copies of each chromosome. They reproduce by sexual reproduction, and during embryogenesis they form distinct layers of cells that develop into specialized tissues. As heterotrophs, animals fulfill their nutritional requirements by eating other organisms or by ingesting organic molecules made by other organisms. They have no cell walls and are capable of movement.

Though animals from various phyla or classes may appear very different, the most striking differences are in size and shape. Animals exhibit a limited number of basic body plans, or programmed patterns of development, with slight variations among lineages. The environment and natural physical laws constrain or place limits on the ranges of overall body structures. For example, active living cells must exist in an aqueous environment, or at least be bathed in such a solution to allow the efficient exchange of nutrients, gases, and other substances. This necessity demands that larger and more complex animals must have extensive invaginations or branches within their bodies to allow for the exchange of nutrients and gases in all tissues. Furthermore, the volume that a given surface area across which exchange occurs can support limits the size of individual cells and ultimately the entire organism. Environmental constraints or the conditions of a particular habitat also influence the degree and types of variations an animal's body plan can tolerate.

HIERARCHICAL LEVELS OF ORGANIZATION

One difference between animal groups is the level of organizational complexity. The earliest life-forms were unicellular organisms, which exhibit the simplest level of organization. All of the functions necessary for life, such as obtaining of nutrition, metabolic processes, elimination of wastes, and reproduction, were carried out by individual cells, as they are today by prokaryotic organisms and unicellular eukaryotic organisms. The next level of organization builds upon the simplest level and is characterized by groups or aggregations of cells that share labor for a collective good. For example, *Volvox* is a green alga that lives in colonies shaped like hollow spheres consisting of hundreds or thousands of individual cells. The individuals must cooperate to reproduce. In asexual colonies, only cells in one region of the colony can divide to form new colonies. For sexual reproduction, male colonies release sperm, and cells within female colonies develop into eggs. The third level of organizational hierarchy is characterized by the presence of tissues, organized aggregates of cells that have similar structures and that perform similar functions. Some biologists consider sponges to have this level of organization, while others believe that sponges do not have any true tissues. Cnidarians like jellyfish do have a nerve net, a specialized tissue, but many of their cells are not organized into tissues. Organs are assemblies of several types of tissues that work together to perform a specific function; for example, the heart is an organ that pumps blood in many animals, such as earthworms, octopuses, and insects. At the highest level of structural organization within an organism, the organ-system level, several organs work together to accomplish a physiological function. For example, the stomach and the intestines are organs that function in conjunction with other organs and tissues as part of the digestive system, a physiological system that carries out digestion and absorption of nutrients.

TISSUE TYPES

After fertilization, the zygote undergoes cleavage, repeated mitotic divisions that result in a cluster of cells called a morula that eventually takes the shape of a hollow ball of cells called a blastula. The enclosed, fluid-filled cavity of a blastula is called the blastocoel. In a process called gastrulation, the cells of the blastula rearrange themselves, usually by first folding inward and then expanding out. Gastrulation results in a structure called a gastrula that comprises an outer layer of cells called the ectoderm and an inner layer called an endoderm. Diploblastic animals such as cnidarians (including hydra, jellyfish, sea anemones, and corals) and ctenophorans (comb jellies) only have these two germ layers. In triploblastic animals, a third layer, mesoderm, forms between the ectoderm and the endoderm. The blind pouch formed during gastrulation and lined with endoderm makes up the archenteron, and the opening where the initial infolding occurred is the blastopore. The archenteron and the blastopore will develop into the animal's digestive tract.

The ectoderm gives rise to the outer layer of skin, the nervous system, and the sensory organs. The endoderm forms the lining of the digestive tract, the respiratory system, the urinary bladder, the digestive organs, the liver, and several glands. The mesoderm develops into most of the skeleton, muscles, circulatory system, reproductive organs, and excretory organs. All of the differentiated tissues that make up the organs and structures of the organ systems are composed of four main tissue types that develop from the primary germ layers: epithelial, connective, muscular, and nervous tissues.

Epithelial tissues are sheets of cells that cover external body parts or line body cavities and ducts. Tight packing of the cells protects the underlying tissues and prevents the loss of fluids and the entry of potentially pathogenic microorganisms. Many cells on the surface of epithelial tissues produce and secrete substances such as digestive enzymes, sweat, or lubricating substances. The number of layers and the shape of the cells distinguish types of epithelial tissue. Simple epithelia have a single layer of cells, and stratified epithelia are multilayered. The shapes of the epithelial cells may be squamous (flattened to permit rapid diffusion), cuboidal (boxlike and often involved in secretion or absorption), or columnar (tall, rectangular prisms that often have microvilli to increase their absorptive surface area). In stratified squamous epithelium, the basal layer continually undergoes mitotic divisions to replace cells from above that are damaged or that have died and sloughed off. Transitional stratified epithelium is adapted to stretching. A supportive basement membrane underlies all epithelial tissues and consists of substances secreted by both the overlaying epithelial tissue and neighboring connective tissue.

Connective tissue, derived from mesoderm, serves a supportive role and consists of a few cells and protein fibers embedded in a matrix, or ground substance. The fibers may be either loosely or densely packed. Loose connective tissue is found throughout the animal body and holds organs together and in place. Fibroblast cells within loose connective tissue form three types of fibers, collagenous, elastic, or reticular fibers. Macrophages, cells of the immune system that phagocytose old cells and foreign materials, wander through the ground substance. Dense connective tissue contains closely packed fibers, making tissues like tendons (that attach bones to muscles) and ligaments (structures that attach bones to other bones) very strong. Blood, lymph, fluids within the body tissues, cartilage, bone, and adipose tissue (which stores fat) are also considered connective tissues.

Muscular tissue, also derived from mesoderm, consists of elongated cells called muscle fibers that contain myofibrils, the units capable of contraction. Vertebrate animals contain three types of this abundant tissue: skeletal, cardiac, and smooth. Both skeletal and cardiac muscle tissues have striations, alternating thick or thin bands that run perpendicular to their length. The distance between these bands decreases during contraction. An animal can voluntarily contract skeletal muscles, resulting in movement, but cardiac muscle, found only in the heart, is under involuntary control. Contraction of smooth muscles surrounding the stomach helps churn the food contents. Smooth muscle also encircles the intestines, esophagus, blood vessels, bronchi, and other passageways where constriction controls the diameter, which in turn affects the rate by which substances move.

Nervous tissue is specialized to transmit, conduct, and receive nervous impulses, electrical signals that travel throughout the body via nerves. In animals that exhibit cephalization, most nerves lead either to or from the brain, which processes the information transmitted by nervous tissue. Nervous tissue consists of nerve cells called neurons and neuroglia, cells that insulate and support nerve cells.

Epithelial, connective, muscular, and nervous tissue make up the organs of all animal bodies, except for sponges, which only exhibit organization up to the tissue level. Different tissues and organs work together to carry out different physiological functions and are dependent on one another as components of the same organism.

BODY PLANS

An animal's body plan refers to its developmentally programmed general shape, type of symmetry, and internal organization. Biological symmetry refers to a balance in the size, shape, and relative proportions of body parts on opposite sides of a median plane or center. With the exception of sponges, which are asymmetrical, animals exhibit either radial or bilateral symmetry. In radial symmetry, more than one plane passing through a central axis divides the animal into symmetric halves. Cnidarians and echinoderms (such as starfish and sea urchins) are radially symmetric. Animals with bilateral symmetry have one plane of symmetry that divides the body into distinct right and left halves. Bilaterally symmetric animals have nonsymmetrical dorsal (top) and ventral (bottom) halves occurring at the ends of the axis, and anterior (front) and posterior (back) sides in addition to right and left sides. The development of bilateral symmetry led to cephalization, the concentration of neural and sensory organs at the anterior portion, the head, of the body. Having eyes, ears, smell, or other sensory organs on the head provides an advantage for organisms as they travel forward: they can be aware of the environment they are entering and move toward food sources or away from danger.

Another evolutionary adaptation found in some bilaterally symmetric animals is a fluid-filled space. In coelomate animals, mesoderm completely lines this body cavity and forms a peritoneum, a sheet of connective tissue that wraps around and suspends the internal organs. This "tube-within-a-tube" arrangement provides extra space for body organs, allows for more body tissues to exchange nutrients and wastes, and allows animals to grow larger. Two different methods give rise to a coelom. In schizocoely, the mesoderm forms from cells near the blastopore, then splits to form the coelom. In enterocoely, the mesoderm forms from pouches of the archenteron. Both methods lead to the formation of a complete coelom with a peritoneum. Acoelomates, such as flatworms, have no body cavity; all of their body cells are in contact with other cells. Pseudocoelomates, such as roundworms, also have the tube-within-a-tube arrangement but do not have a peritoneum. In other words, their body cavity is merely a blastocoel left over from embryogenesis.

Bilateral animals can also be divided into groups based on the fate of the blastopore, a characteristic related to the method by which the coelom forms. After the formation of the archenteron during gastrulation, another opening forms at the end of the embryo opposite the blastopore, where the infolding began. One of the holes becomes the mouth and the other becomes the anus in a one-way digestive tract. In coelomate protostomes, the coelom forms by schizocoely, and the mouth develops from or near the blastopore. In deuterostomes, which are all

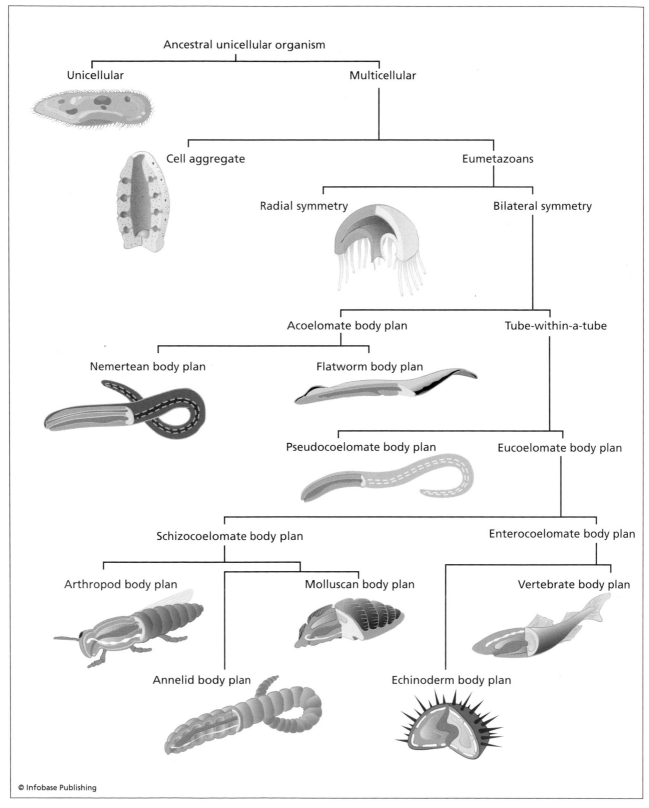

Ancestral unicellular organism

Unicellular — Multicellular

Cell aggregate — Eumetazoans

Radial symmetry — Bilateral symmetry

Acoelomate body plan — Tube-within-a-tube

Nemertean body plan — Flatworm body plan

Pseudocoelomate body plan — Eucoelomate body plan

Schizocoelomate body plan — Enterocoelomate body plan

Arthropod body plan — Molluscan body plan — Vertebrate body plan

Annelid body plan — Echinoderm body plan

Though animals have a variety of shapes and sizes, all are variations of a few basic body plans.

coelomates, the coelom forms by enterocoely, and the anus forms from or near the blastopore. The division of bilateral animals into protostomes and deuterostomes is somewhat controversial because some phyla exhibit qualities of both groups.

Metamerism, or segmentation, is the condition of having the body divided into a linear series of metameres, or repeated segments. Annelids (including earthworms), arthropods (including insects), and chordates (including vertebrate animals) all possess segmented bodies, though in some animals the segmentation is not as apparent as in others. For example, evidence of segmentation exists in vertebrate embryos and in insect larvae, but segments are not as obvious in the adult forms. In animals such as earthworms, many of the internal organs within each segment also repeat.

See also ANATOMY; ANIMAL FORM; CIRCULATORY SYSTEM; DIGESTIVE SYSTEM; EMBRYOLOGY AND EARLY ANIMAL DEVELOPMENT; ENDOCRINE SYSTEM; EXCRETORY SYSTEM; HOMEOSTASIS; HOST DEFENSES; HUMAN REPRODUCTION; INTEGUMENTARY SYSTEM; INVERTEBRATES; MUSCULOSKELETAL SYSTEM; NERVOUS SYSTEM; REPRODUCTION; RESPIRATION AND GAS EXCHANGE; SENSATION; VERTEBRATES.

FURTHER READING

Hickman, Cleveland P., Jr., Larry S. Roberts, Susan L. Keen, Allan Larson, Helen I'Anson, and David Eisenhour. *Integrated Principles of Zoology.* 14th ed. New York: McGraw-Hill, 2008.

antimicrobial drugs Antibiotics are substances that are made by one microorganism that inhibit the growth of or kill another microorganism. Semisynthetic antibiotics are antibiotics derived from a substance produced by a microorganism. Many diseases result from infection by a pathogenic microorganism. The use of chemicals in the treatment of disease is called chemotherapy, the goal of which is to destroy the responsible microorganism while causing minimal harm to the host. In the early 1900s, the German physician Paul Ehrlich conceived of a "magic bullet," an agent that would selectively target and kill a disease-causing organism. Ehrlich's research led to his discovery of salvarsan, the first chemotherapeutic agent, used to treat syphilis. Though salvarsan did serve as an effective chemotherapeutic agent for syphilis, it is not considered an antibiotic since a microorganism does not produce it. In 1928 Sir Alexander Fleming, a British bacteriologist, serendipitously discovered the first antibiotic, penicillin, when he observed that a substance produced by the mold *Penicillium notatum* inhibited bacterial growth in culture. Sir Howard Florey and Ernst Chain from Oxford University performed the first successful clinical trials of penicillin to treat an infection in 1940. These events led to the age of antibiotics, during which microbiologists have discovered thousands of antibiotics. Many potentially fatal infectious diseases became easily treatable, leading to a significant decline in the mortality rate from infectious diseases. Antibiotics are not without problems, however; many induce negative side effects and can be toxic, and the growing resistance of many microorganisms to different antibiotics is cause for alarm.

MECHANISMS OF ACTION OF ANTIBACTERIAL ANTIBIOTICS

Most antibiotics are only effective against bacteria. Infections by fungi, protozoa, helminthes, or viruses necessitate different types of chemotherapeutic agents. Antibiotics work by either inhibiting the growth of a microorganism or killing the microorganism. If an antibiotic prevents further growth of a pathogen, the body's immune system can often eliminate any microorganisms that are already present. The major mechanisms of action for antibiotics include inhibition or destruction of the cell wall, inhibition of protein synthesis, destruction of the cell membrane, inhibition of nucleic acid synthesis, and inhibition of synthesis of other essential metabolites.

The purpose of the bacterial cell wall is to give shape to the cell and protect it against adverse changes in the external environment, such as a decrease in solute concentration, which could result in cell lysis from a buildup of water pressure inside the cell. The main component of most bacterial cell walls is peptidoglycan, composed of repeating disaccharide subunits made of *N*-acetylglucosamine (NAG) and *N*-acetylmuramic (NAM) acid. Polymers of these repeats are linked by cross-bridges consisting of tetrapeptides, chains of four amino acids, forming a sheetlike layer. Prokaryotic cells are grouped into two main categories, gram-negative or gram-positive, based on differences in how they respond to a particular staining method called the Gram stain. Gram-negative cells have as little as a single layer of peptidoglycan, but the cell wall of gram-positive cells can contain as many as 40 stacked layers. Some antibiotics, such as penicillin and its derivatives, prevent the synthesis of peptidoglycan, and therefore the construction of cell walls, in actively growing cells. These antibiotics will not, however, kill any cells that already have completely formed cell walls and that are not growing. Because gram-positive cells contain much more peptidoglycan, these antibiotics are most effective against gram-positive bacteria. Antibiotics that only work on limited types of bacteria, such as penicillin, are said to have a narrow spectrum of activity. Other antibiotics that work by preventing cell wall synthesis are the cephalosporins, bacitracin, and vancomycin. Because human cells do not have cell walls, this class of antibiotic does not harm host cells.

Many antibiotics destroy bacteria by interfering with the process of protein synthesis. Cells that are actively growing must constantly synthesize new proteins. Ribosomes perform the function of reading a messenger ribonucleic acid (mRNA) transcript and linking the correct sequence of amino acids to construct a new polypeptide chain that may or may not combine with other polypeptides to form a complete protein. Ribosomes consist of two subunits, one large and one small. In prokaryotic cells, the subunits are 50S and 30S, which join to form a 70S ribosome; in eukaryotic cells, the subunits are 60S and 40S, which together make an 80S ribosome. (The *S* refers to the way the subunits sediment when centrifuged, a characteristic dependent on their size and overall shape.) Though prokaryotic cells and eukaryotic cells synthesize proteins by a very similar process, the differences in the structure of the ribosomes are significant enough for antibiotics to target prokaryotic ribosomes without affecting eukaryotic protein synthesis. By inhibiting the synthesis of new proteins, the antibiotics prevent cells from growing and dividing and from maintaining sufficient levels of so-called housekeeping proteins. Some antibiotics inhibit ribosome function by binding to the 50S portion of the ribosome—for example, chloramphenicol and erythromycin. Others, such as tetracyclines, streptomycin, and gentamicin, interact with the 30S subunit. Most antibiotics that interfere with ribosome function affect a wide range of bacterial species; in other words, they have a broad spectrum of activity.

Because the construction of gram-negative cell envelopes differs from construction of gram-positive cell envelopes, even though some antibiotics can bind to the ribosomes of all prokaryotes, they cannot penetrate gram-negative cell walls and thus are ineffective for treating infections caused by gram-positive bacteria. Although 80S ribosomes carry out most of the protein synthesis in eukaryotic cells, the mitochondria contain 70S ribosomes that synthesize proteins from mitochondria genes. Antibiotics that interact with prokaryotic ribosomes also interact with mitochondrial ribosomes and therefore can cause negative side effects such as fatigue.

Another group of antibiotics works by damaging the cell membrane. Disruption of the integrity of the cell membrane results in the loss of control over what enters and leaves the cell. If the injury is very large, the cell lyses and releases all its contents, causing immediate cell death. Antibiotics such as polymyxin B attach to the phospholipids that compose the cell membrane, making it more permeable and allowing important nutrients to escape.

Antibiotics that inhibit nucleic acid synthesis either interfere with deoxyribonucleic acid (DNA) replication or transcription, the process of making an mRNA transcript from a gene on the DNA. Transcription must occur prior to translation, the synthesis of polypeptide chains by ribosomes, as a crucial step in new protein synthesis. The processes of DNA replication and transcription are similar in prokaryotic and eukaryotic cells; thus antibiotics intended to harm the pathogenic bacteria may also harm the host. Rifampin, quinolones, nalidixic acid, nofloxacin, and ciprofloxacin are examples of DNA replication inhibitors. Rifamycins inhibit transcription.

The last group of antibiotics includes those that are competitive inhibitors for enzymes that catalyze steps of important metabolic pathways. A competitive inhibitor is a chemical that binds to an enzyme and thus prevents the natural substrate from binding. If the substrate cannot bind, then the enzyme cannot catalyze the reaction at that step in the metabolic pathway. For example, para-aminobenzoic acid (PABA) is a metabolite that bacteria chemically convert into the vitamin folic acid. Bacteria need this vitamin in order to make components of nucleotides, the building blocks of the nucleic acids DNA and RNA. If a cell cannot synthesize these components, transcription and DNA replication will cease. Sulfanilimide, a sulfa drug, mimics the structure of PABA and competes for binding to the enzyme that normally binds PABA. When this happens, the cell cannot make folic acid. Even though humans also need folic acid to remain healthy, they obtain it through diet rather than synthesize it, so the host is not affected. Sulfones and trimethoprim also work by preventing the synthesis of important metabolites.

ANTIFUNGAL, ANTIPROTOZOAN, AND ANTIHELMINTH DRUGS

Most antibacterial antibiotics are useful because they selectively target characteristics of prokaryotic cells such as peptidoglycan or 70S ribosomes; thus they only harm the bacteria in or on the host's body—in other words, they are selectively toxic. Finding chemotherapeutic agents that are selectively toxic for pathogenic fungi, protozoa, or helminths (parasitic worms) is more difficult since they are eukaryotic. Chemicals that adversely affect parasitic eukaryotes often also harm the host cells. Antifungal drugs such as polyenes and azoles often work by targeting sterols, a type of lipid present in fungal cell membranes. Their therapeutic value is limited, however, because human cells also have sterols in their cell membranes. Polyenes, such as amphotericin B, bind to sterols in the membrane and make it more permeable. Azoles (for example, imidazole, triazole, clotrimazole, miconazole, and fluconazole) inhibit the synthesis of sterols.

Quinine, a chemical originally isolated from the Peruvian cinchona tree, and its synthetic derivatives are used to treat the malaria caused by the protozoan *Plasmodium*. Some drugs used to treat tapeworm infestation act by inhibiting adenosine triphosphate (ATP) synthesis or changing the membrane permeability. Other helminth infections are treated by chemicals that cause the worms to have muscular spasms, disrupting microtubule action and interfering with mobility, or by paralyzing the worms.

ANTIVIRAL DRUGS

Viruses are not living entities; thus no drug can kill a virus. Any drugs used to treat viral infections must work by preventing further viral replication. Because viruses depend on host machinery to replicate, finding or designing drugs that inhibit viral DNA replication or transcription without also affecting the host cell activities is sometimes challenging. Antiviral drugs can target replication of the viral nucleic acid or the transcription of its genes when the virus carries specific genes required for these processes. Antiviral drugs often work by interfering with attachment of the viral particle to the host cell, preventing penetration, inhibiting uncoating of the viral nucleic acid once it enters the cell, or preventing the assembly of new viral particles. Some antiviral medications, including acyclovir, ribavirin, and zidovudine (AZT), mimic the structure of nucleotide bases. The DNA polymerase incorporates these base analogs into the viral nucleic acid, then cannot proceed with synthesis because the analog has no attachment site for the next nucleotide. Ideally, the virus encodes its own DNA polymerase, which is selectively inhibited by the base analogs. Retroviruses, such as human immunodeficiency virus, are viruses that have an RNA genome and depend on an enzyme called reverse transcriptase to synthesize DNA from RNA. Some drugs interfere with this enzyme's activity. Drugs called protease inhibitors inhibit enzymes necessary for the assembly of new viral particles and are also used to treat human immunodeficiency virus (HIV) infections. Enzyme inhibitors are used to treat influenza also. Another antiviral chemical is interferon, a natural substance produced by cells in response to infection by a virus that inhibits the virus from infecting neighboring cells. Physicians often prescribe interferon for viral hepatitis.

RESISTANCE

When bacteria do not respond to treatment with a particular antibiotic, the bacteria are said to be resistant, or insensitive, to that antibiotic. In some cases, the mechanism of action is simply not effective for a particular type of bacteria, but in other cases, bacteria that once were sensitive have developed a resistance to the antibiotic. The global increase in antibiotic resistance is a major current public health concern. As resistant populations replace the normal sensitive populations, infections grow increasingly difficult to treat.

Resistance results from the phenotypic expression of a gene, so the development of resistance in bacteria that previously were sensitive results from a change in genotype, the genetic makeup of an organism. Alterations in genotype occur by spontaneous mutations or by acquiring of a new gene through the processes of transformation, conjugation, or transduction. The gene product can lead to antibiotic resistance by one of several different mechanisms. Point mutations usually affect the cellular target of the antibiotic, and acquired genes usually encode enzymes that either inactivate the antibiotic or affect its ability to enter and stay in the cell. One mechanism of resistance works by destroying the antibiotic. Penicillin and its derivatives all contain a unique structure called a β-lactam ring. The enzyme β-lactamase cleaves the β-lactam ring, inactivating it. Expression of these enzymes, generally referred to as penicillinases, is the most common mechanism of resistance to this class of antibiotics. Another common means of resistance involves a mutated membrane transport protein that prevents the antibiotics from gaining entry into the cell. A slight change in the structure of the target of an antibiotic can prevent the drug from recognizing and binding it. For example, a substitution of an amino acid in one of the ribosomal polypeptides can prevent an antibiotic from binding to that ribosomal subunit and inhibit protein synthesis. Alternatively, a cell could up-regulate the expression of the target molecules, so that the concentration of the target increased, bound up all of the available antibiotic, and still had enough left over to carry out the normal function. Depending on the mechanism of action of an antibiotic, down-regulation of a bacterial protein can also lead to resistance. For example, if an antibiotic makes use of a bacterial protein to gain entry into the cell, a decrease in the amount of that protein can decrease the antibiotic's effectiveness. A bacterial cell can become resistant by developing a means for ejecting the antibiotic from the cell before it can interfere with the bacterial cell processes. Enzymes called translocases pump the drugs out of the cell and are relatively nonspecific; thus they cause bacteria to be resistant to several drugs.

Numerous factors have contributed to the alarming increase in resistance to antimicrobials, the most significant being indiscriminate use. Simply put, the presence of antibiotics selects for bacterial strains that are resistant, so frequent use increases the numbers of resistant bacteria. Viruses are the most common cause of colds and sore throats, but many patients

Penicillin and its derivatives share the black-and-white portion of this structure and have variable R groups. The enzyme β-lactamase cleaves the β-lactam ring structure at the covalent linkage indicated by the arrow. Bacteria that synthesize the enzyme β-lactamase destroy penicillin and its derivatives, making them resistant to that class of antibiotics.

demand and many physicians prescribe antibiotics to treat such maladies, though antibiotics do not affect viruses in any way. Viruses do not have cell walls, 70S ribosomes, or prokaryotic molecular machinery, common targets of antibiotics. Hospitals harbor many drug-resistant bacterial strains, an unfortunate fact since many hospitalized patients are already immunocompromised or have wounds that facilitate the entry of pathogens into the host's body. Since the 1950s, strains of *Staphylococcus aureus* found in hospitals went from almost all being sensitive to almost 100 percent being resistant. Methicillin-resistant *S. aureus* (MRSA), a strain that is resistant to most antibiotics, is prevalent in hospitals and other health care facilities and a frequent cause of potentially fatal staph infections. The powerful antibiotic vancomycin was the last resort drug to treat such infections until recently, when vancomycin-resistant strains started appearing. Vancomycin-resistant *Enterococcus* (VRE) is another danger in health care facilities. Though VRE exists as part of the normal human flora, in immunocompromised individuals (infants, the elderly, and very ill people), VRE can overgrow, resulting in diarrhea.

To limit the continued spread of resistant strains, physicians should only prescribe antibiotics when appropriate, and patients should follow instructions regarding dosage and duration of treatment. The simultaneous treatment with more than one antibiotic decreases the chance that a resistant strain will survive. Researchers should continue to search for new antimicrobials with structures that do not lend themselves to easy destruction or avoidance. Measures to decrease the use of antimicrobials in animal feed worldwide will also help to curb the spread of resistance.

See also BACTERIA (EUBACTERIA); ENZYMES; FLEMING, SIR ALEXANDER; FUNGI; GENE EXPRESSION; INFECTIOUS DISEASES; MICROBIOLOGY; PROKARYOTIC CELLS; PROTOZOA; VIRUSES AND OTHER INFECTIOUS PARTICLES.

FURTHER READING
Centers for Disease Control and Prevention, Department of Health and Human Services. "General Information about Antimicrobial Resistance." Available online. URL: http://www.cdc.gov/drugresistance/general/index.htm. Accessed January 13, 2008.
Tortora, Gerald J., Berdell R. Funke, and Christine L. Case. *Microbiology: An Introduction.* 9th ed. San Francisco: Benjamin Cummings, 2007.
U.S. National Library of Medicine, National Institutes of Health, Department of Health and Human Services. "Medline Plus: Drugs, Supplements, and Herbal Information." Available online. URL: http://www.nlm.nih.gov/medlineplus/druginformation.html. Updated August 28, 2008.

Archaea   According to the five-kingdom classification scheme proposed by Robert Whittaker, the kingdom Monera comprised all prokaryotic organisms. In 1977 Carl Woese from the University of Illinois at Urbana-Champaign performed an extensive molecular analysis of several types of prokaryotic organisms and obtained surprising results. Comparison of the sequences for the 16S ribosomal ribonucleic acid (RNA) genes showed that some of the prokaryotes differed tremendously from others, enough so that they warranted their own superkingdom on the tree of life. He proposed the formation of three domains of living organisms: Eukaryota, which contains the protists, fungi, plants, and animals; Bacteria, or Eubacteria, which includes the more familiar prokaryotic organisms; and Archaea, which includes the most recently recognized forms of prokaryotic life.

The domain Archaea is named for the Greek word meaning "ancient," because archaea live in conditions similar to those on Earth 3.5 billion years ago. Most of the planet was covered in water that contained harsh chemicals and often reached boiling temperatures. Microbiologists do not yet know nearly as much about archaea, which have only been recognized for a few decades, as they do about bacteria. Archaea often live in extreme environmental conditions; such organisms are called extremophiles, and their unusual growth requirements make it more difficult to study them in the laboratory. Not all archaea are extremophiles, and not all extremophiles are archaea. Thermophiles are extremophiles that live at very high temperatures, and psychrophiles live in extreme cold. Acidophiles and alkalinophiles live in acidic or basic environments. The halophiles ("salt lovers") live in very salty environments. Another major group of archaea are the methanogens, which produce methane gas (CH_4) as a by-product of their metabolism.

ARCHAEAN CHARACTERISTICS

Archaea are unicellular, prokaryotic organisms that have a variety of shapes, such as rods, cocci, and other unusual forms including triangles. Members of this domain share similarities with both bacteria and eukaryotic organisms. As prokaryotes, archaea lack internal compartmentalization. The single chromosome that makes up their genome is not bound by a nuclear envelope. Like bacteria, archaea can be gram-negative or gram-positive, but the composition of their cell walls differs. Bacterial cell walls consist of peptidoglycan, made of chains

of alternating subunits of *N*-acetylglucosamine and *N*-acetylmuramic acid and linked together by short peptide bridges. Gram-negative archaea have a proteinaceous layer associated with their cell membranes. In some gram-positive archaea, the thick cell wall consists of pseudomurein, which contains *N*-acetylalosaminuronic acid in place of *N*-acetylmuramic acid in addition to different types of amino acids. Other archaea contain different complex polysaccharides. Because of this, antibiotics that are harmful to bacteria, such as penicillin, and chemicals, such as lysozyme, that target peptidoglycan are ineffective against archaea.

The membrane lipids of archaea are also distinctive. The lipids contain branched isoprenes rather than fatty acids, and the lipids are linked to glycerol through ether bonds rather than ester bonds as in bacteria and eukaryotic organisms. Sometimes the

COMPARISON OF ARCHAEA, BACTERIA, AND EUKARYA

	Archaea	Bacteria	Eukarya
nuclear envelope	absent	absent	present
membrane-bound organelles	absent	absent	present
cell wall	composed of proteinaceous subunits (in most), in some made of pseudo-murein	made of peptidoglycan	if present, composed of a variety of substances such as cellulose or chitin
lipids in the cell membrane	branched carbon chains made of isoprene derivatives, ether-linked	straight chain fatty acids, ester-linked	straight chain fatty acids, ester-linked, contain sterols
initiator tRNA	methionine	N-formyl-methionine	methionine
introns	absent	present in some genes	present
sensitive to the antibiotics streptomycin, kanamycin, and chloramphenicol	no	yes (for most)	no
DNA bound by histones	yes	no	yes
chromosome	single, circular	single, circular	linear, number varies
microtubule cytoskeleton	no	no	yes
ribosomes	70S	70S	80S
plasmid DNA	present	present	rare
ribosomes sensitive to diphtheria toxin	yes	no	yes
polycistronic mRNA	present	present	absent

branched portions cyclize, providing more rigidity to the membrane.

Archaea contain a single, circular chromosome, but the sequence of the archaean genome is genotypically distinct from that of bacteria and eukaryotes. Histonelike proteins bind the chromosomes of archaea in a manner similar to the way they bind to eukaryotic chromosomes. As bacteria do, archaea have polycistronic genes that contain no introns, or intervening sequences between coding regions of messenger RNA (mRNA). The initiator transfer RNA (tRNA) is methionine, as in eukaryotic organisms. Their ribosomes are 70S, as in bacteria, but the ribosomes' shape differs and they are insensitive to antibiotics that inhibit bacterial ribosomes. Also, the RNA polymerases are more similar to eukaryotic RNA polymerases than to bacterial RNA polymerases.

Archaea exhibit a variety of means for obtaining their energy and nutrients. Scientists have identified both aerobic and anaerobic species, and autotrophs and heterotrophs. Hydrogen gas (H_2), carbon dioxide (CO_2), or sulfur provides energy for some archaea, and some are photosynthetic. Their metabolisms vary greatly among the different groups of archaea.

ARCHAEAN CLASSIFICATION

Sequence similarities of 16S rRNA genes group the archaeans into four main clades: Euryarchaeotes, Crenarchaeotes, Korarchaeotes, and Nanoarchaeotes. The Euryarchaeota include the methane producers and the halophiles (salt lovers). Some extreme thermophiles are also in Euryarchaeota, but most thermophiles belong to Crenarchaeota. Methanogens, one type of Euryarchaeota, oxidize hydrogen gas (H_2) for energy and are the most common of the Archaea. Molecular oxygen (O_2) is toxic to methanogens; instead they use an inorganic substance such as CO_2 as the oxidizing agent, in the process reducing it to methane gas (CH_4), which they release into the environment. In addition to CO_2, methanogenic species convert other substances, such as formate, methanol, acetate, carbon monoxide, and methylamine, into methane. Methanogens inhabit marshes, producing swamp gas; live as endosymbionts of cattle and termites; or reside in the human gut. Sewage treatment facilities utilize methanogens to help decompose organic matter in wastewater, and some industrial plants harvest the gas given off to utilize as a source of energy. At least 17 methanogenic archaean genera and 93 species have been identified.

The extreme halophiles, which grow in very salty conditions, also belong to Euryarchaeota. Organisms that live in environments that have a higher concentration of solutes than inside the cell must have special adaptations that prevent the cell from dehydrating. Natural tendencies cause the water to diffuse from inside the cell to the outside, but the loss of too much water causes cell death. Some halophiles overcome this obstacle by transporting other solutes, such as potassium, into the cell to balance the osmolarity with the external environment and prevent excessive water loss. Extreme halophiles require a high percentage of salt, a minimum of 9 percent but on average 12–23 percent (for comparison, normal seawater contains 0.9 percent salt). One type of halophile is photosynthetic. The purple pigment bacteriorhodopsin, found in its cell membranes, harvests light energy from the Sun and uses it to generate adenosine triphosphate (ATP). Scientists have identified 20 species of extreme halophiles so far.

The Crenarchaeota includes most of the organisms that live in extreme temperatures, both hot and cold, and the organisms that can tolerate extreme acidity. Most bacteria and eukaryotic organisms are mesophiles, meaning their optimal growth temperature ranges from 68°F to 113°F (20°C–45°C). Bacteria that live in symbiotic relationships with humans or that are pathogenic to humans must be mesophiles, since normal body temperature, 98.6°F (37°C) falls within this range. Thermophiles thrive at a range of 104°F–176°F (40°C–80°C). Extreme thermophiles (hyperthermophiles) prefer temperatures even higher, with an optimum around 212°F (100°C), the temperature at which water boils at standard pressure. Thermophiles and hyperthermophiles have special enzymes and molecules that prevent denaturation at such high temperatures. Thomas D. Brock at the University of Wisconsin at Madison first discovered thermophiles in 1966 in hot springs at Yellowstone National Park. Since then, scientists have discovered microbes that can withstand surprising conditions. In 2003 Derek Lovley and Kazem Kashefi of the University of Massachusetts at Amherst identified a spherical, flagellated archaean species that can thrive at 250°F (121°C), the current record held by a living organism, and, interestingly, the temperature at which autoclave ovens sterilize the equipment and media used in most microbiological research. They discovered these archaea, named Strain 121, near some black smokers 200 miles from Puget Sound and one and one-half miles deep in the Pacific Ocean. Strain 121 respires using iron as the final electron acceptor, forming magnetite in the process. This microbe "breathes" iron, in comparison to aerobic organisms, which "breathe" oxygen. At the other end of the spectrum, psychrophiles can grow at temperatures as low as 32°F (0°C), but they grow best between 68°F and 86°F (20°C–30°C). These organisms can spoil food quickly, even when it is kept in a refrigerator. Some have been found living in an

Antarctic lake of ice. Not very much is known about the psychrophiles.

Most organisms grow best at a neutral pH, around 7.0. Acidophiles, also in Crenarchaeota, have the rare quality of growing best in acidic conditions, and alkalinophiles grow best in basic conditions. Molecules inside the cell, such as the DNA, cannot withstand a pH of this level; thus the cells have mechanisms for maintaining a neutral internal pH despite the harsh external conditions that the cell surface molecules must tolerate.

Korarchaeota is a relatively new clade, first recognized in 1996 when an organism living in a hot spring in Yellowstone National Park was found to be very different from the Euryarchaeota or the Crenarchaeota. Korarchaeota includes members that are more distantly related to the other archaea. Scientists believe that the korarchaeote species more closely resemble the ancient ancestral life-form common to prokaryotes and eukaryotes than the other archaea. The newest clade, Nanoarchaeota, discovered in 2002, contains organisms that are tiny, even by prokaryotic standards. Approximately 1.57×10^{-5} inch (0.4 μm) in diameter, nanoarchaeote genomes only contain 500,000 base pairs, smaller than any other known organism.

See also BACTERIA (EUBACTERIA); BIOGEOCHEMICAL CYCLES; BROCK, THOMAS; MICROBIOLOGY; PROKARYOTIC CELLS; WOESE, CARL.

FURTHER READING

American Society for Microbiology. "Microbe World." Available online. URL: http://www.microbeworld.org. Accessed January 13, 2008.

Dyer, Betsey Dexter. *A Field Guide to Bacteria.* Ithaca, N.Y.: Cornell University Press, 2003.

Madigan, Michael T., and John M. Martinko. *Brock Biology of Microorganisms.* 11th ed. Upper Saddle, N.J.: Prentice Hall, 2006.

Margulis, Lynn, and Karlene V. Schwartz. *Five Kingdoms: An Illustrated Guide to the Phyla of Life on Earth.* 3rd ed. New York: Henry Holt, 2002.

Aristotle (384–322 B.C.E.) Greek *Philosopher* Aristotle was an ancient Greek philosopher who contributed to numerous broad fields, ranging from politics and cosmology to physics and life science. He has been called the first true scientist and his school the first research institution with the first academic library. Though many of his ideas have been disregarded, his work formed the basis of many scientific fields, including the life sciences.

Aristotle was born in 384 B.C.E. in Stagira, in northern Greece. His father, Nichomachus, was a physician for the royal family in Macedonia. Both of his parents died when he was a boy, and at age 17, he went to Athens to study philosophy at Plato's Academy. Plato was a student of Socrates, and both men were brilliant Greek philosophers; thus Aristotle's educational pedigree was of the highest quality. During the 20 years he spent in Athens, he became known as Plato's brightest pupil. Whereas Plato was a renowned philosophical thinker, Aristotle was more of a natural philosopher, who saw reality in objects formed of matter.

After Plato died in 347 B.C.E., Aristotle left the academy, possibly because he was disappointed that he was not named its head, a position given to Plato's nephew. He started his own school in Assos, in northern Asia Minor (now Turkey). While there, Aristotle married Pythias, the niece of the city's ruler, Hermias, who was also his school's benefactor. They had one daughter, whom they named Pythias, after her mother. In 344 B.C.E. Hermias was overthrown, and Aristotle moved to the island of Lesbos. Two years later he returned to Macedonia, where he tutored the future Alexander the Great, son of Philip II, the king of Macedonia. In 339 B.C.E. Aristotle moved home to Stagira, accompanied by a group of devoted students. Some historians believe Aristotle met and married Herpyllis during this time. Whether they married or not, together they had a son, named Nichomachus, after Aristotle's father. In 335 B.C.E. he moved back to Athens and started his own school, the Lyceum, also known as the Peripatetic School because he often lectured while walking along the covered walkways (called peripatus) on the grounds.

In Assos, Aristotle began to work on *Historia animalium (The History of Animals)*, a venture that served as a reference for his other works in zoological science. At the Lyceum, Aristotle continued work on this treatise but also branched out into other sciences, the humanities, and history. His student followers wrote essays on these subjects, and over the years newer students updated them. These documents became a valuable source of historical information, and Aristotle's systematic collection of observations and information became the basis of natural philosophy, which evolved into modern science, as well as mathematics, astronomy, and medicine. Most of the manuscripts that have survived are not Aristotle's own writings, but interpretations or revisions of his notes or lectures that others have arranged and edited. This has led to some inconsistencies in his works and makes it difficult to follow the evolution of his thoughts and ideas. Because he did collect knowledge in the form of written books and maps, the Lyceum was the first school to resemble a modern academic institution.

Aristotle's greatest contribution to science may have been his attempt to classify and organize into hierarchies all the organisms known at the time. He started by dividing life-forms into plants and animals, and then further categorized them on the basis of shared physical characteristics, such as whether or not they had blood. (He mistakenly thought insects did not have blood, so his categories resembled the modern groups of invertebrates and vertebrates.) Other features that defined Aristotle's classification scheme included the distinction between aquatic and terrestrial habitats, ecological roles such as predator and prey, mode of reproduction, number of legs, and other shared anatomical characteristics.

Believing observation was key to understanding the natural world, he dissected animals to look at their insides, something most philosophers at the time saw as dirty work that was beneath the role of an intellectual. Aristotle systematically examined the development of chick embryos by incubating several eggs of the same age and dissecting one each day to see what new features had formed. By dissecting cattle, he found that their digestive tracts contained more than one stomach chamber. He noted observations such as that no single-hoofed animal had horns, and no horned animal also had tusks. He discovered that beehives contained a single queen, whom all the other bees in the hive assisted. His observations demonstrated that hyenas existed as separate sexes—male and female—in contrast to the belief at the time that they were hermaphrodites. Sea life particularly fascinated Aristotle, who was the first to recognize that dolphins nurtured their developing young through a structure called a placenta, something no fish did. Because of this unique feature, Aristotle sagely grouped dolphins with terrestrial animals that had the same structure (now called placental mammals). His books included the description of many marine invertebrates, such as octopuses, cuttlefish, and crabs. His comparison of sea urchin mouths to a five-sided lantern called a horn lantern led to the term *Aristotle's lantern* to portray the arrangement of five teeth around the circular mouth. For almost 2,000 years, natural philosophers relied upon and added to the system Aristotle developed for classifying living organisms. The system worked until the 1700s, when the Swedish naturalist Carl Linnaeus proposed and implemented a new scheme to accommodate the large increase in the number of species discovered as exploration to other lands became more common.

Historia animalium includes observations and descriptions of about 500 different species, including their anatomy, habitats, and behaviors. Collecting the data for this intended reference spanned several years, and it probably was not written in complete form until after many of his other zoological treatises, whose content was based on the observations recorded in *Historia animalium.* When Alexander ruled as king of Macedonia, he sent men far away to obtain all sorts of animals (and plants and rocks), which he gave to Aristotle, whose scientific collections may have formed the first zoo and museum of natural history. The main goals of zoology at the time were to identify and classify animals and to define their purpose or function in nature. Aristotle's early zoological treatises examined the design of tissues and anatomical structures and discussed physiological processes such as movement, respiration, aging, and death. He also explored inheritance, inessential characteristics such as color, and the relationship between composition or structure and reproduction, realizing that living organisms were more than a collection of parts. He observed that offspring often resembled distant relatives more than they resembled their parents and that parents who had suffered injuries or lost limbs did not pass on these malformations to their children. From this he surmised that parents passed on the potential to exhibit certain characteristics rather than the characteristics themselves. He also mistakenly believed that inherited characteristics were passed on through the blood.

In *Historia animalium,* Aristotle described the heart and blood vessels in some detail and even divided animals into two general classes, based on whether they had blood or not. This led to categories similar to those with a backbone and those without. He was fascinated by the beating heart of a chick embryo and observed that all the blood vessels originated at the heart; he did not differentiate between arteries and veins. He studied the vessels by starving the animals so they would become so thin that he could visualize the vessels of the living organism, then sacrificed them by strangulation so the blood would remain inside the vessels for more accurate observations. This method led to some artifactual structural abnormalities, however, such as a swollen right side of the heart, with the atrium and ventricle appearing as one continuous chamber, into which the apparently widened vena cava delivered blood. Though he was a keen observer, he did foster several erroneous beliefs about animal physiology. Aristotle thought that the heart was the center of life and that it warmed the blood, whereas the brain served as a cooling organ for the blood. He also believed that heat generated from the heart caused the lungs to expand, drawing in air, which then cooled the lungs before exhalation. Aristotle was a proponent of preformation, the idea that reproduction occurred when a male deposited a complete, "preformed" miniature individual into a female, in whose body it grew until birth. He thought some animals spontaneously formed from mixtures of mud and water. Physicians and physiologists propagated many of Aristotle's incorrect ideas for hundreds or thousands of years before being challenged. Some of his other zoological works include *De generatione animalium (On the Generation of Animals)*, *De motu animalium (On the Motion of animals)*, *De incessu animalium (On the Progression of Animals)*, and *De partibus animalium (On the Parts of Animals)*.

Having spent so much time comparing animals and observing slight structural differences, he envisioned a progression in their form, a precursor notion to evolution. He placed different types of animals on an imaginary ladder, with higher rungs representing increasing ranks of perfection. The ladder consisted of plantlike animals (corals and anemones) and mollusks at the bottom, followed by worms, egg-laying animals, and eventually mammals, with humans at the top. Aristotle believed in teleology—the philosophy that life-forms have end goals, and their forms and body processes are means of achieving that purpose. Today scientists know evolution is a nondirected process; thus Aristotle's belief in teleology was contrary to modern evolutionary theory, but his system of classification based on natural characteristics and his collection and organization of his numerous observations did, provide a framework upon which future biologists could structure their thinking. Charles Darwin, the British naturalist who proposed the theory of evolution by means of natural selection in the mid-1800s, praised Aristotle's intellect and foresight.

Though Aristotle's inclusion of humans with other animals was revolutionary, he did believe humans were special. In accordance with teleological principles, he thought all living things had a purpose, and therefore the potential to achieve that goal. For example, a seed's purpose was to develop into a complete new plant, but this was the maximum of its potential, as plants had the most primitive of "souls." Animals, on the other hand, had souls with greater potential, as they could move and use their senses to perceive events in the surrounding world. Humans had the ability to think and reason, and Aristotle believed this gave them the capacity to become divine. Through seeking and gaining knowledge and using logic to understand that knowledge, which he believed was the aim of science, one's soul could become eternal.

Another contribution Aristotle made to the development of modern science was his overall approach toward studying natural philosophy, which led to the development of science as a discipline distinct from logic or politics. His reliance on universal scientific principles and formulation of hypotheses based on observable data and facts helped establish scientific knowledge as true or real knowledge. This empirical means of achieving scientific knowledge in a particular field led to an explosion of knowledge in subjects such as botany, medicine, and astronomy. The artificial manipulation of variables during scientific experiments occurred later.

In addition to biology, Aristotle significantly influenced the field of cosmology by attempting to explain how the stars and planets moved. He proposed that a series of concentric spheres surrounding the Earth carried the celestial bodies. The spheres were composed of a transparent substance called ether, and the vapor form of this substance filled empty space. He called the outermost sphere the prime mover. To explain the pathways of the planets, he had to add more spheres to his system, which became more and more complex. In contrast, modern science seeks the simplest possible explanations for natural phenomena.

Aristotle thought that everything in the Earth consisted of different amounts of four basic substances: earth, fire, water, and air. Elements had a tendency to return to their natural state; for example, solid objects fell to the Earth, and bubbles floated to the surface of a body of water, where it contacts the atmosphere. Aristotle also delved into the topic of motion. Just as his ideas about planetary motion had to wait until the 17th-century German astronomer Johannes Kepler provided a better explanation, Aristotle's ideas about projectile motion had to wait for the Italian scientist Galileo Galilei.

After the death of his former pupil, Alexander the Great, anti-Macedonian sentiments forced Aristotle to retire to his mother's former estate in Chalcis. He died the following year, in 322 B.C.E., at the age of 62. After his death, his lecture notes and writings were collected, assembled, and edited in multiple volumes.

FURTHER READING

Adler, Robert E. *Science Firsts: From the Creation of Science to the Science of Creation*. Hoboken, N.J.: John Wiley & Sons, 2002.

Cooper, Sharon Katz. *Aristotle: Philosopher, Teacher, and Scientist*. Minneapolis: Compass Point Books, 2007.

Shields, Christopher. *Aristotle*. New York: Routledge, 2007.

assisted reproductive technology The 2005 Assisted Reproductive Technology (ART) Report, published by the Centers for Disease Control and Prevention (CDC) of the U.S. Department of Health and Human Services, states that in 2005 more than 52,000 children were born as a result of successful ART treatments. The term *ART* refers to any procedure during which both eggs and sperm are handled to help a couple conceive, such as in vitro fertilization (IVF) or IVF-based procedures.

INFERTILITY

According to the 2002 National Survey of Family Growth (the most current data available at the time of this writing), published by the National Center for Health Statistics (of the CDC), an estimated 12 percent of women of reproductive age had difficulty either in becoming pregnant or maintaining a pregnancy in 2002. The number of couples seeking treatment for infertility, medically defined as the inability to conceive after one year of unprotected intercourse or the inability to carry a pregnancy to term, has steadily increased since ART was introduced in the United States in 1981. In order to achieve a successful pregnancy without medical intervention, at least one egg must fully mature and escape from a woman's ovaries, the egg must travel unimpeded down the oviduct (also called the fallopian tube), a mature sperm must fertilize the egg, the developing embryo must implant into the lining of the uterus, and the uterus must maintain the pregnancy for nine months while the embryo develops into a fetus and then a child capable of sustaining life outside the mother. This complex series of events offers much opportunity for conditions to go wrong.

Infertility in males can result from a low sperm count, a condition in which the number of sperm is lower than the number typically required to conceive. In other cases the sperm do not develop or mature properly, resulting in deformed or immotile sperm that are incapable of traveling up the woman's reproductive tract and fertilizing an egg. Having low sperm numbers or poor-quality sperm can result from structural abnormalities of the male reproductive organs, general poor health, a specific disease, or exposure to certain environmental factors such as cigarette smoke, alcohol, toxins, or radiation. In females, poor general health, physical problems with the uterus or oviducts, or a history of sexually transmitted diseases can lower fertility, but the most common cause is ovulation problems. As a woman ages, she ovulates less frequently and eventually stops ovulating altogether. Because more women are waiting until later in life to have children, age is a common factor contributing to infertility.

An evaluation performed by a physician can help determine the cause of infertility. Semen analysis will reveal whether the sperm have any structural abnormalities, whether they do not move properly, or whether the numbers are simply too low. In a woman, hormone levels in the blood in combination with an ultrasound examination of the ovaries might indicate problems with ovulation. Injection of a special dye through the vagina allows a physician to examine the uterus and the oviducts for abnormalities or blockages that could prevent the sperm and egg from finding each other or prevent the embryo from implanting in the uterus. Laparoscopic examination of the pelvic cavity can help diagnose diseases such as endometriosis that interfere with reproduction. Finding a likely cause for infertility helps the physician and the couple to determine the most appropriate and potentially effective course of treatment.

TYPES OF ART

Though 85–90 percent of infertile couples may be treated successfully by the administration of drugs or by surgical repair of reproductive organs, others require more complicated treatment to conceive and deliver a live baby. Assisted reproductive technology (ART) generally refers to procedures that involve the handling of both eggs and sperm, so procedures such as the administration of medicine by itself or simple artificial insemination, in which sperm are injected into a woman's cervix, are not included. ART typically involves removing the eggs from the woman's body, combining the egg or eggs with sperm, and placing the eggs and sperm or the fertilized egg or eggs back into the woman's body. Four common ART procedures are IVF, gamete intrafallopian transfer (GIFT), zygote intrafallopian transfer (ZIFT), and intracytoplasmic sperm injection (ICSI).

IVF, a procedure pioneered by Dr. Patrick Steptoe and Robert G. Edwards in the 1970s, forms the basis of all four techniques and accounts for 98 percent of ART procedures. Since 1978, when the first child conceived by IVF was born, the procedure has benefited thousands of women who have blocked or absent oviducts and men who have low sperm counts. In a typical IVF procedure, the woman is given ovulation-inducing drugs to stimulate the ripening or maturation of multiple eggs within her ovaries. After two intense weeks of carefully orchestrated treatments with fertility drugs, the physician retrieves the eggs by aspiration through a needle guided by ultrasound. The father provides a fresh sperm specimen that is washed and added to a dish containing the harvested eggs in a specially designed culture medium, and the gametes are incubated together for 14 to 18 hours. By this time, the sperm have penetrated the eggs, and the zygotes are placed in a new culture medium optimized for cell division. After one and one-half to five days, depending on the clinic, the physician uses a microscope to select the healthiest embryos and places them inside the uterus using a catheter inserted through the vagina and cervix. Administration of the steroid hormone progesterone helps prepare the uterine lining for implantation.

GIFT begins the same way as IVF; the woman is given drugs to induce ovulation, and the eggs are retrieved and combined with a prepared sperm sample. The difference is that in GIFT, immediately after egg retrieval, the eggs and sperm are combined and placed together in the oviduct, where fertilization naturally occurs, using a laparoscope. Of course, the woman must have at least one open oviduct for this to work. Using GIFT, a physician cannot know whether fertilization has occurred unless the woman becomes pregnant.

In ZIFT, also known as tubal embryo transfer, fertilization occurs in the laboratory, and then the egg is returned to the oviduct before undergoing cell division. ZIFT differs from IVF in that the zygote is returned to the oviduct rather than the uterus, usually earlier than with IVF. If the procedure does not result in a pregnancy, the physician cannot know whether the zygote developed into a normal blastocyst.

When lack of a sufficient number of mature sperm hinders conception, a physician might recommend ICSI. The sperm are examined closely under a microscope, and one that looks normal and healthy is gently drawn into a pipette and then injected into the cytoplasm of a mature egg. After cell division begins, the embryo can be returned to either the oviduct or the uterus.

When the woman's ovaries do not ovulate, donor eggs can be retrieved from another woman. Donor sperm can also be used. Gamete donation is sometimes used when the man or woman carries a gene that causes a disease.

Miscarriages occur in more than 50 percent of pregnancies in women over the age of 42, most resulting from chromosomal abnormalities. After age 42, the likelihood that a woman will have a child born with a chromosomal abnormality reaches one in 39. Preimplantation diagnosis (PGD) reduces this risk. In PGD, a single cell is removed from a three-day-old embryo that has between eight and 10 cells total. The chromosomal makeup of the removed cell is examined. Normally, a human cell contains 46 total chromosomes. Chromosomal analysis will reveal whether too many or too few chromosomes are present or apparent structural abnormalities affect any of the chromosomes. After PGD, only embryos that appear to have a normal set of chromosomes will be implanted. This procedure reduces but does not eliminate the risk of genetic birth defects.

Since the first successful IVF procedure, an estimated 1.2 million children have been born worldwide as a result of ART. The success rate of ART depends on the age and general health of the parents, the condition of the embryos used, the ART procedure, and the particular clinic. Clinics must report their success rates to the CDC. For 2005, the average percentages of ART cycles that resulted in the birth of a healthy baby were 37.3 percent for women under the age of 35, 29.5 percent for women between the ages of 35 and 37, 19.7 percent for women between the ages

of 37 and 40, and 10.6 percent for women between the ages of 41 and 42. Because ovulation-inducing drugs stimulate the maturation of multiple eggs, and because physicians often implant several embryos into a woman undergoing an ART procedure in order to increase the chance for success, multiples such as twins, triplets, or higher numbers occur more frequently than in pregnancies that occur without any medical intervention.

See also HUMAN REPRODUCTION; SEXUAL AND REPRODUCTIVE HEALTH.

FURTHER READING

Edwards, R. G., and Patrick Steptoe. *A Matter of Life: The Story of a Medical Breakthrough.* New York: William Morrow, 1980.

Falcone, Tommaso, with Davis Young. *Overcoming Infertility: A Cleveland Clinic Guide.* Cleveland, Ohio: Cleveland Clinic Press, 2006.

Sher, Geoffrey, Virginia Marriage Davis, and Jean Stoess. *In Vitro Fertilization: The A.R.T. of Making Babies, Assisted Reproductive Technology.* 3rd ed. New York: Facts On File, 2005.

U.S. National Library of Medicine and the National Institutes of Health. Infertility. "MedlinePlus." Available online. URL: http://www.nlm.nih.gov/medlineplus/infertility.html. Updated November 18, 2008.

Vliet, Elizabeth Lee. *It's My Ovaries, Stupid!* New York: Scribner, 2003.

Avery, Oswald (1877–1955) *Canadian-American Bacteriologist* Oswald Avery helped pave the way for the molecular biological revolution when, with Colin MacLeod and Maclyn McCarty, he demonstrated that deoxyribonucleic acid is the molecular basis for the transmission of genetic information. This scientific development was one of the most important of the 20th century.

Oswald Theodore Avery was born on October 21, 1877, in Halifax, Nova Scotia. His family moved to New York City in 1887 when his father accepted an invitation to serve as pastor of a church in the Lower East Side. Avery attended Colgate University, where he made good grades and won several oratorical competitions. After earning a bachelor of arts degree in the humanities in 1900, he proceeded to Columbia University College of Physicians and Surgeons and received his medical degree in 1904. After practicing as a general physician for three years, Avery switched his career focus to bacteriological and immunological research. He took one part-time job with the Board of Health and another performing bacterial counts on milk samples. In 1907 he joined the Hoagland Laboratory, a private bacteriological research institute in Brooklyn, where he became associate director of the division of bacteriology. At Hoagland, Avery researched many topics, including the bacteriology of yogurt and fermented dairy products, immunological proteins, vaccines, and secondary infections in pulmonary tuberculosis. His research on bacterial strains that caused pneumonia, a disease in which the lungs become inflamed and fluid-filled, leading to coughing and difficulty breathing, impressed Rufus Cole, the director of the Hospital of the Rockefeller Institute for Medical Research, who hired Avery in 1913.

At the Rockefeller Institute, Avery concentrated on identifying and characterizing bacterial strains that caused pneumonia. A landmark study performed in 1917 by Avery and his coworker Alphonse Dochez described the presence of an immunologically specific soluble substance in the culture medium of pneumococcus. They subsequently identified this substance as a polysaccharide, a surprising result because chemists generally believed the antigenic agent had to be a protein on the surface of the bacterial cells. Though some scientists suspected Avery's findings resulted from protein contamination, Avery proved otherwise and went further to show that the form of this antigenic polysaccharide varied among pneumococcal types, results that led to the development of immunochemistry as a new field of research. This research led to Avery's nomination for a Nobel Prize in physiology or medicine almost annually from the 1930s until his death in 1955.

For nearly three decades, Avery contributed significantly to the fields of bacteriology and immunology, collaborating with many reputable scientists. His research led to an understanding of how specifically to treat pneumonia cases on the basis of the specific causative bacterial strain. He became a member of the Rockefeller Institute for Medical Research in 1923. The National Academy of Sciences elected Avery to membership in 1933, but the extraordinary findings he published in 1944 earned him a most distinguished place in the history of genetics.

Colin MacLeod, a physician who joined Avery's lab in 1934, and Maclyn McCarty, who joined the lab in 1941, collaborated with Avery. They were interested in pneumococcal transformation, a phenomenon discovered by the British researcher Frederick Griffith in the 1920s. Accumulated evidence showed a correlation between the presence of a smooth capsule surrounding one strain of bacteria and its virulence, or its ability to invade and multiply within host tissues. The capsule prevented host white blood cells called phagocytes from ingesting and destroying the invasive bacteria. Griffith was studying similar phenomena. He had successfully converted a nonvirulent, nonencapsulated strain (designated R) of *Streptococcus pneumoniae* into a virulent encapsulated strain (designated S), by adding heat-killed virulent bacteria to a culture of the nonvirulent strain and injecting mice with the mixture. Though the heat-killed bacteria alone did not make the mice ill, when they were injected simultaneously with the live, harmless, nonencapsulated strain, the mice became sick and died. Avery and his team duplicated Griffith's experiment, showing that a chemical substance from the dead S strain of bacteria was able to confer the ability to produce a smooth coating to a living R strain of bacteria. They prepared extracts from the bacterial cells by repeatedly freezing and thawing them and achieved similar results in vitro.

For 15 years, Avery systematically tried to identify the chemical substance from the extracts responsible for inherited variation in bacteria cells, or, more broadly, to identify the hereditary material of life. Assuming it would be a protein, they obtained many negative results through the years. Surprisingly, a nonprotein substance purified from the crude extracts did show promise in its ability to transform R into S strain bacteria. Avery, MacLeod, and McCarty found the ratio of nitrogen to phosphorus in the substance similar to that of deoxyribonucleic acid (DNA). They wondered whether DNA from the virulent S bacteria could be the transforming material. When they treated the extract with enzymes that destroyed proteins, polysaccharides, or ribonucleic acid, the ability of the material to transform R into S types was not hindered. However, treatment with enzymes that specifically attacked DNA decreased the transformation efficiency. From these results, the trio concluded that the active transforming factor responsible for the formation of polysaccharide capsules in the previously rough strain of bacteria was DNA. Being not only cautious, but modest, Avery waited until he had convincing evidence before publishing the results. Even then he downplayed the enormity of their discovery in the conclusions of their paper, "Studies on the Chemical Nature of the Substance Inducing Transformation of Pneumococcal Types: Induction of Transformation by a Desoxyribonucleic Acid Fraction Isolated from Pneumococcus Type III" published in the *Journal of Experimental Medicine* in 1944. He preferred simply to report his findings objectively and to let others draw their own conclusions.

The report provoked much opposition from scientists who believed that DNA, generally thought to be a simple tetranucleotide, was too basic a molecule to have such profound influence. Once again, other scientists including colleagues at his own institution believed that Avery's extracts were contaminated with protein and that those proteins were the true transforming factor. Some scientists made it their goal to prove Avery wrong, attacking his methods for purifying the transformation material and his chemical analyses, but DNA-dependent transformations performed by others eventually confirmed their findings. The revolutionary conclusion that DNA was the hereditary material was later supported by experiments performed by Alfred Hershey and Martha Chase using bacteriophage (a virus that infects bacteria) in 1952.

Avery became a naturalized citizen of the United States in 1918. He had enlisted as a private in the U.S. Army the year before and was promoted to captain soon after becoming naturalized. Though Avery became a professor emeritus of the Rockefeller Institute in 1943 at age 65, he continued his research there until 1948, when he retired to Nashville, Tennessee, to be closer to his family. The Royal Society of London awarded Avery their prestigious Copley Medal in 1945, and the Lasker Foundation gave him the Albert Lasker Award for Basic Medical Research in 1946. He died on February 20, 1955, of liver cancer.

See also BIOMOLECULES; DEOXYRIBONUCLEIC ACID (DNA); GRIFFITH, FREDERICK; HERSHEY, ALFRED; MacLEOD, COLIN MUNRO; McCARTY, MACLYN.

FURTHER READING

Avery, Oswald T., Colin M. MacLeod, and Maclyn McCarty. "Studies on the Chemical Nature of the Substance Inducing Transformation of Pneumococcal Types: Induction of Transformation by a Desoxyribonucleic Acid Fraction Isolated from Pneumococcus Type III." *Journal of Experimental Medicine* 79 (1944): 137–158.

Dochez, A. R. "Oswald Theodore Avery (October 21, 1877–February 20, 1955)." In *Biographical Memoirs*. *National Academy of Sciences*. Vol. 32. Washington, D.C.: National Academy of Sciences, 1958.

Dubos, René. *The Professor, the Institute, and DNA*. New York: Rockefeller University Press, 1956.

U.S. National Library of Medicine. *Profiles in Science*. Available online. URL: http://www.profiles.nlm.nih.gov/CC/. Accessed January 13, 2008.

Bacteria (Eubacteria) The group of organisms referred to as bacteria once encompassed all the prokaryotic organisms, until phylogenetic analysis demonstrated that prokaryotes consisted of two distinct groups, the Bacteria, also called the Eubacteria or true bacteria, and the Archaea. These groups are now recognized as domains or superkingdoms, a classification level higher than kingdom. Bacteria includes the microorganisms with which most people are familiar—the types that scientists use for research, the strains that inhabit the human intestinal tract, the ones used in food production, pathogenic bacteria, photosynthetic bacteria, and the rest of the Bacteria that inhabit moderate environments. Archaea are prokaryotes that often live in extreme environments such as those characterized by very high or very low temperatures, extreme acidity, or high solute concentrations and methane-producing strains. The names of the domains *Bacteria* and *Archaea* are capitalized, whereas the terms *bacteria* and *archaean* are not capitalized when referring to microorganisms within the domain. Bacteria differ structurally, biochemically, and physiologically from archaeans.

GROWTH AND ADAPTATION

As prokaryotic cells, bacterial cells lack compartmentalization. They reproduce by binary fission, in which one cell divides into two, each containing a copy of the parental chromosome. During optimal growth conditions, replication of this chromosome often limits the generation time, the period required for a population of cells to double. Under optimal conditions, generation times vary between species from 20 minutes to a few hours. In a natural setting, conditions never remain optimal as nutrients become depleted and metabolic waste accumulates, poisoning the microbes. Some bacteria can form endospores, tough dormant structures that can withstand harsh environmental conditions and then germinate when the conditions improve.

Bacterial growth is measured by the number of cells or the size of the population rather than looking at an individual organism. Bacteria growing in a liquid broth culture or in an aquatic environment cloud the liquid as the numbers increase. On a solid growth medium, a single bacterial cell grows and divides repeatedly, piling up to form a raised, circular collection of cells called a colony. All of the cells in a colony contain copies of the same chromosome.

Though clonal populations of bacteria all receive identical copies of the chromosome, bacteria do have mechanisms for creating genetic variation that allows evolutionary adaptation to occur. Growth conditions such as temperature, nutrient availability, pH, oxygen concentration, and the presence of certain chemicals can select for spontaneous or induced mutations, causing them to become fixed. Prokaryotes can also obtain new genes by conjugation, a process in which a hollow tubular structure called a pilus forms a bridge between two cells, allowing for the transfer of deoxyribonucleic acid (DNA). In transformation, bacterial cells uptake pieces of DNA from the environment through their cell membranes. Viruses that infect bacteria, called bacteriophages, can also carry genes from one bacterial cell to another, a process called transduction.

DIVERSITY

The domain Bacteria includes tremendously diverse organisms, making them a very successful group. One area of variation is their nutritional mode—

how they obtain their energy and what they use as a carbon source. Some organisms are phototrophs, meaning they obtain their energy from sunlight, and others are chemotrophs, meaning they obtain their energy from chemicals. Organisms that use carbon dioxide (CO_2) as their sole carbon source are called autotrophs, and organisms that utilize organic molecules as their main carbon source are called heterotrophs. These terms can be combined to give more detailed information about an organism's nutritional requirements. For example, a photoautotroph uses light as its energy source and CO_2 as its carbon source. Bacteria exist that represent all the major nutritional modes: photoautotrophs, chemoautotrophs, photoheterotrophs, and chemoheterotrophs. Bacterial requirements for oxygen (O_2) also vary: obligate aerobes require oxygen to survive; facultative aerobes use oxygen if it is present but can ferment in its absence; and oxygen is toxic to obligate anaerobes. Nitrogen is a necessary nutrient for synthesizing amino acids of proteins and nitrogenous bases in nucleic acids, but bacteria have different capabilities with respect to nitrogen metabolism. Some can fix nitrogen, meaning they can convert nitrogen gas (N_2) to ammonia (NH_3), a form that can be readily incorporated into organic molecules. Bacteria capable of nitrogen fixation play an important role in the cycling of this nutrient in ecosystems. Bacteria with unique metabolic requirements often live in close association with organisms that have complementary metabolisms.

BACTERIAL CLASSIFICATION

One means for categorizing bacteria is by evolutionary relationships. Because organisms that are closely related have fewer mutations between them compared with organisms that diverged a long time ago, the molecular differences provide insight into the evolutionary distance between types of organisms. Analysis of the nucleotide sequence of certain ribosomal ribonucleic acid (rRNA) genes divides Bacteria into five major groups (and many more phyla) that share a common evolutionary ancestor: the proteobacteria, chlamydias, spirochetes, cyanobacteria, and gram-positive bacteria. Morphologies or physiological characteristics within groups can be extremely variable since the groups are defined by molecular relatedness.

The phylum Proteobacteria, the largest of the Bacterial clades, includes more than 1,600 identified species of gram-negative bacteria of all nutritional modes and oxygen requirements. Bacteria are termed gram-negative if they do not retain the crystal violet stain during the Gram staining procedure, and gram-positive if they do retain it. The differential reaction depends on the structure of the cell envelope. Some

proteobacteria are pathogenic, some are free-living, and others live in symbiotic relationships. Proteobacteria are divided into five subgroups, named for the first five letters of the Greek alphabet. The class Alphaproteobacteria includes *Rhizobium* and *Agrobacterium* species. *Rhizobium* lives in a symbiotic association with legumes; it resides in nodules of the roots and fixes nitrogen. The plants, belonging to the pea and bean family, benefit from the supply of usable nitrogen and the bacteria benefit from the water and nutrients taken in through the plant's root system. *Agrobacterium* causes tumor formation in its hosts. Because of its known ability to transfer DNA between itself and the plant, plant geneticists use it to improve crops. Many *Agrobacterium* species have recently been reclassified as *Rhizobium* species. The rickettsias, including the human pathogen *Rickettsia*, are alphaproteobacteria that can only survive as endosymbionts of other cells. Scientists believe the organisms that evolved into eukaryotic mitochondria by endosymbiosis originated from alphaproteobacteria.

The class Betaproteobacteria includes many soil and wastewater species, but also some human pathogens such as *Neisseria*, which causes gonorrhea and a form of meningoencephalitis. Many betaproteobacteria are facultative aerobes, a few are phototrophs, and some have unique metabolisms. The soil microbe *Nitrosomonas* plays an important role in the cycling of nitrogen by oxidizing ammonium (NH_4^+) for energy and producing nitrite (NO_2^-) as a waste product.

Gammaproteobacteria includes many medically significant species as well as the sulfur bacteria. Many familiar species belong to this group. *Escherichia coli*, one well-known species of gammaproteobacteria, is part of the normal human flora and is used extensively in research. The causative organisms for cholera (*Vibrio cholerae*), the foodborne illness salmonellosis (*Salmonella enteriditis*), typhoid fever (*Salmonella typhus*), the plague (*Yersinia pestis*), and an opportunistic human pathogen (*Pseudomonas aeruginosa*) that often infects the pulmonary tract, the urinary tract, or burns are all members of the class Gammaproteobacteria. The purple sulfur bacteria are capable of photosynthesis and live in hot sulfur springs or stagnant water. They oxidize hydrogen sulfide (H_2S) rather than water as plants and algae do and produce granules of elemental sulfur as a product.

Members of Deltaproteobacteria include the unusual myxobacteria, sulfate- and sulfur-reducing bacteria, and other anaerobic bacteria. Myxobacteria live in the soil, move by gliding in swarms, and produce fruiting bodies when growth conditions are unfavorable. The fruiting bodies release resistant spores that germinate when conditions improve. Sulfate-reducing bacteria use sulfate (and sometimes other oxidized sulfur compounds) as an oxidizing agent, reducing it to sulfide, but do not incorporate it in organic compounds. This anaerobic process is called dissimilatory sulfur reduction. Sulfur-reducing bacteria obtain energy by reducing elemental sulfur to H_2S with hydrogen or organic compounds. Some delta-proteobacteria reduce other oxidized inorganic compounds, such as ferric iron.

Members of the class Epsilon Proteobacteria live in animal digestive tracts and sometimes are pathogenic. Examples include *Campylobacter*, which can cause gastroenteritis in humans, and *Helicobacter pylori*, which causes stomach ulcers.

Bacteria that belong to the phylum Chlamydiae are nonmotile obligate parasites, meaning they cannot complete their life cycle without an animal or protozoan host. Their metabolic capabilities are extremely limited, and their cell walls lack peptidoglycan. One example is *Chlamydia trachomatis*, a bacterial species that causes the common sexually transmitted disease nongonococcal urethritis as well as trachoma, a leading cause of blindness in humans.

The phylum Spirochaetes includes bacteria that are characteristically long and helical and possess axial filaments. These flagellalike filaments run the length of the cell between the wall and the cell membrane and twist, causing a bacterium to move by rotating like a corkscrew. Most spirochetes are anaerobic and free-living, but a few are parasitic. *Treponema pallidum* causes the sexually transmitted disease syphilis, *Borrelia burgdorferi* causes Lyme disease, and *Leptospira* causes leptospirosis.

The cyanobacteria are unique in that they are the only prokaryotic organisms that undergo oxygenic photosynthesis, meaning that water serves as the electron donor and oxygen is produced as a by-product. Evidence strongly suggests that chloroplasts in plants and algae evolved from endosymbiotic cyanobacteria. Also called blue-green algae because they were once thought to be algae, these aquatic prokaryotes can be unicellular, filamentous, or colonial. Some cyanobacteria can also fix nitrogen, reducing it to NH_4^+, which can be incorporated in cellular metabolism.

The fifth group of Bacteria, the gram-positive bacteria, includes all the bacteria that retain crystal violet when stained by the Gram procedure. The cell walls contain as many as 40 layers of the carbohydrate-protein complex peptidoglycan, and teichoic acids are present in the cell membrane. Gram-positive bacteria include free-living and parasitic forms and consist of two major phyla: the Firmicutes and the Actinobacteria.

The largest phylum of gram-positive genera is the Firmicutes, whose members have a low percentage composition of the nucleotides guanine (G) and cytosine (C) in their nucleic acid. Firmicutes include two spore-forming genera—*Clostridium*, which includes the species that cause gas gangrene and botulism, and *Bacillus*, which includes the species that causes anthrax—in addition to beneficial species that serve as a source of antibiotics and that are used as natural pesticides. Firmicutes also comprise many non-spore-forming genera: *Staphylococcus*, a normal inhabitant of human skin that can also be pathogenic; *Streptococcus*, including *Streptococcus pyogenes*, which causes strep throat and rheumatic fever; *Lactococcus*, which produces lactic acid as an end product of fermentation; and *Enterococcus*, which can cause urinary tract infections, bacterial endocarditis, diverticulitis, and meningitis. The last group of Firmicutes is the mycoplasmas, tiny cell wall–less bacteria that evolved from gram-positive bacteria. *Mycoplasma pneumoniae* causes pneumonia.

The other major phylum of gram-positive bacteria is Actinobacteria, whose members have a high G-C content. Soil-dwelling actinomycetes play an ecologically important role in the decomposition of organic matter such as cellulose and chitin and are also the source of numerous antibiotics. Some actinobacteria are filamentous and resemble mold. Another type of actinobacteria, the coryneform bacteria, exhibit unique shapes, sometimes resembling the letter Y or V. *Corynebacterium diphtheriae* can cause the disease diphtheria. Members of the genera *Mycobacterium*, characterized by the presence of mycolic acids in the cell membrane, cause tuberculosis and leprosy. *Propionibacterium* species are used in the production of swiss cheese, and some cause acne.

See also ARCHAEA; BIOGEOCHEMICAL CYCLES; INFECTIOUS DISEASES; LEEUWENHOEK, ANTONI VAN; MARGULIS, LYNN; MICROBIOLOGY; PASTEUR, LOUIS; PHOTOSYNTHESIS; PROKARYOTIC CELLS; SPONTANEOUS GENERATION.

FURTHER READING

American Society for Microbiology home page. Available online. URL: http://www.asm.org. Accessed January 13, 2008.

Dyer, Betsey Dexter. *A Field Guide to Bacteria*. Ithaca, N.Y.: Cornell University Press, 2003.

Madigan, Michael T., and John M. Martinko. *Brock Biology of Microorganisms*. 11th ed. Upper Saddle, N.J.: Prentice Hall, 2006.

Margulis, Lynn, and Karlene V. Schwartz. *Five Kingdoms: An Illustrated Guide to the Phyla of Life on Earth*. 3rd ed. New York: Henry Holt, 2002.

Singleton, Paul. *Bacteria in Biology, Biotechnology, and Medicine*. 6th ed. New York: John Wiley & Sons, 2004.

Banting, Sir Frederick G. (1891–1941) Canadian *Physician* Diabetes is a disease characterized by the body's inability to produce or utilize insulin. Without insulin, an excess of glucose collects in the bloodstream and is excreted in the urine, and the body does not obtain the energy it needs, leading to a slow death by starvation. For the victims of diabetes whose bodies cannot make insulin, injections of the hormone help regulate the levels of sugar in the blood. The Canadian physician Sir Frederick Banting is venerated for his discovery of insulin and its utility as a treatment for diabetes.

CHILDHOOD, EDUCATION, AND EARLY CAREER

Frederick G. Banting was born on November 14, 1891, in Alliston, Ontario. He was the youngest of five children and grew up in a deeply religious household on a farm. An average student and a decent athlete, he spent his childhood exploring around the farm and its riverbank. When he was a teen, one of his childhood friends became gravely ill, became very thin, lost all of her energy, and died of a then-mysterious illness called diabetes at the age of fourteen. This memorable event tremendously impacted the course Fred would pursue in his future.

After graduating from the local public school in 1910, Banting enrolled at Victoria College, a liberal arts institution in Toronto. To please his parents, Banting planned on majoring in theology and becoming a minister, but he was fascinated with medicine. In the autumn of 1912, he registered as a medical student at the University of Toronto. While a student, Banting saved to purchase a microscope for $57.50, a considerable sum in those days. In his free time he studied his own blood under the microscope, perfected his tissue preparation skills, and conducted experiments in the laboratory. At the nearby Hospital for Sick Children, Banting specialized in orthopedic surgery, the surgical correction of skeletal deformities. After World War I broke out in 1914, licensed doctors were scarce in the city. The school accelerated the medical students' courses, and they graduated six months early.

In December 1916, Banting entered the Canadian Army Medical Corps as a lieutenant. Banting first went to England, and then France, where he witnessed the suffering of many wounded soldiers and gained extensive surgical experience. During action in 1918, a piece of shrapnel seriously injured Banting's forearm and severed an artery, but he continued providing medical assistance to other soldiers for 17 hours. Later, when the doctors wanted to amputate, he refused to let them and determinedly strove to rehabilitate his arm. He received the Military Cross for his brave conduct.

In 1919 Banting returned to Toronto, where he accepted an orthopedic surgery residency at the Hospital for Sick Children. He specialized in the mechanical correction of childhood deformities such as clubfeet and twisted limbs. After one year, he attempted to start his own surgical practice in London, Ontario, but was unable to attract enough patients, so he accepted a part-time instructorship in anatomy, physiology, and clinical surgery for the medical school at Western University (now the University of Western Ontario). Banting was popular with students and delivered meticulously prepared lectures, but he missed performing medical research. He often joined the chief of physiology, Dr. Frederick R. Miller, in his neurophysiologic investigations. Together the physicians showed that the cortex of the brain was sensitive to outside stimulation.

THE PANCREAS AND HORMONE "X"

Banting spent a lot of time reading medical journals in order to include the latest reports and discoveries in his lectures. In autumn 1920, he began preparing for an upcoming lecture on the pancreas. The pancreas is a large abdominal gland that produces digestive enzymes that travel through a duct to the small intestine, where they chemically break down proteins, lipids, and carbohydrates into simpler molecules that the body can readily absorb. Removal of the pancreas leads to increased levels of sugar in the blood and urine, and death results. While searching the medical literature to learn more about this gland, Banting found descriptions of diabetes symptoms dating back 4,000 years. Victims suffer unquenchable thirst and hunger, high sugar levels in the blood and urine, an acetone-like odor on the breath, tiredness and depression, extreme weight loss, and eventually a coma leading to death. Though the disease had been recognized since ancient times, no treatment or cure had been discovered. Banting wondered why so little was known about treating diabetes.

In 1869 a medical student named Paul Langerhans identified groups of cells in pancreatic tissue, later named islets of Langerhans, that looked different from the regular pancreatic cells that secreted digestive enzymes and did not lead to the small intestine via a duct. In 1889 German researchers, Josef Von Mehring and Oskar Minkowksi, removed the pancreas of a dog, which developed acute diabetes mellitus and died within two weeks. The islets of Langerhans in the pancreatic tissue from deceased diabetics appeared atrophied. Some scientists thought these cells produced an undiscovered hormone that helped the body burn sugar for energy. Several physicians, including Dr. John James Richard Macleod, the head of physiology at the University of Toronto, claimed that there was no proof for the existence of this unknown "hormone X."

On the evening of October 30, 1920, the day before Banting's lecture on the pancreas, he visited the library to search for additional material to include in his lecture. That morning, a new issue of *Surgery, Gynecology, and Obstetrics* had arrived. The journal contained a 12-page article by Dr. Moses Barron titled "The Relation of the Islets of Langerhans to Diabetes, with Special Reference to Cases of Pancreatic Lithiasis." The article said that sometimes an autopsy revealed gallstones blocking the pancreatic duct. In these cases, the pancreatic cells that produce digestive juices had disintegrated, but the Langerhans cells all looked normal and healthy, and the patients showed no symptoms of diabetes. Barron also reported that this effect could be recreated in dogs by surgically tying off the pancreatic duct. After several weeks, the entire pancreas shriveled up except the Langerhans cells.

The information that the Langerhans cells were somehow associated with diabetes interested Banting, who believed that the cells made an unknown hormone X that helps the body burn sugar. Past attempts using pancreatic extracts to relieve diabetic symptoms had been unsuccessful, however. Banting thought digestive enzymes made by the pancreas destroyed the unknown hormone during extraction and thought that tying off the pancreatic duct to destroy the enzyme-making cells before extract preparation would preserve the hormone's activity during extraction.

When Banting shared his idea for a method of obtaining active hormone X with Miller, Miller suggested that Banting speak with the well-known endocrinologist Macleod at the University of Toronto. Several other physicians also deferred to Macleod as the leading expert on blood sugar chemical processes. Banting knew that Macleod did not believe in the existence of hormone X but scheduled an appointment and drove to Toronto anyhow.

Macleod politely listened to Banting but was unimpressed with Banting's lack of research experience on blood chemistry and turned away the discouraged young doctor. Worried that his nervousness interfered with his ability to present a strong case for proceeding with the anticipated research clearly, Banting spent that night typing up a written proposal. The next morning he returned, and Macleod agreed to provide Banting with 10 dogs, an assistant proficient in biochemistry, and laboratory space for eight weeks.

STUDIES DIABETES IN DOGS

A few months later, Banting moved back to Toronto and was joined by an assistant, Charles H. Best, a recent physiology and biochemistry graduate who had previous research experience using chemical procedures, to measure sugar levels in blood and urine. On May 16, 1921, Banting began surgery on the dogs, tying off the pancreatic ducts in hopes of destroying all the pancreatic tissue except the islets of Langerhans. The following week, he attempted to remove the pancreas of one dog using a two-step procedure, but the dog died of shock and infection. Using his surgical experience, he refined a technique to remove the pancreas completely in one operation. As expected, the dog developed diabetes. During the six- to eight-week waiting period for the pancreas of the duct-tied dogs to atrophy, Banting named the unknown hormone that they hoped to find "isletin."

On July 6, 1921, the men opened up two dogs whose pancreatic ducts had been tied and were dismayed to find healthy pancreas glands inside. Examination revealed that Banting had tied the ducts too tightly, and new pathways had formed around the ligature. To correct for this, Banting retied the ducts more loosely than before, but in three different places to ensure digestive juices could not flow through them. In a few dogs, degeneration was occurring, but the men decided to let it progress for two more weeks. Banting was worried that they soon would hear from Macleod, who was vacationing in Scotland for the summer. He had originally promised them eight weeks in his lab, and their time was up. They were also broke, so Banting sold his car to buy more dogs.

They removed a pancreas from another dog, which promptly became diabetic. As it approached the coma stage, they cut open a duct-tied dog and removed its now degenerated pancreas. The islets of Langerhans still appeared healthy, so they crushed the gland in a chilled, buffered saline solution. After filtration, the extract was injected into the neck vein of the dying dog. In an hour, the dog began to lift its head. Within a few hours it was sitting up, wagging its tail, and its blood sugar level had dropped to almost normal. Within five hours, the urine was completely void of sugar. These were exactly the results Banting and Best had anticipated. Isletin had been used successfully to treat diabetes.

Unfortunately, the next morning the dog was dead. Isletin was a treatment, but it was not a cure. They removed the pancreas from a second dog and waited for it to become ill. Then they made more pancreatic extract from another duct-tied dog, but this time they also made extracts from the liver and spleen to demonstrate the previous success was due to a substance specifically from the pancreas. When they injected the liver and spleen extracts into the sick dog, nothing happened, but when they injected the pancreatic extract, again the dog perked up, and within hours the urine contained no sugar. They managed to keep this dog alive for three days.

Though these results pleased Banting, he was disturbed by having to kill healthy animals to obtain extracts that only treated diabetic animals for a short time. He wondered how he could maximize the amount of extract produced while minimizing the number of animals that had to be sacrificed. One technique they tried was exhausting the pancreas by overstimulation with another hormone, but this still yielded limited amounts.

Macleod returned from vacation, and while he was not overly impressed with their progress, he allowed Banting and Best to continue using his laboratory facilities. To ease financial burdens, Banting assumed a position as a demonstrator in the pharmacology department. His responsibilities were minimal, and the small salary was just enough to allow him to continue his studies.

DISCOVERS INSULIN

One day Best came across a paper that said that the pancreas of newborns was richer in Langerhans cells than that of adults. Since fetuses do not digest their own food in utero, they also would not be producing digestive juices. Banting thought slaughterhouses ought to have a sufficient supply of calf embryos from which they could isolate the pancreas. By noon the following day they had obtained nine embryonic calves, from which they extracted isletin. The isletin from calf embryos also reduced the blood sugar levels to normal when given to diabetic dogs. While this method provided more extract than using duct-tied dogs and did not require the sacrifice of otherwise healthy animals, the supply was still limited.

Banting and Best devised a chemical extraction method from adult cattle pancreas involving a combination of acid and alcohol. To ensure it would not cause any undesirable side effects in sick patients, they injected the extract into each other and observed no harmful effects, but of course, neither of them was diabetic. They had a potent extract and were ready for a real human trial.

The opportunity for a human trial presented itself on January 11, 1922, when a 14-year-old boy was admitted to Toronto General Hospital with a severe case of diabetes. His body had wasted away to a mere 65 pounds and death was imminent. A dose of isletin reduced the boy's blood sugar. They worked to purify the extract further and optimize the dosage, miraculously restoring his health. This surprising success attracted the attention of Macleod, who promptly stopped work on his own research and dedicated his entire staff to assisting in the isletin research. He suggested a name change to *insulin*, since it was easier to pronounce. A biochemist by

the name of Dr. J. B. Collip and a recent graduate named E. C. Noble joined Best in the perfection of a technique called fractional alcoholic precipitation to purify the insulin from pancreatic extracts.

In February 1923, a former classmate of Banting's from medical school, Joe Gilchrist, visited Banting. He had developed diabetes during the war and volunteered to act as a human guinea pig for their new extract preparations. Respiration tests showed Gilchrist's body was not burning any sugar. They injected him with insulin, and a few hours later he was producing sugar-free urine. Once he accidentally overdosed with insulin but recovered after drinking a nearby beaker of glucose solution. A few months later, Banting obtained permission from the Canadian government to use Toronto's Christie Street Hospital for Returned Soldiers to begin clinical trials. Later they expanded their testing to Toronto General Hospital.

All the clinical tests taught them that although insulin worked wonders, diet was still an important factor in treatment. They determined proper doses of insulin and observed that injections worked best if administered 20 to 30 minutes before a meal. They learned to recognize the signs of insulin overdose and discovered that administering glucose could prevent insulin shock. Before these tests, six of every 10 diabetics died of coma, and every child was doomed. With the availability of insulin, the death rate dropped considerably.

Banting and Best published their findings, "The Internal Secretion of the Pancreas," in the November 1921 issue of the *Journal of Laboratory and Clinical Medicine*. The results were announced publicly at a medical meeting in New Haven, Connecticut, in late 1922. Banting was not an experienced or polished speaker, but Macleod, who was chairing the meeting, spoke next. He did a much better job telling the story but failed to emphasize who actually performed the work; as a result, many believed that Macleod headed the research. The American Association of Physicians in Chicago asked Macleod to speak. Again, he failed to clarify who led the research that led to the discovery of insulin.

Continued clinical trials gave excellent results, and soon commercial drug companies were given the instructions on how to prepare insulin extracts from bovine pancreas, but people flocked to Toronto to see the "miracle physician." Banting temporarily opened an office to treat diabetic patients, only charging minimal fees. With Best and Collip, Banting patented the process for insulin production and used all the profits to fund continued diabetes research. He received numerous medals, awards, and honorary degrees from several nations. In 1923 Banting was appointed the first full professor of medical research in the history of the University of Toronto. The chair was named the Banting and Best Chair of medical research. The greatest honor bestowed on Banting was the 1923 Nobel Prize in medicine or physiology. The Nobel committee named Macleod and Banting corecipients but did not include Best. This unjust omission angered Banting so much that he initially planned to refuse the award. Banting did accept the award to honor his nation of Canada, and he publicly acknowledged Best's contribution by pledging half of his monetary award to his colleague. This act motivated Macleod to share his award with Collip. Banting also insisted the order of names be changed to Banting and Macleod.

On June 4, 1925, Banting married Marion Robertson, a radiology technician whom he had met briefly before being discharged from the army. They had one son, William Robertson Banting, in 1929, but his marriage ended in divorce in 1932. Banting received custody of Bill.

RESEARCH ON SILICOSIS

After spending so much time touring, lecturing, and treating diabetic patients, Banting was eager to return to laboratory research. He now had a secretary, a laboratory technician, two graduate students, and his own cramped laboratory. Banting was a good teacher in the lab. He forced his students to think for themselves and taught them to speak simply and to get to the heart of the matter. His research interests included cancer, chemical treatments of mental disorders, and royal jelly (the food of the queen honeybee), but he made the most progress examining silicosis.

Silicosis is a lung disease that affected mostly miners. Symptoms included shortness of breath and a persistent cough, and the disease often resulted in total disability. Banting and his workers identified the cause as inhalation of silicon dioxide, which dissolved in the lungs to form silicic acid. The acid irritated the lining of the lungs and caused hardening, or fibrosis, to occur. Filtering the dust from the air was too cost prohibitive, so the Banting team explored other methods to prevent silicosis. They found that dispersal of a fine dust of aluminum powder into the air successfully prevented the formation of silicic acid in the miners' lungs.

HONORS AND LATER CAREER

The world continued to bestow new honors and responsibilities on Banting. In 1930 the University of Toronto opened a new research center, the Banting Research Institute. In 1934 he was created a Knight Commander of the Civil Division of the Order of the British Empire. As chairman of the Medical Research Committee of the National Research Council of Canada, he surveyed the national medical facilities and recommended the formation of a committee of Aviation Medical Research, which he chaired.

In 1937 he married Henrietta Ball, who worked at the Banting Research Institute on chemotherapy and tuberculosis research. In 1939 right before Canada declared war, Banting rejoined the army as a major. The government gave him the task of organizing and administering a major research program that included studies on decompression, the development of an antidote for mustard gas, and the invention of a protective flight suit for airmen. In February 1941, Banting was flying to England to present findings on the newly developed flight suits. The plane's engine failed during flight, and Banting died after a crash landing over Newfoundland. Thousands visited as Banting lay in state at the University of Toronto's Convocation Hall. Sir Frederick Banting was given full military honors at his funeral.

In recognition of Banting's major contributions to developing treatments for diabetes, the International Diabetes Federation established the Banting and Best Memorial Lectureship, and the American Diabetes Association established the Banting Medal and Memorial Lectureship. The Banting Research Foundation continues to commemorate Banting's discovery of insulin by supporting medical research for young Canadian scientists. The Banting and Best Diabetes Centre at the University of Toronto supports and advances diabetes research, education, and patient care.

As a result of Banting's efforts, millions of people afflicted with diabetes are living healthy, enjoyable lives. Though so-called experts on sugar metabolism believed the disease was hopeless, the dedicated, unpaid medical researcher toiled away in a hot, cramped, borrowed attic lab. Because Banting maintained hope, the miracle hormone insulin was discovered.

See also BIOTECHNOLOGY; DIABETES; DIGESTIVE SYSTEM; HOMEOSTASIS; RECOMBINANT DNA TECHNOLOGY.

FURTHER READING
Bliss, Michael. *The Discovery of Insulin.* Toronto: University of Toronto Press, 2000.

The Nobel Foundation. "The Nobel Prize in Physiology or Medicine 1923." Available online. URL: http://nobelprize.org/nobel_prizes/medicine/laureates/1923/index.html. Accessed January 13, 2008.

Beebe, William (1877–1962) American *Marine Biologist* William Beebe was one of the first men to venture into the ocean depths and record the remarkably diverse array of marine organisms. He pioneered the use of the diving helmet and the bathysphere for biological research, paving the way for the exploration of oceanic life.

BECOMES ZOO CURATOR

Charles William Beebe was born on July 29, 1877, in Brooklyn, New York, and moved to East Orange, New Jersey, during his early childhood. Before entering East Orange High School, he dropped the *Charles* from his name. He completed four years of Latin as well as two years of German, languages that would help him later in his career. In addition, he took several courses in the natural sciences. He was a strong student and very physically active. In his spare time he watched and listened to birds, memorized the local wildflowers, collected butterflies, and built up a bird skin collection. Before he even graduated, he had his first scientific publication, a letter to the editor of *Harper's Young People* about a bird, the brown creeper.

After high school Will matriculated as a special student at the University of Columbia, where he took many classes and attended several lecture series. Though he never received his degree, he did make the acquaintance of people who helped advance his career. One noteworthy man was the paleontologist Henry Fairfield Osborn, who was a professor at Columbia and was the curator of the American Museum of Natural History. Osborn was a founder of the New York Zoological Society (today called the Wildlife Conservation Society) and in 1895 became its first president. The society opened a Zoological Park (the present-day Bronx Zoo) in 1899.

In October 1899, the zoo hired Beebe as the first assistant curator for birds. Though he lacked formal training, popular magazines had already published several of his articles, and scientific publications such as *Science, The Auk,* and later *Zoologica* were beginning to do the same. In 1902 the zoo promoted Beebe to curator, and he piloted campaigns to build a spacious bird house and an enormous flying cage.

TRAVELS AND WRITES

Mary Blair Rice became Beebe's wife in 1902. She traveled with Beebe and collaborated on many of his writing projects. Together they traveled to Mexico during the winter of 1903 and 1904 with the goal of identifying and collecting Mexican bird specimens, especially those not indigenous to the southern United States. Mary was well educated and a talented writer in her own right. She assisted in writing Will's first published work, *Two Bird Lovers in Mexico* (1905).

Field research dominated Beebe's interests, and writing occupied his time. In 1906 Beebe published two books: *The Bird, Its Form and Function,* an introduction to ornithology, and another popular book intended to inspire amateur naturalists, *The Log of the Sun.* The poetic manuscript included essays on topics ranging from the life sciences to meteorology and was composed of 52 chapters, one for each week of the year. In 1910 Will and Mary wrote *Our Search for a Wilderness,* describing two expeditions they took together, one to northeastern Venezuela in 1908 and another to British Guiana (present-day Guiana) in 1909. Beebe observed the exotic wildlife and took home some birds for the Zoological Park. He captured 40 birds of 14 different species in Venezuela and 280 birds of 51 species in British Guiana. One interesting encounter was with the hoatzin, a strange bird whose young have claws on the back of their wings for climbing trees.

Beebe traveled with his wife to eastern Asia from 1909 to 1911 to study over 20 different pheasant genera. They visited 20 countries in over 17 months. They divorced shortly after returning, and Beebe took a privately funded five-year leave to pursue museum research and to complete the major scientific publication of his career, *A Monograph of the Pheasants.* Only 600 sets of this very expensive book were printed. The series was as popular among painters for the beautiful photographs and sketches of pheasants as it was among naturalists for the extensive knowledge it contained concerning pheasants. General information about each species, its distribution, description, and life history, was presented. Unusually for scientific writing, Beebe used the first-person singular and very colorful prose. He vividly shared his own personal adventures in searching for the birds.

The first volume of this project was published in 1918, but World War I delayed publication of the remaining three volumes. Almost 40 years old, Beebe volunteered for service during the war, though the nature of his service is somewhat unclear. He served through the French Aviation Service rather than the United States. While enlisted, he learned to fly and instructed other volunteers. After one year, he returned to the United States, but tastes of his wartime experiences peppered his future writings.

In 1926 and 1927, less scientific abridged versions of the four-volume *Monograph* were published. These editions, titled *Pheasant Jungles* and *Pheasants, Their Lives and Homes,* were aimed at a more general audience, and the latter contained fictionalized accounts of his actual experiences. Some scientists scoffed at Beebe's popular writings, claiming they detracted from his reputation as a respectable scientist, and they accused him of exaggerating many of the adventurous claims he recorded in his popular texts. Beebe continued to publish objective scientific accounts of his field research as well as write successful creative books for the general population.

After returning from his five-year pheasant sabbatical in 1915, Beebe traveled to Brazil to collect bird specimens for the thriving Zoological Park. While there, he was amazed at the number and variety of organisms located within a small region underneath one huge cinnamon tree. He pioneered the method of studying one small designated location for an extended period. Significantly, he discovered 76 different types of birds and over 500 total organisms within a few square feet. Beebe's interests switched from birds to tropical research.

In 1916 Beebe established the New York Zoological Society's first tropical research station at Kalacoon, near Georgetown, British Guiana, in the northeastern region of South America. The staff shared their quarters with scorpions, tarantulas, and vampire bats. Beebe found and studied 281 bird species while at Kalacoon. The Zoological Society published many of his findings and those of two other scientists from this research station in *Tropical Wildlife in British Guiana* in 1917. This book included observations on the bright-billed toucans, the reptile-like hoatzins, and the ground-dwelling tinamous. Beebe added to this account in *Jungle Peace* (1918), for which the former president, Theodore Roosevelt, wrote the introduction.

After returning from his war service, Beebe was given the title of honorary curator of the department of birds. In addition, in 1918 the society created a department of tropical research, which Beebe directed until he retired. When his staff returned to Kalacoon, the ecology of the region had changed as a result of the number of rubber trees that had been cut down for war supplies. They moved the research station to nearby Kartabo, at the junction of the Cuyuni and Mazaruni Rivers. Beebe wrote several scientific papers describing the flora and fauna of the jungle there. *Edge of the Jungle* (1921) and *Jungle Days* (1925) were both inspired by Kartabo. They focused on the ecology of jungle life. Beebe's professional interests were expanding once again. A 1926 issue of *Zoologica* included a paper he wrote on the three-toed sloth.

THE GALÁPAGOS ISLANDS

In spring 1923, Beebe journeyed to the Galápagos Islands. He spent two and one-half months at sea on the *Noma* and 100 hours on the islands themselves. The tameness of all the wildlife there, including the mockingbirds that ran up to welcome him rather than flying away, enamored the naturalist. Immersed in birds, sea lions, and iguanas, he pondered the irregular variations among the island species. The expedition spent some time anchored in Darwin Bay, which they named, where they were surprised to find bits of coral on the beach. Beebe was harmlessly

attacked there by a two-foot (0.6-m) moray eel, and they took pleasure in watching birds fight over prime nest-building sticks.

Soon after returning to New York with plant and animal species gathered for the Zoological Park and the American Museum of Natural History, Beebe ventured out again, this time on a steam yacht called *Arcturus*. His main focus now was oceanography, in particular creatures of the sea. He also planned to study the Sargasso Sea south and east of Bermuda and the Humboldt Current, which moves up the Pacific coast of South America toward the Galápagos. The *Arcturus* departed Brooklyn in February 1925 for a six-month trip. Though storms stirred up the Sargasso Sea too much for study to be useful and the Humboldt Current was unexpectedly absent, they collected much valuable information. After five weeks out, the ship needed some repairs, so they rested at Fort Sherman, Panama, for a while. Beebe took pleasure in examining the wildlife there.

Next they anchored at their previous lodging, Darwin Bay, and continued exploring and collecting. Beebe began to use a copper diving helmet as an integral part of his field research. The helmet allowed him to remain submerged for long periods. A leather tube ran from the helmet to a vessel above the surface, and a person hand-pumped fresh air down the tube to the diver. Using a helmet, he was able to collect specimens that had never before been identified and to view marine life in the natural environment. Animals were taken above the surface for further live study in his aquariums or by dissection.

While they were stationed at Darwin Bay, the crew happened to observe volcanic fires from Albemarle, the largest island of the Galápagos archipelago. They set out in that direction. When Beebe crazily decided to explore up close with his foremost assistant, John Tee-Van, the gas and smoke made them nauseous, and Beebe temporarily lost some sight and speech. After stumbling back to the ship, he was severely dehydrated and exhausted, but he recovered. When passing by again nine weeks later, they were amazed to witness the red hot lava flowing into the ocean waters. The hot waters killed many fish that swam too close. Animal scavengers became ill from unknowingly approaching the gaseous exhalations and died as well. The crew watched one sea lion tragically jump high out of the hot water right into the lava.

When Beebe left the Galápagos in June 1925, he had plenty of research material. As usual, he shared highlights of his experiences as well as scientific information through his writing. *Galápagos: World's End* and *The Arcturus Adventure* were published in 1924 and 1926, respectively. At the time, there was still much debate over the origin of the islands. Some believed, as scientists do today, that the archipelago was formed from a volcanic hotspot that spewed out hot magma that piled up over time to form the individual islands. In *Galápagos: World's End,* Beebe stated his belief that the islands were originally one continuous landmass that sank, leaving portions above the sea surface, thus creating the archipelago. This would explain the similarities between species on the islands, yet allow for evolution of unique characteristics by adaptation over time. Beebe also believed that a land bridge formerly existed connecting the Galápagos Islands to the Cocos Ridge. This would explain how terrestrial animals and insects and spiders originally arrived on the islands. Others who believed in the idea of a former land bridge thought it probably connected with Ecuador, which would have been closer. However, Beebe always remained open-minded if proof otherwise was presented. Two years later, on the *Arcturus* expedition, Beebe himself made observations that led him to believe that terrestrial creatures could have entered and inhabited the islands in the absence of a land bridge.

During his *Arcturus* expedition, Beebe set up a sea station halfway between the Galápagos Islands and Central America, 60 miles (97 km) south of Cocos Island. A sea station was a temporary designated area where the boat remained stationary or slowly circled in order to gather data. While in that one spot, the staff hauled up nets, dredged, took bottom samples, recorded temperatures, and made other observations. At this one station, named number 74, they captured a remarkable 136 fish species and more than 50 crustacean species in a 10-day period. In this location, Beebe took surface hauls every 30 minutes for an entire day. This allowed him to observe that some fish only surfaced during the daytime, while others surfaced only at night. This was useful information for marine biologists so that if they wanted to study a particular type of fish, they knew when they were most likely to find it. He also figured out that data on luminescent fish were best gathered at nighttime, and he was able to make rare observations of living luminescent fish.

HAITI AND BERMUDA

By July the expedition had returned to New York, leaving Beebe to sort through his data and numerous specimens. Hooked on marine biology, the following year he set out for Haiti on a schooner chartered by the New York Zoological Society. A biological station was set up on board the *Lieutenant* in the Bay of Port-au-Prince, where they stayed for four months. The goal was to identify fish in Haitian waters and explore the coral reefs. From this expedition he wrote *Beneath Tropic Seas* (1928). A list of 270 species was published in *Zoologica* in 1928 and increased to 324

species when the list was supplemented in 1934. To study coral reef life Beebe depended on his diving helmet. In more than 300 dives Beebe observed diverse life-forms of the coral reef, then classified them by their ecological niche.

While in Haiti, Beebe met Elswyth Thane, a writer for newspapers and motion picture studios. They married in 1927. The two traveled together and separately, both of them involved in their own research.

In 1928 Beebe obtained permission from the British government to carry out studies on the island of Nonsuch in Bermuda. The region of focus was eight miles (12.9 km) in diameter and ranged from 6,000 to 8,000 feet (1,829 to 2,438 m) deep. Most of Beebe's research over the next 11 years was carried out here, including his most famous deep sea dives. Of course, the goal was to study fish from the deep sea as well as from the shores. Beebe and Tee-Van published *Field Book of the Shore Fishes of Bermuda* (1933). They used many of the common research methods that had been practiced for decades, including trawling, dredging, and hauling silk nets, but Beebe found these means limiting.

SETS RECORDS IN BATHYSPHERE

Beebe had been the first to use a diving helmet as an integral part of his field research rather than just for pleasure, but its utility was limited. Though the helmet was fine for dives between 15 and 50 feet (4.6 to 15.2 m) deep, by 100 feet (30.5 m) it became unsafe. Beebe contemplated alternative means of underwater exploration. In 1928 he teamed up with a young man named Otis Barton, a trained engineer, who drew up blueprints for a sphere-shaped vessel and funded its construction. One major feat of the vessel design was to withstand the enormous pressure from the ocean depths. The proposed round bathysphere was well suited to distribute the extreme external pressure evenly.

The bathysphere weighed 5,000 pounds, the outer diameter of the newly named "bathysphere" was four feet nine inches (145 cm), and the walls were one and one-half inches (3.8 cm) thick. The circular door was a mere 14 inches (35.6 cm) in diameter, barely enough for a grown man to wedge himself through. The door was fastened by 10 bolts. There were three window ports, two of which were filled with eight-inch (20.3-cm) fused quartz disks, three inches (7.6 cm) thick. The third was plugged. There were four legs on the bottom to which wooden skids were attached. Inside were pans of calcium chloride to absorb moisture and soda lime for absorbing excess carbon dioxide. Of course, oxygen tanks supplied air for breathing. To circulate air inside the chamber, the two men carried palm-leaf fans.

The bathysphere was specially outfitted for deep underwater exploration of marine life. (© Wildlife Conservation Society)

An electrical line and a communications line were wrapped together in a cable one and one-half inches thick and fed into the top of the sphere. The 3,500 feet (1,067 m) of steel cable necessary to lower the bathysphere into the water weighed 4,000 pounds. Two steam winches on deck moved the enormous hollow ball.

Unmanned test descents commenced in early June 1930. The first test resulted in a tangled mess of communication lines. After adjustments and another unmanned descent, the first manned descent occurred on June 6. Despite a minor leak and a pop from an electric switch at around 300 feet (91.4 m), Beebe and Barton achieved a depth of 800 feet (244 m). Beebe's most amazing observation concerned the colors viewed below the surface. Whereas the water began a light greenish color, as they descended it turned bluish green, then a pale blue, then a blackish

blue. On June 10, Beebe and Barton made another attempt, but at 250 feet (76.2 m) the communications line went out. Without being able to hear the human voice from the surface through Beebe's headphones, the men felt very isolated. The crew pulled them up immediately.

After cutting off 300 feet (91.4 m) of cable, they went down again the following day. They descended slowly, making verbal observations every foot of the way. Beebe's experience in ichthyologic identification qualified him for this task. He was thrilled to observe many specimens that had previously only been seen dead in net hauls. However, many were brand new, and Beebe and Barton relished the opportunity to observe them swimming in their natural environment. At 1,426 feet (435 m), they paused and returned to the surface. Barton donated the bathysphere to the New York Zoological Society that fall.

Two years later history was made again, but this time the world was invited along. The National Broadcasting Company arranged a live radio broadcast of a dive one Sunday afternoon in September 1932. The weather had caused delays, and the sea was still rougher than would normally be acceptable, but the world was waiting. Beebe and Barton descended on their 20th deep dive into the sea, and their eager reports to Gloria Hollister aboard the barge were relayed to America. The British Broadcasting Corporation was also connected by short-wave radio, increasing the listening audience.

In the bathysphere all light had disappeared by 1,700 feet (518 m), but as they continued descending, the numbers of luminescent fish increased. They turned around after dangling momentarily at 2,200 feet (671 m), and on the way back up Beebe spotted two six-foot- (1.8-m-) long fish that he named *Bathysphaera intacta* (untouchable bathysphere fish). He claimed their teeth were luminous, and a linear formation of lights ran along their sides. Later, others doubted Beebe really saw these, believing perhaps they were a few fish swimming end to end. On this trip, a spiny lobster had been tied to the outside of the bathysphere. They expected it to be crushed and act as bait to attract fish to the submerged bathysphere for observation. Astonishingly, the lobster survived the thousands of tons of pressure and went on to live in Beebe's aquarium.

The following year, the bathysphere was displayed at the Century of Progress Exposition in Chicago. The president of the National Geographic Society asked Beebe to consider one more bathysphere expedition. The National Geographic Society would sponsor it, and they did not stipulate an attempt at a new depth record. This sounded attractive to Beebe, who later stated that it was the lack of a request for a new record that made him determined to set one.

The new record dive occurred on August 15, 1934. They reached 3,028 feet (923 m), over one-half mile deep. This record was unbeaten for 15 years. Actually, Beebe and Barton had descended to 2,510 feet (765 m) a few days prior, but Beebe felt that the second dive was totally different despite the exact same location. Several new species were named. Again, Beebe noted the increased number of luminescent fish in deeper regions, as well as that larger creatures were more prevalent.

The bathysphere experiences were invaluable not only because they set records and revealed undiscovered species, but because they challenged oceanographers to develop better methods for undersea studies. Beebe emphasized that what was viewed directly was so much different from what was inferred from deep trawling or dredging or net hauls. Observing marine life in the natural environment was much more informative. Many brand new species were identified and others seen live for the first time. Organisms such as siphonophores (including the Portuguese man-o-war) could be viewed in their true form, rather than as a tangled-up mess of debris pulled to the surface. Information on vertical distribution and relative abundance of different species of fish could also be obtained.

After this season, the bathysphere was retired. Just as a diving helmet did, the sphere had to remain tethered to a surface vessel. Though it had descended over 3,000 feet (914 m), its depth were still limited, as was its lateral mobility. Beebe continued studying oceanography on a yacht named *Zaca*. He continued to use his diving helmet and was as zestful and enthusiastic at the age of 60 as he was at 25 years old. His last sea voyage departed in April 1938.

RETIREMENT

During World War II, Beebe was unable to continue his research off Bermuda, so he returned to jungle research at Caripito, Venezuela. Beebe's last sea book, *Book of Bays,* was published in 1942. In it he expressed concern over man's threat to the world's ecosystems. In 1945, he established another research station, at Rancho Grande in Venezuela, then yet another in 1950, at Simla, in the Arima Valley of Trinidad. He personally purchased this land and lived there during the wintertime. He retired in 1952 and eventually sold this land and its 200 associated acres to the New York Zoological Society for one dollar. In 1955 Beebe made one last trip to Asia to check on the pheasant populations he had studied 45 years before.

Though his spirit remained robust, his health failed during the last three years of his life. He was no longer able to ride his bike around the Zoological Park commanding visitors to go check out the new bird exhibit as he had in his younger days. Unable to tolerate the cold of New York, he spent the months of October through May at Simla. He grew physically weaker and his speech sporadically was slurred. Though he expressed hopes of dying of heart failure from viewing an unexpected amazing natural phenomenon, Will Beebe died of pneumonia on June 4, 1962. He was buried in Trinidad.

Simla was renamed the William Beebe Tropical Research Station, but it was functional for only a few more years. The football field–sized bird house at the Zoological Park was replaced by a newer complex in 1972. The bathysphere is on display at the New York Aquarium, which is managed by the Wildlife Conservation Society. During his lifetime, Beebe was criticized for making up such fantastical, outlandish tales of extravagant undersea creatures. Not until years later did advancements in underwater photography enable others to verify his claims. Vindicated, he was ultimately recognized for discovering hundreds of new species.

Beebe's scientific articles and notes are recorded in scores of *Zoological Society's Bulletin* issues and other scientific journals. He conveyed his passion for deep sea life in his two dozen popular books of naturalist adventures, inspiring a new wave of scientists and naturalists including Sylvia Earle and Rachel Carson.

See also CARSON, RACHEL; EARLE, SYLVIA; MARINE BIOLOGY.

FURTHER READING

Barton, Otis. *The World beneath the Sea*. New York: Crowell, 1953.
Beebe, William. *Adventuring with Beebe: Selections from the Writings of William Beebe*. New York: Duell, Sloan and Pearce, 1955.
———. *Half Mile Down*. New York: Harcourt, Brace, 1934.
Welker, Robert Henry. *Natural Man: The Life of William Beebe*. Bloomington: Indiana University Press, 1975.
Yount, Lisa. *Modern Marine Science: Exploring the Deep*. New York: Chelsea House, 2006.

Bigelow,   Henry (1879–1967)   American *Zoologist* Henry Bigelow was a pioneering ocean researcher of the 20th century. As the first person to perform a comprehensive study of the Gulf of Maine, not only did he collect vast amounts of data during his research, but as a result, he recognized and emphasized the importance of physical, chemical, and biological unity in studying the sea. His recognition of the need to study the complex interdisciplinary nature of the sea led to the establishment of the esteemed Woods Hole Oceanographic Institution, which has supported leading oceanographic research for 75 years.

EDUCATION AND TRAINING

Henry Bryant Bigelow was born to a banker, Joseph Smith Bigelow, and Mary Cleveland Bryant Bigelow on October 3, 1879, in Boston, Massachusetts. As a youth, he often traveled to Europe with his family and developed a love for the outdoors and for sports. The family spent summers at the quaint harbor town of Cohasset, Massachusetts. Henry graduated from the Milton Academy in 1896 and then took courses at the Massachusetts Institute of Technology while working at the Boston Museum of Natural History. He enrolled at Harvard University in 1897 and graduated cum laude four years later. Though in his memoirs he reported that he did not have much of a social life during college, he made contacts with influential individuals that shaped his career.

Bigelow's early scientific interest was birds. He went on a trip to the Canadian province of Newfoundland and Labrador in 1900. His first scientific publication on the American eider (a northern sea duck with soft down), "A Virginia Record for the American Eider," was published in 1901 in the respected ornithology journal *Auk* while he was still an undergraduate. The following year he published a more substantial article, "Birds of the Northeastern Coast of Labrador," also in *Auk*.

In 1901 he was invited to accompany a Harvard professor who was also the director of Harvard's Museum of Comparative Zoology (MCZ), Alexander Agassiz, on an expedition to the oceanic island group the Maldives, located in the Indian Ocean, southwest of Sri Lanka. Bigelow's responsibility was caring for the jellyfish and the siphonophores they collected. He enjoyed the fieldwork during this experience, which sparked an interest in marine invertebrates and taught him the basics of taxonomy, the classification of species. He published papers on the

Henry Bigelow, shown here on the deck of the USS *Grampus* while exploring the Gulf of Maine, was a systematic zoologist who merged the physical and life sciences in his pioneering studies of the ocean. *(National Oceanic & Atmospheric Administration/Department of Commerce)*

medusas of the Maldive Islands in 1904 and 1909, establishing himself as a knowledgeable marine biologist. During the period 1904–05, Bigelow traveled to the eastern tropical Pacific with Agassiz, and in 1907, to the West Indies.

In 1904 Bigelow earned his master's degree from Harvard and in 1906, his doctorate. The topic of Bigelow's doctoral dissertation was the nuclear cycle of *Gonionemus vertens,* a hydrozoan that attaches to eelgrass or sea lettuce using adhesive disks on its tentacles. This experience studying cellular structure and function impressed upon Bigelow the necessary discipline for laboratory research.

Bigelow married Elizabeth Perkins Shattuck in 1906, and they eventually had four children together. His wife often accompanied him on his research travels.

GULF OF MAINE STUDIES

After obtaining his doctorate, Bigelow was given a position as an assistant at Harvard's MCZ, where he catalogued specimens. A visit by John Murray, a Scottish oceanographer who specialized in studying the ocean bottom, prompted Bigelow to investigate the virtually unknown Gulf of Maine. In 1912 he borrowed the schooner *Grampus* from the U.S. Bureau of Fisheries and began an extensive study of the Gulf of Maine that lasted for 12 years and was supported jointly by the MCZ and U.S. Bureau of Fisheries. This work was revolutionary because it was so comprehensive—Bigelow studied everything related to the sea in the area. He collected over 10,000 net hauls of marine organisms, sent out more than 1,000 drift bottles to study the water's currents and flow, and set up hundreds of stations to measure water temperature and salinity levels. He became an expert on fishes and coelenterates, which are aquatic invertebrates with a radially symmetric saclike body and a single internal cavity. Examples include jellyfish and hydras.

The extensive data and results from these studies were published in 1924 in three monumental monographs: *Fishes of the Gulf of Maine, Physical Oceanography of the Gulf of Maine,* and *Plankton of the Offshore Waters of the Gulf of Maine.* In the latter, Bigelow described his fortune in having a "veritable *mare incognitum* lay before us" as they set out on the first oceanographic cruise in the gulf to examine not only the pelagic fauna (the animals living in the open sea), but their ecological role, geographical variations, seasonal successions and migrations, and temperature preferences. According to Alfred C. Redfield, the author of Bigelow's memoir for the National Academy of Sciences (NAS), Bigelow's research on the Gulf of Maine made it the most thoroughly studied body of water of comparable size in

the world. During these years, Bigelow switched his research focus from cytology and zoology to oceanography and developed an appreciation for the need to understand all of the natural sciences in order to comprehend the complex cycle of the sea. Throughout his career, he would impress this conviction onto his colleagues and students, paving the way for modern oceanography.

HYDROGRAPHY AND OCEANOGRAPHY WORK

In 1919 Bigelow took a temporary break from his Gulf of Maine studies to teach navigational skills and to serve as a navigation officer aboard the U.S. Army Transport *Amphion.* He also assisted the U.S. Shipping Board and was a consultant for the U.S. Coast Guard for the International Ice Patrol. From examinations of the plankton drifting, surface temperatures, and salinity, he drew conclusions about drifting icebergs. The hydrography knowledge he gained ultimately benefited his research. Bigelow accepted a teaching appointment at Harvard in 1921.

As secretary for the Committee on Oceanography of the NAS, Bigelow composed a report titled "On the Scope, Problems, and Economic Importance of the Oceanography, on the Present Situation in America, and on the Handicaps to Development, with Suggested Remedies" in 1929. This well-received, influential report was made public in the form of a book, *Oceanography, Its Scope, Problems, and Economic Importance* (1931). In his report, Bigelow portrayed oceanography as a youthful field and defined it in terms of three chief subdivisions:

- geological
- physical-chemical
- biological

He recommended that oceanographers be grounded in the principles of all three disciplines, for which he provided overviews summarizing the current knowledge. In portions of this report that were not published, Bigelow made specific recommendations for how best to resolve the current deficiencies in the state of oceanographic research in the United States. To substantiate the recommendations, Bigelow described the applications of oceanography and discussed the economic ramifications of improved research in the areas described. The Woods Hole Oceanographic Institution (WHOI) was established on the basis of Bigelow's report and the Rockefeller Foundation donated $2.5 million. In addition, the Scripps Institution, the University of Washington, and the Bermuda Biological Station received financial assistance.

Bigelow served as the first director of WHOI from 1930 to 1939. He made daily rounds, making himself accessible to his scientists, and he requested that his staff perform field research at least once a year. The vessel *Atlantis* was made available for them to do so. Bigelow successfully recruited eminent biologists, chemists, and physical geologists without worrying whether or not they had already studied oceanography. He was more concerned with attracting clever, creative scientists who were willing to apply their talents to oceanography. Today, WHOI remains dedicated to research and education in the marine sciences and is the largest independent oceanographic institution in the world. Although Bigelow resigned as director of WHOI in 1939, he maintained a close association with the institute by serving as president of the trustees (1940–50) and then chairman of the board (1950–60).

In the early 1940s, Bigelow researched Georges Bank from the *Atlantis*. Fishermen depended on this region for haddock; thus scientists were exploring the cause of its long phytoplankton season compared to that of the Gulf of Maine. In collaboration with W. T. Edmonson, Bigelow published *Wind Waves at Sea, Breakers, and Surf* in 1947. The popular book summarized the physical nature of wind waves, wave dimensions and contours, and the effect of waves on small vessels, and it served as an introduction to waves for the U.S. Navy, whose use of small vessels and other amphibious craft required such detailed knowledge.

Bigelow served as editor in chief of the series *Fishes of the Western North Atlantic* (1948–64), a cooperative publication written for ichthyologists as well as general naturalists. In collaboration with William C. Schroeder, Bigelow contributed more than 40 papers on ichthyology between 1948 and 1965.

CAREER HONORS AND ACHIEVEMENTS

At Harvard, Bigelow had been appointed a lecturer (1921), associate professor of zoology (1927), professor of zoology (1931), and the Alexander Agassiz Professor of Zoology (1944). Bigelow retired as a professor emeritus from Harvard in 1950, but he continued serving on the faculty for the MCZ until his death. He had served as curator for coelenterates (1913–25), research curator (1925–27), and curator of oceanography (1927–50). When Bigelow jokingly suggested, in 1960, that the university show their appreciation for his lengthy tenure, the president of Harvard presented him with a bottle of bourbon whiskey.

Bigelow had served on several influential committees and did not shy away from administrative functions: the National Research Council Committee on Oceanography (1919–23); the National Research Council Committee on Submarine Configuration, for which he served as vice-chairman (1930–32); and the NAS Committee on Oceanography, for which he was secretary (1928–34) and chairman (1934–38).

Bigelow was the distinguished recipient of numerous medals and honors and was elected to membership of numerous academic organizations. WHOI established a chair in oceanography in his name in 1958. WHOI's board of trustees established the Henry Bryant Bigelow Award, WHOI's highest honor, for those who make significant inquiries into the phenomena of the sea. The first recipient of the medal and cash prize was Bigelow himself in 1960. He also received the Alexander Agassiz Medal of the NAS. Several universities awarded Bigelow honorary doctorate degrees during his lifetime.

Henry Bigelow died on December 11, 1967, at his home in Concord, Massachusetts. A laboratory at WHOI bears his name, as does a marine research institution in West Boothbay Harbor, Maine, which was established in 1974. In 1970 the U.S. Department of the Interior named a bay in the Gulf of Maine, located between Cape Ann and Cape Small, Bigelow Bight.

In addition to the vast amount of information he collected concerning various forms of marine life and for his comprehensive study of the Gulf of Maine, Bigelow is considered a pioneer in oceanography because of the direction in which he led the field. Before the 1930s, the field was merely an assemblage of facts and data—lists of identified fauna, maps of ocean depths at different positions, and locations and directions of currents. Bigelow encouraged oceanographers to synthesize all the information gathered and look for relationships, to embrace the different fields of biology, chemistry, and physical geology to find the connections. This effort for unification gave oceanography an ecological aim, which has persisted into the 21st century. Redfield, in his memoir written for the NAS, summarized Bigelow's contributions to oceanography, "Not only did a man emerge who had prepared himself, perhaps unwittingly, for leadership at a time when men of influence sensed that something should be done to improve the status of marine science in America, but new ideas were in the air, wafted across the ocean from a multitude of general scientific advances. Henry Bigelow, though trained in the classical tradition, was sensitive to these breezes, wise enough to grasp their implication, and bold enough to act on their meaning."

See also BIOMES, AQUATIC; ECOSYSTEMS; MARINE BIOLOGY.

FURTHER READING
Bigelow, Henry Bryant. *Memories of a Long and Active Life.* Cambridge: Cosmos Press, 1964.
———. *Oceanography: Its Scope, Problems, and Economic Importance.* Boston and New York: Houghton Mifflin, 1931.

Bigelow Laboratory for Ocean Sciences home page. Available online. URL: http://www.bigelow.org. Accessed January 17, 2008.

Redfield, Alfred C. "Henry Bryant Bigelow (October 3, 1879–December 11, 1967)." In *Biographical Memoirs: National Academy of Sciences*. Vol. 48. Washington, D.C.: National Academy of Sciences, 1976.

Woods Hole Oceanographic Institution home page. Available online. URL: http://www.whoi.edu/home/. Accessed January 17, 2008.

biochemical reactions A chemical reaction occurs when substances react to form new substances with different chemical properties. Biochemical reactions are simply chemical reactions that naturally occur within or are caused by living organisms, and they typically require enzymes, biological catalysts. Most enzymes are proteins, and all enzymes are highly specific for a particular substrate. Enzymes act by increasing the rate of a biochemical reaction, but they are not altered or consumed during the reaction. The different types of biochemical reactions that occur in living systems can be grouped into six main classes: oxidation-reduction, reactions that involve the transfer of a functional group, hydrolysis reactions, ligation reactions, elimination or addition reactions, and isomerizations.

OXIDATION-REDUCTION REACTIONS

Cells obtain the energy they need to live, grow, and reproduce by oxidizing organic molecules such as carbohydrates. Chemically, to oxidize a compound means to remove one or more electrons from it. When cells burn organic molecules for energy, they are removing the electrons—the electrons that atoms of the molecule share in covalent bonds. Oxidizing or breaking down biomolecules releases the energy stored in the shared electron pairs that make up covalent linkages. Because the element oxygen is prevalent in the Earth's atmosphere and because it has a high electronegativity, oxygen often participates in the process of oxidation (hence its name), though it is not required. Any biochemical reaction involving oxidation must couple with a reduction reaction, because if something is oxidized, something else must be reduced. In other words, if something gives up an electron, something else must take it. The same number of electrons present in the reactants of a chemical reaction must be present in the products of the reaction. The atom or molecule that accepts the electron lost by oxidation becomes reduced when it accepts that electron. When a molecule accepts an electron, it often also picks up a proton (H^+) also, a process called hydrogenation because the molecule has gained a hydrogen atom (which is simply a proton combined with an electron). Hydrogenation reactions are reductions, and the converse, dehydrogenation reactions, involving the removal of a hydrogen atom, are oxidations.

The gradual oxidation of an organic molecule allows the cell to extract its energy more effectively than if all of the electrons were lost at once. (Consider a boy playing catch. If someone throws balls to him one at a time, he can catch more total balls than if someone throws him a dozen balls at once.) During cellular respiration the cell uses molecular oxygen (O_2) to oxidize organic molecules such as glucose ($C_6H_{12}O_6$) gradually, uses the energy to make adenosine triphosphate (ATP), and releases carbon dioxide (CO_2) and water (H_2O). For a single molecule of glucose, the overall process of respiration can be summarized as follows:

$$C_6H_{12}O_6 + 6O_2 \rightarrow 6CO_2 + 6H_2O + {\sim}30\text{--}30\ ATP$$

As O_2 oxidizes the carbon atoms of glucose, the electrons (and accompanying protons) are transferred to oxygen, which becomes reduced, forming water molecules in the process. Notice that the carbon atoms do not completely give up any electrons; they have not ionized. The carbon atoms traded covalent linkages with other equally electronegative carbon atoms in order to form covalent linkages with oxygen, which is much more electronegative than carbon; the new covalent bond is a polar covalent bond. The carbon atoms gave up more than their fair share of electrons and acquired a partially positive charge; thus this is still considered oxidation.

Enzyme names should indicate their activity, and biochemists have given enzymes a variety of names to indicate the catalysis of oxidation-reduction reactions. The names often include the term *dehydrogenase, oxidase, peroxidase, hydroxylase, reductase,* or *oxygenase*. For example, malate dehydrogenase catalyzes the transfer of two electrons from malate to NAD^+ to form oxaloacetate and $NADH + H^+$ during the citric acid cycle. ($NADH + H^+$ then carries the electrons to the electron transport chain for the synthesis of ATP.)

TRANSFER OF FUNCTIONAL GROUPS

Organic molecules often have hydrophilic side groups that increase their solubility in aqueous environments and confer unique chemical and physical characteristics on the molecule. Common side groups, called functional groups (-X), in biomolecules are hydroxyl groups (-OH), carbonyl groups (-C=O), carboxyl groups (-COOH), amino groups ($-NH_2$), phosphoryl groups ($-PO_3$), sulfhydryl groups (-SH), and alkyl groups ($-C_nH_{2n+1}$). Enzymes that catalyze the transfer of a functional group from one molecule to another often have names that include *transferase* or a more specific term that indicates the type of group it transfers, such as transmethylase, which catalyzes the transfer of a methyl group.

$$AX + B \xrightarrow{\text{transferase}} A + BX$$

An example of an enzyme that performs a transfer is aspartate aminotransferase, also called aspartate transaminase, which moves an amino group from aspartate to α-ketoglutarate, resulting in the formation of oxaloacetate and glutamate.

HYDROLYSIS AND LIGATION REACTIONS

Hydrolysis reactions, named so because they involve the lysis or splitting of a water molecule, catalyze the breakdown of polymers by cleaving C-O, C-N, or C-C bonds (and sometimes other types). The enzyme essentially splits the water molecule and adds its components to either side of a covalent bond, breaking it and releasing two smaller molecules and energy in the process. The class of enzymes that facilitate hydrolysis reactions are the hydrolases, including enzymes such as esterases, amidases, glycosidases, peptidases, and phosphatases, named for the type of bond cleaved. For example, amylase is a glycosidase that cleaves glycosidic linkages of starch molecules to release single glucose subunits.

Dehydration syntheses, also called condensation reactions, are the reverse reactions of hydrolysis reactions. They join two molecules by removing a water molecule (an OH from one and an H from the other) and require energy input, usually obtained by the cleavage of an ATP molecule. Because dehydration synthesis reactions are anabolic, meaning they build large molecules from smaller ones, the common names of the enzymes that catalyze these reactions often include the term *synthetase*. As a result of confusion with enzymes that biochemists commonly called synthases (which belong to the class of enzymes called lyases and that catalyze elimination or addition reactions), newer enzyme names use the term *ligase* rather than *synthetase*. Ligases are enzymes that catalyze the formation of ester, thiol ester, or amide linkages. For example, glutamate-ammonia ligase, formerly known as glutamine synthetase, catalyzes the condensation of an ammonium ion and the amino acid glutamate to synthesize the amino acid glutamine using energy released by the hydrolysis of ATP Deoxyribonucleic acid (DNA) ligase forms phosphodiester linkages to connect adjacent nucleotides along a single strand of DNA in a duplex molecule, such as in between the Okazaki fragments formed as a result of discontinuous replication of the lagging strand during DNA synthesis.

ELIMINATION OR ADDITION REACTIONS

Elimination reactions are biochemical reactions in which two substituents are removed from a molecule, typically resulting in the formation of a double bond or a ring. The reverse addition reactions involve adding groups across a double bond. Lyases are enzymes that remove a small molecule, such as an ammonia, water, or carbon dioxide, from another molecule to form a double bond during an elimination reaction, or that catalyze the cleavage of C-C, C-O, or C-N bonds during addition reactions, converting a double bond to a single bond in the process. These reactions are often catalyzed by enzymes with names including the terms *aldolase, synthase, deaminase, hydrase, decarboxylase,* or *cyclase*. One example of such a reaction is the conversion of the four-carbon sugar fumarate to malate by the addition of water to its double bond. A step in the central metabolic pathway the citric acid cycle, this reaction is catalyzed by fumarase, which also catalyzes the reverse reaction, formation of the double bond by the expulsion of a water molecule.

ISOMERIZATIONS

Isomers are molecules that have the same number and types of atoms but different structural arrangements of those atoms. Because many biological molecular interactions are highly specific, two molecules that have the same chemical formula may have very different biochemical functions. Some isomers differ in the location where a functional group attaches; others differ in the geometric arrangement of atoms or side groups. Large biomolecules such as DNA can form loops or twists that alter its overall topology; these different forms are called topoisomers. Isomerases move around the atoms within a molecule to form different isomers. Enzymes that include the

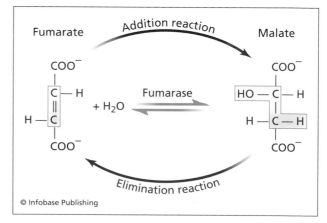

© Infobase Publishing

Fumarase catalyzes the addition of water to the double bond of fumarate, forming malate, and the elimination of water from malate to form fumarate.

terms *isomerase, racemase, epimerase, mutase,* or *tautomerase* are all isomerases. A specific example of an isomerization reaction is the conversion of 1,3-bisphosphoglycerate to 2,3-bisphosphoglycerate (BPG), catalyzed by bisphosphoglycerate mutase. In red blood cells, the molecule 2,3-bisphosphoglycerate binds hemoglobin and lowers its affinity for oxygen. This phenomenon plays an important role in regulating the transfer of oxygen from maternal to fetal circulation and in the adjustment to higher altitudes, where the partial pressure of oxygen in the atmosphere is decreased.

See also BIOCHEMISTRY; BIOENERGETICS; BIO-MOLECULES; CHEMICAL BASIS OF LIFE; ENZYMES.

FURTHER READING
Berg, Jeremy M., John L. Tymoczko, and Lubert Stryer. *Biochemistry.* 6th ed. New York: W. H. Freeman, 2006.

Champe, Pamela C., Richard A. Harvey, and Denise R. Ferrier. *Lippincott's Illustrated Reviews: Biochemistry.* Philadelphia: Lippincott, Williams, & Wilkins, 2005.

Purich, Daniel L. *The Enzyme Reference: A Comprehensive Guidebook to Enzyme Nomenclature, Reactions, and Methods.* San Diego, Calif.: Academic Press, 2002.

biochemistry Biochemistry is the study of chemical compounds and processes that occur in or are caused by living organisms. Considered a subdiscipline of both biology and chemistry, biochemistry bridges the two natural sciences. At the molecular level, all life-forms are similar because they all descended from a common ancestor. The molecules that make up the cells and tissues of living organisms, called biomolecules, are the same in prokaryotes and eukaryotes, in unicellular and multicellular organisms, and across kingdoms. Biochemistry focuses on these molecules and the biochemical reactions that form them and convert them into other substances.

OVERVIEW

Biochemistry involves the study of the structure and function of biomolecules, the major classes of which include carbohydrates, proteins, lipids, and nucleic acids, though other types that play important roles in the biochemical processes carried out by living cells, such as vitamins, are also of interest to biochemists. Most biomolecules consist of many repeated subunits called monomers. The monomeric subunits differ among the major types; carbohydrates consist of monosaccharides, proteins consist of amino acids, nucleic acids consist of nucleotides, and lipids, which are more structurally variable, contain hydrocarbon chains or fatty acids of different lengths. One characteristic all biomolecules have in common is that they are organic molecules, meaning their structure is based on the element carbon. Organic chemistry is the branch of chemistry concerned with carbon-based compounds; thus biochemists must have a strong background in organic chemistry. Though all biomolecules are organic molecules, not all organic molecules are considered biomolecules. For example, benzene, synthetic rubber, and fossil fuels such as petroleum and coal are organic compounds but do not naturally occur in living organisms so are not considered biomolecules. Biomolecules assemble to form macromolecular structures, which compose cells. For example, lipids and proteins combine to form cell membranes, nucleic acids and protein join to form chromatin, ribonucleic acid and proteins assemble into ribosomes, and DNA or ribonucleic acid (RNA) and proteins form viral particles.

Biochemists study all aspects of biomolecules including their basic structure, how they are synthesized, how they are broken down, and their function. Most chemical process that take place within living cells require special catalysts called enzymes, which are usually proteins, but some RNAs also have catalytic properties. Catalysts are substances that speed up the rate of a chemical reaction. All biochemical reactions that occur in living cells could potentially occur without the aid of a catalyst, as catalysts cannot force energetically unfavorable reactions to proceed, but in the presence of the catalyst, the reaction proceeds at 1,000 to 100,000,000 times faster. Enzymes mediate the reactions that convert biomolecules from one form to another; they join monomers, modify molecules, and digest large molecules into smaller ones. Biochemists study how enzymes work and the reactions they catalyze. The reactions are often part of a series of reactions called a metabolic pathway. Metabolism involves all of the anabolic and catabolic activities of an organism. Anabolic pathways are responsible for constructing macromolecules and require energy input, whereas catabolic pathways perform the breakdown of macromolecules into smaller molecules and release energy for use by the cell. Biochemical pathways can be linear, in which one enzyme acts on a substrate, leading to conversion of the substrate into a product, which then acts as the substrate for a second enzyme, and the product of the second reaction becomes the substrate for a third enzyme, and so on. Some biochemical pathways are cyclical, meaning that the initial substrate is regenerated after several steps. Many biochemical pathways, or at least portions of them, function in anabolism and catabolism, and many enzymes catalyze reversible reactions. The regulation of metabolic pathways is also an important topic in biochemistry.

Though only four chemical elements—carbon, hydrogen, oxygen, and nitrogen—compose more than 96 percent of the total weight of living matter, the manner in which they join to one another and their different arrangements create tremendous diversity in the structure of biomolecules. Specific interactions between different molecules play key roles in many biological processes. Biochemists study structure-function relationships at the molecular level to understand better biological phenomena such as antibody-antigen binding, receptor-ligand specificity, enzyme-substrate recognition, allosteric effects, and membrane transport channels and pumps.

HISTORY OF BIOCHEMISTRY

The origins of biochemistry date back to 1828, when the German chemist Friedrich Wöhler synthesized urea ($CO(NH_2)_2$), in vitro, from inorganic ingredients, the first biochemical synthesis. Before his success, people thought that only living organisms could synthesize organic molecules, a concept held by the vital hypothesis, which stated that the mysterious essence of life itself was necessary to synthesize the "stuff of life." The significance of Wöhler's work lay in the realization that organic molecules were simply chemical compounds. The French chemists Anselme Payen and Jean-François Persoz published their work on the isolation of an enzyme complex that they called diastase (the enzyme is now called amylase) from barley malt in the journal *Annales de Chemie et de Physique* in 1833. The Swedish chemist Jöns Jakob Berzelius, who coined the terms *protein* and *catalysis,* wrote a paper on chemical catalysis in 1835, demonstrating that malt extract efficiently digested starch in vitro. In 1860 the French chemist Louis Pasteur demonstrated that the process of fermentation, responsible for the production of alcoholic beverages and pickling, was biological rather than chemical, laying the groundwork for the investigation of other metabolic processes. In 1896 the German chemist Eduard Buchner used extracts from yeast cells to ferment sugars in vitro, supporting the finding that live cells were not required to carry out complex chemical reactions. This finding opened the door for more intensive analysis of biochemical reactions. Buchner received the 1907 Nobel Prize in chemistry for his research on fermentation. Several biochemists unraveled the basic mechanisms by which enzymes catalyze biochemical reactions at the turn of the 19th/20th century.

After Buchner debunked the vital hypothesis, metabolism became a popular research focus among chemists. By 1940 the contributions of several pioneering biochemists led to the elucidation of glycolysis, a central metabolic pathway linking the metabolism of carbohydrates, amino acids, nucleo-tides, and fatty acids. Because of this, glycolysis, one of the first studied, is the best understood metabolic pathway.

The discovery of the double-helical structure of DNA by James Watson and Francis Crick in 1953 initiated a revolution in the biochemistry of nucleic acids and of gene expression. Research along these lines led to the birth of a new field, molecular biology, which encompasses the processes of DNA replication, transcription (the synthesis of RNA), and translation (the synthesis of proteins). The tools and technology used to probe molecular processes, such as recombinant DNA technology and the polymerase chain reaction, have greatly advanced knowledge in biochemistry.

BIOCHEMICAL RESEARCH

Research in the field of biochemistry depends on many techniques such as chromatography, X-ray diffraction, nuclear magnetic resonance (NMR) spectroscopy, electron microscopy, radioisotope labeling, gel electrophoresis, centrifugation, and recombinant DNA technology. One major goal of biochemistry is the purification and characterization of biomolecules, especially enzymes. Biochemists often start by preparing an extract from a piece of tissue, a sample of cultured cells, or even a bodily fluid, such as blood or urine. Separation techniques such as chromatography and centrifugation generate fractions that exclude certain types of molecules on the basis of size, electrical charge, or chemical affinities. The fractions are then subjected to further purification techniques until the sample quality is sufficient for characterization, which includes the determination of size, composition, structure, and function. Molecular biological techniques and computer simulations have facilitated the study of protein structure and function. Using recombinant DNA technology, researchers can replace single amino acids and examine the resulting effect on the protein's structure and therefore function. Computer modeling allows biochemists to compare molecules with similar structures; such comparison can lead to inferences about function. Software programs that predict the overall structure of a protein given its amino acid sequence are not perfect but provide valuable information that can be reconciled with real data from techniques such as X-ray crystallography, electron microscopy, and NMR spectroscopy that reveal molecular structures.

Another aim of biochemistry is to understand bioenergetics, the energy transformations and exchanges carried out by living organisms. All organisms ultimately receive energy necessary to carry out life's processes from the Sun. Primary producers can convert radiant energy from the Sun into chemical energy via photosynthesis, and consumers obtain

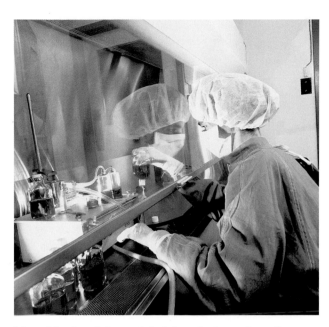

Many biochemists work in laboratories, where they carry out basic research with the aim of gaining a better understanding of biomolecules and biochemical processes that occur in living organisms. *(Publiphoto/ Photo Researchers, Inc.)*

their energy by ingesting organic material either from primary producers or from other organisms that have ingested primary producers. Enzymes act as energy transforming devices by facilitating the formation of covalent bonds in organic molecules, which store chemical energy, and catalyze their breakage, releasing the stored energy. Biochemists study the flow of energy through living systems, which must follow the laws of thermodynamics, stating that energy must be conserved and that processes only occur spontaneously if the total entropy of the system increases.

Additional current avenues of exploration in biochemistry research include the nature of sex determination, regulation of developmental pathways in early embryogenesis, the effect of genes and biochemical basis for behaviors, signal transduction pathways (by which cells receive and communicate information), and apoptosis (programmed cell death). Computational biochemistry has presented opportunities for explorations using molecular graphics, three-dimensional modeling, and computer simulations of biochemical processes. Research relating metabolism to obesity has also gained support recently as a result of the increased incidence of this health concern.

APPLICATIONS

While many biochemists perform basic research with the goal of attaining a better understanding of biomolecules and biochemical processes, knowledge gained from their research can be applied to many fields. Biochemistry is applicable to medicine, as many genetic disorders result when an individual produces defective enzymes or insufficient quantities of certain enzymes. For example, glycogen storage diseases affect carbohydrate metabolism and cause symptoms including muscle weakness, poor growth, and liver problems in children. Many pharmaceuticals act by inhibiting or stimulating biochemical pathways. Understanding the mechanisms for regulating these pathways can lead to the development of novel drug therapies. In addition, biochemists help determine whether the mechanisms by which the body will break down and eliminate drugs and chemicals are safe and do not result in the production of toxic by-products. Cancer often results from the loss of regulation of cell-signaling pathways, which depend on numerous specific interactions among proteins and other biomolecules; thus many biochemists are involved in cancer research.

Many industries hire biochemists to research the biochemical aspects of the products they sell. The commercial production of numerous chemicals exploits biochemical processes carried out by microorganisms. For example, many chemical industries grow microorganisms on a large scale to produce a variety of alcohols, organic acids like citric acid, enzymes that are added to laundry detergents, and vitamins and amino acids used as food supplements. The food and wine industries also depend on the biochemical process of fermentation to produce alcoholic beverages, pickles, sausage, soy sauce, and many dairy products such as sour cream, cheeses, and buttermilk. Factories use biochemical methodologies to purify certain chemical reagents, enzymes, vitamins, pharmaceuticals, monoclonal antibodies, recombinant proteins, and acetic acid or other organic acids from cell cultures, tissue extracts, or other mixtures. Knowledge of biochemical processes helps scientists to determine evolutionary relationships, as biochemists can compare similarities in metabolic pathways and mechanisms between species. Greater degrees of differences in mechanisms or structures of the molecules involved in processes such as photosynthesis, cellular respiration, and DNA replication imply a greater evolutionary distance between species or lineages.

While biochemists have extensively structurally and functionally characterized numerous macromolecular components, none of the macromolecular assemblies per se is "living." Biochemists and other life scientists are still far away from understanding how the assembly and congregation of biomolecules into macromolecular assemblages lead to living cells and organisms.

See also BIOCHEMICAL REACTIONS; BIOENERGETICS; BIOMOLECULES; CENTRIFUGATION; CHEMICAL

BASIS OF LIFE; CHROMATOGRAPHY; ELECTROPHO-
RESIS; ENZYMES; MOLECULAR BIOLOGY; ORGANIC
CHEMISTRY, ITS RELEVANCE TO LIFE SCIENCE; RECOM-
BINANT DNA TECHNOLOGY; WATER, ITS BIOLOGICAL
IMPORTANCE; X-RAY CRYSTALLOGRAPHY.

FURTHER READING

Berg, Jeremy M., John L. Tymoczko, and Lubert Stryer.
Biochemistry. 6th ed. New York: W. H. Freeman,
2006.

Champe, Pamela C., Richard A. Harvey, and Denise R.
Ferrier. *Lippincott's Illustrated Reviews: Biochemistry*. Philadelphia: Lippincott, Williams, & Wilkins,
2005.

Daintith, John, ed. *The Facts On File Dictionary of Biochemistry*. New York: Facts On File, 2003.

Gilbert, Hiram F. *Basic Concepts of Biochemistry: A Student's Survival Guide*. 2nd ed. New York: McGraw
Hill, 2000.

Lennarz, William J., and M. Daniel Lane. *Biological Chemistry*. 4 vols. Amsterdam; Boston: Elsevier, 2004.

biodiversity Biodiversity, or biological diversity,
is the variation in living organisms and living systems within a given geographical area. Typically, this
translates into the number of different species, but
it also refers to genetic variation within a species, or
variation within and between higher levels of biological organization such as communities and ecosystems.
The United Nations Convention on Biological Diversity defines *biodiversity* as "the variability among living organisms from all sources, including, inter alia
[among other things], terrestrial, marine, and other
aquatic ecosystems and the ecological complexes of
which they are part: this includes diversity within
species, between species, and of ecosystems." Biodiversity is one measure of the biological health of
an area; generally, a greater total number of species
indicates a healthier habitat or region.

Estimations of the total number of species present on Earth today range from 2 million to 10
million, though fewer than 2 million have been
catalogued. Genetic variation, differences within
the genetic makeup between individuals of a species or between populations of a species, may affect
fitness, the relative capacity of an organism to survive and pass on its genes to offspring. Through
the natural selection of phenotypes and genotypes
associated with increased fitness, new species have
emerged, and others have become extinct. In the
past, natural factors such as climate, geography,
and interspecific interactions largely determined the
degree to which certain adaptations affected an
organism's fitness and thus controlled the number
and types of species. The effect of human activities,
amplified as the population expands, has altered
the climate, landscape architecture, and distribution of species more rapidly than has ever occurred
by natural means in evolutionary history. This has
led to the present crisis, the sixth major extinction
event in Earth's history. This human-induced mass
extinction is occurring at a rate higher than that of
any past mass extinction event; estimates of species lost each year range from hundreds to thousands. In addition to aesthetic, recreational, ethical,
and utilitarian reasons, scientific reasons to attempt
thwarting this recent severe reduction in species
numbers include maintenance of ecosystem stability, prevention of human-induced environmental
disasters such as drought and flooding, increased
ability to recover from such human-induced and
natural environmental disasters, untapped genetic
resources, and potential medicinal treatments, some
of which have not been discovered yet.

IMPORTANCE OF BIODIVERSITY

All species are connected to each other and to the
environment in which they live through food chains,
the biogeochemical cycling of nutrients, the effects
of living organisms on the climate and physical geology of their habitats, and vice versa. Continued life
depends on the diverse activities and interrelationships of the variety of life-forms that sustain the
planet and share its resources. Society depends on
other living organisms for material resources such
as food, clothing, medicines, and fuel, but a reduction in biodiversity is not only a concern for those
who directly harvest goods from other organisms or
those who are immediately affected by the loss of a
particular species or damage to one small area. All of
human society depends on Earth's natural resources,
including the ecosystems that maintain healthy living
conditions, sustain the atmosphere, provide drink-
able water, temper global climate, control erosion,
naturally control pests, maintain soils for growing
food, and recycle chemical nutrients. Measuring the
services (such as the removal of wastes, filtration of
water, or capture of carbon from the atmosphere)
that an ecosystem provides is one method of gauging
the health of an ecosystem or its ability to function.
The productivity of an ecosystem is another. If the
production of biomass, the amount of living matter,
remains high, then the organisms in the community
are living, metabolizing, cycling nutrients, and performing their tasks in sustaining the environment.
Biologists do not know the lower limits of biodiversity necessary for the maintenance of an ecosystem's health or whether the critical factor is the total
number of species present or the functional types of
organisms present. The recent increase in extinction
rates has turned prompt attention to these questions.

A relationship exists between biodiversity and ecosystem maintenance. An ecosystem can be described as stable if the composition of its community members remains essentially the same, with very few new species introduced and very few species disappearing from the community. Stability can also refer to the sizes of the populations within a community, that is, the number of individual members of a species in a population. One observation is that diverse ecosystems containing many different species are more stable. The greater number of species present, the greater the number of pathways through which energy and nutrients can flow through the ecosystem. If one pathway is disrupted, the energy can still flow, allowing the ecosystem to remain productive. Less biodiversity means less overlap in the functions of the species in the ecosystem and therefore a decreased ability to recover from disturbances. Predicting the effect of the loss of one species is difficult; such a loss can cause a chain reaction since all species depend on other species for resources. Ecologists have shown that in ecosystems with high biodiversity, the population sizes of individual species fluctuate a lot, but the combined productivity of the ecosystem remains high and stable, a good indicator of ecosystem health. Thus, diverse ecosystems are more productive, even though the stability of distinct populations fluctuates. One current research goal is to reveal the mechanisms by which biodiversity contributes to ecosystem stability.

Not all species are ecologically equal. Keystone species are species that play crucial roles in the normal function of an ecosystem. Their removal from an ecosystem may result in the subsequent disappearance of many other species, leading to a reduction in biodiversity. Indicator species provide useful information on the health of an ecosystem. For example, lichens, a symbiosis between fungi and a photosynthetic organism, are particularly sensitive to pollution. Rather than simply measuring quantities of chemicals present in the environment, sampling of indicator species to monitor environmental conditions provides a more accurate picture of what effects pollution may have on an ecosystem, as they reflect the integrated effects of the pollutants and the environment. Sometimes more than one species can fill a niche in an ecosystem, a factor that contributes to the stability of an ecosystem. If the population of one organism decreases, the so-called redundant organism can play the same role, preventing any negative effects on the function or productivity of the ecosystem. This concept can also be applied to diversity within a species and between ecosystems. In the case of genetic variation, the higher the diversity, the more likely that other members of the same or another population can perform similar ecological functions. Likewise, if one ecosystem were destroyed or rendered nonfunctional by a cataclysmic event, the presence of other numerous diverse ecosystems might be able to compensate for the lost one.

BIODIVERSITY THREATS AND CONSERVATION

Human pressures on ecosystems and global climate change have contributed to the serious loss in biodiversity over the past few decades. Expanding human population size and a disproportionate increased demand for natural resources have destroyed habitats, decreased the quality of the environment, and increased the number of extinct and endangered species. The purposeful and inadvertent introduction of nonnative species into new habitats also threatens many species as the nonnative species competes for food and resources and preys on or parasitizes the native species. Human activities that have altered landscapes and water bodies of different areas, contaminated the environment with pollutants, introduced alien species to ecosystems, and made hunting profitable have led to an alarming decrease in biodiversity over the past few decades.

Recognizing the decrease in biodiversity as a global problem requiring international cooperative efforts to turn around, the United Nations (UN) called a meeting in 1992 in Rio de Janeiro. The UN Conference on Environment and Development, also called the Earth Summit, addressed this and other related issues. Representatives from 172 governments met to discuss issues such as the production of toxic substances, alternative energy sources to reduce the effects of burning fossil fuels on global climate, public transportation systems to reduce air pollution, and water conservation. One result of the Earth Summit was the Convention on Biological Diversity, a treaty that outlined international strategies for preserving biodiversity.

Individual choices regarding decisions about what foods and products to buy, water usage, and energy consumption also contribute to reducing the negative impact human society has had on the environment. By decreasing one's personal contributions to pollution, the burning of fossil fuels, and hunting, an individual's actions compounded over time and collectively as a society can make a difference by reducing or eliminating the activities that lead to habitat destruction and fragmentation.

Feeding the world's 6-billion-plus population of humans has a large impact on biodiversity, and protecting biodiversity is important in food production. While some living organisms can make their own food using energy from the Sun and inorganic chemicals from the environment, humans require energy and nutrients in the form of organic compounds made by other living organisms. The food supply is obtained from plants and animals and even fungi and algae that share the Earth's resources. The primary producers, mostly photosynthetic plants and plankton, form the foundation of food webs. Plants require fertile soil, produced as a result of the metabolic activity of diverse microorganisms that decompose organic matter and recycle the nutrients that feed future crops. Bacteria fix inorganic nitrogen from the soil into a form that plants can use, improving their growth. Birds and insects that some consider pests pollinate plants that produce fruits and vegetables that form the staples of the human diet. Though agriculture depends on cultivated species that have been bred specifically to increase yields or improve the quality of products, crop and domestic animal breeding programs still require wild relatives to maintain sufficient levels of genetic variability so they will be able to survive environmental challenges such as exposure to a new pathogen and to increase their overall vigor. Individual actions, such as purchasing organic farm products, buying locally grown products and products in season to reduce chemical treatments necessary to help the product tolerate transportation over long distances and to reduce the associated energy costs, eating more plant products to reduce the extra water and energy required to produce food products from organisms higher up in the food chain, not using plastic or Styrofoam packaging, and reducing paper and food waste, all reduce negative environmental effects and thus contribute to maintaining biodiversity.

The water supply is also connected to biodiversity. Efforts to preserve biodiversity are important in protecting water as a natural resource, and protecting water as a natural resource will help prevent a decrease in biodiversity. As the global population increases, so does the demand for freshwater. A healthy, abundant water supply depends on the actions of numerous diverse organisms including microorganisms, plants, and animals. Watersheds funnel water into lakes and streams, and wetlands absorb many metals and toxic substances that enter the water supply. Many endangered species live on or are supported by watersheds, and the watersheds and organisms that reside in them act as natural filtration systems, helping to purify the water. Landscaping near watersheds, deforestation, and damming waterways all negatively impact the water supply and contribute to the reduction in the number of aquatic species and to the function of aquatic ecosystems. Agricultural and industrial practices add chemicals to the water reserves, reducing the usable supply and requiring more energy to clean up. Individuals can help by reducing water use at home and using fewer chemicals and pollutants in household products.

Human use of fossil fuels for energy also has contributed to the decrease in biodiversity. Burning the fuels releases gases such as carbon dioxide, sulfur dioxide, and nitrogen oxide into the atmosphere, contributing to global climate change and leading to acid rain. Other pollutants released directly kill or inhibit plant growth. Coal mines and oil drilling cause dangerous chemicals and toxins to leach into the environment, destroy habitats, and kill off species.

Global climate change is among the most serious threats to biodiversity. The increased release of carbon dioxide into the air, in combination with the decrease in photosynthetic life that naturally removes it, traps heat in Earth's atmosphere. The radiant energy that is reflected off the Earth's surface cannot escape. Over the last century, the average temperature has risen about 1.3°F (0.74°C), and scientists predict the temperature may rise a few additional degrees over the next 100 years. Global warming causes melting of glaciers, and a resultant sea-level rise, which can flood wetlands, erode coastlines, affect food supplies, change migration patterns, and decrease northern forests.

Protecting biodiversity is crucial to maintaining the ecological health of the planet. The reduction in biological variation within a species, between species, and among ecosystems poses a serious threat to the ability of the biosphere to support many current life-forms, including humans. Scientists continue to explore the relationships among biodiversity, ecosystem stability, and the far-reaching impact of human activities on the environment in order to understand better how to sustain living conditions for all the life-forms that share the planet's resources.

See also CONSERVATION BIOLOGY; ENDANGERED SPECIES.

FURTHER READING

American Museum of Natural History. Center for Biodiversity and Conservation home page. Available online. URL: http://research.amnh.org/biodiversity/. Accessed January 17, 2008.

Anderson, Anthony B., and Clinton N. Jenkins. *Applying Nature's Design: Corridors as a Strategy for Biodiversity Conservation.* New York: Columbia University Press, 2006.

Evans, Kim Masters. *Endangered Species: Protecting Biodiversity.* Farmington Hills, Mich.: Thomson Gale, 2006.

Gaston, Kevin J., and John I. Spicer. *Biodiversity: An Introduction.* Malden, Mass.: Blackwell Science, 2004.

Hulot, Nicolas. *One Planet: A Celebration of Biodiversity.* New York: Abrams, 2006.

Wilson, Edward O. "Threats to Biodiversity." *Scientific American* 261, no. 3 (September 1989): 108–116.

bioenergetics Bioenergetics is the study of energy transformations and energy exchanges within and between living organisms and their environments. Life is a highly structured phenomenon, from the molecular and cellular levels to the community and ecosystem levels. The characteristic order of living systems is contrary to the natural tendency to proceed from a more ordered to a more disordered state. Maintenance of the organization necessary for life requires the constant input of energy. Thermodynamics describes the path of energy flow through living systems, and knowledge about the specific conditions of a system allows one to track the quantities of moving energy.

ENERGY IN LIVING SYSTEMS

Energy is the capacity to do work or make a change such as perform an energy transfer. Several types of work are necessary to support life's processes: synthetic, mechanical, concentration, electrical, heat, and bioluminescence. Synthetic work results in the production of new molecules that assemble into cellular structures, tissues, and whole organisms. Most energy within cells is stored in the form of chemical bonds of organic molecules, and the creation of these bonds requires the input of energy. Mechanical work includes processes such as beating cilia or moving a flagellum for locomotion. Sometimes energy exists in the form of a gradient that is used to perform work such as transporting substances across membranes. The creation of a gradient requires energy, while the dissipation of a gradient releases energy. Electrical work is a special type of work that involves a gradient of charged particles. Electrical gradients are necessary to transmit neural impulses through the nervous system and to signal muscular contraction. Homeothermic (warm-blooded) organisms use heat generated by cellular metabolism to maintain body temperature. Bioluminescence is the production of light by a living organism such as a firefly or fish that lives in the deep sea.

Cells expend energy to perform all of these types of work. The most immediate source of energy inside the cell is adenosine triphosphate (ATP), a high-energy molecule consisting of a ribose sugar attached to a nitrogenous base called adenine and three linked phosphate groups. Cells have the capacity to transform the chemical energy stored in the bonds between the phosphate groups, particularly the second and third phosphate groups, into the types of work described.

The major source of energy found in living organisms is radiation from the Sun. Phototrophs are organisms that have the ability to capture light energy from the Sun and transform it into chemical energy for cellular work. Plants, algae, and some bacteria can carry out this process called photosynthesis. Chemotrophs obtain their energy from chemical sources and can be split into two groups. Chemoorganotrophs oxidize organic compounds such as carbohydrates, fats, and proteins. Animals, protists, and fungi are chemoorganotrophs. Some microorganisms are chemolithotrophs, meaning they obtain their energy from the oxidation of inorganic compounds such as hydrogen sulfide (H_2S), thiosulfate ($S_2O_3^{2-}$), or molecular hydrogen (H_2). Organisms that are photosynthetic are also chemotrophs; the difference is that they can use organic molecules that they have synthesized by using energy obtained from sunlight rather than depend on other sources to produce them.

THERMODYNAMICS OF LIVING SYSTEMS

The energy of living systems is constantly changing forms. Thermodynamics is the study of the principles that govern this energy flow. Bioenergetics is the specialized study of thermodynamics in biology. The practical study of thermodynamics can be applied to living systems to explain energy flow and to predict the direction and extent to which chemical reactions will proceed. Thermodynamics does not, however, give any insight as to the rate of a reaction. Kinetics addresses that issue.

When considering energy flow, one must consider the system and whether it is open or closed. A closed system is separated from its environment so that no energy can enter or leave the system, whereas an open system does permit energy to enter into or be released from the system. As open systems, individual cells or organisms not only allow energy to enter, they require energy input in the form of sunlight or food to sustain life. As soon as the organism is denied a source of energy, the organization fails and the organism dies.

The state of the system refers to the variable properties, such as temperature, pressure, or volume. As long as one of these variables is held constant, one can determine useful information about the changes in energy of the system. Changes in energy usually accompany biochemical reactions, which for all practical purposes occur in relatively constant environmental conditions. Within the fraction of a second it takes for a reaction to proceed, the three variables most important to a physical chemist—temperature, pressure, and cell volume—do not change. This allows one to make predictions about the energy balances and direction of biochemical reactions.

Three laws govern energy flow through systems, but only the first two are particularly useful to a biologist. The first law of thermodynamics, the law of conservation of energy, states that the energy in the universe is constant. Simply put, energy can be transformed, but neither created nor destroyed. One can follow the energy flow by examining enthalpy, or heat content (represented by the symbol H). The change in enthalpy, ΔH, is related to the change in internal energy (energy within the system) and the change in pressure, which is negligible. Thus, ΔH equals the change in energy. If the heat content of the products is less than the heat content of the reactants, ΔH is a negative number and the reaction is termed exothermic, meaning heat is released. If the heat content of the products is more than the heat content of the reactants, ΔH is positive, and the reaction is termed endothermic.

Consider the process of cellular respiration, during which one molecule of glucose is oxidized to carbon dioxide and water. Energy is released as the glucose molecule is oxidized; the chemical bonds between the carbon atoms are broken and carbon dioxide is formed. The reaction can be summarized as follows:

$$C_6H_{12}O_6 + 6O_2 \rightarrow 6CO_2 + 6H_2O + energy$$

In this reaction, ΔH is negative, this reaction is exothermic, and heat is liberated. So, the first law of thermodynamics allows one to determine the energy changes that accompany a biochemical reaction or process.

The second law of thermodynamics, the law of thermodynamic spontaneity, states that every reaction or energy transfer increases the entropy of the universe. Entropy (represented by S) is a measure of disorder or randomness of a system. The second law allows one to predict the direction in which a biochemical reaction will proceed and the amount of energy that will be released or consumed. In order for a reaction to proceed, the change in entropy, ΔS, must be positive; in other words, entropy must increase. An energy transfer such as freezing water involves a decrease in entropy as the hydrogen bonding becomes more structured to form the three-dimensional ice crystals. However, when looking at the universe as a whole, the overall entropy still increases because reactions that are associated with an apparent decrease in entropy of the system accompany other transfers that are associated with even greater increases in entropy of the surroundings. This complicates thermodynamic calculations for biological reactions, which work together to decrease entropy locally in order to maintain the organization that supports life.

Gibbs energy (represented by the symbol G) is a more useful measure for determining whether a reaction will proceed within a specific system. The utility of G, formerly called Gibbs free energy, for biological systems with stable temperature, pressure, and volume is demonstrated by the following equation, which relates free energy to enthalpy and entropy:

$$\Delta H = \Delta G + T\Delta S$$

or, rearranged,

$$\Delta G = \Delta H - T\Delta S$$

with T being temperature in Kelvin units. Reactions characterized by a decrease in Gibbs energy will have a negative value for ΔG. These reactions are exergonic, are considered spontaneous, and can occur. Reactions with a positive value for ΔG are endergonic and cannot occur unless they are coupled to another reaction that has a large enough negative value for ΔG to ensure that when combined, the overall ΔG is still negative. The fact that a biochemical reaction has a negative value for ΔG and can occur does not mean that it will occur. Biochemical reactions require catalysts called enzymes to get them actually to take place.

EQUILIBRIUM

The change in Gibbs energy (ΔG) can be calculated if the specific reaction conditions are known. First one must examine how close the reaction is to equilibrium. At equilibrium, the forward and reverse rates of a chemical reaction are equal, and the concentrations of reactants and products remain constant. A value called the equilibrium constant (K_{eq}) can be determined by dividing the concentration of the products by the concentration of reactants. To illustrate, the equilibrium constant for the following chemical reaction

$$a\text{A} + b\text{B} \rightleftharpoons c\text{C} + d\text{D}$$

is

$$K_{eq} = \frac{[\text{C}]^c \, [\text{D}]^d}{[\text{A}]^a \, [\text{B}]^b}$$

where [A] and [B] represent the concentrations of reactants, [C] and [D] represent the concentrations of products at 25°C, and a, b, c, and d represent the coefficients of each.

The equilibrium constants for most common reversible biochemical reactions have been experimentally determined. One can predict the direction in which a reaction will proceed by plugging in the measured concentrations. If the ratio of the actual concentrations of products to reactants does not equal the known K_{eq} for that reaction, the system is not in equilibrium, and the reaction will proceed in whatever direction is necessary to reach equilibrium. If the ratio is less than K_{eq}, then the reaction will proceed to the right, and if the ratio is greater than K_{eq}, then the reaction will proceed in the reverse direction, to the left.

Calculating ΔG shows how far from equilibrium a particular system is and how much energy will be released when the reaction proceeds toward equilibrium. The following equation is used

$$\Delta G = RT \ln \frac{[C]^c \, [D]^d}{[A]^a \, [B]^b} - RT \ln K_{eq}$$

where R is the gas constant (1.987 cal/mol-K) and T is the temperature in Kelvin (298 K). The result, ΔG, will be expressed in cal/mol and should be a negative number. This confirms that the reaction is spontaneous or thermodynamically possible but, again, does not mean that the reaction actually will occur, or, if it does, indicate the rate. The field of kinetics examines these issues, which are covered under the discussion of enzymes in this text.

See also BIOCHEMICAL REACTIONS; CELLULAR METABOLISM; CHEMICAL BASIS OF LIFE; ECOSYSTEMS; ENZYMES; PHOTOSYNTHESIS.

FURTHER READING

Chang, Raymond. *Physical Chemistry for the Biosciences.* Sausalito, Calif.: University Science Books, 2005.

Murray, Robert K., Daryl K. Granner, Peter A. Mayes, and Victor W. Rodwell. *Harper's Illustrated Biochemistry.* 26th ed. New York: McGraw Hill, 2003.

Myers, Richard. *The Basics of Chemistry.* Westport, Conn.: Greenwood Press, 2003.

Nicholls, David, and Stuart Ferguson. *Bioenergetics.* 3rd ed. Burlington, Mass.: Academic Press, 2002.

biogeochemical cycles The exchange of nutrients, chemical substances required for the normal development, maintenance, and reproduction of organisms, between organisms and the environment is an important feature of a functioning ecosystem. The process of nutrient cycling involves the use, exchange, transformation, and movement of nutrients through an ecosystem. Living organisms require a source of energy and a supply of nutrients. Whereas the Sun constantly inputs solar energy into an ecosystem, nutrients have limited availability. Not only is the quantity of a nutrient in an ecosystem important, but the form of the nutrient affects whether or not organisms can uptake and utilize it in metabolic processes. Autotrophic organisms are capable of fixing elements such as carbon or nitrogen, meaning they can incorporate the inorganic form of an element into organic molecules. Heterotrophic organisms must ingest their nutrition from preformed organic molecules. Because living organisms change the form of certain nutrients, or elements, as part of their normal metabolic processes, the organisms are called biotic components, and they play a significant role in the cycling of nutrients through ecosystems. Abiotic components of nutrient cycling include the processes that are not directly related to living organisms; for example, weathering, erosion, the formation of sedimentary rock, and other geological processes, are abiotic components. Because both biological and geochemical events participate in the movement of nutrients through ecosystems, the process is called biogeochemical cycling.

Chemical nutrients cycle at a local level as well as a global level. Gases in the atmosphere travel great distances, whereas elements in the soil usually stay within the vicinity. A reservoir is a place where a nutrient is kept in storage and serves as a supply. Two characteristics that describe reservoirs involve whether the nutrients they contain are organic or inorganic and whether the nutrients are available or unavailable. The nutrients move between reservoirs by several different processes: assimilation, photosynthesis, fossilization, burning of fossil fuels, respiration, decomposition, excretion, weathering, erosion, and formation of sedimentary rock.

Many nutrients are already tied up in living organisms, as part of cells, tissues, and body fluids. A living organism is thus considered a reservoir made up of organic materials that are available as nutrients to other organisms if the organism is eaten. As living organisms utilize the nutrients and extract energy from the organic compounds, metabolic processes produce inorganic by-products that leave the body by respiration or excretion. Organisms that have died also contain organic material that is available as nutrients to other organisms. For example, many microbes feed off detritus and leaf litter on the forest floor. Decomposition of organic material by detritovores, organisms that consume nonliving organic matter, also returns inorganic nutrients to the environment. Fossilization occurs when sediment buries dead organisms and they are transformed by physical processes into coal, oil, and peat over millions of years. These fossil fuels are composed of organic nutrients, but in a form unavailable to organisms. Burning them releases the elements into the atmosphere, soil, and water as inorganic nutrients that are available for use by organisms capable of photosynthesis or fixation and assimilation. Inorganic materials also contribute to the formation of sedimentary rock; as minerals in rocks they are unavailable, until weathering or erosion wears away the rock and they become available once again. Thus, numerous different biological and geochemical processes participate in the cycling of nutrients through ecosystems.

NUTRIENT CYCLES

Organic material, molecules that living organisms synthesize and the molecules from which they are composed, consists mostly of the elements carbon, oxygen, hydrogen, nitrogen, phosphorus, and sulfur. Though

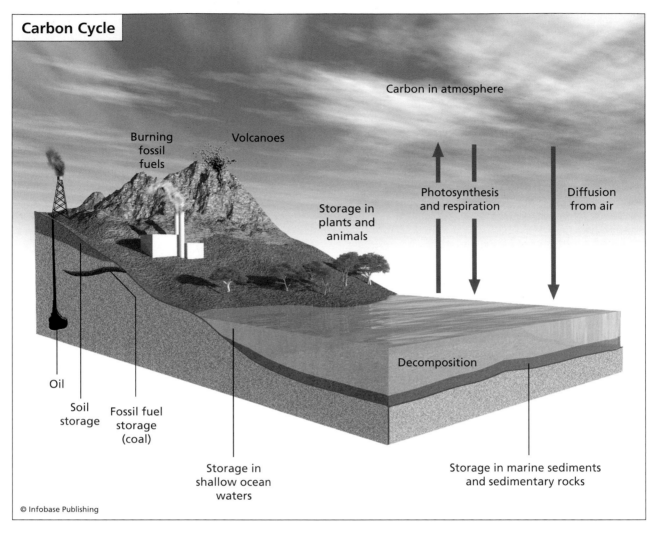

Carbon Cycle

Carbon in atmosphere

Burning fossil fuels

Volcanoes

Storage in plants and animals

Photosynthesis and respiration

Diffusion from air

Oil

Soil storage

Fossil fuel storage (coal)

Storage in shallow ocean waters

Decomposition

Storage in marine sediments and sedimentary rocks

© Infobase Publishing

Numerous biological and geochemical processes drive carbon through the carbon cycle, which includes both inorganic and organic forms of the element. Carbon enters seawater from many sources, including the air, respiration by living things, erosion of carbon-containing rock, and combustion of fossil fuels. Carbon is removed by processes such as photosynthesis, the creation of limestone, and storage in plants and animals.

all nutrients must be recycled, these elements are the most important biologically, as they are the main components of the biomolecules proteins, nucleic acids, carbohydrates, and lipids. The continuity of life depends on the recycling of carbon, nitrogen, phosphorus, and sulfur in addition to water, since water is a major source of hydrogen and oxygen.

Carbon forms the backbone or molecular framework of all organic molecules. Carbon exists in the atmosphere as the gas carbon dioxide (CO_2), though the atmosphere consists of less than 0.1 percent CO_2. This form of carbon is only available to autotrophic organisms, organisms that are capable of incorporating inorganic carbon into organic molecules via photosynthesis or chemosynthesis. Plants, algae, and some bacteria undergo photosynthesis, during which they convert radiant energy from sunlight into chemical energy that they utilize to synthesize carbo-

hydrates. Only specific prokaryotic species are capable of chemosynthesis, in which inorganic chemicals serve as the energy source for synthesizing organic molecules from carbon in CO_2. Primary consumers ingest the autotrophic organisms to obtain their nutrients (already in an organic form) and energy, and secondary and higher-level consumers feed off the organisms lower in the food chain to fulfill their nutritional requirements. For example, algae are autotrophic, and snails are primary consumers that eat the algae to obtain their nutrients. Some fish, such as loaches, eat the snails and are therefore considered secondary consumers. Any organism that eats the fish would be a higher-level consumer. All of these organisms carry out cellular respiration, the process by which cells break down organic chemicals into smaller components to release energy and metabolites that can be used to synthesize other biomolecules.

The cells use the energy to synthesize molecules of adenosine triphosphate (ATP), and they give off CO_2 as a by-product. After living organisms die, microorganisms decompose the organic material left behind as part of their normal metabolic processes, converting some of the carbon back into inorganic forms. When dissolved in water, CO_2 participates in an equilibrium reaction with bicarbonate (HCO_3^-) and carbonate (CO_3^-), which can combine with calcium and precipitate out of solution to join the largest carbon reservoir, carbonate rocks. Geological processes also convert the carbon in the remains of organisms that lived millions of years ago into fossil fuels such as coal, oil, and natural gas. Just as the biological burning of organic materials via metabolic processes releases CO_2 into the atmosphere, burning of fossil fuels and volcanic activity also release CO_2 back into the atmosphere, completing the carbon cycle.

Nitrogen is an important component of nucleic acids, both deoxyribonucleic acid (DNA) and ribonucleic acid (RNA), and of amino acids, the building blocks of proteins. An insufficient quantity of

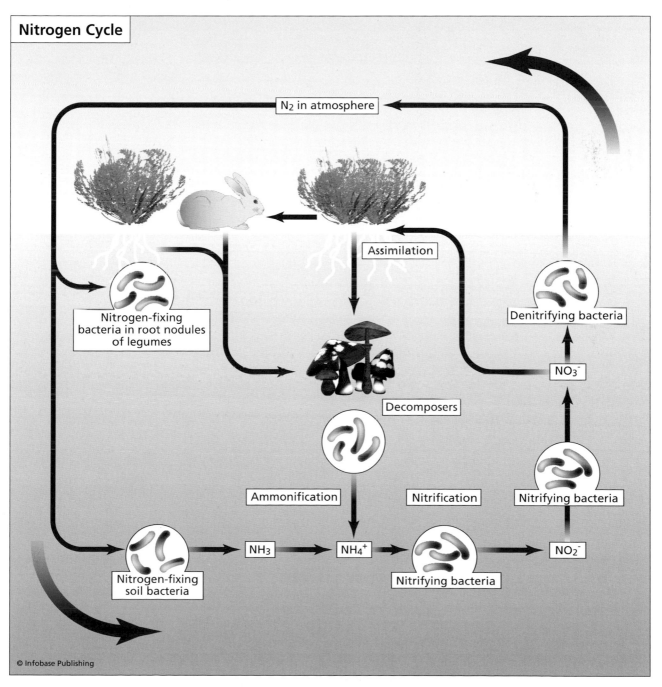

Nitrogen Cycle

© Infobase Publishing

Specialized bacterial species play an important role in the cycling of nitrogen.

nitrogen in the soil often limits the growth of plants, and thus nitrogen is a major ingredient of fertilizers. The atmosphere consists of more than 78 percent nitrogen gas (N_2) and thus serves as the major reservoir of nitrogen, but this form is unavailable to most organisms. The capability of nitrogen fixation is limited to specialized species of bacteria that live in the soil, in water, and as symbionts in the roots of leguminous plants. Genera containing species that can perform nitrogen fixation include *Azotobacter, Azospirillum, Clostridium, Anabaena,* and *Nostoc.* These bacteria can break the strong triple bonds between the two nitrogen atoms in a molecule of N_2, an energetically expensive reaction that results in the reduction of nitrogen, usually to ammonia, which in the presence of water readily transforms into ammonium ions (NH_4^+). Bacteria including *Rhizobium, Bradyrhizobium,* and *Azorhizobium* infect the roots of plants such as soybeans, alfalfa, and peas, causing them to form

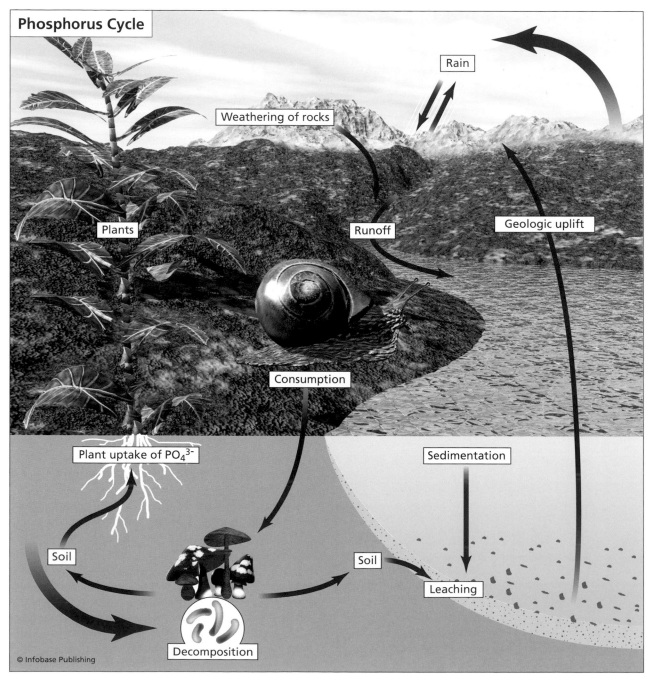

Phosphorus Cycle

Rain

Weathering of rocks

Plants

Runoff

Geologic uplift

Consumption

Plant uptake of PO_4^{3-}

Sedimentation

Soil

Soil

Leaching

Decomposition

© Infobase Publishing

Phosphorus cycles through ecosystems mostly in the form of phosphate (PO_4^{3-}), which is soluble.

nodules. These bacteria fix nitrogen and provide the plants with a constant supply of NH_4^+. Lightning also provides sufficient energy and physical conditions to fix nitrogen. Other bacteria (*Nitrosomonas, Nitrosococcus, Nitrobacter,* and *Nitrococcus*) carry out nitrification, the oxidation of NH_4^+ to nitrite (NO_2^-) or nitrate (NO_3^-). Most plants, algae, and other organisms can readily use NO_3^- to synthesize amino acids and nucleotides for incorporation into proteins and nucleic acids, respectively. Bacteria such as *Clostridium* and *Proteus* decompose the nitrogenous organic material of detritus, releasing NH_4^+ that can either be used by some organisms or oxidized first by nitrifying bacteria and then used. The cycle is completed when nitrogen returns to the inorganic pool of atmospheric nitrogen after denitrification, during which denitrifying bacteria (certain species of *Bacillus, Pseudomonas, Spirillum,* and *Thiobacillus*) reduce nitrates in a stepwise process, creating N_2 gas.

Phosphorus is a major component of nucleic acid, which contains a backbone made of alternating sugar and phosphate moieties. ATP also contains phosphorus in the form of phosphate (PO_4^{3-}) moieties. The major reservoirs for phosphorus are mineral deposits in rocks, marine sediments, and soil. The form of phosphorus found in rocks, fluorapatite, is not soluble, but interaction with acids solubilizes and releases it from rock in a more usable form, PO_4^{3-}. Autotrophs can fix this form into organic compounds, from which heterotrophs can fulfill their phosphorus requirements by consuming the autotrophs. Fertilizers often contain phosphates to compensate for insufficient amounts of phosphate naturally present in the soil. When excess phosphate from fertilizer runoff enters streams and lakes, eutrophication can occur; algae, cyanobacteria, and other surface organisms overgrow and shut off the supply of dissolved oxygen to the rest of the water body. Aerobic microorganisms involved in decomposition of the overgrowth further contribute to the lack of oxygen available to the aquatic life, causing fish and invertebrates to die. The phosphorus continues to cycle as terrestrial organisms die and decompose, and the phosphorus enters the soil as phosphate ions and leaches into bodies of water, where it sediments and geological processes form rock. The geological uplift of marine sedimentary rock transports phosphorus to land, where weathering and erosion of the rocks release fluorapatite, completing the cycle.

Sulfur is a component of the amino acids cysteine and methioine and is therefore present in many proteins. The sulfur cycle shares many similarities with the nitrogen cycle in that specialized bacteria perform many of the transformations of different forms of sulfur. The many forms include elemental sulfur (S), hydrogen sulfide gas (H_2S), sulfate (SO_4^-), and thiosulfate (S_2O_3). SO_4^- is the form that most organisms can utilize. Members of the bacterial genera *Thiobacillus* are lithotrophic, meaning inorganic chemicals, such as sulfur compounds, fulfill their nutritional requirements. They often inhabit swamps, mud, sewage, and acidic environments that other organisms cannot inhabit because of the lack of available organic nutrients or low pH but that do contain plenty of inorganic sulfurous compounds. After other organisms incorporate the oxidized sulfur compounds in their metabolic processes, sulfur-reducing bacteria such as *Desulfovibrio* and *Desulfuromonas* regenerate reduced sulfur in the form of hydrogen sulfide (H_2S) or metal sulfides, completing the sulfur cycle. Springs or areas that contain these sulfur-reducing bacteria often smell like rotten eggs as a result of the abundance of H_2S present.

WATER CYCLE

Water is composed of two parts hydrogen and one part oxygen (H_2O); thus the hydrologic cycle, or water cycle, is responsible for the cycling of these elements. The importance of water is obvious for aquatic species, but water is just as crucial for terrestrial species, as the major constituent of all cells is water. The liquid form is most available to organisms, but some organisms can obtain water from water vapor, though not from ice, or frozen water. The oceans contain approximately 97 percent of the biosphere's water. Polar ice, then lakes and groundwater compose the remainder. Evaporation, condensation, and precipitation are the major physical processes involved in water cycling through ecosystems, and transpiration is the major biological process. Solar radiation drives the cycle by causing evaporation, the conversion of liquid water into the gaseous form, water vapor, through the input of heat energy. Approximately 90 percent of the water in the atmosphere arrived there through evaporation. As a vapor, water enters the atmosphere, where the wind can carry it to new locations. As it rises, the vapor encounters cooler air, causing it to condense into a denser liquid form. When liquid droplets become heavy enough, they fall as precipitation in the form of rain, drizzle, sleet, snow, freezing rain, or hail. All forms of precipitation transport the water back into the oceans or onto land, depending on the location. When water falls onto land, it percolates through the soil and runs off into lakes and streams, some of which empty into the oceans. Plants absorb water through roots embedded in the soil and return it to the atmosphere via transpiration, evaporative water loss, which is

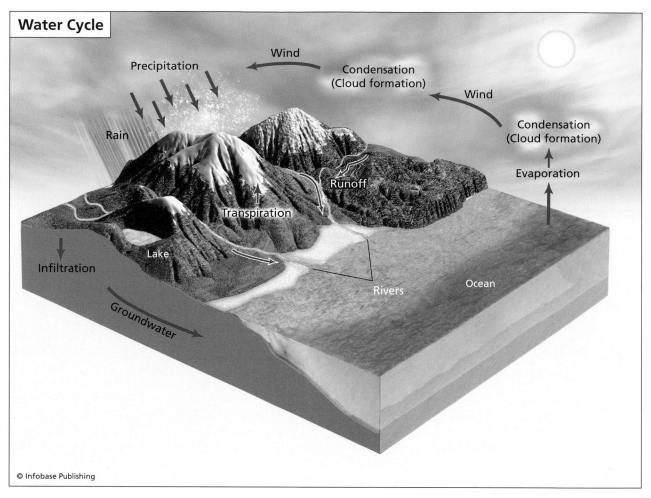

Water Cycle

Wind

Precipitation

Condensation
(Cloud formation)

Wind

Rain

Condensation
(Cloud formation)

Runoff

Evaporation

Transpiration

Lake

Infiltration

Rivers

Ocean

Groundwater

© Infobase Publishing

The major processes that drive the water cycle, or hydrological cycle, include evaporation, transpiration, and precipitation.

responsible for roughly 10 percent of the vapor in the atmosphere.

See also BACTERIA (EUBACTERIA); BIOMOLE-CULES; ECOSYSTEMS; FUNGI; MICROBIOLOGY; PROTO-ZOA; WATER, ITS BIOLOGICAL IMPORTANCE.

FURTHER READING

Moles, Manuel C. *Ecology: Concepts and Applications.* 3rd ed. New York: McGraw Hill, 2005.

Sigee, David C. *Freshwater Microbiology: Biodiversity and Dynamic Interactions of Microorganisms in the Aquatic Environment.* Hoboken, N.J.: John Wiley & Sons, 2005.

biogeography Geography is the science of the description, distribution, and interaction of the physical, biological, and cultural features of the Earth's surface. Biogeography is a subdiscipline of geography that deals with the spatial and temporal distribution of plants and animals. The temperature, precipitation, type of soil, salinity of the soil or water, degree

of Sun exposure, landscapes, and other physical features of an area influence the types of flora and fauna that live there, as do other members of the biological community. This allows for the depiction of general biological patterns on a map, such as geographical features, such as mountain, lakes, and deserts, or meteorological features, such as temperatures, isobars, and precipitation patterns. Because organisms interact with other living organisms and with their physical environment, and because these relationships influence the course of biological evolution, the fields of ecology and evolution are interconnected with biogeography, which also draws on other disciplines such as geology and climatology.

CONCEPTS OF BIOGEOGRAPHY

The Earth's biosphere contains several different biomes, ecological associations characterized by the types of vegetation present and the general climate. These biotic and abiotic factors affect the flora and fauna that typically inhabit a particular region. Dispersal also influences the distribution of a species. Plants and animals have a variety of means for dispersal, or extending the range that they inhabit: swimming, flying, walking, water and wind currents, seeds carried in the feces of animals, spores blown by the wind, parasites on the bodies of other animals, and so on. Geographic barriers, such as mountain ranges, deserts, or oceans, can limit an organism's dispersal.

Islands provide natural isolated model systems for studying biogeographical principles, which can then be applied and examined in ecosystems that are not as clearly defined by sharp physical boundaries. The concept of a biological island is not limited to a tract of land surrounded by water, but mountain ranges, lakes, rivers, altitudes, sharp climate boundaries, and forest fragments all create isolated, inland islandlike habitats, though the degree of isolation may differ as a result of a combination of geographical or biological factors. For example, some lakes (aquatic islands) may be completely isolated, while others may receive drainage or input from streams or have streams that carry water out of them. Birds and insects can fly varied distances to find food or to mate. Some animals are fine swimmers or can travel over mountains. Some plants produce spores that

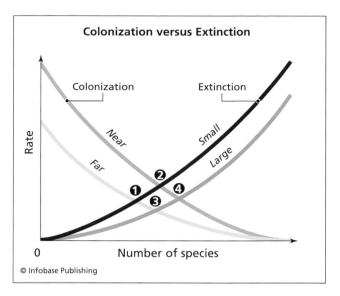

According to the equilibrium model of island biogeography, the maximal number of species for an island is maintained by a balance of colonization and extinction rates, both of which are affected by the island size and remoteness.

Biogeography explores the distribution of species in different geographic locations. Coastal dune forest ecosystems, such as this one in Riga, Latvia, develop under the shelter of the sand dunes, in sand with very little soil development, and where annual precipitation exceeds 27.5 inches (70 cm). *(AJE, 2007, used under license from Shutterstock, Inc.)*

the wind can blow great distances or have seeds that other animals ingest and spread to distant locations.

One main concept of biogeography is that area and isolation affect the species richness of islands and habitat patches on continents. Biodiversity increases with increasing area and decreases with greater isolation. In other words, a smaller island located far away from other ecosystems will contain fewer species than a larger island located nearer to other islands or to shore. Species proliferate and diversify at a faster rate on bigger biological islands. The effect of isolation needs to be examined individually for each species, since dispersal mechanisms and therefore dispersal rates vary widely among species.

Another concept of biogeography is that biodiversity on islands is maintained by a balance of immigration and extinction. This equilibrium model of island biogeography shown above illustrates a relationship between the number of different species on an island and immigration rates and between the number of species on an island and extinction rates. As the number of species present increases, the rate of arrival of new species decreases. Conversely, as the number of species present increases, the rate of extinction increases. The reality is more complex than this simplified model suggests: because the size and isolation of an island affect the species richness, these factors also influence the rates of immigration and extinction. While some islands reach equilibrium in the number of different species, others do not. The species composition, however, is dynamic; this change in composition is known as species turnover.

Species richness generally increases from high and middle latitudes toward the equator. Tropical habitats such as rain forests contain many more species than habitats closer to the poles. Though this observation is well documented, the reasons for it are unknown. Several hypotheses attempt to explain this gradient in species richness, which biogeographers continue to explore. The "time since perturbation" hypothesis states that tropical habitats have been around longer and have not been perturbed frequently by the advance and retreat of glaciers; thus extinction rates are lower. Other hypotheses link the species richness to high productivity, environmental heterogeneity, lack of extreme fluctuations in weather, breadth of niches, or degree of interspecific interactions such as symbioses. The latitudinal gradient in richness could also be due to an increased area. Since species number changes as a result of extinction and speciation events, tropical environments could simply have higher speciation rates or lower extinction rates. Because of the shape of the planet Earth, tropical latitudes, located between 23.5 degrees north and south of the equator, contain more land area and water area than other latitudes. The temperature is also more uniform, giving species that prefer that temperature range a wide region across which they can spread. Larger area allows for a larger population size, which is less prone to extinction than a smaller population. The larger area could also allow for more speciation events, since there is more area over which geographic barriers may exist and lead to allopatric speciation events.

Because the Earth is dynamic, the structure of many ecosystems results from significant historical events or regional processes. Throughout the Earth's history, certain geographic areas have experienced periods of intense volcanic activity, orogeny, flooding, or glacial advancement and retreat. The conditions of a region at a particular time place unique selection pressures on the biota (all the flora and fauna) that are living there or that can colonize there afterward. For example, more species of trees inhabit temperate forests in North America and Asia than in Europe. This can be explained by the fact that temperate forests occur in areas that experienced glacial advancement and retreat during different geological periods. When glaciers advanced into these regions during the last ice age, they forced the trees to migrate farther south, where the temperatures were warmer. In Europe, mountain ranges that run east and west prevented the southward migration of trees; thus they became extinct. Asia has no such mountain barriers and eastern North America only has the Appalachians, which run north and south, so tree populations in these regions could retreat toward the equator during the ice age, only to migrate north again when global temperatures increased.

TOPICS IN BIOGEOGRAPHY

The goal of biogeography is not simply to describe biogeographic patterns, but to examine the processes that determine the distribution and diversity of organisms globally, within a continent, on a marine island, on a mountain island, or on another patch of land habitat. Biogeographers study the geological and ecological processes that affect the origin, spread, distribution, and diversification of biota inhabiting a particular region. For example, one current topic of research is the effect of global climate change on the distribution of plants and animals. As temperatures rise and rainfall patterns change, vegetation that once thrived in a particular area may suffer, and, as a result, the animal life dependent on those plants for food, shelter, or shade will migrate to find a more suitable environment. The rise in sea level resulting from the melting of glaciers and polar ice affects coastal regions, estuaries, and wetlands and the life-forms that those habitats support. Biogeographers are researching the effects that have already occurred and will use this information to try to predict future consequences. In collaboration with ecologists, conservation biologists, and environmental scientists, they can attempt to reduce the potential negative impact.

Biogeographers seek to understand how distribution patterns came about and what phenomena maintain or alter those patterns. These questions are addressed by biogeographers in conjunction with paleobiologists, paleoclimatologists, and evolutionary biologists. Studying the evolutionary history of populations in an area along with the past climates and geological features will provide insight as to the evolution and ecological interactions among species, populations, and communities. This helps biogeographers understand the natural forces that affect extinctions, migrations, and alterations of different habitats, information that can be applied to biological conservation efforts.

A biogeographer might ask why a species lives in one area but not in another. In order to answer this, he or she might examine the other flora and fauna in the two regions and consider the types and abundance of food sources or predator/prey relationships. Alternatively, the explanation could be historical. Perhaps the species previously inhabited both regions but became extinct in one of them. If so, what changed and caused the species to die out in one region but not the other? Or perhaps the species originated in one area. Did the climate change in the one area and select for adaptations that over time led to the formation of the new species? As a scientist, the biogeographer must formulate a hypothesis based on observations and then test its validity.

See also BIOMES, AQUATIC; BIOMES, TERRESTRIAL; ECOLOGY; EVOLUTION, THEORY OF.

FURTHER READING

Biogeography Specialty Group of the Association of American Geographers home page. Available online. URL: http://people.cas.sc.edu/kupfer/bsg.html. Updated July 27, 2007.

International Biogeography Society home page. Available online. URL: http://www.biogeography.org. Updated December 4, 2007.

Lomolino, Mark V., Dov F. Sax, and James H. Brown, eds. *Foundations of Biogeography: Classic Papers with Commentaries.* Chicago: University of Chicago Press, 2004.

MacArthur, Robert H., and Edward O. Wilson. "An Equilibrium Theory of Insular Zoogeography." *Evolution* 17, no. 4 (1963): 373–387.

———. *The Theory of Island Biogeography.* With a new preface by Edward O. Wilson. Princeton, N.J.: Princeton University Press, 2001.

bioinformatics   The term *bioinformatics* can be defined numerous ways: as a field of research, as a collection of methods for analyzing data, or as a synonym for computational molecular biology. Broadly defined, bioinformatics is the use of computers to analyze any biological data. Here, the term *bioinformatics* will refer to the use of computers for storing, retrieving, analyzing, or predicting the composition or structure of biomolecules, specifically nucleic acids and proteins.

BACKGROUND INFORMATION

Nucleic acid includes both deoxyribonucleic acid (DNA) and ribonucleic acid (RNA), molecules made from the assembly of numerous nucleotide subunits, which each contain a sugar, a nitrogenous base, and a phosphate group. In the laboratory, biologists use and manipulate segments of nucleic acid as small as 10 nucleotides and up to thousands of nucleotides in length, though some chromosomes contain millions of nucleotides. DNA consists of four different nucleotides, which differ on the basis of the nitrogenous base they contain: adenine (A), guanine (G), cytosine (C), or thymine (T). RNA consists of four different nucleotides also, but the sugar moiety is slightly different, and the nitrogenous base uracil (U) replaces thymine in RNA. The sequence of the nucleotides in DNA indirectly encodes for the synthesis of proteins. Segments of DNA that contain information on how to make a protein are called genes; genes are located on chromosomes. During the first step in protein synthe-

sis, the enzyme RNA polymerase uses the DNA template to make a molecule of messenger RNA (mRNA) in a process called transcription. The sequence of nucleotides on the mRNA is based on the sequence of nucleotides in the DNA; in other words, the two are said to be complementary. For every A in the DNA, RNA polymerase incorporates a U in the mRNA; for C in the DNA, it adds G to the mRNA; for G in the DNA, it adds C in the mRNA; and for T in the DNA, it adds A in the mRNA. After transcription, ribosomes scan the mRNA and assemble specific amino acids in an order based on the sequence of nucleotides in the mRNA during a process known as translation. The ribosome reads the mRNA three nucleotides at a time. Each set of three is called a codon and specifies one of 20 different amino acids. One triplet codon (AUG) indicates the start site of translation, and three other codons (UAA, UAG, and UGA) cause termination of translation to occur. The remaining codons specify a single amino acid, though in some cases, up to six different codons encode for the same amino acid, a phenomenon known as degeneracy. Although the genetic code is degenerate (redundant) in that multiple codons encode for the same amino acid, each given codon only encodes for one specific amino acid. The chain of amino acids that results from translation is called a polypeptide, and after folding into the correct conformation, or in some cases, when combined with other polypeptides, the protein is complete.

Molecular biologists have sequenced the entire genome of hundreds of viruses and organisms, including numerous bacteria, yeast, fruit flies, mice, and humans. They have isolated and identified many individual protein-coding genes from these organisms and countless others. Identification of a gene allows researchers to study the expression of that gene, determine what turns the gene on and off, the protein it encodes, the structure of that protein, and the function of that protein. Geneticists can mutate the gene, or make changes to the DNA sequence, and observe the effects of the mutation on the protein function. Genetic engineers can clone the gene into other organisms, such as bacteria, and produce large quantities of the protein, which biochemists can purify and then characterize.

ROLE OF BIOINFORMATICS

Bioinformatics facilitates the collection, storage, retrieval, and analysis of all the information gained by sequencing genomes, identifying genes, characterizing proteins, and performing other applications. Without the use of computers, researchers would not have completed the sequencing of the human genome, which contains more than 3 billion nucleotides, or possibly any other organism's genome. Since 1953,

(continues on page 87)

BIOINFORMATICS: SEQUENCE ALIGNMENT ALGORITHMS

by Jason S. Rawlings, Ph.D.
St. Jude Children's Research Hospital

Sequence alignment algorithms are tools that enable life science researchers to compare protein or deoxyribonucleic acid (DNA) sequences with one another. These are powerful tools because of the biological relationship between the information in these sequences and the resulting protein function. According to the central dogma of molecular biology, a gene's DNA sequence will determine the amino acid sequence of the corresponding protein. The amino acid sequence determines the overall three-dimensional structure of the protein, and the structure determines function of the protein. Thus, the DNA sequence of a gene contains valuable information about the function of its resulting protein. An investigator uses a sequence alignment algorithm to determine whether two or more DNA or protein sequences are similar to one another. This essay will present the different types of alignments, and how basic alignment algorithms work and give examples of how scientists use these algorithms in their research.

The true power of sequence alignment algorithms lies in their application; following are but a few examples. They can be used to retrieve the sequence of a specific gene or protein from a database, akin to searching the Internet for a specific Web page. They can also be used to identify an unknown sequence by comparing the unknown to all known sequences in a public database. They can be used to compare two sequences for overall similarity or to find just the regions within each that are similar to one another. Researchers can also determine relationships among several sequences by aligning multiple sequences to identify members of a larger family of genes or to compare proteins from different species in an evolutionary analysis.

The different types of sequence alignments all have different algorithms. Each type of alignment algorithm has its own advantages and disadvantages in terms of the accuracy of the alignment, the information it produces, and the time it takes to complete the alignment. The simplest alignment is between two sequences; these algorithms will provide a detailed, accurate analysis of the similarity of the two sequences. The time to complete the alignment is dependent on the length of the two sequences, and given the speed of today's computers, the time required is negligible.

One can also align three or more sequences together in a multiple sequence alignment. Many available programs can perform these types of alignments. The most popular are derived from the Clustal program. Multiple sequence alignments are advantageous for the amount of information they produce and for the types of downstream analyses that they allow one to perform. However, the time to complete the alignment becomes significant. In the Clustal algorithm, not only the length of each sequence is a factor, but also the number of sequences, because each sequence must first be compared to all of the others to determine the pairwise similarity of all the sequences. This requires a total of

$$\frac{X(X-1)}{2}$$

sequence comparisons, where X is the number of sequences to be aligned. Then, the multiple sequence alignment is progressively built by combining a growing number of subalignments, starting with the most similar pair of sequences. Thus, thousands of comparisons may be required for aligning a large number of sequences, making for a computationally intensive task.

Finally, one can align a single sequence against an entire database of sequences using the Basic Local Alignment Search Tool (BLAST) program at the National Center for Biotechnology Information (NCBI) Web site. The user submits a sequence to the database and the BLAST program returns the sequences that are most similar. Because this process may involve comparing the input sequence to millions of sequences in the database, sacrifices in accuracy and completeness must be made in exchange for speed. Specifically, the program looks for "words" within the submitted sequence that can be matched in the database. These "words" may consist of just 10 nucleotides or amino acids (a typical gene consists of thousands of nucleotides; a typical protein consists of hundreds of amino acids). When the BLAST program finds matches in the database, it analyzes adjacent sequences for similarity, building an alignment in the process. This is repeated until either adjacent sequences are no longer similar or the entire sequence is aligned to the match found in the database. Those sequences that do not have adjacent similarity are no longer considered. The program returns to the user the sequences in the database that aligned best. In the end, only a portion of the submitted sequence may be aligned to sequences in the database.

NCBI offers several variations of the BLAST program; each is optimized for different types of input sequences, the database to be searched, and purpose of the query. Users can submit a DNA sequence in a search against a DNA database. Alternatively, a DNA sequence can be compared to a protein database; the query is translated into the corresponding amino acid sequence by the program before the search is done. Conversely, protein sequences can be compared to protein databases or compared to a DNA database that has been translated in all possible reading frames. Advanced versions of the BLAST algorithm allow for more specialized searches, such as iterative database searches or searches optimized for very short input sequences. Additionally, the user can choose to search the entire public database of sequences or limit the search to a specific genome or subset of sequences. The BLAST algorithm is extremely versatile and fast. The time required to complete a BLAST search depends mainly on the size of the database to be searched as well as the number of other searches in the queue at the time the search is initiated. A typical nucleotide search against the entire public DNA database can be completed in seconds.

At the core, a sequence alignment algorithm must align two DNA or two protein sequences, but how do they work? A typical algorithm such as the Needleman and Wunsch algorithm (Needleman and Wunsch, 1970) uses a predefined set of rules to line up the two sequences dynamically using a point system. For DNA sequences, the rules are simple. Only one of the strands of the DNA needs to be aligned from each of the submitted sequences because of the rules governing complementary base pairing. There are four possible nucleotides at any given position in a DNA sequence: guanine (G), adenine (A), thymine (T), or cytosine (C). At each position in the alignment, the nucleotides are compared; if they match, then points are awarded, while mismatches receive a penalty. The algorithm may introduce gaps, at a great penalty, and extend those gaps for a further penalty, in order to achieve an overall alignment with the best possible score. Overhanging sequences caused either by created gaps or simply by submission of sequences of different lengths are not penalized. The example that follows illustrates how an alignment algorithm might find the best possible alignment between two sequences.

```
1:   TTCGCGAAGATTACGCGAA
2:   TTCGTTACGCGAAGGGGTA
```

Mismatches in the first alignment are bolded below.

```
1:   TTCGCGAAGATTACGCGAA
2:   TTCGTTACGCGAAGGGGTA
```

Matches: 10(10) = 100
Mismatches: 9(−5) = −45
Gaps: 0(−50) = 0
Extensions: 0(−50) = 0
Score: 100 − 45 = 55

The second alignment involves a gap that eliminates any mismatched nucleotides. The introduction of such a gap in sequence 2 leads to an overhang.

```
1:   TTCGCGAAGATTACGCGAA
2:   TTCG........TTACGCGAAGGGGTA
```

Matches: 13(10) = 130
Mismatches: 0(−5) = 0
Gaps: 1(−50) = −50
Extensions: 6(−3) = −18
Score: 130 − 50 − 18 = 62

The creation of a gap in the second alignment results in a better score—62 compared with 55 for the first alignment, which contained only mismatch penalties. In this example, matches are given 10 points, mismatches are penalized five points. Creating a gap incurs a 50-point penalty, with an additional penalty of three for each position in the gap. Even though creating a gap is costly, it results in an overall better alignment score in this example. Gaps incur such a large penalty because of the genetic repercussions they present. Inserting or deleting a single nucleotide in a gene sequence introduces a frameshift mutation, which will drastically alter the resulting protein sequence downstream of the mutation. The specific rules used for an alignment can be modified, providing optimization for the length of the sequences to be aligned and the overall goal of the alignment.

Protein sequence alignments work in a similar fashion; however, they present some additional challenges. There are 20 possible amino acids to consider, compared to just four for nucleotides. Also, many amino acids share similar structural and chemical properties, thus invalidating a simple match/mismatch approach. To tackle these issues, protein alignment algorithms utilize a substitution matrix. Instead of a simple match/mismatch scenario, the matrix provides a similarity score for comparing two amino acids. This score is based on several factors, including amino acid structure, chemical composition, and evolution. For example, tyrosine and phenylalanine would receive a favorable similarity score because they

are structurally very similar: both contain an aromatic ring. Histidine and arginine would also receive a favorable score, even though they are structurally different, because both amino acids are basic. Since amino acid sequence is determined by the codons in DNA, certain amino acids are considered similar. For example, the codon UGU codes for cysteine, a relatively small amino acid containing a sulfur group that can uniquely influence overall protein structure via participation in disulfide bonding. A single base change, to UGG, results in tryptophan, a very bulky amino acid that does not contain a sulfur group. From analyzing many protein sequences, molecular biologists have determined that certain amino acid changes occur more frequently than others during evolution; these also receive favorable scores. As in DNA alignments, protein sequences are aligned using gap and extension penalties, with the goal of achieving the highest possible overall score.

Scientists can exploit alignment algorithms to complete complex tasks or aid in scientific research. The Human Genome Project is a prime example; in this project, scientists used DNA sequencing technology to determine the sequence of the entire human genome. While the entire genome consists of approximately 3 billion nucleotides, a typical DNA sequencer can reliably read between 700 and 1,000 nucleotides per reaction. Thus, sequencing the entire genome required the researchers to perform millions of sequencing reactions. Then, these individual sequences needed to be assembled to form the completed genome, much like putting puzzle pieces together. The strategy for assembling the entire genome relied on sequence alignment algorithms that could identify and align overlapping portions of each sequencing reaction. Both the forward and reverse strands of the genomic DNA are sequenced; the alignment program can recognize this and reverse complement the sequence reads as needed. As more sequencing reactions are added,

(continues)

(continued)

the resulting sequence assembly reflects a progressively larger portion of contiguous genomic sequence. Combined with genetic maps, this approach was used to construct the entire genome sequence.

Scientists can also use alignment programs to guide their research in the characterization of protein function. As an example, consider the *Drosophila melanogaster* protein, Socs36E, analyzed by a team of scientists from the University of Kentucky and published in a paper titled "Two *Drosophila* Suppressors of Cytokine Signaling (SOCS) Differentially Regulate JAK and EGFR Pathway Activities," in *BMC Cell Biology* in 2004. First, the authors submitted the protein sequence in a BLAST query to identify and retrieve similar sequences from other organisms. In mammals, there are eight

suppressor of cytokine signaling (SOCS) proteins, and Socs36E was aligned to all of them using a multiple sequence alignment algorithm. From this, they discovered that *Drosophila* Socs36E is most similar to mammalian SOCS5 and used this information to retrieve the sequences of all SOCS5 orthologues from other species. (Orthologues are versions of the same gene or protein in two or more different species.) Alignment of these sequences using a multiple sequence alignment algorithm allowed for the identification of regions, or domains, of the protein sequence that have been conserved throughout evolution (see the figure titled Multiple Sequence Alignment). All SOCS proteins contain two known functional domains, a centrally located SH2 (Src homology 2) domain and a carboxy-terminal SOCS domain. From the alignment, they identified these domains

in the Socs36E sequence. Further analysis using more sequences (from other proteins containing these domains) might allow for the discovery of the critical amino acid residues required for protein function. A scientist could potentially use this information to construct mutant versions of the protein by deleting or altering these amino acids, permitting detailed study of the protein's function.

Sequence alignment algorithms form the staple bioinformatics tools used by life science researchers. Although the first sequence alignment algorithms were introduced nearly 40 years ago, new alignment programs continue to emerge. Many of them focus on increasing the speed and accuracy of the algorithm, making multiple alignments consisting of many sequences feasible. Other algorithms incorporate secondary information into an alignment. For example, if the

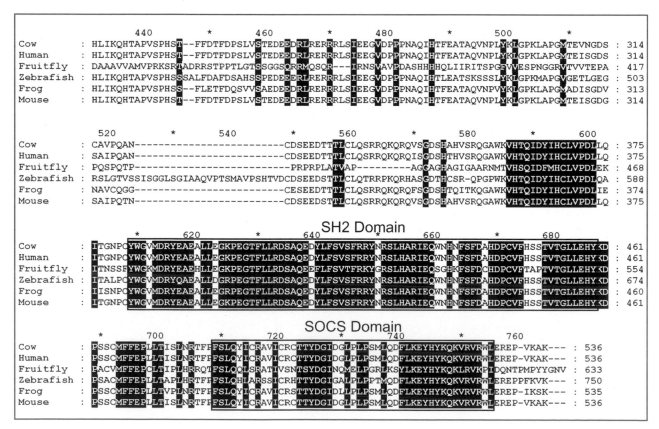

Multiple sequence alignment of an orthologous protein from several species. Identical amino acid residues are shaded in black. The SH2 and SOCS domains are indicated. Only the carboxy-terminal portion of the protein alignment is shown. *(Jason Rawlings)*

three-dimensional structures of some of the proteins in a multiple sequence alignment have been solved, this information can be used to assist in aligning these proteins with those that do not have structural information. As scientists learn more about protein structure and function and how they relate to amino acid and DNA sequence information, these algorithms will continue to improve.

Note: The BLAST program is freely available and can be accessed online. URL: www.ncbi.nlm.nih.gov/BLAST. Accessed April 8, 2008.

FURTHER READING

Altschul, Stephen F., Warren Gish, Webb Miller, Eugene W. Myers, and David J. Lipman. "Basic Local Alignment Search Tool." *Journal of Molecular Biology* 215, no. 3 (October 5, 1990): 403–410.

Altschul, Stephen F., Thomas L. Madden, Alejandro A. Schäffer, Jinghui Zhang, Zheng Zhang, Webb Miller, and David J. Lipman. "Gapped BLAST and PSI-BLAST: A New Generation of Protein Database Search Programs." *Nucleic Acids Research* 25, no. 17 (September 1, 1997): 3,389–3,402.

Higgins, Desmond G., and Paul M. Sharp. "CLUSTAL: A Package for Performing Multiple Sequence Alignment on a Microcomputer." *Gene* 73, no. 1 (December 15, 1988): 237–244.

Needleman, Saul B., and Christian D. Wunsch. "A General Method Applicable to the Search for Similarities in the Amino Acid Sequence of Two Proteins." *Journal of Molecular Biology* 48 (1970): 443–453.

O'Sullivan, Orla, Karsten Suhre, Chantal Abergel, Desmond G. Higgins, and Cédric Notredame. "3DCoffee: Combining Protein Sequences and Structures within Multiple Sequence Alignments." *Journal of Molecular Biology* 340, no. 2 (July 2004): 385–395.

Rawlings, Jason. S., Gabriela Rennebeck, Susan M. W. Harrison, Rongwen Xi, and Douglas A. Harrison. "Two *Drosophila* Suppressors of Cytokine Signaling (SOCS) Differentially Regulate JAK and EGFR Pathway Activities." *BMC Cell Biology* 5 (October 15, 2004): 38.

Wallace, Iain M., Gordon Blackshields, and Desmond G. Higgins. "Multiple Sequence Alignments." *Current Opinion in Structural Biology* 15, no. 3 (June 2005): 261–266.

(continued from page 83)

when James Watson and Francis Crick revealed the structure of DNA, scientists have been studying genes one at a time. Genomics, the study of entire genomes, allows scientists to study all the genes or all the mRNAs in an organism at the same time. Similarly, proteomics, the analysis of the proteins produced by an organism, allows researchers to study whole sets of proteins that may be related by function, tissue specificity, regulatory mechanisms, or other aspects. Then, a researcher can focus in on individual genes or groups of genes for more intense study. One revelation from the vast amount of sequencing information gathered over the past few decades is that the genetic code, the way DNA sequences encode for the synthesis of proteins, is universal. The basic mechanisms are the same and triplet codons generally encode for the exact same amino acids whether in an *Escherichia coli* cell or a human being. Because of this, once a gene has been identified in one species, computer programs can locate similar genes in other species whose genomes have been sequenced. If the structure of a protein has been determined, then as researchers discover proteins with similar amino acid sequences, they can immediately make inferences about the new protein's structure and function. For example, zinc fingers are a common structural motif found in proteins that bind DNA. If an unknown protein has a region that resembles a zinc finger domain, a scientist may hypothesize that the unknown protein binds DNA, a characteristic that can be easily tested in the lab. From further research of information available to the public in bioinformatics databases, the scientist might learn that proteins with zinc fingers often play a role in the regulation of gene expression and design future experiments accordingly.

The structure of a molecule determines its function; ability to predict the three-dimensional structure of a protein on the basis of its amino acid sequence, which can be determined from the DNA sequence, would be extremely valuable. Bioinformatics specialists have developed computer programs that will suggest likely secondary structures given a particular sequence and other programs that will "thread" the amino acid backbone of an unknown protein through the backbone of a protein that has the same domain. For example, one could thread an unknown protein with a zinc finger domain through the structure of a known zinc finger protein to see a possible three-dimensional structure for the unknown.

GENOME AND PROTEIN DATABASES

After completing the sequencing of a gene or of an entire genome, scientists deposit the information into a database. Examples of widely used databases include GenBank in the United States, DNA Databank of Japan (DDBJ), and the European Molecular Biology Laboratory Database (EMBL). These databases and others are freely accessible online through processing centers such as the National Center for Biotechnology Information (NCBI).

One common application of bioinformatics is to retrieve protein sequences or gather other information about a protein of interest. The Expert Protein

Analysis System (ExPASy) proteomics server is one of the most popular servers for protein information and analysis. The Web site contains links to several databases including the UniProt Knowledgebase. One can begin by entering a type of protein or a protein name, such as adenosine triphosphatase (ATPase), an enzyme that breaks apart molecules of ATP to release energy. The server will return information such as a biochemical description, references in the scientific literature related to the protein, and the actual amino acid sequence of the protein. Depending on the type of search, the server may return the sequences of several related proteins. In order to use the amino acid sequence for further analysis, the sequence must be in the FASTA format, which is simply a specific way to type out the sequence so bioinformatics tools can recognize and read the information. All of the retrieved information can be downloaded and saved to a personal computer for future use.

Because protein structure determines its function, the amino acid sequences are usually well conserved between organisms. Two proteins do not need to have identical amino acid sequences to carry out the same function. When amino acid sequences differ, the nature of the amino acids is often conserved. For example, one small hydrophobic amino acid may substitute for another hydrophobic amino acid and not affect the ability of the protein to function. The site of the mutation also matters. Regions of a protein that do not play an active role in the protein's function are more likely to tolerate mutations than regions that are crucial to the protein's function. To illustrate, consider a screwdriver. Whereas the width and thickness of the end of a screwdriver that fits into the depressions in the head of a screw are critical in the screwdriver's ability to rotate the screw, the shape or structure of the handle grip is not as important; it may or may not contain ridges, can be different lengths, or may be made of wood or plastic and still function effectively. The DNA sequence that encodes the protein will be even less conserved between organisms than the amino acid sequence for several reasons. Some parts of a gene do not code for amino acids—regulatory regions, untranslated regions occurring before the translation start site and after the translation stop codon, and introns (the noncoding intervening sequences located within the coding region) in eukaryotic genes—and are therefore under less selection pressure. Also, many codons are redundant; thus the DNA sequence might change but the amino acid sequence does not.

One can use databases such as GenBank, EMBL, or DDBJ to retrieve the DNA sequence for a protein of interest. One simply would use a pull-down menu to select for protein and then enter the name of the protein, or the type of protein, such as HSF1, which is a protein that regulates transcription of a family of genes called heat shock genes. The returned information page will contain a listing of all the related proteins and their source organisms, such as *Saccharomyces cerevisiae* (a yeast) or *Arabidopsis* (a plant). After selecting one of the choices, more detailed information on that particular protein is provided, including the number of amino acids, links to related literature given as references in PUBMED (a literature search system used by the National Library of Medicine), information about the protein's function or subcellular location, recognizable features within the sequence such as common structural motifs or experimentally determined functions of specific regions, and finally, the actual amino acid sequence. Again, the retrieved information can be saved to a computer's hard drive, and then one can use a server like ExPASy to explore the possible structure and function and perform other analyses. To learn more about a nucleotide sequence, one selects *nucleotide* from the pull-down menu before searching. The information returned is similar to that returned for a protein query in that it includes links to related publications about the sequence or gene product as well as general information about the gene product's function. If known, recognizable features such as promoter elements, ribosome binding sites, exons and introns, and polyA sites are also indicated.

BLAST, which stands for "Basic Local Alignment Search Tool," is another NCBI server that facilitates the analysis of sequence information. A researcher can use BLAST to compare a protein or nucleotide sequence with all the other sequences inputted by scientists from around the world. If other similar sequences do exist, BLAST will also offer a report on the degree of similarity and will give rough information about differences in structure or organization between the two genes. For a more accurate and detailed comparison, one can use an alignment program such as CLUSTAL to analyze the conserved regions, examine the degree of similarity, and compare the structure and organization of the two genes. Multiple alignments consist of the simultaneous comparison of more than two sequences. Alignment allows for the identification of regions of the protein that contain amino acids crucial to the protein's function or identification of sequences that are characteristic of specific protein families or secondary structures. The information may also help a researcher predict the three-dimensional structure or the function of a protein.

Comparison of a protein sequence with other protein sequences often reveals conserved features. These regions of the protein might play an important role in the folding or stabilization of the protein, interact with other molecules, or directly participate in the protein's function. Knowing where certain amino acids exist within the folded protein helps in predicting the function of that segment. The sequence of amino acids is referred to as the primary structure of a protein, and it determines the secondary structure the chain of amino acids assumes. Alpha helices and beta sheets are examples of common secondary structures. Random coils are regions of the polypeptide that do not form either. Servers such as PSIPRED, maintained by the Bioinformatics Unit of the University College in London, can predict the secondary structure of a protein on the basis of the amino acid sequence. Other servers, such as Predict-Protein, maintained by Columbia University, offer more complex structural predictions including secondary structures, accessibility of certain regions to solvent, presence of transmembrane helices, globular regions, disulfide bonds, and more. Structural biologists deposit their three-dimensional structural coordinates, obtained by techniques such as X-ray crystallography, into the Protein Data Bank (PDB). Researchers can access this information and view the three-dimensional structures of many proteins. Comparison with sequences of their own protein of interest can reveal information about structural similarities.

See also BIOMOLECULES; BIOTECHNOLOGY; DNA SEQUENCING; GENE EXPRESSION; GENETICS; GENOMES; HUMAN GENOME PROJECT; MOLECULAR BIOLOGY.

FURTHER READING

Batiza, Ann Finney. *Bioinformatics, Genomics, and Proteomics: Getting the Big Picture.* Philadelphia: Chelsea House, 2006.

"The Expert Protein Analysis System (ExPASy) Proteomics Server." Available online. URL: http://www.expasy.org. Updated December 11, 2007.

Krane, Dan E., and Michael L. Raymer. *Fundamental Concepts of Bioinformatics.* San Francisco: Benjamin Cummings, 2003.

National Center for Biotechnology Information home page. Available online. URL: http://www.ncbi.nlm.nih.gov. Updated October 17, 2007.

biological classification   One major goal of life science is to discover and catalogue all the living organisms in the biosphere. The Greek philosopher Aristotle was the first to attempt this never-ending endeavor in the fourth century B.C.E., when he recorded descriptions of as many animals as he could, close to 500 species. He organized them into groups based on observable characteristics, such as feathers, number of legs, or presence of horns. As more species became known, scientists gave them long, descriptive names in their native languages, which hindered the passage of information to other scientists, especially those from countries having different native languages. In the 18th century the Swedish botanist Carl Linneaus introduced the system of binomial nomenclature to facilitate identification of organisms and communication among scientists about various life-forms, the numbers of which naturalists believed approached approximately 10,000. Scientists across the world quickly adopted binomial nomenclature, which is still used today. Linnaeus also proposed a hierarchical system for classifying organisms, with each level becoming more encompassing.

The name for the discipline of life science concerned with identifying, classifying, and naming the diverse life-forms is *taxonomy.* In the 250 years since Linnaeus, the number of species identified and given a latinized name has grown to more than 1.78 million. Biologists suspect that the unknown biosphere holds at least as many as 10 times that number, and some estimates approach 100 million. The job of taxonomists is enormous and consists of more than simply naming new organisms. The world-renowned biologist Edward O. Wilson, of the Harvard University Museum of Comparative Zoology, has called taxonomy "the pioneering exploration of life on a little known planet." Based on evolutionary principles, modern taxonomy is more specifically referred to as systematics, a science with the goal of revealing the evolutionary history of all the organisms on the planet. This approach has increased biologists' understanding of different organisms. The work of systematists has led to the development of the tree of life and provides the foundation for ecological studies and conservation efforts.

DATA USED FOR CLASSIFICATION

Early taxonomists relied on studies of comparative anatomy and embryology, as morphological characteristics were easily observable and the current preferred basis for classification had not yet been proposed. Most naturalists believed species were static, created by a supernatural being in their existing forms. In the middle of the 19th century, the British naturalist Charles Darwin proposed the theory of evolution by means of natural selection, suggesting that new species emerged by descent with modification. Ever since Darwin's ideas gained widespread acceptance, systematists have attempted to arrange organisms phylogenetically, that is, according to their evolutionary histories. Today, biologists still rely on morphological and behavioral characteristics but also have molecular information by which they formulate classification decisions. For example, molecular methods such as nucleic acid and protein sequencing provide much information regarding the evolutionary history of a species or group of species. While molecular approaches reveal information about the relatedness of living species, the fossil record provides information on how those relationships came to be.

A fossil can be either a remnant of an organism that lived in the past or evidence that an organism lived in the past. Fossils form when an organism or part of an organism becomes buried in sediment and either mineralizes or leaves impressions that are preserved in the Earth's crust. Remnants include structures, such as bones, teeth, or shells, that were once part of the organism but have become mineralized. Other types of fossils include impressions, molds, or casts of the original organism or its parts. Impressions form when the shape and texture of the organism are preserved in the sediment that settles around the remains. The remains themselves may decompose, but the indentations left in the surrounding sediment are preserved as it turns into rock. If an empty space remains after the organic matter decomposes, the structure left behind is called a mold. If minerals that crystallize fill the empty space after the organic matter decays, the structure left behind is called a cast. Trace fossils, such as hardened excrement or indentations such as footprints or burrows, also provide useful details. Fossils not only provide morphological information, such as an organism's size, approximate weight, and number and types of external structures, but also indicate the geological period during which the organisms lived. The material within a layer of sediment is all derived from the same period, and the layer below it is older. Thus one can follow the appearance or disappearance of different life-forms through studying the changes in successive layers of sediment.

Anatomical information, both from fossil evidence and from present-day organisms, helps biologists reconstruct phylogenies. Generally, organisms that are closely related share similar structures. For example, the phylum Vertebrata is defined by the common presence of a backbone, but members of the phylum Vertebrata can exhibit very diverse overall morphologies. Consider the appearance of a representative organism from three classes of vertebrates: a human is a mammal, a sparrow is a bird, and a bullfrog is an amphibian. The forelimbs of these animals all look different, but a closer examination of the arrangement of bones within the forelimbs will reveal that all these animals have a humerus, a radius, and an ulna. Developmentally, these bones are all derived from the same embryonic structures. Despite the very different outward appearances of the forelimbs, structures such as these are said to be homologous, meaning they are similar as a result of shared ancestry. Homologous structures in different species are derived from one structure in a common ancestor.

Gross morphological information can be misleading. Combining the structural information with molecular data will reveal a more complete picture. Sometimes organisms look similar but are not closely related. For example, bats and dragonflies both have wings used for flight. Bats, however, are mammals and are more closely related to other nonwinged organisms such as humans and fish than they are to any insect. In this case, the wings are called analogous structures, meaning they are similar as a result of convergent evolution rather than of shared ancestry. Wings for flight evolved separately in insects and in bats. Molecular data can reveal such misconceptions that result from organisms' adapting to similar conditions in the same way.

Biochemical data in the form of sequence information from deoxyribonucleic acid (DNA), ribonucleic acid (RNA), and proteins enable biologists to determine phylogenies even when ancestral species are extinct or fossil evidence is not available for examination. Regions of DNA that encode for functional gene products are highly conserved compared to those that do not encode for functional gene products. Even within a gene that encodes a protein, differences in the nucleic acid and even the amino acid sequence of the protein exist; they are degeneracy in the genetic code and the fact that some amino acids are simply space fillers within a polypeptide, meaning they do not play an active role in the function of the protein. Because of this, comparison of sequences between organisms provides information about their evolutionary relatedness. The concept of homology can be applied to molecular sequences just as it is applied to anatomical structures. When organisms share a high proportion of sequences within a gene or region of DNA, one may conclude that the sequences or genes are homologous, meaning those regions of DNA are derived from the same original sequence belonging to a common ancestor. Organisms will have more sequence information in common with organisms that share a recent common ancestor compared to organisms with more distant common ancestry. Taken alone, molecular sequence information only reveals relative relationships (i.e., these two organisms have more sequence information in common, thus are more closely related than two others who share less homology). The same is true for chronological information. Events can only be ordered relative to one another if fossil evidence from different geological ages is available for similar organisms. If it is, the degrees of difference in sequence can serve as a molecular yardstick.

Biochemical evidence has helped biologists understand the evolutionary relationships among plants, algae, and bacteria that are capable of photosynthesis, the process by which organisms convert radiant energy from the Sun into chemical energy in the form of carbohydrates. While all of these groups of organisms are capable of photosynthesis, they have different types of pigments that absorb the light energy, with the groups of organisms that are more closely related sharing the same kinds. In another example, scientists have examined the genetic divergence in hemoglobin genes to elucidate the phylogeny of hominids, including humans, apes, chimpanzees, gorillas, and orangutans.

Genomics has facilitated classification efforts through the construction of genome maps that can be used to distinguish species. Complete genome sequences are available for hundreds of species from a variety of taxonomic groups, and advancements in biotechnology and bioinformatics have increased the pace at which biologists complete the sequencing of new genomes. The availability of information via the Internet eases communication of raw data and its analysis. High-quality photographs and databases are readily available for comparison with identified organisms within and between phylogenetic groups. Because of currently available methods and new technology, many biologists believe that with an effort like that put into the Human Genome Project, it is not unrealistic to think that the undiscovered biosphere could be completely explored within a few decades—if the degree of funding and the number of personnel currently performing active research toward this goal increase.

PHYLOGENETIC TREES

Taxonomists today have much more information by which they can classify organisms. Whereas 250 years ago Linnaeus resorted to similarities in basic appearance to classify plants and animals, now taxonomists group organisms on the basis of phylogenies. The resulting hierarchical classification system therefore also reveals their evolutionary relationships to other groups of organisms. Phylogenetic trees conveniently depict these evolutionary implications by branching. Branch points, or nodes, represent common ancestors that diverged by splitting and developing into two new separate lineages. The deeper the common branch points for two groups of organisms, the more distant their most recent common ancestor is. Trees may have scaled or nonscaled branches. The length of scaled branches represents the time or degree of evolutionary change. Trees may be rooted or not; unrooted trees do not reveal the position of the common ancestor, only patterns of relatedness, and many different rooted trees may be derived from a single unrooted tree. Because classification is hierarchical, each branch point leads to groups that are more exclusive. Groups of organisms are called monophyletic if they consist only of an evolutionary ancestor and all of its descendents.

Cladograms are branched depictions that represent shared characteristics. Their patterns may or may not reflect the true evolutionary history; if the shared characteristics are homologous, then they do. The term *clade* refers to a group of organisms that consists of all the organisms on all the stems derived from a single branch point. Cladistics is the mathematical method of grouping organisms into clades based on statistical data on the degree of relatedness of different species. Computers have facilitated cladistic analyses.

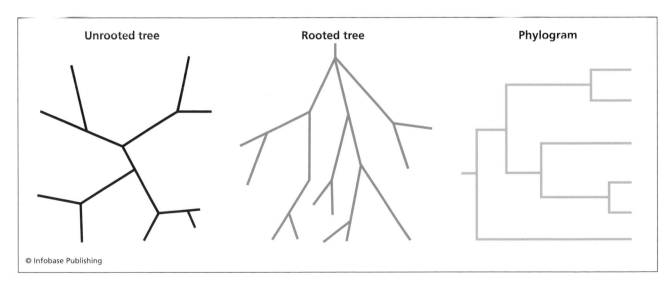

© Infobase Publishing

Phylogenetic trees are branched diagrams that depict evolutionary relationships. Rooted trees indicate the position of a shared ancestor, whereas unrooted trees do not. Phylograms have scaled branches that reflect the degree of evolutionary change.

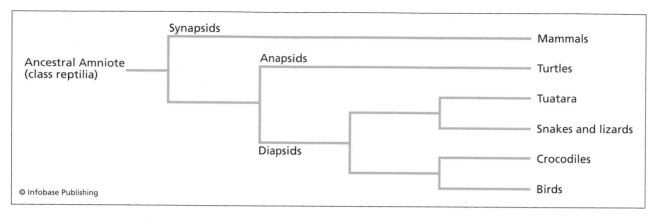

This phylogenetic tree of the class Reptilia shows that birds are more closely related to crocodiles than other reptiles, and that mammals are derived from the same common ancestor as the ancestor that gave rise to anapsid and diapsid reptilian lineages.

Systematists consider monophyletic groups the only valid form of taxonomic category (also called a taxon), as they have a sound theoretical basis and lead to combinations of groups that naturally share a greater number of characteristics. The node that defines a taxon may be deep or shallow, reflecting more encompassing and more exclusive categories. For example, one can refer to the clade of amniotes; or, within that, the smaller clade of diapsid reptiles; or, even more specific, the crocodilian clade. A paraphyletic group of organisms contains an ancestor and only some of its descendants. For example, a grouping that consisted of the ancestral amniote, turtles, tuataras, snakes, lizards, and crocodiles but did not include mammals or birds would be paraphyletic. If the group contained mammals, turtles, tuataras, snakes, lizards, crocodiles, and birds, but not their common ancestor, the group would be polyphyletic.

The National Science Foundation supports an effort called Assembling the Tree of Life (ATOL), the goal of which is to outline the history of organismal evolution by describing the relationships among the estimated 1.7 million species described to date. The simple notion of such a tree serves as a reminder that all life on Earth is connected through the passage of genetic information. The development of such a comprehensive tree of life would organize all biological knowledge and provide a framework for future investigations.

HIERARCHICAL ORGANIZATION AND MODERN CLASSIFICATION

In order to identify new species, biologists must study organisms exhaustively and compare them to descriptions of previously identified organisms to determine whether an organism has already been identified. The scientist begins by grouping it with other organisms that exhibit the same general characteristics, then focuses on more distinguishing characteristics until reaching the most defined taxonomic group, the species, a group of reproductively isolated groups of organisms.

The separation of a species is predominantly natural: that is, species are principally defined by a biological mechanism, the ability to reproduce with one another. The higher levels of classification are man-made. The category above species is genus. In modern scientific literature, organisms are referred to by Latin names for their genus and species, as well as a third designation, a letter representing the person who named the organism. In order of increasing generality, additional levels of classification include families, orders, classes, phyla, kingdoms, and domains. Some of the categories are divided further into subcategories. For example, human beings belong to the following taxonomic categories: domain of Eukarya, kingdom of Animalia, phylum of Chordata (subphylum of Vertebrata), class of Mammalia, order of Primata (suborder of Anthropoidea), family of Hominoidea, genus of *Homo,* and species of *sapiens.* In short, humans are referred to scientifically as *Homo sapiens* L. (the *L* stands for Linnaeus, who first named humans).

Many species have common names that translate differently in different languages, but the scientific Latin names are universal. By using these names in scientific communication about different species, scientists can be sure they are referring to the same organisms. In this manner, they can build on the research of others and advance knowledge in their fields more quickly and efficiently.

In 1735 Linnaeus published *Systema naturae* (System of nature), a treatise that firmly established his scientific reputation and went through 13 editions. In it, Linnaeus strove to arrange the entire natural world into three kingdoms: *Regnum ani-*

male (the animal kingdom), *Regnum vegetabile* (the plant kingdom), and *Regnum lappideum* (the mineral kingdom), which consisted of nonliving matter such as rocks and minerals. He included fungi with plants and categorized different microorganisms, which had been discovered and described by the Dutch draper Antoni van Leeuwenhoek in the 17th century, as either plants or animals. Despite some obvious problems, most scientists did not question Linnaeus's separation of life into two main kingdoms. A century later, the German biologist Ernst Haeckel proposed a third living kingdom consisting of single-celled organisms, Protista, further divided into several phyla including algae, protozoa, and monerans, which he noted did not contain nuclei. The monerans included bacteria and blue-green algae, a prokaryotic organism that today is called cyanobacteria. Haeckel also was the first to construct a phylogenetic tree based on clades.

The American biologist Herbert F. Copeland divided Haeckel's monerans into two kingdoms: Monera, which consisted of all the prokaryotic organisms, and Protoctista, which consisted of the algae and protozoans. In 1969 the American ecologist Robert Whittaker instituted the five-kingdom system that became widely used and that clearly distinguished fungi from plants. Whittaker's five kingdoms were Monera, Protoctista (or more simply, Protista), Fungi, Plantae, and Animalia.

Though biologists still refer to Whittaker's five-kingdom system, the most modern widely accepted classification system, proposed by Carl Woese in 1990, recognizes that the prokaryotic organisms traditionally placed in the kingdom Monera include organisms from two distinct lineages. By studying and comparing the ribosomal RNA genes of various life-forms, Woese concluded that sufficient molecular evidence existed to warrant the proposal of three domains of life: Archaea, Bacteria, and Eukarya. Archaea and Bacteria both consist of prokaryotic organisms, but the two groups are as distinct from each other as they are from eukaryotic organisms. Woese believed the differences were so significant that the simple addition of a sixth kingdom was not sufficient to convey the extent of the differences.

Molecular evidence has also eliminated the kingdom Protoctista as a valid taxon, though many biologists still refer to it for convenience when discussing single-celled eukaryotic organisms other than yeasts. Single-celled eukaryotes are now found scattered throughout the eukaryotic kingdoms. A few decades ago, all of eukaryotes were included in one of four kingdoms: Protoctista, Fungi, Plantae, and Animalia. Taxonomists now believe as many as 21 kingdoms may be required. For example, whereas biologists once grouped seaweeds with algae within Protoctista,

they now have separated brown seaweeds (kelp) and red seaweeds into distinct clades, and many botanists consider green seaweeds as plants. Various slime molds, formerly associated at different times with animals, plants, protozoa, and fungi, now have their own kingdoms. While many more eukaryotic kingdoms than prokaryotic kingdoms exist, this does not reflect the presumed relative numbers of existing species belonging to each domain, but rather a greater effort to identify organisms within the eukaryotic domain and particularly those in taxonomic groups including humans. Though biologists have created many eukaryotic kingdoms, one must remember that the eukaryotic organisms all are closely related, having shared a common ancestor slightly more than 1 billion years ago.

The following outline summarizes one modern classification scheme. The three domains each contain several kingdoms.

1. Bacteria
   a. Proteobacteria (purple bacteria)
   b. Planctomyces and Chlamydiae
   c. Spirochaetes
   d. Bacteroides, Flavobacteria, and relatives
   e. Green Sulfur Bacteria
   f. Gram-Positive Bacteria with high G-C
   g. Gram-Positive Bacteria with low G-C
   h. Cyanobacteria
   i. Green Nonsulfur Bacteria
   j. Thermotogales
   k. Hydrogenobacter/Aquifex

2. Archaea
   a. Euryarchaeota
   b. Crenarchaeota
   c. Karyarchaeota

3. Eukarya
   a. Diplomonads
   b. Microsporida (sporozoa)
   c. Parabasalids
   d. Myxomycota (plasmodial slime molds)
   e. Euglenozoa
   f. Naegleria
   g. Entamoeba
   h. Acrasiomycota (cellular slime molds)
   i. Rhodophyta (red seaweed)
   j. Ciliata
   k. Dinoflagellata
   l. Apicomplexa
   m. Labyrinthulids (slime nets)
   n. Oomycota
   o. Xanthophyta

p. Chrysophyta
q. Phaeophyta (brown seaweed)
r. Diatoms
s. Plantae
t. Fungi
u. Animalia (metazoa)

See also ALGAE; ARCHAEA; BACTERIA (EUBACTERIA); BIOINFORMATICS; CUVIER, GEORGES, BARON; EUKARYA; EVOLUTION, THEORY OF; FUNGI; GENOMES; HAECKEL, ERNST; HISTORY OF LIFE; INVERTEBRATES; LINNAEUS, CARL; PLANT DIVERSITY; PROTOZOA; SLIME MOLDS; VERTEBRATES; WOESE, CARL.

FURTHER READING

Hodkinson, Trevor R., and John A. N. Parnell, eds. *Reconstructing the Tree of Life: Taxonomy and Systematics of Species Rich Taxa.* Boca Raton, Fla.: CRC Press, 2007.

Maddison, D. R., and K.-S. Schulz, eds. "The Tree of Life Web Project, 1996–2006." Available online. URL: http://tolweb.org/tree/. Accessed January 22, 2008.

Nielsen, Claus. *Animal Evolution: Interrelationships of the Living Phyla.* 2nd ed. New York: Oxford University Press, 2001.

Tudge, Colin. *The Variety of Life: A Survey and a Celebration of all the Creatures That Have Ever Lived.* New York: Oxford University Press, 2000.

biological membranes   A cell is defined by the cell membrane, also called the plasma membrane or cytoplasmic membrane, the structure that separates the contents of the cell from the external environment. The main functions of the cell membrane are to contain the cytoplasm and to restrict the passage of substances into and out of the cell. Because membranes have a high lipid content, small uncharged particles can diffuse through membranes, but ions and larger molecules such as proteins or nucleic acid cannot traverse membranes without assistance. The nuclei of eukaryotic cells are bound by a double membrane that extends outward into the cytoplasm, forming a network that encloses the endoplasmic reticulum and Golgi apparatus, structures that are involved in the synthesis, modification, and transport of macromolecules. Other eukaryotic membrane-bound organelles include mitochondria, chloroplasts, lysosomes, and vacuoles. Some prokaryotic organisms have extensive invaginations of their cell membrane that increase the surface area across which membrane-dependent biochemical reactions such as photosynthesis and cellular respiration occur.

STRUCTURE

In general, biological membranes consist of a phospholipid bilayer with numerous embedded proteins. Phospholipids contain a polar head and two nonpolar fatty acid tails. In a cell membrane, the phospholipid molecules arrange themselves so the nonpolar tails align with one another and the polar heads align with one another. Because the cytoplasm of the cell and, in most cases, the external environment is aqueous, the aligned phospholipids form a bilayer, with the nonpolar regions hidden from the water and the polar heads exposed to the external environment and to the cytoplasm of the cell. This phospholipid bilayer is selectively permeable, meaning it allows some substances but not others to pass through it. Small molecules such as water or nonpolar molecules that can dissolve in lipids can pass through the membrane. Charged particles or larger molecules such as proteins, carbohydrates, and nucleic acids cannot move through the nonpolar region; thus the membrane serves as an obstacle for these substances.

Though a phospholipid bilayer forms the basis of the cell membrane, proteins make up about 50 percent or more of the total composition. Some of the membrane proteins completely span the bilayer and have ends protruding into the cytoplasm and the cell's exterior. Carbohydrate side chains linked to the membrane proteins often extend from the protein, providing additional specificity for molecular recognition. Other membrane proteins such as some enzymes are only embedded in the interior layer. Either way, the membrane proteins are held in place by hydrophobic (nonpolar) interactions between the protein and the nonpolar center of the bilayer. Proteins are simply chains of amino acids, and the amino acids can be either hydrophilic (having a strong affinity for water) or hydrophobic (lacking an affinity for water). Membrane proteins often have at least one long hydrophobic stretch that helps position it in the bilayer. The many different types of membrane proteins function as receptors for hormones or other chemical signals, as enzymes, and as transport proteins.

According to the fluid mosaic model of membrane structure, the phospholipids and the membrane proteins can diffuse laterally throughout the bilayer. This is similar to moving through a crowded subway station. As one person moves, the people he approaches slide over just enough so he can pass through, then they close the distance as soon as he bypasses them. The model is called *fluid* because of the free movement of molecules within the membrane, and *mosaic* because of the occasional globular proteins that decorate the exterior surface.

TRANSPORT ACROSS MEMBRANES

Different mechanisms exist for moving substances through biological membranes. As was already stated, despite their polarity, water molecules can diffuse through membranes by a process termed osmosis. In

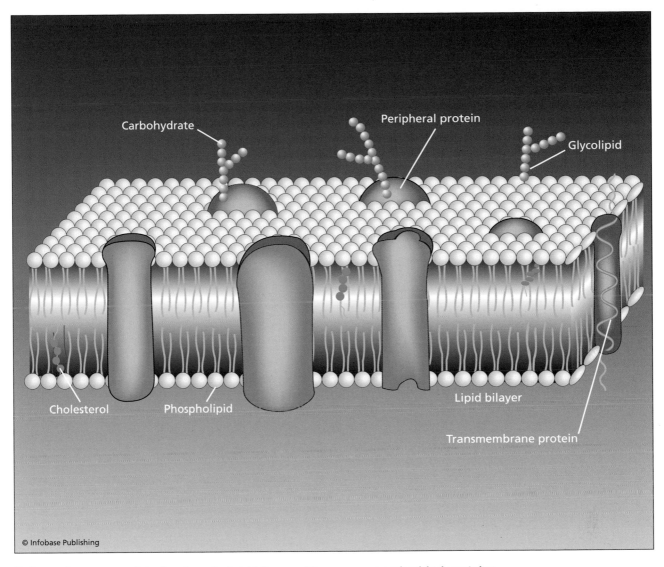

Cell membranes consist of a phospholipid bilayer with numerous embedded proteins.

the presence of solute, water molecules will surround the solute molecules to dissolve them, effectively lowering the availability of free water molecules. Because of this, the water will move from an area with a low solute concentration to an area with a higher solute concentration without additional energy input. Cells in an isotonic environment, a situation in which the solute concentration inside the cell equals the solute concentration outside the cell, will have equal rates of water movement into and out of the cell. In a hypotonic environment, the solute concentration outside the cell is lower than in the cell's cytoplasm. In this situation, water will diffuse into the cell, causing it to swell. Cell walls prevent the lysis or bursting open of cells that live in hypotonic environments by providing structural rigidity. Other adaptations for survival in hypotonic environments include mechanisms for transporting dissolved solutes out of the cell or con-

tractile vacuoles, structures that collect and remove excess water from the cytoplasm.

The concentration of solute is higher outside than inside the cell in a hypertonic environment. Water will move out of the cell, causing it to dehydrate and shrivel up.

Other small uncharged molecules can also freely diffuse through biological membranes. Nonpolar substances such as steroid hormones are soluble in lipids, thus can dissolve in the nonpolar portion of the bilayer to pass through. Nonpolar substances such as the gases molecular oxygen (O_2) and carbon dioxide (CO_2) are small enough to slip between the lipids of the bilayer. Even polar substances that are small enough, such as ethanol or urea, can diffuse through membranes. Many lipid-soluble drugs are designed specifically so they can easily traverse membranes to penetrate cells.

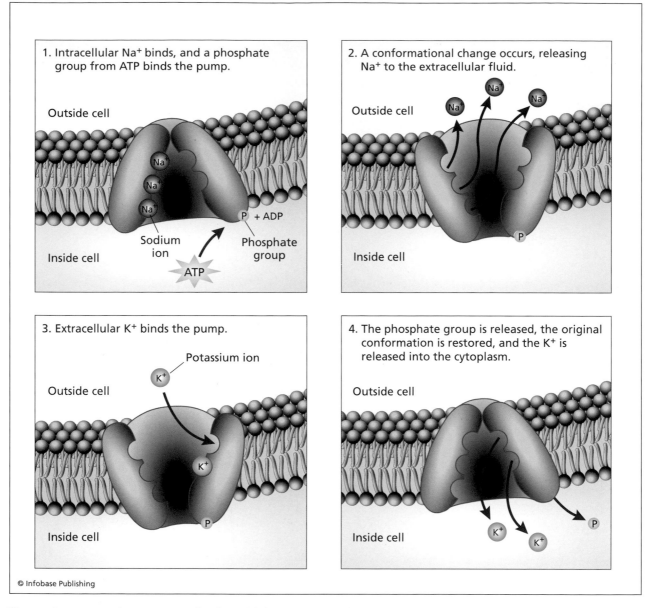

1. Intracellular Na⁺ binds, and a phosphate group from ATP binds the pump.

Outside cell

Na⁺
Na⁺
Na⁺

Sodium ion

P + ADP

Phosphate group

ATP

Inside cell

2. A conformational change occurs, releasing Na⁺ to the extracellular fluid.

Outside cell

Na⁺
Na⁺
Na⁺

P

Inside cell

3. Extracellular K⁺ binds the pump.

Potassium ion

K⁺

Outside cell

K⁺

P

Inside cell

4. The phosphate group is released, the original conformation is restored, and the K⁺ is released into the cytoplasm.

Outside cell

P

K⁺
K⁺

Inside cell

© Infobase Publishing

The sodium-potassium pump maintains a higher concentration of potassium inside the cell and a higher concentration of sodium outside the cell.

Most substances cannot diffuse through biological membranes because they either are too large or are charged and therefore cannot penetrate the hydrophobic interior of the phospholipid bilayer. Specialized proteins span the membrane to facilitate the transport of these substances. When transport proteins, also called carrier proteins, are required to move a solute across a membrane even though the solute is moving down its concentration gradient, the process is called facilitated diffusion. Because physical laws are still controlling the movement, no additional energy is required to power the transport. The transport proteins are specific for the molecules they translocate, and they can transport the solute in either direction depending on the concentration gradient. The solute first binds the carrier protein on the side of the membrane with the higher concentration. The binding causes a conformational change, a slight change in the shape of the carrier protein that exposes the solute to the other side of the membrane where the concentration is lower, and the carrier protein releases the solute. Amino acids and sugars are examples of molecules that move across membranes by facilitated diffusion.

Other mechanisms exist to transport solutes from areas of lower to higher concentration. Because this sort of transport requires the expenditure of energy, it is called active transport. If solutes could only move through biological membranes down their concentration gradients, the internal environment of the cell and of all the cellular organelles would be identical to that of the extracellular fluids. The work performed by the cell requires unique conditions for a variety of tasks, and active transport makes those conditions possible.

Living cells have voltages across their membranes, meaning the electrical charge inside the cytoplasm differs from the charge of the extracellular fluid. This unequal distribution of charges is called a membrane potential, and it is a form of potential energy. More anions (negatively charged ions) are present inside the cell, giving it a more negative charge than outside the cell. In combination with the chemical concentration gradient of the ions, the electrical gradient can be used by the cell to perform work, such as transmitting a neural impulse. Cells also employ membrane potentials to perform work of cellular respiration and photosynthesis.

The sodium-potassium pump demonstrates a specific example of active transport. The concentration of sodium ions (Na^+) is lower inside the cell than outside the cell, and the concentration of potassium ions (K^+) is higher inside the cell than outside the cell. The binding of intracellular Na^+ to the pump

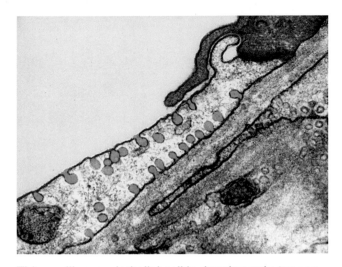

This capillary endothelial cell is drawing substances into the cell by pinocytosis, a specialized form of endocytosis. Pinocytotic vesicles are shown in blue. *(Copyright Dennis Kunkel Microscopy, Inc.)*

protein stimulates the phosphorylation of the protein by adenosine triphosphate (ATP). A resulting conformational change releases the Na^+ to the extracellular fluid, and then K^+ binds the protein. The phosphate group is released from the pump, restoring the original conformation, which releases the K^+ to the cytoplasm and prepares the protein for binding new Na^+ ions. In this manner, three Na^+ ions are exchanged for two K^+ ions for each ATP that is hydrolyzed.

Cotransport is a specialized form of active transport that involves the coupling of a solute that moves down its concentration gradient with the transport of another solute against its concentration gradient. Often, the gradient used to power this type of transport is derived from a proton gradient. The hydrolysis of ATP drives the transport of hydrogen ions (H^+) across a membrane, creating an electrochemical gradient. As the H^+ flows back through the membrane down its gradient, amino acids, sugars, or other nutrients "piggyback" through the membrane carrier protein.

Solutes that are too large to move through carrier proteins are transported through biological membranes by the complementary processes of endocytosis and exocytosis. In endocytosis, the cell membrane reaches out and around, forming an invagination that entraps the substances contained in the enclosed fluid. This forms a membrane-bound vesicle inside the cytoplasm that can fuse with lysosomes, organelles that contain digestive enzymes that break down the substances taken into the cell. Phagocytosis is the endocytosis of solid particles such as debris from a blood clot or a bacterial cell. In receptor-mediated endocytosis, receptors on the surface of a cell bind specific particles, and the cell folds inward, forming a sac around

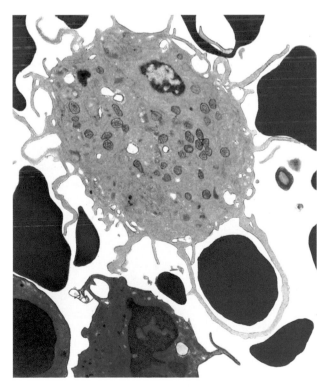

The macrophage in this transmission electron micrograph is engulfing a red blood cell by phagocytosis. Note the membrane of the macrophage has completely encircled the red blood cell. *(NIBSC/Photo Researchers, Inc.)*

the attached particles. Pinocytosis is the uptake of droplets of extracellular fluid by endocytosis. In exocytosis, substances are packaged into vesicles that fuse with the cell membrane and are released to the cell's exterior. Secreted proteins are transported out of the cell by exocytosis. Membranes are also generated and membrane proteins are moved to the surface by this mechanism.

See also BIOMOLECULES; CELL BIOLOGY; CELL COMMUNICATION; EUKARYOTIC CELLS; PHYSIOLOGY; PROKARYOTIC CELLS.

FURTHER READING

Alberts, Bruce, Alexander Johnson, Julian Lewis, Martin Raff, Keith Roberts, and Peter Walter. *Molecular Biology of the Cell.* 5th ed. New York: Garland Science, 2007.

Campbell, Neil A., Jane B. Reece, Lisa A. Urry, Michael L. Cain, Steven A. Wasserman, Peter V. Minorsky, and Robert B. Jackson. *Biology.* 8th ed. San Francisco, Calif.: Pearson Benjamin Cummings, 2008.

Sztul, Elizabeth. "Cell membranes." In AccessScience@ McGraw-Hill. Available online. URL: http://www.accessscience.com, DOI 10.1036/1097-8542.116500. Accessed January 14, 2008.

biological weapons Biological weapons are biological agents, including pathogenic organisms or toxins produced by an organism, that are used for the purpose of inflicting harm or incapacitating others. Ideal biological weapons are highly infective, potent, and easy to deliver and have available vaccines directed against them. A major disadvantage of biological weapons is the possibility of backfire, since containment is difficult once a contagious disease has begun spreading. Biological weapons are usually easy to produce in mass quantities but difficult to deliver effectively to the enemy. Aerosols are the most common form of delivery. The aerosol must be either inhaled or ingested and usually must incubate for a few days before inducing symptoms.

The use of biological weapons against an enemy for hostile purposes or during armed conflict is called biological warfare. In 1972 the Biological and Toxin Weapons Convention made it illegal to develop, produce, stockpile, or acquire biological weapons. As of June 2000, more than 144 countries recognized by the United Nations, including the United States, the United Kingdom, and the Russian Federation, had signed the treaty. Military analysts agree that since biological warfare does not immediately incapacitate an enemy, other weaponry is more effective to prevent an immediate threat. Bioterrorism is another serious concern, since biological weaponry requires minimal expertise, economic outlay, and

equipment. The threat of bioterrorism, the use of biological weapons as a means of terrorizing or coercion, has gained more recognition since the events following the attacks on the United States of America on September 11, 2001. Less than one month later, mail containing spores from the bacterial pathogen anthrax caused 22 confirmed cases of anthrax and five deaths. Fortunately, effective defenses are available against most biological weapons. As soon as the biological agent was identified as anthrax, people who might have been exposed were treated with the antibiotic ciprofloxacin, a precaution that may have prevented additional cases.

The Centers for Disease Control and Prevention (CDC) lists approximately 27 biological agents that could serve as potential biological weapons. The list includes bacteria, viruses, toxins, and other biological products. The CDC categorizes the agents into three categories (A, B, or C) based on the ease with which they spread and the severity of illnesses they cause.

The most serious threats, those belonging to category A, include the pathogens responsible for anthrax, smallpox, plague, viral hemorrhagic fevers, botulism, and tularemia. Anthrax is caused by the bacterium *Bacillus anthracis*. Characteristics that make it ideal as a biological weapon include its ability to form resistant spores that can easily be spread in the form of an aerosol, its high infectivity, and the availability of a preventative vaccine. The manifestation of anthrax depends on the means by which someone is exposed. Inhalation of anthrax spores is the likely mode of transmission if it is used as a biological weapon. Inhalation anthrax causes coldlike symptoms to occur within a week, but treatment is most effective post exposure prior to the establishment of infection. The mortality rate is 50 to 90 percent. Smallpox is caused by a very contagious variola virus and also has a high mortality rate (30 percent), and it can be prevented by vaccination. Because of the success of a worldwide vaccination effort during the middle of the 20th century, smallpox has been eradicated. The last documented U.S. case occurred in 1949, and the last known naturally occurring case worldwide was in Somalia in 1977. The only known specimens of this virus exist at high-security labs in the U.S. CDC and in the former Soviet State Research Center for Virology and Biotechnology. Routine vaccination for the virus that causes smallpox ended in 1972 since it is no longer present in the population; that means the general population is vulnerable to the use of variola virus as an agent for bioterrorism. The plague is caused by the bacterium *Yersinia pestis* and is normally transmitted by bites from fleas carried on rodents. Infection can lead to either the bubonic or the pneumonic form of the plague, but only the pneumonic form can be transmitted from person to person, and an aerosol containing *Yersinia* would cause the pneumonic form. Treatment with antibiotics can cure pneumonic plague if administered within 24 hours of symptom development. Viral hemorrhagic fevers (VHFs) cause profuse bleeding from the mucous membranes and include viral diseases such as Ebola, Marburg, and Lassa fever. They can be extremely lethal, and, with a few exceptions, there is no established cure or drug treatment for VHFs. Fatality rates range from approximately 50 to 90 percent, a factor that limits the spread of these diseases. Botulism results from the ingestion or inhalation of a neurotoxin produced by the bacterium *Clostridium botulinum*. The toxin can cause muscle paralysis, including paralysis of the muscles involved in breathing, a condition that can be fatal without mechanical assistance. If antitoxin is administered early, the symptoms of botulism can be reduced. Most patients do recover within weeks or months. Tularemia is also caused by a bacterium, *Francisella tularensis*. Though nonfatal, a tularemia infection causes severe weight loss, fever, headaches, weakness, and sometimes pneumonia. *Francisella tularensis* is not transmitted from person to person but if used as a weapon would probably be spread by aerosol. One can contract the disease by inhaling or ingesting as few as 10 to 50 bacterial cells. Antibiotics can treat tularemia.

Other toxins that pose a threat include ricin and staphylococcal enterotoxin B, both of which belong to category B. Ricin is a poison that can be made from the waste of ground-up castor beans of *Ricinus communis*, a plant whose oil is used by many industries, including in paints, textiles, and cosmetics. The toxin is very stable and can be produced in the form of a powder or a mist or dissolved in a small quantity of liquid, making it easy to deliver. Even in small doses, ricin is potent; it works by inhibiting cells from synthesizing proteins. Symptoms vary depending on whether the poison is ingested or inhaled, and there is no antidote. In 2003 and 2004 mail containing ricin powder was sent to the White House and the U.S. Senate, but security officials intercepted the letters before any harm was done. Staphylococcal enterotoxin B (SEB) is a toxin produced by *Staphylococcus aureus* and a common cause of food poisoning. As a biological weapon, SEB would be aerosolized and then inhaled by victims. Symptoms after ingesting SEB include nausea, vomiting, cramps, and diarrhea, but if it is inhaled, symptoms would include sudden high fever, chills, headache, muscle aches, and a dry cough. Another class B agent is *Coxiella burnetii*, a bacterium that causes Q fever, a disease characterized by headaches, fever, malaise, coughing, and pneumonia. The mortality rate for Q fever is low, but the incidence would probably be high if it were used as a biological weapon.

Other forms of weaponry (besides firearms, missiles, bombs, and other artillery) that are related to biological weapons are chemical weapons, nuclear weapons, and radiological weapons. Chemical weapons include substances such as inflammatory or combustible mixtures, smokes, or gases that can irritate, burn, incapacitate, poison, or asphyxiate. After inhalation or absorption through the skin, chemical weapons can take effect immediately. Nuclear weapons include those whose destructive power is derived from uncontrolled nuclear reactions. Atomic bombs derive their force from chain nuclear fission reactions, in which neutrons are injected into atomic nuclei, causing them to split, releasing more neutrons, which split additional nuclei, and so on. Another type of nuclear weapon is the hydrogen bomb, which obtains its incredible amount of energy from nuclear fusion, the forced fusion of multiple nuclei to form a single, heavier nucleus. Radiological weapons disperse radioactive material through the use of conventional explosives, by fire, or otherwise by dilution. Radioactive material spontaneously emits dangerous energetic particles as their atomic nuclei disintegrate and can cause radiation sickness, the symptoms of which include tiredness, nausea, vomiting, loss of teeth and hair, and damage to the bone marrow, which can result in a decrease in red and white blood cells.

See also BACTERIA (EUBACTERIA); BIOMOLECULES; INFECTIOUS DISEASES; RADIOACTIVITY.

FURTHER READING

Judson, Karen. *Chemical and Biological Warfare.* Tarryton, N.Y.: Benchmark Books, 2004.

Langford, R. Everett. *Introduction to Weapons of Mass Destruction: Radiological, Chemical, and Biological.* Hoboken, N.J.: John Wiley & Sons, 2004.

Naff, Clay Farris, ed. *Biological Weapons.* Farmington Hills, Mich.: Greenhaven Press, 2006.

Zubay, Geoffrey. *Agents of Bioterrorism: Pathogens and Their Weaponization.* New York: Columbia University Press, 2005.

biology Biology is the branch of natural science that deals with living organisms. The study of life addresses many varied questions: What is life? How do new species originate, and how do they reproduce? What do all life-forms have in common, and how do they differ? How do particular environments affect a species, and how do species interact with one another? The quantity of information that biology encompasses is enormous, but the information can

(continues on page 102)

STRUCTURE-FUNCTION RELATIONSHIPS IN LIFE SCIENCE

by Lisa M. Dorn
University of California at Davis

There exists an interaction so fundamental to the natural world that it can be found at every level of organization, from atoms to molecules, organs, individuals, populations, communities, biomes, and even global climate patterns. This interaction is the intricate relationship that inevitably exists between structure and function. Structure is the way in which something is organized or put together, while function is the activity, intended or otherwise, performed by that something. These two share a reciprocal relationship—it does not make much sense to use a screwdriver to hammer in a nail; nor would it make much sense to design a hammering device that resembled a screwdriver. In the first case, structure dictates the function of an object. When skipping stones across a lake, one looks for the flattest stone because the aerodynamic qualities of its flat shape enable the stone to skip effectively across the lake's surface. A round stone would simply fall to the bottom. Conversely, the intended function of an object will often shape its structure. For example, the structures of airplane and helicopter wings differ because each functions for a different type of flight. The expectation that structure will dictate function is a product of observation, the observation that flat stones are more likely to skip than round stones. In biology, however, the converse expectation, that function shapes structure, is an extension of the theory of natural selection. Natural selection suggests structures that are best suited to a function that improves the organism's fitness will survive and multiply over future generations. Examples of the interaction between structure and function abound in the natural world, at every level of organization. Understanding the role this relationship plays in the development and functioning of the natural world can help illuminate the world in a new light.

At the atomic level, the relationship between the structure and function of a carbon atom helps explain why carbon is so important to the life of our planet. One of the most important aspects of the structure of a carbon atom is that it has four bonding electrons, a fairly boring statistic until you realize just how much four bonding electrons can do. These four electrons allow carbon to form combinations of single, double, or triple bonds with an almost endless combination of atoms. Carbon can even bond to itself in very different ways, forming some of the hardest and softest minerals in the world. A diamond is pure carbon with each carbon atom bonded to four other carbons, forming a three-dimensional crystal structure. This structure makes diamond the hardest mineral in the world. Another common form of carbon is a series of two-dimensional crystals called graphite. In graphite each carbon atom is bonded to only three others, forming flat sheets of carbon that resemble chicken wire. These sheets are only loosely bonded to each other, causing them to slide across each other quite easily. This sliding is what makes graphite one of the softest minerals in the world. Because graphite, the recognizable material at the tip of a pencil, is so soft, it leaves a little of itself behind every time a pencil tip touches paper.

When carbon combines with other elements, it can form gases, crystals, or chains. Carbon gases include carbon dioxide and carbon monoxide. Calcium carbonate, the main ingredient in seashells, marble, and classroom chalk, is a carbon-based crystal. In carbon chains, each carbon atom forms a bond with the carbon atoms directly above and below it, leaving two free electrons on every carbon atom to bond with other elements. As a result, carbon forms the core of thousands of important biological and other molecules including fuels, plastics, and carbohydrates.

Remarkably, all of carbon's amazing shapes and functions can be traced back to its atomic structure of four bonding electrons. In a sense, life is an eventuality of the structure of the carbon atom—an example of the fundamental relationship between structure and function. In the carbon example, function was a direct result of structure at the atomic level. However, in biology the structure-function relationship often results when structure and function affect each other across generations, a process called natural selection.

At the molecular level, the molecule cellulose elegantly demonstrates the structure-function relationship. As a major component of plant tissue, cellulose is one of the most important biological molecules on Earth. At any given time 50 percent of all of the carbon on Earth is tied up in cellulose molecules. Cellulose is an important food source for many organisms and is the major component of paper and cotton. However, the main function of cellulose is as a plant's skeleton. Plants construct cellulose out of small, energy-rich sugar molecules called glucose. Plants can afford to use these potential energy sources as construction materials because plants are experts at turning energy from sunlight into the chemical energy plants and animals need to survive, a process called photosynthesis. When plants capture this energy, they store it in the form of glucose, which is then converted into many other molecules. Some of the glucose is stored in the form of starch, which plants and animals use for food, but much of it is bound together in long strands to make cellulose. The glucose molecules used in cellulose have a unique structure, distinct from the structure of the glucose subunits of starch. This slight structural difference causes cellulose chains to lie flat (unlike starch, which has a coiled structure). The flat shape of cellulose allows the hydrogen atoms at the edges of the glucose molecules to bind to each other, reinforcing the flat shape and creating a strong thin fiber with a structure similar to that of graphite. All of this strength is essential for cellulose to function as plant skeletons. Without the rigid structure provided by cellulose, plants would be amorphous green puddles, and the world would not look anything like it does today.

While the strength of cellulose is important to its function in plants, it creates enormous problems for everyone else. Everybody needs energy to survive, but plants belong to a unique group of organisms called phototrophs that are capable of converting light energy into chemical energy. The rest of us must therefore rely on plants for energy. Because much of the energy originally harvested from sunlight is bound up within cellulose, other organisms (including humans) must have a means of accessing that energy. There are three main strategies for getting the energy out of cellulose: use an enzyme to break down cellulose, cooperate with an organism that has this enzyme, or eat organisms that use one of the first two strategies. An enzyme is a protein specialized for catalyzing biochemical reactions, such as breaking down other molecules. Each enzyme is responsible for working on one particular molecule. The relationship between an enzyme and its target molecule is dependent on the shape or structure of the two molecules (much like two adjacent pieces of a jigsaw puzzle). Only a few organisms—most notably several unicellular microorganisms—have developed the enzyme capable of breaking down cellulose, cellulase. Herbivores rely on cooperation with these microorganisms to obtain plant resources. The larger herbivore provides a safe habitat and food for the microorganisms while the microorganisms provide the herbivore with energy. Some herbivores, such as cows, deer, and antelope, create this habitat inside their own guts. These herbivores, collectively called ruminants, have a complex digestive system with four stomachlike chambers for food digestion. The first of these chambers houses microorganisms and serves as a large fermentation vat, much like the oak barrels used to make wine. Antelopes and other ruminants feed their microorganisms by eating grasses and other plant materials. Once the microorganisms have broken down the cellulose, the products are moved through the remaining chambers, where they are churned and eventually absorbed into the animal. Many of the microorganisms are moved through the digestive tract and digested with the cellulose products. Antelopes will use these microorganisms as a source of protein and vitamins. The structure of the antelope's gut has evolved into a digestive system that functions efficiently for retrieving energy from cellulose.

The antelope's relationship with cellulose also affects the structure of antelope populations. The same is true for other members of the ruminant group. A population is a group of members of the same species that live in the same area and reproduce with each other. The functions of populations include getting food, reproducing, and protecting individuals from predators. These three functions have a large effect on the structure of populations. The East African dikdik, a small gray-brown antelope with large eyes and a strange pointed muzzle, exemplifies this relationship. Dikdik populations are structured into monogamous pairs in small home territories, a direct result of the functions of the population. Dikdiks live in arid habitats and feed on small fruits and new plant growth. These food items are rare and do not last long, so the dikdiks are constantly moving around their home territory to find food. They also need to be experts on their home territory so that they know when and where new food is available. As a result, dikdiks prefer small home territories. Dikdik food and habitats have also caused the dikdiks to be very antisocial. Because their food is scarce, females prefer to avoid competition with other females by living alone. Males, on the other hand, do not need as much food as females (because they do not have to feed offspring), so they do not really care whether they share a home range. What they do care about is finding a mate when the time comes. But female dikdiks do not make this task easy. Females stay so far apart from each other it would take far too much energy for a male to keep running from female to female to see who is ready to mate. Instead, males hedge their bets by sticking with the same female all of the time. The female tolerates the male because he helps keep other dikdiks away from her territory. Finally, the small size of the dikdiks and their tendency to live in pairs help them remain hidden from predators. This is important because dikdiks are too small to run very fast. If they lived in herds on open plains, they might become a predator's favorite snack food. What dikdiks eat, how they find mates, and how they protect themselves from predators determine their population structure.

Populations of dikdiks are surrounded by populations of many other species: the plants they eat, their predators, and many other organisms. In any given area there are all manner of populations living together. These collections of populations living in the same area and interacting are called communities. Even communities exhibit a strong relationship between structure and function. The structure of a community refers to the number of species present and the way in which the species interact with each other. The function of communities is related to their ability to respond to change, a characteristic related to the structure of the community. Communities with complicated structures are better equipped to avoid potentially devastating changes. A complicated community structure is one in which there are many different connections among species and overlaps in species' functions (eat the same prey, etc.). Community structure is often depicted as a food web diagram. Food webs depict how each species in a community uses other species as a resource such as food or nesting material.

Some communities are susceptible to change on the basis of the presence or absence of a single species. These species are referred to as keystone species because they have a large effect on all other species in the community. Removing a random species from the community is unlikely to have a large effect,

(continues)

(continued)

but removing the keystone species can radically alter the structure and function of the community. In desert communities, kangaroo rats are a common keystone species. Kangaroo rats dig burrows into the soil to keep cool and safe from predators. By disturbing the soil they make it easier for small annual plants to establish themselves and compete with perennial grasses. When kangaroo rats are removed, the perennial grasses take over the plant community, the annual plants die out, and the birds and rodents relying on the annual plants soon die out as well. The presence or absence of kangaroo rats determines the structure of the community and, in turn, affects the community's stability (ability to function).

Communities themselves are organized into biomes. A biome consists of all communities in the world with similar climate and similar plant and animal adaptations. Alternatively, one can think of a biome as all communities with an equivalent structure and function. The structure of a biome is a combination of the physical environment with the food web of interactions among species. The function of a biome relates to the way energy moves between species in the food web. Deserts, rain forests, and grasslands are all examples of biomes.

While there are many different types of deserts in the world, all deserts share a similar structure: low rainfall, large daily temperature fluctuations, not much in the way of ground cover or tree canopies, plants that store water, and small burrowing animals. The Earth is patterned in a mosaic of biomes, each with a unique structure and function. The structure of the global climate determines the location of biomes across the Earth. Climate results from wind patterns arranged in large loops of air alternating between the Earth's surface and the upper atmosphere. There are six doughnut-shaped air loops called cells circling the Earth, three in the Northern Hemisphere and three in the Southern. As the winds turn, they alternately heat and cool and move water across the planet in predictable ways. Deserts are created by dry winds picking up water from the surface of the Earth. The first and second set of cells in each hemisphere meet at 30 degrees latitude, where air descends toward the Earth. The air approaching Earth is cold and dry, so as it gets closer to the warm Earth, its temperature rises and it soaks up water from the Earth's surface. The heat causes the air to start rising again, completing the second loop of air. The combination of warm and thirsty rising air leaves the Earth with hot, dry deserts. Looking at a map of the world reveals that

great deserts like the Sahara, the Arabian, the Sonoran, the Kalahari, and the Australian are all found near 30 degrees latitude. Much as the structure of the carbon atom leads to carbon's functioning as the base of our most important biological molecules, so the structure of wind loops over the surface of the Earth leads to the pattern and function of biomes across Earth's surface.

No matter what level of organization one examines, amazing examples of structure-function relationships abound. The intimate association between structure and function explains why the world functions as well as it does and why studying the living world is so exciting.

FURTHER READING

Abbot, E. Stanley. "The Causal Relations between Structure and Function in Biology." *The American Journal of Psychology* 27, no. 2 (April 1916): 245–250.

Brashares, Justin S., Theodore Garland, and Peter Arcese. "Phylogenetic Analysis of Coadaptation in Behavior, Diet, and Body Size in the African Antelope." *Behavioral Ecology* 11, no. 4 (2000): 452–463.

Jarman, P. J. "The Social-Organization of Antelope in Relation to Their Ecology." *Behaviour* 48 (1974): 215–266.

McCann, Kevin Shear. "The Diversity-Stability Debate." *Nature* 405 (May 11, 2000): 228–233.

(continued from page 99)
be categorized by several convenient mechanisms. One can begin to break the subject down on the basis of the type of living organism being studied. For example, botany is the study of plants, zoology is the study of animals, and microbiology is the study of microorganisms. Those branches are still rather large, but they can each be broken down further. They can be subdivided according to more specific types of organisms—microbiology can be split into bacteriology (the study of bacteria), mycology (the study of fungi), protozoology (the study of protozoa), and so on. Another way biological subdisciplines can be further broken down is by focusing on one physiological system or process. For example, a botanist might focus on plant reproduction or plant genetics, a microbiologist might specialize in microbial

ecology, or a zoologist might research ant behavior. Many of the numerous fields of study are listed in the table Branches of Biology.

CHARACTERISTICS OF LIFE

Though only 1.4 million species have been identified, biologists believe there could be as many as 50 million different species on the Earth. Species are particular types of organisms that have similar anatomies and are capable of interbreeding. While the diversity of life-forms that have adapted to exist in practically all Earth's habitats is tremendous, all living organisms share several common properties.

- Living things must have order. According to the second law of thermodynamics, the natural tendency of things is to move toward

BRANCHES OF BIOLOGY

Branch	Is the Study of
anatomy	the structure of organisms
arachnology	spiders
bacteriology	bacteria
behavioral ecology	the ecological and evolutionary basis for animal behavior
biochemistry	chemistry applied to living systems
biogeography	the distribution of a species
biophysics	physical theories and methods applied to living systems
biopsychology	the biological basis of behavior and mental states
botany	plants
cell biology	cellular structures and their functions
ecology	the interactions between organisms and their environment
endocrinology	hormones, the endocrine system, and its associated disorders
entomology	insects
ethology	animal behavior
evolutionary biology	the origin and descent of species and their change over time
genetics	heredity and variation of organisms
herpetology	reptiles and amphibians
ichthyology	fishes
marine biology	organisms that live in the ocean
microbiology	microorganisms
molecular biology	biomolecular structures and processes such as DNA replication, transcription, and protein synthesis
mycology	fungi
ornithology	birds
phycology	algae
physiology	the functions and activities of living matter
population biology	populations of organisms, especially in terms of biodiversity, evolution, and the environment
protozoology	protozoa
reproductive biology	how organisms reproduce
sociobiology	the social organization and behavior of animals
zoology	animals

disorder, but living organisms are highly structured and must maintain this structure to carry out physiological processes associated with life. One way that all organisms exhibit structural order is that they are composed of cells. Even the simplest single-celled organisms compartmentalize certain functions to specialized areas of the cell.

- Living organisms take in energy from the environment in the form of food or light energy and transform the energy to perform work associated with life. Death results when the organism is no longer able to bring in and utilize energy successfully.
- Living things grow and develop according to patterns determined by their genetic information. Chromosomes made of deoxyribonucleic acid contain the blueprints for building and assembling new cells and for the function of those cells.
- Life-forms create new life-forms similar to them. Every individual of a species might not produce offspring, but in a species there must be a mechanism in place for reproduction.
- Organisms can detect stimuli from the external environment and respond accordingly. This could be as simple as a plant's growing toward the sunlight or as complex as a bat's using sound waves to navigate its way through a dark cave.
- Despite changes in the external environment, a living organism must maintain optimal conditions inside its own cells and body. For example, waste products of metabolism must not build up, salt concentrations must remain within tolerable levels, and body temperature must be maintained.
- Life evolves, or changes, over time. Organisms adapt to changing environments, and adaptations that result in increased survival and reproductive efficiency are passed on to offspring.

Objects or things that fulfill some but not all of the characteristics listed are not considered living. For example, fire takes in components from its environment, seems to grow and develop, changes and responds to its surroundings, and even seems to create offspring, but it does not exhibit order; nor does it have a genetic blueprint.

COMMON THEMES OF THE BIOLOGICAL SCIENCES

Several unifying themes emerge when studying the biological sciences. One is that as a science, biology is a process. New information is gained by scientific investigation. Biologists formulate testable hypotheses, tentative explanations for an observed phenomenon, then plan and carry out controlled experiments to examine the validity of the hypothesis. Theories are much broader in scope, explaining multiple phenomena and observations. Before becoming a theory, an idea must be supported by an abundance of evidence and be widely accepted by scientists in that field. Two theories that dominate biology are natural selection and the cell theory. The theory of natural selection explains that individuals or species that are best adapted to their environment have better survival and reproductive success, leading to the perpetuation of those genetic qualities that suited the organism to that environment. The cell theory states that cells are the fundamental structural and functional unit of all living things.

Another common theme of biology is that life is characterized by a hierarchy of order with each level of organization building on the previous level. At each level, structure determines function, beginning with the architecture of an atom all the way through the structure of biological communities. As components from one level join to develop a higher level of organization, new properties emerge and new functions can be carried out.

All matter, including living organisms, is made up of the chemical elements, and atoms are the smallest particles of an element that retain that element's properties. Chemical linkages called covalent bonds hold atoms together to form molecules. Living organisms are composed mostly of biomolecules, which are large, complex molecules, including proteins, carbohydrates, lipids, and nucleic acids. Biomolecules assemble to form cellular structures such as chromosomes, biological membranes, ribosomes, and other organelles, each of which has a unique function that contributes to the growth or maintenance of the cell, the smallest functional unit of life.

An organism can be unicellular, meaning a single cell of that species carries out all of the activities necessary for life, or multicellular, containing up to trillions of individual cells. In complex multicellular organisms, the cells are the functional units of a tissue, a group of cells that have a common structure and work together to perform a common function. Several tissues join to form an organ that is part of a physiological system. For example, the mouth, pharynx, esophagus, stomach, small intestine, colon, gallbladder, and pancreas are all components of the human digestive system. Alone, each performs a specific function—the stomach churns food, the pancreas secretes digestive enzymes, food is chemically digested in the small intestine, and so on. Together, all the organs work to harvest energy from food to provide the body with energy for life's work and material for building new biomolecules.

An organism, then, can comprise a single cell or multiple body systems, each of which cannot support life alone, but together with the others allows a multicellular organism to carry out all of life's necessary processes. All the individual organisms of a species that live in a defined geographic area make up a population. A community comprises all the populations of all the different species that live in the same habitat. The combination of all of the different populations in a community and the abiotic (nonliving) factors make up an ecosystem. The world's largest communities, called biomes, are characterized by specific types of vegetation and climates. Temperature, precipitation, soil types, and wind conditions all affect the types of organisms that inhabit a biome. For example, dry grasslands called savannas exist in tropical areas that have little rainfall. The prolonged dry seasons support the growth of large, scattered trees amid drought-resistant undergrowth. Because of this, large grazing herbivores inhabit savannas. Since no life has been found elsewhere in the universe, the final level of the biological hierarchy of organization is the biosphere, which is simply the entire portion of the Earth that is inhabited by living organisms.

Another unifying theme of the biological sciences is evolution, a process of change. The physical structure of the Earth has evolved over the 4.5 billion years since its formation. Fossil evidence suggests that life first appeared 1 billion years afterward, beginning with Archaea, prokaryotic organisms that could live in the harsh, acidic, hot conditions of the nascent Earth. Just as the conditions of the planet affected the type of organisms that lived on it, the presence of life in turn affected the conditions of the planet. Most notably, molecular oxygen produced by photosynthetic bacteria accumulated in the atmosphere, setting the stage for aerobic life-forms. Species changed, and the diversity of life-forms increased. Between 1.7 and 1.8 billion years ago, eukaryotic life-forms first appeared, and over the next billion years, some evolved into multicellular beings. The first soft-bodied invertebrate animals appeared, and then vertebrate animals. Algae arose, leading to the development of aquatic plants. Approximately 475 million years ago, terrestrial plants and fungi colonized the land, creating an environment that led to the evolution of terrestrial animals.

Throughout each major episode in the history of life on Earth, organisms that were particularly well adapted to a particular environment survived and passed on their characteristics to offspring by natural selection, perpetuating the genetic variations that gave a species reproductive success in that environment. Both abiotic factors, such as temperature, light, water, and nutrient availability, and biotic factors, the presence of other life-forms, affect the organisms that live in a particular environment. Organisms interact with other organisms of the same species by competing for food, territory, and mates. They also interact with other species through predator-prey relationships, symbiotic relationships, and competition for ecological niches. These interactions not only contribute to the evolution of species, but shape populations and entire communities. The resultant changes then select for variations that make an organism better adapted to the new environment.

APPLICATIONS OF BIOLOGY

New knowledge and advances in biology have impacted and continue to impact profoundly the way people live. Understanding the complex organization of the human body has led to improvements in health care and medicine. Humans now have longer life expectancies than ever before, with the average life expectancy of people in the United States reaching 77 years at the turn of the last century, compared with 49 years in 1901. Advances in microbiology have led to the decrease in the spread of infectious diseases and the development of antibiotics and vaccines to treat and prevent infections that were once life-threatening. Production in agriculture has increased as a result of selective breeding programs, genetic engineering, and a better understanding of the pests that destroy crops. Technological advances in biology have permeated forensics by helping criminal investigators to determine the time elapsed since death by the insects associated with a human corpse and to identify suspects on the basis of deoxyribonucleic acid (DNA) evidence they have left behind. Learning about Earth's history and the previous life-forms it supported can help society envision the future, to understand, anticipate, and work to correct detrimental situations such as the depletion of natural resources and extinction of species.

What humans have learned by studying life has shaped populations, affected how people treat one another, changed how humans care for themselves, impacted governmental spending, changed how wars are fought, and revealed the impact humans have had on Earth and the other species with whom humans share the planet. The study of biology has affected almost every aspect of society and will continue to do so as scientists make new and exciting discoveries.

See also AGRICULTURE; ANATOMY; BIOCHEMISTRY; BIOGEOGRAPHY; BIOINFORMATCS; BOTANY; CELL BIOLOGY; CONSERVATION BIOLOGY; ECOLOGY; ENVIRONMENTAL SCIENCE; ETHOLOGY; EUKARYOTIC CELLS; EVOLUTIONARY BIOLOGY; GENETICS; MARINE BIOLOGY; MICROBIOLOGY; MOLECULAR BIOLOGY; ORGANIC CHEMISTRY, ITS RELEVANCE TO LIFE SCIENCE; PHYSIOLOGY; PROKARYOTIC CELLS; SCIENTIFIC INVESTIGATION; SOCIOBIOLOGY; ZOOLOGY.

FURTHER READING
Campbell, Neil A., Jane B. Reece, Lisa A. Urry, Michael L. Cain, Steven A. Wasserman, Peter V. Minorsky, and Robert B. Jackson. *Biology.* 8th ed. San Francisco: Pearson Benjamin Cummings, 2008.
Lewis, Ricki, Bruce Parker, Douglas Gaffin, and Marielle Hoefnagels. *Life.* 6th ed. New York: McGraw Hill, 2007.

biomes, aquatic Many factors profoundly influence the ability of an organism to survive in a particular environment. Biotic factors, the other organisms living in the same area, and abiotic factors, the surrounding chemical and physical conditions, both affect the degree of success to which an organism can find and obtain food and reproduce and raise young and therefore affect the distribution and abundance of a particular species. The biosphere comprises all of the portions of the planet inhabited by life and can be divided into ecosystems that consist of all the organisms and the physical environment in which they live. A biome is the largest useful ecological association that occupies a large geographic region and can contain several ecosystems. Characterized by the types of life-forms that inhabit them, biomes can be largely divided into aquatic and terrestrial types.

Aquatic biomes include marine and freshwater environments and cover approximately 75 percent of Earth's surface. Because the oceans consume so much area, they have a great impact on global climate and weather patterns in addition to playing a major role in the hydrological cycle. Physical factors, including light, temperature, and the movements of waters, and chemical factors, including salinity and dissolved oxygen concentration, influence the biology of aquatic biomes. Divisions of both marine and freshwater biomes into zones based on the amount of light present, the distance from the shore and depth of water, and open water versus bottom waters provide a variety of physical and chemical environments that support diverse life-forms. Marine biomes are typically defined by salt concentrations exceeding 3 percent and are classified according to physical features of the environment, such as zonation, mechanical forces of waves or currents, and range of temperature. Intertidal zones, oceanic pelagic, coral reefs, and marine benthic zones are the major marine biomes. Because estuaries occur where freshwater rivers flow into oceans, they have variable salt concentrations. Freshwater biomes typically have salt concentrations below 1 percent and include lakes and ponds, wetlands, streams, and rivers.

MARINE BIOMES

Marine habitats can be divided into zones dependent on various physical factors. Biological communities are distributed in the ocean on the basis of the adaptations particular organisms have to the conditions provided by the different zones. Because water absorbs light from the sun, the amount of light that penetrates a water body decreases with increasing depth. Organisms that depend on light must therefore reside near the surface of the water. As autotrophs, photosynthetic organisms make their own food from inorganic carbon sources, such as carbon dioxide (CO_2), using radiant energy from the Sun to incorporate the carbon into organic molecules such as carbohydrates. Organisms that undergo photosynthesis are called producers, and they form the basis of food chains. They reside in the photic zone, near the water's surface, where sufficient sunlight penetrates to support photosynthesis. The aphotic zone lies below the photic zone and is characterized by very little light. The benthic zone is the bottom of the ocean floor, where all the debris and dead organisms eventually settle. The abyssal zone is the region located at the ocean's deepest points. The pelagic zone is composed of the open waters, making it the largest, in terms of volume, of the marine biomes. The continental shelf is the submarine tract of land that surrounds a continent and leads to a steep downward slope into the deep ocean. With respect to water depth, the neritic zone encompasses the shallow waters over the continental shelf. The intertidal zone, or shoreline, is the area where land and water intersect. The waters beyond the steep slope that leads into the deep ocean make up the oceanic zone.

Intertidal Zones

As the gravitational actions of the Moon and the Sun cause displacement of the sea surface, the waters alternately submerge and recede from the shores, or intertidal zones, creating a unique sunny environment for marine life-forms. The lower boundary of a shore occurs where the seabed is exposed to the air during the lowest of tides, and the upper boundary is the area wetted during the highest tides. The climate of the geographical area and the nature of the seabed, rocky or sandy, influence the types of life that inhabit an intertidal zone. Different conditions exist even within the boundaries of an intertidal zone, as some areas are exposed to air for longer periods than others. The mechanical action of the waves on the shorelines and the often twice-daily turn of the tides help maintain relatively high nutrient levels. Intertidal zones with sandy seabeds lack a firm substrate to which seaweeds or sessile animals can attach and are completely exposed to the hot Sun and the cold wind. Animal inhabitants include sand fleas, crabs, and clams that burrow in the wet sand. These animals depend on the tides to deliver food in the form of

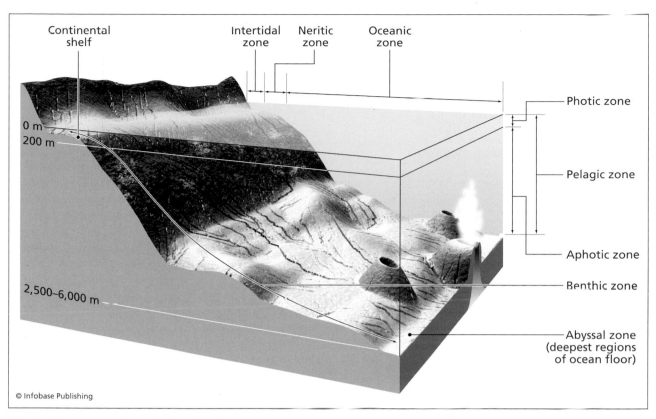

Continental shelf

Intertidal zone Neritic zone Oceanic zone

Photic zone

Pelagic zone

Aphotic zone

Benthic zone

Abyssal zone (deepest regions of ocean floor)

0 m
200 m

2,500–6,000 m

© Infobase Publishing

Physical factors such as the depth, distance from the shore, and light define different zones of marine biomes.

algae, seaweed, or dead marine creatures. Protected sandy areas such as bays or lagoons contain sea grass and algae, producers that can support a food web. Sea turtles and some birds nest in the intertidal zones of beaches. Rocky shores include both steep cliffs and gentle slopes that support the growth of marine algae, lichens, and other organisms that can withstand the harsh action of the waves and periods of no water. The demanding conditions select for organisms with structural adaptations that allow them to attach to a rocky surface and prevent desiccation. Animal life-forms that inhabit rocky shores include barnacles, limpets, marine snails, mussels, crabs, sea urchins, and starfish.

Oceanic Pelagic Zones

The oceanic pelagic zone consists of the vast open waters making up more than 70 percent of the Earth's surface that can be further divided in terms of the gradient of physical features along the depth of the water column. Ocean depths average around 13,123 feet (4,000 m) but extend to more than 32,808 feet (10,000 m) in some places. Light penetrates and the temperatures are warmer at the upper surface, but traveling deeper presents conditions characterized by cooler temperatures, less light, and higher pressures. A thermocline in which the temperature rapidly decreases with depth separates the warmer

upper layer from the cooler deeper waters. Most of the sea's biomass exists in the photic zone as plankton. Phytoplankton, the photosynthetic, microscopic life that floats near the surface, serves as the basis for the pelagic food chain. Zooplankton, animallike microscopic life that floats or swims weakly, such as protozoans, krill, worms, copepods, and larvae of some invertebrates, feed on the phytoplankton, and small fish and larger invertebrates then eat the zooplankton. Nutrient levels are higher in the surface waters as a result of mixing. Moving deeper down the water column, the light gradually fades, and some organisms exhibit bioluminescence, the emission of light from living organisms. Bioluminescence serves different functions in different species: to confuse predators, to recognize other members of one's own species, to attract mates, or to attract prey. Deeper than 3,300 feet (1,000 m), the ocean is completely dark, temperatures approach freezing, and the water pressure is 100 to 1,000 times greater than at the surface. Very little food reaches these depths. To compensate for lack of vision, fish that inhabit this region of the pelagic zone often have lateral line systems, groups of cells that detect vibrations due to disturbances in the water, warning them when predators or prey are near. Creatures that live at these depths also often have an acute sense of smell. Some organisms move up and down the water column daily. They feed near the surface, then migrate to deeper levels to hide in the darkness to be safe or to conserve energy by slowing their metabolism in the cooler temperatures. Some fish migrate across the ocean rather than up and down the water column in order to breed. Other animals that inhabit the oceanic pelagic biome include numerous fishes, squids, sea turtles, and marine mammals.

Coral Reefs

Reefs are massive skeletal structures formed by the secretions of reef-building cnidarians in warmer waters, typically greater than 68°F (20°C). Corals are the most common reef builders, though other invertebrates also secrete calcium carbonate, which precipitates to form a reef structure. Photosynthetic protists called zooxanthellae live within the coral tissues, provide nutrition, and remove carbon dioxide from the water. Because of the dependence of the corals on these photosynthetic organisms, reefs are limited to shallow waters where light can penetrate. When sediment from construction or highways

clouds the waters, photosynthesis is inhibited, and the reefs suffer. Individual sessile polyps sit within a cuplike structure, from which they extend tentacles to feed on small fish, crabs, or other small animals at night, when photosynthesis does not occur. Oceanic currents continuously wash the reefs, introducing nutrients and removing waste products. Unusually strong waves, such as are present during hurricanes or tropical storms, may destroy entire reefs. The polyps of living corals form massive colonies that serve as the basis for unique ecosystems. The diversity within coral reefs exceeds that of any other aquatic ecosystem. The corals themselves are the dominant animals, but the nooks and crannies of the reef structure provide shelter for small fish and numerous invertebrates, fish and sea urchins feed off algae that grows on the coral, and shrimp or smaller fish feed off the parasites living on larger fishes. Starfish and fish such as parrot fish prey on chunks of the living coral tissue. After the corals die, their skeletons remain, and new polyps attach to and grow on the remains. Fringe reefs surround volcanic islands and

Brilliantly colored coral reefs serve as the foundations for unique ecosystems rich in biodiversity. *(National Oceanic & Atmospheric Administration/Department of Commerce)*

enclose a narrow section of water between the reef and the island. Barrier reefs result when the central island begins to sink into the ocean, forming a lagoon inside the reef, which protects the interior from the harsh oceanic swells. When the volcanic island becomes completely submerged, the resulting structure is called a coral atoll.

Marine Benthic Zone

The marine benthic zone is the region of seafloor below the neritic zone, the offshore pelagic zone, extending along the ocean bottom to the abyssal zone, the deepest, coldest regions. The physical and chemical conditions vary tremendously at different depths and distances from the shore in the benthic zone. Sediment formed from the erosion of continental rocks, the calcareous and siliceous oozes and phosphate-rich minerals secreted by marine creatures, and other minerals that have precipitated from seawater, in addition to the remains of former marine life, covers the benthic zone. Bacteria on the ocean floor decompose this and other fallen organic matter, which serves as food for unusual marine benthic creatures. Some areas of seafloor are covered by oceanic crust from recently erupted volcanoes or underwater mountain ranges. Photosynthetic organisms live in the shallow benthic regions, where sufficient sunlight penetrates, but most of the benthic zone receives no sunlight. Unique ecosystems exist near deep-sea hydrothermal vents, cracks in the ocean floor near midoceanic ridges that expel superheated mineral-rich fluids. Near these vents, communities consisting of diverse microbial and animal life thrive. Chemosynthetic prokaryotic organisms live within giant tube worms and use the reduced inorganic compounds such as hydrogen sulfide formed from the heat and sulfates as an energy source to fix carbon and serve as the primary producers for the community. Clams, mussels, copepods, crabs, shrimp, octopi, and starfish also live around hydrothermal vents.

Estuaries and Coastal Wetlands

Estuaries occur where freshwater rivers join the saltwater sea. The water in an estuary flows up the channel back into the river during high tides, and then in the reverse direction out into the sea during low tides. This water movement transports organisms, draws in nutrients, and removes wastes. Because the high salinity of the seawater makes it denser than the freshwater flowing down the river, the saltier seawater often occupies the lower level of the estuary, and the upper layer has a lower salt concentration. This creates a dynamic chemical environment that varies temporally with the coming and going of the tides and along a spatial gradient, both vertically and horizontally. The flow of waters from the river and the incoming sediment create a variety of physical structures such as channels and islands. Mangroves and salt marshes on sea-level coasts and sandy shores transition between land and sea rather than river and sea, but these coastal wetlands also create a unique chemical and physical environment for a variety of marine fish and invertebrates to inhabit. Grasses and algae dominate salt marsh vegetation, and different genera and species of mangrove trees inhabit mangrove forests, which also contain crocodiles and alligators. Fish are abundant in estuaries and salt marshes, and many marine fishes use these grounds to breed and raise young. They lay their eggs on the sediment and the larvae feed off the abundant phytoplankton and zooplankton. Waterfowl and mammals such as muskrats commonly feed and breed in these transitional biomes.

FRESHWATER BIOMES

Freshwater biomes are usually found inland and are characterized by low salinity, typically below 1 percent. Life-forms that inhabit these biomes have adapted to the low-salt conditions and could not survive in marine environments. Freshwater biomes are especially important to people because they provide water used for drinking, energy, transportation, and recreation and support many occupations. The major freshwater biomes include lakes and ponds, wetlands, rivers, and streams.

Lakes and Ponds

Lakes are inland bodies of standing water. Smaller lakes are called ponds and may only cover a few square yards (a few square meters), but some lakes cover thousands of square miles (or square kilometers). Lakes and large ponds share many of the same physical features with respect to zonation as marine environments. A photic zone at the water's surface supports photosynthetic life-forms, and an aphotic zone lies deeper, where sunlight cannot penetrate. The photic zone can be further divided on the basis of the nearness of the waters to the shore of the lake; the littoral zone lies closest to shore, and the limnetic zone lies farther from the shore. The littoral zone is shallow and warm and contains rooted vegetation, phytoplankton, and sometimes floating aquatic plants. These producers support animals including snails, clams, insect larva, crustaceans, fishes, and amphibians. The vegetation and small animals support other animals such as turtles and ducks. The limnetic zone contains most of a lake's phytoplankton, which maintain the health of the lake ecosystem by providing food for other organisms and by oxygenating the water. This zone only extends to the point where light cannot penetrate. Below that point lie the cold, open waters called the profundal zone. The lack of light limits algal or plant growth; thus oxygen content is low. The fish that live in the profundal zone depend on food produced in the limnetic or littoral zone. Snails, clams, crayfish, worms, and insect larvae live on the lake bottoms, the benthic zone.

Lakes exhibit diverse conditions with respect to temperature, salinity, oxygen concentration, and nutrient availability. Variation in the types of organisms present exists between geographic regions, from season to season within a lake, and between the zones within a lake. The biological zones of life are related to the physical structure. Thermally stratified lakes consist of three layers: the epilimnion, the metalimnion, and the hypolimnion. The epilimnion is the wind-mixed top layer that constantly exchanges gases with the atmosphere. The middle layer of the water column, the metalimnion, is also called the thermocline because of its steep temperature gradient. The depth of the thermocline changes during the day. The bottom layer is the hypolimnion. In geographic areas where the temperatures are cold enough to freeze in the winter, the hypolimnion is warmer than the water near the surface immediately underneath the ice. As the ice melts, the thermal stratification shifts. By springtime the waters may be completely mixed, but as the temperatures continue to increase, the epilimnion becomes warmer and less dense, the hypolimnion becomes colder and denser, and the thermocline becomes steeper. The lake remains stratified until fall, when the temperatures begin to decline and the epilimnion cools and becomes denser. Eventually the cold air causes the surface waters to become colder than the bottom waters. After ice forms on the surface and blocks the wind, the layers remain stratified with the hypolimnion warmer than the epilimnion until the coming of spring. Such seasonal changes in temperature within different zones of a lake cause water to mix, as does wind when the surface is not covered with ice. This mixing changes the water chemistry by moving nutrients from the bottom up to the surface waters and carrying dissolved oxygen from the surface waters to the bottom waters. In tropical climates, the surface waters remain warmer than the bottom waters, though at higher elevations the waters cool enough to cause mixing at night.

© Infobase Publishing

Lakes can be divided into four general zones based on depth and distance from the shoreline.

In eutrophic lakes, high primary productivity in surface waters due to high nutrient content leads to oxygen depletion as the biomass decomposes. *(Michael P. Gadomski/Photo Researchers, Inc.)*

Oligotrophic lakes have low nutrient content, high oxygen concentrations, and low biological productivity. The diversity of phytoplankton and the benthic organisms is great, and the fish that inhabit oligotrophic lakes generally require higher oxygen concentrations. Trout and whitefish are commonly found in such environments. On the other hand, eutrophic lakes have high nutrient content, low oxygen concentrations, and high biological productivity. Particularly in warm conditions, the high nutrient concentration and sunlight stimulate primary productivity in the surface waters. The quantities of phytoplankton increase, and as the microorganisms die and sink to the bottom, the action of decomposers depletes the oxygen that is present. These conditions support fish such as carp, catfish, and bowfins and invertebrates that can tolerate low oxygen concentrations and higher temperatures. Vascular aquatic plants are abundant in the littoral zone of eutrophic lakes.

Wetlands

Wetlands, including marshes and swamps, are areas of land that are covered intermittently with shallow water and that can support the growth of aquatic plants. In addition to serving many important ecological functions such as removing nutrients from the surface and groundwater, performing denitrification (the conversion of organic nitrogen back into nitrogen gas), and protecting surrounding terrestrial habitats from storms and floods, wetlands provide environmental intermediate conditions between a terrestrial and an aquatic ecosystem. Because the water that fills the basins is not aerated and the soil is saturated with water, plant species characteristic of wetlands have adaptations that allow them to survive in anaerobic soils. Called hydrophytes, such plants have thin cuticles, less rigid structures since the water supports them, flat leaves that enable flotation, small feathery roots into which water can directly diffuse, and numerous open stomata since water loss from transpiration is not a concern. Examples of hydrophytes include water lilies, buttercups, duckweed, and hornworts. The abundance of plants and algae supports many herbivorous creatures such as crustaceans, muskrats, and aquatic insect larvae, which in turn support a variety of birds. Carnivorous animals including dragonflies, otters, alligators, and owls are also common.

Rivers and Streams

Rivers and streams are running bodies of water that serve to drain landscapes. A watershed is the region drained by a stream or river. Water from rains runs off either above or below the ground, picking up dissolved materials along the way, and collects in small creeks, sometimes called the headwaters. The creeks merge into streams and then rivers that eventually lead to the sea or to large inland basins. The flow of water through rivers and streams is constantly changing as a result of precipitation events, melting snow and ice, evaporation, transpiration, sediment, discharge of water from aquifers, and many human-induced events such as irrigation, levee construction, or the draining of wetlands. Gravity moves the water from higher to lower altitudes, carrying with it sediment, nutrients, minerals, and life-forms. The concentrations of nutrients and salts change as the flow nears the mouth, where the water enters the ocean or other large water body. As in marine biomes and lakes, the degree to which light penetrates influences the type and abundance of organisms near the surface. Light can penetrate deeper in clearer streams than in rivers downstream, but clear streams are still cloudier than clear lakes because of the dynamics of river systems. The chemical composition of a river depends on the climate and the geological features of the watershed. In tropical areas where the annual rainfall is high, the soil has been washed clean of dissolved materials; thus rivers in these regions contain less salt, whereas rivers in drier areas are saltier. The oxygen content depends mostly on temperature. Colder waters carry more dissolved oxygen, and warmer waters have lower oxygen content. Because rivers have a large surface area relative to their volume, and because the water is always moving, oxygen is absorbed and mixed into the waters regularly and does not have a major impact of the types of life that inhabit the waters. The distribution of organisms along a river continuum seems to follow a general pattern. The headwaters usually contain higher amounts of oxygen but low amounts of nutrients and salts, and the bottoms of the channels are typically rocky. Fish such as trout inhabit these oxygen-rich, cold streams. The nutrient and salt concentration increase as the water travels toward its downhill destination, covering river bottoms with silt and sediment. Leaves and pieces of plants that live on the banks of the headwaters and other biomass such as insects and animal feces that are washed into streams during storms serve as the major food source for animals that live in streams. Aquatic microbes attack these food fragments, and stream invertebrates shred the organic material and feed on the fine particles of food. The types of invertebrates that inhabit streams and rivers are mostly benthic, meaning they live on the bottom in the sediments. The content of the sediments affects the composition of the invertebrate communities. Moving down the river continuum to the middle headwater streams, the width typically increases, and because the water is less shaded, algae and rooted aquatic plants are more abundant. These, in addition to the fine particulate matter carried in by the small headwater streams, feed the fish and invertebrates. The water is slightly warmer, moves more slowly, and is less oxygenated—conditions preferred by fish such as bass. Fine particulate matter and increased phytoplankton populations are the main food sources in the larger rivers, contributing to the differences in the type of fish that inhabit larger rivers compared to the headwaters.

See also BIOMES, TERRESTRIAL; BIOSPHERE; ECOLOGY; ECOSYSTEMS; HYDROTHERMAL VENTS; MARINE BIOLOGY; PHOTOSYNTHESIS.

FURTHER READING

Day, Trevor. *Lakes and Rivers.* New York: Chelsea House, 2006.

———. *Oceans.* New York: Chelsea House, 2006.

Moore, Peter D. *Wetlands.* New York: Chelsea House, 2006.

Sigee, David C. *Freshwater Microbiology: Biodiversity and Dynamic Interactions of Microorganisms in the Aquatic Environment.* Hoboken, N.J.: John Wiley & Sons, 2005.

Weigel, Marlene, and Julie L. Carnagie, ed. *Encyclopedia of Biomes.* 3 vols. Farmington Hills, Mich.: UXL, 2000.

Whitfield, Philip, Peter D. Moore, and Barry Cox. *Biomes and Habitats.* New York: Macmillan Reference USA, 2002.

Woodward, Susan L. *Biomes of Earth: Terrestrial, Aquatic, and Human-Dominated.* Westport, Conn.: Greenwood Press, 2003.

biomes, terrestrial Ecology is the study of the relationships between organisms and their environment. Environmental factors include biotic, those relating to other living organisms, and abiotic factors, those relating to nonliving components of the environment, such as physical, chemical, and geological features. A biome is a large region of characteristic ecological associations, characterized by certain types of vegetation and climatic conditions. The biosphere, the portion of the planet that contains life, consists of several major terrestrial and aquatic biomes; scientists have not reached a general consensus about the exact number of types of biomes. While different biomes contain characteristic combinations of climate and types of fauna and flora, individual species or particular environmental conditions are

World Biomes

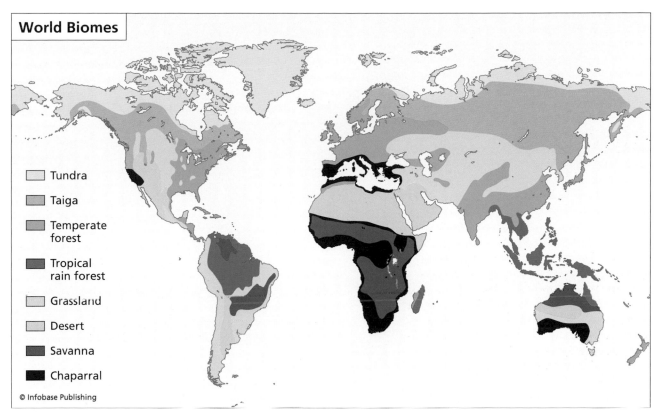

- Tundra
- Taiga
- Temperate forest
- Tropical rain forest
- Grassland
- Desert
- Savanna
- Chaparral

© Infobase Publishing

Differences in climate affect the distribution of life-forms on Earth, forming characteristic communities called biomes that cover large geographic areas.

not necessarily distinguishing, and biomes overlap physically and biologically. Areas within one biome may resemble those of another type of biome, and the same biome at distant geographic locations may contain different resident species. With this in mind, major terrestrial biomes include the tropical rain forest, savanna, desert, chaparral, temperate grassland, temperate broadleaf forest, coniferous forest, tundra, and mountains.

TROPICAL RAIN FOREST

Situated in equatorial and subequatorial regions, tropical rain forests experience year-round warm weather, with temperatures averaging 77°F–84.2°F (25°C–29°C) and 78.7–157.5 inches (200–400 cm) of precipitation annually. Almost half of the precipitation is due to water released into the air by transpiration from the vegetation. Three major geographical regions host tropical rain forests: Central America in the Amazon River basin, the Zaire basin in Africa, and Indo-Malaysia. The warm, moist climate makes this biome the most diverse in the world; an estimated 1.5 million different species live in the rain forest. The different life-forms in the tropical rain forest are distributed into four main layers. The uppermost emergent layer consists of scattered trees with heights between 100 and 240 feet (30 and 73 m), requiring

widespread buttresses to support them. Their bark is thin since water loss is not a concern, and their lowest branches are quite high, since at lower heights sunlight cannot penetrate through the dense vegetation that is characteristic of the tropical rain forest. The leaves are exposed to the Sun and wind, but they are small so they do not dry out. The roots of these trees do not penetrate deeply. The upper canopy, which consists of trees ranging from 60 to 130 feet (18 to 40 m), is rich in food sources and supports most of the animal life. Light can penetrate the upper portion of this layer, but not all the way to the bottom portion. Plants called epiphytes (such as orchids) grow on trees in the upper canopy where sunlight is accessible, and they derive their moisture and nutrients from the air. Leafy vines also cling to the trees for support and as a means to reach the sunlight in the upper canopy. Leaves on the upper canopy trees are sometimes oily and are shaped in a manner that allows the water to drip off them rather than collect in pools where mold can grow. Trees in the third canopy, the understory, reach approximately 60 feet (18 m) tall. This layer also consists of smaller trees, shrubs, and numerous plants that have broad, large leaves to collect as much sunlight as possible. These organisms thrive when a taller tree that has been blocking the sunlight falls down. The air does not

Dense vegetation, trees with giant buttresses, vines, and epiphytes are characteristic of tropical rain forests, such as this one in Oahu, Hawaii. *(Dhoxax, 2007, used under license from Shutterstock, Inc.)*

circulate well in the understory, light barely penetrates this far down, and the humidity remains very high. The forest floor is completely shaded; thus few plants grow there. The topsoil is nutrient poor, as the frequent rains wash away all the minerals and excess nutrients. As soon as the insects, earthworms, fungi, and other microorganisms decompose the leaf litter, the plants take up the recycled nutrients. Insects make up the largest group of animal life in tropical rain forests. Resident insects, such as butterflies and beetles, and birds often display bright colors. Mammals contain adaptations that facilitate living in trees, such as monkeys' having prehensile tails for hanging from branches. Many reptiles and amphibians also inhabit the rain forest.

SAVANNA

Savannas, or tropical grasslands, are plains that have shrubs, a few scattered trees, and drought-resistant undergrowth. Located in equatorial and subequatorial regions, with temperatures that range from 68°F

to 86°F (20°C–30°C) and average 75.2°F–84.2°F (24°C–29°C). The warmth and humidity make for very wet summers, when the majority of the annual rainfall, which averages 11.8–19.7 inches (30–50 cm), occurs. Likewise, the winter is very long and dry and frequently includes several months with no rainfall. The dry conditions select for plants that have adaptations such as smaller leaves, deep roots that penetrate to the water table, and thick bark that protects against fires, which are common and help maintain this biome by controlling the density of woody plants. Grasses serve as food for large herbivorous mammals such as elephants, wildebeest, zebras, and giraffes. Carnivores such as lions and hyenas prey on the grazers. During the wet season, the vegetation is lush and the rivers flow freely, but during the dry season the plants die and the animals must wander in search of water and food.

DESERT

Occupying about one-fifth of the Earth's land, desert biomes receive less than 11.8 inches (30 cm) of precipitation per year, with an average of about 5.9 inches (15 cm) per year. Deserts generally occur in rings that circle the globe near 30° north and south latitude, and most deserts are hot, but some in the arctic are cold. The temperatures are extremely variable, with hot deserts experiencing temperatures that occasionally near 122°F (50°C) and cold deserts as low as -22°F (-30°C). Animals that live in deserts burrow to escape the heat or the cold, depending on the type of desert. The hot and dry conditions of the desert limit the vegetation that can survive there; wind and infrequent flash floods mold the mostly bare landscape. Succulent plants such as cacti and shrubs with roots that penetrate deeply are the most common forms of plant life. Adaptations include tolerance to heat and low moisture, ability to store water, and reduced leaf area to prevent water loss by transpiration. The photosynthetic mechanisms differ in a manner that reduces photorespiration, allowing the stomata to remain closed or to limit their opening to nighttime, thereby reducing water loss. Desert soil is salty, making it even harder for plants to access any available water from it, and it lacks organic matter, with the exception of underneath shrubs. Cold deserts may have some lichens, grasses, and mosses. Because of the extreme weather, many fauna are nocturnal. Snakes, lizards, scorpions, ants, beetles, rodents, and birds represent the animal life found in deserts.

CHAPARRAL

The chaparral is sometimes referred to as the Mediterranean woodland and shrubland after the location of its main area of distribution. Chaparral also occurs along midlatitude coasts of North America,

Desert flora, as shown in this Nevada yucca forest, have adaptations that help the plant tolerate long periods of drought, such as having reduced leaf surface area to prevent evaporative water loss and the ability to store water. *(John and Karen Hollingsworth/U.S. Fish and Wildlife Service)*

Australia, Chile, and South Africa. The landscape is varied—sometimes flat, sometimes rocky, or along mountainsides. Weather changes with seasons; as in the savanna, rainfall ranges between 11.8 and 19.7 inches (30 and 50 cm), but in contrast, chaparral summers are dry and the winters are rainy. Fall, winter, and spring experience temperatures around 50°F–53.6°F (10°C–12°C), but in the summer the temperature may reach 104°F (40°C) and in drought conditions fires may occur. Flora and fauna possess adaptations similar to those of desert life-forms for tolerating hot and dry conditions. Woody plants are evergreen and have small, tough leaves. The trees often have thick, fire-resistant bark, whereas the shrubs are often oily and burn easily, but to compensate, they regrow quickly. Depending on the geographical location, chaparral residents include mammals that eat twigs and buds, such as deer, wild sheep and goats, antelope, kangaroos, and hares. Carnivores such as puma and aardwolves and omnivores such as foxes, skunks, and jackals inhabit this biome, which also serves as a winter home for many migratory birds and insects.

TEMPERATE GRASSLAND

Like the savanna, temperate grasslands are biomes characterized by vast plains dominated by grasses with scattered trees growing alongside streams. Precipitation is greater than in the savanna; rainfall is seasonal and averages between 11.8 and 39.3 inches (30–100 cm) per year. As the name suggests, moderate temperatures typify this biome—20°F (-6.7°C) in the middle of winter and 70°F (21.1°C) in the middle of summer. Humid, rainy grasslands grow tall grasses and are called prairies, and drier grasslands, which experience greater temperature extremes, grow shorter grasses and are called steppes. Though no geological or biological features block the wind, the deep roots of the grasses prevent the soil from blowing away. The soils are naturally nutrient-rich and fertile from the growth and decay of deep grass root systems, though humans have taken advantage of this and converted much of this land for agricultural purposes, an act that has depleted much of the organic matter from grassland soil. In addition to grasses, common plants of temperate grasslands include flowers such as asters, coneflowers, sunflower, goldenrod, and clover. Near rivers, where sufficient water is present, trees such as cottonwoods, oaks, and willows grow. Large mammalian grazers such as horses, pronghorn, antelope, and bison inhabit temperate grasslands alongside many burrowing animals such as prairie dogs and mice. Predators include coyotes, bobcats, and wolves, and birds include wild turkey, eagles, Canada geese, and the endangered prairie chicken.

TEMPERATE BROADLEAF FOREST

Temperate broadleaf forests, also called deciduous forests because they contain trees that shed their leaves before winter each year, occur at midlatitudes in the Northern Hemisphere, as well as in southern Australia and New Zealand. Precipitation in this biome averages 27.6–78.7 inches (70–200 cm) annually, and temperatures range from around 32°F (0°C) in the winter up to 86°F (30°C) in the summer. The soil on the forest floor is usually fertile and contains a lot of organic matter. Because of this, humans have cultivated much of the land that formerly supported these forests in their natural form; many of the temperate forests present today were planted by humans. Biodiversity is not as high as in tropical rain forests, but biomass is. Temperate forests contain several layers or zones, with the highest including tall trees like the common lime, birch, maple, oak, and beech. Smaller trees and saplings grow under the highest canopy, followed by shrubs and ferns, and then herbs. The forest floor contains lichens, club mosses, true mosses, mushrooms, and microorganisms, which play an important role in breaking down old wood and organic matter and in recycling the nutrients within the forest ecosystem. The forest provides shelter, shade, and protection for many birds, insects, and other animals such as squirrels, chipmunks, mice, deer, bears, raccoons, foxes, frogs, salamanders, and snakes. In the winter, some of the animals migrate south, and others hibernate.

CONIFEROUS FOREST

The coniferous forest, also called the taiga, is the largest terrestrial biome. It is located north of the temperate forest, south of the tundra, and extending across northern North America and Eurasia, and its dominant plants are conifers. The Russian taiga plays an important ecological role as a carbon sink, helping to reduce global climate change by incorporating carbon from carbon dioxide into organic material at a rate faster than it emits carbon dioxide back into the atmosphere. Annual precipitation averages about 15.7 to 39.3 inches (40–100 cm), but some coastal coniferous forests receive up to 118 inches (300 cm)—these are considered temperate rain forests. The winter temperatures of the taiga are harsh, dipping as low as -94°F (-70°C) in Siberia. Summertime highs can approach 86°F (30°C), but summers are short—the growing season only lasts about 130 days. In the northern forests of North America, the ocean moderates the temperatures, extending the length of the growing season there. Evergreen trees such as pine, fir, hemlock, and spruce have waxy needlelike leaves and cones as reproductive structures. The trees themselves have a cone shape to prevent snow from accumulating on them and weighing down the flexible branches. Because the leaves do not fall off the trees, they are ready to photosynthesize as soon as the weather permits the growing season to begin. The trees are tall and thin and have thick bark to resist occasional fires during droughts. Deciduous trees such as larches, birches, and aspens also grow in the taiga. Logging companies are destroying many of the trees of coniferous forests, though conservation efforts are attempting to limit this process. The soil is nutrient-poor and acidic, conditions that limit the plant diversity, though shrubs, herbs, and mosses can tolerate the taiga, as do lichens. Berry-bearing shrubs are an important food source for many animal taiga residents. Animal life includes herbivores such as hares, squirrels, voles, deer, beavers, caribou, buffalo, and moose and predators such as Siberian tigers, fox, lynx, wolverines, and bobcats. Black bears eat whatever they can find, including berries, tubers, insects, small mammals, and fish. Birds such as wood warblers and woodpeckers that migrate to the taiga to breed keep the insect population under control. Seed-eating birds such as finches and sparrows and predatory birds such as hawks and owls live in the taiga year-round.

TUNDRA

Many animals, such as caribou, that spend winters in the taiga spend part of the year in the tundra, the biome that covers the northernmost lands of the Arctic Circle. The tundra is typically cold and dry. With the Arctic Ocean nearby to moderate the climate, the temperatures are not as extreme as in the taiga, averaging -22°F (-30°C) in the winter and about 50°F (10°C) in the summer when the sunlight shines on the tundra 24 hours a day, but the growing season is short. Because precipitation only reaches about six to 10 inches (15.2–25.4 cm) a year, plant life is mostly herbaceous, or nonwoody, and includes lichens, mosses, grasses, herbs, low-growing shrubs, some flowers, and very few birch trees in southern areas of the tundra. As the taiga does, the tundra serves as a carbon sink, taking in more carbon from the atmosphere than it releases. The cold temperatures inhibit decomposition by microorganisms; thus organic matter accumulates in this superficial layer of soil that freezes and thaws each year. When it thaws, decomposition occurs. One can observe cracks on the Earth's surface from an aerial view, just as cracks form on concrete from repeated freezing and thawing. Below the uppermost layer of soil, the ground remains frozen. Water and plant roots cannot penetrate through this layer called permafrost, which may extend from 10 inches

Tundra flora typically lack woody tissue and only persist for one growing season; tundra fauna include large grazing animals such as the caribou shown here. *(U.S. Fish and Wildlife Service)*

(25.4 cm) to three feet (30.5 cm) down, so when the surface snow melts in the summer, temporary lakes form. A global increase in temperatures has melted some of the permafrost, increasing the rate of decomposition of the dead plant matter in the upper layer of soil and releasing more carbon back into the atmosphere, which contributes to global warming—forming a vicious, self-feeding cycle. Resident animal species include large mammalian grazers such as musk oxen, and others such as caribou (also called reindeer) migrate to the tundra during the short summers. Polar bears, wolves, wolverines, foxes, and snowy owls prey on these creatures, and on smaller squirrels, rabbits, lemmings, fish, seals, and birds like harlequin duck sandpipers and plovers that migrate from the taiga to nest in the tundra and feed off insects.

MOUNTAINTOPS

The Antarctic is too cold to support life-forms similar to those of the Arctic tundra. The climate on mountaintops, however, does resemble that of the tundra and is sometimes referred to as the alpine tundra or the alpine biome. Vegetation and soil conditions similar to the Arctic tundra's are found at these altitudes of about 10,000 feet (3,048 m). The growing season is short, and the winters are long and cold. Two main differences are that alpine soil can drain well, and

the mountain life-forms must deal with high exposures to ultraviolet radiation from the Sun. Because the atmosphere is less dense, the partial pressures of carbon dioxide, which plants need for photosynthesis, and oxygen, which animals need for respiration, are low; thus plants and animals have special adaptations to compensate. Moving from the top of a mountain down toward its base, one may encounter several distinct ecological associations unique to the particular altitude.

See also BIOMES, AQUATIC; BIOSPHERE; ECOLOGY; ECOSYSTEMS; ENVIRONMENTAL CONCERNS, HUMAN-INDUCED; PLANT FORM AND FUNCTION.

FURTHER READING

Allaby, Michael. *Deserts*. New York: Chelsea House, 2006.

———. *Grasslands*. New York: Chelsea House, 2006.

———. *Temperate Forests*. New York: Chelsea House, 2006.

Day, Trevor. *Lakes and Rivers*. New York: Chelsea House, 2006.

———. *Taiga*. New York: Chelsea House, 2006.

Moore, Peter D. *Tundra*. New York: Chelsea House, 2006.

Weigel, Marlene, and Julie L. Carnagie, eds. *Encyclopedia of Biomes*. 3 vols. Farmington Hills, Mich.: UXL, 2000.

Whitfield, Philip, Peter D. Moore, and Barry Cox. *Biomes and Habitats.* New York: Macmillan Reference USA, 2002.

Woodward, Susan L. *Biomes of Earth: Terrestrial, Aquatic, and Human-Dominated.* Westport, Conn.: Greenwood Press, 2003.

biomolecules Organic molecules that compose cells are known as biomolecules. Often called macromolecules, most biomolecules are rather large and are composed of numerous monomers, repeating subunits covalently linked to form polymers. The joining of monomers is usually accompanied by the loss of a water molecule in what is called a condensation reaction. One monomer contributes a hydrogen atom to the water molecule that is removed and the other, a hydroxyl group. Conversely, biomolecules can be broken down into monomeric subunits by hydrolysis reactions wherein a covalent linkage is broken by the addition of one hydrogen atom from a water molecule to one monomer and a hydroxyl group attaching to the adjacent monomer. Cells break down polymers into monomeric subunits and then reassemble them to create the biomolecules they need. There are four main classes: carbohydrates, proteins, nucleic acids, and lipids. Despite the fact that each of these types of biomolecules mostly consists of only four major elements (carbon, hydrogen, nitrogen, and oxygen) and two minor elements (sulfur and phosphorus), they exhibit a remarkable diversity of structure and function.

CARBON

The element carbon is the major component of all biological molecules. Because it has a valence of 4, it tends neither to gain nor to lose electrons to complete its outer shell. Instead, a carbon atom participates in strong covalent linkages with other carbon atoms (C), oxygen (O), nitrogen (N), and hydrogen (H) to form a variety of large and complex macromolecules. Double and triple bonds can also form between carbon atoms. Hydrocarbon chains form the basis of organic biomolecules, vary in length, and can be straight, circular, or branched. Because carbon-hydrogen bonds are nonpolar, however, simple hydrocarbons are not common in living organisms, which consist of cells that contain 70–95 percent water.

FUNCTIONAL GROUPS

The addition of hydrophilic side groups to the nonpolar hydrocarbon skeletons acts to increase the solubility of biomolecules in an aqueous environment, such as inside a cell or body tissue. In combination with the basic structure of the carbon skeleton of an organic compound, the side chains attached to the carbon atoms help determine that molecule's

Functional Group	Formula	Compound Name
Hydroxyl	— OH	Alcohols
Carbonyl	$-C\overset{\displaystyle\parallel O}{\underset{\displaystyle H}{<}}$	Aldehydes
	$-\overset{\displaystyle\parallel O}{C}-$	Ketones
Carboxyl	$-C\overset{\displaystyle\parallel O}{\underset{\displaystyle OH}{<}}$ *nonionized* or $-C\overset{\displaystyle\parallel O}{\underset{\displaystyle O^-}{<}}$ *ionized*	Carboxylic acids
Amino	$-N\overset{\displaystyle H}{\underset{\displaystyle H}{<}}$ *nonionized* or $-{}^+N\overset{\displaystyle H}{\underset{\displaystyle H}{<}}H$ *ionized*	Amines
Sulfhydryl	— SH	Thiols
Phosphate	$-O-\overset{\displaystyle\parallel O}{\underset{\displaystyle O^-}{P}}-O^-$	Organic phosphates

© Infobase Publishing

The functional groups attached to an organic compound contribute to that molecule's chemical and physical properties.

chemical and physical properties. Side chains that commonly participate in the chemical reactions of a molecule are called functional groups. Six play a crucial role in biochemistry: hydroxyl groups, carbonyl groups, carboxyl groups, amino groups, sulfhydryl groups, and phosphate groups.

CARBOHYDRATES

Carbohydrates include sugars and related compounds that have the general formula $(CH_2O)_n$, with n usually being 3, 4, 5, or 6. For example, the molecular

Cellulose consists of thousands of glucose molecules linked in an unbranched chain.

formula for glucose is $C_6H_{12}O_6$. Also known as simple carbohydrates, monosaccharides such as glucose, fructose, and galactose contain a single aldehyde or ketone group. They can be drawn as open chain structures but in aqueous solutions exist as closed five- or six-carbon ringed structures. Disaccharides form by a condensation reaction between two monosaccharides, forming a special type of bond called a glycosidic linkage. Sucrose, common table sugar, consists of a molecule of glucose linked to a molecule of fructose. Maltose and lactose are two other common disaccharides formed from two glucose molecules or a glucose and a galactose molecule, respectively. When between three and 50 subunits join together, the resulting molecule is called an oligosaccharide. The term *polysaccharide* refers to much larger molecules that contain hundreds or thousands of monosaccharides that are often cross-linked and branched to form more complex structures such as starch, glycogen, or cellulose.

Glucose serves as the main source of cellular fuel, and other carbohydrates are often converted into glucose when energy is needed. When catabolized by aerobic respiration, a single molecule of glucose yields up to 38 molecules of adenosine triphosphate (ATP). Glucose is also the primary sugar dissolved in blood that circulates throughout the body of animals. Plants store surplus energy in the form of starch, a helical glucose polymer consisting of hundreds to thousands of subunits. During digestion, animals hydrolyze, or break, starch into individual glucose subunits that can immediately be utilized for energy or stored in the form of glycogen, a highly branched polymer of glucose subunits. Other carbohydrates perform structural roles. For example, plant cell walls are made primarily of cellulose, unbranched chains consisting of thousands of covalently linked glucose molecules. The configuration of the glucose monomers in cellulose differs from that of those in starch, causing the linkages to differ slightly. Because of this, separate enzymes are required to digest starch and cellulose, and many animals including humans do not synthesize the latter. Undigested cellulose, or fiber, is still an important part of a healthy diet.

Another polysaccharide that performs a structural role is chitin, the major component of arthropod skeletons and fungal cell walls.

PROTEINS

Proteins are the main macromolecular constituent of cells. Over half of a human body is made of protein. Proteins consist of one or more polypeptides that are constructed by linking 20 different amino acids in specific sequences. All amino acids consist of a central carbon bound to four side groups: a hydrogen atom (–H), an amino group (–NH$_2$), a carboxyl group (–COOH), and a variable side chain (denoted by R). At a neutral pH, the amino group and the carboxyl group are ionized to –NH$_3^+$ and –COO$^-$. The chemical composition of the variable group is what gives an amino acid its unique properties. The acidic amino acids include aspartic acid and glutamic acid; the basic amino acids include lysine, arginine, and histidine; amino acids with uncharged polar side chains include asparagine, glutamine, serine, threonine, and tyrosine; and the nonpolar amino acids are alanine, valine, leucine, isoleucine, proline, phenylalanine, methionine, tryptophan, glycine, and cysteine. Amide linkages called peptide bonds form by condensation reactions between amino acids to synthesize polypeptide chains.

After the synthesis of a polypeptide chain, a protein must fold into its final, or native, configuration before it is considered active, or ready to perform its main function, which usually involves the specific interaction with another molecule. The order or sequence of amino acids in a chain is referred to as the protein's primary structure. The secondary structure that a protein assumes depends on hydrogen bonds between hydrogen atoms of the amino groups and oxygen atoms of the carbonyl groups of different amino acids. The alpha-helix and the beta-pleated sheet are two common secondary structures. An alpha-helix resembles an old-fashioned telephone cord. The beta-pleated sheets involve two or more regions of the polypeptide chain lying parallel to each other and have a zigzagged appearance as if one had taken a sheet of paper and folded it back and forth at

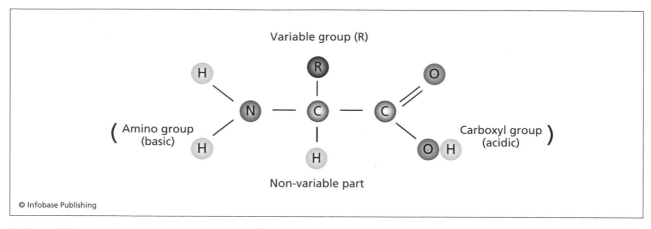

Variable group (R)

Amino group (basic)

Non-variable part

Carboxyl group (acidic)

© Infobase Publishing

Amino acids all have a central carbon with four side groups: a hydrogen atom, an amino group, a carboxyl group, and a variable R group.

regular intervals, then opened it slightly. At the tertiary level of protein folding, the polypeptide chains adopt their three-dimensional conformations that are stabilized by a combination of hydrophobic interactions, hydrogen bonds, and ionic interactions. Disulfide bridges, covalent linkages between the sulfhydryl groups of two cysteines, also reinforce a protein's tertiary conformation. Some proteins consist of more than one polypeptide chain. In these cases, the final quaternary structure results when one or more poly-peptide chains join and aggregate in a purposeful arrangement with one another. Most proteins have one of two general shapes, either fibrous or globular. Fibrous proteins include actin and myosin, components of muscle fibers. Globular proteins such as hemoglobin have a roughly spherical form.

Heating or treatment with certain chemicals denatures proteins, or destroys their conformation by disrupting the chemical bonds and interactions formed between atoms of the molecules. Because the structure of a protein is crucial to its ability to function properly, denaturation results in the loss of function and is sometimes used to destroy cell populations such as in the process of boiling water potentially contaminated with bacteria or other microorganisms.

The enormous architectural diversity of proteins allows them to serve many diverse functions. Some act as structural building blocks for cells or tissues such as that found in skin, ligaments, and muscles. Many play enzymatic roles, and others such as insulin are hormones involved in communication between cells. Receptors located on the surface of cell membranes and antibodies, a component of specific immunity, are proteins. The movement of substances within a cell, across biological membranes, and for cellular motility is also accomplished by proteins.

The final three-dimensional structure of a protein includes a combination of secondary structures arranged in a unique conformation. In this depiction of a CD4 receptor protein molecule, the alpha-helices are shown as cylinders and the beta-pleated sheets are shown as flat ribbons with arrowheads on the ends. *(Dr. Tim Evans/Photo Researchers, Inc.)*

NUCLEIC ACIDS

Nucleic acids are composed of chains of nucleotides bonded together by phosphodiester linkages. Nucleotides consist of three components: a nitrogenous base, a five-carbon sugar, and a phosphate group. There are two main types of nucleic acid, deoxyribonucleic acid (DNA) and ribonucleic acid (RNA).

One major structural difference between DNA and RNA is the sugar molecule they contain. DNA is a polymer of deoxyribonucleotides, which, as the name implies, have deoxyribose sugar molecules.

RNA contains ribose sugar molecules that differ in that they have an extra oxygen atom attached to the 2' carbon. A nucleoside is formed by the link-age of a nitrogenous base to the pentose sugar. The bases are divided into two general categories, purines and pyrimidines. The purines consist of two fused

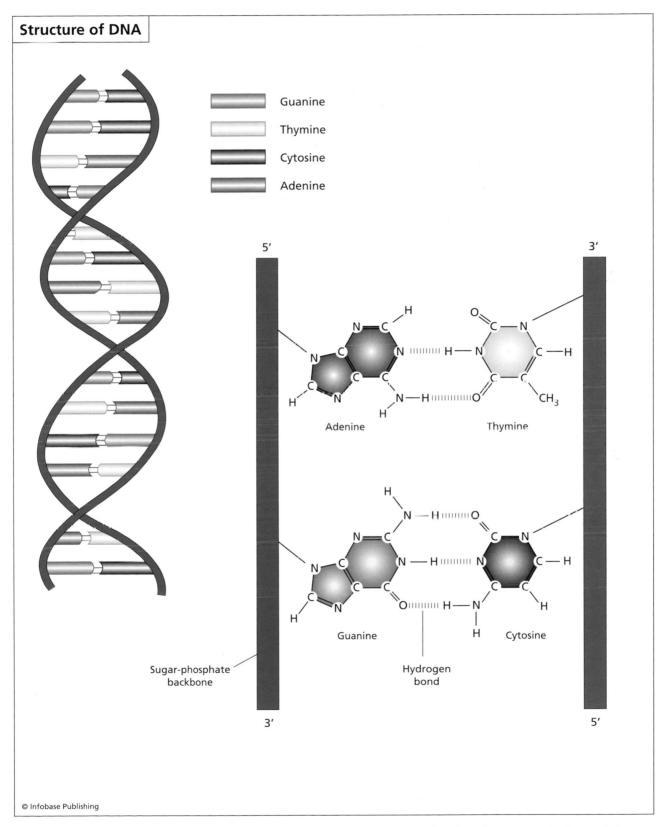

Structure of DNA

Guanine
Thymine
Cytosine
Adenine

Adenine   Thymine

Guanine   Cytosine

Sugar-phosphate backbone

Hydrogen bond

© Infobase Publishing

The two nucleotide chains of DNA are linked by specific hydrogen bonds that form between adenine and thymine, and between cytosine and guanine.

rings and include either adenine or guanine. The pyrimidines have a single ring and include cytosine, thymine, and uracil. In general, thymine only exists in DNA, and uracil is only in RNA. The addition of a phosphate group to the 5' carbon atom of the sugar in a nucleoside results in a nucleoside monophosphate, also known as a nucleotide. A condensation reaction between the phosphate group of one nucleotide and the 3' hydroxyl group of another creates a phosphodiester linkage. Several of these linked in succession form the sugar phosphate backbone of a single nucleotide chain. In DNA, two chains are linked by hydrogen bonds that form between specific base pairs, creating a double-stranded molecule. Adenine always pairs with thymine, and cytosine always pairs with guanine, giving the strands the characteristic of complementarity. The two strands are antiparallel to one another, meaning they run in opposite directions. If the phosphate group attached to the 5' sugar carbon of a nucleotide points upward, on the other strand the phosphate group of the complementary nucleotide will be pointing downward, and the hydroxyl group attached to the 3' sugar carbon will be pointing upward. The regular pairing of a purine with a pyrimidine between complementary chains maintains a consistent width along the length of the double-stranded molecule. The two polymers twist around one another at a repeat length of 10 nucleotides, giving it an overall helical shape.

DNA carries the genetic information in living organisms. Genes are simply segments of a linear DNA molecule that encode for a specific protein or RNA molecule. Inside cells, DNA exists as part of chromosomes that duplicate before the cell divides to ensure that each daughter cell receives all the genetic information. RNA performs several roles during the process of protein synthesis. RNA is also a structural component of ribosomes, and the ribonucleotide adenosine triphosphate (ATP) is the main molecule used by cells when they expend energy.

LIPIDS

Lipids are biomolecules that are insoluble in water but soluble in nonpolar organic solvents. Unlike other biological macromolecules, lipids are not polymers. This diverse class of biomolecules includes triglycerides (commonly called fats), phospholipids, and steroids.

The major class of lipids, triglycerides, contains both fats and oils. Fats are solid at room temperature, whereas oils are liquid. Triglycerides consist of one molecule of glycerol, a chain of three carbons with attached hydroxyl groups, to which three fatty acid chains are attached by ester linkages. Fatty acids have carboxyl groups with long unbranched hydrocarbon tails that give fats their hydrophobicity.

Examples include stearic acid ($CH_3(CH_2)_{16}COOH$), which is found in beef tallow; butyric acid ($CH_3(CH_2)_2COOH$), which is found in butter; and oleic acid ($CH_3(CH_2)_7CH=CH(CH_2)_7COOH$), which is found in olive oil. A fatty acid is said to be saturated if it contains only single bonds and the maximal possible number of hydrogen atoms. Monounsaturated fatty acids have a single double bond. Polyunsaturated fatty acids have more than one double bond per molecule, and therefore a lower melting point, making them liquid at room temperature. Most animal fats are saturated, whereas fats from fish and plants are unsaturated and are referred to as oils.

Phospholipids are structurally similar to triglycerides, except they have a phosphoric acid and an amino alcohol group in place of the third fatty acid. Because the end containing phosphorus is soluble in water but the other end is nonpolar, phospholipids make good emulsifiers, substances that mix with both lipids and water. When placed in an aqueous solution, phospholipids will aggregate to form a micelle, a spherical structure in which the hydrophobic tails all point inward and the hydrophilic heads face outward. In cell membranes, phospholipids form bilayers with the hydrophilic heads facing toward the cell's exterior and the cytoplasm while the hydrophobic tails point inward toward one another, to the exclusion of water molecules.

Steroids share a unique structure consisting of three six-carbon rings fused with one five-carbon ring. Different functional groups are attached to the fused ring structure to give a variety of steroids including cholesterol and the substances synthesized from it including bile salts, the sex hormones, and vitamin D.

Lipids serve a variety of functions in living organisms. Fats are a major source of stored energy that can be utilized when glycogen levels are low, they are the major component of biological membranes, they form a waxy coating on plants that protects against water loss, and they play hormonal roles.

See also BIOCHEMICAL REACTIONS; BIOCHEMISTRY; CELLULAR METABOLISM; CHEMICAL BASIS OF LIFE; DEOXYRIBONUCLEIC ACID (DNA); GENE EXPRESSION; NUTRITION; ORGANIC CHEMISTRY, ITS RELEVANCE TO LIFE SCIENCE.

FURTHER READING

Champe, Pamela C., Richard A. Harvey, and Denise R. Ferrier. *Lippincott's Illustrated Reviews: Biochemistry.* Philadelphia: Lippincott, Williams, & Wilkins, 2005.

Chang, Raymond. *Physical Chemistry for the Biosciences.* Sausalito, Calif.: University Science Books, 2005.

Daintith, John, ed. *The Facts On File Dictionary of Biochemistry.* New York: Facts On File, 2003.

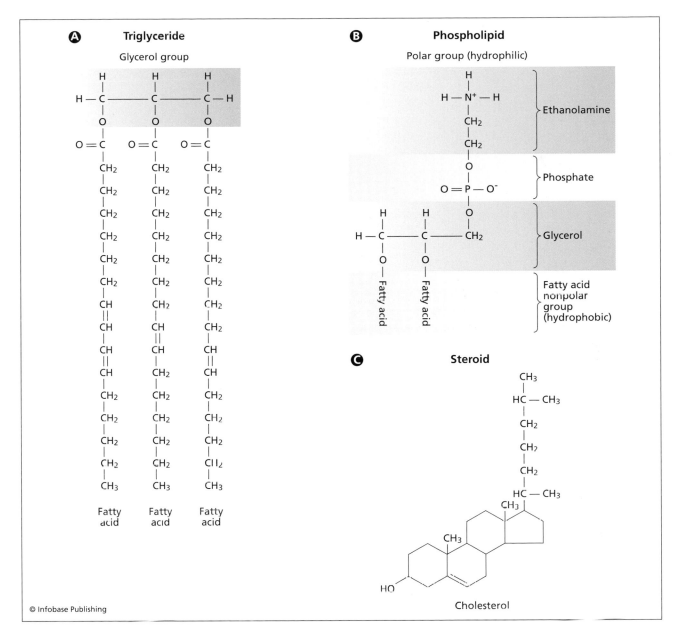

The structures of lipids vary tremendously. A) Triglycerides consist of a glycerol molecule, to which three fatty acids are attached. B) Phospholipids differ from triglycerides in that the third carbon of the glycerol backbone is linked to a phosphoric acid linked to an amino alcohol group. C) Steroids such as cholesterol have a fused-four-ring structure.

Karukstis, Kerry K., and Gerald R. Hecke. *Chemistry Connections: The Chemical Basis of Everyday Phenomena.* Boston: Academic Press, 2003.

Lennarz, William J., and M. Daniel Lane. *Biological Chemistry.* 4 vols. Amsterdam; Boston: Elsevier, 2004.

Murray, Robert K., Daryl K. Granner, Peter A. Mayes, and Victor W. Rodwell. *Harper's Illustrated Biochemistry.* 26th ed. New York: McGraw Hill, 2003.

Widmaier, Eric P. *The Stuff of Life: Profiles of the Molecules That Made Us Tick.* New York: Times Books, 2002.

bioremediation Bioremediation is the use of biological processes to remedy or resolve an environmental problem such as to clean up an oil spill or contaminated groundwater. Microorganisms including bacteria, fungi, and protists have diverse metabolic abilities and naturally utilize a variety of substances to fulfill their energy and nutritional needs. Chemicals and other substances that harm the environment or are toxic to some life-forms serve as food for others. Though microorganisms present in soil and water have been actively decomposing and recycling human waste since the beginning of civilization, the first formal use of bioremediation to treat synthetic compounds and spills was in 1975, when a group of scientists observed that growth of hydrocarbon-degrading bacteria was enhanced by the addition of nutrients to the soil. Increasing the metabolism of the bacteria increased the rate at which the hydrocarbons were consumed. Since then the use of bioremediation has been improved and expanded to remove metals, acidic waters produced by coal mining, toxins, and other synthetic organic compounds, in addition to cleaning up hydrocarbons from oil spills.

Bioremediation is a specific application of biodegradation, the biological breakdown of organic substances. Something that is biodegradable can be broken down by microorganisms that decompose the material into smaller components that feed into either biogeochemical cycles or into smaller molecules that other organisms can take in for nutrition. Because carbon makes up a significant mass of organic compounds, decomposition of natural and human-made organic compounds results in the conversion of much of the carbon into carbon dioxide, which is liberated. The microorganism also incorporates some of the carbon into its own biomass. Bioremediation is the specific use of microorganisms to biodegrade harmful pollutants such as petroleum products, benzene, and toluene, into innocuous products. The bacteria *Pseudomonas* and *Bacillus* and fungi that break down toxins have been most useful, but plants are also used in bioremediation. Some plants can uptake and concentrate toxic metals from diluted samples in soil or groundwater.

In situ bioremediation (ISB) is carried out at the site of contamination and can be used to treat contaminated soils, aquifers, lakes, or other bodies of water. Natural bioremediation refers to the simplest form of bioremediation used today. Microorganisms that are already present in the environment are left to perform the task without any assistance, although a worker usually monitors the levels of the contaminant to ensure the microbes are removing it. One must also make sure the contaminant does not leak away or is not washed away before being metabolized. In many cases, assistance is needed to biodegrade or to remove the pollutant more quickly. Two different strategies can accomplish this: biostimulation or bioaugmentation. The addition of chemicals or nutrients to accelerate the process of bioremediation is known as biostimulation. As the bacterial growth rate increases, the organisms metabolize the polluting substance at a faster rate. This method is more common, but, as natural bioremediation does, biostimulation depends on the previous existence of pollution-eating microorganisms at the contaminated site. Optimizing the conditions with respect to the concentrations of other nutrients such as phosphates,

nitrogen-containing compounds, and oxygen simply encourages the growth of the native pollutant-eating microbes. Hydrogen peroxide, which decomposes to oxygen and water and is more soluble in water than oxygen gas, is often added to meet the high demand for oxygen when the metabolizing bacterial population is large. When using ISB to treat contaminated water, one difficulty is that water movement carries the nutrients away. Because of this, ISB works better for spills on beaches or soil than on open bodies of water. Another method involves seeding the polluted environment with microorganisms known to metabolize the specific contaminating compound. This method, called bioaugmentation, may involve the introduction of a naturally occurring bacterial strain or a genetically engineered strain. In the United States, the Toxic Substances Control Act gives the Environmental Protection Agency the authority to regulate the release into the environment of microorganisms that have been genetically engineered for the purpose of breaking down chemical pollutants.

Despite pumping air, supplying nutrients, or adding microbes to a contaminated site, in some situations, the optimal conditions for the microorganisms to grow and multiply cannot be achieved underground. In these cases, it is appropriate to dig up the contaminated soil and place it in containers called bioreactors. The advantage of bioreactors is that the soil can be heated to speed up metabolism, the concentration of different nutrients can be controlled, the soil can be mixed or aerated efficiently, or anaerobic conditions can be created, and they can be used to treat solid, liquid, or gaseous contaminants. To treat groundwaters, wells are drilled and the groundwater is pumped into tanks, where nutrients and air are mixed into the water to stimulate microbial growth, and then the treated water is returned into the ground. Alternatively, the nutrients and air can be added to the wells.

In addition to being inexpensive, the advantage of bioremediation is its use of natural biodegradation processes; thus it does not require the introduction of any additional dangerous chemicals into the environment. The pollutant is converted to harmless gases and water. With ISB, the contaminated soil or water does not need to be removed and taken to a different location for cleanup, so costs are significantly lower. A disadvantage is that the microorganisms might not be able to tolerate the conditions where the contamination exists. Other chemicals toxic to the pollution-eating microorganisms might be present at the site or the concentrations might be too high. Another disadvantage is the length of time required to bioremediate a site—months or even years depending on the type of pollutant and the conditions of the contaminated site.

Petroleum hydrocarbons were the first type of compounds targeted by bioremediation. Anaerobic bacteria, bacteria that do not use oxygen as an electron acceptor in cellular respiration, have proved efficient at metabolizing other substances, such as chlorinated hydrocarbons, that scientists previously believed were recalcitrant to bioremediation. Since they undergo anaerobic metabolism, oxygen does not need to be supplied. The development of bioremediation techniques for the removal of metals, gasoline additives, and chemicals generated during the synthesis of explosives has also progressed. Researchers have identified microorganisms that can reduce metals to less toxic valence states. Microbiologists continue to seek new bacterial strains that utilize potential pollutants as growth substrates. To isolate them, they collect samples of soil or other environmental samples and culture the microorganisms that are present in media containing mineral salts and the desired growth substrate, in other words, the contaminant to be biodegraded. By adding the desired substrate but not adding other organic compounds, the medium selects for organisms that are capable of using that specific substrate to meet all of their metabolic needs.

Cometabolism refers to the situation when an organism does chemically transform a substance but does not use it as an energy source or for synthesis of other molecules. When a microbe has one enzyme that has a broad specificity and binds and transforms a substrate, but its other enzymes that act at a later stage in the same metabolic pathway do not recognize the transformed compound as a substrate, then the by-product accumulates unless another bacterial species is present and can metabolize it further. For example, several types of bacteria can metabolize aromatic hydrocarbons. Environmental contaminants such as dichlorodiphenyltrichloroethane (DDT) and polychlorinated biphenyls (PCBs) are chlorinated aromatic hydrocarbons that the bacteria begin to metabolize but cannot carry the process through to completion, so chlorinated by-products accumulate. Some bacteria that can utilize biphenyl cometabolize PCBs to chlorobenzoates. Though the bacteria have enzymes that can further metabolize benzoate, the enzymes do not recognize chlorobenzoates, the product of PCB metabolism. Other types of bacteria can metabolize the chlorobenzoate, however, so they perform the next step in the breakdown process. Cometabolism is a slow process but can be accelerated somewhat by the addition of other organic compounds to encourage growth. Scientists are learning that many synthetic organic compounds like DDT that were previously thought to be nonbiodegradable, or recalcitrant to bioremediation, can be biodegraded by microbial consortia, an association of several microorganisms that cooperate to complete the task. Different types of bacteria perform sequential steps in biodegradation of the compound. Though this is good news, the process is complicated by the fact that some of these bacteria are aerobic and others are anaerobic. Using genetic engineering techniques, researchers are trying to create strains of bacteria that contain and express the genes necessary to carry out the entire metabolic pathway to completion within one organism. Researchers have also identified bacteria that can remove the radioactive element uranium from contaminated groundwater and convert it into insoluble uraninite.

Landfarming has been useful in getting rid of petroleum refinery wastes. In this method, a farmer spreads the waste products over a field and allows indigenous oil microbes to metabolize the pollutants. Initial studies suggest that fields used for landfarming will still grow crops, but the quality of the crop product with respect to nutritional and chemical content is still being investigated.

See also ECOLOGY; ENVIRONMENTAL CONCERNS, HUMAN-INDUCED; ENVIRONMENTAL SCIENCE.

FURTHER READING

A Citizen's Guide to Bioremediation. Washington, D.C.: U.S. Environmental Protection Agency, Office of Solid Waste and Emergency Response, 2001. Available online. URL: http://www.clu-in.org/download/citizens/bioremediation.pdf. Accessed January 17, 2008.

Crawford, Ronald L., and Don L. Crawford, ed. *Bioremediation: Principles and Applications.* New York: Cambridge University Press, 2005.

Lovely, Derek R. "Cleaning Up with Genomics: Applying Molecular Biology to Bioremediation." *Nature Reviews/Microbiology* 1 (October 2003): 36 44.

Mulligan, Catherine N. *Environmental Biotreatment: Technologies for Air, Water, Soil, and Wastes.* Rockville, Md.: Government Institutes, 2002.

Singh, Ajay, and Owen P. Ward, eds. *Applied Bioremediation and Phytoremediation.* New York: Springer, 2004.

U.S. Environmental Protection Agency, Technology Innovation Program. "Hazardous Cleanup Information." Available online. URL: http://www.clu-in.org/. Accessed January 17, 2008.

biosphere The biosphere is the zone of the Earth that contains life and consists of the lower part of the atmosphere (air), the hydrosphere (water), and part of the lithosphere (rock and soil). All organisms live within the biosphere, which spans about 13 miles (21 km), or 1/600 the diameter of Earth. Most life exists within a much narrower band close to the surface of the land or oceans, where sunlight and water are readily available. The biosphere can be divided into smaller units called ecosystems, comprising all the living organisms and the physical environment within a defined area. A decaying log on the forest floor and all the life-forms that live in or on it provide an example of an ecosystem, as is the Sahara in northern Africa. The biosphere can be considered the largest, most inclusive ecosystem on the Earth. Within the biosphere, all of the life-forms on the planet and the many varied environments interact with and affect one another.

STRUCTURE OF THE BIOSPHERE

The biosphere encompasses parts of the atmosphere, the hydrosphere, and the lithosphere. The boundaries of the biosphere are not well defined. At one extreme, winds can carry dormant microbial spores higher up in the atmosphere than can support active, metabolizing life-forms. At the other extreme, scientists have discovered microbial life in the subsurface continental sedimentary basins of North America at depths of about two miles (near 3.2 km), and isolated thermophilic prokaryotes from 3.3 miles (5.3 km) in a borehole drilled in granite gneiss below the country of Sweden. Life

also exists near hydrothermal vents on the floor of the deep ocean, found at an average depth of 1.33 miles (2.1 km) and with temperatures reaching 750°F (400°C). Despite these extreme examples, the majority of life exists in the more moderate conditions of the upper 330 feet (100 m) of the lithosphere and hydrosphere. Geographic areas within this crowded region do exist that cannot support much life and are practically barren.

The atmosphere is the blanket of air that surrounds Earth. The major constituents are 78 percent nitrogen (N_2) and 21 percent oxygen (O_2). Carbon dioxide (CO_2) is an important constituent, though its concentration is only about 385 parts per million. Life in the biosphere helps maintain the chemical composition of the atmosphere, which is in a highly oxidizing state. The amount of water (H_2O) vapor is variable and forms clouds in the troposphere, the lowest layer of atmosphere, which contains about 80 percent of the atmospheric mass and extends five to nine miles (8–14.5 km) from Earth's surface. The troposphere is responsible for most of the weather that is observed from the ground and is the only layer of atmosphere known to support life. Birds and insects live in the troposphere.

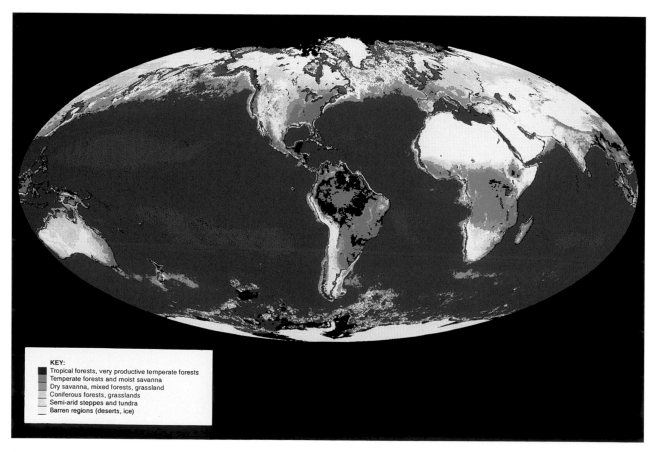

KEY:
- Tropical forests, very productive temperate forests
- Temperate forests and moist savanna
- Dry savanna, mixed forests, grassland
- Coniferous forests, grasslands
- Semi-arid steppes and tundra
- Barren regions (deserts, ice)

This satellite photograph shows differences in vegetation throughout the biosphere; dense vegetation is represented by purple and green, sparse vegetation by various shades of brown. *(NASA)*

The lithosphere is the solid, rocky crust consisting of inorganic minerals covering the entire planet. Most terrestrial organisms live between 10 feet (3 m) underneath and 100 feet (30 m) above Earth's surface.

The hydrosphere comprises all the water on the Earth including the oceans, lakes, rivers, ice sheets and glaciers, and moisture present in the air as vapor. The oceans contain 97 percent of Earth's water, and most oceanic life inhabits the upper 656 feet (200 m).

Although all of the components of the biosphere can be clearly defined, they are not necessarily separated physically. Most habitats contain components of all three, as the air, water, and earth all contribute essential ingredients to life. For example, in a rocky coastal biome, the rocks and dislodged sediment represent part of the lithosphere, the water and moisture sprayed into the air represents the hydrosphere, and the oxygen and other gases mixed into the water by the aerating action of the crashing waves represent the atmosphere.

ABIOTIC FACTORS AFFECTING THE BIOSPHERE

Abiotic factors are nonliving elements that influence whether life can exist, and, if so, what type of life-forms may exist in a particular region of the biosphere. Climate is the major abiotic factor affecting life in the biosphere. The distance from the Earth to the Sun is small enough to warm the entire planet but great enough so it is not too hot. Even slight variations in this distance affect the climate of a geographic region, as demonstrated by the seasonal climate changes due to the Earth's tilt toward or away from the Sun. Other influential abiotic factors include daily weather, erosion, earthquakes, and chemical reactions that occur to the landscape and in the atmosphere. Geological, chemical, and biological processes contribute to the cycling of elements necessary for life, mainly carbon, nitrogen, oxygen, and phosphorus. Climate and weather drive water through the hydrologic cycle, upon which all life depends.

The shape of the planet impacts the biosphere because it affects the path sunlight, the biosphere's ultimate energy source, must take to reach Earth's surface. At the equator, the sunlight hits the Earth's surface at a right angle, delivering the most intense radiation to equatorial areas. As a result of this great amount of energy input, biomes that occur in this region, such as tropical rain forests, are highly productive. The farther away one moves from the equator and toward either pole, the lower the angle of the incoming radiation; thus the sunlight is less direct: in other words, more radiation has already been absorbed by the time it reaches Earth's surface. The intensity of light that reaches the surface also varies seasonally as a result of Earth's tilt. The planet tilts 23.5° relative to its axis, causing the Northern Hemisphere to experience summer while the Southern Hemisphere experiences winter, and vice versa. The regions of the planet that lie between 23.5° north and 23.5° south latitude, known as the Tropics, receive the most sunlight year-round, and the least seasonal variation, since these regions are always closest to the Sun. Because light intensity is seasonal, so are many life cycles. Some animals only mate or breed in the spring, some animals hibernate in the winter, and plant life often becomes dormant before the winter.

Because solar radiation is most intense near the equator, more water evaporates in those areas. The warm moist air rises and then falls as precipitation. Some of the air, now dry, absorbs moisture from the Earth's surface as it moves away from the equator. At latitudes of about 60°, the air becomes moist enough to drop more rain, then is dry again over the poles. The rotation of the planet about its axis as the air flows near its surfaces also creates predictable wind patterns. These winds are named for the location where they originate; for example, the westerlies flow from the west to the east in temperate zones, and the northeast trade winds blow from the northeast. Wind shapes landscapes and amplifies temperature effects; thus organisms living in windy areas of the biosphere must be able to tolerate increased heat and water loss.

Geological features such as mountain ranges or nearby large bodies of water create regional or local effects on climate. Mountains affect sunlight, temperature, and rainfall on an area. For example, in the Northern Hemisphere, south-facing slopes receive more sunlight than north-facing slopes, causing different types of vegetation to grow on different sides of a mountain range. Mountains also direct air upward, where it cools and can cause rainfall in an area. Nearness to water bodies affects life not only by providing water, but by affecting local climate. Because water has a high heat capacity, it can absorb large amounts of heat energy from the Sun, with minimal effects on its temperature. This aspect of the hydrosphere moderates the climate of nearby terrestrial environments, keeping the temperatures lower than in regions farther inland. After the Sun sets and the water cools, the water body releases warmth into the surrounding air, preventing the nighttime temperatures from dipping very low.

The incoming solar radiation warms the Earth's surface; matter on Earth's surface absorbs some of it and reflects some of it back into the atmosphere as heat. Greenhouse gases in the stratosphere trap this heat within Earth's atmosphere. Some of the Sun's warmth serves to evaporate water from the oceans and Earth's surface. Once the water vapor rises into

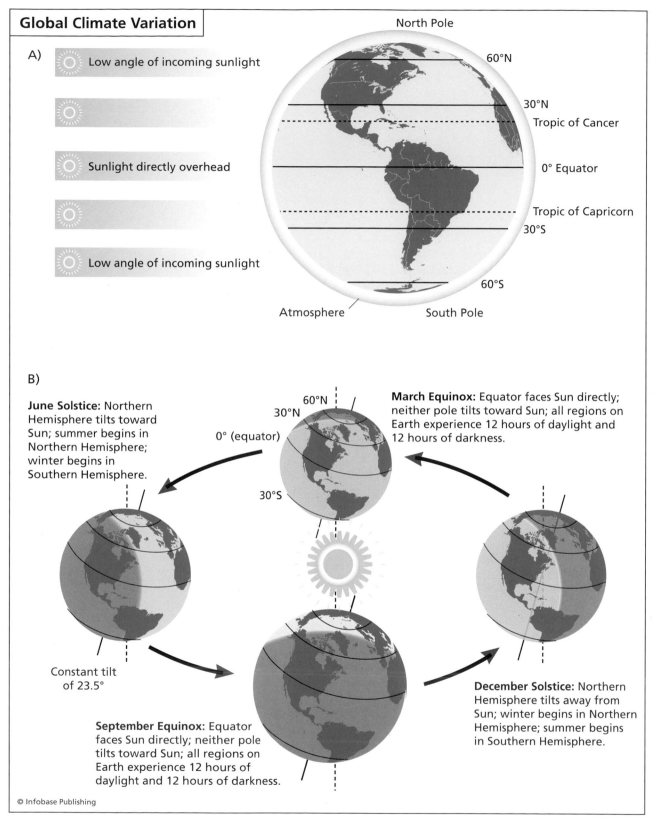

Global Climate Variation

A)

Low angle of incoming sunlight

Sunlight directly overhead

Low angle of incoming sunlight

North Pole

60°N

30°N

Tropic of Cancer

0° Equator

Tropic of Capricorn

30°S

60°S

Atmosphere   South Pole

B)

June Solstice: Northern Hemisphere tilts toward Sun; summer begins in Northern Hemisphere; winter begins in Southern Hemisphere.

March Equinox: Equator faces Sun directly; neither pole tilts toward Sun; all regions on Earth experience 12 hours of daylight and 12 hours of darkness.

60°N
30°N
0° (equator)
30°S

Constant tilt of 23.5°

September Equinox: Equator faces Sun directly; neither pole tilts toward Sun; all regions on Earth experience 12 hours of daylight and 12 hours of darkness.

December Solstice: Northern Hemisphere tilts away from Sun; winter begins in Northern Hemisphere; summer begins in Southern Hemisphere.

© Infobase Publishing

Climate varies globally. A) Because Earth is curved, sunlight that reaches Earth's surface at higher latitudes has penetrated a greater length of atmosphere, thus is more diffuse than sunlight hitting the Earth's surface near the equator. B) Earth's tilt causes the regions near the equator to receive more sunlight annually than regions nearer to the poles.

the atmosphere, winds carry it over the continents, and it falls as precipitation.

The amount of moisture present influences the types of life-forms that can inhabit an area. The simple presence of water is not sufficient; it must be available to organisms. Even marine organisms must have mechanisms for preventing water loss due to osmosis because of the hypersaline conditions, and freshwater organisms must have adaptations that prevent their cells from swelling and bursting as a result of water diffusing into them. All terrestrial life must resist desiccation or dehydration, and organisms living in dry areas must have special adaptations for enduring long periods of drought or for storing water.

The composition of the soil also affects the biosphere. The mineral composition and acidity or alkalinity may limit the type of vegetation that can grow. The physical nature of the landscape is also important, and processes such as erosion and weathering can shape and mold terrestrial and aquatic habitats. The chemical composition of the atmosphere and dissolved gases in bodies of water allows organisms suited for certain environments to respire in the conditions of that specific environment.

See also BIOMES, AQUATIC; BIOMES, TERRESTRIAL; ECOLOGY; ECOSYSTEMS.

biotechnology The term *biotechnology* is sometimes used interchangeably with *genetic engineering* or *recombinant DNA technology*, but the three have distinct meanings. Biotechnology is simply the practical application of knowledge from biological science, and it often refers to the use of living organisms or materials they generate to make commercial products or accomplish specific useful tasks. The products may be pharmaceuticals, cloned mice, or a genetically altered crop of corn. Examples of tasks facilitated by biotechnological advances include bioremediation, the use of microorganisms to clean up an oil spill, and deoxyribonucleic acid (DNA) fingerprinting to identify a suspect in a crime. Genetic engineering is the artificial manipulation of the genome of an organism, a task that is accomplished using tools of recombinant DNA technology. Biotechnology is more encompassing and includes the application of biological knowledge, biological processes, or organisms that may or may not be genetically modified in order to accomplish a task or goal. The benefits of biotechnology to basic research are numerous. New techniques and methods have led to the rapid advancement of knowledge in a variety of diverse fields including cell and molecular biology, developmental biology, applied and environmental micro-

biology, evolutionary biology, infectious disease, human health and medicine, genomics, and microbiology. This article discusses several applications of biotechnology to commercial industries, human health, and agriculture.

People employed knowledge of life sciences for their own benefit long before the term *biotechnology* was coined. One of the oldest applications of biotechnology is fermentation, the process used to make alcoholic beverages such as wine and soured foods such as pickles and buttermilk. Until the mid-1800s, when the French chemist Louis Pasteur demonstrated that microorganisms called yeast were responsible, people thought fermentation was a chemical rather than biological process.

Another traditional use of biotechnology is plant tissue culture, the propagation of plants under sterile conditions. By growing seeds or plants on specialized media, a horticulturist can generate clones of a plant, better control the environmental conditions, and nurture plants that are particularly tricky to grow successfully. Creating clones, identical genetic copies, of plants is easier than creating other organisms because many plant cells are totipotent, meaning they have the ability to develop into a complete new organism under the right conditions. If plant cells have been transformed, genetically engineered to exhibit a particular characteristic, tissue culture allows for the totipotent transformed cells to develop into a whole plant.

Microorganisms perform a variety of beneficial biotechnological services. Certain types of bacteria can be utilized to extract minerals from ores in a biotechnological process called biomining. Traditional methodology involves digging the ores from the earth, mechanically crushing them, and then subjecting the ores to harsh chemicals or physical treatments. Bioprocessing uses bacteria such as *Thiobacillus ferrooxidans* that oxidize inorganic compounds, such as copper sulfides, and in the process produce acid and ferric ions that leach out the copper. *Thiobacillus* is endogenous to the environments that are mined for copper, and the simple addition of sulfuric acid will encourage their growth and increase their metabolism. Two factors that slow the process are the generation of heat and the presence of heavy metals that are toxic to the bacteria. Using bacterial strains that are thermophilic, meaning strains that not only tolerate but prefer higher temperatures, will help solve this problem. Finding genes that help some microorganisms resist harm caused by heavy metals and cloning them into bacteria used in biomining will also improve efficiency.

Bacteria with unusual metabolic capabilities can also be used to digest organic matter into methane and carbon dioxide gases. Biotechnology companies are researching how to take advantage of this by creating financially profitable bioenergy plants. Sewage and wastewater treatment facilities also employ microorganisms to digest organic material before chemical treatments. Bioremediation efforts utilize microorganisms to break down oil, chlorinated solvents, petroleum products, or other hazardous pollutants into inert substances.

Forensic investigations have come to depend on biotechnology. DNA fingerprinting is a method for identifying an individual on the basis of his or her unique DNA. An investigator obtains a sample of a person's DNA from cheek cells, body fluids such as blood or semen, or hair. Biotechnology is used to analyze the sample, taking advantage of the less than a fraction of 1 percent of the differences in DNA sequences between individuals. The resulting DNA fingerprint can be compared with that of related individuals to determine familial relationships or to identify a suspect in a crime. Methods similar to those used for DNA fingerprinting can be used to diagnose certain genetic disorders, which are caused by mutations or alterations in a person's DNA that result in loss of function of a certain gene.

Biotechnology also affects the manner in which governments fight wars or terrorists incite fear in people. Biological weapons employ microorganisms or substances they produce to cause harm. For example, anthrax is a disease caused by *Bacillus anthracis,* a bacterium that forms endospores. These structures are tough, and easily dispersed, and when they make their way inside a host, they convert back into metabolically active, growing bacterial cells that secrete toxins that cause illness and potentially kill the host. In 2001 anthrax was intentionally spread through the U.S. mail, causing the death of five people and illness in 22 others.

As plant tissue culture does, some modern uses of biotechnology involve culturing or manipulating living animal cells in vitro. Assisted reproductive technology (ART) encompasses the different methods and technology during which both eggs and sperm are handled in order to help a couple conceive. In vitro fertilization involves the removal of eggs from a woman's body, combining the eggs with sperm to achieve fertilization, the subsequent culture of the zygote and embryo, and the return of the early embryo into the woman's body for gestation. The success of ART is due to the decades of research on human anatomy, reproductive physiology, hormone structure and function, cell growth and division, and early embryonic development in addition to the perfection of skills, laboratory techniques, and specialized equipment.

Stem cell technology is another branch of biotechnology that has received much recent attention.

As a multicellular organism develops, cells become specialized to perform unique functions in a process called differentiation. For example, an adipose cell functions to store fat, and a muscle cell functions in contraction. After differentiation, a cell cannot switch roles; in other words, an adipose cell cannot turn into a muscle cell. Stem cells are cells that are capable of self-renewal and that have the ability to develop into a number of different cell types, on the basis of signals they receive from other cells or environmental cues. Biologists study stem cells to learn about the processes of development and differentiation, in hopes of developing technology for replacing tissues such as injured spinal cord nerves, brain tissue destroyed by strokes or degenerative diseases such as Parkinson's, or cardiac tissue damaged during a heart attack. Stem cell technology also offers hope for a cure for diabetes and treatment for leukemia. Other potential uses of stem cells include preclinical drug testing, screening chemicals for potential toxicity, and developing gene therapy methods.

Another medical use of biotechnology involves tissue and organ transplantation. In order to be successfully transplanted, a donor organ must share certain molecular characteristics with the recipient's tissues and organs. Because the donor organ must meet these specific criteria, many patients die while waiting for organs that are compatible. Medical researchers have learned to grow some types of human tissues in vitro, including skin, ligaments, and blood cells. A patient's own skin can be cloned, ensuring a perfect molecular match, by removing a small piece, separating the cells, and placing them in a medium designed to nurture them and induce them to multiply. The cloned skin can then be grafted onto the patient. Because current technology limits the number of cell divisions to about 20, only small pieces of skin can be created in this manner. Researchers are also trying to learn how to stimulate stem cells to differentiate into specialized tissue types and someday whole organs for transplantation purposes.

The creation of organs from cells in vitro falls in the realm of bioengineering. One recent success of this application of biotechnology is brain-driven prostheses. Computerized limbs debuted in 2001, and in 2005 scientists created the first robotic arm controlled by thoughts alone.

The Human Genome Project utilized advances in DNA sequencing technology to determine the sequence of the 3 billion nucleotide pairs that make up the human genome. Researchers are currently using bioinformatics, computer technology to analyze and store biological data, to assign functions to the more than 20,000 genes believed to encode proteins. This information will assist medical scientists in determining the underlying cause of numerous genetic disorders, aid in their diagnosis, and lead to potential treatments. One novel approach to treating genetic disorders is through the introduction of normal copies of the gene to tissues that are expressing mutant, nonfunctional protein. This strategy, called gene therapy, is still in development.

Epidemiologists utilize biotechnology to follow the spread of infectious diseases such as avian influenza, the bird flu. Viruses mutate rapidly, resulting in the creation of new strains. By examining the genetic composition of viruses causing different outbreaks, epidemiologists can determine the origin and follow the transmission of a particular strain. Biotechnology also helps physicians diagnose different infectious diseases, thus expediting initiation of the proper treatment. Antibodies, proteins naturally made by the immune system that specifically recognize other proteins, are also used in home pregnancy tests.

Some biotech companies produce antibiotics, hormones, enzymes, or chemicals that are useful to humans by using microorganisms. In some cases the microorganisms have been genetically modified to synthesize the product, but in other cases the microorganisms naturally make the valuable substance. The companies grow the microorganisms in large quantities and extract and purify the substance for market. Plants and animals can also be used to generate proteins or other useful products. Pharming is a technology that combines agriculture and biotechnology. Scientists genetically modify plants or animals so they produce useful proteins and secrete them into their milk, blood, or eggs. Examples of pharming include genetically modified cows that make human serum albumin, sheep that produce blood clotting agents and anticoagulants, and corn that synthesizes a type of plastic that is biodegradable.

The genetic engineering of crops can increase their yields and make them more resistant to drought, extreme temperatures, or pests. Other purposes for genetically modifying food crops include improving the nutritional value by increasing the content of certain vitamins. Pigs have been genetically altered to produce healthier unsaturated fats. Foods may also be engineered for use as edible vaccines, eliminating the need for injections.

Biotechnology has also led to the cloning of several types of animals. While the success of cloning mammals has stimulated debate and controversy about the ethics of cloning, it has also opened doors for many possible applications. Farmers could clone cattle that produce large quantities of high-quality meat, dairy cows that yield more milk, and sheep that generate fine wool. Horse breeders could clone moneymaking thoroughbreds. Conservationists could clone animals in danger of becoming extinct.

During manipulation of the early embryos, scientists can insert genes that will make a cloned animal's organs more suitable for transplantation or genes for biopharming.

The numerous applications of biotechnology have changed lives, saved lives, saved ecosystems, solved crimes, solved genomes, changed genomes, improved crops, and done more. Newly gained knowledge will lead to additional successes, rapidly advancing biotechnology beyond society's ability to adjust. Members of society need to educate themselves about the latest scientific developments in order to make wise decisions about the smartest way to use the available biotechnology.

See also BIOINFORMATICS; CLONING OF DNA; CLONING OF ORGANISMS; DEOXYRIBONUCLEIC ACID (DNA); DNA FINGERPRINTING; DNA SEQUENCING; FORENSIC BIOLOGY; GENE THERAPY; GENETIC ENGINEERING; RECOMBINANT DNA TECHNOLOGY.

FURTHER READING

Biotechnology Australia. "Biotechnology Online." Available online. URL: http://www.biotechnologyonline.gov.au/. Accessed January 23, 2008.

Council for Biotechnology Information home page. Available online. URL: http://www.whybiotech.com/. Accessed January 23, 2008.

Grace, Eric S. *Biotechnology Unzipped: Promises and Realities.* Rev. 2nd ed. Washington, D.C.: Joseph Henry Press, 2006.

Spangenburg, Ray, and Kit Moser. *Genetic Engineering.* New York: Benchmark Books, 2004.

Steinberg, Mark L., and Sharon D. Cosloy. *The Facts On File Dictionary of Biotechnology and Genetic Engineering.* 3rd ed. New York: Facts On File, 2006.

Walker, Sharon. *Biotechnology Demystified: A Self-Teaching Guide.* New York: McGraw Hill, 2007.

botany Botany is the scientific study of plants, organisms that are photosynthetic, are multicellular, have cell walls made of cellulose, and typically lack locomotor abilities or sensory and nervous organs. Historically fungi and algae were included in the domain of botany as well, though these organisms belong to different kingdoms. Through photosynthesis, plants carry out vital processes necessary for sustaining the conditions that make the biosphere habitable by other life-forms: they manufacture organic molecules that many other organisms ingest as food, they produce molecular oxygen (O_2) that many organisms require for cellular respiration, and they remove carbon dioxide (CO_2) from the air. Plants also provide many practical products such as fibers for cloth, fuel for energy, and wood for construction. Because of these essential ecological and utilitarian functions that plants provide, botany is a very important discipline of the life sciences.

WHY STUDY BOTANY?

Early interest in plants stemmed from the practical products they provided: food, wood for fuel and construction, fibers for cloth, and medically useful substances. After the scientific revolution, which ended roughly around the beginning of the 18th century, natural philosophers became interested in plants for the sake of science itself. Clearly living organisms, but so different from animal life, plants were mysterious. Scientists began to explore their structure, physiology, evolutionary history, and means of reproduction. In the early 1770s an English chemist named Joseph Priestley burned a candle in a jar that also contained a sprig of mint and found that plants changed the composition of air. After a while, the air within the jar could no longer support the burning of the candle, but 27 days later, he was able to relight the candle in the jar. The mint had altered the composition of the air, allowing it to support the burning of the candle once again. Priestley also placed a live mouse in a closed jar, and it died. When he placed a mouse in a jar with a plant, the mouse survived. In 1779 the Dutch-born British physician and plant physiologist Jan Ingenhousz expounded on Priestley's experiments and showed that light was necessary for the plant to perform the process that changed the air composition. These early experiments led to the eventual description of photosynthesis, the mechanism by which organisms use light energy to make organic compounds from CO_2, creating O_2 as a by-product.

Photosynthetic organisms have the ability to capture light energy from the Sun, convert it into chemical energy, and then utilize the energy to fix carbon, or to incorporate inorganic carbon from CO_2 into organic food molecules like carbohydrates. Organisms that have the ability to manufacture food by photosynthesis are called producers, and they form the foundation of food webs in all ecosystems. Living organisms that do not have the ability to fix carbon must ingest producers, material made by producers, or other organisms that have fed on producers. Thus all life-forms that are not producers are ultimately dependent on them for food. Though plants may be the most familiar producers on the planet, algae and some prokaryotic organisms can also produce their own food using either sunlight or inorganic chemicals as a source of energy. The plant products wheat, rice, corn, millet, and sorghum supply about 70 percent of all the food energy and about 90 percent of the protein consumed by the world's population. Fruits and vegetables are also plant products, and foodstuffs such as grasses that are not directly consumed by people are necessary to feed cattle or other livestock from which meat and dairy products are obtained. Thus, agriculture depends on knowledge of plants to feed the world.

Photosynthesis is not only important in that all life on Earth depends on the food molecules it supplies, but also in that oxygenic photosynthesis (the type carried out by plants, algae, and most bacteria) removes CO_2 from the atmosphere and releases O_2 as a by-product. (This explains the change in the composition of the air in Priestley's experiments.) Most life-forms, including humans, require oxygen to extract energy from food molecules, and until photosynthesis released enough oxygen into the early atmosphere, the only life-forms on Earth were anaerobic. A complete understanding of photosynthesis is therefore crucial to understanding how plants help sustain the current conditions of the biosphere.

Studying plants is also important as they are responsible for supplying the majority of energy for society. The burning of fossil fuels satisfies more than 85 percent of energy demands to power activities such as transportation and the generation of electricity. Coal, natural gas, and petroleum-based products are produced by natural geological and chemical processes acting on the organic material produced by former life, mostly plants. Most fossil fuels originated during the Carboniferous, the geological period that occurred from approximately 354 to 290 million years ago, from the remains of the abundant forests and vegetation growing in and near swampy lowlands and microorganisms that accumulated at the bottom of water bodies. After these life-forms died, natural chemical and geological processes buried their remains and subjected them to extreme pressures and temperatures over millions of years, eventually converting them into the fossil fuels. Because it took hundreds of million of years to create the fossil fuels currently burned for energy, they are considered a nonrenewable resource. Biomass, the organic material produced mainly by plants that does not go directly into food or consumer products, can be converted to biofuels and is considered a renewable source of energy. The world population fills approximately 7 percent of its energy needs with biomass.

Botany is relevant to the manufacture of many practical and commercially valuable products. Cloth is manufactured from cellulose fibers obtained from plant cell walls. Flexible fibers such as those from cotton or flax plants are used to make clothing, and stronger, tougher fibers are used to make upholstery, ropes, or other materials. Even some synthetic fibers like rayon and acetate are produced by treating cellulose with different chemicals. Wood used as lumber for making furniture and as a raw material for making paper is also taken from plants. Many beverages are derived from plants. Coffee, tea, and cocoa are made by steeping plant products in hot water, and juices are obtained by crushing and straining fruits and vegetables. For thousands of years people have used plants for a variety of medicinal purposes ranging from treating nausea to dulling pain. Researchers continue to seek new potential pharmaceuticals from plant products for treating diseases such as cancer, Alzheimer's disease, and cardiovascular diseases. In addition to these few specific uses, botanical species provide society with numerous other goods. The industries that manufacture these products all require an in-depth knowledge of botany to produce high-quality products efficiently, cheaply, and safely.

Botanical knowledge is also crucial for maintaining a healthy environment. Many botanists have devoted their careers to researching aspects of environmental concerns that involve plants. Understanding the effect that human population growth and the accompanying industrial expansion have on plants could help in preventing severe long-term ecological consequences. Plants may also become part of the solution. For example, during photosynthesis plants remove atmospheric CO_2, a gas that exacerbates the natural greenhouse effect and contributes to global warming. Plants also perform many ecosystem services necessary for maintaining vital functioning ecosystems such as recycling nutrients. Continued research in plant physiology and ecology may lead to solutions to problems such as global warming and chemical pollution or at least reduce the negative consequences.

SUBDISCIPLINES

Plants provide diverse products and perform many functions within ecosystems. Because of this, in addition to the fact that one can examine plants at any level of biological organization, the field of botany encompasses numerous, far-reaching subdisciplines. Plant anatomy is concerned with the structure of plants, which provides useful information about evolutionary relationships and about past climates. For example, analysis of the size and shape of fossil leaves yields information about the mean temperature during a specific period of history. Plant physiology focuses on plant function, including how plants obtain their nutrition, how they distribute water and nutrients throughout their bodies, how they reproduce, and how they develop and grow. Plant anatomy and physiology are often studied in concert since the structure of a plant is closely related to its function, a theme common to all living organisms. To illustrate this connection, consider that some desert plants grow very long roots to help them find moisture deep down in the water table, if present, and others have few or no leaves to reduce

water loss from transpiration. Biochemical properties, such as the chemical processes involved in photosynthesis, are also linked to the unique structure and function of a particular plant.

Plant ecology is the study of how plants interact with other members of the community in which they live as well as their physical environment. In addition to providing ecosystem services related to the events of photosynthesis (carbon fixation, O_2 release, and CO_2 removal), plants perform numerous other diverse functions for ecosystems. Many plants live in symbiotic relationships with other organisms; for example, nitrogen-fixing bacteria may reside in and on plant root systems, and insects that feed off plant nectar, in turn, pollinate organisms farther away than the wind could carry pollen grains. Tall vegetation provides cool, shady areas for other organisms and offers protection by hiding them. Root systems of trees slow the rate of erosion of sloped areas. Transpiration, the loss of excess water through the leaves, helps maintain the humidity of tropical rain forests. Even after dying, plants continue to serve an important role in their community. The leaf litter on a forest floor provides nourishment for saprobes (organisms that obtain nourishment from dead or decaying organic matter), insulates the ground, prevents moisture from escaping the ground, supplies materials for nest building, fertilizes the upper layer of soil, and provides a habitat for many arthropods, worms, fungi, protists, and bacteria. A tree stump or fallen branch can be home to hundreds of other species by providing a variety of microhabitats. In addition to examining the ecological functions such as the listed examples, a plant ecologist may study the impact of human activities on the plants in a particular environment, or how the introduction of a nonnative plant to a new habitat affects the native species.

Plant taxonomy, or plant systematics, involves the naming and classification of plants. Historically, plant classification depended on easily observable but artificial characteristics such as the number or length of stamens. Modern classification schemes depend on evolutionary relationships.

Economic botany is the study of the relationship between people and plants. As already mentioned, people depend on many plant products to fulfill the basic needs of food, shelter, and clothing, but also to obtain other commercially valuable and useful products such as herbs and spices, dyes, beverages, and medicines. Examples of medicines extracted from plants or made from plant products include digoxin from foxgloves used to treat congestive heart failure and arrhythmia, the anticancer drug taxol from Pacific yew trees, quinine from the bark of the South American cinchona tree used to treat malaria, and the decongestant pseudoephedrine from ephedra species such as Ma Huang. Another aspect of economic botany is horticulture, growing fruits, vegetables, flowers, and ornamental plants. Horticulture is considered both a science and an art, depending on one's purpose—food production or aesthetics. Substances from plants are not always beneficial, and economic botany covers these also. Poison ivy, poison oak, and poison sumac produce urushiol, which causes many people to develop itchy rashes. Because people depend on the ecological functions of plants, economic botanists are also helpful in conservation efforts and in development of strategies for the sustainable use of plant resources. Many common plants can cause severe problems if ingested: raw rhubarb leaves can cause convulsions, coma, and death; daffodil, hyacinth, and narcissus bulbs cause nausea, vomiting, and diarrhea; and mistletoe berries can be fatal.

Other subdisciplines of botany include bryology, the study of mosses and liverworts, and pteridology, the study of ferns and related plants. Paleobotany is the study of fossil plants, and palynology is the study of pollen and spores, both modern and fossilized.

BRIEF HISTORY OF BOTANY

Throughout history, humans have been interested in plants for food and for the medicinal products they provide. The ancient Greek philosopher Theophrastus is considered the founder of botany. He authored two manuscripts in approximately 300 B.C.E., *De causis plantarum* (About the reasons for vegetable growth) and *De historia plantarum* (A history of plants). For 1,500 years the focus of botany was related to issues of farming and gardening. One of the first botanical references leading to the development of botany as an empirical science was *De plantis libri* (Book of plants), written by the Italian botanist Andrea Cesalpino in 1583. Cesalpino classified plants into three main groups—trees, shrubs and herbs, and seedless plants—which he further divided on the basis of structures of the fruits. The Swiss naturalist Gaspard Bauhin published the first significant work that attempted to describe all of the approximately 6,000 known species in 1623, titled *Pinax theatric botanici* (Illustrated exposition of plants). In the 18th century, the Swedish botanist Carl Linneaus went further by classifying all the known plants according to the structures of their reproductive organs and introduced binomial nomenclature, a system for naming plants that facilitated communication about different species among scientists. Linnaeus published numerous works; two of the most influential were *Systema naturae* (System of nature), initially published in 1735, and *Species plantarum* (Plant species) in 1753.

Microscopy allowed scientists to examine plant morphology at the cellular level, a procedure that led to a greater understanding of plant physiology, including the means by which plants reproduce. The British scientist Robert Hooke had first visualized cells while examining cork under the microscope in the late 17th century, and the English physician and botanist Nehemiah Grew and the Italian physician and biologist Marcello Malpighi founded the field of plant anatomy with their microscopic studies of plants. A better understanding of plant anatomy led the way to plant physiology, pioneered by the English botanist and clergyman Stephen Hales, who investigated the movement of water throughout plants, transpiration, plant respiration, and other topics in the early 18th century. A few decades later, Priestley and Ingenhousz contributed to the knowledge of the process of photosynthesis. The French botanist Charles-François Mirbel observed that each plant cell was enclosed by a cell membrane in 1809, and by 1830, most of the cellular structures of plants and the main plant tissues had been described.

The 19th century saw the emergence of cell biology, genetics, and evolutionary biology, all of which greatly influenced the development of botany. Famous botanists from this time whose work strongly influenced all life sciences included Matthias Schleiden, a German botanist who is considered a cofounder of the cell theory, and the Swiss botanist Karl Wilhelm von Nägeli, who classified tissues and proposed a mechanism by which plant cells form. The Scottish botanist Robert Brown distinguished between gymnosperms and angiosperms and discovered the cell nucleus in 1840, and in 1875 the German plant cell biologist Eduard Strasburger elucidated the process of nuclear division in plants. During the 20th century, many advances and discoveries further developed botanical science: the process of energy transfer during photosynthesis, the identification and the action of plant hormones, the ecological roles of plants, knowledge of genetics to improve agriculture and horticulture, causative agents of many plant diseases, and the biochemistry and metabolism of plants.

See also CALVIN, MELVIN; ENVIRONMENTAL CONCERNS, HUMAN-INDUCED; INGENHOUSZ, JAN; LINNAEUS, CARL; PHOTOSYNTHESIS; PLANT DIVERSITY; PLANT FORM AND FUNCTION; PRIESTLEY, JOSEPH; SCHLEIDEN, MATTHIAS.

FURTHER READING

Botanical Society of American home page. Available online. URL: www.botany.org. Accessed January 18, 2008.

Levetin, Estelle. *Plants and Society*. 4th ed. New York: McGraw Hill, 2006.

Stern, Kingsley R., James Bidlack, and Shelley Jansky. *Introductory Plant Biology*, 11th ed. New York: McGraw Hill, 2008.

Boveri, Theodor (1862–1915) German *Cytologist* Theodor Boveri was the first to describe chromosomes as independent entities and to establish their continuity throughout the cell cycle. His observations of the cell divisions that immediately followed fertilization in sea urchin embryos elucidated the function of centrosomes in cell division and demonstrated that individual chromosomes contributed different hereditary features to the offspring. He also studied different aspects of early development and showed that the egg and sperm were equivalent with respect to hereditary input to the embryo.

PERSONAL LIFE

Theodor Boveri was born in Bamberg, Germany, on October 12, 1862. He had one older brother and two younger brothers. He attended the Realgymnasium in Nuremberg from 1875 to 1881 and entered the University of Munich. Though he had a talent for painting and playing the piano, he planned to study history and philosophy. After only one semester, he switched to natural science. He became an assistant to Carl von Kupffer at the Anatomical Institute in Munich and conducted his dissertation studies on the structure of nerve fibers.

After receiving his doctorate in 1885, he accepted a fellowship, the Lamont-Stipendium, and moved to the Zoological Institute in Munich. The institute had a new director, the German zoologist Richard Hertwig, who established his reputation in the 1870s by describing the process of fertilization as the fusion of a sperm cell within the membrane of an egg cell. In Munich, Boveri obtained his habilitation (an academic achievement following a doctorate) in zoology and comparative anatomy in 1887. Hertwig took on Boveri as his assistant from 1891 to 1893; during that time Boveri became interested in cell biology.

In 1893 Boveri became a professor of zoology and comparative anatomy at the University of Würzburg and director of the Zoological Institute. In 1897 he married one of his Ph.D. students, Marcella O'Grady, an American biologist, with whom he had one daughter, Margaret.

Boveri suffered from depression and mental illness. His first episode occurred in 1890, initiated by the news that his father had financial difficulties and his mother was ill. Boveri became severely depressed and could not work for several months. For the remainder of his life he suffered recurrences of neurasthenia, a psychological disorder characterized by lack of motivation, extreme tiredness, and low self-esteem. His general health was also poor, and around 1912 he experienced an illness that caused slight paralysis on one side. The outbreak of World War I distressed him further, and he died in 1915 at the young age of 53.

CHROMOSOMAL RESEARCH

Boveri's major researches began in 1885, after arriving at the Zoological Institute in Munich. By the time Boveri began his studies, biologists had learned much about cell division and chromosomes, but no one had firmly established that chromosomes carry the genetic material or explained how they are reproduced and pass on to daughter cells. Hertwig had already described the process of zygote formation by fusion of the egg from the mother with a sperm from the father. The nucleus of the fertilized egg divided numerous times to form the nuclei of all the body cells; therefore all body cells contained nuclear matter from both parents, and the substance of this matter was likely to be the hereditary material. The German anatomist Walther Flemming had described the movement and distribution of chromosomes during mitosis in 1882. While studying chromosomes in roundworm eggs, the Belgian embryologist and cytologist Edouard van Beneden determined that the number of chromosomes is constant within a species, the number is halved when eggs and sperm are formed, and fertilization restores the normal number of chromosomes found in body cells. Today the process of nuclear division that results in half the number of chromosomes is called meiosis, and it is well characterized.

These studies all contributed to the foundation for Boveri's research on chromosomes in the roundworm *Ascaris megalocephala*. He published a series of brilliant papers in 1886–90 laying the foundation for and demonstrating the physical basis of several observed phenomena. Boveri began by simply describing the process of egg maturation up to the point of fertilization, including the formation of polar bodies, small cells that remain unfertilized but contain "leftover" chromosomes.

He next established the individuality of chromosomes, a property retained during cellular division. The chromosomes are only visible under the microscope during mitosis, when they are condensed and thickened enough to be seen. Biologists were therefore uncertain whether they were present at other times. If chromosomes did play such an important role as carrying the genetic material, which provided continuity during the cell cycle and from generation to generation, then one would expect the chromosomes to be permanent organelles. Boveri had observed finger-shaped lobes in the *Ascaris* nuclei during early cleavage. Using these as landmarks, he collected morphological evidence demonstrating that chromosomes were continuously present; they simply condensed during mitosis and dispersed between mitotic events. He went further to propose that the chromosomes were independent units and that they were organized, stable structures. The establishment of these concepts was necessary before chromosomes could be considered carriers of the genetic information, the main tenet of the developing chromosome theory of inheritance.

As an extension to the studies demonstrating the individuality of chromosomes, Boveri confirmed the equal contribution of the egg and the sperm to the chromosomal composition of the fused nucleus of the zygote, as proposed by van Beneden. Shaking unfertilized sea urchin eggs resulted in nucleated and nonnucleated fragments. Fertilization of both types could result in normal development, and occasionally nonfertilized, nucleated fragments developed. The fact that both the maternal and the paternal nucleus could direct normal development demonstrated their equivalence. These studies also provided strong evidence that the chromosomes are the nuclear factors of heredity.

GENETICALLY DIFFERENT CHROMOSOMES

In 1889 Boveri had found that during normal fertilization the egg cell engulfs a centrosome from the midpiece of the sperm in addition to the sperm head. While observing the development of *Ascaris* eggs, he occasionally came across tetrasters, eggs containing four rather than the normal two mitotic poles from which microtubules of the spindle apparatus grow. These structures formed when an egg engulfed two sperm midpieces. From the arrangement of the chromosomes and the four poles, he concluded that the resulting four nuclei would contain different chromosomal elements. For example, in one such tetrafoil cleavage he observed that two nuclei would only receive one chromosome, the third nucleus would receive four, and the fourth would receive two.

Normally one sperm cell fertilizes one egg cell, forming a diploid zygote that divides to form two cells that subsequently divide to form an embryo with four total cells. Boveri demonstrated that the centrosome is the organizing center for cell division. After fertilization, it divides, and each half nucleates the growth of microtubules and serves as a pole competing for chromosomes in the nucleus of each of the resulting cells. Each chromosome duplicates to form two connected equal copies that a normal bipolar spindle apparatus divides into two equal sets. Boveri also determined that a normal bipolar spindle resulted from the fusion of two half-spindles growing from the two different division centers that meet at a duplicated chromosome. All of this laid the ground-

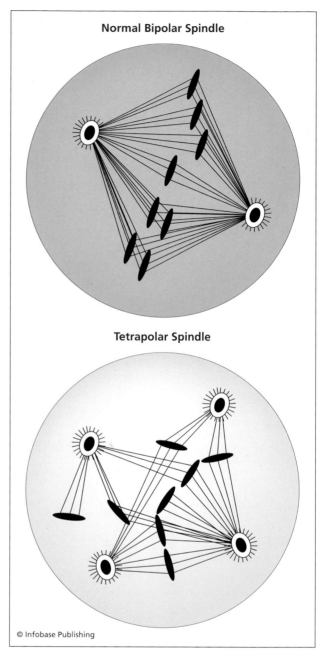

Theodor Boveri examined normal (bipolar) and multipolar cleavages, such as the imminent tetrafoil cleavage shown here, which resulted in unequal distributions of chromosomes. These studies allowed Boveri to draw conclusions regarding the role of chromosomes in inheritance.

work for Boveri's next major discovery—that chromosomes are genetically different.

Scientists did not know whether each chromosome contained all the hereditary material or each chromosome contained different portions of the complete genetic complement such that the entire set of hereditary elements was divided among the individual chromosomes. Boveri was aware that in sea urchin eggs one could artificially produce the tetrafoil cleavages that he previously examined in *Ascaris* by performing artificial fertilization in the presence of excess sperm, conditions that encouraged the double fertilization of one egg with two sperm. A single round of cell division in the dispermic zygotes produced a four-cell embryo. (Both of the sperm cells carry a centrosome that divides to produce two centrosomes for a total of four mitotic poles.) The dispermic embryos seemed to develop normally until gastrulation, when they died. Boveri ingeniously figured out that he could exploit these induced tetrafoil cleavages in fertilized sea urchin eggs to examine the effect of the composition of chromosomes and distribution of the cytoplasm on inheritance and development. The number of cells after division corresponds to the number of poles, and cleavage planes form midway between the poles. Equal distribution of the content within only two sets of chromosomes into four cells is therefore impossible. Because of this, tetrafoil cleavage results in cells with abnormal numbers of chromosomes. Examination of the development of these cells would give information about the composition of the individual chromosomes. If the chromosomes carried different genetic properties, then as the cells with abnormal numbers developed, defects would develop in portions of the embryo that arose from the abnormal cells.

In 1895 the American cytogeneticist Thomas Hunt Morgan had observed that shaking dispermic eggs after insemination caused trefoils (resulting in three cells) to form instead of tetrafoils. The agitation probably disturbed one of the centrosomes. Because the chromosomes would only divide three ways rather than four, trefoils result in nuclei with a correct chromosomal distribution more often than tetrafoils. Boveri hypothesized that if each chromosome contained all the genetic material, then development should proceed normally, but if each chromosome only contained a portion of the genetic material, that is, each chromosome were qualitatively different, then each cell would need a complete set of equally distributed chromosomes in order to develop normally. Probability predicted that approximately 11 percent of triasters would develop normally. Boveri performed the experiment and raised the cells of the tetrafoils and trefoils in isolation after separating them by treatment with calcium-free seawater. The data showed that trefoils developed normally with a higher frequency than tetrafoils—about 11 percent (as expected) in comparison with only 1 percent of tetraster eggs that developed normally. Comparing the abnormal chromosomal distributions with the developmental abnormalities, he concluded that not only was the total number of chromosomes important, but so was the mix of chromosomes. Around

the same time in 1902, Walter S. Sutton, a cytologist from Columbia University in New York, published his findings from studies of grasshopper chromosomes that suggested chromosomes were individual units, they occurred in pairs, each parent contributed one member of a pair to the offspring, and the paired chromosomes separated during nuclear division in gamete formation. Together, Sutton and Boveri's work demonstrated that chromosomes are the physical basis of the laws of inheritance as described by Gregor Mendel.

ROLE OF CYTOPLASM IN INHERITANCE AND DEVELOPMENT

Though nuclei clearly played the major role in inheritance, Boveri was also interested in the role, if any, of the cytoplasm of the egg in inheritance. Though the egg and sperm contributed equally to the formation of the nucleus, the egg contributed almost all of the cytoplasm to the zygote. Boveri knew that differential distribution of the chromosomes affected the normal development of the tetrafoils and trefoils, but the possibility existed that differential distribution of the cytoplasmic contents also played a role. Dispermic embryos turned out to be useful for examining this scenario. Boveri separated the cells of normal four-cell embryos by placing them in calcium-free seawater and observed that the development of all four proceeded normally and equally to produce quadruplets; thus all of the cellular contents (with respect to cytoplasm and chromosomes) must have been distributed equally. The cytoplasmic material was also evenly distributed for trefoil and tetrafoil cleavages, and, when isolated, their cells usually developed normally until the blastula stage, when most of them died. The rest died during the gastrula stage. In general, cells from isolated trefoils seemed to proceed normally until later stages of development than cells from tetrafoils, but this must have been due to higher frequencies of normal chromosomal distribution, as mentioned previously. These experiments suggested that differential cytoplasm probably does not play a role in the abnormal development of trefoils and tetrafoils.

In another study examining the potential role of cytoplasm in inheritance, Boveri attempted artificial fertilization of nonnucleated fragments of eggs from one species of sea urchin with sperm from another species. The two species displayed different larval skeleton shapes, making it easy to determine whether the sperm nucleus was responsible for the inherited traits. Some of the resultant larvae resembled the species of the mother and some resembled the species of the father. Since no original nuclei were supposedly present, the observation that some larvae resembled the father's species reinforced the importance of the nucleus in inheritance, but the fact that some resembled the mother suggested the cytoplasm played a role in inheritance. Unfortunately, Boveri had technical difficulties repeating the experiment and later retracted his published findings. He concluded that the larvae resembling the mother must have resulted from the persistence of some maternal chromosomes in the egg fragment that cooperated with chromosomes supplied by the sperm to allow development to proceed. Though technical limitations at the time prevented him from exploring this further, other researchers later reached a similar conclusion. However, Boveri remained open-minded about the possibility that cytoplasmic factors did play a role in inheritance.

Boveri was more successful in demonstrating that the cytoplasm affected development. During embryogenesis in *Ascaris*, a process called chromatin diminution eliminates certain chromosomes or parts of certain chromosomes from some cells that form somatic tissues. He showed that positioning of the nuclei during the first cleavage of dispermic eggs affected whether or not diminution occurred. Thus, components of the cytoplasm must interact with the nucleus in directing this process.

TUMOR RESEARCH

In 1914 Boveri published a paper on the origin of malignant tumors, suggesting that tumors resulted from abnormal numbers of chromosomes, a condition called aneuploidy. In 1890 the German cytologist David Hansemann had reported that chromosomes failed to separate properly in the cells from human skin cancers. This led to daughter cells with abnormal numbers of chromosomes. Decades later, Boveri proposed that individual cells with incorrectly combined sets of chromosomes could develop into tumors; thus tumors could originate from any condition that gives rise to aneuploid cells, including both unbalanced division of the chromosomes during cell division or from divisions involving more than the normal two centrosomes. Boveri was one of the first scientists to propose that tumors resulted from irregularities at the cellular level. Today cancer researchers continue to observe the chromosomal chaos in cancer cells—whereas human cells normally have 46, cancer cells may have fewer or even hundreds of chromosomes. This phenomenon of aneuploidy in cancer cells probably is a result of the uncontrolled cell divisions that follow transformation rather than a cause of tumor formation.

At the time of his death on October 15, 1915, the linear order of genes along chromosomes had not yet been established by Morgan and his coworkers. Several decades passed before the extent and significance of Boveri's discoveries were appreciated. His scientific publications were all in German and few

English translations are available, but his name is frequently mentioned as a pioneer in genetics. His work merged the fields of cytology, genetics, and embryology and provided the foundation for the interpretation of genetic phenomena at the chromosomal and cellular levels and led to a better understanding of the cycle of cell division.

See also CELLULAR REPRODUCTION; CHROMOSOMES; GENETICS.

FURTHER READING

Baltzer, Fritz. "Theodor Boveri." *Science* 144 (May 15, 1964): 809–815.

Moritz, Karl B., and Helmut W. Sauer. "Boveri's Contributions to Developmental Biology—a Challenge for Today." *International Journal of Developmental Biology* 40 (1996): 27–47.

Oppenheimer, Jane. "Theodor Boveri." In *Dictionary of Scientific Biography*. Vol. 2. Edited by Charles Coulston Gillispie. New York: Scribner, 1,970–1,976.

Sander Klaus. "Theodor Boveri on Cytoplasmic Organization in the Sea Urchin Egg." *Roux's Archives of Developmental Biology* 202: 129–131.

Brock, Thomas (1926–) American *Bacteriologist* Thomas Brock is a bacteriologist who discovered bacterial life thriving at temperatures higher than scientists previously imagined any life-form could exist. His discovery of prokaryotic organisms growing in hot springs at Yellowstone National Park led to the proposal of a whole new domain of life, the Archaea.

BECOMING A MICROBIOLOGIST

Thomas D. Brock was born in Cleveland, Ohio, on September 10, 1926, to Thomas Carter Brock, a power engineer, and Helen Sophia Ringwald, a former nurse. When he was age 15 the family moved to Chillicothe, and within a few months his father passed away. Although the older Thomas had only received an eighth-grade education, he encouraged his son in educational pursuits such as backyard chemistry experiments and building electrical equipment. After graduating from high school in Chillicothe, Brock entered an electrical program in the U.S. Navy. In 1946, after travels to Chicago and Alaska, he enrolled at Ohio State University (OSU) on the G.I. bill. His initial plan was to become a writer, but he switched his major to botany and graduated with honors in 1949. OSU offered Brock a graduate assistantship, and he performed research in mycology (the study of fungi) for a master's and a doctoral degree. By the time he earned a Ph.D. in 1952, Brock had developed an interest in ecology, in particular, microbial ecology.

After graduating, however, postdoctoral positions and academic jobs in general were scarce. The Ohio Agricultural Experiment Station in Wooster, Ohio, hired Brock as a summer research associate to work on soil fungi, a position that prepared him for a position in the Antibiotics Research Department at Upjohn, a pharmaceutical company. He moved to Kalamazoo, Michigan, with his new wife, Louise.

Brock only had two microbiology courses at OSU, but while working for Upjohn, he learned a lot about microorganisms, especially bacteria, since most antibiotics treat bacterial infections. In his free time he also researched a local railroad and taught himself German, a skill that would be useful later in his career. Antibiotics research became boring after five years, and he left Upjohn to join the Biology Department at Western Reserve University (now Case Western University), in his hometown of Cleveland. Part of his job responsibility was teaching general bacteriology, nursing microbiology, mycology, medical microbiology, and advanced microbiology. Having learned enough microbiology to begin feeling like a "real microbiologist," but realizing that the teaching load was too heavy to develop the research program he desired, Brock gave up his assistant professorship and became a postdoc for L. O. Krampitz in the Department of Microbiology at the medical school. With a project on the M protein of group A *Streptococci*, this position provided the opportunity to learn biochemistry, immunology, and a bit of clinical microbiology.

During the summers of 1958 and 1959, Brock vacationed in northern Ontario on Lake Memesagamesing, where he learned boating, fishing, and other aquatic activities. Meanwhile, Brock put his German skills to good use by translating many scientific papers of historical microbiological significance. He included these in a book about the history of microbiology, *Milestones in Microbiology*, the first edition of which was published in 1961.

MICROBIAL ECOLOGY

After only one year of postdoctoral studies, Brock accepted a job as an assistant professor of bacteriology at Indiana University, in Bloomington, in 1960. His teaching responsibilities and research initially focused on medical microbiology, but Brock began thinking about microbial ecology again.

In summer 1963 Brock made plans to study marine microbiology at the Friday Harbor Laboratories of the University of Washington. He focused his research on *Leucothrix mucor*, a widespread marine microorganism, and found that this filamentous organism formed knots under certain culture conditions. Brock was the first to observe these sorts of structures. Not only did this research lead to a

cover story in the journal *Science* and a feature article in the *New York Times* in 1964, but it also marked the beginning of his research on microbial ecology of sulfur hot springs.

Thiothrix, an organism related to *Leucothrix,* lives in sulfur springs. This led Brock to visit Yellowstone National Park, in Montana, in 1964. The developments of microorganisms present in the run-off channels of the Yellowstone hot springs surprised him, and he collected some samples. At the same time, he was working on authoring a text, *Principles of Microbial Ecology* (1966), and his research on ecosystems for the book stimulated his view of the springs as steady-state ecosystems. During the following summer, while vacationing with his wife, Brock planned to examine the chlorophyll levels (in cyanobacteria) in the thermal gradient of the outflow channels. He observed pink, gelatinous, stringy masses of biological material at high temperatures (131°F–140°F or 55°C–60°C). The sample contained large amounts of protein but no chlorophyll, leading to his conclusion that the material was bacterial. On the basis of his findings, he wrote a proposal and received a grant from the National Science Foundation to fund further research on microorganisms in the Yellowstone hot springs. He established a temporary laboratory facility at West Yellowstone and set to work.

DISCOVERY OF *THERMUS AQUATICUS*

Initial studies concentrated on photosynthesis measurements in the thermal mats of cyanobacteria. As part of an undergraduate honor's thesis project, one of Brock's students, Hudson Freeze, attempted to culture the pink bacteria from one of the springs. Though that attempt was unsuccessful, they did grow yellowish bacteria at 158°F (70°C). They named it *Thermus aquaticus* and found similar strains from several other sources.

Over the next two years, Freeze worked on determining the deoxyribonucleic acid (DNA) base composition and measuring the growth rates of *T. aquaticus* at different temperatures. With the assistance of his technician Pat Holleman, Brock worked on the taxonomy of the hyperthermophilic microorganism. He had observed the pink bacterium growing in hot springs with temperatures of 176°F (80°C) but suspected they might grow at even higher temperatures. He placed a string in the source pool, Octopus Spring, which had a temperature greater than 194°F (90°C), and observed pink growth on it. Photomicrographs taken from slides immersed and incubated for one day in the spring appeared in a *Science* paper published in 1967, "Life at High Temperatures." Brock and some of his graduate students also immersed slides in boiling pools, with temperatures at or greater than 198°F (92°C). Because of the high altitudes, water boils at lower temperatures in Yellowstone than at sea level. Amazingly, they discovered bacteria actively growing and dividing at boiling temperatures as well.

The article received much attention, and Brock soon found himself traveling across the globe to search for hyperthermophiles in geothermal regions in Italy, Iceland, New Zealand, Japan, Central America, and the Caribbean. He found bacteria in boiling springs of neutral or alkaline pH at all locations, even at the lower altitudes, where the water boiled at 212°F (100°C).

IMPACT OF THE DISCOVERY OF HYPERTHERMOPHILES

The discovery in 1977 of hydrothermal vents deep in the ocean broadened the scope of thermophilic bacterial research. These fissures in the ocean floor spew superheated, mineral-rich fluid. Temperatures can reach 750°F (400°C) where the water shoots out of the vent. Hydrothermal vents support diverse communities of organisms, with thermophilic bacteria forming the bottom of the food chain. As chemoautotrophs, they synthesize organic compounds using reduced inorganic iron and sulfur compounds as their energy source. The discovery of the hydothermal vent communities gave insight into the earliest life-forms on Earth, as the conditions produced by the vents resemble the presumed conditions of the early Earth. Thus the interest in hyperthermophiles increased.

In 1983 Kary Mullis, a biochemist from California, conceived of the idea of the polymerase chain reaction (PCR). This molecular biological technique functions to generate large quantities of a specific segment of deoxyribonucleic acid. PCR involves repeated cycles of replication, interrupted by brief periods of heating, to denature double-stranded DNA. The enzyme DNA polymerase, like most protein enzymes, is sensitive to heat. At high temperatures the enzyme denatures and loses functionality. The power of PCR lies in its ability to amplify DNA exponentially: each cycle can theoretically double the amount of DNA already present. But the heating step of each cycle destroyed most DNA polymerases, requiring the researcher to add new enzyme after each cycle. The fact that *T. aquaticus* grew at such high temperatures meant it possessed a thermostable DNA polymerase. Such an enzyme would mean the cycles could run repeatedly without the need for adding fresh enzyme. After biochemists isolated and purified the enzyme, called Taq (for *T. aquaticus*) polymerase, the popularity of PCR took off, and Kary Mullis won the Nobel Prize in chemistry in 1993 for his invention of PCR. *Science* magazine created a new award in 1989, "The Molecule of the Year,"

and named Taq polymerase the first awardee. Today all modern molecular biology laboratories employ PCR as an integral part of DNA-based research.

Carl Woese, a professor of microbiology at the University of Illinois at Urbana-Champaign, proposed a new biological classification system that includes a taxonomic level higher than kingdom, that of domain. He suggested that all life-forms belong to one of three domains: Eubacteria, now simply called Bacteria; Eukarya, including all eukaryotic organisms; and another prokaryotic category, Archaebacteria, now simply called Archaea. Woese's phylogenetic research examining ribosomal ribonucleic acid from bacteria and archaeans suggested members of the two domains were sufficiently different to warrant two distinct domains. Woese placed many of the organisms Brock discovered in Yellowstone in the domain Archaea. The fact that Brock had already researched and characterized these microorganisms facilitated Woese's analysis.

WISCONSIN, BOOKS, AND LAKES

In 1971 Brock joined the Department of Bacteriology at the University of Wisconsin-Madison as the E. B. Fred Professor of Natural Sciences. He had married Katherine Middleton earlier that year. After the move to Wisconsin, Brock maintained his productivity. His Yellowstone research continued for five more years, and he branched out into investigating the bacterial genera *Sulfolobus* and *Chloroflexus*. *Sulfolobus* species belong to Archaea and grow optimally at a pH of 2 or 3 and temperatures around 167°F–176°F (75°C–80°C). *Chloroflexus* species are photosynthetic thermophiles belonging to the group of green, nonsulfur bacteria.

The Wisconsin lakes attracted Brock, whose interest in ecology led to a desire to learn more about limnology, the study of freshwater lakes. His major course of investigation during the late 1970s was the study of cyanobacterial populations of Lake Mendota, though he also studied other lakes. In 1985 he published the book *A Eutrophic Lake: Lake Mendota, Wisconsin*.

In addition to *Principles of Microbial Ecology* and *Milestones in Microbiology*, Brock has published numerous scientific papers and articles concerning microbiology written for a more general audience. During 1967–69 Brock spent his evenings and weekends working on another college textbook, *Biology of Microorganisms*, originally written for nonmajors, but its newer editions are intended for undergraduate microbiology majors. The fifth edition, published in 1986, became the first full-color microbiology textbook on the market. The latest edition, the 11th, titled *Brock Biology of Microorganisms* (2006), was written by Michael Madigan and John Martinko.

Brock wrote the first full-length English biography of the German physician and microbiologist Robert Koch. This successful book, *Robert Koch: A Life in Bacteriology and Medicine,* was published in 1988. In 1989 Brock published an updated and revised version of *Milestones in Microbiology,* and in 1990 he published a history of bacterial genetics, *The Emergence of Bacterial Genetics.* According to Brock's own account, the University of Wisconsin did not view his historical research and publications as valid research, and Brock was pressured to retire.

Brock's discovery of life at high temperatures impacted the life sciences in many ways. Before the 1970s biologists thought the upper temperatures at which life could exist approached 163.4°F–170.6°F (73°C–77°C), but Brock showed that some bacteria thrived in boiling springs at temperatures greater than 212°F (100°C). His research on extreme thermophiles led to better understanding of ancient life-forms and eventually to the proposal of a whole new domain of life, Archaea, consisting of many prokaryotic organisms that display other unique extreme abilities. Some survive in acidic environments, some require high salt concentrations, and some produce methane as a by-product of their metabolism—giving new meaning to the limits of biological diversity.

See also ARCHAEA; BACTERIA (EUBACTERIA); HYDROTHERMAL VENTS; MICROBIOLOGY; POLYMERASE CHAIN REACTION; PROKARYOTIC CELLS.

FURTHER READING

Brock, Thomas D. "Knots in *Leucothrix mucor.*" *Science* 144 (1964): 870–872.

———. "Life at High Temperatures." *Science* 158, no. 3804 (1967): 1,012–1,019.

———. "Life at High Temperatures." *Science* 230, no. 4722 (1985): 132–138.

———. *Milestones in Microbiology 1546–1940.* Washington, D.C.: American Society for Microbiology Press, 1999.

———. "The Road to Yellowstone—and Beyond." *Annual Reviews of Microbiology* 49 (1995): 1–29.

———. "The Value of Basic Research: Discovery of *Thermus aquaticus* and Other Extreme Thermophiles." *Genetics* 146 (August 1997): 1,207–1,210.

Buffon, Georges-Louis Leclerc, comte de (1707–1788) French *Naturalist*

Georges-Louis Buffon was a prominent 18th-century naturalist, who authored a mammoth 44-volume treatise, *Histoire naturelle* (*Natural History*), which summarized everything that scientists knew about the natural world at the time. His observation that organisms living in similar but distant environments were distinct became one of the founding principles of the field of biogeography. Buffon's work influenced other renowned biologists for generations, including the famous English evolutionary biologist Charles Darwin.

Georges-Louis Leclerc was born on September 7, 1707, in Montbard, France. His father, Benjamin-François Leclerc, married Anne-Christine Marlin, who had rich relatives, making them members of the French aristocracy. George-Louis had four younger siblings.

In 1717 Georges-Louis entered the Collège des Jésuites in Dijon. From 1723 to 1726 he studied law, and then he traveled around, studied medicine and botany, and finally settled back in France in 1732. His mother had died, and he inherited her family's fortune and the title comte de Buffon. From 1734 on, he was known simply as Monsieur de Buffon.

After he engaged in some engineering and mathematical pursuits, the French Royal Academy of Sciences admitted Buffon as an assistant member of the section on mechanics in 1734. He translated two well-known works into French: *Vegetable Staticks,* by Stephen Hales, and *The Methods of Fluxions and Infinite Series,* by Sir Isaac Newton. Buffon began his botanical researches in the 1730s, and after becoming an associate member in 1739 he transferred his appointment to the botanical section. That same year he also became keeper of the Jardin du Roi (the royal botanical garden). Under Buffon's direction, the garden doubled in area, enlarged its buildings, increased its collections, and developed into a major scientific research center.

At the request of the minister of the navy, Jean-Frédéric Phélypeaux, comte de Maurepas, Buffon began cataloging the royal natural history collections that eventually formed the basis of the Musée National d'Histoire Naturelle (the [French] National Museum of Natural History), which was formally founded in 1793. This project developed into the ambitious endeavor that resulted in the comprehensive work *Histoire naturelle, générale et particulière* (*Natural History, General and Particular,* 1749–1804). Though Buffon believed that humans could not hope to understand nature completely, his writings attempted to convey all that could be understood with respect to natural history, geology, and anthropology. *Histoire naturelle* was translated into many languages and distributed worldwide. Of the 50 proposed volumes, Buffon was only able to complete 36 before his death.

At the time, the widely accepted explanation for the diversity of life was the biblical account stating that a supernatural divine being created all life-forms in their existing forms approximately 6,000 years ago. Buffon bravely suggested that species changed and that many species had died out. After observing that different geographical regions had unique flora and fauna, he concluded that life originated in one central region, and then species either grew more complex or degenerated into the existing forms. As species spread out, some could survive in the particular conditions of a specific habitat, while others died out in that area. The well-known evolutionary biologist Charles Darwin later praised Buffon for his emergent conception of species evolution.

The first 15 volumes appeared in 1749–67 and covered topics such as the history and theory of Earth, the formation of the planets, some general biology, development and reproduction, the natural history of humans, and the natural history of animals. The next seven (1774–89) supplemented the first 15. The rest of the work consisted of nine volumes on birds (1770–83) and five on minerals (1783–88). The remaining eight volumes on reptiles, fishes, and cetaceans were written by comte de Lacépède after Buffon died. One section of the fifth supplementary volume was particularly famous. Published in 1778, *Les epoques des la nature* (Epochs of nature), described his theory of the Earth and attempted to merge geology and biology. His theory is summarized as follows: the planet originated from a piece of the Sun that broke off; some solidification occurred as the primitive Earth cooled and vapors and other substances condensed to form a vast ocean covering the surface; marine organisms appeared; sediment formed from physical and chemical action in the ocean; water burst through barriers in the ground that led to underground caverns, leading to a lowering of the water level; volcanoes, earthquakes, and the force of the waters formed geological features; animals emerged; continents separated; and finally humans dominated Earth. Buffon's chronology of the planet's history suggested that Earth was much older than the biblical estimate of 6,000 years. Experiments he performed on cooling rates of globes made of different materials and of different sizes led him to propose 75,000 years as the age of Earth, but sedimentation observations led him to consider an age of around 3 million years. The Catholic Church condemned his books for these statements.

In *Histoire des animaux*, Buffon discussed animal nutrition, development, and reproduction. Biologists did not know how new organisms came to be. Many thought that miniature beings existed in the gametes (the sperm or the egg), a notion termed preformation, and then development consisted of growth and unfolding of parts on the preformed beings. Though preformation was a commonly accepted theory, Buffon rejected it, for one reason, because it did not explain heredity. Also, nobody had yet observed an egg from an animal that gave birth to living offspring (as opposed to an egg that hatches after being laid). His more complex theory of reproduction involved the ingestion of nutritive matter that subsequently took on characteristics of each internal organ, then passed these on to the embryo during development.

Buffon noted that animals did exhibit a wide variety of physiologies that corresponded to their habitats, and that different forms of animals were spread around the world. The species that inhabited different geographic locations were distinct even when the climates of the regions were similar. This observation, which came to be known as Buffon's law, was one of the founding principles of the field of biogeography. Regarding the concept of species, Buffon subscribed to the belief that a species was defined reproductively by the ability of two members successfully to create fertile offspring as a result of mating. More encompassing groups were arbitrary, he claimed, and invented for the convenience of classification. This insight resembled more modern thinking, as biologists over the last 250 years have been gradually reclassifying organisms on the basis of phylogenetic histories rather than seemingly random but conveniently observed traits.

While Buffon did assert that humans are subject to the same scientific analyses and natural processes as other animals, he did believe that they are superior. He reasoned that human beings became superior because they are social animals and that the development of language as necessitated by their social characteristics led to the development of the ability to reason, which set humans apart from other animals. This intellect allowed human beings to dominate all climates and environments, whereas animals were limited in this capacity. Buffon also considered that humans and apes shared a common ancestor.

In addition to *Histoire naturelle*, Buffon presented a variety of topics to the Academy of Sciences, including works on astronomy, mathematics, physics, forestry, physiology, and pyrotechnics. These contributions are preserved in the academy's *Memoirs* (1737–52).

Scientists around the world respected Buffon, and many academic organizations elected him to membership: the French Royal Academy of Sciences, the French Academy, the Royal Society of London, and the academies of Berlin and St. Petersburg.

In 1752 Buffon married Françoise de Saint-Belin-Malain, and they had one son together. Buffon died on April 16, 1788, in Paris. In addition to his historical scientific treatise on the natural world, his bravery in contradicting the popular beliefs regarding the age of the Earth and the immutability of species paved the way for future naturalists to explore scientific truths about the natural world.

See also DARWIN, CHARLES.

FURTHER READING

Roger, Jacques. *Buffon: A Life in Natural History.* Translated by Sarah Lucille Bonnefoi. Ithaca, N.Y.: Cornell University Press, 1997.

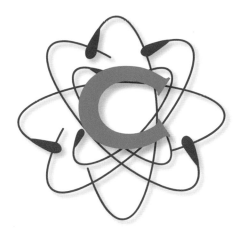

Calvin, Melvin (1911–1997) American *Chemist*
Melvin Calvin was a chemist at the University of California at Berkeley who unraveled the details of photosynthesis, a biochemical process whose importance to life science cannot be overstated. All life-forms require energy in the form of organic molecules such as carbohydrates that they oxidize during cellular respiration. Heterotrophic organisms must take in these organic molecules through the ingestion of other living organisms or organic matter made by other living organisms. Autotrophic organisms, also called primary producers, can synthesize their own organic molecules if provided a source of inorganic carbon, such as carbon dioxide (CO_2), and an alternate source of energy. Many autotrophs are photoautotrophs, meaning they obtain their energy from sunlight. Plants, algae, and certain types of bacteria have specialized cellular structures and biochemical pathways that allow them to use this radiant energy to build carbohydrates. A few types of prokaryotic organisms can use energy stored in reduced inorganic molecules, such as hydrogen sulfide, in order to make organic molecules. Such organisms form the foundation of biological communities found near hydrothermal vents on the deep ocean floor. The overwhelming majority of ecosystems depend on photosynthetic primary producers as the basis of their food chains. For his research on the assimilation of carbon dioxide into organic compounds during photosynthesis, Calvin received the Nobel Prize in chemistry for 1961.

CHILDHOOD AND EDUCATION

Melvin Calvin was born on April 8, 1911, in St. Paul, Minnesota. His father had immigrated to the United States from what is now Lithuania and worked as a cigar maker, an auto mechanic, and then a grocer. His mother was a seamstress who emigrated from the Georgian region of Russia. Melvin; his younger sister, Sandra; and his parents moved to Detroit before Melvin entered high school. He became interested in chemistry after learning about the periodic table of the elements and recognizing the beauty in its organized and explanatory nature. While working in a grocery store as a high school student, he realized the importance of chemistry in everyday life, as it was related to the composition of foods, to the ink on the printed labels of cans, and to everything else that surrounded him.

Calvin enrolled at the Michigan College of Mining and Technology (now Michigan Technological University) in 1927. For financial reasons, he took some time off after his sophomore year and gained practical chemistry experience working as a quality control analyst for a brass factory. After receiving his bachelor of science degree in chemistry in 1931, he studied the electron affinity of halogens for his doctoral thesis under the guidance of George Glockler at the University of Minnesota. After receiving a Ph.D. in chemistry in 1935, he researched the catalytic behavior of coordinated metal compounds as a postdoctoral fellow in the laboratory of Michael Polanyi at the University of Manchester, in England. During this time Calvin became interested in metalloporphyrins, large organic molecules composed of four rings with a metal atom in the center. Calvin examined the activation of molecular hydrogen (H_2) by porphyrins with and without metal present. His later work would involve the oxidation-reduction properties of another well-known porphyrin—chlorophyll, a pigment that absorbs sunlight for use in photosynthesis.

Gilbert Lewis, an American physical chemist whose name is associated with dot structures used to depict chemical bonds, hired Calvin as an instructor at the University of California at Berkeley (UCB) in 1937. Calvin was the first non-UCB graduate the Chemistry Department had hired in 25 years. Within 10 years he became a full professor of chemistry and also served as professor of molecular biology in 1963–80. He was director of the Laboratory of Chemical Biodynamics and associate director of Lawrence Berkeley Laboratory from 1967 to 1980.

DELINEATING PHOTOSYNTHESIS

Calvin's early research at Berkeley focused on hydrogen activation and hydrogenation reactions. He also assisted Lewis in writing a review on the color of organic molecules, a characteristic related to electron excitation. This endeavor helped prepare him for his future work on photosynthesis, as did coauthoring with Gerry Branch *The Theory of Organic Chemistry* (1941), a textbook credited with launching the field of theoretical organic chemistry in the United States. After the war, Ernest O. Lawrence, director of the Radiation Laboratory at Berkeley that is now named in his honor, suggested to Calvin that he find a useful application of carbon 14, as the laboratory had an available supply of this radioisotope after the war ended. Calvin proposed an interdisciplinary research endeavor for a bioorganic chemistry group that would draw on knowledge from biology, physics, and chemistry. Calvin's team used the radioactive carbon 14 isotope to follow the pathway of carbon during photosynthesis, the process by which photosynthetic organisms convert carbon dioxide (CO_2) and water (H_2O) into carbohydrates $(CH_2O)_n$ and oxygen (O_2) using energy obtained from sunlight. Using the unicellular green alga *Chlorella pyrenoidosa*, Calvin and his colleagues exposed the cells to CO_2 that contained radioactive carbon 14, killed the cells, made extracts from the cell contents, and then analyzed the molecules that contained carbon 14 using paper chromatography. After identifying the labeled components, Calvin was able to puzzle together the steps of the pathway through which the carbon traveled. In this simple, yet elegant manner, Calvin delineated the biochemical process of photosynthesis.

$$CO_2 + H_2O \xrightarrow{\text{light}} (CH_2O)_n + O_2$$

Calvin and others published 23 papers and two books (and numerous others later) directly related to the research performed during the period 1946–56 on the path of carbon in photosynthesis. One surprising finding of Calvin's research was that the CO_2 is initially assimilated into a familiar organic compound, phosphoglycerate, rather than being reduced, as scientists assumed. Scientists believed that photoexcited chlorophyll led to the transfer of hydrogen atoms from water to the CO_2, forming formaldehyde, a two-carbon sugar that could polymerize to form larger sugars, releasing oxygen in the process. In reality, the carbon atom from CO_2 combines with something else to form phosphoglycerate, which is subsequently reduced, and the light energy from the Sun is used to regenerate cofactors used in the assimilation, rather than in the initial reduction, as scientists thought occurred. Calvin outlined all the intermediate products that are involved in the regeneration of the assimilation products.

The 1940 discovery of the isotope carbon 14 by Samuel Ruben and Martin Kamen, also at the University of California at Berkeley, provided Calvin the tool he would need to trace the steps leading to the synthesis of sugar from CO_2 during the dark reactions. The following year, Kamen and Ruben showed that the O_2 released during photosynthesis was derived from water, not from CO_2, as scientists assumed.

THE EXPERIMENTAL PROCESS

When Calvin began his studies, biochemists knew a few things regarding photosynthesis. The generation of O_2 was a separate physical and chemical event from the reduction of carbon dioxide. The term *light reactions* referred to the reactions that occurred during the illumination of the chloroplasts, including the production of oxygen gas. Scientists named the reactions that resulted in the reduction of CO_2 the *dark reactions*, though they may occur in the presence or absence of light.

In one of his earliest experiments, Calvin illuminated some plants and stored them in the absence of CO_2 to let the plants synthesize plenty of the high-energy intermediate compounds. When they reintroduced CO_2 to the plants, even in the absence of light, the plants were able to incorporate radiolabeled CO_2. This initial result confirmed that CO_2 incorporation was part of the so-called dark reactions.

Calvin switched to using the green alga *Chlorella*, so they could grow large quantities quickly and reproducibly in a continuous culture. An apparatus called a lollipop because it was shaped like a flat circle held a suspension of the algal cells, and carbon 14–labeled CO_2 was streamed through the suspension during exposure. Lights on each side of the clear lollipop provided illumination. Dropping the suspension into alcohol stopped the enzymatic reactions and killed the cells. After extracting the organic compounds, they used ion exchange chromatography to separate the extract components. Because anion exchange columns (chromatography columns that bind negatively charged ions) bound the radioactive material tightly, they concluded that the initial products were acidic. The strength of the ionic interaction suggested that the early compounds included phosphate esters rather than ordinary carboxylic acids (molecules containing –COO⁻), which would have eluted easily. Additional tests suggested that the compound was phosphoglycerate, a compound already known to have a role in glycolysis, the breakdown of the sugar glucose.

Calvin and various students and collaborators spent the next 10 years trying to identify all the compounds containing carbon 14. A key technique in their procedure was two-dimensional paper chromatography. After extracting the radiolabeled organic compounds, they allowed the sample to diffuse in one direction through the paper filter. Then they turned the paper 90 degrees and allowed the compounds to diffuse in a second direction. Compounds with different sizes and properties diffused at different rates, resulting in a spotted pattern on the filter paper, with each spot representing one specific organic compound. After exposing X-ray film to the filter paper and developing it, they could line up the black spots with areas on the paper that contained the carbon 14–containing compounds. The number of spots, their positions on the filter paper, and their relative radiolabel intensities contained all the information Calvin needed to follow the pathway of carbon during photosynthesis except the identity

of each black spot on the chromatogram. Identifying the compounds turned out to be a complex task. Sometimes fluorescence or ultraviolet absorption of the paper itself was informative; other times, they had to elute the material from the paper, perform chemical manipulations, and run another chromatography experiment to gain information about the compound's identity. They found that within 30 minutes the cells had incorporated the radiolabeled carbon in several carbohydrates, amino acids, and numerous intermediates. The labeled compounds included alanine, aspartic acid, citric acid, glutamic acid, glycine, malic acid, phosphoenolpyruvic acid, phosphoglyceric acid, serine, sucrose, triose phosphate, uridine diphosphoglucose, sugar phosphates, and sugar diphosphates.

By exposing *Chlorella* to radioactive CO_2 for shorter periods, they figured out which compounds were produced early after incorporation and which were produced later on. By briefly increasing the span of time between the CO_2 exposure and dropping the algal suspension into alcohol, they obtained different relative intensities of radioactivity in the compounds, reflecting the sequential reactions in which the carbon 14–labeled compounds participated. Exposure for only a fraction of a second led to one distinctly dominant radiolabeled compound—phosphoglycerate, the first product of photosynthesis. The subsequent appearance of hexose phosphates suggested

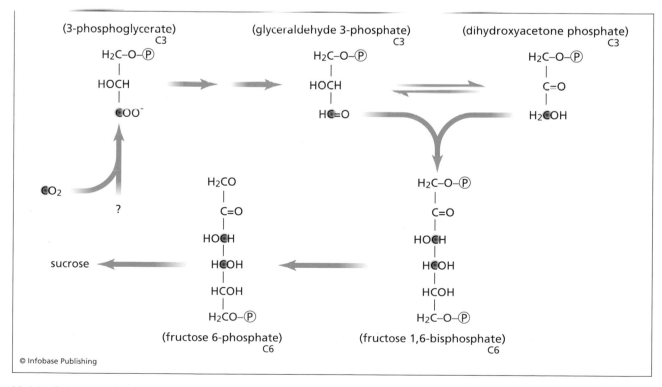

Melvin Calvin and his colleagues systematically determined the steps that occurred between the incorporation of the carbon 14 isotope (represented in red) from carbon dioxide and the formation of a hexose phosphate.

that two phosphoglycerate molecules combined to form a hexose phosphate.

Next they chemically dismantled the radioactive phosphoglycerate to see which one of its three carbon atoms was the radioisotope. By doing the same for the hexose phosphate, they were able to follow the path of the carbon 14 from phosphoglycerate through the stepwise formation of the hexose phosphates and ultimately into sucrose.

The fact that phosphoglycerate contained three carbons suggested that the single carbon atom from CO_2 combined with a two-carbon intermediate; thus their next step was to identify that presumed two-carbon compound. This turned out to be more complicated than expected, as no two-carbon intermediate existed. To build up the levels of the unknown precursor molecule, they restricted the CO_2. While searching for a potential carbon acceptor, rather than increased levels of a two-carbon substance, they found large quantities of five-carbon (pentose) and seven-carbon (heptose) sugars. The relationship of these compounds was not immediately clear. In order to determine the sequential order in which these molecules participated in the photosynthetic carbon pathway, they identified which carbon atoms of the various sugars were radiolabeled and the relative percentage of each that was labeled. Five-carbon sugars were labeled at one end of the molecule, while the six- and seven-carbon sugars had labels at the central carbons. They concluded that the labeled CO_2 combined with the five-carbon sugar ribulose 1,5-bisphosphate to form a six-carbon sugar that subsequently split into two three-carbon phosphoglycerate molecules. From these studies they were also able to determine the pathways that explained the origin of the ribulose 5-phosphate, which could be converted to ribulose 1,5-bisphosphate by the transfer of a phosphate group from an adenosine triphosphate (ATP) molecule. Transferring a two-carbon fragment from the seven-carbon fragment sedoheptulose to glyceraldehyde would result in two five-carbon compounds, both of which can be converted to ribulose 5-phosphate in a single step.

C_7 (sedoheptulose 7-phosphate) + C_3 (glyceraldehyde 3-phosphate) $\rightarrow$ C_5 (ribose 5-phosphate) + C_5 (xylulose 5-phosphate)

They explained the origin of the seven-carbon sedoheptulose by the joining of a three-carbon derivative of glyceraldehyde 3-phosphate with a four-carbon compound.

C_4 (erythrose 4-phosphate) + C_3 (dihydroxyacetone phosphate) $\rightarrow$ C_7 (sedoheptulose 1,7-bisphosphate)

This leads to the next question: where did the four-carbon compound come from? Calvin explained the origin of the four-carbon compound with a reaction that simultaneously explained the seemingly unbalanced radioisotope intensities of the different carbons in the pentose phosphates.

C_6 (fructose 6-phosphate) + C_3 (glyceraldehyde 3-phosphate) $\rightarrow$ C_4 (erythrose 4-phosphate) + C_5 (xylulose 5-phosphate)

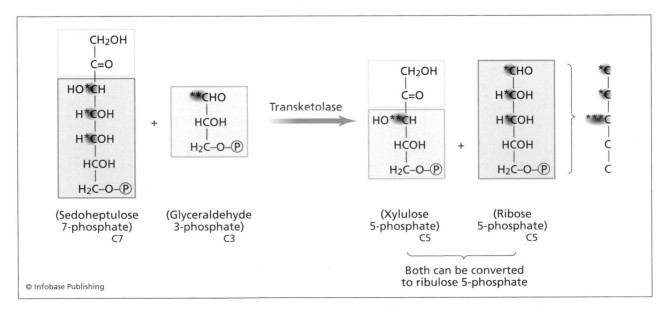

A reaction involving sedoheptulose and glyceraldehyde explained the asymmetric radiolabeling scheme observed in ribulose 1,5-bisphosphate. The red asterisks indicate the relative intensities of the radiolabel.

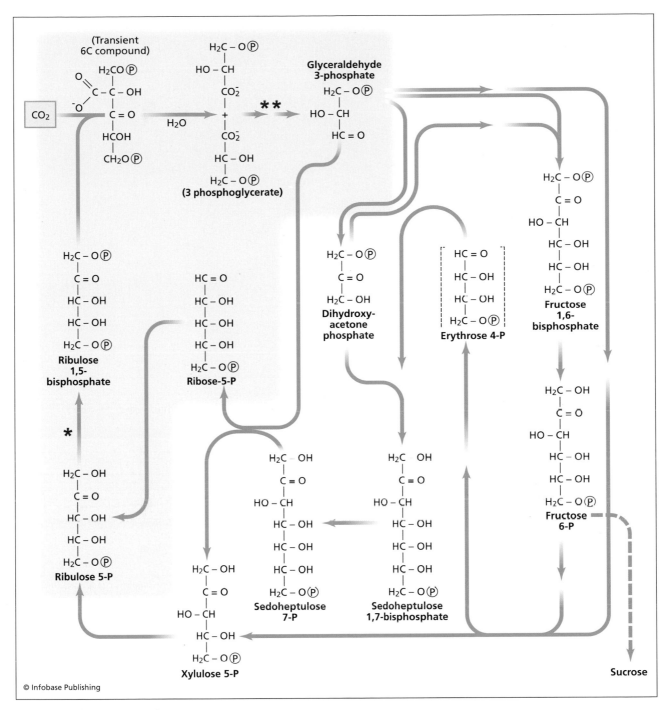

The first product of photosynthesis is 3-phosphoglycerate. The remainder of the cyclical photosynthetic carbon pathway serves to regenerate ribulose 1,5-bisphosphate to allow continued incorporation of carbon from CO_2. The red asterisks indicate the steps in the pathway that require high-energy compounds formed during the light reactions of photosynthesis. The highlighted (yellow) portion of the diagram indicates the reactions that textbooks typically depict when summarizing the Calvin cycle.

After identifying the sources of all the radiolabeled sugars revealed by their chromatography experiments, Calvin was able to puzzle them together in a sequence by following the position of the carbon 14 label. The result was a cyclical pathway driven by the energy ultimately obtained from sunlight. Each com-plete turn of the pathway fixed an additional carbon atom and regenerated the ribulose 1,5-bisphosphate. When they turned off the light source, glyceraldehyde 3-phosphate continued to accumulate, but sucrose did not. From this Calvin concluded that the conversion of glyceraldehyde 3-phosphate into sucrose requires high-energy compounds (ATP [adenosine triphosphate] and NADPH [nicotinamide adenine dinucleotide phosphate]) generated during the light reactions. The CO_2 combined with ribulose 1,5-bisphosphate during the dark reactions. The regeneration of ribulose 1,5-bisphosphate also requires high-energy compounds made during the light reactions.

The Royal Swedish Academy of Sciences awarded Calvin the 1961 Nobel Prize in chemistry for this body of work.

OTHER RESEARCH, AWARDS, AND SERVICE

Calvin contributed his expertise in bioorganic compounds to many other questions related to life science. He performed research on the chemical evolution of life by simulating presumed conditions of the early Earth in order to see what types of organic molecules could form. He also examined ancient rocks for molecular fossils, the presence of organic compounds that might support the existence of life. Applying similar methods, he examined meteorites in search of evidence of extraterrestrial life. He found none. Using knowledge and skills that he developed while studying photosynthesis, he sought plants capable of producing hydrocarbons that could be used as an alternative, renewable energy resource and attempted to create artificial chloroplasts for capturing solar energy.

In addition to the Nobel Prize, Calvin received the Davy Medal of the Royal Society in 1964, the Priestley Medal of the American Chemical Society in 1978, and the National Medal of Science in 1989. He belonged to the National Academy of Sciences, the American Academy of Arts and Sciences, the Royal Society of London, and more. He served as chairman for the Pacific Division of the American Chemical Society (1951), president of the national American Chemical Society (1971), and president of the American Society of Plant Physiology (1963–64). Numerous national panels and committees sought Calvin's expertise, including the Joint Committee on Applied Radioactivity (now the Atomic Energy Agency), the President's Science Advisory Committee for Presidents John F. Kennedy and Lyndon Johnson, and the Energy Research Advisory Board. He consulted for and then joined the board of directors of the Dow Chemical Company in 1963 and served until 1981, when his age forced him to retire. Several universities awarded Calvin honorary doctorate of science degrees. In 1997 Calvin's alma mater, Michigan Technological University, established the Melvin Calvin Nobel Laureate Lecture in his honor. The building that Calvin helped design to replace the old Radiation Research Laboratory at Berkeley was renamed after him when he formally retired in 1980.

In 1942 Calvin married the former Genevieve Jemtegaard, a juvenile probation officer whom he met through a friend. She later assisted Calvin in his research and is a coauthor with him on several scientific papers. The couple had two daughters, Elin and Karole, and one son, Noel. Calvin died of a heart attack on January 8, 1997.

Today the carbon pathway of photosynthesis is referred to as the Calvin cycle. Melvin Calvin's work was revolutionary not only in that it explained the mechanism that supports almost all life on Earth, but also in that it demonstrated how to use radioisotopes to trace carbon through a biochemical pathway. Since carbon forms the backbone of all organic molecules, this pioneering technique led to an explosion in biochemical research and a much greater understanding of cellular metabolism.

See also BIOMOLECULES; CHROMATOGRAPHY; PHOTOSYNTHESIS.

FURTHER READING

Calvin, Melvin. *Following the Trail of Light: A Scientific Odyssey.* Washington, D.C.: American Chemical Society, 1992.

———. "The Path of Carbon in Photosynthesis." *Science* 135, no. 3507 (March 16, 1962): 879–889.

Calvin, Melvin, and Andrew A. Benson. "The Path of Carbon in Photosynthesis." *Science* 107 (1948): 476–480.

The Nobel Foundation. "The Nobel Prize in Chemistry 1961." Available online. URL: http://nobelprize.org/nobel_prizes/chemistry/laureates/1961/index.html. Accessed January 23, 2008.

Seaborg, Glenn T., and Andrew A. Benson. "Melvin Calvin (April 8, 1911–January 8, 1997)." In *Biographical Memoirs: National Academy of Sciences.* Vol. 75. Washington, D.C.: National Academy Press, 1998.

cancer, the biology of Cancer develops when cells lose the ability to control their proliferation. Several checkpoints throughout the cell cycle monitor the cell's size, damage to deoxyribonucleic acid (DNA), and readiness to proceed through mitosis. If the conditions are not met, progression through the cell cycle will be arrested until the unfinished steps are completed. The immune system has mechanisms for recognizing and destroying abnormal cells, such as those that are precancerous, but sometimes precancerous cells escape recognition and continue to grow and divide, producing more abnormal cells. Cancer is considered a genetic disease, meaning mutations in certain genes lead to cancer. Usually several mutations must accumulate before cancer develops.

One characteristic of cancer cells is their lack of differentiation. During development and maturation, cells undergo a process called differentiation, in which they become specialized to perform a particular function dependent on the tissue type. Tissues are composed of conglomerations of cells that perform a common function in the body. The following are four main tissue types:

- connective tissue
- muscle tissue
- nervous tissue
- epithelial tissue

Connective tissue functions to bind other tissues together and to provide support. Some connective tissues are fibrous, such as tendons and ligaments, and others are fluid, such as blood and lymph. Muscular tissue functions to move the body and its parts. Nervous tissue composes the brain and spinal cord and is specialized for receiving stimuli and transmitting neural impulses between the brain and spinal cord and the rest of the body. Epithelial tissue forms a protective cover for the body and lines body cavities. Many specialized epithelial cells also secrete substances, absorb nutrients, and play a role in excretion of waste products. Cancer cells lose their specialized function and no longer contribute to the tissue's function.

To the trained observer, cancer cells look different from normal cells under the microscope. They have larger nuclei that sometimes contain an abnormal number of chromosomes, and their cytoskeleton is disorganized, making the cell shape appear irregular. Normal cells stop growing after completing a limited number of cell cycles, but cancer cells continue to grow and divide indefinitely. This conversion is called transformation, and the transformed cells are termed immortal. Whereas normal cells stop growing once they have spread out and completely covered the bottom surface of a culture dish, cancer cells pile up when grown in culture. In healthy cells, the presence of chemical signals called growth factors either stimulates cell growth or inhibits it, depending on the conditions. Cancer cells do not require the presence of growth factors to undergo cell division; nor do they respond to instructions to stop dividing.

Cells arise from other cells. When a so-called parent cell grows large enough, it undergoes mitosis and cytokinesis, a series of highly regulated events including the duplication and partitioning of a cell's genetic material and cytoplasmic contents, resulting in the production of two identical daughter cells. Those daughter cells each grow and eventually divide into two more daughter cells, and so on. Several complex mechanisms regulate the cycling of a cell through this progression of events. One checkpoint within the cycle ensures that any mutations in the DNA are repaired prior to the synthesis of new DNA. The development of cancer begins with a single cell containing a mutation that interferes with the regulation of cell division. If the mutation is not repaired and the cell escapes destruction by the immune system, the daughter cells will inherit the same mutation, and all the cell descendants will also divide uncontrollably. Without properly functioning control mechanisms, the cell ignores the checkpoints as it moves through the cell cycle, increasing the chance that new mutations will become permanent. New mutations can promote the development of a tumor, or neoplasm, a mass of tissue that has lost its original physiological function and arises from uncontrolled cell division. In order for cancerous cells to continue growing and dividing, particularly once they pile up to form dense masses of cells, they need nourishment. Angiogenesis is the formation of new blood vessels that penetrate into the tumor to supply all the cells with necessary nutrients and oxygen. Benign tumors do not have the ability to invade tissues and remain localized. Additional mutations can give the tumor cells the ability to invade underlying tissues, a condition called malignancy. If malignant tumors invade lymphatic or blood circulation, their cells can metastasize, or travel to distant body parts, and initiate the growth of secondary tumors throughout the body.

CAUSES OF CANCER

Scientists have identified several genes associated with the development of cancer. These genes fall into two main groups: oncogenes are genes that cause cancer, and tumor suppressor genes are genes that stop cancerous growth. Protooncogenes are genes that encode proteins that stimulate progression through the cell cycle and prevent apoptosis, the programmed self-destruction of abnormal cells. Apoptosis is a normal physiological process designed to prevent mutated or damaged cells from dividing. The mutation of protooncogenes can transform them into oncogenes, genes that convert normal cells into cancerous cells. For example, many protooncogenes encode growth factors, chemical signals that normally regulate mitosis and cellular differentiation. Other types of protooncogenes include those encoding proteins that are involved in signal transduction pathways such as enzymes called kinases, membrane receptors that bind hormones or other chemical signals, and transcription factors that bind to DNA and activate genes.

Tumor-suppressor genes encode proteins that normally function to inhibit the cell cycle or to promote apoptosis. Mutations in these genes can result in the loss of this control measure and lead to cancer. The molecule p53 is a tumor-suppressor protein that halts the cell cycle and promotes the repair of damaged DNA. The p53 protein can also stimulate apoptosis. Some protooncogenes interfere with p53 function. If p53 loses its ability to inhibit the cell from progressing through the cell cycle, either because its gene is mutated or because some other malfunctioning protein inactivates it, the cell can become cancerous. This gene is associated with more than half of all cancers.

Some types of cancer run in families. Two different genes associated with breast cancer, *BRCA1* and *BRCA2*, encode tumor-suppressor proteins. Humans have two copies (called alleles) of every gene, one from the father and one from the mother. If only one allele of the breast cancer gene is mutated, the normal protein encoded by the other allele will compensate. Cancer develops when someone has two faulty versions. If a child inherits a mutated form of one of the breast cancer genes from either parent, the risk for cancer increases because only one more mutation must occur. The tumor-suppressor protein encoded by the *RB* gene is linked with retinoblastoma, a cancer that originates in the retina. As with the *BRCA* genes, both copies of the *RB* gene must be mutated for cancer to develop. A third type of cancer for which someone can inherit a predisposition is thyroid cancer. Only one copy of the associated gene, *RET,* needs to be present for someone to have an increased risk for this type of cancer.

Exposure to environmental agents such as chemical mutagens found in tobacco smoke, benzene, asbestos, pesticides, and herbicides also increases the risk of cancer. Mutagens are agents that cause mutations, and carcinogens are agents that induce unregulated cell division. If the right combination of genes is affected, mutagenesis can result in carcinogenesis. Some chemical agents induce mutations by substituting organic chemical groups for hydrogen atoms on DNA. Other chemicals, such as polycyclic hydrocarbons found in tobacco, are not carcinogenic until they are inside the body cells, where natural biochemical processes alter them, making them highly reactive and able to cause breaks or mutations in DNA. Phorbol esters are not mutagenic but cause cancer by activating a biochemical pathway that encourages cell division.

Viruses cause cancer by carrying oncogenes into cells, inhibiting tumor-suppressor genes, or inhibiting the function of tumor-suppressor proteins. Research has demonstrated associations of several viruses with certain types of cancers: hepatitis B and C cause liver cancer; Epstein-Barr virus causes non-Hodgkin's lymphoma and nasopharyngeal cancer; human immunodeficiency virus (HIV) is linked with Kaposi's sarcoma and non-Hodgkin's lymphoma; human papillomavirus can cause cancer of the cervix, vulva, and penis; and human T-cell leukemia virus (HTLV-1) is associated with adult hairy cell leukemia and lymphoma.

Radiation from ultraviolet light (as from the sun), X-rays, radon gas, and radioactive isotopes such as uranium 235 has been shown to cause cancer. Ultraviolet radiation induces the formation of covalent linkages between adjacent nucleotides that, if left unrepaired before the next round of DNA synthesis, can lead to permanent mutations. The other three types of radiation listed are ionizing radiation because they promote the formation of very reactive substances that can directly damage the DNA or other biomolecules.

TYPES AND TREATMENTS

The type of tissue cells from which a tumor originates defines the cancer. Malignant neoplasms that arise from connective tissue, bone, cartilage, or muscle are called sarcomas. Epithelial cells and tissues, such as the skin, mucosal membranes, or glandular tissues, give rise to carcinomas. Leukemia, cancer of the blood, is characterized by a marked increase in the number of white blood cells, and lymphoma is cancer of the lymph tissue. People often refer to carcinomas by the location in the body where the cancer originated. For example, breast cancers are carcinomas derived from epithelial cells of the ducts or the lobes of the breast tissues. Colon cancers are carcinomas that originate in the innermost lining of the wall of the large intestine.

Different types of cancer progress at different rates and respond best to different types of treatment. The size and location of a tumor and whether or not it has spread also help a physician determine the best course of treatment, and sometimes a combination of methods is used. Such methods include surgery, chemotherapy, radiation therapy, hormone therapy, and biologic therapy. The goal of surgery is to remove the tumor. Chemotherapy involves potent pharmaceuticals that aim to destroy the cancer cells while causing minimal negative side effects. To target the cancer cells, the drugs exploit the fact that cancer cells are rapidly growing, synthesizing DNA, and dividing, whereas normal cells are mostly quiescent. A few mechanisms of action for anticancer drugs include binding the enzyme that synthesizes DNA (DNA polymerase), preventing DNA synthesis by mimicking nucleotides, and preventing the formation of microtubules of the spindle apparatus. A disadvantage of chemotherapy is that some healthy body cells also actively grow and divide, and these cells suffer damage from the treatment. Affected cells include the bone marrow, resulting in anemia and a weakened ability to fight infection; the lining of the gastrointestinal tract, resulting in nausea, vomiting, and diarrhea; the hair follicles, resulting in hair loss; and the kidneys, resulting in the buildup of toxins in the blood due to kidney failure. Radiation therapy involves using ionizing radiation to destroy the tumor site with minimal injury to healthy tissues. The ionizing radiation probably kills the cells by breaking chemical linkages in DNA and other biomolecules. Cancers that depend on certain hormones for growth respond to hormone therapy. The goal of this type of treatment is to stop the body from producing the necessary hormone or to prevent it from acting on the cancer cells. Biologic therapy, also called immunotherapy, aims to stimulate the immune system to fight and destroy the cancer cells. Interleukins and interferon are examples of agents used in biologic therapy. The immune system naturally produces these substances, but the doses administered are higher than natural levels.

See also ANIMAL FORM; CELL BIOLOGY; CELLULAR REPRODUCTION; GENE EXPRESSION.

FURTHER READING

American Cancer Society homepage. Available online. URL: http://www.cancer.org. Accessed January 14, 2008.

Hammar, Samuel P. "Cancer (Medicine)." In AccessScience@ McGraw-Hill. Available online. URL: http://www.accessscience.com, DOI 10.1036/1097-8542.105800. Accessed January 14, 2008.

Keinsmith, Lewis J. *Principles of Cancer Biology.* San Francisco: Benjamin Cummings, 2006.

National Cancer Institute homepage. National Institutes of Health. Available online. URL: http://www.cancer.gov. Accessed January 14, 2008.

Panno, Joseph. *Cancer: The Role of Genes, Lifestyle, and Environment.* New York: Facts On File, 2004.

Carson, Rachel (1907–1964) American *Conservation Biologist* Rachel Carson was a marine biologist and writer who launched the environmental movement after writing one of the most influential books of all time, *Silent Spring,* which warned readers about the irreversible damage caused by the indiscriminate use of pesticides. Carson had a talent for making scientific information accessible to general readers, and this book educated and alarmed the public into taking action.

CHILDHOOD AND EDUCATION

Rachel Louise Carson was born on May 27, 1907, in Springdale, Pennsylvania. She grew up on a 65-acre farm, spending time with her older brother and sister listening to the songbirds, strolling through the apple orchard, and exploring nature. Rachel attended the local school but did not have many childhood friends.

She spent most of her free time reading or with her mother, with whom she shared an especially close relationship throughout her life.

Rachel especially enjoyed reading books about nature and animals by writers such as Ernest Thompson Seton and Henry Williamson. In 1918 *St. Nicholas,* a literary magazine that published works of young authors, accepted her story "A Battle in the Clouds," which told of a young pilot's struggle after being shot by German gunfire. Two others were accepted over the next year. Though at the time most children only attended public school until the 10th grade, Rachel's parents agreed that she should continue her schooling, so they enrolled her at Parnassus High School, just across the Allegheny River. After graduation in 1925, Rachel enrolled at the Pennsylvania College for Women (PCW, now Chatham College). Because she enjoyed writing, she decided to major in English.

Rachel's family struggled to pay for her college even though she received some aid in the form of a scholarship. After an initial adjustment to being away from home, she participated in extracurricular activities including basketball, field hockey, the student newspaper, and the literary magazine. During the summertime, Rachel tutored students to earn money. Though she was an English major, in order to fulfill general education requirements, as a sophomore, Rachel enrolled in a biology course that changed her life. She was captivated by learning about nature and the plants and animals that she had admired and that had given her pleasure since childhood. After taking several more biology courses, Rachel changed her major to biology and graduated with high honors in 1929.

Carson spent the summer following graduation as an intern at the Marine Biological Laboratory at Woods Hole, Massachusetts, where she saw the sea for the first time. Examining tissue specimens from ocean organisms under the microscope and studying the nervous system of turtles, she yearned to learn more about marine biology. That fall she entered Johns Hopkins University on a full tuition scholarship to study marine zoology. To help support her family, Carson worked as a laboratory assistant and taught a lower-level zoology course at Johns Hopkins and other courses at the University of Maryland in College Park, all the while working on her master's thesis, titled "The Development of the Pronephros during the Embryonic and Early Larval Life of the Catfish." The pronephros is a temporary kidney that only functions in catfish embryos for 11 days before being replaced by a permanent kidney. After obtaining a master's degree in 1932, she continued teaching at Johns Hopkins and the University of Maryland. She began the doctoral program at Johns Hopkins,

Rachel Carson's book *Silent Spring* warned the public about the dangers of pesticide overuse. *(U.S. Fish and Wildlife Service/NOAA)*

but in 1935 her father died, leaving her financially responsible for her mother.

JOINS THE BUREAU OF FISHERIES/FISH AND WILDLIFE SERVICE

Carson contacted Elmer Higgins, whom she had met at Woods Hole in 1929. He was the head of the division of scientific inquiry at the Bureau of Fisheries in Washington, D.C., and in charge of writing a series of radio scripts for a program called "Romance under the Waters." He invited Carson to help him with this task, and her entertaining and informative scripts exceeded his expectations. In 1936 she took the civil service test required by all government employees, and after she obtained the highest score, the bureau hired her as a junior aquatic biologist. When the radio broadcasts were to be published as a booklet, Carson's boss asked her to compose a general introduction. She did, but her boss felt it was too literary for the government booklet, and he suggested she submit it to The *Atlantic Monthly*. After rewriting the introduction, she eventually submitted her piece, "Undersea," which was accepted for publication. She enjoyed writing very much and upon receiving such positive feedback, she began writing natural history articles for the *Baltimore Sunday Sun Magazine*.

BECOMES BEST-SELLING AUTHOR

Upon reading "Undersea," an editor at the Simon and Schuster publishing company invited Carson to expand her article into an entire book about the ocean and marine life. Carson loved biology and she loved writing. The idea of not having to choose between two exclusive careers thrilled her. Though still working at the bureau, she accepted the proposal and went to work.

Carson was a perfectionist, and that trait made writing *Under the Sea-Wind* a major task. She wanted to instruct her readers about sea life and keep them interested, so she wrote from the animals' perspective. She researched, wrote, and rewrote, carefully choosing every word. The book was divided into three parts: life on the shore, life in the open sea, and life at the sea bottom. Though the book was nonfiction, she wrote it as a story about seabirds and marine animals, naming them (on the basis of their scientific names) and giving them human characteristics. She described their habits and daily activities—a pair of sanderlings named Silverbar and Blackfoot sought sand bugs and crabs in between waves rolling onto the shore, a mackerel named Scomber escaped prey and haul nets, and an eel named Anguilla journeyed out to sea. Her efforts were rewarded when the book was published in 1941, and reviewers praised her ability to make science understandable for the average reader. The timing of the book's release was unfortunate, however, since Japan attacked Pearl Harbor in December 1941, and the United States entered World War II. People were too preoccupied to read a fantastical book about sea life.

Carson became discouraged and felt that writing books was not worth the time it took. She now was supporting her mother and her two nieces since her sister had died, and she was dealing with her own health problems including appendicitis and shingles. In 1940 the Bureau of Fisheries had merged with the U.S. Biological Survey to form the U.S. Fish and Wildlife Service (FWS). At work, Carson contributed to the war effort by writing conservation bulletins to educate people about fish and encouraging them to consume fish as an alternative source of protein in times of scarcity. She also continued submitting brief nature articles to popular magazines. By 1949 Carson was promoted to chief editor for all the FWS publications, and she moved to Silver Spring, Maryland, her home base for the remainder of her life.

Though she remembered the feeling of frustration from the poor sales of her first book, after working on a series of FWS pamphlets, Conservation in Action, about the nation's wildlife refuges, she found herself yearning to write another book, a biography of the ocean for the average reader. She was awarded a Eugene F. Saxton Memorial Fellowship so she could afford to take time off from her job at the FWS to immerse herself fully in her research. She started interviewing oceanographers, reading technical papers, and converting scientific jargon into understandable prose.

Her goal was to depict the wonders of the ocean while melding the fields of marine biology and physical oceanography. She began by summarizing the origin of the planet Earth and its oceans and progressed into the history of the ancient Earth and the emergence of the first life-forms. She profiled the field of marine geology, described the seasonal changes in the sea, and outlined the divisions of life within the different oceanic zones. The book contained a simplified summary of relatively new information regarding tides, waves, and currents that was gathered during the war. The text emphasized the importance of all life-forms in the oceans, from the microscopic protozoa to the larger fish, introducing to readers ecology, the concept of food chains, and the interdependence of organisms within a biological community on one another. She also pointed out the potential economic wealth in the forms of minerals and petroleum found within the sea. While showing why all nature was worth preserving, she was priming her readers for the acceptance of the ideas she would later propose in her masterpiece, *Silent Spring*.

The *Yale Review, Science Digest, Nature,* and the *New Yorker* all published portions of her manuscript before it was released. *The Sea around Us* was published by Oxford University Press in 1951, jumped onto the *New York Times* best-seller list within two weeks, and remained there for a record 86 weeks. (It was number one for 39 weeks.) This instantly successful book received the National Book Award for nonfiction in 1952 and was voted Outstanding Book of the Year in the *New York Times* Christmas Poll. Carson received the John Burroughs Medal for writing a natural history book of outstanding literary quality. PCW and Oberlin College awarded Carson honorary doctorate degrees, and the National Academy of Arts and Letters elected her to membership. *Under the Sea-Wind* was rereleased and hit the best-seller list this time. Royalties from both of her books permitted her to resign from her government job in 1952 so she could concentrate on writing full-time. She bought land in West Southport, Maine, and built a summer cottage with a breathtaking view of the ocean. There she spent summers with her mother and nieces, wading in the waters and looking for marine life.

Having published a book that explored the lives of sea creatures and another on the physical aspects of the oceans, she began toying with the idea of writing a field guide for animals that lived on the Atlantic shores of America. She had applied for and received a fellowship from the Guggenheim Foundation so she could take time off again from FWS but ended up returning a portion of the money, since sales of her other books now allowed her to quit her job. She struggled for a few years on how to arrange the manual and finally decided to organize it by ecosystem. One section was devoted to life on the rocky shores of New England, another to life on the sandy beaches of the mid-Atlantic, and the third to life in the coral reefs and mangroves of the South. Titled *The Edge of the Sea* and published in 1955, her third book was also a best-seller. The National Council of Women of the United States named it Outstanding Book of the Year, and the American Association for University Women gave Carson an Achievement Award.

The following year Carson wrote an article providing suggestions for how to teach children to appreciate nature. The article, "Help Your Child to Wonder," was published in *Woman's Home Companion* (1956). This text was adapted into a book, *The Sense of Wonder,* published in 1965, after Carson's death.

WRITES INFLUENTIAL BOOK *SILENT SPRING*

Though Carson never married, she developed several close relationships with women, from whom she received support, encouragement, and advice. Much of her correspondence with these women is preserved. One friend, Olga Owens Huckins, mailed Carson a copy of a letter to the editor of the *Boston Herald* complaining about the mosquito control program in Massachusetts. She wrote that in her backyard bird sanctuary, she had discovered as many as 14 lifeless songbirds that she believed died from pesticide spraying. The lifeless birds had their claws clutched to their chests and their bills gaping open as if they had died in agony. Carson viewed this dreadful scenario as a call for action.

For more than a decade, companies had been synthesizing a chemical called dichlorodiphenyltrichloroethane, a chlorinated hydrocarbon more commonly known as DDT. The Swiss chemist Paul Hermann Müller was awarded the Nobel Prize in physiology or medicine in 1948 for his discovery of DDT as a contact poison against insects. DDT was found to be effective against many types of annoying insect pests, including flies, lice, gnats, beetles, and mosquitoes. After studies showed it was safe for human use at insecticidal doses, it was used to combat typhus and malaria during World War II by killing the insects that transmitted the diseases. At the time, people said Müller was lucky to have discovered a substance so beneficial to medicine while researching it as an insecticide for moths. Slowly, however, scientific reports began to find its use was not as safe as initially believed. Carson had been following these reports since the 1940s, when she first encountered them while working at the FWS.

The American Association of Economic Entomologists published an article concerning DDT's potentially damaging side effects in 1944. The article pointed out that DDT killed not only pests but also insects that were beneficial to humankind: for example, insects that feed on crop-destroying pests were killed, as were insects that play an important role in plant pollination. Carson dug into the scientific literature and explored the effects on her own, learning that many fertilized bird eggs never

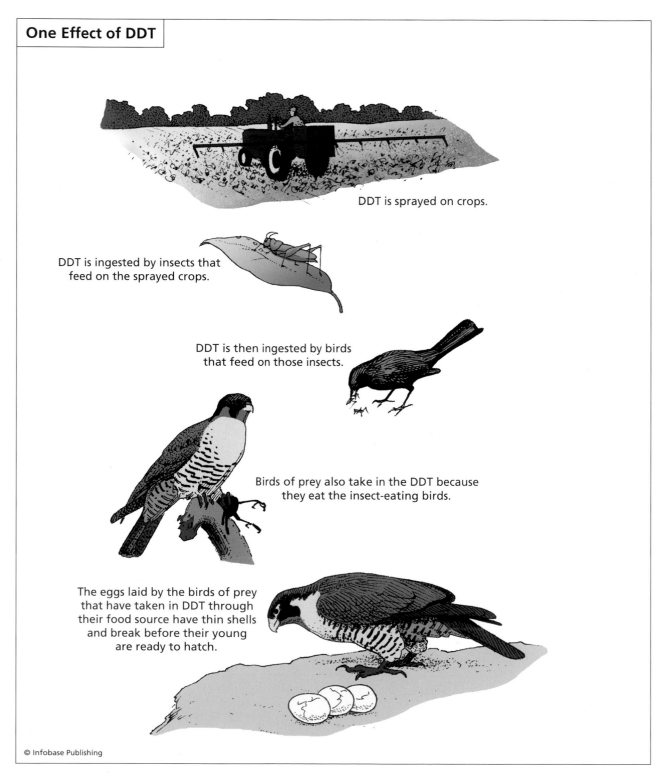

One Effect of DDT

DDT is sprayed on crops.

DDT is ingested by insects that feed on the sprayed crops.

DDT is then ingested by birds that feed on those insects.

Birds of prey also take in the DDT because they eat the insect-eating birds.

The eggs laid by the birds of prey that have taken in DDT through their food source have thin shells and break before their young are ready to hatch.

© Infobase Publishing

As pollutants such as DDT move up through the food chain, their concentration increases.

developed or hatched and others hatched malformed animals. The bird population was steadily declining, and Carson believed this was due to the spraying of DDT. She searched for a magazine willing to publish an article about the dangers of DDT, but despite the fact that Carson was a best-selling author, the magazines were all afraid to publish something so controversial.

Carson decided to write another book, one that pointed out the dangers of this chemical that industries were hailing as the miracle solution to all pests. In a letter to her editor at Houghton Mifflin, she pledged to expose the dangers of supposedly safe pesticides and to provide substantial scientific evidence. This job would require familiarity with the scientific method and scientific literature, and knowledge of cell biology, physiology, ecology, agronomy, organic chemistry, and biochemistry. Though she was recognized as a writer, she was trained as a scientist. Carson convinced her editor that she was up to this task, and she began accumulating evidence.

At first, the government and scientific establishment readily supplied Carson with answers to he requests for information. After her reason for gathering the evidence became known, however, her requests were often blocked or left unanswered. She examined congressional testimony and interviewed countless medical and agricultural experts. Her motivation was fueled in 1957, when residents of Long Island lost a suit against the state of New York to stop the spraying of DDT for gypsy moths.

Her book, *Silent Spring,* opened with an imaginary scenario of a picture-perfect town in middle America. Suddenly, the farm animals all became sick, crops suffered, and birds and chirping insects were no longer heard, all as a result of pesticide and chemical fertilizer overuse. She exposed the truth about the poisonous effects of lingering pesticides and herbicides (chemicals that kill weeds) in the soil and water. Carson did not deny the major benefit of pesticide use in agriculture—namely, the increase in crop yield and therefore an increased food supply. She simply asked, "At what cost?" She wanted the use of most pesticides to be evaluated carefully and controlled accordingly, but she passionately believed that DDT had to be banned altogether. Carson squashed the argument that chemicals lingering in the soil do not directly affect humans by explaining that the dangerous chemicals accumulate in all creatures, beginning with the tiny organisms at the bottom of the food chain, such as plankton and small fish, and working their way up to humans. After entering the food chain, the DDT, which cannot dissolve in water, accumulates in the fatty tissues. Those organisms at the top of the food chain were at the greatest risk, since the organisms they ate contained increasingly concentrated amounts of the stored residues, and as a result, they could develop cancer and have shortened life expectancies. Birds were particularly susceptible; DDT interferes with their calcium levels, leading to weakened egg shells. Humans, as an integral part of the Earth's ecosystem, were not immune. Carson demonstrated DDT toxicity to living cells and said that DDT spraying programs were pointless since the insects developed resistance anyhow. None of this information was brand new, but all the previous reports were hidden in scientific journals and focused on one small aspect of the problem. She made this information accessible and meaningful by presenting a clear, complete overview in plain language.

Not only were the reports Carson uncovered appalling, the years she spent performing her extensive research were difficult personally. Her mother and her niece had both passed away, and at age 50, she adopted her five-year-old grandnephew, who required more attention than Carson's work permitted. In addition, she was diagnosed with breast cancer and was undergoing radiation treatment. In January 1962, she finally submitted the completed manuscript for *Silent Spring* to Houghton Mifflin, complete with 55 pages of references supporting her claims.

Silent Spring reached number one on the bestseller list within two weeks. The release of her critical book infuriated the billion-dollar chemical and agricultural industries, which were at risk of losing money. Unable to retaliate by discounting the overwhelming scientific evidence presented against pesticide use, they resorted to personal attacks, calling Carson a hysterical woman and assaulting her credentials. But the public bought her book, read it, and took her message to heart. They wrote to their elected representatives in Congress and to government agencies in protest. President John F. Kennedy set up a special panel of the President's Science Advisory Committee to evaluate the positive and negative effects of pesticide use. By May 1963, Rachel Carson was vindicated. She testified to Congress, imploring them to develop new policies to protect the environment. The committee recommended eliminating the use of persistent toxic pesticides. As a result of Carson's educating the public and the government on this scientific matter, within one year, more than 40 bills were passed through state legislatures regulating the use of pesticides. Within 10 years, the federal government followed suit. The effects of her discoveries reverberated around the world. In 1972 the use of DDT was banned in the United States.

CARSON'S IMPACT

Rachel Carson succumbed to cancer on April 14, 1964, in Silver Spring, Maryland. She had received numerous awards and recognition prior to her passing, including the Conservationist of the Year Award from the National Wildlife Federation, the National Audubon Society Medal, and the Schweitzer Medal of the Animal Welfare Institute. In 1969 the Department of the Interior changed the name of the Coastal Maine Refuge to the Rachel Carson National Wildlife Refuge. In 1980 she was posthumously awarded the Presidential Medal of Freedom by Jimmy Carter. Carson's greatest reward, however, was knowing her work resulted in the salvation of countless voiceless animals and in turn, of humankind.

Not only did Rachel Carson educate the general public regarding marine science and effect change in the way pesticides were used, but she also pioneered movements in environmental conservation and ecology. Ecology is the study of the relationships between organisms and the environment; both affect one another, and ecologists aim to understand how. One avenue of investigation for modern ecologists is a direct consequence of Carson's speculation on the effect of human activities on the environment. Though her content was scientific, Carson's poetic style of writing appealed to general audiences, making it an effective means to educate the public about technical matters. She called attention to the irony that though the goal of pesticide use was to benefit humans, it actually harmed biological communities, including humans. Though the idea of conservation had been around for almost a century, the public as a whole did not take action to preserve nature until Carson revealed and explained how current practices caused contamination of the food supply with substances that caused cancer and genetic damage and could lead to the extinction of species. She scared the world into caring.

In 1965 friends of Carson founded a nonprofit organization, the Rachel Carson Council, to educate the public about chemical contaminants and alternatives. On April 22, 1970, Americans celebrated the first Earth Day, now recognized annually as a means to protest practices damaging to the Earth and to contribute positively to protecting our environment.

See also ECOLOGY; ENVIRONMENTAL CONCERNS, HUMAN-INDUCED; ENVIRONMENTAL PROTECTION AGENCY; ENVIRONMENTAL SCIENCE.

FURTHER READING

Byrnes, Patricia. *Environmental Pioneers*. Minneapolis: Oliver Press, 1998.

Carson, Rachel. *Silent Spring*. Boston: Houghton Mifflin, 1962.

Lear, Linda. *Rachel Carson: Witness for Nature*. New York: Henry Holt, 1997.

———. "The Life and Legacy of Rachel Carson." Available online. URL: http://www.rachelcarson.org/. Accessed January 14, 2008.

Cech, Thomas (1947–) American *Molecular Biologist* Among molecular biologists, the name Thomas Cech is permanently linked with ribozymes, the term given to ribonucleic acid (RNA) molecules that catalyze biological reactions, a task that scientists previously believed only proteins were capable of performing. His discovery of catalytic RNA earned him the 1989 Nobel Prize in chemistry, shared with Sidney Altman. Their groundbreaking research demonstrated that RNA played a more critical role in cellular processes than previously imagined. Since 2000 Cech has served as president of the Howard Hughes Medical Institute (HHMI), the largest private supporter of biomedical research.

BECOMES A BIOCHEMIST

Born on December 8, 1947, in Chicago, Illinois, Thomas grew up in Iowa City with his father, who was a physician; his mother, who was a homemaker; and his sister and brother. In elementary school he showed an interest in geology and claims to have knocked on the doors of professors at the University of Iowa to ask about meteorites and fossils as a young teenager.

In 1966 he enrolled at Grinnell College, where he met his future wife, Carol. Cech was drawn to physical chemistry and performed undergraduate research at the Argonne National Laboratory and at Lawrence Berkeley Laboratory. After graduation in 1970, Cech entered the University of California at Berkeley as a graduate student in chemistry; there he researched chromosome structure and function in the laboratory of John Hearst. Thomas and Carol moved to Cambridge, Massachusetts, in 1975. He performed postdoctoral research at the Massachusetts Institute of Technology under the direction of Mary Lou Pardue.

NOBEL PRIZE–WINNING RIBOSOMAL RNA STUDIES

The Department of Chemistry and Biochemistry at the University of Colorado (CU) at Boulder hired Cech in 1978. Though as a doctoral student and postdoctoral fellow he studied chromosomal structure and organization in the mouse genome, when he set up his own lab at CU, he switched to using the single-celled, ciliated freshwater protozoan *Tetrahymena thermophila*, planning to look at a single gene. *Tetrahymena* has two nuclei, a micronucleus that functions in reproduction and a macronucleus that contains the expressed genome. Not only is the macronucleus transcriptionally active, but it contains numerous copies of each gene, including about 10,000 copies of the large ribosomal RNA (rRNA) gene. These copies exist extrachromosomally in the nucleoli as small deoxyribonucleic acid (DNA) molecules that resemble minichromosomes in that they have their own telomeres. The gene encodes a large rRNA precursor that is posttranscriptionally processed to become the 17S, 5.8S, and 26S rRNAs. The portion of the gene that encodes the 26S rRNA contains an intron, an intervening noncoding segment that must be removed to create mature, functional 26S rRNA. Having a high copy number and being extrachromosomal made the large rRNA gene an attractive candidate for structural and functional analysis, since these features would facilitate the isolation of the complete gene and all of its associated proteins. One of Cech's first goals was to examine the proteins that regulated the transcription of this gene.

In vitro transcription of isolated nuclei in the presence of α-amanitin, a toxin that inhibits transcription of messenger RNA (mRNA) and transfer RNA (tRNA), yielded several rRNA products. One product of about 9S accumulated posttranscriptionally, and sequencing performed by his student Art Zaug showed that it was the intervening sequence within the 26S rRNA. Excited to have achieved such clean splicing of the intron in the in vitro transcription system, Cech decided to attempt isolation of the splicing enzyme. The notion of introns was still new, having been discovered in 1977 independently by Phillip Sharp at the Massachusetts Institute of Technology and Richard Roberts at Cold Spring Harbor Laboratory, who shared the 1993 Nobel Prize in physiology or medicine for their finding. Cech believed the splicing enzyme must be present in high concentration in the *Tetrahymena* nuclei since so many copies of the gene were transcribed and processed.

To set about isolating the enzyme, Zaug first isolated the unspliced precursor RNA to use as a substrate. They set up two groups of tubes with conditions similar to their in vitro transcription conditions and added substrate RNA to both groups. To the experimental group they added nuclear extracts, and to the other group they did not, planning for that group to serve as the negative control. With no nuclear extract added, they thought no enzyme would be present, and no splicing would occur. Surprisingly, splicing occurred in both sets of tubes.

At the time, scientists believed the sole functions of RNA were to serve as an intermediate for carrying genetic information from the DNA to the protein building machinery and to assist in protein synthesis. Proteins performed the crucial functions of catalyzing biochemical reactions (enzymes) and controlling gene expression by serving as regulatory proteins. So Cech assumed Zaug had made a mistake when setting up the experiment. After numerous repeated attempts using precursor RNA that had been treated in a manner that denatures proteins and destroys their ability

Thomas Cech discovered that RNA had catalytic abilities while he was studying splicing of an intron from an rRNA precursor molecule from the single-celled protozoan *Tetrahymena*. *(University of Colorado)*

to function, such as adding detergent and phenol, digesting with proteinases, and boiling, they obtained the same result. Cech then suspected that the precursor RNA had already been spliced before they set up the in vitro splicing reactions. While examining this possibility, Cech noticed that the spliced segment contained an extra guanosine residue that was not present in the original transcript. Other experiments showed that while the other three nucleotides were not necessary for the splicing to occur, small quantities of guanosine triphosphate (GTP) were necessary to release the intron.

Cech secretly performed an experiment in which he added radiolabeled GTP to unlabeled purified precursor rRNA to see whether it was being added during excision. He did not want his students or colleagues to think he was so naïve as to believe that a covalent bond could form between the GTP and the RNA without a protein enzyme present to catalyze the reaction. The results clearly demonstrated what Cech himself was resistant to accept, that in the absence of protein, the RNA catalyzed its own cleavage and, in the process, added a guanine to the removed segment. Though they had numerous negative results suggesting no protein was involved, in order to obtain a result that would positively demonstrate RNA was responsible, Cech and his lab group used brand new recombinant DNA technology. By making the precursor RNA in vitro from recombinant DNA and using purified RNA polymerase, they eliminated the possibility that a small amount of protein remained tightly bound during RNA extraction and purification from the *Tetrahymena* nuclei. In 1981 and 1982 Cech published several papers that broke a paradigm in life science. RNA could function as an enzyme; they named such RNAs *ribozymes*. The precursor RNA catalyzed the removal of its own

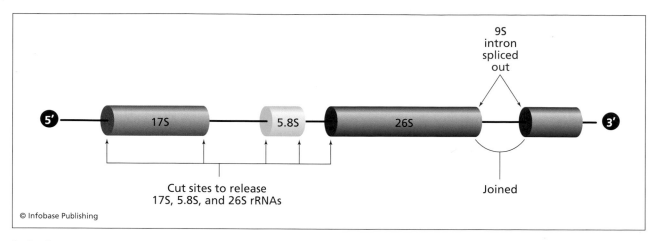

9S
intron
spliced
out

5′ 17S 5.8S 26S 3′

Cut sites to release
17S, 5.8S, and 26S rRNAs

Joined

© Infobase Publishing

A single precursor RNA transcript encodes three rRNAs in *Tetrahymena*.

intron through breaking one bond and reforming another.

Over the next few years Cech and his research group unraveled the mechanism by which RNA catalyzed its own splicing. The first step was the addition of the "extra" guanosine to the phosphorus atom of the 5′ splice site by a transesterification reaction. Then the newly formed 3′ hydroxyl attacked the 3′ splice site in a second transesterification.

Enzymologists argued that self-splicing was different from catalysis because true enzymes did not change during the course of the reaction, but by removing part of itself, the precursor RNA was changed. Cech and Zaug responded by removing some of the precursor RNA's nucleotides and showing that it could then function as a multiple turnover catalyst. Though Cech's results seemed at first unbelievable, other groups soon had similar results. In 1983 Sidney Altman at Yale University published data showing that a stable RNA molecule had all the classic characteristics of an enzyme. The catalytic activity of RNase P, which cleaves precursor transfer RNA molecules, cleaved RNA in the absence of any protein. Cech and Altman shared the Nobel Prize in chemistry in 1989 "for their discovery of catalytic properties of RNA."

HOWARD HUGHES MEDICAL INSTITUTE

In 1988 Cech became an HHMI investigator. HHMI is the largest private supporter of basic biomedical research, providing financial resources for researchers at U.S. universities and other research organizations. With $781 million in disbursements for 2006, HHMI is second only to the federal government in dollars spent on scientific research. In January 2000, Cech became the president of the HHMI. With an endowment near $17.5 billion, HHMI supports six major areas of research: cell biology, genetics, immunology, neuroscience, computational biology, and

structural biology. As president, Cech has addressed ethical issues such as stem cell research, cloning, and the relationship between academic scientists and biopharmaceutical companies. Two of Cech's main priorities at HHMI have been bioinformatics and bioethics. Cech has announced his plans to step down as president of HHMI; Robert Tjian, a molecular biologist from the University of California at Berkeley, will assume the office on April 1, 2009.

Structural biologists in Cech's lab have made significant progress toward determining the structure of the *Tetrahymena* ribozyme in order to gain further insight into the mechanism by which the active site catalyzes splicing. Others in the Cech lab are trying to obtain atomic-resolution structures of group I introns, a common type of intron found in mRNA, rRNA, and tRNA that share the same folding pattern. Another current focus of Cech's laboratory are telomeres, the structures at the ends of chromosomes. Scientists believe that maintenance of these structures is related to cellular senescence and therefore important in aging and cancer research. Telomerase, an RNA-containing enzyme, prevents the deterioration of telomeres caused by shrinking during successive rounds of replication. The Cech lab has cloned the gene for the catalytic center of the telomerase enzyme and found that it shares motifs in common with reverse transcriptase, an enzyme that synthesizes double-stranded DNA from RNA and that was thought to be limited to retroviruses. Studying the telomerase reverse transcriptase (TERT) will help scientists better understand how telomeres replicate themselves.

In addition to his Nobel Prize, Cech has received the National Medal of Science (1995), the Heineken Prize from the Royal Netherlands Academy of Arts and Sciences (1988), the Award in Molecular Biology from the National Academy of Sciences (1987), and the Pfizer Award in Enzyme Chemistry from the American Chemical Society (1985). Most recently,

In 2007 the Chemical Heritage Foundation awarded him the Othmer Medal. The National Academy of Sciences elected Cech to membership in 1987, and the American Academy of Arts and Sciences in 1988. Cech also holds an appointment as a professor of biochemistry, biophysics, and genetics at the University of Colorado Health Sciences Center in Denver.

Cech and his wife, Carol, have two daughters. Allison was born in 1982, and Jennifer was born in 1986.

See also ALTMAN, SIDNEY; BIOMOLECULES; GENE EXPRESSION.

FURTHER READING

Cech, Thomas R. "RNA as an Enzyme." *Scientific American* 255, no. 5 (November 1986): 64–75.

Cech, Thomas R., Arthur J. Zaug, and Paula J. Grabowski. "In Vitro Splicing of the Ribosomal RNA Precursor of Tetrahymena: Involvement of a Guanosine Nucleotide in the Excision of the Intervening Sequence." *Cell* 27, no. 3 (1981): 487–496.

Doudna, Jennifer A., and Thomas R. Cech. "The Chemical Repertoire of Natural Ribozymes." *Nature* 418, no. 6894 (July 11, 2002): 222–228.

Jacobs, Steven A., Elaine R. Podell, and Thomas R. Cech. "Crystal Structure of the Essential N-Terminal Domain of Telomerase Reverse Transcriptase." *Nature Structural and Molecular Biology* 13, no. 3 (March 2006): 218–225.

Kruger, Kelly, Paula J. Grabowski, Arthur J. Zaug, Julie Sands, Daniel E. Gottschling, and Thomas R. Cech. "Self-Splicing RNA: Autoexcision and Autocyclization of the Ribosomal RNA Intervening Sequence in *Tetrahymena.*" *Cell* 31, no. 1 (1982): 147–157.

Lundmark, Cathy. "More Roles for RNA." *Bioscience* 54, no. 12 (December 2004): 1,162.

The Nobel Foundation. "The Nobel Prize in Chemistry 1989." Available online. URL: http://nobelprize.org/nobel_prizes/chemistry/laureates/1989/index.html. Accessed January 23, 2008.

cell biology As the name suggests, cell biology is the subdiscipline of biology that explores life at the cellular level, specifically, the activities, functions, and properties of cells. All life has a cellular basis, and the cell is the smallest fundamental unit of life. Below the cellular level, only lifeless molecules and ions are present; thus an appreciation for cell biology is important to all aspects of life science. In order to answer the question "What is life?" scientists have resorted to reductionism, examining the components of living systems to understand how they work together. Cells differ in size, shape, composition, and structure. While in some cases a single cell makes up an entire organism that carries out all of life's processes, in other cases, a cell is one of billions or trillions that make up a single organism. Cells may be specialized to carry out unique functions within a multicellular being, but all cells have in common a few fundamental properties. Cell biology encompasses all that cells have in common, and the differences between individual types. By studying different types of cells and making comparisons, biologists can better define life and its limitations. Over the last few decades, the convergence of cell biology, biochemistry, genetics, and molecular biology has advanced knowledge in all those fields synergistically.

BRIEF HISTORY OF CELL BIOLOGY

Although some individual cells are visible to the naked eye, the observation of most cells requires the assistance of a microscope to magnify them. Cell size is most often expressed in terms of micrometers; one micrometer (1 µm) equals one-millionth of a meter (10^{-6} m), the equivalent of 3.94×10^{-5} inch. Prokaryotic cells are typically a few micrometers in diameter or length, whereas plant or animal cells average 20–30 µm. Subcellular structures, sometimes called organelles, are measured in nanometers (nm); one nanometer equals one-billionth of a meter (10^{-9} m), or one-thousandth of a micometer. Because of the size of typical cells, the discipline of cell biology lagged behind the invention of the compound microscope.

A microscope is an optical instrument used to magnify an image of an object. Near the end of the 16th century, in Holland, a man named Zacharias Janssen made a living grinding lenses for eyeglasses. He discovered that when he put two lenses together, the effect was much greater than using a single lens alone. Compound microscopes have two or more lenses in succession that enlarge the image of an object, resulting in a greater total magnification. Common compound microscopes today achieve magnifications of approximately 1,000 times, putting most cells in the observable range.

Two 17th-century scientists who paved the way for cell biologists were Robert Hooke and Antoni van Leeuwenhoek. Hooke was an accomplished British scientist who served as the curator of experiments and secretary for the Royal Society of London. In 1665 he published a best-selling book titled *Micrographia,* which contained drawings of specimens and descriptions of objects viewed under a microscope, including insects, bird feathers, sponges, and more. Biologists credit Hooke for the discovery and naming of cells. The boxlike pores or perforations he observed in thinly sliced cork reminded him of the cells or rooms of a monastery. At the time, Hooke did not realize the connection between life and these structures, which he named *cells,* which were actually the remains of cell wall tissue from dead plant cells. Leeuwenhoek was a Dutch shop owner who sold clothes and buttons. He used ground glass lenses to examine carefully the quality of cloth sold in his shop. As was common at the time, he ground his own lenses, and he became remarkably skilled at this process. Though he had no training in science, his interest in it, his ability to create lenses of high quality, and his keen observational skills led to a decades-long communication with the Royal Society about his microscopic observations. He first examined rainwater under his microscope and was shocked to see numerous tiny creatures, which he called animalcules. Leeuwenhoek had discovered microorganisms, and for the remainder of his life he observed different microbiological specimens including peppercorn and hay infusions, scrapings from his teeth, bodily fluids and excrement, and other objects such as hair and insect parts. He kept very detailed descriptions of his observations in the form of correspondence with the Royal Society, and though his prose, grammar, and style were not scientific, the information contained within his letters was invaluable and led to the founding of microbiology. His letters are the first record of living organisms composed of a single cell.

By 1831 the resolving power of microscopes had improved sufficiently for the English botanist Robert Brown to recognize the ubiquity of a structure that he named the nucleus from his microscopic examinations of germ cells of orchids. Leeuwenhoek had observed these round structures but never elaborated on their importance, whereas Brown determined they were crucial to the process of fertilization. In 1838 the German botanist Matthias Schleiden concluded that all plant tissues were composed of cells and that new plants developed from single cells. The following year, the German physiologist Theodor Schwann concluded that all animal tissues were composed of cells. Independently, Schleiden and Schwann formulated the basis of the cell theory, purporting that all organisms are composed of one or more cells and that the cell is the basic structural unit of life. In 1858 the German pathologist Rudolf Virchow added another tenet to cell theory, stating that all cells originate in preexisting cells. The cell theory has both defined and shaped biology to the same degree that the atomic theory has influenced chemistry.

Modern cell theory is founded on the same principles as originally proposed by Schleiden and Schwann and supported by Virchow. In summary, the theory states the following:

- Cells are the fundamental structural and functional unit of life.
- All organisms consist of one or more cells (viruses are not considered cells).
- All cells arise from preexisting cells (with the exception of the first cells formed when life originated).
- All cells are composed of the same basic substances.
- Metabolism is a cellular function.

As the use of dyes and the development of staining techniques improved the level of detail seen when using the microscope, scientists learned more about the structures within the cell.

Meanwhile, in the late 1800s, the French chemist Louis Pasteur tied living organisms to what are now referred to as biochemical processes when he demonstrated that living organisms carried out fermentation, the chemical conversion of sugars to alcohols. Later biochemists performed biochemical reactions outside living organisms using extracts, which facilitated the understanding of cellular metabolism. In the 1920s and 1930s German biochemists, including Gustav Embden, Otto Meyerhof, Otto Warburg, and Hans Krebs, delineated pathways that further explained cell function. Gycolysis and the Krebs cycle explained how cells extracted energy from organic molecules and how cells synthesized many of their building blocks. During the same 50-year period, other scientists were pioneering the field of genetics. The Austrian monk Gregor Mendel established the basic laws of inheritance in the late 1800s. Soon afterward Johann Friedrich Miescher discovered nucleic acid, and then Walther Flemming discovered chromatin, which he noted separated into threadlike strands during cell division. Wilhelm Roux and August Weismann proposed the germ plasm theory of inheritance, suggesting that chromosomes carried the genetic information in sperm and egg cells to the next generation. (Other aspects of the germ plasm theory were incorrect, such as the claim that only germ cells contained all the hereditary information, and that somatic [nongametic] cells only contained certain portions relevant to their specific function.) Today biologists know that all normal nucleated cells contain all the deoxyribonucleic acid (DNA) of an organism, but different somatic cell types express different portions of the genome. Walter Sutton and Theodor Boveri proposed the chromosomal theory of heredity, stating that the behavior of chromosomes during meiotic division explained Mendel's laws of inheritance.

The invention of electron microscopy in 1932 led to many new breakthroughs in cell biology. For the first time, biologists could directly observe details of subcellular organelles and even distinguish shapes of macromolecules. In place of light, as is used in light microscopy, a beam of electrons illuminates the specimen. Since electrons have wavelengths much shorter than wavelengths of photons of visible light, the resolution is much greater, allowing the observer to distinguish between objects 0.1–0.2 nm apart, approximately 1,000-fold greater than the resolution of light microscopes.

By the middle of the 1900s, biologists had a decent appreciation for cells, how energy flowed in and out of them, and how they reproduced. Cell biologists, molecular biologists, and biochemists continued to work out the details of these processes. Oswald Avery, Colin MacLeod, and Maclyn McCarty demonstrated that DNA mediated genetic transformation in bacteria, and then Alfred Hershey and Martha Chase showed that DNA was the carrier of genetic information in bacteriophage. In 1953 James Watson and Francis Crick solved the structure of DNA, a discovery that led to a rapid explosion of knowledge in the molecular life sciences.

SUBDISCIPLINES AND CURRENT RESEARCH IN CELL BIOLOGY

Cell biology encompasses a variety of research and often overlaps with the subdisciplines of molecular biology and biochemistry. Microbiologists, botanists, or zoologists may study cell biology from the perspective of their organismal specialty. One can also approach cell biology from an evolutionary standpoint, or the reverse—describe the origin and evolution of life at the cellular level.

Anatomy at the cellular level is just as important as it is at the organismal level. A biologist needs to explore the biological system as a whole first, then examine the individual components, and ultimately determine how a component contributes to the working system. Beginning at the cellular level, a scientist might observe and describe the function of a liver cell, specifically the fact that the hormone glucagon stimulates the catabolism of glycogen into glucose. How does this happen? One must be familiar with the structural characteristics of the cell and its membrane to answer this question. The hormone never enters the cell, but it causes the release of glucose from the cell into the external environment, where in the body it would find its way to blood circulation. First, glucagon is a protein, and structure and composition of the membrane that surrounds the cell do not permit the movement of proteins across this barrier to the cell's interior. Proteins are too large, and even if the molecule were smaller, glucagon is soluble in aqueous solutions, so it could not make its way through the lipid bilayer of the membrane. Cells have receptors that are embedded in the membranes and extend into the cell's interior, the cytoplasm, and to the exterior. The exterior portion contacts the hormone, which initiates a conformational change in the transmembrane receptor, an event that activates a nearby enzyme to produce another molecule, which stimulates a cascade of events that occur within the cell and ultimately leads to the synthesis of glucose. As with glucagon, glucose cannot simply diffuse through the cell membrane but can be transported across by a carrier protein, another type of structure embedded in the cell membrane. The receptor protein embedded in the membrane functions in communication between the external environment and the interior of the cell, and the structure embedded in the membrane functions in the transport of molecules across the membrane. One can only understand the cell's response to glucagon after knowing how the cell membrane is constructed as well as the structures embedded within it. Thus an understanding of cellular physiology requires a prior appreciation of cell anatomy.

The preceding example relates to one current hot topic in cell biology, the subject of cell signaling and communication between cells. Signaling includes the recognition of a stimulus at the exterior portion of the cell membrane and the transfer of that message to the cytoplasm of the cell. One hallmark of life is the ability to perceive and respond to changes in the external environment, and this often involves the communication of information across a cell membrane. In multicellular organisms, the external environment includes everything that is happening in other parts of the body. An appropriate response often involves the coordination of multiple events. For example, consider two seemingly simple steps in the digestion of a meal by a mammal. As food reaches the animal's stomach, the nervous and endocrine systems begin instructing the stomach and small intestine to secrete digestive enzymes to break down the food. At the same time, the smooth muscles surrounding the stomach and small intestine must alternately contract and relax to move the food through the digestive tract. What would happen if the small intestine secreted digestive enzymes but never underwent peristalsis to move the food through the tract? Or what if the food moved through the digestive tract without ever being catabolized? In either case, the animal would not efficiently meet its nutritional needs from that meal. Cell signaling is necessary for the secretory cells and the muscle cells to work together to achieve a goal.

Cell biologists have also found cell signaling to be important in regulation of the cell cycle. When cells lose the ability to control their own division, they may become cancerous. Cancer research is another major topic of cell biological research. In addition to mechanisms of cell signaling, cancer researchers study other aspects of mitosis and cytokinesis in order to find new ways to treat cancer. Any type of drug that interferes with cell growth, division, DNA replication, or formation and function of the spindle apparatus for moving chromosomes could be a candidate for cancer therapy.

Differentiation and development are also popular research topics among cell biologists. During development of a multicellular organism, cells become differentiated, or specialized, to perform a particular function for the organism. Once a cell completes differentiation, it maintains its general character, meaning it cannot change into another type of cell. Stem cells are cells that have not yet fully differentiated and have the ability to continue dividing indefinitely. In an adult organism, certain types of stem cells are responsible for renewing tissue that has been damaged or that has a limited life span. Embryonic stem cells are unique in that they have the ability to develop into any type of specialized cell. By studying both stem cells and differentiated cells, cell biologists can look for similarities and differences. Embryonic stem cell research will help cell biologists understand the natural processes that occur during development. This knowledge has the potential to cure patients with serious diseases and conditions, such as diabetes, Parkinson's disease, and spinal cord injuries.

METHODS IN CELL BIOLOGICAL RESEARCH

Cell biologists utilize a collection of methods and techniques that allow them to explore the structure and function of cells. A few standard methods used in modern cell biology laboratories are microscopy, cell fractionation, and cell and tissue culture. Many experiments also rely on biochemical and molecular biological methods, such as protein separation and detection techniques, enzyme assays, and recombinant DNA technology, including DNA cloning and mutation analysis.

The utility of microscopy in cell biology research has been invaluable. Structural analysis typically precedes functional studies at any level of biological organization. Cells are too small to observe with the naked eye; different types of microscopy allow biologists to observe various aspects of cellular anatomy. Treatment with dyes or specialized staining techniques may highlight subcellular structures or allow a biologist to examine the intracellular location of certain biomolecules or cellular structure. For example, Gram staining is a technique that distinguishes between types of prokaryotic cells on the basis of the composition of the cellular envelope, a characteristic that is important in identification of microorganisms and that has medical diagnostic significance. Another staining protocol, Masson's trichrome, helps histologists (scientists who study tissues) distinguish cells from surrounding connective tissue. The incorporation of fluorescent labels linked to antibodies designed to recognize and bind specific proteins highlights areas within a cell where the protein of interest is found. For example, if an antibody recognizes histone proteins that compact DNA into chromatin, the chromosomes will fluoresce a certain color, depending on the type of label. If the antibody specifically recognizes receptors on the surface of the endoplasmic reticulum (ER), the membrane of the ER will fluoresce. Specialized methods used to prepare specimens for electron microscopy are designed to preserve or accentuate certain subcellular structures. For example, freeze-fracture replication is a procedure in which small pieces of tissue are rapidly frozen, placed within a vacuum chamber, and then split apart using the edge of a sharp knife. As the fracture plane spreads from the point of contact, bilayered membranes separate, leaving behind impressions and elevations where integral membrane proteins existed. After coating the exposed fractured plane with a heavy metal, one can observe the specimen using an electron microscope and study the topography. Transmission electron microscopy is best for examining internal structures of cells, and scanning electron microscopy is best for examining the surface structure of three-dimensional specimens.

Just as biologists can gain a better understanding of multicellular organisms by studying the structure and function of isolated tissues and cells, cell biologists can learn more about the structure and function of whole cells by studying their subcellular structures in isolation. Cell fractionation is the separation of different organelles from cells. Most often the starting material is a tissue sample, which the researcher homogenizes in a buffered solution by chopping, grinding, subjecting to repeated cycles of freezing and thawing, or exposing to hypotonic conditions. The homogenate may need to be filtered, depending on the type of tissue. Differential centrifugation then separates the different organelles according to their different densities. Generally, centrifugation of the cell homogenate at increasing force causes the nuclei to pellet out of solution first, followed by the mitochondria and lysosomes, and finally microsomes (vesicles formed from the endoplasmic reticulum and the Golgi apparatus). The fractions containing different organelles can then be used for further analysis.

The capability of studying cells in isolation allows cell biologists to examine the effects of certain physical conditions or of certain chemical substances on cell function. Whole cells can be removed from living organisms and grown in vitro in the laboratory under optimized conditions. Certain cell types are more tolerant of this than others and can grow for longer periods. Scientists have discovered a few cell lines that are immortal: in other words, they can grow and multiply in vitro indefinitely. By studying cells from tumors that have acquired this characteristic, scientists have learned much about what controls the normal process of cell division and what goes wrong in transformed cells, cells that have lost the ability to regulate this process. Treatment of cultured cells with carcinogenic chemicals, ultraviolet radiation, or tumor viruses has resulted in their transformation, leading to the creation of new cell lines for use in cellular, biochemical, or medical research.

See also CELL COMMUNICATION; CELL CULTURE; CELLULAR METABOLISM; CELLULAR REPRODUCTION; CENTRIFUGATION; CHROMATOGRAPHY; CHROMOSOMES; DUVE, CHRISTIAN DE; EUKARYOTIC CELLS; HOOKE, ROBERT; LEEUWENHOEK, ANTONI VAN; MICROSCOPY; PROKARYOTIC CELLS; RECOMBINANT DNA TECHNOLOGY; SCHLEIDEN, MATTHIAS; SCHWANN, THEODOR; VIRCHOW, RUDOLF.

FURTHER READING

American Society for Cell Biology (corporate author), Joseph G. Gall and J. Richard McIntosh, eds. *Landmark Papers in Cell Biology: Selected Research Articles Celebrating Forty Years of The American Society for Cell Biology.* Woodbury, N.Y.: Cold Spring Harbor Laboratory Press, 2000.

Bechtel, William. *Discovering Cell Mechanisms: The Creation of Modern Cell Biology.* New York: Cambridge University Press, 2006.

Hine, Robert, ed. *The Facts On File Dictionary of Cell and Molecular Biology.* New York: Facts On File, 2003.

Ling, Gilbert N. *Life at the Cell and Below-Cell Level: A Hidden History of a Fundamental Revolution in Biology.* New York: Pacific Press, 2001.

Rothman, Stephen. *Lessons from the Living Cell: The Limits of Reductionism.* New York: McGraw Hill, 2002.

cell communication One of the hallmarks of life is the ability to respond to the environment. As the basic unit of life, a cell must have a means for transmitting information from the outside of the cell to the inside of the cell. In multicellular organisms, the cells that make up different tissues and organs must communicate with other tissues and organs to coordinate the life processes necessary for growth, survival, and reproduction. Information is carried from the inside of a cell to the outside (and vice versa) through its membrane via molecular signals. Cells secrete chemicals that bind to specific receptors on other cells and initiate a series of intracellular reactions. The process of signal transduction involves pathways that are stimulated by the specific binding of a small molecule called a ligand to a receptor and that result in a variety of cellular responses.

Just as people can communicate by having a face-to-face conversation or over long distances through the use of telephones or e-mail, cellular communication also can occur over a range of distances. Short-distance communication occurs when a cell secretes chemicals such as growth factors into the extracellular fluid, affecting the neighboring cells. This is called paracrine signaling. Transmission of neural impulses across chemical synapses is another example of cell communication over short distances. A neuron releases chemicals called neurotransmitters into a space called a synapse between the neuron and another cell. The neurotransmitter binds specific receptors on the second cell and initiates a response in that cell. The cells of many tissues are connected to their neighbors by gap junctions, where the membranes of the adjoining cells are separated by a narrow gap, across which channels connect the cytoplasms of the two cells. Gap junctions allow solutes such as ions and water-soluble organic molecules in the cytoplasm to diffuse freely between the two cells. Sometimes cells communicate directly by the binding of a molecule exposed at the surface of one cell to a molecule on the surface of an adjacent cell.

Hormones work over greater distances. In animals, hormonal signaling is a function of the endocrine system. Cells of endocrine glands produce and secrete hormones into blood circulation, through which they travel all over the body until they reach their target cells, cells that have receptors that specifically recognize and bind the hormone. Plant hormones can travel through vessels, through cells, and through the air as a gas.

COMMON CELL-SIGNALING MECHANISMS

Many different molecules work together through several different complex mechanisms and pathways to communicate information between cells. Though different pathways use different molecular intermediates, they share several common features. Some common characteristics include the binding of a ligand to a specific receptor on a cell surface, the use of G proteins as molecular switches for biochemical pathways, second messengers to amplify signals and stimulate different intracellular activities, phosphorylation as a mechanism for activating proteins, and biochemical cascades. These factors work together to generate a variety of responses.

Communication between cells involves chemicals called ligands, small signal molecules that specifically bind to receptor molecules on a cell's surface. Ligands are usually water-soluble and therefore cannot diffuse through cell membranes. Though a ligand may contact numerous types of cells throughout the body or in the local extracellular medium, only cells that express the specific receptors will be affected. A single cell may have up to 100,000 receptors on its surface, though one-tenth of that is more common. The number and combination of receptors are dynamic and change as a cell grows or develops. The receptors are mostly glycoproteins that are embedded in the phospholipid bilayer and span the entire width of the membrane so portions of the receptor are in contact with both the extracellular environment and the cell's interior. Ligand binding induces a conformational change in the receptor glycoprotein that is transmitted through the membrane and initiates a specific response inside the cell.

Cell-signaling pathways often employ guanosine triphosphate (GTP) binding proteins (G proteins) as switches to turn pathways on or off. G proteins can bind either GTP or guanosine diphosphate (GDP). When GDP is bound, the G protein is inactive, and when GTP is bound, the G protein is active and can interact with other proteins to alter their activity. The G protein cannot bind other proteins when the GTP has been hydrolyzed to GDP; the switch is turned off. G proteins are also called heterotrimeric G proteins, because they consist of three different subunits. Another class of G proteins, referred to as small monomeric G proteins, also bind guanine nucleotides but function in signaling mediated by a different type of receptor.

Second messengers play a role in relaying the information sent to the cell membrane by a signal molecule to the cell's interior. These small, non-protein, water-soluble molecules or ions can diffuse through the cell rapidly, to carry information throughout the cytoplasm. They amplify the original signal and mount a complex coordinated response that results in a change in cellular activity. Cyclic adenosine monophosphate (cAMP) and calcium ions (Ca^{2+}) are common second messengers.

Proteins often carry the information in a cell-signaling pathway. Usually, the proteins are already present in the cytoplasm of the cell but are inactive. Phosphorylation, the covalent addition of one or more phosphate groups, is a common means of changing the activity of a protein. Protein kinases are enzymes that catalyze the transfer of a phosphate group from an adenosine triphosphate (ATP) molecule to a protein, either stimulating or inhibiting that protein. The kinase activity may be incorporated in the receptor itself, but more commonly kinase exists as a separate enzyme that is either associated with the membrane or floats freely in the cytoplasm. Phosphatases, enzymes that remove phosphate groups, work in conjunction with kinases to regulate the activity of cellular proteins. Phosphorylation affects several types of proteins including different enzymes, channel proteins, ribosomal proteins, regulatory proteins, and receptors.

CELL SURFACE RECEPTORS

Three main classes of membrane receptors include G protein–linked receptors, tyrosine-kinase receptors, and ion-channel receptors. G protein–linked receptors play a variety of roles and are present across kingdoms. They are important in development, disease, and sensation, and many pharmaceuticals act by influencing G protein pathways. As the name suggests, G protein–linked receptors work in conjunction with G proteins. The binding of a signal molecule to a G protein–linked receptor causes a conformational change that activates the receptor, which in turn activates the G protein by exchanging the bound GDP with a GTP. Once activated, the G protein then transfers the signal to an intracellular target, either another nearby enzyme or an ion channel. At the same time, the G protein hydrolyzes the GTP to GDP and an inorganic phosphate and dissociates from the enzyme. The G protein will remain inactive unless another signal is received by the G protein–linked receptor.

Several G protein–linked receptors have been cloned and their structures determined. A common structural motif involves a single long polypeptide chain that loops back and forth to cross the membrane seven times. The amino terminus of the polypeptide extends into the extracellular fluid and the carboxyl terminus protrudes into the cytoplasm, with three loops occurring on each side of the membrane. The last exterior loop usually binds to the signal molecule, and the last interior loop interacts with the G proteins.

Another class of receptors, the tyrosine-kinase receptors, are responsible for mediating cell responses to growth factors, signaling molecules that exert a positive effect on cell growth or differentiation. The cytoplasmic portion of these receptors has an enzymatic activity that transfers a phosphate group from a molecule of ATP to the amino acid tyrosine on a protein. In contrast to the G protein–linked receptors, tyrosine-kinase receptor proteins only span the cell membrane once. When bound by growth factors at the exterior sites, two receptor monomers aggregate to form a dimer. Each monomer then transfers a phosphate group from an ATP to a tyrosine residue on the cytoplasmic side of the other receptor protein. When phosphorylated, the receptor protein is activated to bind certain cytoplasmic proteins, causing conformational changes that result in the initiation of several signal transduction pathways.

Receptors that transport ions through the membrane are called ligand-gated ion channels. Ligand binding can open the channels or close them. The channels are specific for the ions they transport, such as sodium or calcium ions. These sorts of receptors play an important role in the transmission of nervous impulses between cells at chemical synapses.

SIGNAL TRANSDUCTION PATHWAYS

Binding of a ligand to its membrane receptor initiates a complex set of reactions that result in altered activity of the target cell. The ultimate goal or change in the cell's activity could be the expression of a new gene product, alteration in the rate of cell division, secretion of a hormone, or turning on a specific metabolic pathway while turning off another. The final result depends on the type of cell and the signal that reaches it, but before any change in activity can occur, the signal must be converted from the form of a signal molecule to a response by the cell, a process called signal transduction. The signal molecule itself remains on the outside of the cell, and the receptor remains embedded in the membrane, but the information is relayed to

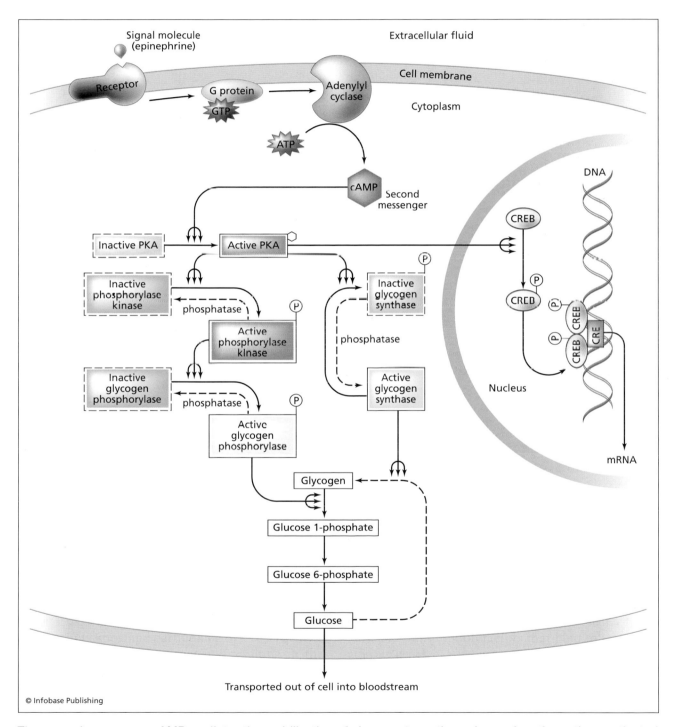

© Infobase Publishing

The second messenger cAMP mediates the mobilization of glucose stores through a series of reactions activated by PKA.

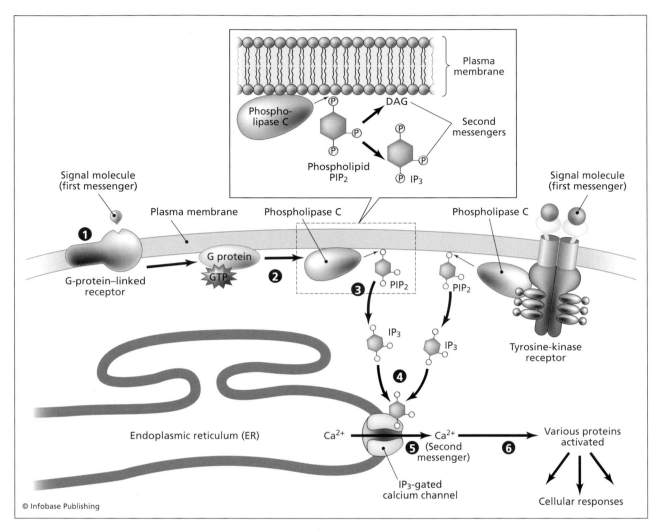

Inositol triphosphate (IP$_3$) and calcium (Ca^{2+}) often function as second messengers in signal transduction pathways.

proteins in the cytoplasm by various intracellular signal transduction pathways.

One example of an intracellular pathway is the cAMP-mediated pathway for glucose mobilization. In times of high stress, the body needs to increase the amount of available glucose in anticipation of muscular activity. The adrenal medulla secretes the hormone epinephrine, and the pancreas secretes glucagon. These hormones bind to their receptors on the surface of liver cells where glycogen is stored. The membrane-associated enzyme adenylyl cyclase converts ATP to cAMP in response to the binding of a signal molecule to its receptor. The cAMP molecules diffuse into the cytoplasm, where they bind a site on the enzyme protein kinase A (PKA). PKA consists of two catalytic subunits and two regulatory subunits. The binding of cAMP to the regulatory subunits causes them to dissociate from the catalytic subunits, activating the enzyme. PKA can then add phosphate groups to serine residues on its

target proteins, phosphorylase kinase and glycogen synthase. The first enzyme, phosphorylase kinase, catalyzes the phosphorylation of another enzyme, glycogen phosphorylase, which breaks down glycogen into glucose 1-phosphate moieties. Another enzyme converts glucose 1-phosphate into glucose 6-phosphate, and yet another converts that into glucose. Glycogen synthase, the other enzyme phosphorylated by PKA, is inhibited by phosphorylation and cannot perform its normal function of converting glucose to glycogen for storage. Because the original hormone binding at the cell surface can activate hundreds of adenylyl cyclase enzymes, leading to the production of numerous molecules of the second messenger cAMP, the signal is amplified. Also, each enzyme can catalyze numerous reactions, further amplifying the original signal.

The signal relayed by the second messenger cAMP extends beyond the cytoplasm. The activated PKA not only phosphorylates cytoplasmic enzymes, but travels to the nucleus, where it phosphorylates a transcription factor called CRE-binding protein (CREB). This transcription factor binds to genes that contain a specific sequence, the cAMP response element (CRE). Binding of CREB turns on the CRE-containing genes, whose products influence processes such as cell proliferation and differentiation. While PKA mediates most of cAMP's effects, cAMP can also directly bind to and open some ion channels, such as in sensory neurons involved in the sense of smell.

Other enzymes called protein phosphatases rapidly reverse the action of protein kinases. Whereas kinases add phosphate groups to proteins, phosphatases remove them. These antagonistic enzymes work together to coordinate activities controlled by phosphorylation.

Another common intracellular signal transduction pathway involves second messengers derived from the membrane phospholipid phosphatidylinositol-4,5-bisphosphate (PIP_2), found on the inner layer of the phospholipid bilayer. Different forms of the enzyme phospholipase C are activated either by G proteins or by tyrosine-kinases after the binding of a signal molecule to its cell surface receptor. Phospholipase C cleaves PIP_2 into diacylglycerol (DAG) and inositol triphosphate (IP_3). Because IP_3 is water-soluble, it quickly diffuses through the cytoplasm and binds to IP_3 receptor channels on the endoplasmic reticulum. The cell normally maintains a low cytoplasmic concentration of Ca^{2+} by actively pumping it out of the cell and into the endoplasmic reticulum (ER), but binding of IP_3 to the channels opens them, releasing Ca^{2+} from the ER into the cytoplasm. Ca^{2+} binds to the protein calmodulin, activating it. The Ca^{2+}-calmodulin complex activates several other proteins, resulting in the stimulation of muscle contraction, activation of transcription factors, opening of ion channels, and regulation of metabolic enzymes. Meanwhile, the DAG remains associated with the membrane and activates the enzyme protein kinase C, which in turn phosphorylates serine and threonine residues on target proteins that are often involved in cell growth and differentiation.

Another pathway initiated by tyrosine-kinase receptors activates the small monomeric G protein Ras, an important regulator of cell growth. As with the heterotrimeric G proteins, Ras can bind either GDP or GTP, but it is only active when GTP is bound. A guanine nucleotide-release protein (GNRP) exchanges a bound GDP with a GTP, triggering a cascade of cellular events. One significant event is the activation of mitogen-activated protein kinases (MAP kinases), a family of kinases that are highly conserved throughout eukaryotic kingdoms and play a role in mating and sporulation in yeast and cell growth and differentiation in multicellular organisms.

Activated MAP kinases enter the nucleus, where they phosphorylate transcription factors of genes involved in cell growth.

The signal transduction pathways described here have been simplified. In reality, signal transduction pathways are not linear progressions from one starting point to a single end result. Completely separate pathways initiated by different signal molecules can converge to activate a shared protein intermediate. Conversely, a single signal molecule can bind its receptor, initiating one pathway, but then diverge or branch out into several different pathways that result in different cellular actions. Even more complex, cross talk sometimes occurs between two different pathways. Cross talk is the web that is formed when information is transferred back and forth between two pathways.

See also BIOLOGICAL MEMBRANES; BIOMOLECULES; CELL BIOLOGY; ENDOCRINE SYSTEM; ENZYMES; EUKARYOTIC CELLS; NERVOUS SYSTEM; PHYSIOLOGY; PROKARYOTIC CELLS; SENSATION.

FURTHER READING

Alberts, Bruce, Alexander Johnson, Julian Lewis, Martin Raff, Keith Roberts, and Peter Walter. *Molecular Biology of the Cell.* 5th ed. New York: Garland Science, 2007.

Brivanlou, Ali H., and James E. Darnell, Jr. "Signal Transduction and the Control of Gene Expression." *Science* 295 (2002): 813–818.

Campbell, Neil A., Jane B. Reece, Lisa A. Urry, Michael L. Cain, Steven A. Wasserman, Peter V. Minorsky, and Robert B. Jackson. *Biology.* 8th ed. San Francisco: Pearson Benjamin Cummings, 2008.

Cooper, Geoffrey M., and Robert E. Hausman. *The Cell: A Molecular Approach.* 4th ed. Sunderland, Mass.: Sinauer Associates, 2007.

King, Michael W. "The Medical Biochemistry Page: Signal Transduction." Available online. URL: http://web.indstate.edu/thcme/mwking/signal-transduction.html#intro. Updated March 22, 2006.

cell culture The ability to maintain living cells in the laboratory is an important aspect of life science research. Though the ultimate goal is to understand how biological units function in natural settings, by being able to control the environment, biologists can study different aspects of organisms, tissues, cells, and molecules. In the case of many microorganisms, a single cell is a whole organism; thus in these cases, a researcher grows and observes populations of the organisms. For multicellular organisms, cell culture usually consists of maintaining a piece of tissue or isolated cells from a specific tissue. Depending on the cell type, the cells may actually divide and multiply in number, or they may simply survive without multiplying when cultured.

In order to grow or maintain cells in vitro, the conditions must generally mimic the in vivo or natural conditions with respect to moisture, salts, pH, and nutrients. The substance that supplies most of these is called the medium, and it may be liquid, solid, or semisolid. The culture is then placed in an incubator that maintains the appropriate temperature and gaseous requirements.

ASEPTIC TECHNIQUE

Controlled experiments on cultured cells demand that only the desired type of cell is present in the culture. Employing aseptic technique ensures that unwanted microorganisms are not accidentally introduced. Aseptic technique is the manipulation of sterile instruments or culture media in a way that maintains sterility. Technicians sterilize the culture medium before it is used, often by autoclaving it. Autoclaves are equipment that use heat and steam to kill any viable microorganisms in culture media or on instruments prior to use. When working with cultures, countertops are often sanitized with bleach, alcohol, or a cleansing solution to reduce the numbers of microorganisms present. Metal instruments such as inoculating loops or forceps can be sterilized by directly flaming them. Other supplies, such as plastic petri dishes, flasks, and pipettes, are often sterilized by treatment with toxic gases or ultraviolet radiation and then wrapped in a manner to preserve the sterility. These types of supplies are often disposable. Briefly flaming the openings of bottles, flasks, and tubes after opening and before closing can reduce contamination as well. During handling, the worker must be careful not to allow nonsterile objects to contact anything that will have contact with the culture itself.

When the preceeding measures are not sufficient, laminar flow hoods provide extra "clean" working conditions. The work area typically consists of a box enclosed on five sides. The sixth side is partly covered, usually with a transparent pane, such as of plexiglass, so the researcher can see inside, but air exhaled during breathing is blocked from entry into the hood. The worker can insert his or her arms through an opening underneath this pane at the front of the hood. The air flow system takes in air through a filter that traps any airborne microorganisms and gently sends it through the hood in a direction leading out the opening in front. The force of the air is just strong enough to prevent any airborne contamination, such as fungal spores or flecks of dust that may be carrying viral particles, from entering through the opening and settling on the work area. Laminar flow hoods also often have an ultraviolet light inside.

Handling and treating cell and tissue cultures within a laminar flow hood reduce contamination. *(James King-Holmes/Photo Researchers, Inc.)*

When nobody is working inside the hood, the light is turned on to kill any microorganisms that might succeed in gaining entry.

BACTERIAL AND FUNGAL CULTURE

Bacteria are cheap and easy to maintain in a laboratory environment. They can be grown either in liquid or on semisolid media containing sugars, amino acids, vitamins, and other necessary nutrients. Broths are liquid media that are used to grow bacteria in test tubes, flasks, or bottles, which are sometimes shaken during growth in order to aerate the media to provide more oxygen. When sterile, most broths are transparent, and as bacterial growth occurs, the medium becomes cloudy. Semisolid media are made by the addition of agar, a gelatinous substance extracted from algae, to broth. When heated, the agar dissolves, and the medium is poured into petri dishes, small circular plates with lids. As it cools, the medium solidifies and bacteria can be spread over the surface to grow. If the bacterial culture is diluted sufficiently before plating, then individual bacterial cells divide and multiply to give rise to single circular-shaped colonies that are usually approximately 0.04 inch (1 mm) in diameter.

To inoculate fresh broth or a fresh agar plate with bacteria, one simply dips a sterilized tool called an inoculating loop (a long metal stick with a wire at one end bent into the shape of a circle with a diameter of about two millimeters) into media containing bacterial growth to pick up a drop of the culture, then dips it into the fresh tube or gently touches the surface of the agar to transfer the bacteria to the new medium. Alternatively, one could remove a specific volume from a broth culture using a sterile volumetric pipette or a micropipette and dispense it into the fresh medium. Whether bacteria are grown in liquid or semisolid media, they are incubated at the temperature optimal for growth of that species. Bacteria that cause diseases in humans usually grow best at 98.6°F (37°C), normal body temperature. When inoculating media with specific bacteria, it is important to maintain a sterile environment so that no undesired contaminating bacteria are accidentally introduced.

Many types of media are specialized. Selective media inhibit the growth of some types of bacteria while allowing for the growth of others. For example, a high salt concentration may be prohibitive to some species but not others. The addition of antibiotics to a medium also makes it selective. Only bacteria containing genes that confer resistance to that antibiotic will be able to grow in its presence. Differential medium allows one to distinguish between different phenotypes after growth. For example, the medium may contain a dye indicator that changes color at a low pH.

Fungi are also cheap and easy to grow in the laboratory. Molds, filamentous fungi, are usually grown on media containing agar as a semisolid support. The media must contain organic compounds such as simple sugars because fungi are heterotrophs. Because fungi multiply by forming spores that can easily become airborne, containing the mold cultures so they do not contaminate other cultures is difficult. Yeasts are unicellular fungi and are often grown in broth but can be grown on petri plates with agar-containing medium as well.

CELL/TISSUE CULTURE

The cells of most plants and animals can be maintained in culture, allowing a researcher to examine them periodically by microscope, to perform biochemical tests on them, or to examine the effect of adding or removing a component of the growth medium. The typical growth medium for mammalian cells contains amino acids, vitamins, salts, glucose, and growth factors. The addition of antibiotics helps to prevent contamination with microorganisms, and a pH indicator dye reveals whether the pH deviates too far from the optimal pH of 7.4. The cells are grown in an atmosphere of 5 to 10 percent carbon

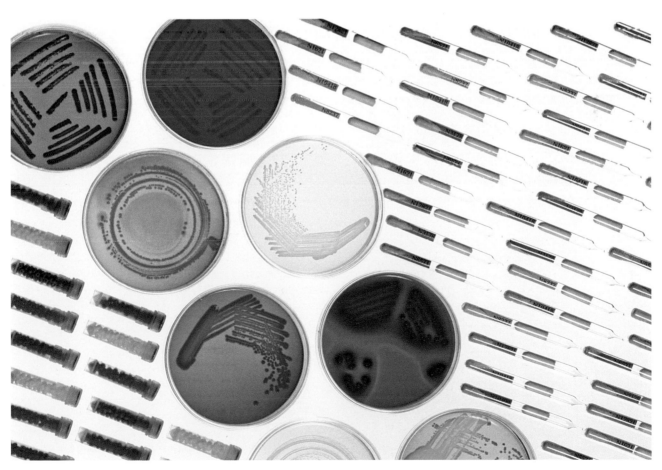

Many types of media are available for growing and storing bacteria. *(Geoff Tompkinson/Photo Researchers, Inc.)*

dioxide, as the medium usually contains a sodium bicarbonate/carbonic acid buffer system. Often, serum is added to supply any undefined necessary growth factors or hormones. When studies involve assigning a specific macromolecule to a particular cell function or the identification of the factors minimally necessary to support growth, then the medium must be serum-free and chemically defined, meaning the investigator knows every component in the medium and its concentration. The availability of chemically defined media has made possible the study of signaling molecules involved in communication between cells.

Cells are usually grown in flasks laid on their sides or in petri dishes so the cells may attach to the bottom surface and spread out in a monolayer as they grow and so the investigator can view them under a microscope without opening the lids. Many tissue culture flasks are coated with extracellular components such as collagen or laminin to help the cells adhere, since a solid support is required for most tissue cells. Transformed cells often lose their anchorage dependence.

Primary cultures are cultures of cells isolated from tissues that have not proliferated in vitro. As the cells proliferate, the technician may subculture them,

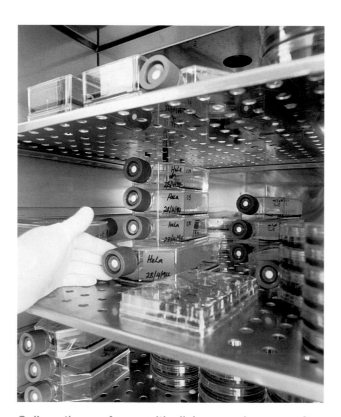

Cells or tissues from multicellular organisms are often grown in flasks that are incubated on their sides so a small quantity of broth covers a larger surface area of growth. *(James King-Holmes/Photo Researchers, Inc.)*

that is, transfer small quantities to new flasks. Most vertebrate cells have a limited number of times they can divide in culture; for example, fibroblast cells undergo 25–40 rounds of cell division before they stop. This phenomenon is called cell senescence, and immortalized cell lines have overcome this limitation. Cancer cells are immortal—they have inactivated the checkpoints of the cell cycle and divide continuously, even in culture. When grown in culture, cancer cells can reach high densities and often pile up and float—they do not require a solid support. One can induce cells to become immortal by the introduction of certain chemicals, oncogenes, or oncogenic viruses, but the transformed cells differ from the normal cells, and this must be taken into consideration when drawing conclusions from data obtained using them. Many cell lines from different animals are commercially available—for example, fibroblast cell lines, epithelial cell lines, kidney cells, ovary cells, and embryonic stem cells. Aliquots of cell lines can be stored in vials kept in liquid nitrogen at –320.8°F (–196°C). One of the most widely used human epithelial cell lines, HeLa cells, originated from a cervical cancer from a woman named Henrietta Lacks in 1951. Though she died eight months later, her contribution has advanced research on viruses, genetics, and cancer more than any other cell line.

Plant tissue culture is often performed as a means to propagate plants. Unlike animal cells, many plant cells are totipotent, meaning they have the ability to develop into a complete organism. Cuttings or sections removed from one plant can be cultured on semisolid medium that contains nutrients and hormones to stimulate growth. For example, auxins stimulate root growth, and cytokinins stimulate shoot formation. As the plant grows, new cuttings can be taken and subcultured in new flasks with fresh medium. Each plant that results from the cutting of another is a clone, a genetically identical copy of the parent plant. This is one method horticulturists use to produce large numbers of desirable organisms or to conserve endangered plant species. In the field of plant breeding, researchers use plant tissue culture to screen for beneficial characteristics, such as resistance to a pathogen.

See also CELLULAR REPRODUCTION.

FURTHER READING

Freshney, R. Ian. *Culture of Animal Cells: A Manual of Basic Technique.* 5th ed. New York: John Wiley & Sons, 2005.

cellular metabolism Organisms must be able to obtain energy from their environment in order to carry out the chemical reactions necessary for life.

Metabolism is the sum of all the chemical reactions that serve to break down and build up molecules used in life processes. Anabolism includes all of the processes that act to construct or synthesize materials needed for cellular functions such as growth, repair, or reproduction. The complementary set of reactions, catabolism, serves to break down components into smaller units. The synthesis or breakdown of chemicals inside the cell occurs via multistep pathways catalyzed by enzymes, specialized proteins that increase the rate of biochemical reactions by lowering the activation energy necessary for the reaction to proceed. Catabolic pathways generally release energy stored in chemical bonds whereas anabolic pathways typically require the input of energy to form new chemical bonds. Many intermediates of metabolism can enter either anabolic or catabolic pathways depending on the needs of the cell at the time. This property, amphibolic metabolism, improves cell efficiency.

CELLULAR RESPIRATION AND ATP SYNTHESIS

Almost all energy in living systems ultimately is derived from the Sun, but most energy in living organisms is stored as chemical energy in the form of organic molecules including carbohydrates, proteins, lipids, and nucleic acids. Organisms capable of photosynthesis can harvest the radiant energy in sunlight and convert it to chemical energy stored in biomolecules. The energy stored in organic compounds can be transferred to other organic compounds through metabolism and to other organisms through the consumption of food. Cellular respiration is the process by which organisms extract the energy from organic compounds to synthesize adenosine triphosphate (ATP), the cell's currency of energy. Once the high-energy compound is made, the cell can use it to provide energy and fuel cellular processes.

ATP is an organic molecule that consists of three components: ribose, a five-carbon sugar; adenine, a nitrogenous base; and a chain consisting of three linked phosphate groups. In order to overcome the natural tendency of the negatively charged phosphate groups to repel each other, the covalent linkages between the phosphates must contain large amounts of energy. When these phosphate bonds are broken, the energy within them is released and can be utilized by the cell for processes such as anabolism, movement, or transport of molecules. During aerobic respiration, the cell produces ATP by two mechanisms, substrate level phosphorylation by the direct transfer of a phosphate group from an organic intermediate to adenosine diphosphate (ADP) and oxidative phosphorylation by the addition of an inorganic phosphate to ADP following electron transport.

GLYCOLYSIS

The six-carbon sugar glucose is the cell's major energy supplier. The first stage of aerobic cellular respiration, glycolysis, occurs in the cytoplasm of cells and involves the breakdown, or oxidation, of glucose into two three-carbon molecules of pyruvate. Eight different enzymes catalyze the eight different reactions in the glycolytic pathway. A ninth enzyme catalyzes the interconversion of two different forms of one intermediate. Though glycolysis produces four ATP molecules by substrate level phosphorylation, two are used during the initial steps, giving a net yield of two ATP molecules for each molecule of glucose that enters the pathway. As glucose is oxidized, some of its electrons and accompanying hydrogen atoms are transferred to the electron carrier nicotinamide adenine dinucleotide (NAD^+), forming the reduced form, $NADH + H^+$, often simply noted as NADH. Each NADH carries a pair of electrons to the electron transport system, where they will play a role in the synthesis of additional ATP by oxidative phosphorylation. During glycolysis, two NADH are formed in addition to a net of two ATP, but most of the energy from the original glucose molecule is now stored in the two molecules of pyruvate. Glycolysis can be summarized as

$$glucose + 2NAD^+ + 2ADP \rightarrow 2 \text{ pyruvic acid} + 2NADH + 2ATP$$

TRICARBOXYLIC ACID CYCLE

The tricarboxylic acid cycle occurs in the cytoplasm of prokaryotic cells but in the mitochondrial matrix of eukaryotic cells; thus in eukaryotes, pyruvate, the end product of glycolysis, must be transported through the mitochondrial membranes to the matrix. Pyruvate cannot directly feed into the next stage of cellular respiration; it must first be decarboxylated, meaning a carbon is removed, in a reaction requiring coenzyme A (CoA) to form the two-carbon molecule acetyl-CoA. During this preliminary reaction, one molecule of carbon dioxide (CO_2) is released, and one molecule of NAD^+ is reduced to NADH.

The acetyl group of acetyl-CoA enters the tricarboxylic acid cycle, the next main phase of cellular respiration. Also called the Krebs cycle after Sir Hans Krebs, an English biochemist who elucidated its steps, or the citric acid cycle after one of its intermediates, this cyclical pathway has the main purpose of continuing oxidation of organic compounds in order to release more electrons for ATP synthesis via oxidative phosphorylation. Many intermediates of the tricarboxylic acid cycle feed into other anabolic and catabolic pathways. In the initial step, the acetyl group of acetyl-CoA joins a four-carbon molecule called oxaloacetate to form the six-carbon molecule citrate, the ionized form of citric acid. In the seven remaining steps to complete one turn of the cycle, three NAD^+ are reduced to NADH, one flavin adenine dinucleotide (FAD, an electron carrier similar to NAD^+) is reduced to $FADH_2$, two CO_2 molecules are released, and one ATP is produced directly by substrate-level phosphorylation. The last reaction in the cycle regenerates the four-carbon oxaloacetate, with which an acetyl group that enters the tricarboxylic acid cycle combines. Because a single molecule of glucose yielded two molecules of pyruvate, the preliminary pyruvate decarboxylation step and the tricarboxylic acid cycle yield a total of eight NADH, two $FADH_2$, two ATP, and six CO_2 for each molecule of glucose that is catabolized. This can be represented as

$$2 \text{ pyruvic acid} + 8NAD^+ + 2FAD^+ + 2ADP \rightarrow$$
$$6CO_2 + 8NADH + 2FADH_2 + 2ATP$$

ELECTRON TRANSPORT SYSTEM AND CHEMIOSMOSIS

At this point in cellular respiration, the catabolism of one molecule of glucose has generated four total molecules of ATP, but most of the remaining available energy is being carried by NADH and $FADH_2$ in the form of electrons. The reduced NADH and $FADH_2$ carry the electrons extracted from the original glucose that entered cellular respiration to the electron transport system, a chain of molecules that participate in a series of oxidation-reduction reactions coupled to the production of ATP in oxidative phosphorylation. Some common molecules that participate as carriers in electron transport are pyridine nucleotides, flavoproteins, quinines, iron-sulfur proteins, and cytochromes. These molecules are embedded in the inner mitochondrial membrane of eukaryotic cells and the cell membrane of prokaryotic cells. NADH and $FADH_2$ hand off the electrons they have been carrying to a member of the electron transport chain. Each molecule in the chain has its own unique position dependent on its relative tendency to donate or accept electrons. After accepting an electron, a carrier in the chain passes the electron to its neighbor, which has a greater tendency to accept electrons, until the electron reaches a final electron acceptor. Molecular oxygen, being very electronegative and having a great affinity for electrons, acts as the final electron acceptor in aerobic respiration. During the final step of the electron transport system, molecular oxygen is reduced to water by the following reaction:

$$2H^+ + \tfrac{1}{2}O_2 \rightarrow H_2O$$

As the electrons pass through this chain to electron transport system members that have progressively higher affinities for electrons, energy released in manageable amounts is used to make ATP by chemiosmosis, the establishment and utilization of a proton gradient across the membrane to generate ATP. $NADH + H^+$ and $FADH_2$ have been carrying not only electrons liberated by the oxidation of glucose but also the accompanying protons (H^+). As the electron carriers transport the electrons down the chain, they simultaneously pump protons from the mitochondrial matrix into the intermembrane space between the inner and outer mitochondrial membranes in eukaryotic cells or exterior to the cell membrane of prokaryotic cells. This creates a chemical gradient as the concentration of hydrogen ions increases on one side relative to the other side of the membrane and also an electrical gradient as the positive and negative charges are separated across the membrane. The phospholipid bilayer of the membrane prevents the immediate dissipation of the gradient, which would result in wasted energy. At certain locations, unique enzyme complexes called adenosine triphosphatases (ATPases) span the membrane and act as doorways, allowing the hydrogen ions to pass through freely. As the gradient dissipates at these sites, the enzyme complexes use the released energy to form a high-energy bond by linking an inorganic phosphate to a nearby ADP, forming ATP.

Because NADH and $FADH_2$ drop off their electrons at different positions along the transport chains, the amount of energy liberated when those electrons are carried down the gradient varies. As a result, approximately three ATP molecules are synthesized for each pair of electrons carried to the chain by NADH, whereas only two molecules of ATP are synthesized for each pair of electrons carried by $FADH_2$. However, in eukaryotic cells, the NADH molecules created during glycolysis must be actively transported into the mitochondria to take their electrons to the electron transport system. Thus, these two NADH generate a net gain of two ATP rather than three. A maximum of 38 ATP can be generated from the oxidation of one molecule of glucose.

FERMENTATION

In the absence of oxygen, cells can still oxidize nutrients by a process called fermentation. Though the electronegativity of oxygen allows for the most efficient oxidation of organic molecules, anaerobic catabolism by fermentation allows the synthesis of ATP to continue. In the initial stages of glucose fermentation, glycolysis breaks down glucose into two molecules of pyruvate. The oxygen requirement of aerobic respiration occurs at the last stage, when molecular oxygen acts as the final acceptor of the electrons extracted from the food. The glycolytic pathway does not require any oxygen but yields a net of two ATP molecules. Glycolysis will continue to break down glucose into pyruvate as long as

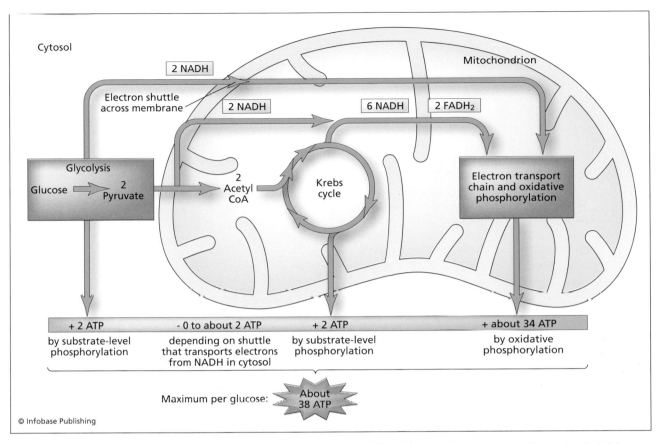

Aerobic respiration yields a maximal output of 38 molecules of ATP, although in eukaryotes the actual yield may be lower due to the "cost" of energy required to transport electrons across the mitochondrial membrane.

the supply of NAD^+ is sufficient. Without NAD^+ available to accept the electrons released during oxidation, glycolysis cannot continue. During fermentation, pyruvate or a derivative of pyruvate acts as the final electron acceptor. Different organisms use different organic molecules, resulting in a variety of end products. Mammalian cells and some other organisms directly transfer the electrons from NADH to pyruvate, forming lactate, the ionized form of lactic acid. Organisms that undergo alcoholic fermentation break down pyruvate to CO_2 and acetaldehyde, which then accepts the electrons from NADH, forming ethanol and regenerating NAD^+, thus allowing glycolysis to continue so that more ATP can be produced. This process, called alcoholic fermentation, is the major step in the process of making beer, wine, and other alcoholic beverages. Fermentation that produces acidic products such as lactic acid or acetic acid is used in the production of certain foods, and fermentation that produces solvents such as acetone or butanol are important in industrial processes.

BIOSYNTHESIS OF MACROMOLECULES

While the function of catabolism is to break down organic nutrients for energy to produce ATP, anab-

olism serves to create macromolecules needed by cells to survive, grow, and reproduce. Cells obtain the building blocks of proteins, carbohydrates, lipids, and nucleic acids—amino acids, monosaccharides, fatty acids, and nucleotides, respectively—by transporting them in directly from the environment, synthesizing them through metabolic pathways, or releasing them from the breakdown of larger macromolecules. Key intermediates such as pyruvic acid and acetyl-CoA play important roles in catabolism and anabolism. As they are central components of metabolism, small modifications can transform them into compounds that feed pathways that accomplish the cell's most immediate needs.

For example, triglycerides are simple fats composed of a glycerol molecule linked to three fatty acids that contain long hydrocarbon chains. The cell can remove the fatty acids from the glycerol and convert it into pyruvic acid, the end product of glycolysis. The pyruvic acid molecule can then move into the mitochondria and feed into the tricarboxylic acid cycle and electron transport system. Enzymes digest the long hydrocarbon chains of the fatty acids by beta-oxidation. During this process, coenzyme A is attached to the end of the chain, then two carbons are broken off at a time, creating numerous molecules of acetyl-CoA that can feed directly into the tricarboxylic acid cycle. One triglyceride molecule with three chains containing 18 carbons each yields approximately 450 molecules of ATP.

Carbohydrates and lipids are the major sources of cellular energy, but proteins can be catabolized for energy in starvation conditions. Their building blocks, amino acids, can be deaminated, meaning the amino group is removed. The product can then enter the tricarboxylic acid cycle at one of the intermediate steps.

The ability of a cell to synthesize complicated molecules from simpler ones depends on its genetic makeup. Some organisms can synthesize almost anything from glucose; others do not have the necessary enzymes and require the availability of certain precursors. Essential nutrients are nutrients that must be obtained in preassembled form because of an organism's inability to make them from raw materials. A nutrient that is essential to one organism is not necessarily essential to another. Of the 20 naturally occurring amino acids used to synthesize proteins, adult humans must obtain eight through their diet but can synthesize the other 12 through metabolic pathways.

REGULATION OF METABOLISM

To conserve energy, the cell employs several methods to regulate its metabolism, the most common being feedback mechanisms. When high levels of an end product of a metabolic pathway are present, that end product inhibits an enzyme that catalyzes an early step in the pathway in a process called feedback inhibition. In this manner, the cell diverts key intermediates into pathways whose end products are in short supply. One method of catabolic regulation involves feedback inhibition by ATP of an enzyme that catalyzes the third reaction of glycolysis. If plenty of ATP is available to satisfy the cell's energy requirements, then glycolysis slows down. As ATP levels begin to drop, that inhibition is lifted and cellular respiration speeds up again until the levels increase sufficiently.

See also BIOCHEMICAL REACTIONS; BIOCHEMISTRY; BIOENERGETICS; BIOLOGICAL MEMBRANES; BIOMOLECULES; CHEMICAL BASIS OF LIFE; ENZYMES; EUKARYOTIC CELLS; NUTRITION; PHOTOSYNTHESIS; PROKARYOTIC CELLS.

FURTHER READING
Harris, David E. "Biological Energy Use, Cellular Processes of." In *Macmillan Encyclopedia of Energy*. Vol. 1. Edited by John Zumerchik. New York: Macmillan Reference USA, 2001.

Kusinitz, Marc. "Metabolism." In *Gale Encyclopedia of Science*. 3rd ed, Vol. 4. Edited by K. Lee Lerner and Brenda Wilmoth Lerner. Detroit: Gale, 2004.

Murray, Robert K., Daryl K. Granner, Peter A. Mayes, and Victor W. Rodwell. *Harper's Illustrated Biochemistry*. 26th ed. New York: McGraw Hill, 2003.

Widmaier, Eric P. *The Stuff of Life: Profiles of the Molecules That Make Us Tick*. New York: Times Books, 2002.

cellular reproduction   The continuity of life depends on the ability of cells to reproduce, or to give rise to new cells. A cell grows, duplicates all its contents and information, and divides into two equal cells termed daughters. Cell division is necessary for replacing damaged or unviable cells, forming multicellular organisms from unicellular zygotes, and increasing the population of unicellular organisms. In multicellular organisms, programmed cell death balances cellular reproduction. This genetically directed process of cell destruction, called apoptosis, occurs as a normal part of development and growth. Cells that are old or injured or that have developed an abnormality self-destruct, and mitosis replaces them with healthy, new cells. While a few cell types are constantly either actively dividing or preparing to divide, most cells spend most of their time engaging in other functions, such as transmitting neural impulses in the case of neurons or synthesizing and secreting hormones in the case of endocrine cells. Cells divide in a highly regulated manner at different rates, depending on the type and the environmental conditions. Cancer results when a cell loses the ability to control its divisions.

THE CELL CYCLE

Cells arise by division of their parent cell (also called mother cell) and give rise to their own daughter cells by cell division. The stages between these divisions compose the cell cycle, which can be divided into two major parts: the mitotic (M) phase and interphase. During mitosis, the nucleus duplicates its contents to form two separate nuclei. In coordination with the nuclear division, cytokinesis partitions the cytoplasmic organelles and material, resulting in the formation of two complete cells that are identical to one another and to the parent cell. Though actual separation of the nuclei and cytoplasm occurs during M phase, the cell performs many preparations for mitosis during interphase.

Most cells spend the majority of their time in interphase, which consists of three stages: G_1, S, and G_2. During G_1, the first gap, the cell grows and carries out the particular functions for which the cell is suited. If the cell is an intestinal epithelial cell, it serves as part of a protective lining, acts as a selectively permeable barrier between the lumen of the intestine and the body's internal tissues, and actively transports materials from the digestive tract into the circulatory system. If the cell forms part of a vegetative fungal hypha, it secretes digestive enzymes into the environment and absorbs nutrients into the fungus. G_1 phase ends in two possible fates: either a cell commits to DNA synthesis and progressing through the cell cycle or the cell exits the cell cycle and enters a quiescent state, called G_0. Most cells in the human body are in G_0. Muscle and nerve cells do not undergo cell division. Other cells, such as liver cells, are typically in G_0 but can reenter the cell cycle under certain conditions. Cells that will ultimately divide continue to experience growth by the production of new proteins and cellular organelles such as mitochondria and endoplasmic reticulum. The cell also accumulates materials necessary for the next phase, the synthesis (S) phase.

As the name implies, synthesis of deoxyribonucleic acid (DNA) occurs during S phase. Chromosomes, located inside the nucleus, contain the DNA. Eukaryotic cells contain linear chromosomes consisting of double-helical DNA wound tightly around proteins called nucleosomes, which package the DNA into manageable units. During G_1 of interphase, the chromosomes are not visible by microscopic examination. The chromosomes exist as extended, dispersed fibers so that the transcription machinery can easily access the genes in order to create templates for the synthesis of new proteins. The uncoiled, extended form also facilitates DNA replication during S phase. A chromatid is a single linear strand of double-helical DNA, and prior to DNA replication, a chromosome consists of a single chromatid. Following replication, chromosomes assume an X shape, due to the physical linkage of the two identical daughter chromatids (called sister chromatids) at a structure called the centromere. Animal cells also duplicate their centrioles during S phase. Centrioles are cellular organelles that play a role in forming the spindle apparatus, which aids in separation of the sister chromatids during mitosis.

The period between the completion of DNA replication and the beginning of mitotic division is termed G_2. During this last stage of interphase, the cell synthesizes materials needed for mitotic division, such as the proteins necessary to build the spindle apparatus.

MITOSIS

Though sometimes the term *mitosis* is used to mean cell division, technically, it refers to nuclear division. Mitosis occurs in four continuous stages: prophase, metaphase, anaphase, and telophase. Cytokinesis, the division of the cell itself, usually accompanies the last stage of mitosis, resulting in two daughter cells, identical to each other and to the parent cell from which they originated. Mitosis makes up a small segment of the cell cycle, taking only 30 or so minutes, compared with the hours, days, weeks, or years that a cell can remain in interphase.

During the first stage of mitosis, prophase, the chromosomes condense up to several thousand-times, forming dense X-shaped structures that one can easily observe on a stained slide using a microscope. Before compaction, the smallest human chromosome extends up to 0.55 inch (1.4 cm) in length, but after condensation it approaches 7.87×10^{-5} inch (2 μm). Packaging the DNA in this manner makes it much easier to move around the cell during mitosis so it can be equally divided in the daughter cells. Also during prophase, the nuclear envelope begins to disappear, the endoplasmic reticulum and Golgi complex fragment, and the microtubules of the cytoskeleton undergo disassembly. Though nonmitotic cells contain one pair of barrel-shaped centrioles (arranged perpendicular to one another as part of a structure called the centrosome), the cell duplicated the centrioles during S phase, so prophase cells have two pairs. Each pair migrates to one end of the cell and nucleates the assembly of spindle fibers, arrays of microtubules that will assist in the movement of chromosomes later in mitosis.

Metaphase can be divided into two subcategories: prometaphase and metaphase. During prometaphase, microtubules attach to the kinetochores, disklike structures on the outer surface of the centromeres of

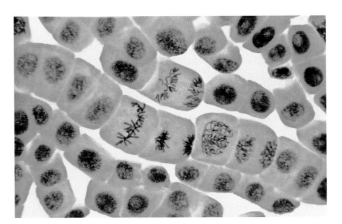

This stained preparation of cells from an onion root tip reveals several stages of mitosis. Most of the cells shown here are in interphase, evidenced by round, densely stained nuclei. The cell in the upper middle region of the photo is in anaphase; a metaphase cell with its chromosomes lined up along the equator appears just left of center; telophase is shown in the centermost cell, having two distinct darkly stained patches of chromosomes; and the cell immediately right of center with chromosomes condensed but not yet lined up at the equator is in prophase. *(M. I. Walker/Photo Researchers, Inc.)*

each chromosome. The chromosomes migrate to the metaphase equator, an imaginary plane that divides the cell into two halves. At metaphase, the centromeres of all the chromosomes are lined up at the equatorial plane. Spindle fibers connect each chromosome to both poles, the opposite ends of the cell containing the centrioles.

During anaphase, the centromeres split and the conjoined sister chromatids separate. Motor proteins fueled by adenosine triphosphate (ATP) propel the chromatids to opposite ends of the cell. The poles also move farther apart.

Telophase occurs when the chromatids reach the opposite poles. If everything has proceeded correctly, the poles should contain equivalent sets of chromosomes. In human cells, each pole should have 46 chromatids, and each chromatid at this stage is an independent chromosome. The chromosomes begin to decondense or disperse, and the nuclear envelope reforms around each set of chromosomes, creating two nuclei. The endoplasmic reticulum and the Golgi apparatus also reform, and cytokinesis, or cell division, occurs. In plants a cell plate is synthesized in the center of the cell, and in animal cells partitioning of the cytoplasm takes place by a narrowing or constriction of the cell membrane around the middle of the cell. The result is two distinct new cells.

REGULATION OF THE CELL CYCLE

Tight control mechanisms regulate the cell's progress through the cell cycle. Geneticists and cell biologists have identified numerous genes involved in cell cycle control over the past decade. Mutations in these genes, referred to as cell division cycle (cdc) genes, lead to unregulated cell proliferation. Many cdc genes encode protein kinases, enzymes that transfer a phosphate group from an ATP molecule to a protein and, in doing so, alter the protein's activity. Cyclins are another key gene product involved in cell cycle control. They are termed cyclins because their concentration increases and decreases as the cell progresses through the cell cycle. Cyclin-dependent kinases, or Cdk proteins, are kinases that are only active when bound to a cyclin protein.

Several feedback mechanisms check to see whether the cell has completed certain events successfully before allowing it to advance to the next stage. Various checkpoints that occur throughout the cell cycle ensure that only cells ready to proceed to the next stage actually do. The most important checkpoint occurs near the end of G_1 and is called the restriction point in mammalian cells. Duplicating the cell's DNA is energetically expensive, and if mutations are present when synthesis begins, they will be perpetuated when the DNA is copied. For these reasons, cells must survey their internal conditions before continuing through the cell cycle. External signals also play a role in controlling cell cycle events through signal transduction pathways. A cell should not enter S phase unless it is committed to completing the cycle and following through until the completion of mitotic division. If the cell has grown sufficiently and the DNA is in good condition, then the concentration of a specific Cdk protein increases, allowing it to combine with its cyclin partner to form a cyclin-Cdk complex that acts as a signal for the cell to proceed into S phase.

Another important protein in making it past the G_1-S checkpoint is p53, the product of a tumor-suppressor gene. The p53 protein binds to DNA and stimulates the production of proteins that inhibit cell growth. Mutations in tumor-suppressor genes are the most common genetic alterations found in cancerous cells. The protein p53 normally functions in apoptosis, a natural process that prevents the proliferation of mutated cells, such as those that are cancerous. While genetic mutations can lead to loss of function of this protein, outside sources can also interfere with p53 function. For example, viral DNA (specifically, from human adenovirus and papillomavirus) can bind to p53 and inactivate it. In some types of cancer, a particular gene (mdm2) is overexpressed, and the gene product binds to and inactivates p53.

A second checkpoint occurs early in G_2, shortly after a cell has completed DNA replication. A Cdk protein binds to a different cyclin to form a complex called maturation-promoting factor (MPF). If a cell has successfully completed DNA replication and repair and is ready to enter mitosis, the concentration of MPF will be high enough to perform its responsibilities such as stimulating the compaction of chromatin, forming the spindle apparatus, and phosphorylating proteins of the nuclear lamina to promote breakdown of the nuclear envelope.

A last checkpoint occurs during metaphase of mitosis. If the spindle apparatus is properly assembled, the chromosomes have all aligned at the equator, and all the chromatids are joined to spindle fibers through their kinetochores, then cell traverses this final checkpoint and proceeds with anaphase. This ensures that chromosomes will not be missing nor be present in duplicate in the daughter cells.

Cancer results when the proliferation of cells goes unchecked. For various reasons, cancer cells do not respond to the mechanisms that regulate progression through the cell cycle. Atypical cells that would normally be eliminated continue to proliferate out of control. Cancer begins when a single cell loses control, or undergoes transformation. If the immune system does not attack and destroy the transformed cell, it will grow and divide numerous times, forming a tumor. If the tumor cells remain at the site of origination, the tumor is called benign and can often be removed by surgery. Problems arise when the tumor becomes malignant and invades neighboring tissues, impairing their function. When this happens, the individual is said to have cancer. Metastasis occurs when cells break off from the original tumor and spread through lymphatic and blood circulation to

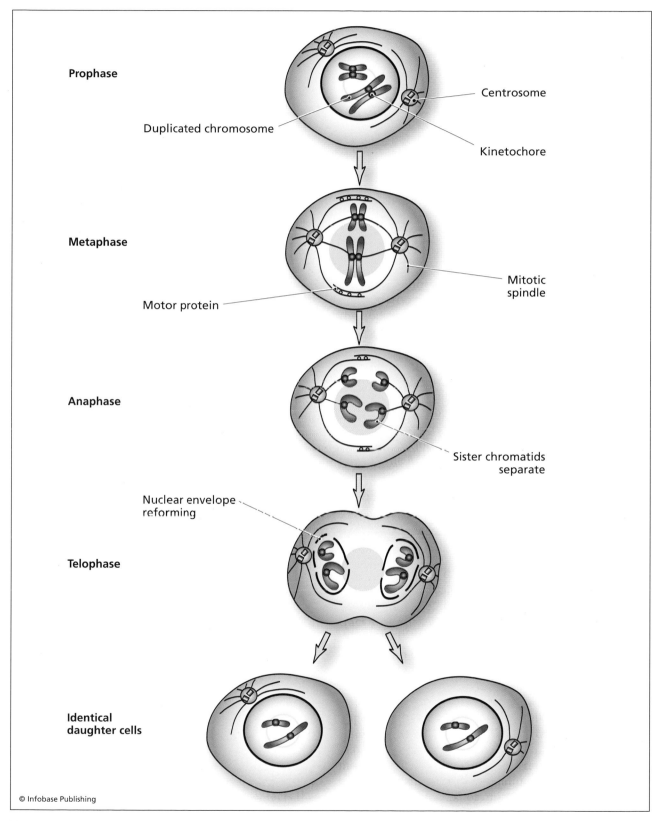

Prophase

Duplicated chromosome

Centrosome

Kinetochore

Metaphase

Motor protein

Mitotic spindle

Anaphase

Sister chromatids separate

Nuclear envelope reforming

Telophase

Identical daughter cells

© Infobase Publishing

Mitosis involves the separation of duplicated chromatids into two newly formed nuclei and, when accompanied by cytokinesis, results in two genetically identical cells.

other parts of the body, where they can develop into new tumors.

The gene that encodes the p53 protein is linked to slightly more than half of all cancers. When p53 function is lost, damaged cells proliferate rather than die. In some families, a mutated form of this gene passes from generation to generation. Exposure to mutagenic chemicals or radiation increases one's risk of developing cancer because they cause mutations in the DNA, potentially to genes that play a role in regulation of the cell cycle.

See also CANCER, THE BIOLOGY OF; CELL BIOLOGY; CHROMOSOMES; DEOXYRIBONUCLEIC ACID (DNA); EMBRYOLOGY AND EARLY ANIMAL DEVELOPMENT; EUKARYOTIC CELLS; MOLECULAR BIOLOGY; REPRODUCTION.

FURTHER READING

Becker, Wayne M., Lewis J. Kleinsmith, and Jeff Hardin. *World of the Cell.* 6th ed. San Francisco: Benjamin Cummings, 2006.

Kleinsmith, Lewis J. *Principles of Cancer Biology.* San Francisco: Benjamin Cummings, 2006.

Klug, William S., and Michael R. Cummings. *Essentials of Genetics.* 5th ed. New York: Prentice Hall, 2005.

Panno, Joseph. *Cancer: The Role of Genes, Lifestyle, and Environment.* New York: Facts On File, 2004.

———. *The Cell: Evolution of the First Organism.* New York: Facts On File, 2004.

centrifugation Alongside microscopy, centrifugation is one of the most important laboratory techniques in life science research. The process of centrifugation allows for the separation of substances of different densities through the use of centrifugal force, the apparent force felt by an object traveling in a curved path that acts to move the object outward from the center of rotation. Centripetal force, the force that acts to pull the object inward (in opposition to centrifugal force), keeps the object moving in a curved path. A centrifuge, the piece of equipment that performs this task, consists of an electric motor that spins a rotor around a fixed axis. The size of centrifuges used for life science research varies with type; some fit on top of a lab bench, while others are the approximate size of a standard washing machine. The rotor holds the tubes that contain the substances to be separated. Different-sized buckets of various rotors hold different-sized tubes. A rotor can have buckets that swing, allowing the tubes to reach a horizontal position while spinning, so that the contents are forced outward along a path parallel to the sides of the tubes. Other rotors have buckets set at fixed angles, so the tubes are maintained at a constant angle relative to the axis upon which the rotor spins.

Acceleration, or the rate of centrifugation, is often reported in multiples of *g*, the acceleration due to gravity at the surface of Earth, equal to 32.174 ft·s^{-2} (9.80665 m·s^{-2}). Acceleration is the product of the radius (from the fixed axis to the end point of the sample tube) and the square of the angular velocity. Reporting centrifugation using *g* allows one to duplicate experimental conditions when using different-sized rotors or centrifuges. Relative centrifugal force (RCF) is a means for reporting the force applied to a sample within a centrifuge.

$$RCF = 0.00001118 \times r \times N^2$$

The variable *r* is the rotational radius measured in centimeters, and *N* is the revolutions per minute.

Using centrifugation, a scientist can separate cells, components of cells, and biomolecules; separate solids from liquids in a mixture; or separate

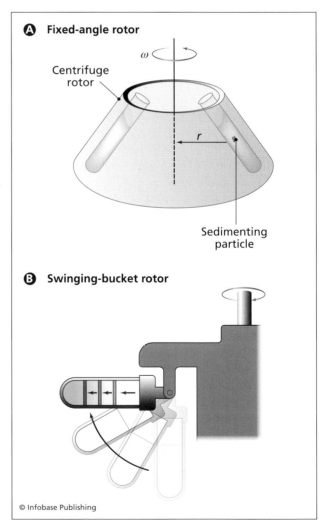

A **Fixed-angle rotor**

ω

Centrifuge rotor

r

Sedimenting particle

B **Swinging-bucket rotor**

© Infobase Publishing

Some centrifuge rotors hold the samples at fixed angles, while others have hinged buckets that swing outward during centrifugation.

liquids of different densities. Microcentrifuges, commonly found in molecular biology or biotechnology laboratories, handle samples with volumes less than 1.5 mL. A much more powerful type of centrifuge, an ultracentrifuge, can achieve very high accelerations and therefore must be operated under a vacuum system to reduce heat due to air friction. Some ultracentrifuges are designed to reach accelerations up to $500,000 \times g$ and therefore can be used to separate biomolecules from one another. The importance of ultracentrifugation technology was demonstrated when the Swedish chemist Theodor Svedberg won the Nobel Prize in chemistry in 1925 for inventing and developing this technique. In his honor, the unit of measurement that describes an object's behavior when centrifuged, the sedimentation rate or sedimentation coefficient, bears his name—the *Svedberg unit*, abbreviated as *S*, and equal to 1×10^{-13} second. In all centrifuges, the samples must be loaded in a balanced manner to prevent a force imbalance at high speeds. If unbalanced, the rotor can damage the spindle, upon which it sits, detach, and cause injury in addition to destroying the samples.

Three types of centrifugation important for life science research are differential centrifugation, density gradient (or rate-zonal) centrifugation, and equilibrium density (or buoyant density) centrifugation. Differential centrifugation separates cellular components on the basis of differences in size, shape, and density. In 1974 two Belgian scientists, Albert Claude and Christian de Duve, and the American scientist George E. Palade shared the Nobel Prize in physiology or medicine for their achievement in pioneering subcellular fractionation by differential centrifugation and their subsequent discoveries related to cell structure and function. To fractionate cellular components, the tissue sample or the cells must first be broken open, or lysed. The investigator can accomplish this by osmotically shocking the cells, using ultrasonic vibrations, or physically grinding them up. Employing gentler procedures helps ensure the organelles or other large structures will remain intact. Because cellular organelles vary so much in size and weight, they will sediment, or travel through the sample tube at different speeds when centrifuged. Particles that are large or dense sediment at a faster rate, and they will have a greater sedimentation coefficient. For example, eukaryotic ribosomes are composed of two subunits, a larger 60S subunit and a smaller 40S subunit. Because sedimentation rate depends on shape and density as well as mass, Svedberg units are not additive—an assembled ribosome has a sedimentation coefficient of 80S, not 100S. After the tissue is homogenized in an appropriate ice cold buffer, subcellular fractions are collected by centrifuging the homogenate, then performing repeated

Centrifugation of whole blood separates the blood cells and platelets from the plasma. *(Klaus Guldbrandsen/Photo Researchers, Inc.)*

cycles of centrifugation of the supernatants, spinning out the particles using greater forces with each cycle. The pellets that form at the bottom of the tubes contain successively lighter or less dense cellular components. The supernatant is the fluid that remains in the centrifuge tube after the particulate matter has formed a pellet. When fractionating cellular components, unbroken cells and nuclei will pellet first; then mitochondria, lysosomes, and peroxisomes, followed by fragments of cell membrane and endoplasmic reticulum; and finally, free ribosomes and large macromolecules.

Density gradient centrifugation, also called rate-zonal centrifugation, is a specialized type of centrifugation in which the investigator layers the mixture containing the particles to be separated over a solution that has a higher concentration of solute at the bottom of the tube and a lower concentration at the top. Particles will move through the gradient of solute as discrete bands traveling at different rates depending on their density. As they move through the gradient, the particles encounter higher solute concentrations, and therefore higher densities. Larger particles with greater sedimentation coefficients will move faster through the gradient. The increasing solute concentration keeps particles that are similar in shape and size in a tight band as they travel through the gradient, and at the end of the spin, particles of different sizes will have migrated to different positions on the gradient.

Equilibrium gradient centrifugation, or buoyant density gradient centrifugation, also uses a gradient over which the solution containing the particles to be separated is layered. Though not necessary in rate-zonal centrifugation, in equilibrium gradient centrifugation, the density of the solution at the bottom of the tube must exceed the density of any of the

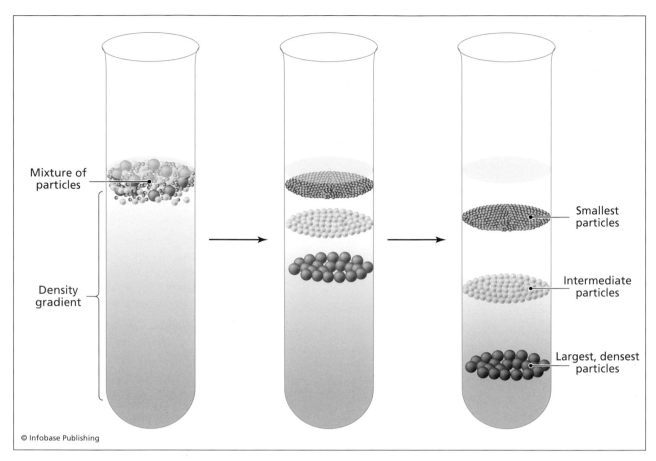

In density gradient centrifugation, particles travel through a gradient in discrete sedimenting bands.

particles to be separated. In other words, the range of densities in the prepared gradient must span the densities of all the components being separated. Sucrose usually serves as the solute for organelle isolation, whereas cesium chloride works well for biomolecules of different molecular weights or structures. After the mixture is layered on top of the gradient, the samples are centrifuged. The particles travel in discrete bands based on density, just as in rate-zonal centrifugation; however, they stop migrating when they reach a position in the gradient at which the density of the solute equals the density of the particles. The particles stop moving at this point because no force acts on the particles when they are surrounded by a solution of equal density. In rate-zonal centrifugation, the particles will all eventually reach the bottom of the tube if centrifuged for a sufficient period, but in equilibrium gradient centrifugation, after they settle to a characteristic position in the gradient, they cease moving.

After both types of density gradient centrifugation, the fractions are collected by puncturing the bottom of the tube with a sharp object and collecting the material that drips out into a series of tubes. After collection, the fractions can be subjected to further separation techniques, examined using microscopy,

or analyzed for various biochemical or biological properties.

See also CELL BIOLOGY; CHROMATOGRAPHY; DUVE, CHRISTIAN DE; EUKARYOTIC CELLS; MICROSCOPY.

FURTHER READING
"Centrifuge." In *Encyclopaedia Britannica.* 2008. From Encyclopaedia Britannica Online. Available online. URL: http://search.eb.com/eb/article-9022102. Accessed January 14, 2008.

Chargaff, Erwin (1905–2002) Austrian-American *Biochemist* Erwin Chargaff is best known for formulating what is known as Chargaff's rules, statements describing the nucleotide composition of deoxyribonucleic acid (DNA). These quantitative relationships were instrumental in the discovery of the double-helical structure of DNA by James D. Watson and Francis Crick.

Erwin Chargaff was born in Czernowiz, Austria, on August 11, 1905. His family was forced to move to Vienna at the outbreak of World War I. He attended the University of Vienna as an undergraduate in chemistry and obtained his doctorate in chemistry in 1928 from Spath's Institute of the University of Vienna, with a thesis on organic silver complexes. During the seven years that followed, Chargaff held several positions including research fellow at Yale University, where he studied lipids of *Mycobacterium tuberculosis;* assistant in the Public Health Department at the University of Berlin; and research associate at the Pasteur Institute in Paris. With 30 published papers to his name, in 1935, he returned to United States to join the faculty at Columbia University, where he remained until his retirement. He started as a research associate in biochemistry (1935), then became an assistant professor (1938), associate professor (1946), and full professor (1952), eventually serving as department chair (1970) and achieving emeritus status in 1974.

When Chargaff settled in at Columbia, chromosomes were known to carry genes and to consist of protein and nucleic acid. Because proteins consist of 20 different amino acids that can be combined in a practically unlimited number of combinations, scientists assumed that proteins were the component of chromosomes that contained the information necessary to direct the expression of an immense amount of variable characteristics. In 1944, after Avery and colleagues published their surprising results showing that DNA encoded for the presence of a polysaccharide capsule in pneumococcal bacteria and that this heritable trait could be transferred from one strain to another via chemically purified DNA, Chargaff's research at Columbia focused on the composition of nucleic acids. Chemists knew that DNA consisted of four different nucleotides. Each nucleotide contained a deoxyribose sugar, a phosphate group, and one of four different nitrogenous bases—adenine (A), guanine (G), cytosine (C), or thymine (T)—but they erroneously believed that the structure of DNA was a nonspecific aggregate of these four subunits, a belief called the tetranucleotide hypothesis. Chargaff isolated DNA from cells of many different types of organisms. He liberated the nitrogenous bases from the nucleotides, then separated them by paper chromatography. Because different nucleotides absorb ultraviolet light of different wavelengths, he was able to determine the quantities of the different nucleotides by measuring how much light each base absorbed. After collecting data from a variety of species including corn, chicken, octopus, rat, and human, he found that, across kingdoms, the amount of A nearly always equaled the amount of T, and the amount of C equaled the amount of G. However, between species, the relative proportions of the two pairs differed. For example, in human DNA, A = 30.9 percent and T = 29.4 percent, whereas G = 19.9 percent and C = 19.8 percent. This information convinced Chargaff that DNA provided sufficient variability to encode genetic information. The regularities in base composition came to be called "Chargaff's rules."

Despite the importance of these results and their implications for DNA structure and self-replication, Chargaff was careful not to overinterpret his data. The results were of key importance in solving the structure of DNA. James D. Watson and Francis Crick interpreted Chargaff's rules to mean that A paired with T by the formation of hydrogen bonds that linked two strands of DNA together, the basis for complementarity between the two strands. Likewise, C specifically formed hydrogen bonds with G. Because A always exists as a member of an A-T base pair in the double helix, the amount of A always equals the amount of T, and because C and G always exist together as a base pair, the amount of C equals the amount of G. When Watson and Crick revealed their stunningly simple double-helical model of DNA to the world, Chargaff was upset that his contributions were not acknowledged. The fact remains, however, that Watson and Crick are the ones who successfully solved the structure, even if they reached their conclusion by incorporating the work of others: the X-ray diffraction data from Rosalind Franklin and Maurice Wilkins, the nucleotide composition observations of Chargaff, the model-building approach of Linus Pauling, the suggestion of their colleague Jerry Donohue that the bases appeared in their keto rather than enol forms, and so on. Chargaff did not recognize the biological significance of his crucial findings concerning the equivalence of the purines and pyrimidines; thus his name is not as well known as those of Watson and Crick.

After the structure of DNA was solved, Chargaff contributed to the understanding of how it encoded for protein construction. Through the years, Chargaff's research also delved into other aspects of biochemistry. He studied blood clotting, lipids and lipoproteins, the metabolism of amino acids and inositol, ribonucleic acid, and phosphotransferases.

Chargaff became an American citizen in 1940. He spoke 15 languages and was respected for his tremendous intellect. He assumed emeritus status at Columbia University in 1974, though he continued to write profusely on a variety of scientific and philosophical topics. Chargaff received many honors and awards during his lifetime including the Pasteur Medal in 1949, the Charles Leopold Mayer Prize from the French Academy of Sciences in 1963, and the Distinguished Service Award from Columbia University in 1982. Erwin Chargaff married Vera Broido in 1928, and they had one son named Thomas. Chargaff died on June 20, 2002.

See also AVERY, OSWALD; CHROMATOGRAPHY; CRICK, FRANCIS; DEOXYRIBONUCLEIC ACID (DNA);

FRANKLIN, ROSALIND; PAULING, LINUS; WATSON, JAMES D.; WILKINS, MAURICE H. F.

FURTHER READING

American Philosophical Society. "Erwin Chargaff Papers." Available online. URL: http://www.amphilsoc.org/library/mole/c/chargaff.htm. Accessed January 14, 2008.

Chargaff, E. "Chemical Specificity of Nucleic Acids and Mechanism of Their Enzymatic Degradation." *Experientia* 6 (1950): 201–209.

———. "Structure and Function of Nucleic Acids as Cell Constituents." *Federation Proceedings* 10 (1951): 654–659.

Chase, Martha (1927–2003) American *Geneticist*
Martha Chase is best known for a landmark experiment performed with Alfred Hershey that confirmed deoxyribonucleic acid was the carrier of genetic information in bacteriophage.

Martha Cowles Chase was born in Cleveland Heights, Ohio, on November 30, 1927. She received a bachelor's degree from the College of Wooster in 1950 and took a job as a lab assistant to Alfred D. Hershey, a microbiologist at the Carnegie Institution of Washington in Cold Spring Harbor, on Long Island, in New York.

In 1944 Oswald Avery and his colleagues Colin MacLeod and Maclyn McCarty at the Rockefeller Institute demonstrated that deoxyribonucleic acid (DNA) was the molecule responsible for transforming a nonvirulent strain of pneumococcal bacteria into a virulent strain. This was a surprising result because most geneticists believed proteins carried genetic information. The structure of DNA was still unknown, but chemists assumed that since it was composed of only four different nucleotides, the molecule was too simple to carry all of the necessary information for a cell to perform the complex biological functions of replication and protein synthesis. Proteins, on the other hand, consisted of 20 different amino acid building blocks and thus was the preferred molecular candidate for the carrier of genetic information.

Hershey and Chase performed a series of elegantly designed experiments with bacteriophage that convinced even the skeptics of the scientific community that DNA was in fact the genetic material. Bacteriophages are viruses that specifically infect bacterial cells. The basic structure of a bacteriophage consists of a protein coat surrounding a nucleic acid core. When a phage attaches to a host bacterium, it injects its genetic material through the cell membrane into the cell but the phage particle itself remains external to the cell. Soon afterward, the bacterial cell starts synthesizing viral components, which are then assembled into new viral particles. The bacterial cell eventually bursts open, releasing the newly synthesized viral particles into the environment, where they seek and infect new bacterial cells. The mysterious substance that the bacteriophage injected into the bacterial cell was the genetic material. The experiments performed by Hershey and Chase addressed the nature of that substance.

To identify the injected material, they infected bacterial cultures with T2 bacteriophage that had been labeled with radioactive sulfur (^{35}S) or radioactive phosphorus (^{32}P). Sulfur is a component in proteins, and phosphorus is a component of DNA. The radioactive labeling of these atoms allowed Hershey and Chase to track, or to follow, the location of the viral proteins and nucleic acids during their experiment. After allowing sufficient time for the radioactive bacteriophage to attack the bacteria, they agitated the cultures in a blender and then centrifuged the cultures to separate the bacteria cells and their contents from viral particles. Though most of the viral particles remained in the supernatant, many had already injected their genetic material into the host bacterial cells. In the culture infected with T2 labeled with ^{35}S, the radioactivity remained in the supernatant, indicating that the protein coats did not enter the bacterial cells. When they examined the ^{32}P-labeled culture, they found a significant portion of the radioactivity in the bacterial cell pellet; thus the nucleic acid must have entered into the bacterial cells during the infection process. They concluded from these results that DNA must encode the information necessary to direct the synthesis of new viral particles within a bacterial host cell.

These experiments were published in an article titled "Independent Functions of Viral Protein and Nucleic Acids in Growth of Bacteriophage" in 1952. This landmark paper convinced biologists that nucleic acid, not protein, was the carrier of genetic information. Though not common practice, Hershey included Chase's name as a coauthor of their manuscript. The following year Chase began working at Oak Ridge National Laboratory in Tennessee, then later at the University of Rochester. In 1959 Chase moved to California to begin work toward a doctorate degree in microbial physiology, which she earned from the University of Southern California in 1964. She returned to the Cleveland area and died of pneumonia on August 8, 2003. Martha Chase's name will forever be associated with the famous blender experiment that demonstrated DNA to be the genetic material of phage.

See also AVERY, OSWALD; BIOMOLECULES; CENTRIFUGATION; DEOXYRIBONUCLEIC ACID (DNA); HERSHEY, ALFRED; MACLEOD, COLIN MUNRO;

McCarty, Maclyn; viruses and other infectious particles.

FURTHER READING

Hershey, A. D., and Martha Chase. "Independent Functions of Viral Protein and Nucleic Acid in Growth of Bacteriophage." *Journal of General Physiology* 36 (1952): 39–56.

chemical basis of life Anything that has mass and occupies space is considered matter—water that fills the oceans, the bark of a tree, snowflakes, dust particles floating in the air, or the fossil remains of a long extinct dinosaur. All of the physical material in the universe is composed of chemical elements, substances that cannot be broken down or chemically converted into other substances. Elements consist of a single type of atom, the smallest particle of an element that still retains its chemical properties. When two or more atoms join by covalent linkages, the result is a molecule, the smallest particle of a substance that retains all the properties of the substance. Compounds are substances made of two or more elements combined in definite proportions.

The same chemical elements make up all living and nonliving matter. The element carbon has the same chemical properties whether it exists in the form of diamond, as a component of carbon dioxide gas, or in starch of a potato. Life scientists must study the chemical nature of matter in order to understand life at all levels. Whereas the reason why a biochemist must understand basic chemistry might be obvious, the reasons why a geneticist, ecologist, or any other life scientist must are just as important. The mechanisms by which one generation transmits its genes to the next and the way genes determine one's physical characteristics are based on the molecular processes of deoxyribonucleic acid (DNA) replication, transcription, and translation. Ecologists aim to understand the interactions between organisms and their environment, both the physical surroundings and the other organisms living in the same location. Without understanding the chemical nature of matter, an ecologist could not appreciate the biogeochemical recycling of nutrients such as nitrogen or why some organisms are autotrophic and can make their own food while others are heterotrophic and must ingest food in the form of organic compounds.

When studying life at the molecular or atomic level, as well as at the whole organism or community level, structure and function are intimately related. The structure of atoms and molecules determines their chemical characteristics, such as how they will react with other atoms or molecules and what types of bonds they may form. Though understanding atomic architecture and how different types of chemical bonds hold molecules together is crucial to the development of an appreciation for life science, the question of how life arises from inanimate matter remains unanswered.

ATOMIC STRUCTURE

Atoms consist of protons, neutrons, and electrons. The protons and neutrons have similar masses and reside in the nucleus. Electrons exist in defined orbitals that surround the nucleus, and they do not contribute significantly to the mass of an atom. Protons have a positive electrical charge, and electrons have a negative charge. The atoms of each element have an atomic number that equals the number of protons in the nucleus. Because atoms are electrically neutral, the number of protons in an atom equals the number of electrons. When different numbers are present, the particle carries an electrical charge and is called an ion. Neutrons do not carry a charge and therefore do not contribute to the net electrical charge of an atom, but they do help stabilize the nucleus. The exact number of neutrons in the atoms of a particular element varies. For the elements with lower atomic numbers, the number is usually close to the number of protons, whereas heavier elements require larger ratios of neutrons to protons to achieve stability. The atomic weight of an atom, its mass relative to a hydrogen atom, essentially equals the number of protons plus the number of neutrons. Atoms of the same element (meaning they have the same number of protons) that have different numbers of neutrons are called isotopes and can be distinguished on the basis of their atomic weights. For example, atoms of the element carbon usually have an atomic mass of 12 due to six protons and six neutrons, but carbon 14 isotopes with eight neutrons also exist in nature. Isotopes with too few or too many neutrons have unstable nuclei and may disintegrate spontaneously, a phenomenon known as radioactive decay. The slow but steady rate at which carbon 14 decays to carbon 12 allows researchers such as geologists and paleontologists to determine the age of fossils and other organic matter.

The periodic table of the elements (see appendix) displays more than 100 chemical elements, but more than 96 percent of the matter in living organisms is due to only four: carbon (C), hydrogen (H), oxygen (O), and nitrogen (N). Phosphorus (P) and sulfur (S) are the next most common, and other elements including magnesium (Mg), sodium (Na), chlorine (Cl), potassium (K), and calcium (Ca) each make up less than a fraction of 1 percent of living organisms. These elements combine in different ways to make up biomolecules that assemble into cellular structures. The major classes of biomolecules are carbohydrates, proteins, nucleic acids, and lipids. The manner in which the atoms interact to form molecules depends on the structure of the individual atoms, especially the electrons in the valance shell, the outermost energy level.

The protons and neutrons of an atom cluster together in a region called the nucleus. Electrons continuously move around the nucleus in discrete orbitals, defined as regions where electrons might be found within an atom. Groups of orbitals form main energy levels or electron shells; those nearest to the nucleus have the lowest energies, and those farther away have higher energies. Atoms fill their energy sublevels from the lowest to the highest level, depending on how many electrons they have.

- The lowest-energy level contains a single orbital that can only hold two electrons.
- The second energy level consists of two sublevels that can hold a total of eight electrons.
- The third energy level contains three sublevels and can hold a maximum of 18 electrons.

Though atoms can have more than three energy levels, because of their lower atomic weights, atoms in biological molecules rarely have more. The valence electrons are the electrons located in the outermost shell of an atom. Because of the order in which the sublevels fill, the maximal number of valence electrons in a ground-state (lowest energy state) atom is eight. Atoms are most stable when the valence shell contains eight electrons, a phenomenon called the octet rule. If the outermost shell is the first energy level, two electrons are sufficient to fill it and obey the "octet rule."

To illustrate, consider the atomic structures for carbon, hydrogen, oxygen, and nitrogen. Carbon has an atomic number of 6 and a mass of 12; the nucleus has six protons and six neutrons. Two electrons fill the first energy level, leaving four electrons for the second level. Hydrogen has an atomic number of 1. Because no neutrons are required to stabilize a nucleus that only has one proton, it also has an atomic mass of 1. A single electron orbits the proton. Oxygen has an atomic number of 8 and a mass of 16; its nucleus holds eight protons and eight electrons. After two electrons fill its innermost shell, six remain and exist in the outer shell. Nitrogen has an atomic number of 7 and an atomic mass of 14. Two electrons fill the first energy level, and the valence shell contains the remaining five.

CHEMICAL BONDS

The valence electrons determine how an atom interacts with other atoms. Atoms are most stable when

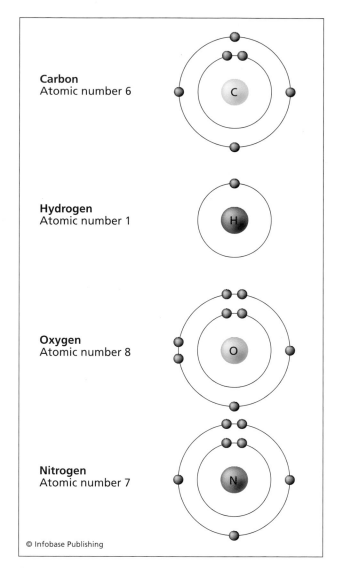

Electron shell diagrams indicate the number of electrons, represented by dots, in each shell of an atom.

their outermost shell is filled; thus they interact with other atoms in a manner that allows them to achieve this. Interactions are considered chemical bonds when their combined product forms a species that has distinct chemical properties.

Covalent bonds form when two atoms share valence electrons so they both have a filled valence shell, containing either a total of two (if the valence shell is the first energy level) or of eight valence electrons. The number of covalent bonds in which an atom can participate depends on the number of unpaired valence electrons. Each shared pair of electrons constitutes one covalent bond. When two or more atoms are covalently bound, the result is a molecule. For example, in a molecule of methane (CH_4), carbon shares each of its four valence electrons with a hydrogen atom. As a result, carbons shares four pairs of electrons with four hydrogen atoms, each of which contributes its lone electron to a shared pair. Each hydrogen atom acquires a configuration with two electrons in its valence shell, and the carbon atom achieves a total of eight electrons in its valence shell. Sometimes, two atoms attain octets by sharing more than one pair of electrons. In a double bond, two atoms share two pairs of electrons, and three pairs are shared in a triple bond. In biomolecules, carbon often participates in double and triple bonds. Oxygen can also form double bonds, since it has two unpaired electrons that can participate in covalent bonds, and nitrogen can form double or triple bonds since it has three pairs of unshared electrons. In molecular diagrams, either a pair of dots or a short dash represents a covalent linkage.

Atoms of each element have characteristic electronegativities, or abilities to attract electrons for participation in a chemical bond. In 1932 the American chemist Linus Pauling developed a scale of electronegativities ranging from 0.7 (francium) to 3.98 (fluorine). Two atoms with a difference in electronegativities of less than 0.4 will generally form covalent bonds; in other words, they will share their valence electrons equally. For example, consider a molecule of CH_4. Carbon and hydrogen have electronegativities of 2.55 and 2.20, respectively. The difference is 0.35, which is less than 0.4; thus carbon and hydrogen form covalent bonds. Diatomic molecules, which are composed of two atoms of the same element (H_2, N_2, O_2, F_2, and Cl_2), have pure covalent bonds. As the same element, both atoms have equal electronegativities. Covalent bonds are strong, take a lot of energy to form, and release energy when broken. In living cells, special molecules called enzymes facilitate the formation and breakage of covalent bonds during the synthesis and degradation of biomolecules.

When the difference in electronegativities of two atoms equals or exceeds 1.7, an ionic bond forms. In this case, one atom has a much stronger attraction for the electrons of another atom, often because it has more protons in its nucleus and therefore a stronger positive charge to attract the electrons orbiting another atomic nucleus; or it has a smaller atomic radius, so the attraction of the positively charged nucleus does not have to extend as far to reach and interact with the electrons of another atom. Generally, the most electronegative elements are positioned in the upper right corner of the periodic table. When two atoms with significant differences in their electronegativities interact to form an ionic bond, the atom with the lower electronegativity gives up one or more electrons to the atom with

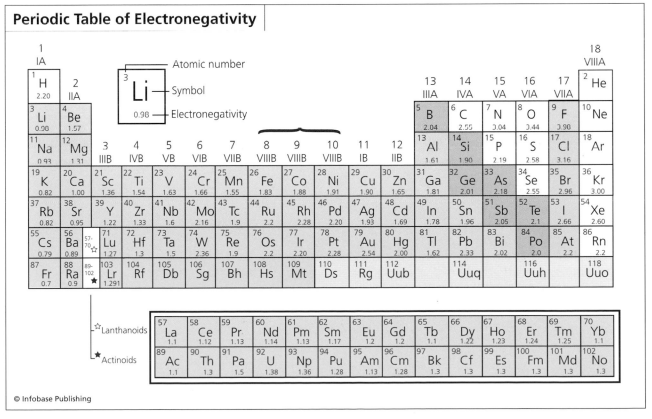

The nature of a chemical bond between two atoms depends on the differences in their electronegativities, shown here as a periodic property.

the greater electronegativity. As in covalent bonding, the driving force is still the tendency of the atom to reach a stable configuration, that is, a configuration with a filled valence shell.

When a neutral atom donates or accepts one or more electrons, it becomes an ion, a charged particle. Ionic bonds are attractions between oppositely charged ions. Sodium chloride (NaCl), also known as table salt, is an example of an ionic compound. Sodium has an electronegativity of 0.9, and chlorine has an electronegativity of 3.0. The difference is 2.1, which is greater than 1.7; thus their interaction results in the formation of an ionic bond. A sodium atom has 11 electrons, so the valence shell has one lone electron. Chlorine has 17 electrons, so its valence shell has seven electrons. By donating its lone electron to chlorine, the sodium atom attains a stable octet in its outermost shell, and because it now has one less electron than proton, it has a positive charge of 1 (Na^{+1}) and is considered an ion.

Likewise, chlorine attains a stable octet in its valence shell by accepting the electron donated by the sodium atom. The chloride ion has one more electron than proton, thus has a charge of negative 1 (Cl^{-1}). The sodium ion and the chlorine ion join to form an ionic bond, an attraction between oppositely charged ions. Whereas compounds bonded covalently are called molecules, compounds held together by ionic bonds are usually referred to as salts and assume regular, repeating, three-dimensional structures (crystals) formed by balancing the charges. The environment of living systems is typically aqueous, and ionic compounds rapidly dissociate in the presence of water. When polar water molecules completely surround the ions, the salt is said to be dissolved. Covalent bonds do not break apart in the presence of water.

While water does not cause atoms held together by covalent bonds to dissociate, the nature of a covalent bond does determine the molecule's solubility in water. Covalent bonds, as defined earlier, form

A **Covalent bond:** equal sharing of electron pairs

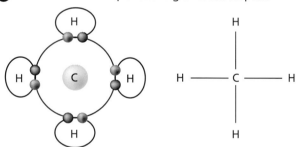

B **Polar covalent bond:** electrons held closer to more electronegative element, causing unequal charge distribution

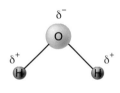

C **Ionic bond:** an atom donates one or more electrons to another atom, and electrostatic attraction holds them together

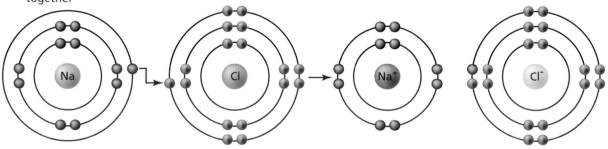

D **Hydrogen bond:** weak attraction between a partially positive hydrogen atom and an electronegative atom

$$\begin{array}{c} \diagdown \\ \diagup \end{array} C = \underset{\delta^+ \quad \delta^-}{O} \text{---------} \underset{\delta^+ \quad \delta^-}{H} - N \begin{array}{c} \diagup \\ \diagdown \end{array}$$

© Infobase Publishing

The characteristic electronegativities of different elements determine the types of chemical bonds that they form with other elements.

between atoms that have a difference in electronegativity of less than 0.4, and ionic bonds form between atoms with electronegativity differences greater than 1.7. When the difference falls in the range of 0.4 to 1.7, the bond formed is termed polar covalent. The atoms still share their electrons, but they do not do so equally. The tendency of one of the atoms is great enough to localize the charges by holding the shared electrons closer to its nucleus than they are to the less electronegative atom's nucleus, but the difference is not great enough to pull the electrons off the less electronegative atom completely. In polar covalent bonds, the more electronegative atom will carry a partial negative (δ^-) charge, while the other atom will have a partial positive charge (δ^+). Polar molecules have dipoles, opposite charges at different ends or sides of the molecule. This separation of charges makes polar molecules soluble in water and other aqueous substances; they are said to be hydrophilic because of their affinity for water. Water molecules, consisting of one oxygen atom and two hydrogen atoms, are very polar. Oxygen is very electronegative and holds the electrons shared with the bound hydrogen atoms more tightly than do the hydrogen atoms. This lopsided distribution of electrons gives the oxygen end of the triangle-shaped molecule a partial negative charge and the hydrogen atoms a partial positive charge. In biomolecules, oxygen and hydrogen form polar covalent bonds, as do nitrogen and hydrogen. Carbon, however, has a slightly lower electronegativity than oxygen or nitrogen, and so it shares electrons relatively equally with hydrogen. When two polar covalent molecules are positioned near one another, the opposite partial charges can form a weak ionic interaction with one another. Though the ionic interactions between polar covalent molecules are weak, several can act jointly to hold molecules together. For example, partially charged atoms in the active site of an enzyme can hold a specific substrate in the correct position and orientation to allow a biochemical reaction to take place and then release it after product formation.

A hydrogen bond is a noncovalent interaction that occurs between two atoms that are already participating in polar covalent bonds with other atoms. As the name implies, one of the atoms is a hydrogen atom that is bound to one electronegative atom, and the other is another electronegative atom, usually a nitrogen or oxygen. The hydrogen atom is basically sandwiched between the two electronegative atoms but is only covalently bound to one of them. Hydrogen bonds can form between different molecules or between different parts of one large molecule. They are very weak, and simple thermal motion causes them to break and reform continuously. Hydrogen bonds hold water molecules together and are responsible for the many unique properties of water. The three-dimensional structures and therefore the proper function of many proteins and nucleic acids also depend on hydrogen bonding. The two strands of a double-helical molecule of DNA are held together by hydrogen bonds between complementary nucleotides. The individual hydrogen bonds are weak enough so the two strands can be easily separated in order for replication or transcription to occur, but in a polymer thousands of nucleotides long, the collective strength of the hydrogen bonds holds the two strands of DNA together to form a stable molecule.

Hydrophobic interaction, another biologically important noncovalent interaction, occurs between nonionic, nonpolar substances. Such molecules are not hydrophilic—they are hydrophobic, meaning they do not have an affinity for water and seem to repel it. Composed of atoms that have similar electronegativities, nonpolar molecules (such as the long hydrocarbon chains that are found in lipids) have their electrons equally distributed; thus hydrophobic molecules do not dissolve in water. Hydrophobic interactions are the forces that cause the nonpolar substances to join when placed in an aqueous solution in order to minimize interaction with the polar water molecules.

One last type of noncovalent interaction found between biomolecules are van der Waals forces, weak interactions that result from the attraction of transient dipoles in neutral atoms or molecules. The electrons in an atom or molecule are in constant motion, and the presence of other nearby particles influences their exact location at any instant in time. Thus transient dipoles are constantly forming around the molecules, and the weak intermolecular forces resulting from the attraction of the local charge fluctuations are called van der Waals forces.

See also BIOMOLECULES; WATER, ITS BIOLOGICAL IMPORTANCE.

FURTHER READING

Alberts, Bruce, Alexander Johnson, Julian Lewis, Martin Raff, Keith Roberts, and Peter Walter. *Molecular Biology of the Cell.* 5th ed. New York: Garland, 2007.

Lennarz, William J., and M. Daniel Lane. *Biological Chemistry.* 4 vols. Amsterdam and Boston: Elsevier, 2004.

Myers, Richard. *The Basics of Chemistry.* Westport, Conn.: Greenwood Press, 2003.

chromatography Chromatography is a technique used for separating mixtures. A mobile phase (such as a liquid or gas) containing the mixture in a solvent passes or moves through a stationary phase (such as a paper or a gellike matrix). Different chemical or physical properties of molecules in the mixture cause them to move through the stationary phase at different rates, and they can be further purified, measured, or analyzed. Different categories of chromatographic methods depend on the type of mobile phase: gas, liquid, or supercritical fluid. Life science research often employs liquid chromatography, in which a solution containing a mixture of biomolecules moves through a medium composed of a variety of possible materials depending on the characteristic used to separate the components. In column chromatography, the stationary phase is typically prepared as a slurry of resin and buffer that is packed into a narrow cylindrical glass or plastic column. In paper chromatography capillary action pulls the liquid up through a strip of paper acting as the stationary phase. An absorbent material spread over a flat glass or plastic plate serves as the stationary phase in thin-layer chromatography.

In liquid column chromatography, gravity pulls the mobile liquid phase through the resin of a column. The resin retains the components of the mixture to different degrees on the basis of different characteristics of the components. The components will thus move through the column at different rates.

In column chromatography, a liquid mobile phase travels through a column packed with a material that retains mixture components to different degrees. *(Maximilian Stock Ltd./Photo Researchers, Inc.)*

The solution is collected in fractions as it leaves, or elutes from, the column. All of the fractions contain equal amounts of solution, but the concentration of the mixture components in each fraction varies as a result of different retention times of the various molecules. To determine what is present in each fraction, the researcher then assays each one by a variety of methods depending on the purpose. For example, the total protein or nucleic acid concentration can be measured, or the ability of a fraction to carry out a specific biochemical reaction can be assessed.

Different types of column chromatography exploit different properties of the molecules in the mixture to achieve separation. The three most common types in life science research are size exclusion, ion exchange, and affinity chromatography.

Size exclusion chromatography, also called gel filtration chromatography, separates components on the basis of their physical dimensions. The column is packed with porous beads resembling tiny spheres with tunnels running through them. Large molecules will not be able to enter into the pores; they are said to be excluded from the porous beads and flow rapidly through the column. They basically travel in a straight downward path and emerge in the void volume, the volume of the buffer that surrounded the beads of the column. The first few fractions of a size exclusion column will contain all the molecules that were larger than the size of the pores; thus a researcher must carefully choose beads that contain an appropriate size pore on the basis of the biomolecules to be purified. The pore size should be slightly larger than the molecule of interest. Molecules whose size resembles that of the pores in the resin will penetrate some of the pores but not others, so the retention time of these intermediate-sized molecules will be longer than that of the larger molecules. As the size of the molecules decreases, the retention time increases because smaller molecules will spend more time inside the beads, traveling a greater total distance through the column. Common uses of size exclusion chromatography are in fractionating proteins from extracts and purifying oligonucleotides from individual nucleotides after radiolabeling them.

Another type of chromatography commonly used in the purification of biological molecules is ion exchange chromatography, which separates molecules on the basis of their charge. Anion exchange columns use a positively charged stationary phase, such as a resin with diethylaminoethyl (DEAE) groups attached. As negatively charged molecules flow through the column, they interact with the DEAE groups, causing an increased retention time. In cation exchange, a negatively charged resin such as phosphocellulose retains positively charged molecules. Once a charged molecule binds to the column,

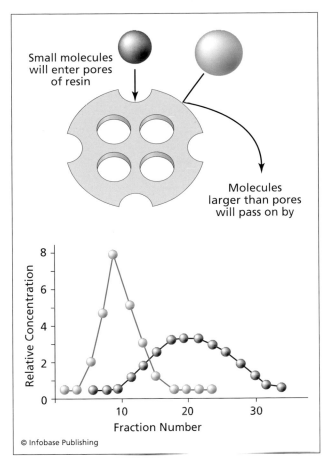

Small molecules will enter pores of resin

Molecules larger than pores will pass on by

© Infobase Publishing

The column resin used in size exclusion chromatography contains pores of a size selected to allow some molecules to flow through easily but slow down others. Larger molecules will elute in earlier fractions, whereas the column will retain smaller molecules for longer periods.

it will not elute until the ionic strength or the pH of the buffer solution is changed. Gradually increasing the ionic strength will allow the ions in the buffer to replace the charged side chains of the molecule bound to the column, releasing it into the upcoming fractions. Altering the pH of a solution will either protonate (add a proton) or deprotonate (remove a proton) ionic side chains of a molecule bound to the resin, lowering its affinity and allowing it to elute. Because this method is highly selective and the resins are inexpensive, ion exchange chromatography is often used at early steps in purification protocols.

Affinity chromatography employs resins to which specific compounds have been attached. This is the most selective form of chromatography and can be used to purify specific enzymes on the basis of their binding to a specific substrate, to purify antibodies on the basis of their interaction with a specific antigen, or purify protein receptors on the basis of their recognition of a particular ligand. The

compound for which the molecule desired has an affinity is chemically attached to an inert matrix, often agarose, a polysaccharide isolated from algal cell walls. As the solution or extract flows through the column, the molecule of interest binds. In order to liberate the purified molecule, a buffer with a high ionic strength is passed over the column, disrupting the intermolecular ionic interactions holding the molecules together.

See also BIOMOLECULES; CENTRIFUGATION; CHEMICAL BASIS OF LIFE; ELECTROPHORESIS.

FURTHER READING
Heftmann, E., ed. *Chromatography: Fundamentals and Applications of Chromatography and Related Differential Migration Methods.* Boston: Elsevier, 2004.
Poole, Colin F. *The Essence of Chromatography.* Boston: Elsevier, 2003.

chromosomes Chromosomes are the threadlike structures that carry the genetic information in living organisms. Consisting of chromatin, a complex of deoxyribonucleic acid (DNA) and proteins, eukaryotic chromosomes are found in the membrane-bound nucleus and contain linear DNA molecules. In contrast, typical prokaryotic chromosomes are closed, circular structures that are not enclosed within a membrane-bound compartment but rather are concentrated in a region called the nucleoid. Prokaryotes have a single chromosome, whereas the number of chromosomes in eukaryotic cells is species dependent. Also, most eukaryotic cells contain two each of several chromosome types, a condition called diploidy, with one member of each pair from each parent. Cells that only contain one of each type of chromosome are termed haploid and play an important role in sexual reproduction. The "Chromosome Number by Species" table gives a sampling of the number of chromosomes found in various eukaryotic species. No consistent relationship exists between the number of chromosomes and the complexity of the organism.

CHROMOSOMAL THEORY OF INHERITANCE
The chromosomal theory of inheritance states that chromosomes are the cellular components that carry genes, the functional units of heredity. Though this concept seems simple, the chromosomal theory revolutionized 20th-century biology. The Austrian monk Gregor Mendel laid the foundation in the 1860s when he presented his research on inheritance in pea plants. He collected data from thousands of offspring that resulted from controlled pollinations between plants with specific traits, and he proposed an insightful explanation regarding inheritance. To summarize,

CHROMOSOME NUMBER BY SPECIES

Species	Diploid Number of Chromosomes
Myrmecia pilosula (an ant)	2
mosquito	6
Canis familiaris (dog)	8
Drosophila melanogaster (a fruit fly)	8
petunia	14
Planaria torva (a flatworm)	16
Aspergillus nidulans (a mold)	16
corn	20
Saccharomyces cerevisiae (a yeast)	32
Homo sapiens (human)	46
Pan troglodytes (chimpanzee)	48
chicken	78
king crab	208
Ophioglossum reticulatum (Indian fern)	1,260

he said that every diploid organism possesses two determinants (today called genes) for each character (trait). The two copies of a gene (now called alleles) may or may not be identical. By definition, a dominant allele will mask the presence of a recessive allele if an individual has one copy of each, meaning the individual will exhibit the phenotype (the observable characteristic) associated with the dominant allele. If an individual has two identical copies of a gene, then the phenotype associated with the alleles the individual possesses will manifest. Mendel also proposed the law of segregation, stating that though every individual has two copies, only one passes from parent to offspring through the gametes; in other words, the two copies of a gene separate during the formation of gametes. At fertilization, offspring receive one allele from each parent and thus have two copies. Random chance determines which allele passes to the offspring through an egg or the sperm cell that fertilizes it. Mendel also put forth the law of independent assortment, which states that the alleles for one gene separate independently of the alleles for other genes. (Scientists later discovered this only holds true for alleles located on different chromosomes or sufficiently distant on the same chromosome.) Mendel

presented his research in 1865, and it was published the following year, but the work was unnoticed until the turn of the century.

Meanwhile, unaware of Mendel's findings, cell biologists sought the hereditary material. Since egg and sperm were thought to contribute equally to the offspring, and the sperm contained very little cytoplasm, the biologists focused on the nuclear contents. In 1882 Walther Flemming at the University of Kiel in Germany described his observations of chromosomes and their movement during cell division, a process he named mitosis. In 1883 the German physiologist August Weismann proposed that chromosomes were the bearers of the genetic information.

In 1900, while performing literature searches related to their own studies, three scientists independently rediscovered Mendel's work: Hugo de Vries, Erich Von Tschermak, and Carl Correns. An American graduate student named Walter Sutton and a German biologist named Theodor Boveri separately but simultaneously had been studying chromosomes and their behavior during mitosis (the process resulting in duplication and separation of the nuclear contents prior to cell division) and meiosis (the process that during the formation of gametes results in nuclei that only contain half of the total number of chromosomes). The attention drawn to Mendel's experimental data, Correns's suggestion that chromosomes might be the carriers of inherited traits, and Sutton's own research on chromosome structure and function in grasshoppers led to Sutton's 1902 *Biological Bulletin* paper, "On the Morphology of the Chromosome Group in *Brachystola magna*." In that paper and one he published the following year titled "The Chromosomes in Heredity," Sutton concluded that genes are located on chromosomes, described how chromosomal behavior during meiosis explained Mendel's laws of heredity, and provided supportive data from his own research. Also in 1903, Boveri drew the same conclusion from his studies in roundworms. Science credits both Sutton and Boveri with formulating the chromosomal theory of inheritance. With a powerful theory to explain the transmission of inherited characteristics, genetics advanced quickly in the early 1900s. The work performed by Thomas Hunt Morgan, in particular, provided a preponderance of confirmatory evidence. Biologists did not determine that DNA, rather than protein, was the component of chromosomes that acted as the molecular carrier of genetic information until the 1940s.

CHROMOSOME STRUCTURE

The molecule of heredity is DNA, which consists of a long polymer of four alternating nucleotides, referred to by the nitrogenous base the nucleotide contains: *A* for adenine, *C* for cytosine, *G* for guanine, or *T* for thymine. Molecules of DNA are double-stranded, and the nucleotides between the two strands pair up in a specific manner. Nucleotides containing the nitrogenous base A always pair with nucleotides containing the nitrogenous base T, and C always pairs with G. The sequence of the four nucleotides carries the information within genes, which exist as distinct segments along the length of a chromosome. All types of living organisms have DNA as their genetic material, but the manner in which the DNA is organized and stored within a cell varies. The genetic material of viruses can be either DNA or ribonucleic acid (RNA), which is very similar in structure to DNA and can be single-stranded or double-stranded, linear or circular, but viruses are not considered living organisms.

Prokaryotic cells include members of the domains Archaea and Bacteria. They usually have only one chromosome, consisting of a single, circular molecule of double-stranded DNA. The length is much shorter than that of the typical eukaryotic chromosome. The chromosome of *Escherichia coli*, the most extensively researched bacteria, is just under 5 million base pairs long, which if straightened and measured, would extend about 0.05 inch (1.2 mm), the approximate width of a letter on this page. Sometimes bacteria also have plasmids, much smaller circular pieces of DNA found in the cytoplasm. Plasmids are considered extrachromosomal pieces of DNA, though they can be considered part of the organism's genome, the entire complement of genes within an organism. An average rod-shaped *E. coli* cell measures 0.5 × 2.0 micrometers (μm, 1 μm = 1 × 10^{-6} m); thus compaction of the chromosome is necessary for it to fit within the cell. Topoisomerase enzymes that twist the DNA, causing it to coil into a tight bundle, package the chromosome, which is attached to the inner cell membrane. Positively charged proteins called HU and H bind the negatively charged phosphate groups of the bacterial DNA, helping it fit within the relatively small volume of a bacterial cell while still allowing it to function as a template for DNA replication and for transcription during protein synthesis.

Eukaryotic chromosome structure is much more complicated than prokaryotic chromosome structure. Except during mitosis, the chromosomes of a eukaryotic cell exist as chromatin dispersed within the nucleus. Eukaryotic chromosomes vary in length; even different chromosomes of one species can differ. For example, human chromosomes range from 51 million to 245 million base pairs long, which translates into 19–73 millimeters (mm, 1 mm = 10^{-3} m) if fully extended. Since the diameter of a typical nucleus is only 5–10 μm, extensive packaging of the chromatin is necessary so it will fit within the confines of the nuclear envelope.

Eukaryotic chromosomes contain a much higher proportion of protein than prokaryotic chromosomes. Histone proteins are positively charged proteins that interact with the DNA of eukaryotic chromosomes. The five major types are H1, H2A, H2, H3, and H4. Two each of the latter four histone proteins combine to form an octamer that serves as the core of a structure called a nucleosome, around which a segment of about 147 base pairs of DNA wraps 1.7 times. The amino terminus of each histone protein extends outward and plays a role in the regulation of gene expression. Linker DNA, stretches of eight to 114 base pairs, exists between the nucleosome core particles. When viewed with an electron microscope, DNA packaged at this level resembles beads on a string, with the beads representing the nucleosomes and string representing the linker DNA. The extended chromatin fibers measure about 10 nanometers (nm, 1 nm = 10^{-9} m) in width at this stage, and the DNA has been compacted about threefold.

The next level of packaging requires the binding of a fifth type of histone protein, H1, which binds to the DNA near the nucleosome and further compacts the DNA. H1 binding causes the chromatin to coil into a thicker and shorter 30-nm solenoid structure. Each turn of the solenoid consists of about six nucleosomes, resulting in compaction of about five-times. At the next level of packaging, the 30-nm fiber forms

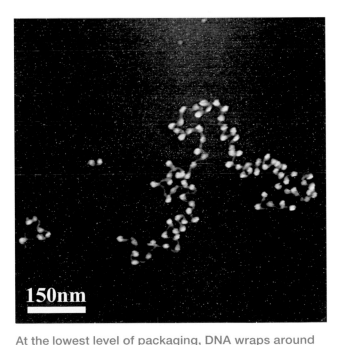

150nm

At the lowest level of packaging, DNA wraps around histone octamers, forming nucleosomes, a structure that resembles beads on a string. This transmission electron micrograph was obtained using purified chromatin fragments from chicken erythroid cells. *(Dr. Zhifeng Shao, University of Virginia)*

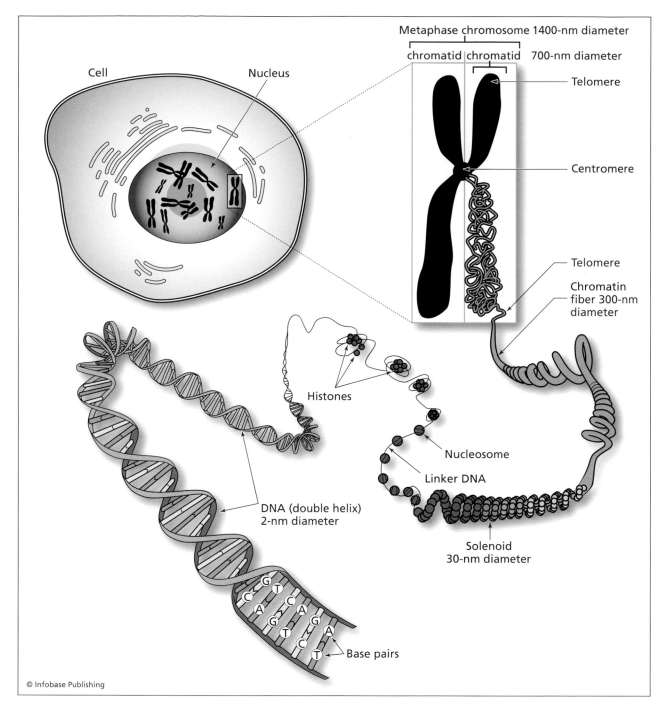

DNA in eukaryotic cells undergoes a series of successive levels of packaging.

looped domains that attach to a protein scaffolding, forming a 300-nm chromatin fiber. When further compaction is required, as when a cell prepares for mitosis, these looped domains and the scaffolding coil and fold, forming the arms of the chromosome and resulting in a fiber of approximately 700 nm in width (this measurement varies). Because the chromosomes of cells undergoing mitosis are duplicated at this stage, the width of the two attached sister chromatids is twice the value. The final degree of compaction varies from approximately 1,000-fold to 10,000-fold.

Prior to mitosis, the chromosomes in a nucleus duplicate, and during mitosis, the chromosomes separate and migrate toward opposite poles of the cell. Cytokinesis ensues, and the two resulting daughter cells each contain a complete set of the chromosomes. Most of the time, the chromosomes are not fully condensed but are long and threadlike and thus are not visible using light microscopy. They do, however, retain an organized higher level of structure. In the absence of scaffolding, looped domains of chromosomes attach to the inside of the nuclear envelope and to fibers of the nuclear matrix, maintaining organization and preventing entanglement with other looped domains and other chromosomes. During mitosis, the chromosomes are fully condensed and, if stained, can be easily observed with a light microscope. Because the chromosomes are still duplicated, each chromosome consists of two sister chromatids attached at a centromere. The position of the centromere and the relative length of the mitotic chromosomes allow cytogeneticists to distinguish among chromosomes types. When performing a karyotype analysis, the cytogeneticist identifies all the chromosomes and arranges them in pairs according to their number. This enables the diagnosis of genetic disorders due to abnormalities in chromosome number or structure. The use of specialized staining techniques results in unique banding patterns on the chromo-

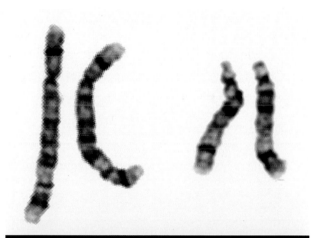

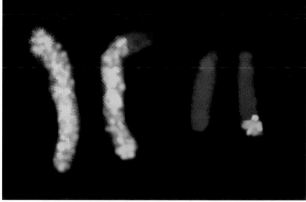

Specialized staining techniques create unique chromosomal banding patterns that aid in identification of each chromosome. The use of fluorescent dyes helps in the detection of chromosomal structural abnormalities, as shown in this translocation between human chromosome numbers 5 and 14. *(Addenbrookes Hospital/Photo Researchers, Inc.)*

somes. These bands facilitate the identification of each type. Fluorescent dyes are useful in detecting abnormalities in chromosomal structure. For example, in translocations, a piece of one chromosome swaps locations with another.

The major function of DNA is to encode for the synthesis of proteins. This occurs in several steps, the first of which is transcription, whereby the two stands of DNA in a region that contains a gene separate, and RNA polymerase reads the template strand of DNA to build a complementary strand of messenger RNA. In its most compact state, the DNA of a chromosome is inaccessible to the enzymes and other proteins that perform this task. In some cells, segments of chromosomes that either do not encode for proteins or encode for proteins those cells do not need to express are turned off. These sections, called heterochromatin, remain tightly compacted and stain darker than sections that are packed more loosely, called euchromatin. Protective structures called telomeres, repetitive sequences of DNA found at the ends of linear chromosomes, and centromeres also remain tightly packaged.

Two unique types of chromosomes are polytene chromosomes and lampbrush chromosomes. The French embryologist Edouard-Gérard Balbiani first observed polytene chromosomes in 1881, but their hereditary nature was not explored until the 1930s. Found in certain tissues of developing fly larvae, polytene chromosomes result from multiple successive rounds of replication without separation. The long DNA molecules remain attached and align with the many other sister chromatids. The condensation of the genetic material at different positions along the length of the chromosome forms distinct bands, sometimes called chromomeres. The interbands, regions between the bands, are uncoiled and genetically active, as evidenced by large puffs called Balbiani rings. Polytene chromosomes are interesting because they are highly visible even during interphase, whereas most chromosomes are not visible unless a cell is undergoing mitosis. Biologists believe that polytene chromosomes serve to increase the nuclear and cellular volume in developing organisms and allow higher levels of gene expression than if the cells were diploid, a factor that may be important for developing larvae. Lampbrush chromosomes are so named for their resemblance to the brushes once used to clean kerosene-lamp chimneys. Found mostly in oocytes but also in some spermatocytes, lampbrush chromosomes are meiotic, meaning they only exist in cells that undergo meiosis, the specialized type of cell division that results in a halving of the genetic information to create haploid sex cells such as eggs and sperm. Lampbrush chromosomes are unique in that they exist in an extended form rather

The banding patterns of polytene chromosomes, which are large enough to observe using light microscopy, are distinctive for each species. *(Andrew Syred/Photo Researchers, Inc.)*

than condensed, as most chromosomes are during nuclear divisions. Their structure is characterized by lateral loops that emanate from each chromomere.

CHROMATIN REMODELING

Histones, the proteins responsible for the first level of packaging, are nonspecific and can bind to any DNA sequence in vitro. The positioning of nucleosomes along a DNA molecule within the nucleus of a cell, however, is influenced by several factors. Certain regions must remain exposed to function properly and the chromosomes in similar cell types are all packaged similarly. The chromatin must also be capable of remodeling, during which the chromatin relaxes its packing temporarily to provide access to replication and transcription enzymes and machinery and then returns to its packaged state when no longer active. In 1993 Bryan M. Turner suggested that epigenetic information resides in histone tail modifications. Epigenetic information is genetic information that is not encoded within the DNA sequence. In 2001 Thomas Jenuwein and C. David Allis proposed the "histone code," a system of posttranslational modifications to histone amino termini that acts as an epigenetic regulatory mechanism for processes that act on chromatin. In 2006 Eran Segal of the Weizmann Institute of Science in Israel, Jonathan Widom of Northwestern University, and their colleagues published a study showing that combinations of DNA sequences direct nucleosome positioning by rendering certain regions of the DNA

more flexible for wrapping around histone octamers. The pattern appears to have very loose requirements; thus the sequences that direct positioning do not conflict with protein coding sequences of genes. The existence of such a nucleosome code explains features of transcriptional control previously not understood.

Three chemical modifications to the amino acids in histone tails, the amino termini that extend outward from the nucleosome core, influence the activity of a chromosome's genes and its chromatin structure: acetylation, methylation, and phosphorylation. An enzyme called histone acetyltransferase adds acetyl groups ($-COCH_3$) to the positively charged amino acid lysine in the histones. This neutralizes the charge and reduces the electrostatic interaction with the negatively charged DNA. Loosening up the histones binding to the DNA makes the DNA more accessible; thus acetylation is a means of activating chromatin. Females have two X chromosomes, but the genes of one are sufficient; males only have one. The extra X chromosome in females becomes inactivated by lack of acetylation, forming a structure called a Barr body. Methyltransferases add methyl groups ($-CH_3$) to the positively charged amino acids lysine and arginine of the histone and, as a result, inactivate genes. The DNA itself also can be methylated, particularly cytosines adjacent to a guanine. Kinases transfer phosphate groups ($-PO_4^-$) to the amino acids serine and histidine, thus making the protein more negative. Though the exact effect is not understood, phosphorylation is related to relaxing and compacting the chromatin before and after DNA replication and in gene activation in nonmitotic cells.

Because histones do not completely dissociate during DNA replication, modifications of chromatin at the level of the histones can be passed on to future generations. Until recently, geneticists believed all heritable traits were due to changes in the nucleotide sequence. A relatively new concept, epigenetic inheritance, describes the transmission of genetic information in the form of reversible changes in DNA rather than alterations in the DNA sequence. Heritable chemical modifications to the histones or to the DNA itself affect the regulation of gene expression without altering the sequence. Epigenetic effects can occur during embryogenesis or during normal cellular reproduction, and recently scientists have shown that environmental factors such as diet can induce changes to one's epigenome that can be transmitted to the next generation.

See also CELLULAR REPRODUCTION; DEOXYRIBONUCLEIC ACID (DNA); EUKARYOTIC CELLS; GENE EXPRESSION; GENETIC DISORDERS; GENOMES; PROKARYOTIC CELLS.

FURTHER READING

Calladine, Chris R., Horace R. Drew, Ben F. Luisi, and Andrew Travers. *Understanding DNA: The Molecule and How It Works.* 3rd ed. San Diego: Elsevier Academic Press, 2004.

Gibbs, W. Wyat. "The Unseen Genome: Beyond DNA." *Scientific American* 289, no. 6 (December 2003): 106–113.

Gregory, T. Ryan, ed. *The Evolution of the Genome.* Burlington, Mass.: Elsevier Academic, 2005.

Olins, Donald E., and Ada L. Olins. "Chromatin History: Our View from the Bridge." *Nature Reviews* 4 (October 2003): 809–814.

Richmond, Timothy J., and Curt A. Davey. "The Structure of DNA in the Nucelosome Core." *Nature* 423 (2003): 145–150.

Wagner, Robert P., Marjorie P. Maguire, and Raymond Stallings. *Chromosomes: A Synthesis.* New York: Wiley-Liss, 1993.

Watters, Ethan. "DNA Is Not Destiny." *Discover* 27, no. 11 (November 2006): 33–37, 75.

Wolffe, Alan. *Chromatin: Structure and Function.* 3rd ed. San Diego: Academic Press, 1998.

circulatory system Living cells must exchange nutrients and waste products across their membranes; thus the surface area to volume ratio restricts the maximal size a cell can reach. As the diameter of a cell increases arithmetically, its volume increases exponentially. The surface area of the cell membrane must be large enough to support the diffusion of substances to meet the metabolic needs of contents within the entire volume of the cell. In single-celled organisms, substances diffuse directly between the cytoplasm and the external environment. The cells of lower invertebrates (animals without a backbone) such as sponges and cnidarians have close enough contact with the environment for nutrients to reach all of the organisms' cells and waste products to leave by simple diffusion and osmosis.

Invertebrates with multiple layers of body cells such as clams, insects, and spiders require special adaptations for nutrient and waste exchange because some of the cells are too far away from the external environment. Circulatory systems connect the organs or tissues of gas and nutrient exchange with all the other body cells when the distance is too great for simple diffusion alone to supply the body cells with the oxygen and nutrients necessary for metabolism. Open circulatory systems consist of a heart that pumps hemolymph, fluid containing essential substances, throughout a network of vessels and open spaces, or sinuses. The hemolymph bathes the body tissues, exchange occurs, and the fluid returns to the heart. Other invertebrates such as starfish, earthworms, and octopuses and all vertebrate animals have more efficient closed circulatory systems that transport substances throughout the body. In closed circulatory systems, the circulating fluid, called blood, travels via blood vessels throughout the body but never leaves the body tissues. Instead, substances diffuse through the walls of smaller branched vessels into the extracellular fluids of tissues, where exchange occurs.

ORGANIZATION OF THE HUMAN CIRCULATORY SYSTEM

In mammals gas exchange occurs in the lungs, nutrients are absorbed through the digestive tract, and nitrogenous waste products exit the body via the excretory system. All the cells in the body need to exchange gases, nutrients, and waste products, but many are too far from the specialized organs for transport by diffusion to achieve this efficiently. The human cardiovascular system comprises a muscular heart, blood, and numerous branched vessels that functionally connect all the body tissues and organs.

Blood circulates through an extensive branched network of vessels, including arteries, capillaries, and veins. Arteries transport blood from the heart to the body's tissues. As arteries approach the tissue they supply, they branch into arterioles, smaller vessels leading to capillaries that infiltrate the tissues. Capillaries converge into venules that converge into veins, the vessels that carry blood back to the heart.

Arteries are thicker and more muscular than other blood vessels because they receive blood directly from the heart; thus the blood flows through them with greater force than through the other vessels. The walls of arteries must be flexible so they can expand and contract to accommodate the force of the blood expelled from the heart and help regulate blood pressure. Arterial walls consist of three layers of tissue. A single layer of cells called the endothelium lines the innermost region of the artery, giving it a smooth interior to minimize resistance during blood flow. The thick middle layer consists of bundles of smooth muscle and elastic fibers that surround the endothelial layer. The outer layer of connective tissue also contains elastic fibers that wrap around and protect the vessels. Capillaries have only a thin epithelial layer and its basement membrane. These very thin vessel walls facilitate osmosis and diffusion of gases, nutrients, hormones, and other molecules. Their diameter is very narrow, barely wider than that of a single blood cell. The construction of veins is similar to that of arteries except the walls are thinner and valves are present. Whereas the force of the heart's contracting propels blood through the arteries at high speeds and great pressure, both the velocity and the pressure decrease as the blood travels through capillaries and then into venules and veins. Flaps of

Major Arteries and Veins

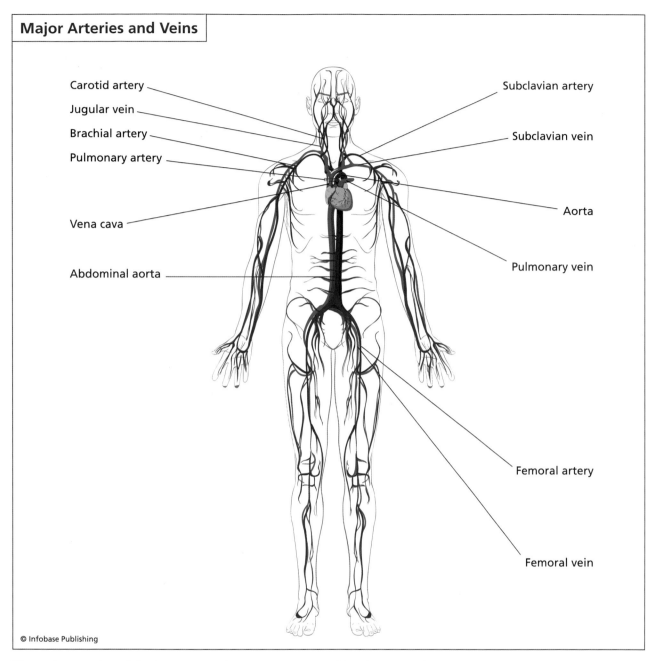

Carotid artery
Jugular vein
Brachial artery
Pulmonary artery
Vena cava
Abdominal aorta

Subclavian artery
Subclavian vein
Aorta
Pulmonary vein
Femoral artery
Femoral vein

The major arteries of the human circulatory system carry blood from the heart to all the body's tissues. The veins return blood to the heart, completing the closed systemic circuit. Note: Not all of the major arteries and veins are shown here.

tissue called valves prevent the backflow of blood in the veins, keeping it moving in the direction toward the heart. Contraction of skeletal muscles squeezes blood inside the veins, forcing open the one-way valves and allowing blood to pass. When the muscles relax, the valves close. This explains how movement assists in circulation.

A separate second system of branched vessels functions to return excess fluid to blood circulation. The fluid that leaks from capillaries into the body tissues is called interstitial fluid. The lymphatic sys-

tem collects the excess fluids from body tissues and dumps the fluid, now called lymph, back into the bloodstream near the junction of the jugular and subclavian veins on each side of the body. Lymph flows through lymphatic vessels mostly by skeletal muscular contraction and nearby arterial pulsing, squeezing the lymph vessels. Valves similar to the valves found in veins prevent backflow of the lymph. Swelling, or edema, occurs when interstitial fluid persists in the tissues, especially during times of inactivity when lack of skeletal muscular contraction reduces movement of lymph through the lymphatic vessels. This is why finger rings that fit fine before going to bed feel tight after waking up in the morning. The lymphatic system also plays an important role in defense against infection.

HEART STRUCTURE AND FUNCTION

The human heart contains four chambers and is similar in size to a closed fist. Made mostly of cardiac muscle, the heart functions to pump blood through the circulatory system in order to transport oxygen and nutrients to the body's tissues and carry carbon dioxide and other waste products away from the body's tissues to organs specialized in their elimination from the body. Three layers make up the walls of the heart: the thin endocardium lines the interior of the heart, the myocardium is the bulky muscular portion, and the epicardium is the thin outer layer that holds the coronary arteries responsible for supplying blood to the heart tissue. A thin sac called the pericardium encases the entire heart, protecting it and separating it from the rest of the internal organs. The septum, a muscular wall, separates the heart into two sides, each consisting of two chambers that fill

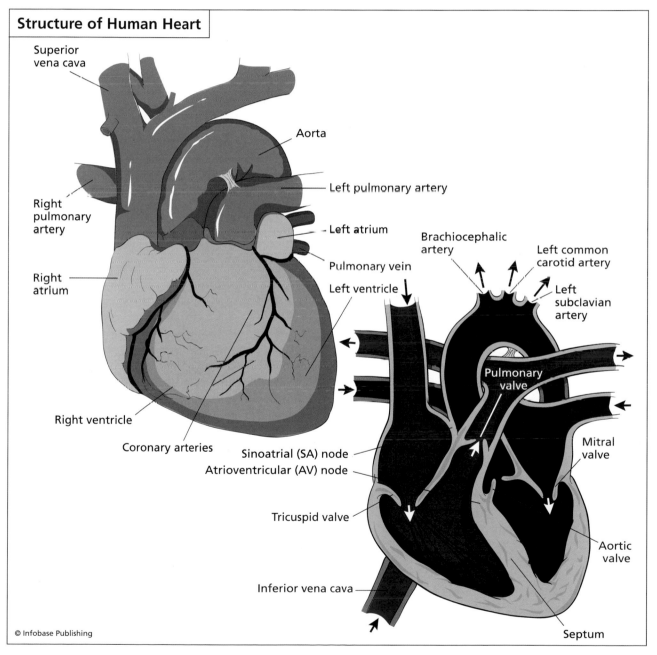

Structure of Human Heart

Superior vena cava

Aorta

Right pulmonary artery

Left pulmonary artery

Left atrium

Pulmonary vein

Left ventricle

Right atrium

Brachiocephalic artery

Left common carotid artery

Left subclavian artery

Pulmonary valve

Mitral valve

Right ventricle

Coronary arteries

Sinoatrial (SA) node

Atrioventricular (AV) node

Tricuspid valve

Aortic valve

Inferior vena cava

Septum

© Infobase Publishing

The human heart contains four chambers, two atria and two ventricles, through which blood flows in a unidirectional manner.

with blood, atria and ventricles. The smaller atria receive blood from different parts of the body via veins, and the ventricles pump the blood through arteries out to the body. Four valves inside the heart prevent the backflow of blood.

The heart pumps blood through a unidirectional circuit. The two largest veins of the body, the superior and inferior vena cava, feed oxygen-poor blood into the right atrium. The blood travels through the tricuspid valve into the right ventricle, which pumps the blood into the pulmonary arteries through the pulmonary semilunar valve. After obtaining a fresh supply of oxygen and releasing carbon dioxide in the lungs, the blood returns to the left atrium of the heart via the pulmonary veins. The blood travels through the mitral valve into the left ventricle, the most muscular chamber, which pumps blood through the aortic semilunar valve into the largest artery, the aorta, which branches into the body's major arteries. After passing through the capillary beds of the tissues, blood moves through the veins to the vena cava, completing the circuit.

Contraction of the heart muscle decreases the cavity size of the chambers, forcing blood along its path. The rhythmic beating of the heart causes the arteries to pulse. The force of blood rushing through the arteries causes them to expand to accommodate the incoming blood with each contraction of the heart and then relax as the fluid moves along. A cardiac cycle is one complete cycle of the heart chambers' filling and draining. The contraction stage is called systole, and the relaxation stage is called diastole. A stethoscope is an instrument that facilitates the detection of a heartbeat. The first sound one hears, the "lub" in the "lub-dub" of each beat, is the recoil of the blood against the closed valves between the atria and ventricles. The second sound, the "dub," is the recoil against the valves that prevent backflow of blood leaving the ventricles. A healthy heart beats an average of 72 times each minute. The heart rate increases during exercise and decreases during rest.

The sinoatrial node, a small cluster of cardiac muscle cells located in the right atrium, acts as a natural pacemaker that controls the regular contractions of the heart. Certain vertebate cardiac muscle cells are unique in that they can contract without receiving a neural impulse. Gap junctions that allow almost instantaneous transmission of electrical impulses separate cardiac muscle cells. When the sinoatrial node initiates an electrical impulse, the signal immediately travels throughout the atria, resulting in simultaneous contraction of all atrial cells. The atrioventricular node, located between the right atrium and right ventricle, delays transmission of the electrical signal for about one-tenth of a second, then relays it to the muscle cells in the ventricles. This delay allows the atria to empty into the ventricles completely before the ventricles contract. Though cardiac muscle cells can contract without external input, other factors such as hormonal influences or body temperature can affect the heart rate.

COMPOSITION OF THE BLOOD

Blood is a type of connective tissue consisting of several types of cellular components (erythrocytes, leukocytes, and platelets) suspended in a liquid matrix (plasma). The average 150-pound (68-kg) adult human circulatory system holds approximately 5.5 quarts (5.2 L) of blood. About 60 percent of the blood volume is plasma, the liquid portion of the blood. Plasma is 90 percent water but contains numerous solutes. Sodium, chloride, bicarbonate, and other ions dissolved in the plasma maintain the osmotic balance and act as buffers that keep the pH of the blood at its optimum of 7.4. Proteins present in plasma also function to maintain osmotic balance as well as transport lipids, contribute to blood viscosity, and aid in clotting. Immunoglobulins, or antibodies, are proteins that fight infection, and protein hormones act as molecular signals that communicate information between cells of distant body parts. Steroid hormones also travel around the body via blood circulation, as do vitamins, amino acids, simple sugars, and dissolved gases such as oxygen and carbon dioxide.

Cellular components make up the remaining 40 percent of the total blood volume. Erythrocytes, also known as red blood cells, are the most abundant cell type present in the blood. Their main function is to

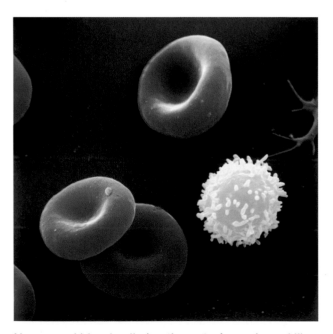

Human red blood cells (erythrocytes) are shaped like biconcave disks and function in oxygen transport. White blood cells (lymphocytes) help fight against infection. (Eye of Science/Photo Researchers, Inc.)

CELLULAR ELEMENTS OF BLOOD

Cellular Element	Major Function	Number (per mm³ of blood)
ERYTHROCYTES (RED BLOOD CELLS)	transport oxygen and also carbon dioxide	5–6 million
LEUKOCYTES (WHITE BLOOD CELLS)	various immune functions	5,000–10,000
neutrophils	phagocytosis	3,000–7,000
eosinophils	defend against parasitic worms, counteract allergic responses	100–400
basophils	secrete histamine and heparin	20–50
lymphocytes	specific immune response, produce antibodies	1,500–3,000
monocytes	phagocytosis	100–700
PLATELETS (CELL FRAGMENTS)	blood clotting	250,000–400,000

carry oxygen from the lungs to all the body tissues. Shaped like a biconcave disk that is thinner in its center than around the edges, a red blood cell contains about 250 million molecules of hemoglobin, an iron-containing protein that binds oxygen. Mature human erythrocytes have no nuclei and therefore cannot repair themselves or undergo mitosis. They also have no mitochondria and cannot aerobically respire; that makes sense, considering that their function is to carry oxygen (that aerobic metabolism consumes) to other body tissues. Because erythrocytes have a life span of only a few months, stem cells in the bone marrow must constantly replenish them.

Approximately 1 percent of all blood cells are leukocytes, or white blood cells, that function in the immune response. Different types of white blood cells perform different tasks in preventing and fighting infection. Some white blood cells such as neutrophils and monocytes are phagocytic and engulf bacteria or other foreign particles. Basophils produce and secrete chemicals such as histamine that play a role in inflammation, and eosinophils help fight infections caused by parasites. Two types of lymphocytes, B cells and T cells, play central roles in the specific immune response.

Platelets are fragments of cells made when pieces of cytoplasm are pinched off large cells in the bone marrow. They circulate in the blood and stick to damaged epithelium of an injured vessel. At the injured site, the platelets secrete a sticky substance and attach to the proteins on the wall, plugging up any holes at the site of injury. A clotting enzyme released by platelets initiates a cascade of reaction that result in the formation of fibrin, a protein that forms a network of fibers that aid in blood clot formation, a process called coagulation.

Red blood cells express unique surface antigens composed of complex branched chains of sugars, or oligosaccharides. These antigens form the basis for one blood typing system used to determine compatibility before performing a blood transfusion. The ABO blood typing system involves two different oligosaccharides called A and B. People who express only the A oligosaccharide on the cell membrane of their red blood cells have type A blood. Individuals with the B oligosaccharide on their cells have type B. People who express both types A and B on their cells have type AB blood, and people with neither form are said to be type O. The Rh factor is another type of antigen found on the surface of some red blood cells. People whose blood cells have the antigen are Rh⁺ and those who do not are Rh⁻.

CARDIOVASCULAR HEALTH

Cardiovascular disease (CVD), afflictions related to the heart and blood vessels, is the leading cause of death of people of all racial and ethnic groups living in the United States. A heart attack occurs when part of the organ stops working and dies. If the coronary arteries that supply blood to the heart become blocked and blood cannot pass through, the tissue supported by those vessels will suffer from oxygen deprivation. The seriousness of the heart attack depends on the size of the area affected. A victim might die immediately or live for decades after a heart attack. During a heart attack some victims experience intense crushing pain in their chest, mild pain in an arm, or nausea or break into a sweat.

A stroke results when part of the brain dies as a result of interruption of its blood supply either by blockage or by rupture of a blood vessel. Body movement, function, and sensation can be impaired after a stroke, and death results in approximately one of three cases. Signs that someone might be experiencing a stroke include sudden numbness or weakness (especially if on one side of the body), confusion, dizziness, loss of sight, loss of balance, or severe headache with no known cause.

Two factors that increase the risk of heart attack and stroke are high blood pressure and atherosclerosis. Blood pressure is related to the amount of blood the heart pumps and the resistance it encounters as it flows through the arteries. High blood pressure often produces no symptoms until it is advanced, but it can be easily measured in a doctor's office. A blood pressure reading consists of two numbers, and normal is 120/80 or below. The first number, called the systolic pressure, indicates the pressure generated by the heart when it contracts and pushes blood through the arteries. The second number, the diastolic pressure, indicates how much pressure is on the arteries when the heart is at rest. Anything that narrows the diameter of the arteries will cause an increase in blood pressure. The body can do this naturally to compensate for a low blood volume or during periods of increased activity by contracting the smooth muscles surrounding the arteries, causing them to constrict and resulting in an increased blood pressure. High blood pressure increases the risk of heart attack and stroke.

Atherosclerosis is a disease in which fatty deposits such as cholesterol build up on the interior surface of arteries forming growths called plaques. The buildup decreases the diameter through which the blood flows, increasing the pressure. When calcium is deposited on the fatty acid buildup, the result is a hardening of the arteries, a condition termed arteriosclerosis. Hardened arteries do not have the elasticity of healthy arteries and cannot expand during the contraction of the heart, so the heart must work harder to push the blood through the vessels.

A thrombus, or clot, is often the cause of blockage in an artery that leads to a heart attack or stroke. The thrombus can develop in a coronary artery or an artery leading to the brain, or it can form elsewhere and travel through circulation in the blood until it becomes trapped in a narrow vessel. A traveling clot, called an embolus, often develops at sites where plaques have formed inside the arteries. Circulating platelets recognize plaques as injured tissue and begin the clotting process.

Many risk factors for CVD are easy to control. Cholesterol travels through the blood bound to proteins. High levels of low-density lipoprotein (LDL) cholesterol in the blood are associated with cholesterol deposition and plaque formation. Smoking, overweight, poor diet, and lack of exercise increase LDL levels. High levels of high-density lipoprotein (HDL) seem to protect against heart attack. Researchers believe HDL either carries cholesterol from the arteries back to the liver or removes excess cholesterol deposits from plaques, slowing their buildup. Exercise increases HDL levels. Not smoking, exercising regularly, maintaining a healthy diet and weight, and having medical checkups all decrease the risk of developing CVD. Hereditary factors cannot be controlled, but awareness of a family history of CVD can help someone develop a plan to decrease other CVD risk factors.

See also ANATOMY; ANIMAL FORM; HARVEY, WILLIAM; HOMEOSTASIS; HOST DEFENSES; INVERTEBRATES; PHYSIOLOGY; VERTEBRATES.

FURTHER READING
American Heart Association homepage. Available online. URL: http://www.americanheart.org/. Accessed January 14, 2008.

Nagel, Rob. *Body by Design: From the Digestive System to the Skeleton.* Vol. 1. Detroit: U*X*L, 2000.

Romaine, Deborah H., and Otelio S. Randall. *The Encyclopedia of the Heart and Heart Disease.* New York: Facts On File, 2004.

Seeley, Rod, R., Trent D. Stephens, and Philip Tate. *Anatomy and Physiology,* 7th ed. Dubuque, Iowa: McGraw Hill, 2006.

Silverthorn, Dee Unglaub. *Human Physiology: An Integrated Approach.* 4th ed. San Francisco: Benjamin Cummings, 2006.

Clark, Eugenie (1922–) American *Ichthyologist* Eugenie Clark is a world-renowned ichthyologist who has discovered numerous new species as well as being an expert on sharks. Her pioneering studies on shark behavior dispelled several falsehoods and demonstrated that sharks are intelligent creatures rather than ferocious manhunters.

CHILDHOOD, EDUCATION, AND TRAINING
Eugenie Clark was born on May 4, 1922, in New York, New York, to Charles and Yumiko Clark. Her mother was a swim instructor, and her father was the manager of a private pool. He died when Genie was only two years old, and by then, she already knew how to swim. They lived with Genie's Japanese grandmother in Queens but often went to the beach on Long Island, where they used chewing gum to plug their ears before plunging into the ocean. Her mother then worked to support the family at a newspaper and cigar stand in the lobby of the Down-

town Athletic Club. Genie often accompanied her on Saturdays and waited for lunchtime at the nearby New York Aquarium, where she amused herself by watching the fish swim in their long tanks. She soon began collecting fish at home in a 15-gallon (68-L) tank, became the youngest member accepted into the Queens County Aquarium Society, and learned to keep methodical records of all her pets. By the time she entered high school, she also kept pet snakes, toads, salamanders, and alligators, and biology was her favorite subject.

After graduation Genie enrolled at Hunter College in New York City. In addition to taking every zoology course offered, she took field courses in zoology and botany at the University of Michigan Biological Station during the summers. After obtaining a bachelor's degree in zoology in 1942, she worked as a chemist for the Celanese Corporation of America at their plastics research labs in New Jersey.

Clark entered graduate school at New York University, where she specialized in ichthyology, the scientific study of fishes. She researched the puffing mechanism of blowfish (the order *Tetraodontiformes* or *Plectognathi* includes the triggerfishes, puffers, filefishes, boxfishes, globefishes, and ocean sunfishes) under the guidance of Dr. Charles Breder, who was the curator of the Department of Fishes and taught an ichthyology class at the American Museum of Natural History. He was so impressed with her work that he published her research results in the museum's scientific magazine, *Bulletin of the American Museum of Natural History,* "A Contribution to the Visceral Anatomy Development, and Relationships of the Plectognathi" (1947).

Clark met Dr. Carl Hubbs, of the Scripps Institution of Oceanography of the University of California, at the 1945 meeting of the American Society of Ichthyologists and Herpetologists in Pittsburgh, Pennsylvania. When she completed her master's degree in zoology at NYU in 1946, she joined him as a part-time research assistant and began research toward a doctorate degree. Hubbs taught Clark to dive with a face mask and to walk on the ocean floor using a metal helmet connected by a long hose to a compressed air supply aboard a ship. Scuba gear was not yet widely available.

In 1947 the United States Fish and Wildlife Service (FWS) hired Clark to study fish in the waters surrounding the Philippines, but en route, she was delayed in Hawaii only to learn that they had changed their mind about hiring her because of her gender. Though disappointed, Clark used her time in Hawaii to explore the waters around the islands and study tiny tropical puffers. She returned to NYU and continued her dissertation studies on the mating habits of platies and swordtails under the supervision of

Eugenie Clark is a renowned American ichthyologist. *(Al Danegger/University of Maryland)*

Professor Myron Gordon. In aquariums, platies and swordfishes mated to create hybrid fish, but hybrids were never found in the wild. Clark described the act of true copulation in platies and determined that in a competition between sperm of the two different species, the same-species sperm had an advantage over sperm from a different species. This meant that even if both species deposited sperm inside a female, the sperm from the same species as the female would successfully fertilize the eggs. Clark also performed the first successful artificial insemination of a fish in the United States.

While working toward her Ph.D. in zoology, in 1949 Clark accepted a job studying the fish of the South Seas for the Scientific Investigation in Micronesia program of the United States Office of Naval Research (ONR). Since the United States had acquired many of the South Pacific islands after World War II, the navy wanted to know whether commercial fishing in the area would be profitable and to determine which fish were safe to eat and which were poisonous. One method Clark used for collecting fish was to add rotenone to tide pools. Rotenone is a chemical that can be extracted from plant roots that stuns the fish, causing them to float to the surface. The fish are still safe for consumption, however, and the vegetation is not harmed. Clark caught many interesting specimens, some of which were normally hidden or too small to be captured by other methods. She shipped the preserved specimens to the American Museum of Natural History. From the islands of Micronesia, Guam, Kwajalein, and Palau, Clark collected hundreds of specimens from the tide pools and also several puffers that she sent back to California for poison analysis. On the Palauan island of Koror, Clark learned how to hold her breath for a long time underwater and to spearfish near the coral reefs. She found many new types of plectognaths and other fish she had never seen before and encountered sharks and razor-toothed barracudas.

THE RED SEA AND CAPE HAZE

Clark earned her doctorate in zoology in 1950 and received a Fulbright Scholarship to study fish of the Red Sea in the Middle East. The Red Sea is very warm and salty, and its name is derived from the reddish appearance caused by tiny red algae that live on its surface. Clark had read an article describing the symbiotic relationship between a sea anemone and a clown fish and found many similarities between the Red Sea and the tropical Pacific Ocean. Many of the tropical plectognaths she studied years earlier had been described originally by ichthyologists from the Red Sea, but no one had scientifically analyzed the Red Sea fish in 70 years. Using the Marine Biological Station in Ghardaqa, Egypt, as a base, she traveled around, collected specimens of more than 300 species, wrote detailed descriptions, discovered three new species, and performed dissections of the many forms of marine life. The only poisonous fish she found were puffers. Several scientific papers and a bestselling autobiography, *Lady with a Spear* (1953), resulted from her year spent exploring the Red Sea.

Back in New York, Clark taught biology at Hunter College and conducted ichthyological research at the American Museum of Natural History. She also lectured at educational institutions around the country.

One reader of Clark's autobiography happened to be a wealthy Floridian with a son who was interested in marine biology. Anne Vanderbilt and her husband William invited Clark to their home on the Cape Haze peninsula. The meeting began a long association between Clark and the Vanderbilts, who funded the Cape Haze Marine Laboratory for her to direct in Placida, Florida, on the Gulf of Mexico. The lab opened in 1955.

The first marine mystery Clark solved at her new lab involved a fish called the belted sand bass from the genus *Serranus*. She observed several egg-carrying fish and assumed they were females, but she could not locate any of the males. Dissection and microscopic examination revealed not only eggs but sperm. The fish were hermaphroditic, meaning they had the reproductive organs of both males and females. The fish could switch from male to female in less than 10 seconds, and their coloring changed at the same time. They could even fertilize their own eggs.

"SHARK LADY"

Prompted by a request for fresh shark livers by Dr. John Heller, a medical researcher at the New England Institute of Medicine, Clark broadened her focus from fish reproduction to sharks. She identified 18 species living off the coast, including hammerheads, blackfins, dog fish, nurse, tiger, bull, lemon, and sandbar sharks. She performed hundreds of dissections and examined their stomach contents to learn about their diet, finding mostly small fish, crabs, eels, octopuses, and other sharks. She built pens at the end of the dock so she could keep live sharks for observation. A shark expert, Dr. Perry Gilbert from Cornell University, taught Clark to spray a chemical at the mouth and over the gills of sharks to render them unconscious for about 10 minutes, during which the workers could haul the sharks safely into the pens. Over the 12 years that Clark directed the Cape Haze Marine Laboratory (which moved to Sarasota in the early 1960s), she became an expert on sharks and earned the nickname "Shark Lady." The National Science Foundation and the ONR provided financial assistance in the form of grants that Clark used to expand the lab, build more shark pens, and execute research.

Though many people believed sharks were dangerous and stupid creatures, no one ever had inves-

Clark helped dispel several myths about sharks, such as this white-tip reef shark from the Northwestern Hawaiian Islands reserve shown here. *(National Oceanic & Atmospheric Administration/Department of Commerce)*

tigated their behavior. Clark observed that lemon sharks swam toward people approaching their tanks, as if they expected to be fed, and believed they were capable of learning. Her research demonstrated that sharks were actually intelligent creatures with the ability to learn and remember. To teach a shark to associate a ringing bell with food, she painted a square wooden target white, hung fish from it, and lowered it by rope into the shark pen. When the shark went for the bait, he bumped into the target and a bell sounded. After three days, she lowered the fish only after the shark bumped the target rather than attaching it, then progressively lowered the fish farther and farther away from the target until the shark had to swim to the other side of the pen to retrieve his food. Clark also performed experiments that showed that sharks can distinguish colors and shapes.

When Prince Akihito of Japan, who was interested in ichthyology and had studied many types of fish, invited Clark to be his guest in 1965, she gave him a nurse shark that she had trained to ask for food by bumping a target. She was surprised to learn that he had never been diving, so a few years later, in 1967, when he stopped by Sarasota on the way from South America back to Japan, Clark met him at 5:30 A.M., before any reporters were around, and taught him to skin-dive.

Though sharks were the main focus of her research, Clark also engaged in other scientific activities. Exploration of nearby freshwater springs led to the discovery of many ancient human bones and other traces of Native American life from over 7,000 years ago. In 1959 Clark broke the record for the deepest dive with compressed air for a woman at 210 feet (64 m). During a trip with her family to the Middle East sponsored by the National Geographic Society in 1964, she investigated a colony of garden eels near Elat, Israel, and identified and named a new species of sandfish found in the Red Sea *Trichonotous nikii* (after her youngest son, Niki).

Clark left the lab in the late 1960s to move back to New York and recommended Perry Gilbert at Cornell to succeed her as director. With financial assistance from a businessman named William Mote, the lab was expanded and renamed the Mote Marine Laboratory. Today scientists continue to research a variety of marine disciplines at the laboratory, which consists of research centers for sharks, marine mammals and sea turtles, fisheries enhancement, ecotoxicology, coastal ecology, aquaculture research and development, and tropical research (located in Summerland Key). Mote also offers a series of educational programs, houses aquarium exhibits, and runs a dolphin and whale hospital.

For two years Clark taught zoology at the City College of New York and was a visiting professor at the New England Institute for Medical Research. She accepted a position in the Zoology Department at the University of Maryland in 1968 and was promoted to full professor in 1973. The following year she published a second autobiography, titled *The Lady and the Sharks*. Many of her discoveries were published in articles that she wrote for *National Geographic* magazine.

In 1972 Clark examined fluid secreted by the Moses sole, a sand-dwelling flat fish that she had first observed a dozen years before. The whitish substance that oozed from pores by the fins made her fingers tingly and numb and killed sea urchins and reef fishes in small doses. When she placed a Moses sole in a shark tank as bait, the mouths of sharks seemed to freeze open as they approached the fish, and the sharks wildly shook their heads back and forth. When she put one on an 80-foot (24-m) line with nine other types of fish and lowered it into the sea, sharks consumed all the fish except the Moses sole. She pulled it up and rubbed the scales with alcohol then lowered it again, and it was immediately eaten. Putting a small shark in a tank with the fluid killed the shark in six hours. She wondered whether the Moses sole produced a substance that could act as a shark repellent. Though initial studies seemed promising, the substance was unstable at room temperature and could not be sold for general use. As Clark learned more about sharks, she did not believe shark repellents were necessary anyhow; people were more dangerous to sharks than sharks were to people. She felt that understanding the creatures' behavior and acting accordingly were a better measures against shark attacks. The Moses sole also produced an antidote for its own poison that was later found also to protect against bee, scorpion, and snake venom, but the antidote had to be injected at the same time as the poison itself to be effective.

In 1972 a friend in Mexico sent Clark photographs of sharks in underwater caves off the Yucatan Peninsula and described some unusual behavior. The sharks seemed to be sleeping or dazed. Clark was interested and in 1975 traveled to Mexico to investigate the phenomenon. Biologists thought that sharks needed to swim constantly in order to survive, but these sharks remained motionless inside the caves for extended periods. To obtain oxygen, they pumped water over their gills while they remained stationary. Clark noticed that freshwater was leaking into the caves, lowering the salt concentration. She suggested the change in salinity caused a trancelike state in the sharks. Her team also noticed that remoras could easily remove all the parasites from the sharks' skin while in the less salty water. The question of whether or not sharks or other fish actually sleep, as defined by a distinctive change in brain waves, has never been resolved.

As the 1970s drew to a close, pollution threatened many of the world's waters. Clark was particularly concerned with the fate of the Red Sea since one of her favorite places to dive was Ras Muhammad, located at the southern tip of the Sinai Peninsula. She initiated efforts to have the site declared a national park, to protect it from the traffic that caused damage to its coral reefs and pollution. Her vision became a reality in 1983, and the park now is often referred to as an underwater Garden of Eden.

In 1981, on the coast of Baja California, Clark took her first ride on a whale shark. These gentle creatures grow up to 40 feet (12 m) long and mostly eat plankton. Later she would discourage others from doing the same, to leave the sea creatures in peace.

From 1987 to 1990, Clark was the chief scientist for the Beebe Project, funded by *National Geographic.* She was in charge of 71 deep ocean dives in deep ocean submersibles, underwater vessels that can travel distances of up to 20,000 feet (6 km). Her longest dive was 17.5 hours and her deepest was to 12,000 feet (3,658 m). She also served as a consultant for several television specials about marine life including "The Sharks" in 1982, a program that was sponsored by National Geographic and received the highest Nielsen rating for a television documentary. She wrote a children's book, *The Desert beneath the Sea,* in 1991, with the author Ann McGovern. That same year, she visited Ningaloo Reef Marine Park in Australia to study whale sharks. She saw 200 in a single month and observed their eating habits. Though she retired as a professor emerita from the University of Maryland in 1992, she continues to travel to exotic locations around the world. Since 1996 Clark has been studying the behavior of a small reef fish off the coast of Papua New Guinea and off the island of Mabul off North Borneo. *Pholidichthys leucotaenia,* called convict fish because of the striped pattern, burrow into a labyrinth of tunnels as adults and stay there. The juveniles, however, leave the tunnels to eat plankton and then return to the tunnels, where the parents take the juveniles into their mouths. The young regurgitate their food into the parent, and the parent releases them from their mouth.

PERSONAL LIFE AND ACCOMPLISHMENTS

In 1942 Clark married a pilot named Hideo "Roy" Umaki, who served in the army and was stationed overseas. They divorced in 1949. The following year she married a Greek orthopedic surgeon named Ilias Papakonstantinou (later shortened to *Konstantinou*). They had two daughters and two sons during the next seven years. Clark divorced Konstantinou and married a writer named Chandler Brossard in 1967. The marriage between Clark and Brossard dissolved, and in 1970 Clark married a scientist from the National Institutes of Health named Igor Klatzo. That marriage also ended in divorce. In 1997 she married a longtime friend, Henry Yoshinobu Kon, who died in 2000.

In a career that has spanned seven decades, Clark has contributed significantly to knowledge in the fields of fish and shark behavior, taxonomy, and ecology. She has discovered 11 new species, authored more than 165 scientific and popular articles about marine science, and consulted for or participated in more than 200 radio and television programs dealing with conservation, marine biology, fish, diving, and career women. She has received numerous medals and awards for her research. The National Geographic Society, the Society of Women Geographers, the Maryland Women's Hall of Fame, the American Society of Oceanographers, and other organizations have acknowledged her. The University of Massachusetts awarded her an honorary doctorate in 1992, and the University of Guelph and Long Island University awarded her honorary doctorates in 1995. Four new fish species bear names in her honor: *Callogobius clarki, Sticharium clarkae, Enneapterygius clarkae,* and *Atrobucca geniae.* As Mote director emerita, Dr. Eugenie Clark remains committed to teaching others to protect and appreciate marine life.

See also MARINE BIOLOGY.

FURTHER READING

Clark, Eugenie. *The Lady and the Sharks.* New York: Harper & Row, 1969.

Dr. Eugenie Clark home page. Available online. URL: http://www.sharklady.com/. Accessed January 18, 2008.

Mote Marine Laboratory welcome page. Available online. URL: http://www.mote.org/. Accessed January 18, 2008.

Reis, Ronald A. *Eugenie Clark: Marine Biologist.* New York: Ferguson, 2005.

cloning of DNA   To clone something means to create an identical copy. Cloning often refers to whole organisms, in other words, making genetically identical individuals. Cloning deoxyribonucleic acid (DNA) means making molecules that are identical copies of an original DNA molecule or portion of one. If the cloned DNA encodes a gene product, then the procedure is referred to as cloning a gene. The process of cloning a gene involves recombinant DNA technology to isolate a foreign gene, insert it into another DNA molecule called a vector, and put it into bacterial cells or the cells of another simple organism such as yeast. When the microorganisms grow and divide to produce new cells (clones), the cloned DNA will also be replicated; thus cloning

DNA allows for the production of large numbers of copies of a gene for further genetic analysis, medical research, or other studies.

In 1972 Paul Berg synthesized the first recombinant DNA, made by using restriction enzymes to cleave DNA from two different organisms and joining the molecules. His hybrid genome contained viral DNA and genes from *Escherichia coli* (bacteria), an achievement that won him the 1980 Nobel Prize in chemistry. The American biochemists Stanley Cohen at Stanford University and Herbert Boyer at the University of California at San Francisco built upon Berg's accomplishment to clone the first gene in 1973. Cohen worked with bacterial plasmids, and Boyer worked on restriction enzymes. Boyer's research group had discovered the commonly used restriction enzyme *Eco*RI in 1968. In a collaborative effort, they used an enzyme to cut a plasmid that already encoded for resistance to the antibiotic tetracycline and, in the cut site, specifically inserted a gene encoding resistance to the antibiotic kanamycin obtained from another bacterium. After sealing the nicks in the plasmid, they introduced the recombinant plasmid into *E. coli*. The bacteria exhibited resistance to both antibiotics. Next, they successfully inserted genes from the toad *Xenopus laevis* into bacteria, the first transplantation of genes from an animal to a bacterial species. In 1980 the Lasker Foundation awarded the Albert Lasker Award for Basic Medical Research to Berg, Boyer, Cohen, and the virologist Dale Kaiser, who also advanced recombinant DNA methodology through his studies of bacteriophage lambda DNA.

Today, cloning a gene often simply means putting a particular gene into a plasmid. Scientists can obtain genes or DNA segments for cloning experiments from gene libraries. Breaking up an organism's DNA into smaller pieces makes it easier to manipulate in the laboratory. Restriction enzymes recognize and cut double-stranded DNA at specific sequences. Most restriction enzymes leave behind single-stranded overhangs that facilitate subsequent attachment of the DNA to other DNA molecules that have been cleaved with the same restriction enzyme. Once the DNA segments are of manageable size, scientists incorporate the fragments into plasmid vectors—small circular DNA molecules that carry DNA between cells of different organisms. A genomic plasmid library is a collection of bacterial colonies that each possess specific plasmid that contains a different DNA segment from a given species, such as human or fruit fly. Collectively, a copy of the organism's entire genome is represented. A complementary DNA (cDNA) library is more specific in that it only contains genes that encode for proteins—noncoding regions of DNA are excluded. The *c* stands for "complementary," as cDNA molecules are molecules of DNA made from messenger RNA (mRNA) molecules using the enzyme reverse transcriptase. Cells synthesize mRNA molecules by reading the DNA and adding ribonucleotides that are complementary to the template strand of DNA to a growing chain of RNA. This process is called transcription, and in eukaryotic cells, the product is a molecule of RNA that contains alternating exons and introns. Exons are the regions of the newly made RNA that encode for protein, and introns are intervening regions that do not encode for protein. Before protein synthesis, splicing occurs; the introns are removed and the exons are joined. The result is a mature mRNA molecule that ribosomes translate into a chain of amino acids during protein synthesis. To make a cDNA library, a researcher extracts and purifies mRNA from cells, then uses that mRNA as a template for reverse transcription, in which the enzyme reverse transcriptase reads the sequence of ribonucleotides on the mRNA and incorporates complementary deoxyribonucleotides into one strand of DNA, which it then uses to make a complementary second strand. The result is a double-stranded segment of DNA that contains only the coding region of a gene that encodes a protein that the original cell type expresses. Because of this, libraries can be specific for a specific cell or tissue type; for example, a pituitary library would only contain the genes expressed by the pituitary gland, which would differ from the set of genes expressed by the testes, for example. The DNA segments are maintained in plasmid vectors for transforming bacteria to generate clones. The library comprises the collection of bacteria containing the DNA from the cells or organisms of interest.

Having access to a premade gene library expedites the process of cloning. The next step is to find the bacterial clone that contains the gene to be cloned. The bacteria are spread out onto semisolid medium called agar to grow. This method of culturing the bacteria allows for individual colonies to be isolated, and each colony contains hundreds of thousands of clones of the original bacterial clone. The researcher must screen the colonies to locate one that contains the gene of interest using one of several different possible methods depending on the goal and on the biological reagents that are available. If the researcher is looking for a gene in one species that has already been isolated and characterized from another species, the researcher can use that DNA in hybridization methods for identifying colonies from the library that contain similar DNA. Hybridization occurs by the formation of hydrogen bonds between nitrogenous bases of complementary nucleotides. Molecules of single-stranded nucleic acid hybridize to other molecules that have

Construction of a Genomic Library

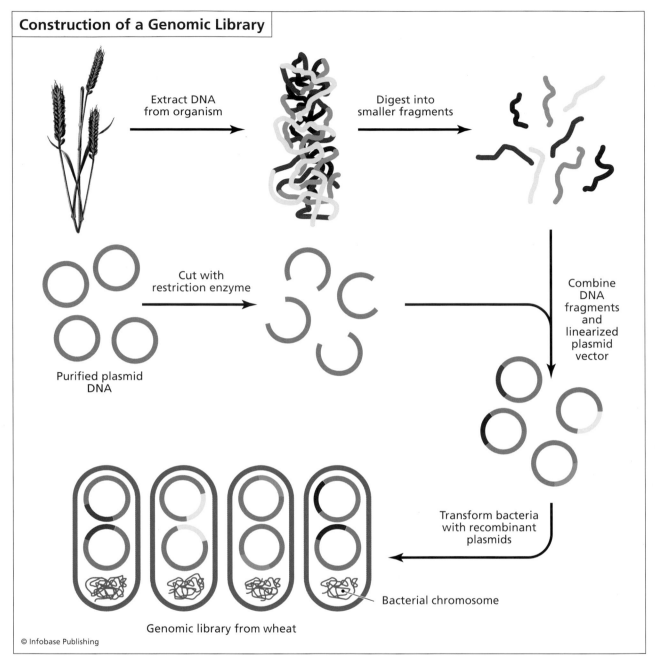

Extract DNA from organism

Digest into smaller fragments

Cut with restriction enzyme

Purified plasmid DNA

Combine DNA fragments and linearized plasmid vector

Transform bacteria with recombinant plasmids

Bacterial chromosome

Genomic library from wheat

© Infobase Publishing

Genomic libraries contain a copy of an organism's entire genome as fragments of DNA inserted into plasmid vectors maintained in bacterial clones.

complementary sequences. A probe is a molecule that is labeled with either a radioactive isotope or a fluorescent dye and is used to search for complementary molecules on a substrate such as a nitrocellulose or nylon membrane to which DNA has been bound. One way to accomplish this is to blot the surface of an agar plate with a membrane so that the bacterial colonies contact the membrane, forming a replica of the pattern of the colonies on the agar plate. The researcher then treats the membrane in a manner that immobilizes the DNA in specific positions on the membrane and denatures the DNA to ensure it is single-stranded and therefore available for hybridization. Subsequent incubation of the probe diluted in a solution with the membrane allows complementary molecules to recognize and bind to, or hybridize with, one another. The radiolabel or fluorescent dye enables detection of the locations on the membrane where the probe hybridized on a piece of X-ray film. Aligning the pattern of dark spots on the film with the colonies on the plate reveals which colonies contain the DNA of interest.

A variation of this method involves the use of antibodies to assay for specific protein products of the genes in the library. Antibodies are proteins made by the immune system that specifically recognize and bind to other molecules including proteins. The antibodies can be tagged with fluorescent labels so the researcher can detect their position on a membrane to which proteins have been immobilized. If the gene product is an enzyme, the researcher can simply assay for the product of the biochemical reaction catalyzed by the enzyme.

Once the investigator has identified the bacterial clones that contain the DNA of interest, he or she can culture the bacteria to grow large quantities. The bacterial cells can be harvested and the plasmid DNA extracted and purified. The researcher may need to move the cloned gene to a different plasmid vector that is more suitable for future investigations. Some vectors are specially designed to express the gene product. Called expression vectors, these plasmids contain a strong bacterial promoter to initiate transcription and an adenine-thymine-guanine (ATG), the triplet codon that initiates translation by ribosomes. The gene might need to be cloned into a different organism in order for the gene product to be properly processed after translation of the mRNA into a polypeptide chain, or the gene might need to be cloned into a different cell type or organism in order to carry out functional studies.

For investigations into the control of gene expression, DNA segments from the regulatory region of a gene may be of more interest than segments from the coding region or the gene product itself. For these studies, the region regulatory elements of a gene may be cloned and inserted into a plasmid in front of a reporter gene, something that is easy to measure. One commonly used reporter gene is β galatosidase, an enzyme that hydrolyzes the disaccharide lactose into glucose and galactose subunits. The activity of this enzyme is often associated with a chemical reaction that causes colonies expressing this protein to turn blue if a certain substrate is included in the medium, making them easy to identify. Green fluorescent protein (GFP), originally isolated from a type of jellyfish, is another popular reporter gene that fluoresces a bright green color when exposed to blue light. One advantage of GFP is that its activity can be measured in living cells, allowing a researcher to monitor the gene's expression under different conditions. When the GFP coding region is attached in frame to the gene of a protein of interest, one can use a microscope to look at cells under blue light and observe the intracellular localization of the protein.

After the cloning of a gene, the potential applications are practically limitless and benefit many people and improve daily life. The advent of cloning has enabled pharmaceutical companies to produce hormones, such as insulin and growth hormone, and blood clotting factors for treating patients whose glands or tissues do not synthesize sufficient quantities. The human genes that encode these proteins have been inserted into bacteria that can be grown in large quantities easily and cheaply. Medical researchers also clone genes from viruses, bacteria, and toxins that bacteria synthesize to make recombinant vaccines to protect people from infectious diseases. Cloning technology has benefited basic research tremendously by allowing biochemists, geneticists, and cell and molecular biologists to study single genes or proteins in controlled conditions. Having the gene inserted into a vector also allows researchers to induce specific mutations and to observe their effects. Commercial industries use cloned genes to manufacture enzymes for use in products such as laundry detergents and biomedical research. Agricultural applications include using cloned genes for genetically engineering crops or livestock that possess traits that improve yields or make them more marketable. The introduction of cloned genes into certain microorganisms increases their efficiency in degrading chemical pollutants for bioremediation purposes.

The polymerase chain reaction (PCR) is a technique used to amplify segments of DNA. Whereas recombinant DNA technology requires several days to generate large quantities of a gene or DNA segment, PCR takes only a few hours. Depending on the future research plans for a particular cloned gene, PCR may eliminate the need to clone the DNA in some cases.

See also BIOTECHNOLOGY; CLONING OF ORGANISMS; DEOXYRIBONUCLEIC ACID (DNA); GENETIC ENGINEERING; POLYMERASE CHAIN REACTION; RECOMBINANT DNA TECHNOLOGY.

FURTHER READING

Birren, Bruce, Eric D. Green, Sue Klapholz, Richard M. Myers, and Jane Roskams. *Genome Analysis: A Laboratory Manual Series.* 4 Vols. Cold Spring Harbor, N.Y.: Cold Spring Harbor Laboratory Press, 1997.
Sambrook, Joseph, and David Russell. *Molecular Cloning: A Laboratory Manual.* 3rd ed. 3 Vols. Cold Spring Harbor, N.Y.: Cold Spring Harbor Laboratory Press, 2001.

cloning of organisms   Biologically defined, a *clone* is a genetically identical copy of an organism. Fragments of genetic material, or deoxyribonucleic acid (DNA), can also be cloned, meaning identical copies have been made and usually inserted into bacteria for maintenance. Molecular biologists, biochemists, geneticists, and other researchers regularly clone genes or pieces of DNA as part of standard laboratory procedures. Cloning organisms, however, is much different and falls under the realm of developmental biology rather than genetics. In this case, cloning means creating a whole organism that is genetically identical to another; the individuals are referred to as clones.

For organisms that reproduce asexually, cloning is no big deal. Asexual reproduction is the creation of new individuals from one parent and is the normal means of reproduction for many bacteria, fungi, protists, and numerous plants. Most unicellular organisms replicate by synthesizing an identical copy of their DNA, then splitting into two cells, with each cell receiving the same genetic information. Some multicellular organisms can also reproduce asexually. For example, fungi can reproduce both asexually and sexually, depending on whether two different organisms contribute genetic material to the spores that develop into new complete organisms. Plants can also naturally create genetically identical but separate organisms by several different methods. For example, strawberries asexually reproduce via stolons, extensions that grow outward from the base of a plant along a horizontal surface. Every so often, the stolon makes contact with the ground and forms an organized structure from which roots grow downward to penetrate the soil and shoots grow upward, forming a new plant that can grow independently if separated from the parent clone. The eyes of potatoes are really buds that can also sprout to form new plants. Gardeners and horticulturists often take cuttings to propagate a favorite plant asexually.

Though in all the preceding examples the new organisms that are formed are indeed clones of their parent, in most discussions, *cloning* refers to cloning animals, a complex process that requires technological intervention. While nature can accomplish this by the spontaneous division of a fertilized egg into two separate embryos that become identical twins, purposeful animal cloning is a feat of cell biology and embryology. Success is due to the hard work of numerous experimental embryologists who each contributed their insight and skills to put into practice today what a century ago was considered the domain of God, the creation of designed life.

BRIEF HISTORY OF ANIMAL CLONING

The German embryologist Hans Spemann performed one of the first vertebrate animal cloning experiments more than 100 years ago, in 1902, when he split apart a two-celled salamander embryo and demonstrated that each embryonic cell contained all of the information necessary to create a new organism. He went on to show that as development progressed and cells became more differentiated, they lost the ability to form entire new organisms. Spemann won the Nobel Prize in physiology or medicine in 1935 for discovering the organizer effect, an important aspect of early embryogenesis by which one group of cells influences the developmental pathway taken by another group of nearby cells, a phenomenon today called embryonic induction. In 1938 Spemann also first conceived of nuclear transfer, the transfer of a nucleus from one cell into another from which the nucleus has been removed. Nuclear transfer later became the basis for successful animal cloning procedures. In 1952 the American biologists Robert Briggs and Thomas J. King cloned tadpoles. Their protocol involved removing a nucleus from a blastula-stage cell (an embryonic stage during which the cells form a fluid-filled sphere but have not yet begun to differentiate into specialized tissues) and inserting it into a fertilized egg from which they had already removed the nucleus. After numerous attempts failed, they eventually produced 27 tadpoles from 104 nuclear transfers.

In 1958 the American biologist F. C. Steward grew whole carrot plants from differentiated root cells. This was a significant event, because in vivo, after a cell becomes completely differentiated, there is no turning back. A liver cell remains a liver cell until it dies; a fat cell cannot convert into a neuron to replace damaged brain tissue. Steward's root cells formed all of the new tissue types necessary to create a whole plant. An animal is different from a plant, however, given that many plants naturally reproduce by creating clones of themselves. Only a few years later, in 1962, John Gurdon cloned frogs from differentiated intestinal cells. This would have been a major milestone, but critics doubted whether what Gurdon thought had happened really did happen. Scientists still do not know whether in fact Gurdon's reported frogs did result from successful cloning using differentiated cells, but the event put cloning in the limelight.

Over the next two decades, recombinant DNA technology took center stage, leading to the creation of genetically engineered bacteria in 1973. Meanwhile, the Danish scientist Steen Willadsen perfected a method for freezing and thawing livestock embryos that resulted in live births. The next real advance in cloning was Willadsen's 1984 success in producing a live lamb by fusing cells from an eight-cell lamb embryo with unfertilized eggs. He coated the embryos in agar for protection, grew them in a sheep oviduct for one week, and then transferred the developing embryos to a surrogate mother sheep, one of which resulted in a live birth, the first cloned mammal. The following year Willadsen birthed a new industry cloning cattle for Grenada Genetics, and in 1986, he had success using nuclei from differentiated cells of a one-week-old cow embryo. A team in Wisconsin, Neal First, Randal Prather, and Willard Eyestone, used a slightly different method to clone cattle. They fused single-cell embryos to fertilized eggs using an electric current, allowed them to develop to the eight-cell stage within oviducts that were removed and kept alive in culture, and then transplanted them into a surrogate.

In 1995 Sir Ian Wilmut and Keith Campbell at the Roslin Institute in Scotland produced two cloned sheep from differentiated embryo cells by nuclear transfer. They discovered that a key step was to synchronize the cell cycles of the nuclear donor and recipient cells. The following year Wilmut and Campbell cloned the first sheep, named Dolly, from differentiated adult udder cells. Though experiments had been under way for years with the goal of cloning animals from differentiated cells, the announcement of Dolly's birth stunned the world. To the general public, this event made the possibility of cloning humans closer to home. One year after Dolly, the Roslin group reported success in creating the first transgenic lamb, Polly, who expressed a human gene

Dolly the sheep was the first mammal cloned from adult differentiated cells. *(AP Images)*

for factor IX, a blood-clotting protein used to treat hemophilia.

Since the successes at the Roslin Institute, numerous other mammals have been cloned. In 1998 Ryuzo Yanagimachi and Teruhiko Wakayama, at the University of Hawaii, reported that they cloned 50 mice, including some clones of clones. The technique, now called the Honolulu technique, was more efficient and involved the injection of nuclei from cumulus cells into an enucleated egg. Cumulus cells are cells that surround eggs during development inside the ovaries. After letting the cells recover for six hours, they added chemicals to the medium to stimulate cell division, and they transplanted the embryos into a surrogate mother after reaching the blastocyst stage. A year later, they cloned the first male, using nuclei from cells removed from the tip of a male mouse's tail. To date, the following mammals have been cloned: mice, sheep, cows, pigs, cats, a rhesus monkey, rabbits, mules, a deer, a dog, a horse, rats, and a guar.

HOW IT IS DONE

Somatic cell nuclear transfer, the method used to create Dolly and many other cloned animals, requires a great amount of skill and time. The goal is to create an egg that contains a nucleus from a donor cell of the individual to be cloned. The term *somatic* means the cell is not from a reproductive cell such as an egg or sperm, but is from another part of the body. A skilled technician uses a micromanipulator to insert a needle into an egg cell and gently sucks out the nucleus, which contains all of the DNA. Enucleation, the process of removing the nucleus, essentially turns an egg cell into a cloning factory. The genetic material is gone, but the rest of the egg contents provide the necessary building blocks for biomolecules: amino acids, sugars, nucleotides, and fatty acids. The egg cell also contains the machinery and enzymes necessary to synthesize proteins and factors that carry out cell division.

The technician then uses another needle to insert a different nucleus from a cell of the animal to be cloned into the enucleated egg. Alternatively, after removing the nucleus from an egg, the egg can be fused with a whole cell from the animal to be cloned. Because the nucleus is from a differentiated cell, reprogramming of its genetic material is necessary in order to create a whole new individual. During normal development, the DNA of differentiated cells becomes chemically modified by the addition of methyl groups ($-CH_3$) to the nitrogenous bases of DNA, an event that leads to the inactivation of genes that a specialized cell will not need to express. Early embryonic cells are totipotent, meaning they each individually have the capacity to give rise to

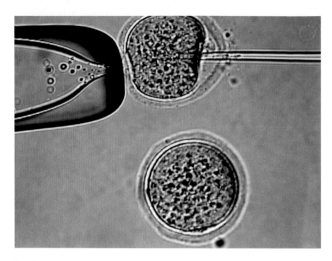

In the process of nuclear transfer, a micropipettor is used to remove the nucleus from one cell and replace it with the nucleus removed from another cell. *(James King-Holmes/Photo Researchers, Inc.)*

a complete individual. As cells become specialized, they lose this ability (in part because of methylation of the chromatin); that is why successful cloning from differentiated cells was such a milestone. The egg cell "reprograms" the donor nucleus by removing the previously added methyl groups and then reestablishing new methylation patterns.

Administration of an electrical shock stimulates the manipulated cell to divide, beginning the process of embryogenesis. Each of the embryo's cells, called blastomeres, contains nuclear DNA identical to that of the parent clone, the organism from which the nucleus was taken: in other words, the embryo is a clone.

For some applications, such as research purposes or therapeutic cloning, the cloned embryonic cells are the final goal. When the goal is to create a complete cloned organism, transplantation of the embryo clone into a surrogate mother follows. If transplantation is successful and the surrogate mother's uterus implants the embryo and accepts the pregnancy, gestation follows the normal course. Despite all the progress by cell and developmental biologists, the success rates are still very low, less than 1 percent on average, with variation depending on the species and the techniques used.

In January 2008 a group of scientists published a report of their success producing cloned human embryos from adult skin cells using somatic cell nuclear transfer (French et al., 2008). While this is an important step in the development of stem cells from somatic cells, the scientific community eagerly awaits the creation of cloned human embryos that mature and produce patient-specific embryonic stem cells from which a stem cell line could be created.

In November 2007 two groups reported success at reprogramming human skin cells into cells that resembled embryonic stem cells. In a study led by James Thomson of the University of Wisconsin-Madison, researchers showed that four factors were sufficient for reprogramming somatic cells to become pluripotent stem cells. Called "induced pluripotent cells," the cells exhibited the essential characteristics of embryonic stem cells and had the developmental potential to differentiate into cells of all three primary germ layers. In the second study, led by Shinya Yamanaka at Kyoto University in Japan, researchers inserted four genes (two of them different from those of the first study) into adult human fibroblasts. Their induced pluripotent cells also exhibited stem cell–like characteristics. Because reprogramming does not utilize eggs or embryos, this newer technique diminishes some of the ethical objections that cloning raises, but reprogrammed cells carry an increased theoretical risk for cancer. John Gearhart, the scientist who discovered human fetal embryonic stem cells, claims this alternative approach "is the future of stem cell research."

APPLICATIONS OF ANIMAL CLONING

While polls show that the majority of people overwhelmingly disapprove of the application of cloning technology for the purpose of reproducing humans, there are many other beneficial applications for cloning. Animal cloning methodology can be used for the commercial production of useful gene products, for cloning farm animals with desirable characteristics, and for therapeutic medical uses and has advanced and will continue to advance research in many different fields, especially developmental biology.

One beneficial application of cloning technology is the bioengineering of animals to produce useful proteins. A transgenic animal is one whose genome has been deliberately modified, such as by the insertion of a gene from another species. The first transgenic sheep, Polly, expressed a human gene that encoded a clotting factor used to treat hemophilia patients. Transgenic cattle have also been or may soon be created to produce and secrete in their milk proteins such as insulin and growth hormone and proteins for treating diseases such as emphysema, cystic fibrosis, or phenylketonuria. Transgenic animals created by cloning can also be designed to exhibit specific disease symptoms, so medical researchers can use them as living disease models to understand the disease better and to explore treatments.

For thousands of years farmers have used selective breeding to improve the yields and quality of agricultural products of animals, such as milk, meat, and wool. Farmers would profit from cloning their

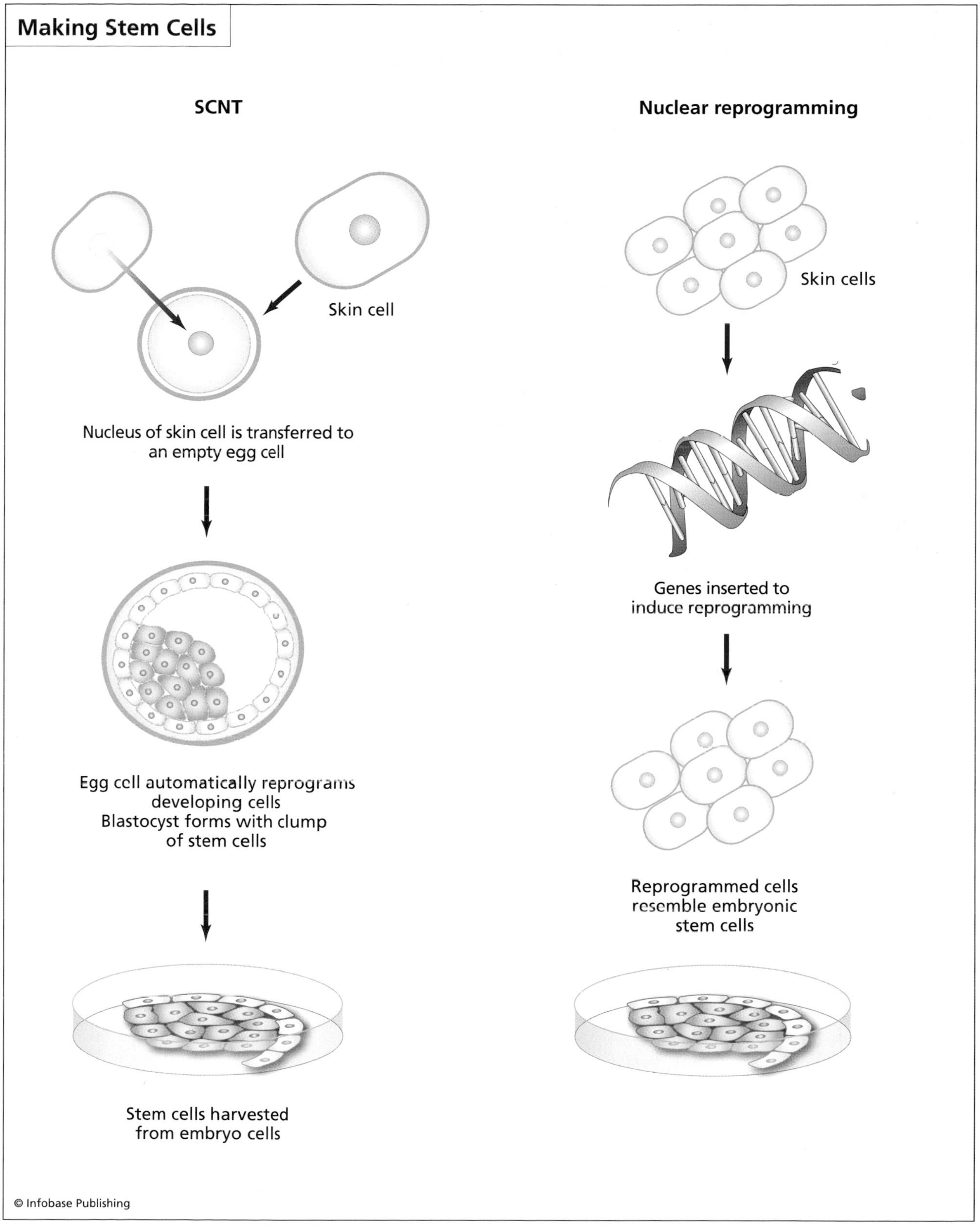

Making Stem Cells

SCNT

Nuclear reprogramming

Skin cell

Skin cells

Nucleus of skin cell is transferred to
an empty egg cell

Genes inserted to
induce reprogramming

Egg cell automatically reprograms
developing cells
Blastocyst forms with clump
of stem cells

Reprogrammed cells
resemble embryonic
stem cells

Stem cells harvested
from embryo cells

© Infobase Publishing

In somatic cell nuclear transfer (SCNT), after transferring the nucleus from the parent clone to an enucleated egg, the resulting embryo cells can be used for research purposes, as in therapeutic cloning. In nuclear reprogramming, genetic factors are inserted into adult cells, causing them to "regress" back into a state that resembles stem cells capable of differentiating into any cell type.

fattest cattle and pigs, and the U.S. Food and Drug Administration preliminarily ruled in late 2006 that milk and meat from cloned cows, goats, and pigs is perfectly safe for human consumption. The general public, however, hesitates to embrace the idea of eating cloned animals, mostly because of lack of understanding of the process of cloning and the perception that cloning makes something unnaturally grotesque or monsterlike. Another desirable characteristic that could be potentially engineered into farm animals is disease resistance.

The potential benefits of therapeutic cloning may force people to overcome their negative feelings about cloning. Therapeutic cloning is the use of somatic cell nuclear transfer and other cloning methodology for the purpose of carrying out stem cell research and studying and curing diseases. One possible application for this methodology is to treat diabetes. To accomplish this, a cell from a diabetes patient is cloned using an unfertilized egg cell. After stimulating the cell to start dividing in vitro, the stem cells can be harvested and transplanted back into the patient. Because the transplanted cells would contain the same genetic information, the patient's immune system would not recognize them as foreign and thus would not reject them by mounting an immune attack. Inside the patient, the cells could differentiate into cells that produce and secrete insulin and respond to the appropriate regulatory control mechanisms, minimizing the need for constant monitoring and adjustments in hormonal administration. Patients with neurodegenerative diseases such as Parkinson's and Alzheimer's diseases as well as spinal cord injuries or brain damage from a stroke may also benefit from therapeutic cloning. Nerve cells do not proliferate after differentiation; thus once they become damaged, the body cannot replace them. By cloning cells from patients with these conditions, scientists could create undifferentiated stem cells with a patient's own genetic material that upon transplantation into the patient could potentially replace the damaged cells and restore nervous system function.

Therapeutic cloning would also allow researchers to study the development of certain diseases, to learn how cells become diseased, and to follow the molecular and cellular events associated with the progression from normal to malfunctioning cells. A better understanding of these processes could lead to novel therapies and treatments, which could also be studied using cloned cells, tissues, and organisms that have been designed to model the disease.

Thousands of patients who await organ transplantation each year could also benefit from the applications of therapeutic cloning research. The number of organs that become available by the death of others falls far below the number needed. Another problem is that when organs are available, they must be a perfect match for molecular markers expressed by the recipient's tissues; otherwise the recipient's immune system will attack and attempt to destroy the transplanted organ. Even when the tissues match fairly well, the recipient must take immunosuppressant drugs for the rest of his or her life, which make him or her susceptible to infections and other maladies. One means to overcome the availability problem is to use organs or tissues from other species, such as pigs. This procedure, called xenotransplantation, still carries the risk of rejection. Using therapeutic cloning to create an organism that expresses the same molecular markers as the transplant recipient could prevent the patient's immune system from recognizing the transplanted tissue or organ as foreign and thus prevent rejection. In the future, perhaps scientists could grow whole new organs from embryonic cells created by somatic cell nuclear transplantation, both eliminating the risk of rejection and reducing the wait time for donor organs.

One can imagine that reproductive cloning could help restore populations of endangered species. In 2001 a biotech company cloned a guar, a type of endangered ox, using a cow as the surrogate mother for gestation. The guar died at two days old of dysentery. In 2001 Italian scientists reported using adult cells to clone a mouflon, an endangered species of wild sheep, which now resides at a wildlife center in Sardinia. One problem with cloning endangered species is that the egg and the surrogate used in the process are different species from the animal being cloned. Also, cloned populations would not have the genetic variability that is necessary for species to survive environmental challenges.

HUMAN CLONING

The cloning of humans raises many complex issues. Many concerns are ethical, but numerous unanswered scientific questions also remain, making it impossible for any well-educated panel to make intelligent, informed decisions regarding human cloning. Scientists do not know how human cloning will differ from animal cloning, which itself continues to be plagued by unresolved questions and difficulties. The success rate of cloning from nuclear transfer is very low because of tremendous stress placed on the nucleus introduced into an egg. Dolly was the single successful live birth resulting from 277 nuclear transfers. How many sacrificed embryos would be necessary to achieve success in cloning a human? Scientists also cannot predict the long-term effects of cloning. Cloned animals have demonstrated symptoms associated with an age much greater than the time since birth should permit. If an animal clone results from the transfer of a nucleus taken from a cell of a five-year-old, is the cellular age at birth already five years? Cell biologists believe that cells may have predetermined life spans, so are clones doomed to shorter life expectancies? Will clones have any acquired characteristics of their clone parent? Would any changes that occurred to the parental clone's DNA, such as rearranged genes, chemical modifications to subunits of the DNA, and the manner of packaging the DNA into chromosomes, be transmitted to cloned offspring? Nuclei from cells at earlier stages of development should have fewer permanent changes.

A few of the ethical questions include, When does life begin? Considering the low success rate, what number of sacrificed embryos is acceptable in order to achieve a live birth? Who should be cloned? Must certain criteria regarding degree of intelligence, health, fertility, lack of known genetic defects, or family history be met? Who makes these decisions? Who pays for the process? What rights do the clones have?

A few international organizations have policies regarding cloning. The United Nations (UN) approved a nonbinding ban on all human cloning in 2005, but the wording does not specifically address somatic cell nuclear transfer. The Royal Society, one of the world's oldest and most respected scientific organizations, denounced the UN for adopting the "ambiguous and badly-worded" declaration. The European Court of Human Rights of the Council of Europe added a protocol to their Convention on Human Rights and Biomedicine that was entered into force in 2001 banning any intervention that seeks to create genetically identical human beings, and though the treaty specifically mentions nuclear transfer, it is unclear whether this procedure is permitted for therapeutic cloning. The United Kingdom passed legislation in 2001 that allows somatic cell nuclear transfer for research purposes, as long as permission is first obtained from the Human Fertilization and Embryology Authority. That same year, U.S. president George W. Bush announced that federal money could not be used in research that involved harvesting new cells from embryonic stem cells (including for therapeutic cloning) but allowed for continued research on cell lines from embryonic cells previously harvested. No federal laws address reproductive or therapeutic cloning using private money, but some states have taken action on this issue. To date 13 states have enacted legislation that prohibits reproductive cloning, and two others prohibit the use of public money for such research. Seven of the 15 states that have laws addressing cloning specifically allow therapeutic cloning. Some fear that the U.S. ban on stem cell development and research is leading to a decline in the country's competitive edge in scientific research, but some very reputable U.S. research institutions including Harvard University and the University of California at San Francisco are getting around the ban by establishing privately funded programs. Other countries have recently lifted bans on therapeutic cloning. Australia lifted its ban on therapeutic cloning for research in late 2006 but requires that embryos be destroyed within 14 days. In 2004 France approved a law allowing stem cell research using human embryos and banning reproductive cloning. A Japanese government panel also approved the use of human clones for research purposes that same year.

See also CELLULAR REPRODUCTION; EMBRYOLOGY AND EARLY ANIMAL DEVELOPMENT; REPRODUCTION; WILMUT, SIR IAN.

FURTHER READING

Edwards, J. L., F. N. Schrick, M. D. McCracken, S. R. van Amstel, F. M. Hopkins, M. G. Welborn, and C. J. Davies. "Cloning Adult Farm Animals: A Review of the Possibilities and Problems Associated with Somatic Cell Nuclear Transfer." *American Journal of Reproductive Immunology* 50 (2003): 113–123.

French, Andrew J., Catharine A. Adams, Linda S. Anderson, John R. Kitchen, Marcus R. Hughes, and Samuel H. Woods. "Development of Human Cloned Blastocysts following Somatic Cell Nuclear Transfer (SCNT) with Adult Fibroblasts." *Stem Cells* published online (January 17, 2008). Available online. URL: http://stemcells.alphamedpress.org/cgi/reprint/2007-0252v1.pdf.

Panno, Joseph. *Animal Cloning: The Science of Nuclear Transfer.* New York: Facts On File, 2004.

Takahashi, Kazutoshi, Koji Tanabe, Mari Ohnuki, Megumi Narita, Tomoko Ichisaka, Kiichiro Tomoda, and Shinya Yamanaka. "Induction of Pluripotent Stem Cells from Adult Human Fibroblasts by Defined Factors." *Cell* 131, no. 5 (November 2007): 861–872.

U.S. Department of Energy Office of Science, Office of Biological and Environmental Research, Human Genome Program. "Human Genome Project Information: Cloning Fact Sheet." Available online. URL: http://www.ornl.gov/sci/techresources/Human_Genome/elsi/cloning.shtml. Updated August 29, 2006.

Yu, Junying, Maxim A. Vodyanik, Kim Smuga-Otto, Jessica Antosiewicz-Bourget, Jennifer L. Frane, Shutan Han, Jeff Nie, Gudrun A. Jonsdottir, Victor Ruotti, Ron Stewart, Igor I. Stukvin, and James A. Thomson. "Induced Pluripotent Stem Cell Lines Derived from Human Somatic Cells." *Science* 318, no. 5858 (December 21, 2007): 1,917–1,920.

community ecology Ecology is the study of the interactions between organisms and their environment, including both living and nonliving components. Populations of different species that live near and interact with one another form biological communities. The boundaries that define a biological community may be small, such as a community of phytoplankton, aquatic plants, zooplankton, invertebrates, and fish that live in a pond. Alternatively, the boundaries may be vast, such as those demarcating the African savannah. The size depends on the research focus of the individual scientist studying the community.

Community ecology is the study of how interactions of living components within an ecosystem affect that community's structure and organization. Within communities, the organisms depend on one another for survival and reproduction. For example, insects, birds, and bats pollinate plants, helping them to reproduce with other individuals that may be located at too great a distance for wind alone to accomplish this. The plants provide food in the form of nectar or pollen for the pollinators. Their green tissues serve as food for plant eaters in a community, and their seeds as food for other organisms. The roots of plants help stabilize the soil from erosion, and they also provide shade and shelter for other animals. The numerous, complex interactions are not always obvious; unraveling them is the major goal of community ecologists.

INTERSPECIFIC INTERACTIONS

The individuals and populations of species within a biological community interact in a variety of ways. In a mature community, each species plays an important role in maintaining the community's health, or stability. This does not mean that all the individuals benefit from interactions with individuals of other species; rather, the community as a whole benefits, as does the ecosystem to which it belongs. In some cases, the link between two or more species is so close that the species coevolve: that is, their dependence on each other limits the extent to which modifications may occur. Relationships between two species (the simplest situation) include competition, predation, herbivory, symbiosis, and disease.

Competition is the demand of two or more organisms for an environmental resource that is in short supply. The 19th-century naturalists Charles Darwin and Alfred Russel Wallace evoked competition as a major factor driving natural selection. They suggested that because environmental resources such as food and space are limited, and because individuals produce more offspring than can survive, species must compete for such resources. In doing so, over evolutionary time species develop adaptations that confer an advantage to them in that particular environment. Intraspecific competition occurs between members of the same species. For example, damsel-fish compete for nooks and crannies in coral reefs that they use for shelter and spawning and from which they obtain food. Interspecific competition is the condition when populations of different species compete.

Each species has its own niche within a biological community. A niche is the sum of the biotic and abiotic environmental factors that affect the survival, growth, and reproduction of a species and includes the relationships between that species and others in its community. In 1934 G. F. Gause formulated the competitive exclusion principle, which states that two species cannot occupy the same niche in a community. Many species have similar needs or have overlapping roles in a community, and this feature of a community helps maintain the function of that ecosystem. The more biodiverse a community is, the more overlapping roles there will be, and the better chance that if the population of one species decreases, the function it performed will still be carried out by another species. Though some overlap in roles is good, the roles they play must differ in some respect, or one of the species (the one more favorably suited for that environment) will eventually overtake the other in the competition for resources. The two species cannot occupy identical niches. For example, the Galápagos have numerous species of finches that differ in their food source. Even though many feed on seeds, the beaks of certain species are adapted to feed on different-sizes of seeds and seeds of varying hardness, thus reducing the competition between the finch species, allowing them to coexist. The degree of niche overlap correlates with the degree of competition that a species faces, and over time competition for a niche can lead to a phenomenon called character displacement, in which species evolve to occupy different niches.

Factors that influence interspecific competition in plants include exposure to sunlight, water availability, and nutrient availability in the soil. Other factors such as composition or acidity of the soil can affect which species will "win" the competition and dominate a region when planted together. This condition can lead to a gradient of species at boundaries where environmental conditions change, such as from a low to a higher altitude or from a coastal region moving inland.

Another type of interspecific interaction is predation. Predators hunt and kill other organisms to obtain their food. In response to hearing the word *predator,* one often thinks of large animals such as mountain lions, sharks, or snakes. Indeed these are examples of successful predators, but so are sea stars that prey on mussels, and so are stoneflies that feed on mayflies, caddis flies, and true flies. As do all interspecific interactions in a community, predation plays

Many leguminous plants, such as the broad bean shown here, participate in symbiotic associations with nitrogen-fixing bacteria that form nodules on the roots. *(Adam Hart-Davis/Photo Researchers, Inc.)*

an important role in maintaining the structure of a community by controlling population size. Because predators are consumers, and often secondary or tertiary consumers, they have a significant impact on all the trophic levels of an ecosystem, particularly considering the amount of biomass and energy necessary to support organisms at higher trophic levels.

Organisms that feed off plants or algae, even though they sometimes kill them by doing so, are usually not referred to as predators but simply herbivores. (Most herbivores feed off living plants.) Like predation, herbivory is a type of interspecific interaction in which one organism exploits another to increase its own fitness at the expense of the fitness of another species. Herbivores in both terrestrial and aquatic environments have unique adaptations that allow them to ingest and grind up plant material, process and digest it (this often involves symbioses, discussed later), and distinguish between poisonous and nonpoisonous food sources.

Symbiotic associations are intimate interspecific interactions. Two or more species live in direct contact to the benefit of at least one of them. Mutualistic symbioses benefit both species, as exemplified by the bacterium *Rhizobium* and broad bean plants. Although nitrogen is abundant in the atmosphere in the form of nitrogen gas (N_2), most organisms cannot use this form because of the strong triple covalent bond that holds the two nitrogen atoms together. Certain species of bacteria, such as *Rhizobium*, are capable of fixing this nitrogen by converting it into ammonia (NH_3), which plants can readily incorporate into organic molecules such as amino acids and nucleic acids. Some nitrogen-fixing bacteria live freely in the soil, but others live in symbiotic associations within the roots of plants, mostly legumes. The plant benefits because it has a built-in source of available nitrogen, which is often a limiting nutrient for plants. Though the bacteria expend much energy fixing more nitrogen than they will use themselves, in return they receive a steady supply of carbon, which the plants fix through photosynthesis.

Another common example of symbiosis involves herbivores and microorganisms that live in their guts. The microbes produce an enzyme that allows them to digest cellulose, the main component of plant cell walls. Herbivores that do not produce this enzyme but harbor microorganisms that do can extract more energy from their food. The microorganisms benefit from the warm, moist environment in the animals and the constant supply of food.

In commensalistic symbiotic associations, one species benefits, but the other is neither harmed nor helped. These sorts of relationships are difficult to document in nature, as close linkages between species typically have some effect, either positive or negative, on both species. The western clown anemone fish and the sea anemone demonstrate one possible commensalistic symbiosis. The fish hides within the tentacles of the sea anemone to escape predators. The anemone protects the fish because its tentacles cause

The emperor shrimp hangs on to the skin of an inedible animal, such as the sea cucumber shown here, in order to hide from predators. Because the shrimp benefits, but the sea cucumber neither benefits nor is harmed, the relationship is called commensalistic. *(Georgette Douwma/Photo Researchers, Inc.)*

Many wasps, such as *Aleiodes indiscretus* shown here, parasitize caterpillars, such as this gypsy moth caterpillar, by drinking their blood, or hemolymph. *(USDA/Agricultural Research Service/Scott Bauer)*

poisonous stings to most species, but the anemone receives no obvious advantage from the fish.

In parasitism, another type of symbiosis, one partner benefits from the association, but the other one suffers from it. The parasite feeds off the host's bodily fluids or tissues. Endoparasites live inside a host; for example, pinworms live inside the intestines of dogs. Ectoparasites feed from an external surface on the host. For example, mosquitoes drink blood from a host, but they do not enter the host's body. Another type of parasitism involves an insect that lays eggs on or in a host, so when the eggs hatch, the larvae have a ready food source.

Pathogens are similar to parasites in that they benefit while causing harm to their host organism. Because parasites depend on a living host for continued nourishment, however, they typically do not kill their host, whereas pathogens cause diseases and often kill their host as a result. Another difference between pathogens and parasites is that pathogens are usually microorganisms such as bacteria, fungi, or viruses, whereas parasites are usually unicellular eukaryotic organisms or multicellular animals such as worms. Examples of pathogens and the diseases they cause are the bacterium *Treponema pallidum,*

which causes syphilis, and the influenza virus that causes flu.

COMMUNITY STRUCTURE

Community structure refers to the sum and the result of all the relationships among populations of different species in a community. Structural characteristics of a community include features such as the composition (which species are present), the abundance (how many of each species are present), and the diversity (how many types of species are present). The interspecific interactions affect these aspects of a community, particularly the trophic structure, or feeding relationships. Food chains depict the feeding relationships in a community. Primary producers, organisms that can fix carbon from inorganic sources to form organic molecules, are found at the bottom of a food chain. These are most commonly photosynthetic organisms but can also be chemosynthetic organisms, as is the case in hydrothermal vent communities. Organisms that feed off primary producers are called primary consumers. Secondary consumers feed off primary consumers, and tertiary consumers feed off secondary consumers, and so on. Because a tremendous amount of energy is lost at each level, the maximum for most communities is four or five links in a food chain. Food webs are more accurate representations of the real feeding relationships in a community. They recognize and depict the fact that consumers feed off several types of food and also can be eaten by more than one type of predator.

Some species have a greater impact on the community structure and function than others. In most communities, one or a few species are more abundant than the others. Called the dominant species, these species usually garner a larger share of the community resources, but in return they contribute most to the ecosystem's productivity. Ecosystems are often named for their characteristic dominant species; for example, a few common C_4 grasses (grasses that undergo a specialized form of photosynthesis as an adaptation of living in hot, dry climates) dominate tallgrass prairies, and the large brown seaweed called kelp dominates the kelp forests of the ocean. Recent research has shown that dominant species of a community play as much if not more of a role than species diversity in maintaining an ecosystem's stability and function.

Keystone species are species that have a great impact on community structure even though they represent a low proportion of the community's biomass or contribute a small percentage to the community's productivity. For example, though the removal of any species within a community will cause a shift or disturbance in the community's structure, the

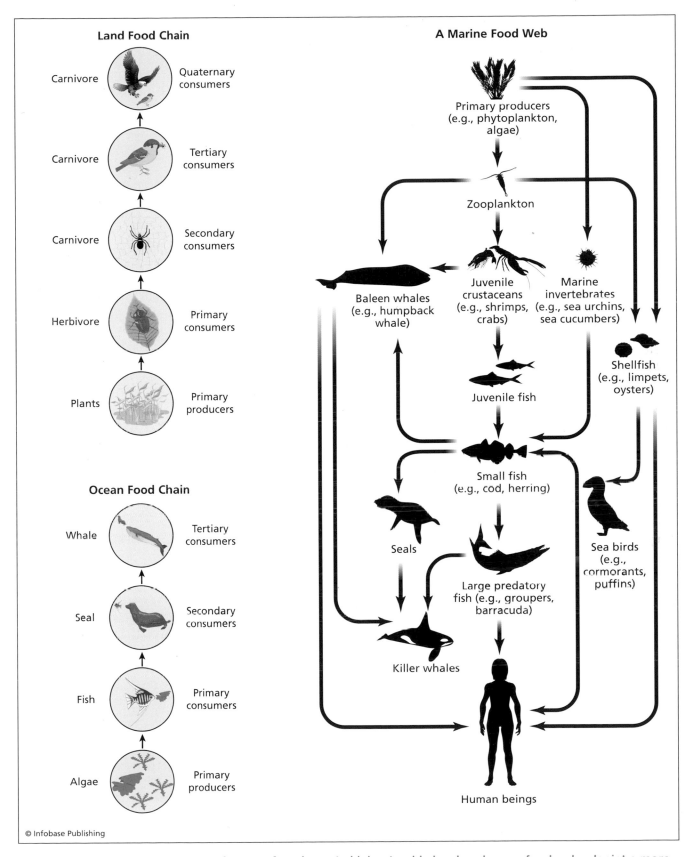

Land Food Chain

Carnivore	Quaternary consumers
Carnivore	Tertiary consumers
Carnivore	Secondary consumers
Herbivore	Primary consumers
Plants	Primary producers

Ocean Food Chain

Whale	Tertiary consumers
Seal	Secondary consumers
Fish	Primary consumers
Algae	Primary producers

A Marine Food Web

Primary producers (e.g., phytoplankton, algae)

Zooplankton

Baleen whales (e.g., humpback whale)

Juvenile crustaceans (e.g., shrimps, crabs)

Marine invertebrates (e.g., sea urchins, sea cucumbers)

Shellfish (e.g., limpets, oysters)

Juvenile fish

Small fish (e.g., cod, herring)

Seals

Large predatory fish (e.g., groupers, barracuda)

Sea birds (e.g., cormorants, puffins)

Killer whales

Human beings

© Infobase Publishing

Food chains depict the movement of energy from lower to higher trophic levels, whereas food webs depict a more complete representation of the numerous complex interactions occurring among the different populations of a community.

removal of a keystone species alters it tremendously. Because of this, keystone species are an important focus of conservation biologists. One example of a keystone species is the African elephant that roams the grasslands of Africa. If the elephants were removed, such as by hunting, woody plants would overgrow because elephants normally pull out young trees by their roots. If allowed to grow, the woody plants would block the sunlight from the grass, and the grasslands would transform into forests or shrub-dominated ecosystems.

Foundation species are those that control the physical structure of a habitat in which a community lives. For example, beavers can convert forests into flooded wetlands by chopping down trees and building dams. When a foundation species is removed or becomes greatly reduced, such as by global climate changes, the introduction of exotic species, or pests, the physical structure changes in ways that disturb the stable local conditions and ecosystem processes. For example, the human introduction of the hemlock woolly adelgid, an insect native to Asia, has reduced the number of eastern hemlock trees in riparian forest communities (those near riverbanks) in the eastern United States. Hemlocks can grow up to 150 feet tall (45 m) and measure up to six feet (1.8 m) in diameter. These insects destroy hemlocks by feeding on the tree sap, disrupting the nutrient flow, and the needles fall off. Without the needles, the tree dies within about five years of infestation. The massive size of the hemlocks provides much-needed shade at the riverbanks or along creeks. This helps moderate the local temperature, crucial to the survival of trout and cold-water species of the water body, which play a key role in the trophic structure of the aquatic ecosystems. Because of this, hemlocks are considered foundation species, or ecosystem engineers.

Ecologists use different models to describe community organization. In bottom-up control, the species at lower trophic levels exert unidirectional influence on the structure of the community. For example, the availability of nutrients in the soil affects plant growth, which in turn affects the numbers of herbivores and other species moving up the trophic levels. Top-down models purport that predators have the largest effect on community organization. For example, reduction in the number of a species at higher trophic levels will allow species at a lower trophic level to overgrow. The species at the next lower level will decline as a result because it serves as food for the now overpopulated species. More realistic models are intermediate, involving controls that exert their influence in both directions, but the simplified models facilitate analysis. Communities probably fluctuate temporally in the degree of bottom-up or top-down control, something that continued research will help ecologists better understand.

DISTURBANCE

Biologists once thought that mature communities reached a point of equilibrium, at which the relative species composition, abundance, and diversity were established. Ecologists are realizing that many communities are as dynamic as the organisms that inhabit them. Communities evolve when challenged, as do species. Any event that alters a community is called a disturbance. Natural phenomena such as a hurricane, a forest fire, an epidemic of a disease, or destruction of a habitat as a result of human influence can all disturb a community. Disturbance is not necessarily bad; it can open up opportunities for new species to join a community or fill a niche previously occupied by a different species. Human-induced disturbances, such as urban development and clearing of vegetation for agricultural purposes, do often cause negative effects, such as loss of biodiversity or reduction in ecosystem services.

Natural disturbances such as a volcanic eruption or glacial formation and action remove all or most of the vegetation in an area. The processes by which the habitat reforms after the volcanic activity ceases or after a glacier retreats are somewhat predictable. Succession is the stepwise process in which communities reform at a habitat that has been dis-

The combined characteristics of the photosynthetic organism and the fungus that make up a lichen allow lichens to inhabit areas that do not contain fertile soil; thus lichens are usually among the first organisms to inhabit relatively lifeless areas during primary succession. *(Robert F. Balazik, 2007, used under license from Shutterstock, Inc.)*

Retreating glaciers, such as this one in Alaska, leave behind relatively lifeless habitats that become recolonized through ecological succession. *(Charles D. Winters/Photo Researchers, Inc.)*

turbed. The species that initially recolonize the area are eventually displaced by a new group of species, and then another group, as a result of the physical effects the living organisms have on the environment. Primary succession is the initial recolonization of an area that has been left lifeless and where no soil is present. When a disturbance clears a community but leaves habitable soil behind, such as occurs when a fire decimates a forest, the process by which a community becomes reestablished is called secondary succession.

The first type of organism to colonize the area are autotrophic prokaryotes, followed by mosses and lichens, which arrive at the new location by wind-blown spores. Lichens are symbiotic associations between a photosynthetic organism and a fungus. The photosynthetic organism, either an alga or a cyanobacterium, fixes carbon and provides organic nutrients for the fungus. The fungus absorbs water and minerals for the phototroph and provides a structure to which the phototroph can adhere. In other environmental conditions, the relationship between the two symbionts may be described as commensal or even parasitic, but under the conditions in which primary succession occurs, the relationship is mutualistic.

Lichens and mosses do not require soil to thrive, as they can grow attached to substrates such as a rock or tree bark, but as they grow, reproduce, and decompose, they contribute to the organic matter of soil. Erosion of rocks also contributes to the formation of soil. Eventually grasses and then shrubs can take root. As they grow, they too enrich the soil, and over time the species that will characterize the community become established. Wind and animals carry in new spores and seeds. The process may take hundreds or thousands of years.

In succession, the organisms that colonize the habitat early alter the environment in a way that facilitates the arrival or success of later-arriving species. As mentioned, one way they do is by forming soil that can support the growth of plants. After soil is formed, its composition continues to change as other plants and soil bacteria establish themselves. For example, the amount of available nitrogen may increase and the acidity may change. Another way that early-arriving species prepare the habitat for later-arriving species is by providing the vegetation sufficient to support primary consumers, or herbivores. Also, as trees grow, they provide shade and microhabitats for smaller animals. The types of organisms that arrive during an early stage of ecological succession may also inhibit or prevent the establishment of other specific types of organisms.

See also BIODIVERSITY; DARWIN, CHARLES; ECOLOGY; ECOSYSTEMS; HYDROTHERMAL VENTS; POPULATION ECOLOGY; WALLACE, ALFRED RUSSEL.

FURTHER READING
Molles, Manuel C., Jr. *Ecology: Concepts and Applications.* 3rd ed. New York: McGraw Hill, 2005.

conservation biology   Conservation biology is the scientific study of processes that affect the maintenance, loss, and restoration of biological diversity, or biodiversity. Though conservation biology just emerged as a scientific discipline in the 1980s, the interdisciplinary field that draws on life sciences such as ecology, evolution, and genetics as well as other disciplines, including economics, sociology, political science, and philosophy, has quickly gained impetus as scientists have recognized that rates of extinction are higher than ever before on the planet. Biologists realized that because of human influence, species were disappearing before scientists had a chance to describe and scientifically study them. *Biodiversity* refers to all the variation present in living systems, including genetic variation within and among populations of a species, the total number of different species, and the varied composition and functions of all ecosystems in the biosphere. Conservation biologists believe that natural diversity is valuable in itself, but preserving biodiversity is also crucial to supporting life and maintaining the ecological health of the biosphere. The biodiversity intrinsic to functioning ecosystems is responsible for cycling nutrients, regenerating water supplies, purifying air, decomposing wastes, dispersing seeds, protecting shorelines and preventing erosion, controlling atmospheric chemistry, regulating climate and weather, pollinating crops, controlling agricultural pests by natural enemies, and making the soil fertile. The aim of biological conservation efforts is to preserve genetic variation, populations, species, biological communities, and entire ecosystems by educating the public about the negative impact of the reduction in biodiversity, identifying existing and potential threats to biological diversity, preventing further reduction of biodiversity, and restoring that already lost.

CONSERVATION GOALS AND EFFORTS
The major goals of research in conservation biology are to attain a better understanding of all life-forms, how they affect one another, and the mechanisms by which biodiversity contributes to the function of ecosystems, including Earth's largest, all-encompassing ecosystem, the biosphere. With this knowledge, conservation biologists can educate the public and the governments about conservation efforts, apply ecological principles to design strategies to preserve biodiversity, and assist in the development of conservation programs and the formulation of legislation and policies related to conservation.

Conservation biologists have identified numerous threats to Earth's biological diversity. The worst threat is habitat destruction, which can result from urban development, agriculture, forestry, mining, and pollution. The global warming trend has led to the physical alteration of many habitats and forced the displacement of many species and endangering of others that remain homeless. The introduction of foreign or nonnative species to new geographic areas threatens the native populations of a biological community. Whether the introduction is purposeful or inadvertent, without the natural population controls such as predators and pathogens that exist in their original habitat, the introduced species may grow uncontrollably and outcompete the native species for resources or even prey on the native species in the new habitat. Overexploitation occurs when humans harvest or kill a species at a rate faster than the species reproduces. The threatened species usually contains or provides a product that has great commercial or medicinal value. Species that play a significant or highly specialized role in an ecosystem are called keystone species. Their removal from a habitat or their extinction can cause a domino effect, impacting many other species of the community and thus the function of the entire ecosystem. For example, ecologists have shown that the removal of large carnivorous animals, the top predators, from ecosystems lifts a natural control on the population size of the prey. This, in turn, decreases the supply of the prey's food. Because food resources are often shared, the resultant increased competition for the food may have a domino effect, reaching other branches of a food web. These major threats to biological diversity are an integral consideration of the strategies aimed to preserve it.

Data collection must precede the development of a conservation management plan. The first step is to identify a species, a population, or an area of conservation interest. Reasons a species becomes of conservation interest extend beyond the scientific; efforts also aim to protect species of symbolic, economic, or cultural importance. After determining which species are in need of conservation, means to monitor success must be defined and specific target goals must be set. Researchers must identify the most important factors in preventing extinction of the species and figuring out how to affect those factors to achieve the desired result. Conservation actions may include taking steps toward restoring lost habitats, establishing protected areas and habitats to prevent further pollution or destruction, stimulating changes in society's attitudes and actions that threaten biodiversity, assisting migration, breeding animals in captivity for eventual release back into the wild, and taking steps to increase the natural success of breeding based on observations concerning mating and breeding behaviors.

Careful landscape planning complements conservation management programs. The boundaries between ecosystems and the structures within them affect the biological richness of the ecosystems. Boundaries can be natural, such as where a grassland meets a forest, or artificial, such as where a city meets the edges of what once was a forest. The species that reside in or near boundaries may require characteristics of both ecosystems, and thus increasing man-made boundaries may increase their populations and thus might negatively affect the balance of the ecosystems. Human actions also act to fragment ecosystems, such as by creating an artificial waterway, building a road, or chopping down parts of forests and leaving patches behind. This affects both the populations that reside on the boundaries and those that live in the interior. Research suggests that smaller areas support fewer species in the interior, disrupting the balance of an ecosystem and potentially contributing to the reduction in biodiversity. Developers and others who alter the natural landscape of an area should consult conservation biologists in the planning stages as a preventative measure. The creation of corridors through which animals can move between fragmented sections of a formerly connected ecosystem is one way to address the issue of fragmentation. These strips or patches of habitat that join fragments allow animals to interbreed with larger populations and thus help maintain genetic variation.

The establishment of protected areas and laws preventing development within national parks and on nature reserves helps preserve biodiversity. For this strategy to be most effective, conservation biologists must identify biodiversity hot spots, areas where numerous species, especially those that are at risk or are endangered, inhabit a small area. Since only a percentage of the Earth's surface is set aside for the purpose of conservation, preserving the hot spots will have the greatest influence. One concern is that selection of targeted areas focuses too much on vertebrates and plants compared to invertebrates and microorganisms that carry out many crucial ecosystem services such as nutrient recycling. The size of the protected area is also important, as the biological boundary necessary to sustain a species may exceed the available area defined by legal or political boundaries; thus disturbances in the surrounding areas can still considerably affect a defined nature reserve.

Efforts to conserve biodiversity and prevent further loss are useful and necessary, but a comprehensive approach to conservation efforts includes the restoration of habitats that human activities have already destroyed and continue to destroy—for example, cleaning up oil spills. One method of restoration favored by environmentalists because it provides a natural means to correct an unnatural problem is bioremediation, which utilizes the natural ability of some microorganisms to metabolize chemicals that are considered pollutants or that are toxic to other species. After inoculation of an oil spill or contaminated water source, the microorganisms break down the undesirable substance and release innocuous substances or at least less harmful ones back into the environment. For example, a microbe might ingest oil, break it down, and release carbon dioxide and water. One avenue of research in environmental microbiology aims to expand the repertoire of substances that microorganisms can safely remove from the environment through recombinant DNA technology and finding and characterizing new species with unique metabolic abilities; thus not only is habitat restoration important to biodiversity, but biodiversity is important to finding new ways to restore habitats.

The purpose of biological augmentation is also to restore habitats, but in contrast to bioremediation, which removes pollutants, the means it uses is adding nutrients in order to assist and accelerate recovery. If a nutrient has been depleted from an ecosystem, any life that depends on that nutrient will be threatened. Adding back the limiting factor can encourage growth of the organisms dependent on that factor and possibly lead to the restoration of the ecosystem's structure and therefore function.

CONSERVATION LEGISLATION AND ORGANIZATIONS

The growing concern over the loss of biological diversity in the 20th century led to the enactment of legislation that helps conservation biologists achieve their goals. Many countries have laws that make it a crime to kill, capture, or harm threatened or endangered species or to damage their habitats or range. The U.S. Environmental Protection Agency (EPA), founded in 1970, is a governmental agency charged with the mission of protecting human health and the environment, a goal consistent with biological conservation efforts. In addition to developing and enforcing environmental laws, the EPA supports education and performs research related to environmental concerns such as acid rain, beaches, pesticides,

(continues on page 224)

PARTNERS IN PROTECTION: COOPERATIVE APPROACHES TO PROTECTING WIDE-RANGING SPECIES

by Rachel Mazur, Ph.D.
Wildlife Biologist
*Sequoia and Kings Canyon
National Parks*

Sequoia and Kings Canyon National Parks are contiguous national parks, found on the western slope of the Sierra Nevada Mountains in California. They are famed for containing both the highest mountain in the continental United States and the largest trees in the world. The parks are also home to more than 450 species of birds and mammals. Not surprisingly, these two parks, along with all of the other parks in the system, are legislatively mandated to "conserve the scenery and the natural and history objects and the wildlife therein." With a total area of 860,000 acres (348,030 ha) between Sequoia and Kings Canyon National Parks, one might think that this would be a simple task. After all, the boundaries of the park are established to allow for conservation, are they not?

The answer is yes *and* no. Yes, in that efforts were made to establish boundaries containing the famous sequoia groves, the high alpine scenery, and the wildlife species. No, in that politics and money also played a role in boundary establishment, thus cutting critical areas of scenery and habitat out of the final designation. Boundaries are human constructs that nature has no obligation to recognize. Wind continues to blow air pollution into the parks. Climate change causes vegetation to shift its distribution into, or out of, the parks.

Then there is the wildlife. Even the largest parks in the world are too small to protect the diverse habitats used by wide-ranging mammals or migratory birds fully. This article presents two case studies demonstrating how biologists at Sequoia and Kings Canyon National Parks reached outside their political boundaries to solve these issues.

STUDY 1: Creation of the Sierra Interagency Black Bear Group

The mountains of the Sierra Nevada are home to a growing population of black bears (*Ursus americanus*). Black bears are found at a range of elevations but spend the majority of their time in the coniferous zone. Here, they find grasses in the meadows during spring, berries and insects during summer, and abundant pine seeds in the fall. When there is a good acorn crop, black bears will move to lower elevations to feast on this high-calorie treat and then return to the coniferous zone to den. Historically, there was little reason for black bears ever to venture into the high country, where scenery is abundant but food is scarce. Their entire range was literally defined by sources of high-calorie foods, which they are adept at finding and accessing with their natural curiosity, intelligence, and strength.

In the mid-1900s, however, backpackers began to visit the high country for its abundant scenery and with them carried high-calorie foods. Bears began to exploit this new food source by stealing it from unsuspecting hikers, who in turn began to sleep with their food to protect it. At night the bears would sneak up quietly and pull it away. When hikers hung their food from trees with ropes, bears shook the trees until the food bags fell. Bears figured out how to foil each new method of food storage almost as soon as it was invented. The situation escalated until bears became so bold that they would simply snatch food from hikers. By the late 1980s, hikers were continually losing food to bears, sustaining injuries from encounters with bears, and leaving a mess (e.g., ropes hanging from trees) in the high country. At the same time, some bears became so bold, destructive, and potentially dangerous that they had to be destroyed.

There did appear to be one solution. A machinist named Richard Garcia in Visa-lia, California, had worked with national park personnel to develop a portable, bear-resistant canister. After many failures he came up with the Garcia canister, which could fit in a backpack, hold a one-week supply of food, and withstand battering by a hungry bear. Its greatest disadvantage was its weight—small price for protecting one's food and keeping bears wild. Because of its great success, one national forest and one national park quickly mandated areas of the high country where canisters were required, but there were three persistent problems. First, regulations were created by each agency without consulting the other, so while the agency boundaries were adjacent, the restricted areas were not, a situation that made little ecological sense. Second, other manufacturers saw the Garcia canister's potential and began producing their own canisters for sale. But some of these newer canisters were easily opened by bears, so users became disillusioned and lost interest in using any canister. Finally, there was no central area where the public could find information about multiple agencies' regulations. Instead, each agency had to be consulted separately—quite a task for a hiker planning a long backcountry trek that would cross the boundaries of different agencies.

In 2000, a particularly bad year for human-bear conflict in the front country, exhausted biologists from Sequoia, Kings Canyon, and Yosemite National Parks and Inyo National Forest decided to get together to talk. The resulting meeting in fall 2000 produced the Sierra Interagency Black Bear Group. The group drafted a "Memorandum of Understanding" with the stated goal of working "to preserve a healthy black bear population free of human influence on a regional-scale by sharing information, techniques, and ideas; coordinating policies and information; and eliminating political barriers to progress."

At first, work by the group was a labor of love. None of the members had funding for meetings. They met during their free time and funded their own gatherings, hoping that together they would make a difference. They created a Web page where visitors and staff could conveniently access information about bear biology and food storage regulations for each agency. More difficult was the creation of a joint protocol for testing and approval of portable food-storage containers. The group agreed to coordinate on any new policies so they would be collaborative, and based on ecological rather than administrative boundaries.

Since its creation, the group has expanded to six agencies, with a seventh expected to join in 2009. As a result of this effort, the public is better informed about food storage restrictions and bear biology, knowing where and how to access information. Bear management specialists have more technical training and are better informed. The increased efficiency of all agencies working together has reduced workloads to the point where they meet as a part of their work schedules. The group is considered an authority on human-bear conflict and is now called upon to advise other national parks and forests across the country. Coordinated food-storage restrictions now cover ecological areas, even if that means crossing agency boundaries. Most importantly, human-bear incidents are down by more than 50 percent.

STUDY 2: Partners in Protection: A Cooperative Approach to Protecting Migratory Birds

Migrant songbirds are another group of well-known residents of Sequoia and Kings Canyon National Parks, although they spend only a portion of their lives within the confines of the parks. They arrive in the spring, set up and defend territories, breed, raise their young, and are gone by late summer. Where do they go? Most fly thousands of miles south to Central America, although some simply migrate farther south in California and others go all the way to South America. There, the birds spend the winter where there are milder weather and abundant food. The following spring, they migrate back north to breed. To conserve these species, habitats at the birds' breeding and wintering grounds must remain intact, as well as those stopover points along the migration route where birds rest and eat.

The challenges of conserving a species that migrates internationally are significant, involving various governmental agencies with different cultures and languages. Biologists in each nation's national parks recognized these difficulties but until recently were restricted to working within their own political boundaries. Then, in 2001, the U.S. National Park Service initiated Park Flight, an innovative program with the goal of working with Central American parks to protect habitats shared by their migratory bird species. The program was funded by generous grants from American Airlines, the National Park Foundation, and the National Fish and Wildlife Foundation.

The program awarded grants to U.S. and Central American parks that were interested in participating. Sequoia and Kings Canyon National Parks were among the first to receive one of these grants and have been involved ever since. The program, "Partners in Protection: A Cooperative Approach to Protecting Migratory Birds," revolves around three main components: education, biological monitoring, and international exchange.

International exchange is the heart of the program. Each year Sequoia and Kings Canyon National Parks host an intern from Central America or Mexico who is training to work in the field of conservation and has great interest in bird biology. They have hosted interns from Guatemala, Nicaragua, Honduras, and Mexico. Some arrive speaking no English; others are bilingual. All have a great love of birds, but their monitoring experience varies widely. Most interns have been working as biologists in their home countries and want to expand their bird experi-

tise and language skills. All have a wealth of knowledge about the birds' wintering grounds, conservation challenges, and monitoring techniques. Park employees have learned an immense amount from them.

The intern is paired with a U.S. biological technician. Both the intern and the technician attend a week of training on bird identification, data collection, and bird safety. Together they run two bird-banding stations. Birds are caught when they fly into large nets that they cannot see. Biologists gently remove the birds from the nets; measure them; determine their species, age, and breeding status; find out whether the bird is already banded and, if not, affix a small metal band around one leg; and then release them. These data are used to monitor trends in species numbers and to assess any declines (or increases) in their breeding or wintering grounds.

Another critical component of the program involves monitoring peregrine falcons, which were once on the brink of extinction. Cliffs that are known to attract both peregrines and human climbers are watched for falcon activity. When a peregrine aerie (nest) is located, nearby climbing routes are temporarily closed until the young successfully fledge from the nest. That way, the birds are not disturbed by climbers, and the climbers are still allowed access to most of the cliff face.

Another type of monitoring involves conducting yearly point-count transects. To do this, one stops at preset intervals along a pathway and records all the birds that are seen or heard. By comparing data over several years, one can determine whether various bird species are increasing or decreasing. All of these monitoring programs may be, and many have been, implemented by the interns when they return to their home parks.

The education component includes the offering of bilingual bird walks to the public, which allows us to reach our many Spanish-speaking visitors. The

(continues)

(continued)

parks celebrate International Migratory Bird Day by teaching people how they can prevent birds from hitting windows, that they should take their birdfeeders inside during the winter and keep cats indoors.

As of this writing, Sequoia and Kings Canyon National Parks have hosted nine international interns through Partners in Protection. Many of these interns have learned new monitoring techniques, improved their English, and, most importantly, become conservation leaders in their countries. At the same time, our employees have learned about conservation issues in the birds' wintering grounds, improved their Spanish, and even gone to Central America to serve as interns for people who were once our interns. Park

employees look forward to hosting a 10th intern during summer 2009.

FUTURE DIRECTIONS

Both the Sierra Interagency Black Bear Group and the Partners in Protection program are now integral to the wildlife management program at Sequoia and Kings Canyon National Parks. Both programs required large initial investments of time and commitment but have reaped enormous rewards. Clearly, the different nations' parks are stronger working together. Around the world, thousands of collaborative projects such as these are under way, all with the ultimate goal of protecting a species, group of species, ecosystem, or natural process. To succeed, these projects first require conservation biologists to conduct sound scientific research into the

problem. They then share their results with a range of specialists, from economists to politicians to social scientists, who work collaboratively to find a solution. This type of cooperation moves the global community closer to realizing the goal of conservation biology—the protection of the biological diversity of the Earth.

FURTHER READING

National Park Service. U.S. Department of the Interior. "Global Conservation: Park Flight." Available online. URL: http://www.nature.nps.gov/globalconservation/parkflight.cfm. Accessed February 2, 2008.

Sierra Interagency Black Bear Group Web site. Available online. URL: www.sierrawildbear.gov. Accessed February 2, 2008.

(continued from page 221)

wetlands, oil spills, the ozone layer, climate change, and clean air—all relevant to habitat preservation and the effects of human impacts on biodiversity. First passed in 1973, the Endangered Species Act (ESA) is one of the major environmental laws forming the legal foundation for EPA programs and actions. The ESA prohibits actions that result in the taking, harming, or killing of threatened or endangered species; actions that adversely affect their habitats; or trade of listed species. Two governmental agencies jointly administer the ESA: the Fish and Wildlife Service, whose mission is to conserve, protect, and enhance fish, wildlife, and plants and their habitats, and the National Oceanic and Atmospheric Administration (NOAA) Fisheries Service, who work to conserve, protect, and manage living marine resources through science-based conservation programs and promotion of healthy ecosystems. Much of the text of the ESA is taken from the Convention on International Trade in Endangered Species of Wild Fauna and Flora (CITES), an international agreement that restricts the trade of more than 28,000 species of plants and 5,000 species of animals, living or dead, as well as their parts. In August 2008 CITES had 173 participating countries.

Recognizing the unique need for international cooperation to address environmental issues effectively, the United Nations (UN) has implemented several programs. The objectives of the United Nations

Environment Programme (UNEP) are to advocate, educate, and promote conservation and sustainable use of all the Earth's natural resources. The UN Convention of Biological Diversity specifically supports the conservation of biodiversity and promotes the sustainable use of biological resources. The UN Convention on the Conservation of Migratory Species of Wild Animals (CMS) protects terrestrial, marine, and avian species whose natural migration patterns cross political boundaries.

The world's largest conservation network is IUCN—the World Conservation Union, whose self-reported mission is "to influence, encourage, and assist societies throughout the world to conserve the integrity and diversity of nature and to ensure that any use of natural resources is equitable and ecologically sustainable." IUCN—the World Conservation Union publishes a comprehensive "red list" that tracks and evaluates the extinction risks for thousands of plants and animals. On the basis of criteria such as population size, rate of decline, geographic distribution, and degree of population fragmentation, organisms are placed into one of nine categories: extinct, extinct in the wild, critically endangered, endangered, vulnerable, near threatened, least concern, data deficient, and not evaluated. Conservation biologists worldwide use this list to develop conservation management plans and recovery programs. IUCN—the World Conservation Union also set up a World Commission of Protected Areas (WCPA), with the goal of selecting, establishing, and managing national parks and protected areas.

Several private organizations have made significant contributions toward conservation efforts. The World Wildlife Fund, the largest privately financed, multinational conservation organization with a membership of more than 5 million, oversees 2,000 projects in 100 countries, all with the aim of nature conservation. Wildlife conservation is also a major goal of the Association of Zoos and Aquariums (AZA), which supports zoos and aquariums that spearhead efforts to protect wild animals and restore threatened populations. Another governmental service that plays a key role in conservation efforts in the United States is the National Park Service, which established the Conservation Study Institute in 1998 in order to stay current with contemporary issues in conservation research and efforts and to develop partnerships, tools, and strategies for future conservation efforts.

Often, the efforts to conserve populations or protect areas where threatened or endangered species live conflict with private property rights or economic ventures. Educating landowners and local, state, and federal officials about conservation biology is imperative for cooperation in conservation efforts. The fact that so many organizations and agencies are working toward a shared goal of biological conservation holds promise for the future of conservation biology. As scientists gain an improved understanding of the ecological principles that govern ecosystems, communities, and populations, perhaps the world will develop better methods for managing and conserving the Earth's biological diversity.

See also BIODIVERSITY; ENDANGERED SPECIES; ENVIRONMENTAL PROTECTION AGENCY.

FURTHER READING
Anderson, Anthony B., and Clinton N. Jenkins. *Applying Nature's Design: Corridors as a Strategy for Biodiversity Conservation.* New York: Columbia University Press, 2006.
Center for Applied Biodiversity Science home page. Available online. URL: http://science.conservation.org. Accessed January 19, 2008.
Groom, Martha J., Gary K. Meffe, and C. Ronald Carroll. *Principles of Conservation Biology.* 3rd ed. Sunderland, Mass.: Sinauer, 2006.
IUCN—The World Conservation Union home page. Available online. URL: http://www.iucn.org. Accessed on January 19, 2008.
Primack, R. N. *Essentials of Conservation Biology.* 4th ed. Sunderland, Mass.: Sinauer, 2006.
Society for Conservation Biology (SCB) home page. Available online. URL: http://www.conbio.org/. Accessed January 19, 2008.
Soulé, Michael E. "What Is Conservation Biology?" *Bioscience* 37, no. 11 (1985): 727–734.

Crick, Francis (1916–2004) British *Molecular Biologist* Francis Crick codiscovered the structure of deoxyribonucleic acid with James D. Watson. He shared the 1962 Nobel Prize in physiology or medicine with Watson and Maurice Wilkins "for their discoveries concerning the molecular structure of nucleic acids and its significance for information transfer in living material."

TRAINED IN PHYSICS
Francis Harry Compton Crick was born on June 8, 1916, in Northampton, England. He received a bachelor of science in physics from University College in London in 1937 and for a time researched the viscosity of water under pressure. His graduate studies were interrupted by World War II, when he went to work for the British Admiralty designing magnetic and acoustic mines. After the war ended, in part stimulated by reading Erwin Shrödinger's *What Is Life?*, a book that emphasized the importance of learning about genes, Crick completely changed his focus to molecular biology. He received a small scholarship to attend Cambridge in fall 1947.

While working at the Strangeways Laboratory at Cambridge, he studied the physical properties of the cytoplasm of chick fibroblast cells and began learning about biology, organic chemistry, and crystallography. He joined the new Medical Research Council Unit at the Cavendish Laboratory of Cambridge University as a doctoral candidate in 1949. His supervisor, Max Perutz, was a well-known protein chemist and soon had Crick examining the structure of hemoglobin using X-ray crystallographic techniques. An enthusiastic graduate student, he was easily distracted from his project and often irritated his coworkers with his incessant chatter.

SOLVES STRUCTURE OF DNA WITH JAMES WATSON
When James D. Watson entered the lab as a postdoctoral fellow in the laboratory of John Kendrew in 1951, the two quickly became friends, partly because of their common enjoyment of scientific discussions, but also their youthful arrogance. Crick needed frequent breaks from his research to talk about science, and Watson enjoyed the chance to offer his opinions of Crick's ideas. Excitement over deoxyribonucleic acid (DNA), which had only recently been identified as the carrier of genetic information, became the focus of their conversations. Both believed genes were composed of DNA and soon became obsessed

with the way genes copied themselves, a problem they hoped to address by solving its structure.

The fact that neither Crick nor Watson had the background training one would expect for someone tackling such a problem did not deter either of them. The problem was that all DNA-related research belonged, by unspoken scientific etiquette, to Maurice Wilkins of King's College in London. Wilkins was not only the first scientist in Great Britain to show an interest and make progress in the subject, but he was also a personal friend of Crick. Sir Lawrence Bragg, head of the Cavendish Laboratory, even directed them to stick to their own research subjects—proteins—but the two eager scientists could not dismiss thoughts of DNA.

Scientists knew two types of nucleic acids existed, DNA and ribonucleic acid (RNA). The structures of each of the five nucleotide building blocks (the purines adenine and guanine and the pyrimidines cytosine, thymine, and uracil) were also known. The Austrian biochemist Erwin Chargaff had determined that, in DNA, the percentage composition of adenine equaled that of thymine and the percentage composition of cytosine equaled that of guanine. X-ray diffraction evidence from photographs taken by Rosalind Franklin in Wilkins's laboratory suggested that DNA was helical.

Using molecular models that resembled toy building blocks, Watson and Crick attempted to recreate the biological structure. Watson attended a lecture given by Wilkins in spring 1951. During the talk, Wilkins showed an X-ray diffraction photograph obtained from crystallized DNA. Though the picture was not remarkable, the fact that DNA could be crystallized and yield X-ray diffraction patterns at all meant that DNA had a regular structure and therefore was solvable. This knowledge excited him greatly, because knowing the structure of DNA would surely lead to an understanding of the way genes worked. Watson attended a colloquium that fall, in which Franklin presented some of her X-ray diffraction data. Though Watson had attempted to learn as much crystallography as possible before hearing her presentation, he failed to remember some important details regarding water content. Unaware, Crick and Watson constructed a model that they believed was the true structure, a triple helix with a sugar-phosphate backbone running down the center. After they eagerly invited Wilkins and Franklin to view their model, Franklin at once saw their structure contained at least 10-fold less water than the molecule had and chastised them. This embarrassing fiasco led to a chewing out by Bragg, who demanded Crick get back to his thesis task of examining the effect of different salt solutions on hemoglobin crystals.

They laid low for a while and limited their DNA discussions to lunchtime. In 1952 Crick and Watson had the opportunity to meet with Chargaff personally. Though Chargaff was unimpressed with their knowledge about nucleic acid chemistry, he discussed his data regarding the composition of DNA, now referred to as Chargaff's rules, with them. Crick recognized the data were of key importance. In December of that year, Linus Pauling wrote to his son, Peter, who had joined Kendrew as a Ph.D. student six months prior. Pauling said that he solved the structure for DNA and would soon send a copy of the manuscript. When it arrived the first week of February 1953, Peter showed the crestfallen Watson and Crick the manuscript in which Pauling described DNA as a triple helix with a sugar phosphate backbone in the center, very similar to their design that Franklin scoffed at 15 months before. To their extreme delight and disbelief, they saw that Pauling had ignored the fact that DNA was an acid and had attached hydrogen atoms to the phosphate groups, neutralizing their negative charge. Furthermore, Pauling had made these hydrogen atoms participate in bonds that held the structure together. The error was unmistakable, and they knew that Pauling would soon recognize it. After trying unsuccessfully to convince Wilkins to work with them to beat Pauling in determining the correct structure, they frantically resumed their model building. Wilkins did assist them by sharing some of Franklin's X-ray diffraction data that suggested the structure was indeed a helix and giving the unit repeat length. They immediately ordered new purines and pyrimidines from the Cavendish machine shop, meanwhile working with cardboard cutouts.

Though Crick knew more about physical science and Watson about biology and genetics, Crick credits himself with figuring out that Chargaff's rules meant that adenine paired with thymine and that cytosine paired with guanine and credits Watson with determining how the nucleotide base pairs fit together chemically. On February 28, 1953, all the puzzle pieces seemed to fall into place, and by March 7, 1953, they constructed a double-stranded model that resembled a twisted ladder, with the rungs being the paired nucleotides. The nitrogenous bases of the nucleotides were linked by specific hydrogen bonds between adenine and thymine and between cytosine and guanine. Strands that pair in this manner are termed complementary. The simplicity of the model was surprising and made apparent a mechanism for replication. Each of the two strands could serve as a template for the synthesis of a complementary strand. The model was biologically as well as chemically attractive.

Coincidentally, Perutz, Crick's supervisor, was serving on a committee whose job was to oversee the biophysics research at King's College. He had access to a comprehensive summary of Franklin's research as part of a report and shared it with Crick and Watson, so they could check their model against the X-ray diffraction data. This event later led to some controversy in the distribution of credit for the eventual discovery of the structure of DNA.

Once again, Wilkins went to Cambridge to examine their proposed model. Surprisingly, he did not act bitter, and within a few days he called to confirm that both his and Franklin's data strongly supported the double helix. Though many were working to solve the structure, Crick and Watson were successful because of their persistence and determination. They chose a biologically important problem, immersed themselves in the necessary background information, and were successful. Their short article "A Structure for Deoxyribose Nucleic Acid" was published in *Nature* in April 1953.

After determining the structure of DNA, in 1953, Crick earned his doctorate from Caius College of Cambridge for a thesis on X-ray diffraction of polypeptides and proteins and went on to serve science in many capacities. He worked in Brooklyn for one year, then returned to Cambridge. After spending a sabbatical at the Salk Institute for Biological Studies in San Diego, Crick joined the faculty in 1977, obtaining an endowed professorship, and became an emeritus president of the institute's Kieckhefer Center for Theoretical Biology. His studies included work on the structures of other biological molecules such as collagen, polypeptides, and polynucleotide chains. He also contributed to the elucidation of how DNA and RNA direct and carry out the synthesis of proteins and how variations in the sequence of nucleotides lead to genetic mutations. In later years he switched to neurophysiology and explored the mammalian visual system.

The Royal Society of London elected Crick to membership in 1959. His contributions to the discovery of the structure for DNA led to his sharing the Nobel Prize in physiology or medicine and the Lasker Foundation Award in 1962 with Watson and Wilkins in addition to many other honors. In 1992 Queen Elizabeth II awarded Crick the Order of Merit.

Crick was married to Ruth Doreen Dodd from 1940 to 1947. They had one son together, Michael. In 1949 he married Odile Speed, with whom he had two daughters, Gabrielle and Jacqueline. Crick died on July 28, 2004.

See also BIOMOLECULES; CHARGAFF, ERWIN; DEOXYRIBONUCLEIC ACID (DNA); FRANKLIN, ROSALIND; PAULING, LINUS; WATSON, JAMES D.; WILKINS, MAURICE H. F.; X-RAY CRYSTALLOGRAPHY.

FURTHER READING

Crick, Francis. *What Mad Pursuit: A Personal View of Scientific Discovery.* New York: Basic Books, 1988.

The Nobel Foundation. "The Nobel Prize in Physiology or Medicine 1962." Available online. URL: http://nobelprize.org/medicine/laureates/1962. Accessed January 14, 2008.

Watson, J. D., and F. H. C. Crick. "A Structure for Deoxyribose Nucleic Acid." *Nature* 171 (April 25, 1953): 737–738.

———. "Genetical Implications of the Structure of DNA." *Nature* 171 (May 30, 1953): 964–967.

Cuvier, Georges, Baron (1769–1832) *French Comparative Anatomist* Georges Cuvier established the reality of the existence and extinction of past life-forms at a time when people hesitated to believe that animals never seen by human eyes had once crawled, walked, hopped, swum, and flown over Earth's surface. As an expert comparative anatomist, he advanced the field of paleontology through his studies of fossil mammals, which he suggested disappeared from the Earth as a result of catastrophic geological events.

BECOMES A NATURLIST

Georges Cuvier was born on August 23, 1769, in Montbéliard, Wurttemberg. At the time, Montbéliard was under German jurisdiction, but it was a French-speaking community. In 1793 the territory was annexed, and Cuvier became a French citizen. He was baptized as Jean-Leopold-Nicholas-Frédéric Cuvier. Later his mother added Dagobert to his name. When his elder brother Georges died as a child, he adopted the name Georges and used it for the rest of his life. As a child Cuvier spent time sketching various animals he read about in a 44-volume encyclopedia about the natural world, *Natural History*, written by Georges-Louis Leclerc, comte de Buffon. Cuvier was a skilled artist and later in life provided many of the drawings for his own published works.

Though his parents wanted him to enter the ministry, Cuvier's teachers did not recommend him for a scholarship to theology school. Georges's father was a soldier, and the family did not have enough money to send him to a university. Fortunately, he gained admission to a school founded by the duke of Württemberg—Caroline University in Stuttgart, Germany. From 1784 to 1788, he studied a variety of subjects ranging from administration and economics to the scientific discipline of zoology and the art of dissection. Georges also mastered the German language.

After Georges's graduation his father helped him obtain a position as a private tutor for an aristocratic family in Normandy, France. This protected him from the immediate effects of the French Revolution and allowed him to indulge in independent studies of natural history during his free time. He collected fish, mollusk, and shorebird specimens from the nearby port of Fécamp. He kept detailed notes of his dissections, observations, and sketches in scientific diaries. During this time, Cuvier regularly corresponded with a friend from Caroline University, Christian Heinrich Pfaff. These letters mentioned many of the scientific ideas for which Cuvier became famous in the early 1800s. In them, he described his scientific endeavors and wrote about topics such as the geology of Normandy and his opinion of the proposed notion that all living beings formed a continuous chain of increasingly complexity. Cuvier and his notebooks attracted the attention of French naturalists, who encouraged him to go to Paris.

In 1795 he found himself teaching animal anatomy at the recently reformed Muséum d'Histoire Naturelle, the largest institution dedicated to scientific research at the time. Having a proper forum, he promptly presented the results of his Normandy researches. Among these was his proposal of a new means to classify invertebrates. Whereas previously zoologists had divided all invertebrates into two groups, worms or insects, Cuvier distinguished among mollusks, crustaceans, insects, worms, echinoderms, and zoophytes.

CAREER AND PERSONAL ACHIEVEMENTS

Cuvier was a gifted teacher and soon obtained the position of professor of zoology at the Écoles Centrales. His abilities and reputation led to several other responsibilities and appointments. In 1796 he became the youngest member of the Class of Physical Sciences at the Institute of France (hereafter referred to as the *Institute*), which partly replaced the Royal Academy of Sciences. In 1800 he was appointed professor at the Collège de France, and he was appointed professor of comparative anatomy at the muséum in 1802. He became permanent secretary of the physical sciences for the Institute in 1803. Napoleon appointed him university counselor in 1808 and sent him to reorganize higher education in Italy, the Netherlands and southern Germany. As compensation, in 1811 Cuvier received the title of chevalier, awarding him the privileges of a low-ranking nobleman. In 1814 Cuvier became councillor of state and head of the Interior Department of the Council of State from 1819 until his death. Cuvier was elected a member of the Académie Française in 1818. He was made a baron in 1819 and given the title of *grand officier* of the Legion of Honor in 1824. In 1831 Cuvier was given the status of peer of France, an honor of high-ranking noblemen.

Cuvier married a widow of a victim of the Revolution, Mme. Davaucelle, in 1804. She already had four children from her previous marriage. Together they had four more children. Tragically, Cuvier was preceded in death by all of his children.

IDENTIFICATION OF ANIMALS FROM FOSSIL REMAINS

For much of his career, Cuvier gathered information on the structure of living beings from fossil remains. The term *fossil* refers to the remains or traces of a formerly living organism. Ancient bones and animal tracks found in sediment are both considered fossils. Fossils from aquatic life are more common than fossils from terrestrial life, as deposition of loose sediment washed away by erosion occurs in bodies of water. Normally microorganisms decompose the remains of dead organisms, but sometimes the remains are preserved. Hard materials such as bones and teeth may be found essentially unaltered. Even softer parts may be found intact if the organism was quickly frozen in a block of ice. Other fossil remains become petrified, meaning that minerals have replaced the organic matter and hardened. This process usually preserves the basic shape. In other instances, holes or cavities may become filled in with hardened mineral deposits. Another common type of fossil occurs when an organism is compressed within the Earth's crust. Over time the organism decomposes, and a thin layer of carbon is left behind. The last type of fossil results when an organism becomes trapped in a layer of sediment and the sediment hardens around it. If acidic liquid gains access to the organism, the organism may dissolve, leaving behind an imprint in the hardened sedimentary rock. Depending on the type and quality of the fossil, an anatomist can extract much information about the former organism's anatomy from the fossil.

The scientific study of fossils is called paleontology. Paleontologists are interested in a variety of topics. For example, some paleontologists use fossil evidence to investigate geologic time. The history of the Earth is divided into a set of geologic periods. Each division of time is characterized by a unique group of fossil remains; thus fossils may be used to determine when the rock layer in which they are embedded was formed. Paleontologists also use fossils to learn about tectonics, the movement of landmasses, throughout the history of the Earth. Cuvier used his knowledge of anatomy to investigate the relationships among organisms, past and present.

In 1796 Cuvier presented to the Institute his treatise *Mémoire sur les éspèces d'elephans tant vivantes que fossiles* (Memoir on the species of elephants, both living and fossils). He gave a detailed description of the osteological (related to the study of bones) features of two known elephant species, African and Indian. He discussed their teeth, skulls, and jaws, among other structures. Then he confidently claimed that fossil elephant remains belonged to a distinct third species, identified as *Elephas primigenius,* an extinct hairy maned mammoth. Cuvier was the first to suggest that comparative anatomy could be used to learn about geological history. For example, one popular geological theory was that the Earth had been gradually cooling since its formation. Scientists assumed that locations where elephant remains were found but where elephants no longer live must have previously been warmer. But Cuvier responded that since the remains belonged to an entirely new species, the extinct animal might have been better adapted to cooler climates than the living species, so the Earth might not necessarily be cooling. Cuvier claimed there must have been a primitive prehuman world that was destroyed by some major catastrophe, but he left the determination of the specific nature of the cataclysmic event to experts in geology.

Later that year Cuvier was sent plates of fossil bones from a large animal found in South America. Again, using careful anatomical comparison, he concluded that this elephant-sized beast that he named *megatherium* was also extinct. He concluded that it was another animal from the ancient world. These studies increased Cuvier's interest in fossil anatomy and set him on a mission to study all fossil animals.

The nearby gypsum quarries of Montmartre and Mesnilmontant contained an abundance of well-preserved fossil remains. Cuvier appealed to a quarrier (someone who excavates stone from a quarry or pit) to give him fossils uncovered during excavation. Cuvier had to draw on his expert skills to examine these fossils since they were embedded in hard plaster stone rather than loose sediment. Many of the fossils were from unknown species. He surmised that the Earth must have been previously crawling with vertebrate animals that no longer existed. In other words, he asserted that animals became extinct.

Cuvier awed other scientists with his ability to identify organisms from only fragments of their skeletal remains. He claimed this was possible because organisms were well-integrated wholes: their parts were not independent of one another. For example, if he found a tooth, he deduced the animal's diet from its construction. If it were a meat eater, the animal's form would have to allow it to move in a manner that would permit it to capture its prey and its jaws strong enough to crush it. Carnivory would require a digestive system that could efficiently extract necessary nutrients from this type of food source. Thus each body part told enough of the full story that an anatomist familiar with the fundamental laws of comparative anatomy could reconstruct the entire organism with astounding accuracy. Cuvier went so far as to claim even the musculature could be

reconstructed from imprints left on the bone. Since organisms are functionally integrated, he depended on their structures to infer their habitat and even the physical history of the Earth at the time that they roamed it.

EXTINCTION AND CATASTROPHISM

Cuvier began to wonder why animals became extinct. By now he realized that the fossils of extinct organisms were not all the same age, and that difference in age pointed to a series of revolutions throughout Earth's history. What happened in the geological past that these animals could no longer survive? Why were their forms no longer sufficient for survival in the habitats that had previously supported their needs? Consequently, Cuvier's interests shifted toward geology in order to learn about the physical environment during the time that the extinct lifeforms lived. He studied the material in which the fossils were found and tried to figure out what was happening at the time the strata were laid down. He was looking for keys to the geological history of Paris, clues to what might have happened that wiped out entire species. He appealed to other natural historians for collaboration. Meanwhile, he continued to publish prolifically on the bones of fossil animals. He identified several new extinct species including mammals similar to present-day otters, gazelles, hares, tapirs, opossums, and others.

Regarding the Earth's history, Cuvier accepted the divisions of periods of time, called epochs. He believed there was a primary, universal, lifeless ocean prior to the formation of continents. Marine life appeared, and then terrestrial life. The lack of human-type fossils and of intermediate fossils convinced him of the reality of extinction and creation of life in its original form. His lectures on geology contained few original ideas except when linked to fossil evidence. Even the idea of several cataclysmic revolutions of the Earth was not new.

Cuvier's training was in anatomy. A new interest in a field does not necessarily qualify one for productive study in that field. Cuvier was simply a biologist with an interest and natural talent in geology. In Alexandre Brongniart he found his complement, a geologist interested in biology. Brongniart had a background in mining engineering. Beginning in 1804, these two men undertook a study of the Seine basin in northern France. They traveled around France examining the succession of strata, paying particular attention to the distinct groups of fossils embedded in each. Each fossil bed demonstrated that the surface of the Earth was not as it always had been.

In 1808 they presented a joint preliminary report to the Institute, *Essai sur la geographie mineralogique des environs de Paris* (Essay on the mineral geography of the Paris region). A fuller version was published in 1811. Cuvier graciously gave the majority of credit to his associate for the efforts that resulted in this essay. This work outlined the principles of paleontological stratigraphy and included a color-coded mineralogical map of the strata and detailed descriptions of nine different successive formations and their corresponding fossils. One goal was to identify the chronology of the Montmartre fossil beds. A significant finding was that there were major differences between the groups of fossils found in different beds. Beds contained significantly different fossils from the beds above and below them. Cuvier and Brongniart also reported finding both saltwater and freshwater organism remains in the same location. They suggested that fossils could be used to determine geological chronologies. For example, they could determine that a particular region first was submerged in salt water, then became dry land, then later was covered by freshwater. The strata may have looked similar, but if the fossils differed, the chronologies did as well. Cuvier also described some of the quadrupeds, or four-legged animals, that he found.

As one ascends a stratigraphical column, sediments near the bottom contain fossils from the oldest periods in geologic time. As Cuvier and Brongniart traveled upward toward more recently laid strata, they noticed that mammal remains suddenly appeared, though they were not remains of extant creatures, that is, creatures still in existence. As they continued moving up the column, remains of current recognizable species were finally observed. The succession was not gradual, but erratic. Cuvier concluded that the breaks represented actual geological breaks and were indicative of major revolutions in Earth's history; he did not believe that gradual geological cycling sufficiently explained his observations. These revolutions wiped out entire species, such that the species alive today do not represent the complete assortment of this planet's animals. Cuvier did not focus on whether new creations occurred after each catastrophe, though others thought this was a logical possibility.

Cuvier was a catastrophist. Catastrophism suggests that certain geological features of the Earth's crust are the result of past cataclysmic events such as volcanic activity or flooding. Cuvier thought mass extinctions might result from such geologic catastrophes. The term *catastrophism* had not yet been used. Instead, Cuvier referred to "global revolutions." He thought periodic revolutions reasonably explained why remains of saltwater and freshwater organisms were found at the same location and why there were apparent breaks in geological time according to rock strata. Moreover, though Cuvier did not specifically identify any revolutions with biblical events, catastrophism did not require that he abandon his Protestant upbringing. The biblical account of creation and the great flood was compatible with a catastrophic viewpoint. The great flood was simply the most recent great catastrophe. However, Cuvier's paleontological findings did convince him that creation must have occurred in many stages.

As far as Cuvier's claim that former life-forms were no longer in existence, doubters questioned why God would create something and then allow it to disappear. Some thought that the fossils were the remains of living organisms, just incorrectly identified. Others asserted that the species to which the fossil remains belonged had simply not yet been observed or identified by humans. Perhaps the species resided in an unexplored part of the world, they speculated.

By 1812 Cuvier's fossil research had almost ended. He collected, reordered, and reissued many of his previous related papers into the four-volume work *Recherches sur les ossemens fossiles de quadrupèdes* (*Researches on Fossil Bones of Quadrupeds*). To this work, he added a *Discours préliminaire* (Preliminary discourse) immediately after the preface. This introductory text, which was technically accessible to the public, summarized the evidence of global revolutions, geological structures and formations, research on fossil bones, their utility in uncovering Earth's history, and the extinction of life-forms. In 1826 this piece was published separately with a new title, *Discours sur les révolutions de la surface du globe, et sur les changemens qu'elles ont produit dans le règne animal* (*A Discourse on the Revolutions of the Surface of the Globe, and the Changes Thereby Produced in the Animal Kingdom*, 1829). It was ultimately reprinted many times in several languages and was considered a masterpiece on its own. By the time Cuvier died it was in its sixth edition.

COMPARATIVE VERTEBRATE ANATOMY

Next Cuvier refocused on his original field of comparative anatomy. In 1817 he published a zoological masterpiece, *Règne animal distribute d'après son organization* (*The Animal Kingdom, Distributed According to Its Organization*, 1834–37). This work included descriptions of the entire animal kingdom. In it, he modified the classification system proposed by the Swedish naturalist Carl Linnaeus. Cuvier recommended four major divisions of animal life: vertebrata, mollusca (including shellfish), articulata (including insects), and radiata (including echinoderms). This system may seem crude today, but at the time it emphasized the diversity of animal life, particularly the invertebrates, and marked a change in the approach to the classification of animals.

Georges Cuvier's classification system considered single-celled organisms (number 1, top left corner) to be the simplest of animals and humans (number 140, bottom right corner) to be the most advanced. *(Sheila Terry/Photo Researchers, Inc.)*

In his 1809 *Zoological Philosophy*, the French naturalist Jean-Baptiste Lamarck suggested that living things transmutated, that is, evolved. Their forms gradually changed to become better adapted to their environment, and these changes were passed on to the next generation. According to Lamarck, animals were becoming more and more complex. At the time it was popular to believe in the stability of life-forms, that each creature existed as God originally had created it. The form of each creature was not subject to mutation. Change would not only violate moral law, but decrease the ability of an animal to survive in the particular environment for which it was divinely suited. Besides, if animals were mutable, then the entire science of taxonomy would have no basis. Étienne Geoffroy Saint-Hilaire, who had initially written to Cuvier inducing him to move to Paris, and who himself believed that animals were subject to

change, gave Cuvier some mummified ibises (a type of bird) from Egypt in 1802. After studying them, Cuvier found that though the mummified birds were more than 3,000 years old, their morphology was identical to that of current ibises. At the time 3,000 years was considered a very long time. Most people believed the Earth itself was only about 6,000 years old. Cuvier thought that if they had not changed in 3,000 years, they were never going to change.

In 1829 two of Geoffroy's followers tried to demonstrate to the academy that a link existed between cephalopods (invertebrates) and fish (vertebrates); Cuvier interfered. This eventually led to a huge public debate between the former collaborators and friends, Geoffroy and Cuvier, at the Royal Academy of Sciences in Paris in 1830. The issue was whether form determined mechanical function or function dictated form. Geoffroy believed that all vertebrates had a common form of basic organization, with slight modifications. He claimed that vestigial organs such as the appendix demonstrated that all vertebrates had originally shared a common ancestral form. He thought that if structures were connected to other structures in the same manner, then differences in size or shape were not so important. Cuvier responded that similarities in form were simply the result of similar functions. He believed in the integration of parts into functional wholes. Today scientists accept both concepts, depending on the structures and species that are being compared. For example, the wings of a bird and the wings of an insect both allow the organisms' function of flight. However, these two types of animals do not share a common structural archetypical ancestor. These structures are considered analogous. On the other hand, the wings of birds and the wings of a bat do share a common vertebrate structural ancestor; they are more closely linked evolutionarily. Being derived from a common ancestral structure, the structures are referred to as homologous.

Another enormous zoological work by Cuvier was *Histoire naturelle des poisons* (Natural history of fish). Written in collaboration with Achille Valenciennes, it summarized all of the content knowledge of the field of ichthyology, the study of fish. The first volume was published in 1828, with eight more being completed prior to Cuvier's death. The 22nd and final volume was published in 1849. The classification system contained in this work remains the basis of modern ichthyologic classification.

CUVIER'S LEGACY

In May 1832, Cuvier suffered an attack of paralysis and died within a few days. His legacies include his library, boasting more than 19,000 volumes and thousands of pamphlets, and an increase in the collections at the muséum from a few hundred to more than 13,000 specimens, all arranged according to his own classification system. Cuvier is most remembered for supporting catastrophism, establishing the reality of extinction of past life-forms, and expanding Linnaeus's classification system of animals. However, his contributions in the form of progress reports on science and scientific biographies submitted in the capacity of permanent secretary to the institute are also noteworthy.

Constantly striving to overcome his humble beginnings, Cuvier was reported to be somewhat arrogant, hurried, and eager to receive flattery. He did achieve prominence during his lifetime and was appropriately honored with many appointments and titles. He was admired for his intelligence, but he was too stubborn to open his mind concerning the variability of species. Because of this, once the theory of evolution by means of natural selection became widely accepted, Cuvier's reputation diminished. However, he will always hold a place in scientific history for bridging the gap between the life sciences and the Earth sciences by founding the science of vertebrate paleontology.

See also ANATOMY; BUFFON, GEORGES-LOUIS LECLERC, COMTE DE; GEOFFROY SAINT-HILAIRE, ÉTIENNE; LAMARCK, JEAN-BAPTISTE; LINNAEUS, CARL.

FURTHER READING

Oldroyd, David. *Thinking about the Earth: A History of Ideas in Geology.* Cambridge, Mass.: Harvard University Press, 1996.

Rudwick, Martin J. S. *Georges Cuvier, Fossil Bones, and Geological Catastrophes: New Translations and Interpretations of the Primary Texts.* Chicago: University of Chicago Press, 1997.

Waggoner, Ben. "Georges Cuvier (1769–1832)." University of California Museum of Paleontology. Available online. URL: http://www.ucmp.berkeley.edu/history/cuvier.html. Accessed January 23, 2008.

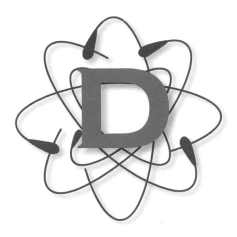

Darwin, Charles (1809–1882) English *Naturalist* Charles Darwin formulated the theory of evolution by means of natural selection following a five-year voyage around the world aboard the H.M.S. *Beagle*. When he published his theory in *On the Origin of Species by Means of Natural Selection, or the Preservation of Favoured Races in the Struggle for Life* in 1859, a controversy began; it would last until the latter part of the 20th century. While Darwin was not the first to propose that life-forms change or evolve, he was the first to propose a scientific mechanism for the process of evolution and to provide an overwhelming amount of organized evidence in support of it.

CHILDHOOD AND EDUCATION

Charles Robert Darwin was born on February 12, 1809, in Shrewsbury, England. His father, Robert, was the son of Erasmus Darwin, a respected physician and nature writer. His mother, Susannah, was the daughter of Josiah Wedgwood, a renowned potter and philanthropist. Both families instilled in their children a high regard for education. This was passed on to Robert and Susannah's own six children, but young Charles was not a promising student.

One year after his mother died, when Charles was nine, his father enrolled him at Shrewsbury School, where his teachers struggled to teach him Latin, the classics, and history. Charles, who preferred to spend his time in a chemistry lab his older brother had fashioned from a tool shed, was terribly bored. Even after the headmaster publicly admonished Charles for wasting his time on scientific pursuits when he should have been studying, Charles did not take a more serious interest in his studies.

When Charles was 16 years old, his father enrolled him at Edinburgh University to study medicine. Though his grandfather and father were both successful physicians, medicine was not to be Charles's destiny. He was disgusted by the animal dissections, and during a mandatory observation of human surgery, he was so repulsed that he ran out of the room.

Two years later Charles Darwin enrolled as a divinity student at Christ's College at Cambridge University, but he continued his personal scientific studies. He also joined the Glutton Club, whose members ate, drank, and played cards frequently. In his spare time he hunted birds and foxes and collected beetles. He successfully completed the requirements for a divinity degree in 1831, but he still needed additional elective credits, so Darwin enrolled in Professor Adam Sedgwick's geology class. For this class he read *A Personal Narrative of Travels to the Equinoctial Regions of the New Continent during the Years 1799–1804* by Alexander von Humboldt. He was fascinated by the voyage of discovery. A botany professor, the Reverend John Stevens Henslow, and Sedgwick both recognized Darwin's scientific mind and encouraged Darwin to pursue natural history as a career. Hesitant to challenge his father's wishes, Darwin earned his bachelor's degree in theology in 1831.

VOYAGE ON THE H.M.S. *BEAGLE*

At the same time, Commander Robert Fitzroy (1805–65) was preparing to depart for South America. Fitzroy was the commander of the H.M.S. *Beagle,* and he was looking for an intellectual companion during an upcoming voyage to explore the coasts of South

The British naturalist Charles Darwin was the first to provide overwhelming evidence in support of biological evolution and to propose a mechanism for evolution. *(Library of Congress)*

America and the Pacific Islands. His friends recommended he take Darwin along as the voyage's naturalist. Darwin's father initially balked at this idea but eventually gave in and provided financial support for his son on what would become one of the most influential scientific expeditions of all time.

The H.M.S. *Beagle* set sail from Plymouth, England, on December 27, 1831. Darwin suffered from severe seasickness and spent the first several weeks in the hammock of his cramped cabin. To pass time on board, Darwin read a recently published textbook by the Scottish geologist Charles Lyell (1797–1875), *The Principles of Geology.* Lyell opposed the popular ideas of the day regarding the history of the Earth. Most scientists believed in a strict biblical account of the creation of the world and the origin of life—God had created the world approximately 6,000 years ago and created all organisms in their present-day form. Fossils told a different story, one that included the prior existence of life-forms no longer present on Earth. Catastrophists believed that huge earthquakes and floods, such as the flood described in Genesis, accounted for the extinction of some species. Lyell argued that the Earth's current

physical form was the result of gradual forces such as erosion and volcanic activity acting over a period of millions of years. Darwin agreed with Lyell. When he voiced his opinions to Fitzroy, Fitzroy was outraged by what he considered blasphemy.

On January 16, 1832, the *Beagle* stopped at the Cape Verde Islands off the northwest coast of Africa. Darwin performed the tasks he was taken along to accomplish—collecting specimens and making detailed notes of his observations. Darwin also observed evidence in the strata that supported Lyell's views on the gradual nature of Earth's change.

After a stop at Tenerife of the Canary Islands, the ship arrived in Brazil on February 28, 1832. Darwin was amazed by the wealth of life in the tropical rain forests and struck by the different types of life-forms. The crew reached Rio de Janeiro in April, and Darwin noticed the nearby rain forests had been destroyed to accommodate the city's growing population. He was also appalled at the treatment of slaves. He mentioned this to Fitzroy, who vehemently disagreed, causing conflict between the two men. Though Fitzroy later apologized, it was apparent the two did not have as much in common as they initially thought.

As they traveled down the east coast of South America, Darwin took time to explore the Punta Alta beach, where on September 23, 1832, he discovered the head of a large animal. The remains were from an extinct toxodon, a rodent the size of an elephant, similar to the present-day capybara, which is about two feet in length. A few days later he found the bones of a 20-foot- (6-m-) tall ground sloth. Darwin wondered why God bothered to create such similar animals. Why did God destroy the larger ones only to replace them with smaller versions?

Darwin also noticed snakes with tail rattles that were not as efficient as those displayed by the North American rattlesnakes. In Patagonia he observed two different forms of unusual ostriches. He also took note of the mountains, valleys, and other geological features of the region. In February 1835, while in Chile, he experienced an earthquake that destroyed villages and killed inhabitants. The earthquake visibly raised the land level and altered other geological formations. Darwin had witnessed firsthand how an ordinary natural disaster affected the Earth's surface and the life it supported.

On September 15, 1835, the H.M.S. *Beagle* landed at the Galápagos Islands. This isolated equatorial chain of over one dozen volcanic islands lay 500 miles (800 km) off the west coast of Ecuador. The islands were home to numerous strange animals, animals that were never observed on the mainland of South America and certainly not in Europe. One of the most legendary species inhabiting the Galápagos

The giant tortoises Darwin observed on the Galápagos Islands seemed prehistoric to him. *(Paul Guther/ U.S. Fish and Wildlife Service)*

Islands is the giant tortoise. These tortoises weigh approximately 500 pounds (227 kg) and have shells with a circumference of eight feet (2.4 m). Darwin and his colleagues were easily able to ride on their backs. The tortoises seemed somehow prehistoric to them.

Another famous native of this archipelago is the finch. Darwin observed and sketched 13 different small birds, some of which resembled finches, along with others that had atypical beaks. He noticed that not only did these birds differ from the mainland birds, but each island seemed to have its own unique species. Darwin found it odd that geologically similar islands would have different species. They all resembled each other, but all varieties had different beak shapes. Some were clearly designed for cracking nuts and seed shells. Others were ideally suited for eating

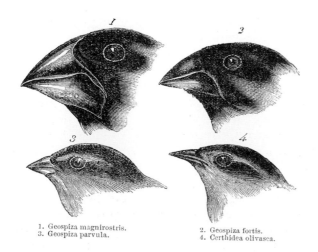

1. Geospiza magnirostris.
3. Geospiza parvula.
2. Geospiza fortis.
4. Certhidea olivasca.

Darwin concluded that the variations in beak shape exhibited by the Galápagos finches were adaptations to the available local food supplies on different islands. *(HIP/Art Resource, NY)*

insects or fruit. One beak type resembled that of a woodpecker; its shape allowed the bird to extract larvae from tree bark. Darwin again wondered why God would create so many creatures that were very similar, yet had discernible differences. He suspected that some natural principle was at work, and that God was not the cause.

Late in the year of 1835, the *Beagle* left the Galápagos. They crossed the Pacific Ocean and stopped at Australia. Darwin wondered why kangaroos, wombats, and wallabies only lived in Australia. Darwin had a lifetime's worth of data to examine and lots of unanswered questions.

EVOLUTION BY MEANS OF NATURAL SELECTION

The ship returned to England on October 2, 1836. Knowing he was not interested in working as a clergyman, Darwin was worried about facing his father. His spirit was in science. Upon his return, he was pleasantly surprised to learn that his father was proud of his work as a naturalist. Professor Henslow had been circulating Darwin's correspondence, and he was respected among intellectuals. Darwin was relieved not to have to become a minister and set to work immediately writing up the narrative of his travels. *The Journal of Researches into the Geology and Natural History of the Various Countries Visited by the H.M.S. Beagle under the Command of Captain Fitzroy, R. N., from 1832–1836* was published in 1839.

Darwin settled back into life in England with many observations and specimens that he still needed to process and evaluate. He returned to the question of why so many subtly different forms of animals such as tortoises, birds, ostriches, and snakes existed. He concluded that the different forms descended from a single common ancestor. Physical variations slowly accumulated, they way geological changes gradually shaped Earth. The idea of evolution was not novel. In fact, during the 1770s, Darwin's own grandfather, Erasmus Darwin, had published a book discussing the concept, but the world was not ready to accept evolution. No probable method had been proposed, and no one had gathered or compiled enough factual evidence to support the theory. An even greater hindrance was that acceptance of evolution required abandonment of strict biblical teaching. Most people believed that God created the world and all of its species in their present form. Darwin knew that in order to convince the world of his theory, he would have to address a persuasive argument posed by those who opposed the idea of evolution. The argument stated that organisms were each perfectly suited for the environment in which they lived, and if they did accumulate modifications, these variations would lead them to become less well suited for their habitat. Darwin had to figure out a way that organisms could change so that it appeared that they were designed that way.

Darwin had his work cut out for him. Fortuitously, he had the benefit of several intellectual colleagues with whom he could discuss his ideas. One such colleague was Lyell, whose geology text had influenced Darwin's thinking during the *Beagle* expedition. The two became good friends. Another friend was John Gould, a respected ornithologist. Gould confirmed for Darwin that the Galápagos finches he took back were all distinct species, not simply slightly different varieties of the same species.

By 1838 Darwin had already spent much time contemplating how offspring differ from their parents. Offspring had subtle yet discernible variations. Darwin thought that the accumulation of enough variations might lead to the formation of a new species over thousands of generations. He thought about artificial selection, the process by which farmers select domesticated animals or cultivated plants for breeding on the basis of their possession of a desirable characteristic. Over several generations, the incidence or degree of the favored characteristic increases. For example, a farmer may choose sheep that yield a superior amount of wool for breeding purposes. In the next few generations, the sheep's offspring also would yield more wool.

Darwin also considered extinction. Organisms were thought to become extinct because a change in climate or environmental conditions meant they were no longer perfectly adapted to their environment. Darwin concluded that if some of the offspring had accumulated enough variations, they might have an advantage over those that had not. Then the variant organisms might be better suited to survive in the new environmental conditions. Nature selected against the offspring that were not able to adapt.

In September 1838 Darwin started reading a popular book to give his mind a rest from thinking about evolution. That book was *An Essay on the Principles of Population* by Thomas R. Malthus, an English economist and clergyman. Malthus described how plants and animals produce more offspring than can survive. He discussed human populations and how poverty, famine, and disease acted to keep population size under control. Darwin recognized the significance of the natural struggle for existence, which may be summarized as follows: Animals produce many more offspring than can possibly survive. Those that do survive face a constant struggle for food and territory. Even if they successfully reach adulthood, they then must compete for mates. Offspring with variations that give them some advantage in their particular environment have a better chance

to survive to reproductive age and to breed. Thus, those individuals who are best suited for survival in their environment are the members of that species that pass on their characteristics to the next generation. Since offspring are very similar to their parents, the likelihood is high that they possess the same characteristics that gave their parent an advantage. In other words, nature selects variations that are advantageous for survival and reproduction in a particular environment, just as farmers artificially select for economically desirable characteristics. Darwin called this process natural selection and believed it was the method by which evolution occurred over thousands of generations. Though this insight would eventually cause a revolution in science, Darwin hesitated to make it public.

Darwin had married his cousin, an intelligent woman named Emma Wedgwood, in January 1839. They moved to London and enjoyed the upper-class lifestyle courtesy of wedding gifts from their parents. They eventually had 10 children, only seven of whom survived infancy. Shortly after his wedding, Darwin became mysteriously ill. He was plagued with headaches, fatigue, and sleeplessness. Modern physicians have suggested that he suffered from Chagas disease, a tropical parasitic infection, but none of Darwin's doctors could diagnose his illness. The family moved to Down House, Kent, in September 1842, and Darwin retreated from the public.

The year he moved to the Kent countryside, Darwin sketched out his theory of evolution by natural selection in a 35-page outline. In 1844 he expanded it to 230 pages, but he delayed publishing it. Instead, he delved into the study of barnacles for the next eight years. Emma suspected this was a way of avoiding the expected controversy. Darwin knew that though he had collected adequate evidence for evolution and had formulated a plausible method, the majority of people would reject his theory on religious grounds. Darwin was a shy and now frail man. He did divulge his ideas to a few friends including Lyell and an English botanist named Joseph Hooker. They encouraged him to continue collecting evidence and developing his theory.

By 1856 Darwin still had not shared his ideas concerning evolution with the rest of the scientific world. Lyell and Hooker exhorted him to publish something before anyone else did. They told him it would be a shame for him to be preempted after two decades of tireless work, so Darwin began slowly writing. Lyell and Hooker urged him to work more rapidly, but Darwin wanted to be thorough.

Then, on June 18, 1858, Darwin received a letter from a young naturalist named Alfred Russel Wallace, who was in the Malay Archipelago at the time. Wallace had developed an idea for how species might change with time, influenced by environmental changes acting to select for advantageous variations in offspring. An essay titled "On the Tendency of Varieties to Depart Indefinitely from the Original Type" was enclosed. Wallace wanted to know whether Darwin thought it worthy of publication. Darwin was stunned to read in Wallace's essay many of the same ideas contained in his own book, which was still in progress. He appealed to Lyell and Hooker for advice. They acted quickly by presenting both Wallace's essay and Darwin's outline to the Linnean Society on July 17, 1858. Hooker helped to establish priority for his friend by asserting that they had discussed the same ideas more than a dozen years prior. Surprisingly, Wallace was very chivalrous about this.

Now Darwin wrote furiously, producing a 200,000 page manuscript by March 1859. That November, *On the Origin of Species by Means of Natural Selection, or the Preservation of Favoured Races in the Struggle for Life* was published. All 1,250 printed copies sold the first day. The book was very detailed and was composed of three main sections. The first section described the process of natural selection of favorable variations. The second section dealt with objections to common arguments against evolution such as the lack of transitional forms and the development of complex specialized organs such as eyes. The third section elucidated how the theory of evolution by natural selection explained many previously unexplained phenomena such as extinction and the slight resemblance of modern and ancient species. *On the Origin of Species* presented immense substantiation for Darwin's theory of evolution by natural selection.

Despite the well-defined supportive arguments and the massive evidence Darwin provided for evolution, his theory engendered a storm of brutal criticism. Famous scientists including his former professor Sedgwick, the English zoologist Richard Owen, and the Swiss-American naturalist Louis Agassiz were all outraged and viciously reviled Darwin and his theory. Darwin had dreaded this outcome and was predictably upset. Luckily, he found equally aggressive and competent defense from his loyal friends. One such ally was Thomas Henry Huxley, a well-known biologist and educator. Huxley was an excellent public speaker and welcomed the challenge of a public debate against Bishop Samuel Wilberforce.

The famous debate took place on June 30, 1860, at Oxford University during the annual meeting of the British Association for the Advancement of Science in front of a crowd sporting more than 700 anxious people. Wilberforce spoke first, denouncing evolution and criticizing Darwin. His speech consisted mostly of personal opinions. The audience applauded loudly and cheered when Wilberforce ended his dialogue by asking Huxley whether it was through his grandmother or grandfather that he was descended from a monkey.

Huxley spoke next. He pointed out that Wilberforce did not state anything new and did not even appear to understand Darwin's theory or arguments. After carefully reviewing Darwin's theory and clearly presenting the arguments in favor of evolution by natural selection, he ended his speech with the statement that he would rather be descended from an ape than be related to a man bestowed with great intellectual gifts who used them to obscure the truth and mock serious scientific debate. The audience was uncontrollable. Some women fainted. Fitzroy was present. He wildly waved his Bible in the air and yelled abominations against Darwin.

Hooker then calmly made his way to the podium. He was disgusted by the behavior of the audience. He proceeded systematically to destroy all of Wilberforce's arguments over a two-hour period. In the end, the force of truth prevailed.

The controversy continued, however. Darwin left the defense of his ideas to his qualified contemporaries and spent his time in the gardens of Down House. Though he had deliberately left out any mention of the human species in *On the Origin of Species*, it had become the focus of the debate between creationism and evolution. In 1867 Darwin tackled this directly by composing *The Descent of Man*, published in 1871. Darwin declared that humans and apes had evolved from a common ancestor, but this idea often is represented incorrectly as that humans descended from apes. Darwin braced himself for more attacks, but this book did not generate the controversy that *On the Origin of Species* did. Most of the scientific world had already dealt with the notion of human evolution and accepted it as part of evolutionary theory.

The remainder of Darwin's life was peaceful. He published other works including *The Expression of the Emotions of Man and Animals* (1872), *Insectivorous Plants* (1875), and *The Movements and Habits of Climbing Plants* (1875). He also wrote an autobiography for his children in 1876 and enjoyed time with his family. His unexplained illness disappeared. In December 1881, he suffered his first heart seizure. Charles Darwin suffered a second heart seizure and died on April 19, 1882. Though he was elected a member of the Royal Society of London and even awarded their Copley Medal in 1864, he never received formal recognition from the British government while he was alive because his work offended the leaders of the Church of England. After his death, Parliament requested that he be buried in Westminster Abbey near Sir Isaac Newton.

Today most people have heard of Charles Darwin and consider Darwinism synonymous with evolution. Fundamentalists still decry evolution and fight to suppress its teachings. Amazingly, the negative feelings the general populace harbors concerning evolution are strong enough to force strict guidelines addressing the manner in which it is taught in public schools. In the scientific community, however, evolution by natural selection is a fundamental unifying theory of all the life sciences.

See also DOBZHANSKY, THEODOSIUS; EVOLUTIONARY BIOLOGY; EVOLUTION, THEORY OF; GOULD, STEPHEN JAY; HUMBOLDT, ALEXANDER VON; WALLACE, ALFRED RUSSEL.

FURTHER READING

The Charles Darwin Foundation for the Galápagos Islands home page. Available online. URL: http://www.darwinfoundation.org. Accessed January 19, 2008.

Darwin, Charles. *On the Origin of Species.* Cambridge, Mass.: Harvard University Press, 1964.

Dobzhansky, Theodosius. "Nothing in Biology Makes Sense Except in the Light of Evolution." *The American Biology Teacher* 35 (1973): 125–129.

Jastrow, Robert, ed. *The Essential Darwin.* Boston, Mass.: Little, Brown, 1984.

Wilson, Edward O., ed. *From So Simple a Beginning: The Four Great Books of Charles Darwin.* New York: Norton, 2006.

data presentation and analysis Data are the objective information collected during the course of an experiment and may take the form of measurements or descriptive observations. Data analysis is the conversion of the raw data into useful information from which a researcher can draw conclusions.

GRAPHS

Charts, tables, and graphs are useful ways to present data. Selecting the most appropriate method for presenting data facilitates analysis by accentuating patterns. Charts are often used for raw data collection; the researcher records the information directly as lists, notes on a diagram or on a map, or on other defined areas of a chart. After recording all the data on a chart, the researcher may compile the information in table format or in a spreadsheet. Tables consist of rows and columns that contain combined data or information. Generating a table facilitates graphing, especially when the researcher uses computer software to generate the graphs or pictorial representations. The use of the many different forms of graphs to present data reveals trends or patterns that might not be as apparent when using a chart or a table to present the information.

Bar graphs depict quantitative data from discontinuous categories. The horizontal axis contains the distinct categories, and the vertical axis shows the number of occurrences. Vertical bars shaped like rectangles extend upward to the demarcation for the number of units associated with that data set. For example, a researcher surveying a defined area of a deciduous forest for different types of fungal species might depict data in the form of a bar graph. One can look at the graph below and immediately see that puffballs were the most common type of fungus in the observed area of forest, and that coral fungi were not observed at all.

Pie charts or pie graphs show the relative frequencies of subdivisions of a set of data. The area of each

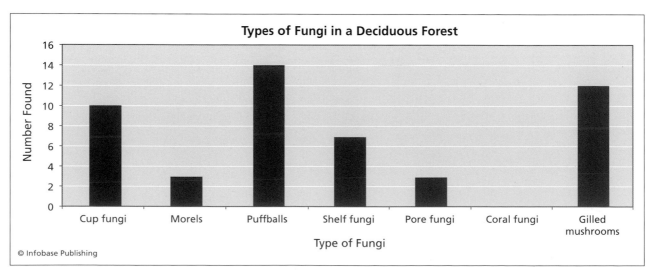

A typical bar graph

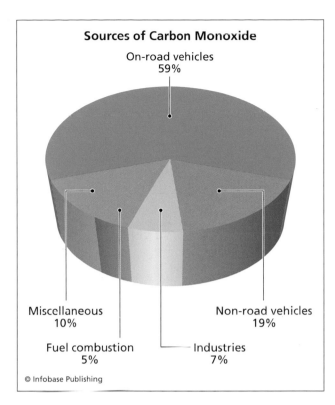

Sources of Carbon Monoxide

On-road vehicles
59%

Miscellaneous
10%

Non-road vehicles
19%

Fuel combustion
5%

Industries
7%

© Infobase Publishing

A typical pie chart

"slice" of the circular pie represents the percentage of the whole that a particular subdivision makes up. For example, an environmental scientist may report the sources of carbon monoxide emissions in a particular city in the form of a pie chart like the one above. This allows one to see at a glance that the major source of emissions is on-road vehicles, with nonroad vehicles (such as boats or construction equipment and machinery) contributing the second highest amount.

Histograms also depict frequencies, but they reveal how the frequency of the data relates to another particular variable. One can study a histogram to see the distribution of the frequencies over the entire range of the variable. For example, a botanist might use a histogram to show the frequency of number of ears per stalk for a new strain of corn. The number of ears on a stalk would be indicated on the x axis, and the rectangular bars would extend up to the number of occurrences for each number. From the histogram depicted at right, one can see that four ears per stalk occurs most frequently and that the numbers ranged from zero to eight ears per stalk.

Line graphs reveal relationships between two sets of continuous data and are especially useful for displaying data over time. The data points are plotted and connected by lines that may reveal informative trends. For example, the line graph on page 240 shows that circulating levels of the hormone progesterone progressively increase during gestation.

One can infer values for unobserved data by extending the line of a line graph beyond the last data point. Scientists often use this process, called extrapolation, to make predictions about the future effect of a trend if it continues along the same pathway. For example, if data show that the number of fish species in a particular lake is declining at a rate of one per year, and the lake currently supports 17 different species of fish, then one could use extrapolation to predict that if the current trend continues, within 17 years the lake will be devoid of fish. Interpolation is the process of determining a value at an interval between two observed data points. For example, if the ichthyologist responsible for counting the number of fish species in the lake recorded data for every year between 1993 and 2007 with the exception of the year 2002, then one could use interpolation to infer the number of fish that were present in the lake in 2002 from the data obtained before and after that time point.

In a scientific experiment, the independent variable is the factor that the experimenter intentionally manipulates, and the dependent variable is what the experimenter evaluates for change. If the experiment is carefully controlled, meaning the only difference between experimental systems is the independent variable, then the general direction and the steepness of the line drawn between data points demonstrate how changing the independent variable affects the dependent variable, if at all. The independent variable is graphed on the horizontal axis, which is referred to as the x axis. The dependent variable is graphed on the vertical axis, which is referred to as the y axis. In the line graph, the dependent variable is the number of mutant bacterial colonies that grow on a plate of medium and is a function of the concentration of the chemical mutagen added to the medium, at least for the range of concentrations tested.

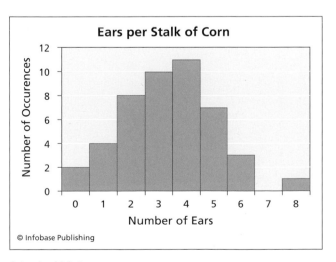

Ears per Stalk of Corn

Number of Occurences

Number of Ears

© Infobase Publishing

A typical histogram

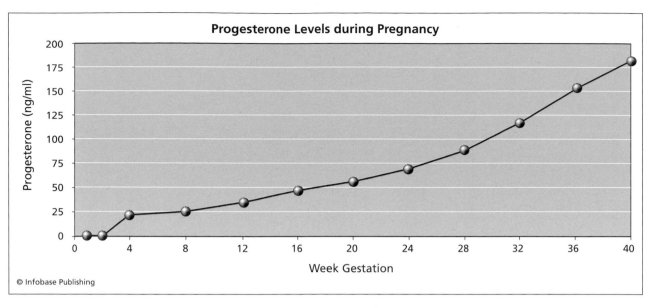

This line graph reveals a trend—progesterone levels increase with time of gestation.

One must be careful, however, in concluding from a line graph that one variable is causing the measured variable to change. Some plots may reveal a trend or relationship without one of the variables directly influencing the other. The variables change together in a way not expected on the basis of chance alone, but one does not necessarily cause the other to change. In some cases the researcher cannot ensure that only a single factor is altered during an experiment. Data collection might involve observing a phenomenon that occurs in nature and cannot be manipulated. For example, the average temperature of the ocean may vary at different times of the day,

but the researcher cannot experimentally control the many factors that cause this effect (e.g., the intensity of the radiation from the Sun as it reaches the water, oceanic circulation patterns). In this case, the time of day does not cause the change in temperature—other factors that also change over time cause the resulting temperature variations—but plotting the data on a line graph will still reveal the temporal patterns in temperature fluctuations.

Scatter plots are similar to line graphs but are used to determine whether and how one variable relates to another. After the data are entered onto the scatter plot, a computer can generate a line of best fit, also

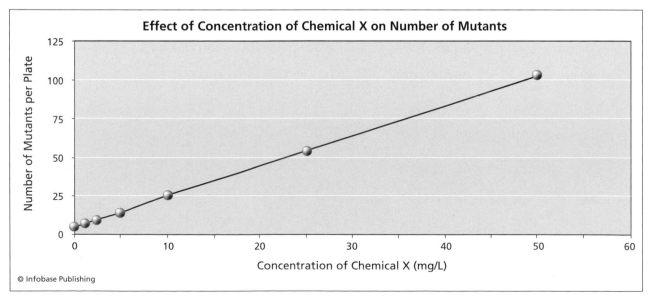

Line graphs illustrate the relationship between independent and dependent variables.

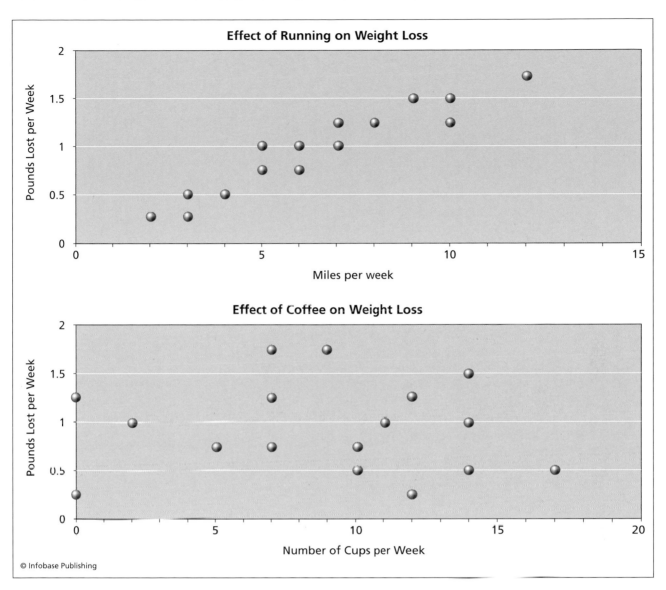

Scatter plots reveal possible correlations between variables.

called a regression line, through the data points. If the data points lie on a reasonably straight line, they are said to be correlated. A widely scattered pattern indicates that no correlation exists between the two factors. If the y values increase as the x values increase, then the variables have a positive correlation, and if the y values decrease as the x values increase, the variables are said to be negatively correlated. To illustrate this, from the scatter plots (generated from hypothetical data), one can conclude that running larger distances correlates with greater weight loss, whereas the number of cups of coffee someone drinks each week does not appear to be correlated to weight loss.

STATISTICAL ANALYSIS

Scientists use statistical analysis to interpret numerical information obtained during the course of an experiment. Statistics is a branch of mathematics devoted to collecting data and extracting meaningful information from those data. Because the data obtained during an experiment must be analyzed afterward, the researcher must carefully design an experiment that will allow one to perform an efficient and meaningful statistical analysis. Statistics helps the experimenter do this, for example, by indicating how many data must be collected in order to draw reasonable conclusions from the results. Another purpose of statistical analysis is to determine the reliability of a data set. Ideally, the information obtained from the observed or tested sample reflects information that is true for a larger population. In order to determine whether a sample accurately depicts the behavior or a result characteristic for the larger population, one must first be able to summarize the experimental data. The mean and standard deviation are two important statistical descriptions of a sample data set.

The mean (χ) is the average of a sample of numerical values and can be calculated by the following formula:

$$\chi = \frac{\sum\limits_{i=1}^{i=n} x_i}{n}$$

Simply, the mean equals the sum of all the numerical values in the data set divided by the number of values (n). As the sample size increases, that is, as the number of values in the data set increases, the mean of the sample approaches the mean of the population (μ) that the sample represents. (Statisticians use Roman letters to indicate sample values, referred to as statistics, and use Greek letters to represent population variables, referred to as parameters.)

Standard deviation (s) and variance (s^2) indicate the spread of the population, in other words, the degree to which the individual data points differ from the mean. A small standard deviation means that all the measured variables are very close in value. The standard deviation of a sample (s) of n data is an estimate of the standard deviation of the entire population (σ). The variance is its square.

$$s = \sqrt{\frac{\sum\limits_{i=1}^{i=n} (x_i = \overline{x})^2}{n-1}}$$

$$s^2 = \frac{\sum\limits_{i=1}^{i=n} (x_i = \overline{x})^2}{n-1}$$

Because the difference between each datum and the mean is squared in the formulae, the value will increase the standard deviation whether the datum is less than or greater than the mean. The standard deviation will have the same units as the mean, but the variance will have the units of the mean squared.

To illustrate the concepts of mean, standard deviation, and variance, consider the following hypothetical data for number of trunk segments in *Geophilus* species (centipedes), presented in the table at right Number of Centipedes with Different Numbers of Segments. The mean equals 54.06 segments (though the actual number of trunk segments in centipedes is always odd). The standard deviation equals 4.77 segments, and the variance is 22.72 segments^2.

The standard deviation is basically the average of the squared differences of the individual data

NUMBER OF CENTIPEDES WITH DIFFERENT NUMBERS OF SEGMENTS

Number of Segments	Occurrences
43	2
45	5
47	6
49	9
51	28
53	36
55	39
57	12
59	10
61	4
63	3
65	2
67	4
69	2

points and the mean, but the sum of the squares is divided by the degrees of freedom rather than n. Degrees of freedom (df) is a mathematical restriction that indicates the number of independent variables upon which a calculation is based and equals $n - 1$, since the mean must be determined before computing standard deviation, and calculating the mean uses up one degree. In other words, the degrees of freedom are the number of values in a calculation that are free to vary, a measure of the amount of precision an estimate of variation has. The more parameters that must be estimated, the fewer the degrees of freedom.

Standard deviation relates to another statistical descriptor called the confidence interval, a range of values about a sample mean that is likely to contain the population mean with a stated probability, such as 95 or 99 percent. Both indicate a degree of reliability for a set of data. Confidence limits are the values at the extreme ends of the defined confidence interval. Remember that when collecting data, a sample is meant to represent the entire population. The mean of the representative sample will probably differ slightly from the mean of the population, but one can define a range of values and declare that the population mean will fall within that range defined by the sample data 95 percent of the time. One can extend the limits slightly and predict with 99 percent certainty that the population mean will fall within the slightly larger range. As the sample size increases, the sample mean will better estimate the population mean and the standard deviation will decrease, allowing for the prediction of a tighter range for the population mean while maintaining the same degree of confidence.

In order to determine how much variation to expect, an investigator calculates the frequencies with which various results could occur for a given experiment. For example, if a geneticist crosses two plants heterozygous for a trait, the expected ratio of offspring that exhibit the dominant characteristic to offspring that exhibit the recessive characteristic is 3:1. Using simple rules of probability and a mathematical expression called a binomial expansion, one can determine the probability that in a sample of four offspring, three would express the dominant characteristic and one would express the recessive characteristic. If one calculates the probabilities for all the possible outcomes (all four offspring dominant, all four recessive, two of each, three dominant and one recessive, or three recessive and one dominant) and plots them as expected frequencies for each outcome on a graph, the result is called a normal frequency distribution. The peak of a normal distribution curve represents the mean, and the spread or dispersion increases or decreases as the standard deviation increases or decreases.

Because the y axis depicts frequencies rather than actual numbers, the distribution curve reflects possible outcomes for all sample sizes. If the investigator repeats the experiment numerous times, most often the results will be close to 3:1, but on occasion, they may deviate significantly. A normal distribution curve shows how often one would expect such deviations to occur on the basis of chance. To illustrate the concept of a normal distribution, consider another example, in which someone flips a coin 10 times. Assuming the probability of landing on heads is the same as that of landing on tails, of all the possible outcomes, the probability that of 10 coin flips, five would be heads and five would be tails is the highest. However, few people would be surprised if the result were six heads and four tails, or even seven heads and three tails. Most people would be surprised, however, if all 10 flips resulted in heads. Though

this is a possible outcome, randomly flipping a coin and obtaining the same result 10 times would be an unlikely outcome. The normal distribution curve created by calculating and plotting the probabilities for

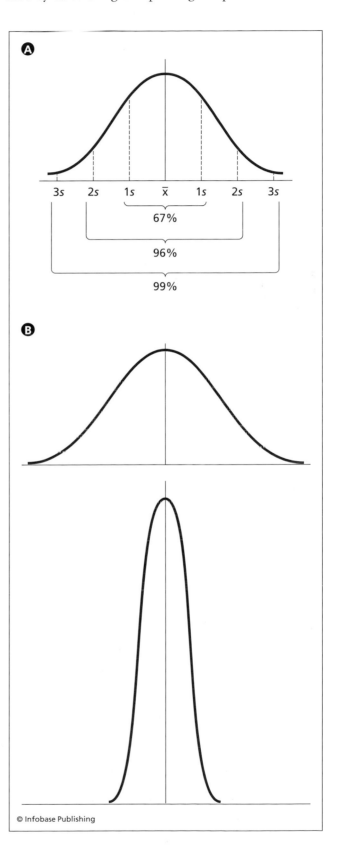

A) The curve of a normal distribution resembles a bell, with 67 percent of the area within 1s of either side of the mean, 96 percent within 2s of the mean, and 99 percent of the area within 3s of the mean.
B) Both of these plots are normal distributions about the same mean, but the sample depicted in the lower diagram has a smaller standard deviation than the data depicted in the upper diagram, as indicated by the difference in the spread of the curves.

all the possible outcomes would peak at the position reflecting the outcome of 50 percent heads and 50 percent tails. The area underneath the curve represents all of the possible outcomes. One can draw vertical lines at each end of the curve that mark off 5 percent of the total area (2.5 percent at each end). The remaining area represents 95 percent of the possible outcomes. Thus, one can predict that 95 percent of the time, a data set will give results that fall under this portion of the curve.

One common application of statistics is to test the validity of a hypothesis. Frequency distributions allow a researcher to determine whether to accept or reject a hypothesis. A null hypothesis is the hypothesis against which the data are tested; it assumes the observed data do not deviate significantly from the expected data; that the differences are due to chance alone. After data collection, the researcher calculates the probability of obtaining those data assuming the null hypothesis to be true. If the probability is high, the data support the hypothesis. If the probability is too low, the researcher rejects the hypothesis. If something is statistically significant, it is unlikely to occur by chance. A researcher must decide the level of significance to use when drawing conclusions from their data. The significance level is a fixed probability of wrongly rejecting the null hypothesis when it is true. Scientists often base their determinations on a 5 percent level of significance. This means that the data will deviate from the expected results enough to reject the hypothesis 5 percent of the time on the basis of random chance alone, and the scientists risk rejecting a true hypothesis under those circumstances.

The chi-square test is one method life science researchers use to see whether their data are consistent with the null hypothesis or whether the data reject the null hypothesis. The formula for calculating a chi-square value (χ^2) is

$$\chi^2 = \sum \frac{(O - E)^2}{E}$$

where O is the observed number for a category and E is the expected number on the basis of the hypothesis being tested. Comparison of the calculated value with a value from an available table of chi-square values for specific degrees of freedom and probabilities tells the researcher whether the data reject or fail to reject the hypothesis. The values from the table are the probabilities of obtaining a chi-square value calculated from experimental data given certain numbers of variable parameters or categories (degrees of freedom) for a normal frequency distribution. One must decide the level of significance to use. Note that as the degrees of freedom increase, the critical values for chi-square increase, as more parameters increase the expected standard deviation, even if the sample data fit well with the expected outcome. Because one must add the calculated chi-square values from each category in order to obtain the total chi-square value, simply increasing the number of categories will increase the critical chi-square value. A calculated total value that is less than the critical chi-square value indicated by the table means that the data support the hypothesis; a calculated value that is higher than the value given by the table means the data differ significantly from the outcome predicted by the hypothesis, and thus, reject the hypothesis. If the data fail to reject the hypothesis, one can conclude that the data support the hypothesis, but not that they prove it.

Imagine a population biologist counting the number of male and female babies born in a particular city. Since fertilization is a completely random event, one may hypothesize that the data will reveal equal numbers of boys and girls. In other words, the null hypothesis would be a 1:1 ratio of boys to girls. Because this example contains two categories (boys and girls), there is one degree of freedom. Only one of the categories can freely vary—if the result in one category is known, so is the other, since there are only two possibilities. For example, if the outcome of one event is not a boy, it must be a girl. Suppose the

CHI-SQUARE (χ^2) VALUES

Degrees of Freedom	Probabilities						
	0.99	0.95	0.80	0.50	0.20	0.05	0.01
1	0.000	0.004	0.064	0.455	1.642	3.841	6.635
2	0.020	0.103	0.446	1.386	3.219	5.991	9.210
3	0.115	0.352	1.005	2.366	4.642	7.815	11.345
4	0.297	0.711	1.649	3.357	5.989	9.488	13.277

researcher collects data on 1,000 total children, 528 of whom are boys and 472 of whom are girls. Given the null hypothesis of 1:1, the expected number of boys is 500, and the expected number of girls is 500. Chi-square for this example is calculated as follows:

$$\chi^2 = \sum \frac{(O - E)^2}{E}$$
$$= \frac{(528 - 500)^2}{500} + \frac{(472 - 500)^2}{500}$$
$$= 3.136$$

The critical chi-square value for one degree of freedom with a 5 percent level of significance is 3.841. This means that given the null hypothesis of a 1:1 ratio of boys to girls, the probability of getting a chi-square value greater than or equal to 3.184 by chance alone is 5 percent. Because 3.136 is less than the critical value of 3.841, the researcher should not reject the hypothesis of 1:1 on the basis of the sample data.

When the sample size is small, one can use a test called the *t*-test to determine whether the sample mean differs significantly from the population mean. This type of test assumes the dependent variable is normally distributed. The *t* value can be calculated and compared to a table of values to determine the probability that the *t* value would be exceeded in a larger number of replicates. The two-sample *t*-test is another common test of significance used to determine whether two independently observed groups of sample data are from populations that have the same mean or whether they are statistically different. After calculating the *t* value, a table of significance will reveal whether the value is significant.

FURTHER READING

Hibbert, D. Brynn, and J. Justin Gooding. *Data Analysis for Chemistry: An Introductory Guide for Students and Laboratory Scientists.* New York: Oxford University Press, 2006.

McCleery, Robin H., Trudy A. Watt, and Tom Hart. *Introduction to Statistics for Biology.* 3d ed. Boca Raton, Fla.: Chapman and Hall/CRC, 2007.

Newman, Isadore, Carole Newman, Russell Brown, and Sharon McNeely. *Conceptual Statistics for Beginners.* 3rd ed. Lanham, Md.: University Press of American, 2006.

deoxyribonucleic acid (DNA) Deoxyribonucleic acid (DNA), the molecular basis for heredity, contains within its sequence of nucleotides all of the information necessary for an organism's growth, maintenance, and reproduction. In eukaryotic cells, the DNA is compartmentalized in the nucleus, and in prokaryotic cells, the DNA exists in the cytoplasm. The unique structure of DNA allows for duplication of the genetic material as well as encoding of all the genetic information for the synthesis of proteins and other macromolecules that result in the outward expression of species-specific and individual characteristic traits.

STRUCTURE

The double-helical molecule of DNA is composed of two individual strands, each a chain constructed from four different nucleotides. The two strands wind around each other and are held together by hydrogen bonds between complementary nitrogenous bases.

The individual nucleotide subunits that compose DNA consist of a deoxyribose sugar, a phosphate group, and a nitrogenous base. The nitrogenous bases of adenine (A) and guanine (G) make up the purines, nucleotides that contain two linked rings—one with five members and one with six members. Cytosine (C) and thymine (T), the pyrimidine nucleotides, each have a nitrogenous base that consists of a single six-membered ring. A phosphate group linked to the 5'-C of the deoxyribose component of one nucleotide forms a phosphodiester linkage with the hydroxyl group extending from the 3'-C of the deoxyribose sugar of the adjacent nucleotide. The nitrogenous base extends from the 1'-C of the deoxyribose sugar.

The two strands that make up a double-helical molecule of DNA are complementary, meaning that the sequence of one strand can be predicted from the sequence of the other. Because A always forms hydrogen bonds with T and G always forms hydrogen bonds with C, the sequence of one strand determines that of the complementary strand. Two hydrogen bonds link A and T, whereas three hydrogen bonds connect C and G. In order for the correct hydrogen bonds to form, the two strands must run antiparallel to one another, meaning the orientations of the two strands are opposite. Following along the backbone of a single strand, the 5'-phosphate group of one nucleotide leads to the pentose sugar, then to the 3'-hydroxyl group of that same nucleotide, which is attached to the 5'-phosphate of the adjacent nucleotide, and so on. Looking at the complementary strand, however, the directionality is reversed. If the one strand runs 5' to 3', then the opposite strand runs 3' to 5'. Alternate arrangements of hydrogen bonding can occur between nucleotides, but the regular pairing of a purine with a pyrimidine maintains the constant width of approximately 20 angstroms in this form of DNA.

The helix formed by wrapping the two strands of DNA around each other is right-handed. In other

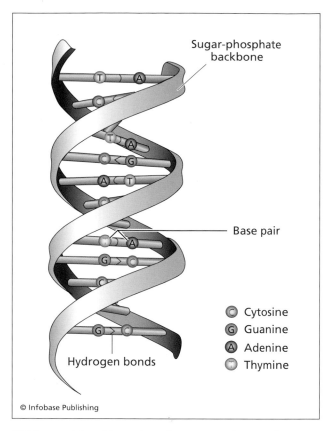

Sugar-phosphate backbone

Base pair

Ⓒ Cytosine
Ⓖ Guanine
Ⓐ Adenine
Ⓣ Thymine

Hydrogen bonds

© Infobase Publishing

DNA consists of two strands, wrapped around one another and connected by hydrogen bonds between specific pairings of nitrogenous bases of the nucleotides.

words, when viewed from the top of its axis down the center, the helix turns in a clockwise direction. If unwound, the double-stranded DNA would resemble a ladder, with the base pairs represented by the horizontal rungs of the ladder and positioned perpendicular to the sugar-phosphate backbone, represented by the vertical side rails of the ladder. The helix completes one turn every 10 base pairs, or 34 angstroms. The form of DNA described thus far is called the B form, but alternate forms of DNA do exist. In the Z form, the helix adopts a left-handed character and a zigzagged backbone. This form could potentially occur naturally in GC-rich areas by the addition of methyl groups to the cytosine moieties, and it might be involved in regulating gene expression. Another form, A DNA, is characterized by slightly tilted base pairs and has more base pairs per turn than B DNA.

DNA REPLICATION

The characteristic of specific pairing between nucleotides supports a semiconservative mechanism for DNA replication, in which each linear chain of nucleotides serves as a template for the creation of a new, complementary strand. Double-helical daughter molecules contain one parental strand and one newly synthesized strand.

Prior to division, the two strands of DNA separate at specific sequences termed origins of replication. First, an enzyme called helicase unwinds the DNA, the hydrogen bonds are broken, and the two strands open up in that region, forming two Y-shaped replication forks from which replication will proceed in both directions. Proteins called single-stranded binding proteins bind to each strand of DNA to prevent the hydrogen bonds from reforming. Replication begins when an enzyme called primase creates primers of approximately 10 ribonucleotides in length at the replication fork. This step is necessary because DNA polymerase, the enzyme that forms covalent linkages between deoxynucleotides, cannot initiate synthesis of nucleic acid without a preexisting free 3'-hydroxyl group. After the synthesis of primers, DNA polymerase reads the parental template strand of DNA, then adds the appropriate complementary deoxynucleotide to the nascent strand, forming an ester linkage between the existing 3'-hydroxyl group and the 5'-phosphate of the incoming nucleotide.

DNA polymerase can only synthesize a new strand of DNA in the 5' → 3' direction. The enzyme scans or reads the template strand 3' → 5' and elongates 5' → 3'. This limitation in conjunction with the antiparallel nature of a double-stranded DNA molecule complicate the progress of replication. The two parental strands of DNA at each replication fork run antiparallel to one another. Thus, only one of them can be scanned continuously in the 3' → 5' direction by DNA polymerase toward the junction of the replication fork. The newly synthesized daughter strand that is complementary to this one is termed the leading strand. The other daughter strand is called the lagging strand because it is synthesized in short, interrupted fragments termed Okazaki fragments. The enzyme primase initiates the synthesis of numerous primers to which DNA polymerase adds nucleotides, progressing away from the replication fork until it reaches the end of another RNA primer. A different DNA polymerase then removes the ribonucleotides from the primer and replaces them with deoxyribonucleotides. The short fragments of DNA are eventually joined together by yet another enzyme, DNA ligase, in order to form one continuous new strand of DNA.

The 5' ends of a DNA molecule present a unique problem for DNA polymerase. Because the enzyme requires a 3'-OH to add nucleotides, as the DNA polymerase approaches the 5' end of a linear DNA molecule, such as a eukaryotic chromosome, the lagging strand cannot be synthesized to the very end by usual mechanisms. The small gap that would be left at the 5' end of the lagging strand would grow larger with each successive round of replication. To prevent this gradual erosion at the telomeres (the physical ends of the chromosomes), eukaryotic organisms have between 100 and 1,000 repeated DNA sequences at the ends of their chromosomes. An enzyme called telomerase adds these terminal repeats to maintain chromosome length over time.

The crucial function of DNA as the carrier of a cell's genetic information demands a high fidelity. The error rate for DNA replication approaches only one in 1 billion nucleotides. Immediately after the initial pairing of an incoming nucleotide with its complementary partner on the original DNA template, DNA polymerase proofreads, or double-checks, to ensure that proper base pairing rules have been obeyed. If an incorrect nucleotide was incorporated, the polymerase excises it and replaces it with the correct complementary nucleotide. Other mechanisms repair mutations on existing DNA.

ESTABLISHMENT OF DNA AS THE GENETIC MATERIAL

In the early 1900s, scientists believed protein was the most likely candidate for the molecular carrier of genes. Composed of chains of 20 different amino acids, proteins seemed the most varied of biological macromolecules in structure and function, making them the probable means of encoding all of a cell's genetic information. Not many suspected DNA, thinking it was a tetranucleotide, that is, a short chain of the four different deoxyribonucleotides linked to one another. It did not seem as if that simple structure could encode the tens of thousands of different proteins made by a cell and allow all the genetic diversity observed among organisms. Thus, the discovery that DNA was the carrier of genetic information was quite a surprise to the scientific community.

The establishment of DNA as the genetic material was the result of the work performed by many individuals. The process started with a British researcher named Frederick Griffith, who studied the different types of *Streptococcus pneumoniae,* a type of bacterium that causes pneumonia. In the late 1920s he was particularly interested in two strains, one that was deadly to humans and one that was harmless. The virulent bacterial strain was encapsulated with a polysaccharide covering that gave its colonies a smooth appearance; thus it was named *S* for "smooth." Because the nonvirulent strain produced rough-looking colonies in comparison, it was referred to as *R*, for "rough." When the S strain was injected into mice, the mice developed pneumonia and quickly died, whereas the R strain had no deleterious effect. When Griffith first killed the S type by heating it, injecting the bacterial cells into mice did not kill them. However, when he injected mice with both living R cells and heat-killed S cells, the mice died. Griffith demonstrated that the R bacteria had acquired polysaccharide capsules as the S bacteria had and that the new characteristic was passed on to progeny of the transformed bacteria. He figured that the gene that encoded the information to make the capsule was lost in the R strain, and the molecule that carried this genetic information was transferred from the S strain to the R strain, transforming it. He published his results in 1928 without having any idea of what molecule was responsible.

This question was painstakingly addressed by Oswald Avery, Colin MacLeod, and Maclyn McCarty at the Hospital of the Rockefeller Institute for Medical Research. Avery's assistants first duplicated Griffith's results in vitro. They made extracts of the S strain by repeated freezing and thawing and used these extracts to transform R strain bacteria. To identify the substance in the extract that was responsible for transformation, they purified different substances from the extracts by treating them with various enzymes known to break down specific types of molecules. Expecting a protein to be the transforming factor, they attempted to perform transformations with the purified substances, but the only substance that successfully transformed R into S bacteria was DNA. The carrier of genetic information was DNA. This result astonished them, but further tests confirmed their results, and they published their data in 1944.

The worldwide scientific community did not immediately embrace the idea that DNA was the genetic material. Four main factors contributed to the lack of enthusiastic reception of the proclaimed biological activity of pneumococcal DNA.

- The timing, during the height of the United States' involvement in the war, was unfortunate.
- The readership of the journal in which Avery published his paper was limited.
- The members of the Rockefeller team were bacteriologists, not geneticists.
- Finally, the erroneous assumption that protein was the genetic material was just too hard for many to surrender.

Though some scientists gradually accepted the enormity of the conclusions of Avery and his colleagues, the next major event that convinced the remaining nonbelievers to accept DNA as the genetic material occurred in 1952. Alfred Hershey, an investigator at Cold Spring Harbor Laboratory, and his assistant, Martha Chase, published their results from some ingenious experiments performed with T2 bacteriophage, a virus that infects bacterial cells. They demonstrated that DNA physically entered bacterial cells and directed the synthesis of new viral particles with their well-known blender experiment. Briefly, they radioactively labeled the bacteriophage by inoculating bacterial cultures growing in the presence of either radioactive phosphorus (^{32}P) or radioactive sulfur (^{35}S) with T2. Phosphorus is a component of nucleic acid, and sulfur is found in proteins. They used the radiolabeled T2 to infect fresh cultures of bacteria and found the ^{32}P label inside bacterial cells and within new viral progeny, but the ^{35}S remained outside the bacterial cells, showing that nucleic acid but not protein entered the bacteria during the infection process.

The common theme that structure determines function pervades the biological sciences, and the molecular level provides no exception. By the time that Hershey and Chase published their results, the race was on to determine the structure of this biological molecule now known to be of utmost importance. The top contenders were the British biophysicists Maurice Wilkins and Rosalind Franklin at King's College in London, the American biochemist Linus Pauling at the California Institute of Technology, and James D. Watson and Francis Crick at Cambridge University. In 1950 the Austrian biochemist Erwin Chargaff found that the amount of the nitrogenous base adenine always equaled the amount of thymine and that the amount of cytosine always equaled the amount of guanine. Another clue to the puzzle was X-ray diffraction evidence obtained by Franklin and Wilkins suggesting the molecule was helical. Watson and Crick employed a model-building strategy, repeatedly constructing and disassembling structures to fit better with all the biochemical information, chemical bonding requirements, and structural data. In early 1953 they proposed their famous double-helical model. Knowing the structure of DNA, biologists were soon able to figure out the mechanisms for its replication and for the coding of genetic information, leaving no room for any doubt as to its profound significance and role as the molecule of life.

See also AVERY, OSWALD; BIOMOLECULES; CELLULAR REPRODUCTION; CHARGAFF, ERWIN; CHASE, MARTHA; CHEMICAL BASIS OF LIFE; CRICK, FRANCIS; FRANKLIN, ROSALIND; GENE EXPRESSION; GENETICS; GENOMES; GRIFFITH, FREDERICK; HERSHEY, ALFRED; MACLEOD, COLIN MUNRO; McCARTY, MACLYN; MOLECULAR BIOLOGY; PAULING, LINUS; WATSON, JAMES D.; WILKINS, MAURICE H. F.

FURTHER READING
Calladine, Chris. R., Horace R. Drew, Ben F. Luisi, and Andrew A. Travers. *Understanding DNA: The Molecule and How It Works,* 3rd ed. San Diego: Elsevier Academic Press, 2004.
Lewin, Benjamin. *Genes IX.* Sudbury, Mass.: Jones and Bartlett, 2008.
Watson, James D., Tania A. Baker, Stephen P. Bell, Alexander Gann, Michael Levine, and Richard Losick. *Molecular Biology of the Gene.* 6th ed. San Francisco: Benjamin Cummings, 2007.

diabetes The common endocrine system disorder diabetes is characterized by inadequate secretion or utilization of the hormone insulin. This results in excessive urine production, thirst, weight loss, and high glucose (sugar) levels in the urine and blood. Insulin, made by the pancreas, helps the body utilize glucose, the major energy source for the body's cells. Diabetes results when the pancreas does not produce sufficient insulin, or when the body's cells do not respond to it as they should. There are several types of diabetes: Type 1, formerly known as insulin-dependent diabetes mellitus; Type 2, previously called non-insulin-dependent diabetes mellitus; and gestational diabetes, which occurs during pregnancy. Other types of diabetes can result from genetic abnormalities, surgery, drugs, or other causes. According to the International Diabetes Federation, diabetes afflicts 246 million people worldwide and kills 3.8 million people each year.

ROLE OF INSULIN

Living organisms must maintain relatively constant internal conditions despite fluctuations in the external environment. For example, most organisms can only survive within a certain range of temperature. A decrease or increase in temperature may stimulate a behavioral or physiological response that functions to counteract the change in order to maintain the internal body temperature within a safe range. Blood sugar level is another factor that the body must maintain within a limited range. After the digestive system breaks down carbohydrates, glucose diffuses into the bloodstream, which then transports the glucose throughout the body. In the tissues, the cells uptake the glucose and utilize it as an energy source to carry out cellular functions.

Insulin helps keep the concentration of glucose in the blood within a healthy range whether someone has just eaten a meal or has not eaten in several hours. It is a protein hormone that facilitates the cellular uptake of glucose molecules. Because glucose is polar, it cannot simply diffuse through the phospholipid bilayer of a cellular membrane; specific carrier proteins allow the glucose to move through the membrane without contacting the interior hydrophobic layer of the membrane. In the presence of insulin, the transporters that carry glucose across the cell membrane are present at the cell surface. In the absence of insulin, the transporters are stored in vesicles inside the cell. In this manner, insulin facilitates the movement of glucose from the blood into the body's cells.

A portion of the pancreas called the islets of Langerhans consists of groups of endocrine cells that produce and release the hormones that regulate blood glucose levels. Beta (β) cells make and secrete insulin, and alpha (α) cells make and secrete glucagon. After a meal, the increased blood glucose levels directly stimulate the release of insulin. This results in glucose uptake into the cells, which decreases the circulating levels once again. When the blood glucose levels are no longer higher than normal, the release of insulin stops. Within a few hours after eating, the blood glucose levels return to normal. This negative feedback system maintains relatively constant levels of blood glucose in healthy individuals.

TYPES AND SYMPTOMS OF DIABETES

Type 1 diabetes results from the destruction of more than 90 percent of the insulin-producing cells of the pancreas. Because the pancreas cannot produce enough insulin, this form of diabetes used to be called insulin-dependent diabetes. Approximately 5–10 percent of all people who have diabetes have Type 1, and most of them develop symptoms before they reach 30 years old. Because of this, Type 1 is also sometimes referred to as juvenile-onset diabetes.

Type 2 diabetes, also called non-insulin-dependent diabetes or adult-onset diabetes, results when a person's cells develop a resistance to insulin. The pancreas still produces and secretes insulin, but as the disease progresses, much more insulin is needed to move the transporters to the cell membrane to uptake glucose from the blood. Type 2 diabetes typically affects adults, and 80–90 percent of people who have this form are obese. People who take corticosteroids or who have high levels of natural corticosteroids may also develop Type 2 diabetes, or if they already have diabetes, steroid injections for other conditions (such as asthma, acute poison ivy, or arthritis) may worsen the diabetes.

Pregnant women who have never had diabetes may develop it late in their pregnancy. This form of diabetes is called gestational diabetes, and approximately 4 percent of pregnant women in the United States develop it. The cause is not known, but hormones naturally synthesized by the placenta during pregnancy are known to interfere with insulin action. The woman may require up to three times the normal amount of insulin to overcome this resistance. Gestational diabetes can result in too much weight gain in the fetus because of the excessive amounts of sugar available in the blood. This may cause difficulties during childbirth, and these children have a higher risk for becoming obese and developing Type 2 diabetes later in their lives. The mother's diabetes usually ends after the pregnancy, but she is at higher risk for developing Type 2 diabetes in her future.

Whatever the type of diabetes or the cause, without sufficient insulin, the body cells starve. Though the person may eat plenty, the glucose cannot enter the cells. Hyperglycemia, or high blood sugar levels, leads to increased urine production. The excess sugar spills into the urine, and the presence of excess solute in the urine filtrate prevents water from being reabsorbed. The volume of urine increases, and the person feels very thirsty from the fluid loss. This dehydration can cause circulatory failure due to decreased blood volume and may even result in death if cerebral blood flow is reduced. Fatigue, hunger, nausea, and blurred vision may also occur.

In the case of Type 1 diabetes, the "starved" cells switch on an alternate metabolic pathway that allows them to obtain energy from fat cells. Compounds called ketones are made as fat stores break down, and acidic by-products accumulate, leading to a condition called ketoacidosis. In addition to extreme thirst and increased urine production, the person may experience nausea, vomiting, fatigue, and abdominal pain. Respirations will be deep and rapid in order to rid the body of extra carbon dioxide in an attempt to maintain a more neutral blood pH. Ketoacidosis causes the breath to have a distinct odor. In the absence of treatment, the patient may die within hours.

Whereas the onset of Type 1 diabetes can occur rapidly, Type 2 usually develops slowly, gradually becoming worse over a period of several years. The person will exhibit similar symptoms to those exhibited by Type 1 diabetics, but symptoms worsen slowly to the point at which the person feels the need to seek medical attention.

DIAGNOSIS AND TREATMENT

When a patient seeks medical attention for symptoms including constant thirst and frequent urination, the physician will look for sugar in the urine by dipping a paper strip that has been treated with special chemicals into a sample obtained from the patient. If the result of this test is positive, then blood will be drawn after a period of fasting and the levels of sugar present will be measured. If the levels are higher than the normal range, the person may have diabetes mellitus. Other symptoms and factors will help the physician diagnose Type 1 or Type 2 diabetes and will determine the course of treatment.

For all types of diabetes, controlling blood sugar levels requires a healthy diet, meaning one high in fiber, low in saturated fats, and low in simple sugars. Treatment of Type 1 diabetes usually involves the administration of insulin, either by injections or by a pump. Eating a healthy diet and a consistent one with respect to amount of total calories consumed and having regular mealtimes makes it easier to determine the proper doses of insulin. Exercise is also beneficial. Because diabetes compounds the negative effects of alcohol and smoking, people who have diabetes should abstain from or limit these activities.

Successful treatment of Type 2 diabetes may not require medication; the disease may respond sufficiently to a controlled diet, weight loss, and exercise. In addition to insulin, medications to treat Type 2 diabetes include drugs that stimulate the pancreas to produce more insulin or to do so more quickly, increase sensitivity to insulin, decrease the amount of glucose produced by the liver, and slow the absorption of ingested starches.

If blood sugar levels fluctuate too much or remain too high, as when the diabetes is untreated, complications may result. High blood sugar level can cause the interior diameter of the blood vessels to shrink by leading to increased deposition of fatty substances on their walls, a condition called atherosclerosis. This prevents good blood flow and inhibits circulation. Because of this diabetics are at higher risk for heart attacks and stroke, and heart attacks are more likely to be severe and lead to death in diabetics.

Diabetes can also lead to neuropathies, disorders in which nerve damage leads to numbness or pain in the extremities. Vascular and nerve damage can reduce one's sensitivity to pain in the feet, cause poor circulation, and slow the healing of foot ulcers. When these problems become serious, amputation may be necessary. Eye problems and kidney disease can also result from high blood glucose level and high blood pressure from diabetes. Improper dosages of medications used to treat diabetes can lead to blood sugar levels that are too low, called hypoglycemia.

DIABETES INSIPIDUS

Diabetes insipidus produces similar symptoms to diabetes mellitus, extreme thirst and excessive urine production, but has a different underlying cause. In this type of diabetes, either the body fails to synthesize or secrete sufficient quantities of antidiuretic hormone (ADH) or the kidneys fail to recognize or respond to it. The hypothalamus normally synthesizes ADH, and the posterior pituitary gland stores it until the hypothalamus chemically stimulates the pituitary to release it during a water deficit. ADH then travels through blood circulation until it reaches the kidney, which has special receptors that recognize and bind the hormone. In response, the kidneys reabsorb more water during urine production. Diabetes insipi-

dus can result from damage to the hypothalamus or pituitary gland, interfering with ADH synthesis or release. Failure of the kidneys to respond to ADH also can cause diabetes insipidus.

See also BANTING, SIR FREDERICK G.; ENDO-CRINE SYSTEM; HOMEOSTASIS.

FURTHER READING
American Diabetes Association home page. Available online. URL http://www.diabetes.org/home.jsp. Accessed August 12, 2007.
International Diabetes Foundation home page. Available online. URL http://www.idf.org. Accessed January 23, 2008.
U.S. Department of Health and Human Services. Centers for Disease Control and Prevention. National Center for Chronic Disease Prevention and Health Promotion. "Diabetes Public Health Resource." Available online. URL: http://www.cdc.gov/diabetes/. Updated December 4, 2008.
"World Health Organization Diabetes Programme." Available online. URL: http://www.who.int/diabetes. Accessed January 23, 2008.

digestive system The purpose of the digestive system is to take food into the body and convert it to a usable form that can be transported throughout the body. Animals are heterotrophs, meaning they must take in food from their external environment as a source of energy and nutrients. One way to classify animals is their usual source of food. Herbivores eat autotrophs, organisms with the ability to synthesize organic molecules from inorganic carbon sources, such as plants or algae. Carnivores consume the flesh of other animals. Omnivores obtain their nutrition from plants or algae and other animals. Each type of feeder has unique adaptations that allow it to obtain, ingest, and digest food from its usual source. Animals also can be classified according to the mechanism used for obtaining food. Suspension feeders filter food particles such as plankton or small invertebrates from the water. Substrate feeders live on or inside the substance they eat. Fluid feeders obtain their nutrition from organic fluids such as blood or nectar obtained from a living host. Bulk feeders consume large pieces of food, a task that often requires adaptations for chasing and killing prey.

STAGES OF DIGESTION

The four processes of digestion are ingestion, digestion, absorption, and elimination. Ingestion is the process by which an animal takes food substances into the body, for example, by tearing pieces of leaves from a plant with mouthparts, sucking nectar from a flower, filtering surrounding water for food particles, or grasping and placing large pieces of food into the mouth. Usually, the body cannot utilize the form of food introduced. Mechanical processes such as chewing or grinding help break food into smaller, manageable pieces that digestive juices can efficiently dissolve. Macromolecules such as proteins, lipids, nucleic acids, starches, or other polysaccharides are too large to be absorbed by body cells. The process of digestion breaks down the food into usable fragments. Enzymes secreted by cells of digestive tissues or accessory glands reduce the macromolecules into individual subunits such as amino acids, monosaccharides, fatty acids, or nucleotides. Absorption is the uptake of these smaller nutrient substances from the digestive compartment. In primitive invertebrates, individual body cells absorb the nutrients directly from a gastrovascular cavity, but in most animals, a separate circulatory system often transports the nutrients throughout the body. The indigestible material remaining in the digestive tract after absorption exits the body as feces through a process called elimination.

DIVERSITY OF DIGESTIVE SYSTEMS

Some animals perform intracellular digestion while others digest food extracellularly. In intracellular digestion, each cell obtains in its own food and digests it. More primitive animals such as sponges and cnidarians digest food in this manner. Extracellular digestion is more complex and often requires a body system with several parts to accomplish each process of digestion. The animal breaks down the food outside cells, inside a body cavity that is continuous with the external environment, and then individual cells absorb the nutrients.

Many animals, such as flatworms and cnidarians, have a single opening, a mouth, through which food enters and travels to a gastrovascular cavity. Because most of the body cells have direct contact with the gastrovascular cavity, no special system or mechanism is necessary to transport the nutrients. Other animals have a digestive tract, a tube with openings at both ends. Food enters at the mouth, then travels through a long tube that has parts specialized in different digestive functions. For example, earthworms have a muscular pharynx that sucks in food through the mouth. The food moves down the esophagus into the crop for storage, before moving into the gizzard, where bits of gravel grind up the food into smaller particles. Chemical digestion and absorption occur mostly in the intestine, and undigested material exits the body through the anus. Birds have digestive tracts similar to those of earthworms but also have a stomach located between the crop and gizzard.

Insect digestive systems consist of three regions: a foregut, a midgut, and a hindgut. In grasshoppers, the foregut includes the mouth, pharynx, esophagus, crop, and gizzard. Mouthparts vary among insect species depending on whether the animal feeds on plant foliage and needs parts adapted for tearing and chewing or obtains nutrition from fluids, requiring mouthparts adapted for piercing and sucking. The labrum and labium are liplike structures surrounding the mouth, and the maxillae and mandibles are jawlike parts for chewing. Some insects have a long tube called a proboscis for sucking nectar from flowers. The salivary glands produce and secrete saliva containing some digestive enzymes. The tubelike esophagus leads from the pharynx to the crop, which stores food until the muscular gizzard pulverizes it. The midgut, where most digestion and absorption occur, includes the gastric cecum and the stomach. Bacteria and protozoa that aid in digestion reside in the gastric cecum, which also produces digestive enzymes. The stomach mixes the digestive enzymes with the ground food particles. Absorbed materials enter the hemolymph, the fluid that bathes the tissues of animals with open circulatory systems. The hindgut contains the openings of the Malpighian tubes, the intestine, colon, rectum, and anus. Hemolymph diffuses into the Malphigian tubes, which serve primarily as excretory organs to remove nitrogenous wastes and in osmoregulation. Water reabsorption occurs in the hindgut, and undigested food and nitrogenous wastes exit through the anus.

HUMAN DIGESTIVE SYSTEM

The digestive tract (gastrointestinal tract or GI tract) of humans consists of a long tube through which food passes as it is broken down, digested, and absorbed or eliminated and several accessory organs and glands. The digestive tract includes the mouth, pharynx, esophagus, stomach, small intestine, large intestine, rectum, and anus. The accessory glands are the salivary glands, the pancreas, the liver, and the gallbladder. The layers of the digestive tract share similar qualities all the way from the esophagus to the anus. Epithelial cells that line the tract transport substances into and out of the lumen and produce and secrete enzymes, mucus, and hormones. Epithelial cells lining the GI tract only live for a few days before sloughing off and being replaced. Layers of smooth muscle function to move substances within the lumen, either to churn it (as in the stomach), increase the flow of substances across the surface area of the epithelial cells for increased contact, or push it through the tract. Contractions force the food forward in successive waves by peristalsis. Connective tissue supplies blood vessels, lymph vessels, and nerves to the wall of the GI tract. The serosa, a thin layer of connective tissue that surrounds the entire tract, holds the digestive organs in place and

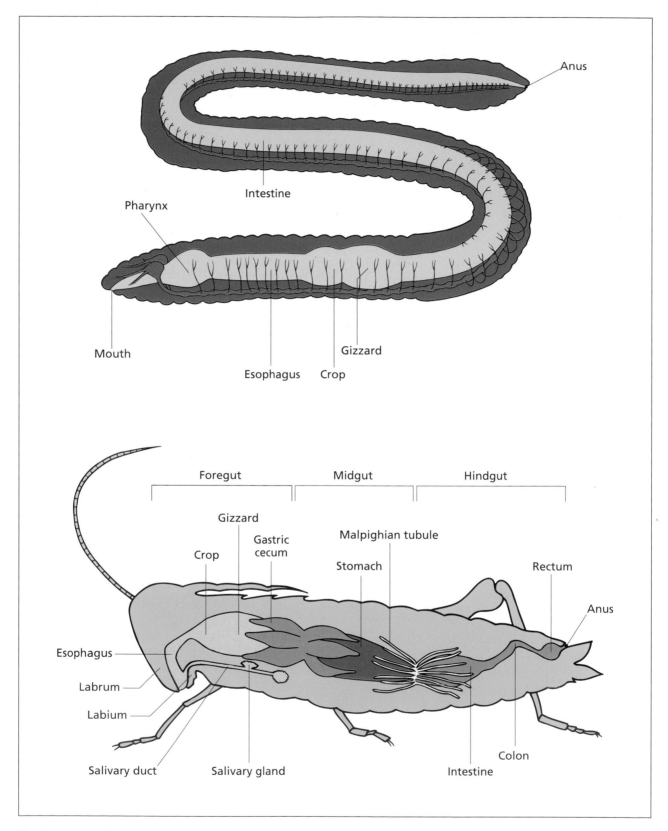

Earthworms and grasshoppers have one-way digestive tracts with regions specialized to perform different digestive functions.

is continuous with the peritoneum, which lines the entire abdominal cavity.

Digestion begins when food enters the mouth, where chewing begins the process of mechanical breakdown. Three pairs of salivary glands secrete saliva into the mouth, lubricating the food and initiating the process of chemical digestion. Digestive enzymes secreted by the accessory organs and glands catalyze the breakdown of macromolecular polymers. The enzyme amylase, present in the saliva, breaks down large starch molecules into maltose, a disaccharide made from two glucose subunits. When a bolus of chewed food or liquid reaches the pharynx, the swallowing reflex is triggered, and a little flap of cartilaginous tissue called the epiglottis moves to its down position and blocks the entrance to the respiratory tract. A muscular ring called a sphincter relaxes, and food moves into the esophagus. Peristalsis moves the bolus down the esophagus into the stomach within five to 10 seconds.

While the stomach churns the food, cells of its gastric glands secrete a variety of chemicals and enzymes that work chemically to digest it. Hydrochloric acid (HCl) denatures proteins, kills most microorganisms that have been ingested, and disrupts the matrix that holds many cells together in meat and plant tissue. The gastric glands secrete pepsinogen, an inactive digestive enzyme. The acidic environment created by the HCl in the stomach causes the cleavage of pepsinogen, converting it to pepsin, an active enzyme that cleaves peptide bonds between amino acids of protein chains. The gastric glands also secrete mucus and bicarbonate, substances that protect the digestive juices from digesting the stomach lining. Contractions of the smooth muscles surrounding the stomach move the contents, now called chyme, toward the pyloric sphincter, the opening to the small intestine.

The pyloric sphincter limits the amount of chyme that enters the duodenum, the first portion of the small intestine. In humans, the duodenum is about 10 inches (25 cm), and the entire length of the small intestine is approximately 20 feet (6 m). The pancreas, gallbladder, liver, and intestinal gland cells all discharge digestive enzymes and other substances into the duodenum. By a combination of hormonal and neural mechanisms, distension of the intestine stimulates the pancreas to secrete pancreatic amylase, lipases that digest lipids, and several types of proteases that digest proteins through a duct into the duodenum. Some enzymes are secreted as zymogens, meaning they are secreted in an inactive form and must undergo a chemical change to become active. The pancreas also secretes bicarbonate, a substance that buffers the acidic pH of chyme entering from the stomach. The liver produces a solution called

bile from bile salts, bile pigments, and cholesterol. The bile travels via hepatic ducts to the gallbladder for storage until food enters the digestive tract, then flows through the common bile duct into the duodenum. Bile contains no enzymes, but the bile salts help emulsify lipids to dissolve them in the fluids flowing through the GI tract. The brush border, the epithelial lining of the duodenum, also secretes many digestive enzymes into the lumen, but other enzymes including peptidases that hydrolyze peptides, disaccharidases that split disaccharides into monosaccharides, and a protease called enteropeptidase remain attached to the surface of the brush border while the chyme and the enzymes it contains move forward by peristalsis and other contractions. By the time the chyme reaches the end of the duodenum, enzymes have digested most of the carbohydrates, proteins, lipids, and nucleic acids into monosaccharides, amino acids, fatty acids and glycerol, and nucleotides, respectively.

The remaining two regions of the small intestine, the jejunum and the ileum, function mainly in absorption of nutrients. To increase the surface area across which absorption occurs, the lining forms large circular folds that are covered with fingerlike projections called villi. Microscopic extensions called microvilli extend from the villi, effectively increasing the surface area of the internal lining and, consequently, the rate of absorption. Though the small intestine of an average adult human is about 20 feet (6 m) long, the surface area approaches 360 square yards (about 300 m^2). Transport from the lumen of the small intestine across the epithelial cells occurs by a combination of passive diffusion and active transport. Inside the epithelial cells, fatty acids and glycerol molecules are reassembled into triglycerdes and combined with cholesterol molecules to form chylomicrons. Exocytosis moves the chylomicrons from the epithelial cells into lacteals, small ducts that drain into lymphatic vessels. The lymphatic system eventually drains into blood circulation. The sugars, amino acids, nucleotides, and vitamins diffuse into tiny capillaries that extend into each villus. The capillaries carry the nutrients to veins that converge into the hepatic portal vein, which transports the absorbed materials to the liver.

Approximately 5.8 quarts (5.5 L) of food and beverages and 3.7 quarts (3.5 L) of secretions pass through the small intestine of a human each day. The small intestine reabsorbs most of this in addition to various ions, but about 1.6 quarts (1.5 L) of chyme moves on to the large intestine. After passing through the ileocecal valve, chyme enters a pouch called the cecum, from which the appendix projects. Chyme then passes through the ascending colon, traverse colon, descending colon, sigmoid colon, and rectum, where undigested material is stored until it is

Human Digestive Tract

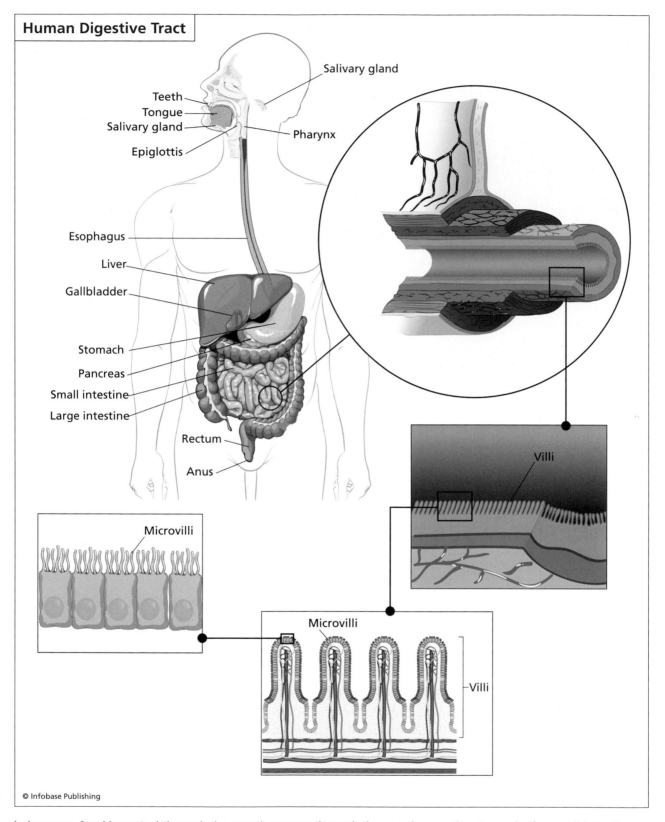

Salivary gland

Teeth
Tongue
Salivary gland
Epiglottis

Pharynx

Esophagus
Liver
Gallbladder
Stomach
Pancreas
Small intestine
Large intestine
Rectum
Anus

Villi

Microvilli

Microvilli

Villi

In humans, food ingested through the mouth passes through the esophagus, the stomach, the small intestine, the large intestine, and the rectum during the process of digestion. Several additional accessory glands including salivary glands, the liver, the gallbladder, and the pancreas secrete substances that aid in the chemical breakdown of the ingested material.

eliminated. The major function of the large intestine, which measures about five feet (1.5 m), is to recover water as the chyme travels its roundabout route. The material solidifies as it moves through the tract, sodium is absorbed, and potassium is added to the waste material. Billions of bacteria inhabit the large intestine and aid in digestion of complex carbohydrates and proteins. Some also produce vitamins, such as vitamin K, that the human host absorbs, and intestinal gas as a by-product of their metabolism. One voluntary and one involuntary sphincter control the movement of waste from the rectum through the anus, the posterior opening of the digestive tract. Peristaltic contractions push the feces toward the anus with the assistance of voluntary abdominal contractions. Constipation, the infrequent discharge of feces, results in dry, hard bowel movements as a result of increased water reabsorption. Diarrhea is characterized by watery stools and results when substances move through the large intestine too quickly, preventing the efficient reabsorption of water. Persistent diarrhea can cause dehydration, a dangerous medical condition.

Hormones control the release of digestive enzymes, ensuring they are only secreted when food is present. The presence of amino acids in the stomach or distension of the stomach stimulates cells in the gastric glands to release the hormone gastrin, which travels through the bloodstream before acting back on the stomach. A pH below 1.5, indicating no food is present, inhibits the release of gastrin. The protein hormone gastrin promotes acid release, and acid triggers somatostatin release, creating a negative feedback loop that inhibits further gastric acid release, gastrin secretion, and pepsinogen secretion. When fatty acids and amino acids are present, the duodenum secretes cholecystokinin (CCK), a hormone that causes the pancreas to secrete digestive enzymes and the gallbladder to release bile. Enterogastrone, secreted by the duodenum, slows peristalsis when a diet rich in fats is ingested, giving the intestine more time to digest the lipids. The duodenum also secretes secretin, a hormone that signals the pancreas to release sodium bicarbonate to neutralize the acidic chyme.

Mammals and other vertebrates with specialized diets have adaptations that meet different needs. Carnivores require sharp teeth that can rip and tear into flesh, while herbivores have teeth with broad surfaces for grinding plant tissue. Omnivores have teeth that possess both characteristics. Herbivores and omnivores also have longer digestive tracts because plant tissue takes longer to digest than animal tissue. Many herbivores also have additional structures or cham-

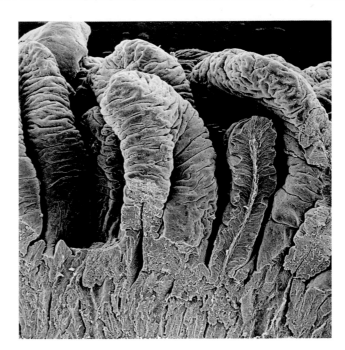

Small fingerlike projections called villi cover the interior surface of the intestinal wall and function to increase the surface area across which nutrient transport occurs. *(David Scharf/Photo Researchers, Inc.)*

bers for housing symbiotic microorganisms such as bacteria and protozoa that aid in the digestion of cellulose, a carbohydrate that animals cannot digest because of the lack of the enzyme cellulase. Because snakes swallow prey whole, they have an adaptation that allows them to unhinge the lower jaw in order to ingest large animals.

See also ANATOMY; ANIMAL FORM; BIOMOLE-CULES; CIRCULATORY SYSTEM; DIABETES; NUTRITION; PHYSIOLOGY.

FURTHER READING

Nagel, Rob. *Body by Design: From the Digestive System to the Skeleton.* Vol. 1. Detroit: U*X*L, 2000.

National Institute of Diabetes and Digestive and Kidney Diseases, National Institutes of Health, U.S. Dept. of Health and Human Services. *Your Digestive System and How It Works.* National Digestive Diseases Information Clearinghouse. Bethesda, Md.: April 2008. Available online. URL: http://digestive.niddk.nih.gov/ddiseases/pubs/yrdd/. Accessed August 2, 2008.

Seeley, Rod, R., Trent D. Stephens, and Philip Tate. *Anatomy and Physiology.* 7th ed. Dubuque, Iowa: McGraw Hill, 2006.

Silverthorn, Dee Unglaub. *Human Physiology: An Integrated Approach.* 4th ed. San Francisco: Benjamin Cummings, 2006.

dissection   Dissection is an important tool for studying plant and animal anatomy. The person performing the dissection carefully separates the specimen into pieces or takes it apart in a manner that exposes the internal structures for detailed analysis. Students often perform dissections on preserved specimens in order to learn about the structure, organization, and relationships of tissues and organs in a body. Grasshoppers, earthworms, starfish, frogs, and fetal pigs are common dissection specimens in high school teaching laboratories. More advanced anatomy classes often use cats, and dental and medical students use human cadavers. Physiologists, pathologists, or medical researchers may dissect animals in order to observe the effects of a drug treatment on certain organs or tissues or to look for internal damage caused by a particular disease. An autopsy is a specialized type of dissection performed by a coroner whose purpose is to investigate the cause of death when there is reason to suspect that it was not due to natural causes. Vivisection is a dissection performed on a living animal for physiological or pathological study.

Groups have expressed differences in opinion as to the ethics of animal dissection. Though labeled illustrations and computer simulations depict the same structures, many educators believe that no alternative adequately provides a similar enough experience to an actual dissection. Schools, universities, and research institutions have guidelines for the proper use and care of animals in teaching and research, and the U.S. Department of Agriculture establishes and enforces procedures used in the procurement of animals. Many educators believe that when these are followed, the advantages to the student—the memorable experience of a hands-on exploration that demonstrates the complexity of the interrelationships and internal workings of the body's organs and structures—far outweigh the disadvantages. Artificial models and diagrams cannot convey or teach characteristics such as texture, the true colors, and individual variations. Other groups believe that dissection is not necessary and that it is unethical to use animals for teaching or research purposes.

Ancient Greek philosophers such as Aristotle and early physicians such as Herophilus and Erasistratus were the first to perform dissections to study animal anatomy and physiology, believing direct observation was the best way to learn about a subject. Because many people believed an intact body was necessary to enter the afterlife, public pressure ended the practice of dissecting human bodies from about 250 B.C.E. until the 16th century, when the Flemish physician Andreas Vesalius drew attention to the fact that human anatomy differed from that of other animals, and the practice resumed. Vesalius published the results of his extensive dissections in what is recognized as the first anatomy textbook, *On the Structure of the Human Body,* in 1543. His detailed instructions on the methods for dissection helped revitalize the field of anatomy.

The most common tools used for dissection are a teasing or dissection needle for pulling apart muscle tissue or gently working connective tissue away from the organs it envelopes, dissecting scissors for cutting through tissues and exposing organs, and a very sharp scalpel, for cleanly slicing thin layers of tissue. Pins are often used to hold structures in place during viewing or further dissection.

See also ANATOMY.

DNA fingerprinting   Humans have 99.9 percent of their genetic makeup in common. Differences in the remaining 0.1 percent of the human genome may be due to naturally occurring alleles (alternate forms of a gene) that change the product of one of the 30,000 genes encoded by the 3 billion nucleotide base pairs of the human genome. Most of the deoxyribonucleic acid (DNA) of the human genome does not appear to encode for anything, and many of those noncoding regions contain repetitive sequences. The differences in these regions are only apparent by DNA testing and, depending on the exact techniques used to examine these sequences, may appear as unique as one's fingerprints. Because of this, medical researchers, forensic biologists, genealogists, and anthropologists can use this information to identify individuals or to determine the relatedness between individuals or groups of individuals. Information gained from DNA fingerprinting can answer questions concerning the structure and migration of human populations, establish paternity, and place a suspect at the scene of a crime.

In 1985 Sir Alec Jeffreys at the University of Leicester invented a technique called DNA fingerprinting, also known as DNA typing, DNA profiling, or DNA testing. Jeffreys and his colleagues were researching a human α-globin gene when they discovered that it contained several adjacent repeats of the same sequence. Further investigation revealed similar repetitive sequences, called minisatellites, elsewhere in the genome. Minisatellites are regions of repetitive DNA consisting of repeated sequences approximately 15 to 100 base pairs in length. Many minisatellites have a GC-rich core sequence and may occur at numerous locations, called loci, scattered throughout the genome. In developing the DNA fingerprinting technique, Jeffreys exploited the existence of these newly discovered minisatellites in the human genome. Variations in the genetic makeup of individuals are called DNA polymorphisms, and they exist as distinct forms within populations. At each minisatellite, the core may repeat a different number of times in different individuals as a result of slippage during DNA replication or of recombination events during meiosis. These polymorphisms make each individual unique.

DNA fingerprinting initially involved restriction fragment length polymorphism (RFLP) analysis. Restriction enzymes cut DNA at locations that contain specific sequences. The digestion releases linear DNA fragments of different lengths, depending on the location and frequency of the recognition sequences in the DNA. Jeffreys used restriction enzymes that did not have a recognition sequence within the repeated sequence of the minisatellite; thus the enzymes cut on either side of the segment containing the repeat. The length of the restriction fragments varies, depending on the number of repeats an individual has at a given locus; accordingly, RFLPs that are polymor-

phic because of differences in the number of tandem repeats are called variable number tandem repeats (VNTRs). Gel electrophoresis separates the fragments on the basis of size, with the smaller fragments migrating faster than the relatively larger fragments, and therefore farther down a gel in a given period. A procedure called Southern blotting transfers the DNA fragments to a nylon membrane while preserving the banding pattern from the gel. The membrane is treated with a radioactively labeled DNA probe that contains minisatellite sequences complementary to DNA sequences on the membrane. The probe specifically binds by the formation of hydrogen bonds to minisatellite sequences on the membrane, and X-ray film detects the resulting pattern, which is the so-called DNA fingerprint. While RFLP analysis of VNTRs was novel and useful, the technique is time-consuming and required large amounts of good-quality DNA.

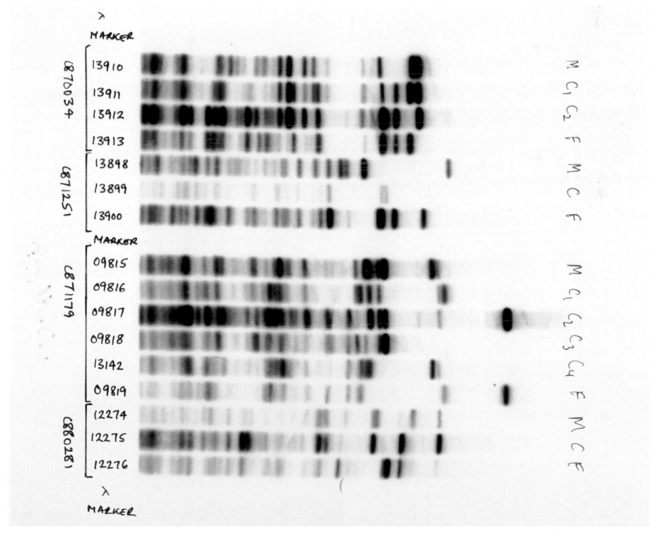

DNA fingerprinting can be used to analyze family relationships. This photo shows an autoradiograph of DNA fingerprints from a father (F), mother (M), and several children (Cₓ). *(David Parker/Photo Researchers, Inc.)*

Kary Mullis, an American biochemist who was working for Cetus Corporation, invented the polymerase chain reaction (PCR) in 1983. This technique generates numerous copies of a specific segment of DNA within a few hours. Oligonucleotide primers, single-stranded pieces of DNA averaging about 20 nucleotides in length, are allowed to hybridize to complementary sequences that flank the site to be

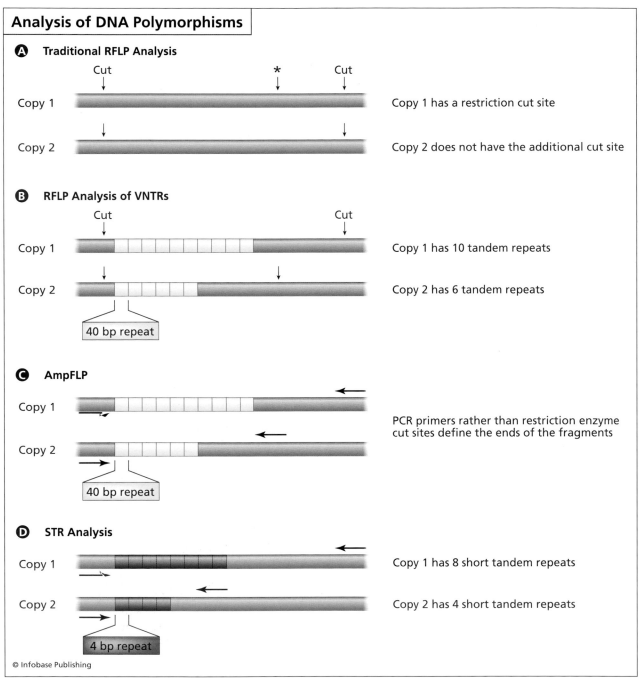

DNA polymorphisms can be detected by several different methods. A) Traditional RFLP analysis reveals the presence or absence of a specific enzyme cut site within a defined restriction fragment. B) Sir Alec Jeffrey's original DNA fingerprinting technique identified RFLPs resulting from differences in the number of tandem repeats in a minisatellite, called VNTRs. C) When PCR performed using specific primers that flank the VNTR amplifies the segment of DNA containing the polymorphism, the technique is called AmpFLP. D) The most common method used for DNA fingerprinting today is STR analysis, which differs from AmpFLP in that microsatellites that contain different numbers of very short repeated sequences, up to five rather than up to 100 base pairs, are examined.

amplified. Then DNA polymerase makes a copy of the template DNA, to which more primers bind during the next round, and the cycle repeats 20 to 30 times. Theoretically, millions of copies of a template can be synthesized within a few hours. PCR was applied to DNA fingerprinting by using primers that amplified polymorphic loci. A major advantage to PCR is the very small amount of DNA is required; thus a tiny sample of blood, semen, saliva, a single hair with the root cells attached, or a few cheek cells is sufficient.

Amplified fragment length polymorphism (Amp-FLP) analysis combines the basis of RFLP analysis with PCR. The DNA is digested with one or more restriction enzymes, and then restriction half-site-specific adaptors are ligated to the ends of the restriction fragments. The primers used to amplify the DNA by PCR anneal to sequence of both the restriction site and the attached adaptor segment. Electrophoresis follows PCR and a specific banding pattern results. Biotechnology companies simplified AmpFLP analysis by developing kits that allowed the detection of single-nucleotide polymorphisms (SNPs) by hybridizing PCR products with cards spotted with different probes.

The most common method used in DNA fingerprinting today is short tandem repeat (STR) analysis. The STRs used for this application are four or five nucleotides long, and they occur in tandem, one directly adjacent to the next. DNA sequences that are between two and five nucleotides long and occur as tandem repeats are called microsatellites. As minisatellites do, microsatellites occur throughout the genome, and individuals possess different numbers of the repeated sequences at different locations throughout the genome. For example, the STR named D7S280 is a locus found on the human chromosome 7, and it consists of the tetrameric repeat GATA. Different alleles of D7S280 contain between six and 15 tandem repeats of this tetramer. In STR analysis, PCR amplifies the desired regions containing the repeated sequence by using specifically designed primers, and then the resulting fragments are separated by either gel electrophoresis or capillary electrophoresis, which provides higher resolution. Fluorescent dyes attached to the PCR primers allow detection. Though polymorphisms at a single locus are not uncommon, simultaneous analysis of several loci results in a specific DNA profile for an individual. Greater numbers of examined loci give more statistically discriminating profiles. Because the polymorphic regions of DNA tested during STR analysis are much shorter than in VNTR analysis, the probability is high that sufficient quantities of useful DNA will be present even in minute samples. Spontaneous chemical degradation often reduces the

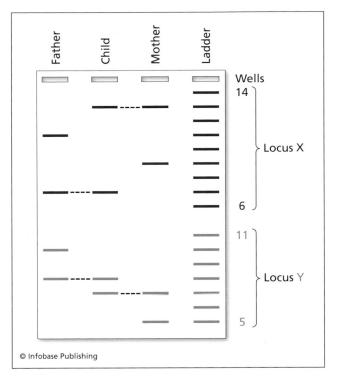

© Infobase Publishing

Gel electrophoresis separates the amplified STRs and reveals an individual's genotype. Since people have two copies of each chromosome, each individual has two alleles for each of the two loci shown here. Each allele is represented by one band. In the diagram shown here, the father has the genotype of 11, 7 for locus X and 10, 8 for locus Y. This means, for example, that one of the father's X STR alleles contains 11 repeats, and his other X allele has 7 repeats. The mother has the genotype 13, 9 for X and 7, 5 for Y. The child must have received allele 13 from the mother and 7 from the father for locus X, and 8 from the father and 7 from the mother for locus Y.

quality of the DNA left behind at a crime scene, but STR analysis only requires very small segments of DNA to be in good shape, meaning readable by DNA polymerase during PCR.

In the United States, the Federal Bureau of Investigation (FBI) maintains a database of the DNA profiles of sex offenders, convicted felons, and others who have previously had DNA fingerprinting analysis performed. The database, called *CODIS* for "Combined DNA Index System," allows law enforcement agencies to search stored profiles for a match to DNA from a crime scene even when they do not have a suspect. CODIS profiles contain information from 13 different STRs, and the odds of two individuals' having the same profile are approximately 1 in a billion.

See also BIOMOLECULES; DEOXYRIBONUCLEIC ACID (DNA); ELECTROPHORESIS; GENOMES; POLYMERASE CHAIN REACTION; VARIATION, GENETIC VARIATION.

FURTHER READING
Jeffreys, Alec J., V. Wilson, and S. L. Thein. "Hypervariable Minisatellite Regions in Human DNA." *Nature* 314, no. 6006 (1985): 67–73.
Kobilinsky, Lawrence, Thomas F. Liotti, and Jamel Oeser-Sweat. *DNA: Forensic and Legal Applications*. Hoboken, N.J.: Wiley-Interscience, 2005.
Rudin, Norah, and Keith Inman. *An Introduction to Forensic DNA Analysis*. 2nd ed. Boca Raton, Fla.: CRC Press, 2002.

DNA sequencing Deoxyribonucleic acid (DNA) serves as the molecular carrier of genetic information. All the genes that instruct cells how to grow, metabolize, multiply, and perform other functions that define life exist on molecules of DNA. The individual building blocks of DNA are nucleotides, subunits that consist of a deoxyribose sugar, a phosphate group, and a nitrogenous base. The nitrogenous bases found in DNA are adenine, cytosine, guanine, and thymine, and the nucleotides are often abbreviated *A, C, G,* and *T,* respectively. The genetic information is contained within the specific order of the four different nucleotides along the length of a DNA molecule, referred to as the sequence. DNA sequencing is the process of determining the composition and order of nucleotides for a segment of DNA, for a gene, for an entire chromosome, or for a whole genome. In the laboratory, molecular biologists sequence DNA molecules by carrying out a set of biochemical reactions called sequencing reactions and then perform gel electrophoresis on the reaction products to analyze, or read, the sequence.

DEVELOPMENT OF THE METHODOLOGY

The earliest DNA sequencing methods depended on the partial hydrolysis of small pieces of DNA and were difficult and time-consuming. In 1975 Frederick Sanger and Alan Coulson at the Medical Research Council Laboratory of Molecular Biology in Cambridge, England, developed an entirely different approach that represented a turning point in DNA sequencing technology. Their method, which they called the plus-and-minus method, used making DNA rather than breaking it down as a strategy for determining its construction. The basis of their approach was the tendency of DNA elongation to stop when one of its nucleotide building blocks is in limited concentration. When the polymerization reactions were carried out in four different tubes, each with a different limiting nucleotide, new DNA chains of different lengths were produced. Sanger employed this method to determine the 5,386-nucleotide sequence of the bacteriophage ΦX174, which led to the prediction of the amino acid sequences for the 10 viral proteins. While this method was much improved over the previous techniques, it still had difficulties; thus Sanger continued to explore ways to improve the method.

Meanwhile, in 1977, Allan Maxam and Walter Gilbert published a different method for determining the sequence of a DNA molecule, the chemical degradation method, which quickly became popular. The Maxam and Gilbert method involves radiolabeling purified DNA fragments, then setting up four different reactions and treating each in a manner that results in the differential cleavage of the DNA at one of the four types of nucleotides. After this step, all of the labeled segments in one tube will terminate at the same type of nucleotide; however, many different lengths will be present. Polyacrylamide gel electrophoresis of the four reactions separates the fragments on the basis of size, and exposure of the gel containing the separated radiolabeled bands to a piece of X-ray film allows the researcher to determine the order of nucleotides in the DNA used.

Meanwhile, Sanger and Coulson made a change to the plus-and-minus technique that led to its replacement. The newer method still relied on DNA polymerase but involved the addition of dideoxynucleotides to the reaction mixtures. Once DNA polymerase incorporated a dideoxynucleotide, DNA synthesis stopped at that position. One drawback was the difficulty in creating and purifying DNA template for use in the sequencing reactions; the physical separation methods were messy and inefficient. Joachim Messing and his colleagues at the Waksman Institute of Microbiology, Rutgers, the State University of New Jersey, contributed to the utility of the dideoxy method when they introduced a cloning technique that facilitated the generation of relatively large quantities of purified single-stranded DNA to use as templates in the sequencing reactions. Cloning technology has greatly improved the mechanisms by which researchers obtain the desired DNA for sequencing, and today researchers simply heat the double-stranded DNA to generate single-stranded templates. Though additional modifications to the specific techniques have improved the results, and automation has improved the efficiency, the dideoxy chain termination method remains the basis for DNA sequencing performed today.

Walter Gilbert and Frederick Sanger received the Nobel Prize in chemistry in 1980 "for their contributions concerning the determination of base sequences in nucleic acids." They shared the award with the American biochemist Paul Berg, a pioneer of recombinant DNA technology. Sanger had previously received the Nobel Prize in chemistry in 1958 for his research on the structure of the protein insulin. DNA sequencing technology has come a long way in the last three decades. Improvements include the use of automated sequencers that read the sequences and the replacement of radioactivity with fluorescent dyes.

DIDEOXY CHAIN TERMINATION METHOD

Dideoxynucleotides have no 3' hydroxyl (-OH) group attached to the deoxyribose sugar. During DNA synthesis, DNA polymerase forms a phosphodiester bond between the 3' OH of the last nucleotide added to the newly synthesized strand of DNA and the 5' phosphate group of an incoming nucleotide. If DNA polymerase incorporates a dideoxynucleotide rather than a normal deoxynucleotide, the 3' OH will be missing, and DNA chain elongation terminates at that position.

To sequence a fragment of DNA using this method, the researcher sets up four separate reactions. Each reaction contains the DNA template, which has been heated to denature the hydrogen bonds between complementary strands and generate single-stranded templates to be sequenced. The reactions also require short pieces of DNA called primers, because DNA polymerase I cannot add new nucleotides without having a 3' OH already present. (In the cell, a separate enzymatic activity carries out this function to initiate DNA replication.) The primers are about 20 bases long and specifically designed to complement a region of the cloning vector that contains the desired DNA. Each reaction also contains a mixture of the four deoxynucleotide (indicated by the lowercase letter d) subunits from which DNA is synthesized: 2'-deoxyadenosine 5'-triphosphate (dATP), 2'-deoxycytidine 5'-triphosphate (dCTP), 2'-deoxyguanosine 5'-triphosphate (dGTP), and 2'-deoxythymidine 5'-triphosphate (dTTP). A fraction of the dATP contains a radiolabeled phosphate group (^{32}P) that facilitates the detection of the DNA after completion of the reactions. Each of the four tubes also contains one dideoxynucleotide (indicated by dd), either ddATP, ddCTP, ddGTP, or ddTTP. Addition of the enzyme allows the reactions to proceed, extending the primers to form a new strand of DNA complementary to the template to which the primer binds. The concentration of the dideoxynucleotide is such that the enzyme incorporates the dideoxynucleotide randomly, so some of the fragments are terminated early on, and others proceed farther before termination. When the DNA polymerase adds a dideoxynucleotide, the reaction stops.

After the reactions are complete, all four tubes contain newly synthesized DNA fragments that have the same 5' end but will terminate in different residues at the 3' end, depending on which dideoxynucleotide was added to that reaction tube. Within a tube, all the fragments will terminate in the same residue but will be of various lengths, depending

on how far the enzyme proceeded before adding a chain-terminating dideoxynucleotide rather than a normal deoxynucleotide. Denaturing polyacrylamide gel electrophoresis on the products of the four reactions, run in parallel in adjacent lanes, separates the fragments within each tube according to their length. Laying a piece of X-ray film over the gel will result in the exposure of the film at the positions of the radiolabeled bands in each lane; after developing the film, the pattern will become apparent. Since the smallest fragments will be at the bottom of the gel and the longest fragments will be at the top, one simply starts at the bottom and notes which lane the lowest band is in, then notes which lane the next lowest band is on, and so on.

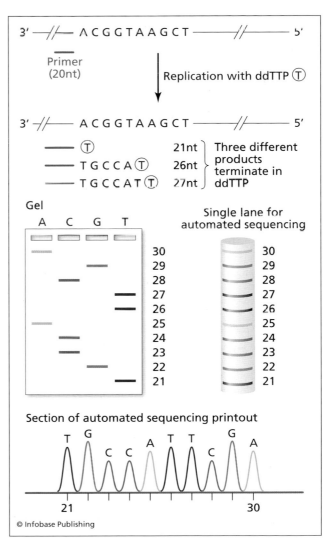

The dideoxy chain termination method for sequencing utilizes dideoxynucleotides to stop DNA synthesis at specific residues. Denaturing polyacrylamide gel electrophoresis separates the fragments on the basis of length, allowing one to read the ordered sequence of nucleotides of a DNA molecule.

Automated techniques are available for sequencing large amounts of material, such as the human genome, which is more than 3 billion nucleotides. One variation involves tagging the primers with one of four different fluorescent dyes. Each color is specific for the dideoxynucleotide used in a reaction tube. For example, green might indicate A, blue might be C, orange might be G, and pink might be T. The different colors allow all four reactions to be loaded into the same lane on a gel. After running a gel to separate the fragments, a laser scans the gel, and as it passes over each band, a detector reads the color emitted by the fluorescent label and sends the information to a computer that records the data.

See also DEOXYRIBONUCLEIC ACID (DNA); ELECTROPHORESIS; ENZYMES; GENE EXPRESSION; GENOMES; HUMAN GENOME PROJECT; RADIOACTIVITY.

FURTHER READING

Church, George M. "Genomes for All." *Scientific American* 294, no. 1 (January 2006): 47–54.

DOE Joint Genome Institute, U.S. Department of Energy, Office of Science. "How Sequencing Is Done." Available online. URL: http://www.jgi.doe.gov/education/how/. Updated March 15, 2006.

Maxam, Allan M., and Walter Gilbert. "A New Method for Sequencing DNA." *Proceedings of the National Academy of Sciences USA* 74, no. 2 (1977): 560–564.

The Nobel Foundation. "The Nobel Prize in Chemistry 1980." Available online. URL: http://nobelprize.org/nobel_prizes/chemistry/laureates/1980. Accessed January 24, 2008.

Sanger, Frederick. "Sequences, Sequences, Sequences." *Annual Reviews of Biochemistry* 57 (1988): 1–28.

Sanger, Frederick, and Alan R. Coulson. "A Rapid Method for Determining Sequences in DNA by Primed Synthesis with DNA Polymerase." *Journal of Molecular Biology* 94 (1975): 441–448.

Sanger, Frederick, Steve Nicklen, and Alan R. Coulson. "DNA Sequencing with Chain-Terminating Inhibitors." *Proceedings of the National Academy of Sciences USA* 74, no. 12 (1977): 5,463–5,467.

Dobzhansky, Theodosius (1900–1975) Russian American *Biologist*

Theodosius Dobzhansky was one of the 20th century's most influential evolutionary biologists. Best known for his famous quotation, "Nothing in biology makes sense except in the light of evolution," Dobzhansky was also the first to define and demonstrate clearly the relationship between inheritance as described by the Austrian monk Gregor Mendel and speciation by descent with modification as proposed by the English naturalist Charles Darwin.

BUTTERFLIES AND LADYBIRD BEETLES

Theodosius Dobzhansky was born on January 25, 1900, in Nemirov, Russia, to Grigory Dobzhansky, a high school mathematics teacher, and Sophia Voinarsky, who home-schooled Theodosius until he was nine years old. A German governess taught him her native language, and later in life he became fluent in English and at least three other languages. Theodosius entered his father's gymnasium (high school) in 1909. The following year, Grigory suffered a head injury that left him paralyzed and forced him to quit work. The family moved to Kiev, and Theodosius entered the local gymnasium, where his interest in science grew. When he and a friend found a copy of Darwin's *On the Origin of Species,* they read it together. While the world fought the First World War, he collected butterflies and studied ladybird beetles of the genus *Coccinella.* After collecting them, he dissected them and examined them under the microscope. He learned to distinguish different species and observed their migration patterns.

The army drafted many of Dobzhansky's friends, but because his birthday was in January and the cutoff was December, the army did not call up Dobzhansky. He enrolled at the University of Kiev, where Professor Kushakevich took him under his wing. Kushakevich was a cytologist, a biologist who studies cells. For three years Dobzhansky researched sexual differentiation in snails, but he also continued his studies on *Coccinella.* His first scientific publication, when he was only 18, described a new species of ladybird beetle, *Coccinella luchniki.* Despite the war, scheming to avoid military duty, the Communist takeover of government, the death of both of his parents, and practically starving to death, Dobzhansky managed to graduate with a degree in biology in 1921. Having already worked at the Polytechnic Institute in Kiev the year before, he obtained a position as an instructor in zoology and soon developed an interest in genetics. With a well-equipped laboratory, he published two more papers in the few years he was there.

From a colleague, Dobzhansky became aware of the work of one of the most famous *Drosophila* geneticists of all time, Thomas Hunt Morgan at Columbia University in New York. As it did among many geneticists at the time, Morgan's research focused on the work of Mendel, whose work regarding the manner by which traits are passed from one generation to the next was rediscovered at the turn of the century, after being ignored for almost 35 years. Dobzhansky had heard about his research on gene linkage, the phenomenon whereby more than one characteristic is passed on together. For example, Morgan found that when he mated the offspring of a white-eyed male fly with a female red-eyed fly, and then crossed their offspring, only males exhibited white eyes—females' eyes were always red. Dobzhansky admired Morgan's work, and when he obtained a collection of mutant *Drosophila*, he excitedly embarked on some studies examining whether genes controlled single traits or multiple traits. The microscopy skills he learned from Kushakevich came in handy. He studied the mutant flies under the microscope and discovered pleiotropy, the phenomenon by which one gene produces a variety of effects or phenotypes. He published his novel findings, and his name attracted the attention of Professor Yuri Filipchencko, the head of the Genetic Department at the University of Leningrad. Filipchencko was looking for an assistant in his lab, and he asked Dobzhansky whether he was interested.

In 1924 Dobzhansky joined Filipchencko's lab, where he continued his studies on pleiotropy in *Drosophila*. Although Filipchencko was a zoologist, at the time the focus of the lab was wheat. Dobzhansky worked in the lab but also continued studying ladybird beetles. During one trip to Central Asia he unexpectedly came across another new species at altitudes where entomologists did not predict ladybird beetles would exist. The same year that he moved to Leningrad, Dobzhansky married an evolutionary biologist, Natalia (Natasha) Sivertzev. In 1933 the couple had one daughter, named Sophie.

The research on pleiotropy impressed Filipchencko, who recommended Dobzhansky for a fellowship from the International Education Board of the Rockefeller Foundation. With this support, Dobzhansky and his wife went to the United States, and in 1927 Dobzhansky joined the lab of none other than his hero, Thomas Hunt Morgan. His wife, Natasha, obtained a position maintaining the fly stocks in Morgan's lab. The following year Morgan accepted a position as head of the new biology division at the California Institute of Technology. Dobzhansky followed. Shortly thereafter he became an American citizen. Cal Tech offered Dobzhansky an assistant professorship in genetics in 1929 and then promoted him to professor in 1936.

CALTECH RESEARCH

Shortly after joining Morgan's group, Dobzhansky came under the tutelage of Alfred H. Sturtevant, one of Morgan's main collaborators, who was responsible for the idea that one could use the data from gene linkage experiments to assign relative positions of genes along the length of chromosomes. While the physical conditions of Morgan's laboratory, one section of which was fondly called the fly room, did not impress Dobzhansky, the quality of Morgan's science did. Dobzhansky was interested in studying genetics in order to understand evolution better. Hermann J.

Muller, a former colleague of Morgan, had shown that hereditary mutations, or alterations in the genes, could be induced by X-rays, thus speeding up the process of collecting mutants. (Muller received the 1946 Nobel Prize in physiology or medicine for his discovery.) As Dobzhansky saw it, the mutations were the basis of evolutionary theory. With Sturtevant, he worked on chromosomal translocation, an event involving the breakage of a segment of a chromosome and its attachment elsewhere. If it attached to the same chromosome from which it separated, but in the opposite orientation, the alteration was called an inversion. If it attached to another chromosome, it was called a translocation. He observed the result of these events by removing a female fly's ovaries, fixing them, embedding them in paraffin, slicing them, staining them, and finally examining them under the microscope. Whereas mutations to individual genes cannot be viewed in this manner, such gross structural changes could. The translocation events were rare, but exposure to X-rays increased their incidence. Dobzhansky found that the offspring of irradiated flies had genes on different chromosomes than normal flies. These translocations were the first to be observed under a microscope, and they provided the physical confirmation of the linkage maps generated mathematically from information on the frequencies of combinations of different traits. Morgan's conclusion that genes occurred in linear arrangements along the length of chromosomes was still being challenged; Dobzhansky's cytological maps furthered that hypothesis as well.

In the early 1930s, when most of the Morgan lab was concerned with the newly discovered giant chromosomes found in the salivary glands of *Drosophila* larvae, Dobzhansky's interests took a different turn—toward the evolutionary aspect of genetics. He believed that the changes he observed in chromosomes had some relation to the evolutionary process, and he wanted to determine exactly what. He also began to wonder about the concept of a species. These unresolved questions led Dobzhansky to embark on research concerning the evolutionary history of *Drosophila* populations in North America. At the time, cytologists were aware that different species possessed different numbers and shapes of chromosomes. For example, *Drosophila melanogaster* had four pairs of chromosomes, but other *Drosophila* species had five pairs. Two so-called races, A and B, of *Drosophila pseudoobscura* rarely mated in the wild, and when they did, the male offspring were sterile. He drove to northern California and collected fruit flies by setting out mashed bananas mixed with yeast. To gain insight as to why the male hybrids were sterile, he examined their chromosomes. He published the first of a series of papers about hybrid sterility in 1933 in the *Proceedings of the National Academy of Sciences*. These studies on the genetic causes of sterility in the flies helped Dobzhansky develop his concept of a species as a reproductively isolated group of like organisms. Scientists today continue to use methods he pioneered and described in these papers, specifically the use of genetic markers to study hybrid sterility.

Though he had been using *Drosophila* as a model system for studying genetics, he still carried out research on ladybird beetles. He combined his new knowledge of cytogenetics with his background in field work in a 1933 paper published in *American Naturalist* describing the geographical variation and evolution in ladybird beetles.

THE MODERN SYNTHESIS OF EVOLUTIONARY THEORY

In fall 1936 Leslie C. Dunn, a geneticist at Columbia University, invited Dobzhansky to return to New York to give a six-week lecture series on evolution. Darwin's theory on the origin of species had inspired Dobzhansky to become a biologist; thus he was enthusiastic, as was his audience. Zoologists, botanists, paleontologists, mathematicians, and geneticists all developed their own theories concerning the evolutionary process as it related to their specific subdiscipline. One of Dobzhansky's greatest contributions to science was integrating all of these theories into one unified, simple biological theory of evolution.

During his lectures, Dobzhansky presented his summary of what came to be known as the modern synthesis of evolutionary theory. In 1859 Darwin had proposed that all life-forms had gradually evolved through the accumulation of minor variations of lower life-forms. Dobzhansky accepted this and the notion that a collection of all the life-forms that ever existed on Earth would exhibit a relatively continuous array of forms. Darwin suggested that ancestral forms diverged again and again; thus more variability exists between extant species and ancestral ones. What Darwin did not explain was the how the changes that led to variation and divergence came about.

Biologists approached resolving this issue in different ways. Some examined evolutionary histories of different types of living organisms; others researched mechanisms of evolution from the genetics perspective. A former colleague of Dobzhansky named Sergei Chetverikov examined genetic similarities and differences in populations of *Drosophila*. J. B. S. Haldane and Sir Ronald A. Fisher (both in England) and Sewall Wright (in Chicago) used theoretical mathematics to describe changes in gene frequencies by creating models of selection, mutation, migration,

genetic drift, and inbreeding in the evolutionary process. Dobzhansky was very familiar with the work of all these founders of population genetics, and by the time he delivered his lectures at Columbia in 1936, he had integrated their work into his unified theory of biological evolution. Dobzhansky published an influential book based on these lectures, *Genetics and the Origin of Species* (1937), defining the relationship between the variations studied by geneticists and the theory of evolution. The book was written in plain language and included discussion of diversity, genetic variation and chromosomal rearrangements, mutation as the source of variation, natural selection, the development of reproductive isolation during speciation, the concept of a species, and differences among members of the same species. He provided examples from his own research and from the scientific literature to demonstrate that such genetic mutations were the underlying cause of the differential characteristics upon which natural selection acted. The book was a remarkable success, considered by some to be the most influential book of evolutionary theory of the 20th century. According to Francisco Ayala, the author of Dobzhansky's biographical memoir written for the National Academy of Sciences, the book "had enormous impact on naturalists and experimental biologists, who rapidly embraced the new understanding of the evolutionary process as one of genetic change in populations. Interest in evolutionary studies was greatly stimulated, and contributions to the theory soon began to follow, extending the synthesis of genetics and natural selection to a variety of biological fields." Dobzhansky revised *Genetics and the Origin of Species* twice under the same title (1941, 1951), and a third time under the title *Genetics of the Evolutionary Process* (1970). Modern texts on evolution follow the same pattern laid out by Dobzhansky.

After returning to Caltech, he decided to look at the degree of variation that was naturally present in a population in order to assess the accuracy and usefulness of mathematical modeling. For his population studies, he chose *Drosophila pseudoobscura*, the fly species in which he had examined hybrid sterility. Without technology that has become available through a better understanding of molecular biology, studying variation in a population was difficult. By nature, recessive genes were masked. Detecting their presence in a population required performing testcrosses with homozygous recessive individuals and making deductions based on observations of the unknown population's offspring. To overcome this difficulty, Dobzhansky studied chromosomal rearrangements of this species in collaboration with Sturtevant. By collecting data on the presence and frequency of specific chromosomal inversion events in different populations from different geographical locations, he reconstructed the evolutionary history of this organism. Wright created mathematical selection models that Dobzhansky then tested by examining wild populations. This line of research led to Dobzhanksy's proposal of the concept and term *isolating mechanisms* to refer to phenomena that prevent different species from exchanging genes, or reproducing. Another important finding derived from these studies was that sterility related somehow to the autosomes, the nonsex chromosomes. Dobzhansky continued his work on *Drosophila pseudoobscura*, in collaboration with Sturtevant and other geneticists, until his death. A series of articles that shared the primary title "The Genetics of Natural Populations" (1938–68) summarized much of his findings related to the geographical and temporal variation of chromosomal arrangements. Others used Dobzhansky's techniques to reconstruct the evolutionary histories of other organisms.

EVOLUTION AS IT RELATES TO HUMANS

In 1940 Dobzhansky left Caltech to accept a professorship in zoology back at Columbia University. His teaching demands were greater, and because of his increased reputation and fame, others sought his expertise more often, leaving Dobzhansky with less time for experimental research. He did continue studying the evolution of *Drosophila* populations but also began to branch into the social sciences and philosophize about the nature of humans, the evolution of the human species, the biological significance of race, and the evolution of human culture.

One public stance taken by Dobzhansky during this stage in his life involved the Soviet agronomist Trofim D. Lysenko, who openly defied classical genetics, including the work of both Mendel and Morgan, by reverting to the belief that acquired characteristics could be inherited. Lysenko claimed that heredity resulted not from genes but from development, and since environmental conditions could affect developmental processes, heredity could also be modified. Josef Stalin succeeded Lenin after he died in 1928, becoming the de facto party leader and dictator of the Soviet Union. Stalin's goal was to advance Russian industry and agriculture rapidly. The Russian government praised Lysenko for his work speeding up the growing season of economically important crops such as wheat. The battle between followers of Lysenko and those of revered scientists such as Mendel and Morgan was mostly political, though also due in part to a misunderstanding of genetics and the faulty assumption that it conflicted with Darwinian evolution, which was by then widely accepted. Because Stalin supported Lysenko's opposition to genetics, Lysenko's power grew. The

ignorant man became the leader of Soviet biology, and, as a result, scientific progress in the Soviet Union came to a screeching halt. Many of Dobzhansky's former acquaintances and colleagues were forced from their jobs, were arrested, and faced the threat of death for opposing Lysenko. In 1946 Dobzhansky translated one of Lysenko's books, *Heredity and Its Variability*, into English to expose Lysenko's ridiculous ideas to American scientists, and he wrote several articles criticizing Lysenko's so-called science. Decades passed before Soviet science showed signs of recovery. Dobzhansky had been ignored after leaving his home country, but now Russia declared the renowned "American" geneticist to be an enemy of the people.

Dobzhansky also became involved with the eugenics movement, as an outspoken critic. The term *eugenics*, referring to the science of heredity or "good breeding," emerged in the 1880s and by the 1930s had grown into a movement aiming to enhance or improve the "quality" of the human species by controlled breeding. Dobzhansky saw this movement for what it truly was—a pseudoscientific façade for racial and social prejudice. Political and social leaders with selfish motives tried to manipulate the masses of society into believing that limiting reproduction of a "higher" class or "favored" race to members of the same class or race would improve the quality and purity of that group. They used similar faulty reasoning to suggest that preventing the reproduction of groups of people considered inferior (because of race, behavioral traits, or certain medical conditions) would improve the human species as a whole. Incensed by the racial bigotry fueled by this movement, Dobzhansky spoke out strongly and publicly against it, declaring that no scientific evidence demonstrated a genetic basis of social or personality traits of humans and, furthermore, certainly no evidence of significant differences between racial groups. More genetic variation existed within the human species than any minor differences between races. In 1946, in collaboration with fellow Columbia geneticist Leslie C. Dunn, he published *Heredity, Race, and Society*, a book that explained the scientific differences with respect to individuals, populations, and races.

In the laboratory, Dobzhansky mixed *Drosophila* populations containing specific chromosome types and recorded his observations on changes in the population with respect to those introduced genetic differences. He observed that the populations evolved different genetic characteristics, even when the parental strains, the temperatures, the food supply, and the exposure to light and darkness were the same. Though he was studying fruit flies, he saw the relevance of these studies to the evolution of human populations. These experiments were extended to attempt the creation of a new species in the laboratory, a successful effort published in *Nature* in 1971 as "Experimentally Created Incipient Species of *Drosophila*." He applied the concept of inherited traits to human nature in a work titled *Mankind Evolving* (1962), which combined knowledge from anthropology, genetics, sociology, and evolutionary biology.

NEVER RETIRED

In 1962 the Rockefeller Institute (now Rockefeller University) in New York City hired Dobzhansky as a professor. For almost a decade, he traveled with his wife on expeditions all over the world, including locations in Brazil, India, and Australia. The purpose of these trips was to find and collect new species of *Drosophila* for study.

In 1969 Natasha suffered a fatal heart attack. Afterward, Dobzhansky tried to keep busy to avoid feeling lonely. During this time he completely revised *Genetics and the Origin of Species* incorporating many new exciting findings such as the discovery of the structure of deoxyribonucleic acid (DNA) and the demonstration that DNA was the molecule of heredity. The new edition, *Genetics of the Evolutionary Process,* published in 1970, was even more successful than his previous editions, as Dobzhansky kept abreast of the scientific literature on the subject and the text reflected decades of progress toward understanding the relationship between genetics and evolution. In 1971 he became professor emeritus at Rockefeller and moved back to California, choosing to serve as an adjunct professor in genetics at the University of California at Davis.

Dobzhansky belonged to the National Academy of Sciences, the American Philosophical Society, and the American Academy of Arts and Sciences. Several foreign scientific organizations elected him to membership as well: the Royal Society of London, the Royal Swedish Academy of Sciences, the Royal Danish Academy of Sciences, the Brazilian Academy of Sciences, the Academia Leopoldina, and the Academia Nazionale dei Lincei. He served as president of the Genetics Society of America (1941), the American Society of Naturalists (1950), the Society for the Study of Evolution (1951), the American Society of Zoologists (1963), the American Teilhard de Chardin Association (1969), and the Behavior Genetics Association (1969). Numerous organizations recognized Dobzhansky's contributions to life science by awarding him medals: The U.S. National Academy of Sciences awarded him the Daniel Giraud Elliot Medal (1946) and the Kimber Genetics Award (1958), the Academia Leopoldina awarded him the Darwin Medal (1959), Yale University awarded him the A. E. Verrill Medal (1966), the American Museum of Natural History awarded him the Gold Medal Award for Distinguished Achievement in Science (1969), and the Franklin Institute awarded him the Benjamin Franklin Medal (1973). In 1964 President Lyndon B. Johnson awarded Dobzhansky the National Medal of Science.

In 1968 physicians had diagnosed Dobzhansky with chronic lymphatic leukemia and gave him only a few months to a few years to live. Seven years later, Dobzhansky died of the effects of lymphatic leukemia, on December 18, 1975, one month shy of his 72nd birthday. He had worked until the day before his death. His students, colleagues, and peers remember him for his generosity, loyalty, exuberance, and willingness to give his time to serve science and teach young scientists. The rest of the scientific community venerated Dobzhansky for his incredible ability to synthesize theories that interwove ideas from numerous fields. Having published nearly 600 papers and books, he was one of the most influential and prolific biologists of the 20th century. Dobzhansky's work continues to manifest its influence in the fields of evolutionary biology, population genetics, and sociology.

See also CHROMOSOMES; EVOLUTION, THEORY OF; MENDEL, GREGOR; MORGAN, THOMAS HUNT.

FURTHER READING

Adams, Mark B., ed. *The Evolution of Theodosius Dobzhansky: Essays on His Life and Thought in Russia and America.* Princeton, N.J.: Princeton University Press, 1994.

Ayala, Francisco J. "Theodosius Dobzhansky (January 25, 1900–December 18, 1975)." In *Biographical Memoirs: National Academy of Sciences.* Vol. 55. Washington, D.C.: National Academy Press, 1985.

Dobzhansky, Theodosius. *Genetics of the Evolutionary Process.* New York: Columbia University Press, 1970.

———. "Nothing in Biology Makes Sense Except in the Light of Evolution." *American Biology Teacher* 35 (March 1973): 125–129.

Land, Barbara. *Evolution of a Scientist: The Two Worlds of Theodosius Dobzhansky.* New York: Thomas Y. Crowell, 1973.

Lewontin, R. C., John A. Moore, William B. Provine, and Bruce Wallace, eds. *Dobzhansky's Genetics of Natural Populations I–XLIII.* New York: Columbia University Press, 1981.

Duve, Christian de (1917–) Belgian *Cell Biologist* The pioneering research of Christian de Duve solidly established cell biology as a subdiscipline of life science. He observed and described the structure and function of two previously undiscovered organelles, the lysosome and the peroxisome. As a result of de Duve's efforts to improve the methodology available for dissecting cells, biologists have identified and elucidated the role of numerous organelles important to cell physiology.

TRAINING IN INSULIN AND BIOCHEMICAL RESEARCH

Christian R. de Duve was born on October 2, 1917, in Thames-Ditton, near London. His parents were Belgian, but the family had escaped to England when the Germans invaded Belgium during World War I. They returned to Belgium in 1920, and Christian grew up in Antwerp.

In 1934 de Duve entered the Catholic University of Louvain, where he planned to study medicine. He obtained research experience in the laboratory of Professor J. P. Bouckaert, a physiologist who studied the role of insulin in glucose uptake. Insulin is a protein hormone that the pancreas synthesizes and releases when blood sugar levels are high, such as after a meal. Cells respond by transporting glucose from the blood into the cells, where they break it down for energy or store it for later use. By the time de Duve graduated with a medical degree in 1941, he had changed his career ambition from practicing medicine to investigating the biochemical mechanism of insulin action.

The Second World War interfered with de Duve's immediate career plans. He served in the army for a while, spent some time in a prison camp, and eventually returned to Louvain. The lack of available supplies hindered research efforts, so de Duve obtained a job at the Cancer Institute and simultaneously enrolled in a program to earn an additional degree in chemical sciences. In 1945 he received the most advanced degree of university level education, *agrégation de l'enseignement supérieur*. His thesis research led to the publication of several articles in the scientific literature and a book titled *Glucose, Insulin, and Diabetes*. He believed that biochemistry was the best approach for determining how insulin worked. To advance his knowledge of biochemistry, de Duve worked for 18 months in the laboratory of Hugo Theorell at the Medical Nobel Institute in Stockholm, Sweden. Theorell received the Nobel Prize in physiology or medicine in 1955 for his studies on oxidation enzymes. Then he went to St. Louis, Missouri, to work with Carl Cori and Gerty Cori, whose research on the metabolic pathways of glycogen metabolism earned them the 1947 Nobel Prize in physiology or medicine. He also collaborated with Earl Sutherland, who received the 1971 Nobel Prize in physiology or medicine for discoveries on the mechanisms of action for hormones. Having received extensive training with such distinguished researchers, de Duve returned to Louvain in December 1947 to make a name for himself.

LYSOSOMES AND PEROXISOMES

De Duve taught physiological chemistry at the medical school and assumed the position of full professor in 1951. After more than 12 years of training and preparation, as a new professor he focused his research efforts on insulin and glucagon action. In addition to making insulin, the pancreas synthesize glucagon, a hormone that promotes the breakdown of the complex carbohydrate glycogen in the liver. Enzymes digest the glycogen into individual molecules of glucose that enter the bloodstream, leading to an increase in blood sugar levels. Thus the action of glucagon is antagonistic to the action of insulin, which acts to decrease blood sugar levels. While examining carbohydrate metabolism in the liver, de Duve became sidetracked by an observation unrelated to insulin action—the latency of the enzyme acid phosphatase. This diversion led to the discovery of two new cell organelles and to the development of better methodology for studying cell components.

Glycogen breakdown occurs by the release of monomers of glucose 1-phosphate, which the enzyme phosphoglucomutase converts to molecules of glucose 6-phosphate. The enzyme glucose 6-phosphatase removes a phosphate group from the molecule glucose 6-phosphate, resulting in the simple sugar glucose. Though both liver and muscle cells store glucose in the form of glycogen, only liver cells express the enzyme glucose 6-phosphatase. Because of this, de Duve thought it played an important role in the effect of insulin on liver cells; specifically, he thought this enzyme blocked the effect of insulin, and he was trying to learn more about it in hopes of gaining insight as to the mechanism of action of insulin on liver cells. The first step was to purify the enzyme for further characterization. Isoelectric precipitation, a standard method for purifying proteins in those days, involves altering the pH of the solution containing the protein so no net charge exists on the protein molecules. At this pH, called the isoelectric point (pI), the amide groups and the carboxyl groups of the amino acids are uncharged, and the protein loses its ability to interact with the water molecules and falls out of solution. The researcher then resolubilizes the precipitated material in a solution with a neutral pH, at which the side chains of the protein can once again readily interact with the solute molecules. When de Duve attempted isoelectric precipitation to purify glucose 6-phosphatase, the enzyme would not redissolve; thus he concluded that it was not simply precipitated, but was associated with some sort of cellular structure. Forced to consider alternate methodology, he decided to purify it by a relatively new technique called cellular fractionation.

Scientists had recently worked out methods for separating cellular components using centrifugation. Albert Claude, a professor at the Rockefeller Institute noted for his studies on animal cells using electron microscopy, had published a procedure for differential centrifugation in the mid-1940s. This technique takes advantage of the fact that cellular components of various sizes and morphologies sediment at different rates when centrifuged. After grinding a tissue such as a mouse or rat liver in a sucrose solution, three successive centrifugations of the homogenate resulted in four different fractions: the nuclear fraction, the mitochondrial fraction, the microsomal fraction (microsomes are fragments of the endoplasmic reticulum), and the remaining supernatant. George Palade, who joined the Rockefeller Institute in 1947, improved upon Claude's methods and described (in collaboration with Keith Porter, who became well known for his electron microscopy work) the structures found in the cell's cytoplasm such as the endoplasmic reticulum, which is a membranous, folded sac that occupies much of the cytoplasm. He also discovered ribosomes, the cellular structures that he demonstrated were responsible for synthesizing proteins, and outlined the pathway by which proteins destined for secretion reached the cell's exterior via the Golgi apparatus, another cytoplasmic structure.

When de Duve utilized differential centrifugation as described in the literature, the glucose 6-phosphatase activity did not separate into a distinct fraction. Activity was distributed such that a portion of the activity was present in the mitochondrial fraction and a portion of the activity was present in the microsomal fraction. They modified the fractionation procedure to generate a small fifth fraction intermediate to the mitochondrial and microsomal fractions. This fraction contained high concentrations of glucose 6-phosphatase.

At the same time, de Duve's group assayed for the distribution of other enzymes, including acid phosphatase, in the cell fractions. At one point when they assayed fractions for acid phosphatase, the enzyme activity was not detectable—it seemed to have disappeared. De Duve assumed they had made a mistake, and they reassayed the fractions after storing them in the refrigerator for five days, at which point the activity had reappeared. This observation intrigued de Duve, and he put his research on insulin action on hold in order to examine it further. What was happening was that the enzyme activity was enclosed within a membrane; thus the enzyme molecules did not have access to the external substrates. The activity was hidden. As the irreversible agglutination that occurred during the attempted isoelectric precipitation suggested, the distributions of the enzymes with specific fractions indicated that the enzymes were associated with specific subcellular particles.

They examined several other enzymes that exhibited similar behavior. These included oxidases, catalase, reductases, phosphatases, nucleases, and others. Latency was observed for some of the enzymes but not others, and one enzyme (urate oxidase) functioned optimally at an alkaline pH rather than an acidic pH as most of the enzymes they had been studying did. These observations suggested that more than one type of subcellular particle contained the enzymes. To separate the types of particles, they turned to density gradient centrifugation. Characterization of the physical properties of the cellular particles in the density gradient fractions revealed three distinct particle populations, with different associated enzymes—now called the mitochondrial, lysosomal, and peroxisomal fractions.

While de Duve was looking at biochemical activities in cellular fractions, others were making advances in the use of electron microscopy. De Duve did not even have a microscope in his lab at the time. In collaboration with Claude in Brussels and Wilhelm Bernhardt in Paris, de Duve was able to examine his fractions. He later acquired his own electron microscope, and, with the assistance of Palade, Henry Beaufay from de Duve's lab became skilled in its use. The lysosomal fractions contained dense bodies surrounded by a membrane, and the peroxisome fractions contained particles referred to as microbodies.

In 1955 de Duve published a paper proposing the existence of these two new types of cellular particles. De Duve's group had studied acid hydrolases. The lysosomes contain these enzymes, which work best at an acidic pH and digest other molecules such as proteins, nucleic acids, and polysaccharides. Because hydrolytic enzymes are capable of attacking and destroying components of the cell's own cytoplasm, he suggested that enclosure in a membrane-bound compartment was necessary to protect the cell from degrading its own parts. When he treated the fraction with substances to dissolve the membrane, the enzymes were released, confirming their enclosure. The activity of the enzymes depends on the acidic conditions maintained within the lysosome; thus if the membrane of the structure ruptures or leaks, the enzymes do not function in the neutral pH of the cytoplasm.

Lysosomes function in intracellular digestion to get rid of old and worn-out cell parts and to destroy foreign substances taken into the cell through phagocytosis. These organelles form from the Golgi complex, another cytoplasmic, membranous structure consisting of sacs and vesicles. Deficiencies of enzymes found in the lysosomes cause medical conditions that result from the accumulation of undigested material built up in the cells. One such example is Hers disease, a hereditary disease caused by a deficiency in the enzyme glycogen phosphorylase and characterized by glycogen accumulation, liver enlargement, and low blood sugar level.

Peroxisomes help rid the cell of toxic compounds, including toxic by-products of oxygen metabolism. They contain oxidases, enzymes that use molecular oxygen to oxidize organic molecules. One of the peroxisomal enzymes that de Duve studied was urate oxidase, which catalyzes the oxidation of uric acid in most mammals. The name of the organelle is derived from *hydrogen peroxide,* a product made by enzymes within the peroxisome. The enzyme catalase further metabolizes the hydrogen peroxide into water and oxygen. Unlike lysosomes, peroxisomes are self-replicating.

THE NOBEL PRIZE, ICP, AND WRITING

In 1962 the Rockefeller Institute (now Rockefeller University) in New York appointed de Duve professor. He split his time between Rockefeller and Louvain, where he retained his professorship. In 1974 de Duve founded a new institute, the International Institute of Cellular and Molecular Pathology (ICP), at the Louvain medical school in Brussels. The mission of the institute is to facilitate the application of basic knowledge in the cell and molecular sciences.

Christian de Duve was awarded the Nobel Prize in physiology or medicine in 1974, along with Claude and Palade, "for their discoveries concerning the structural and functional organization of a cell." Together, the three are credited with the creation of modern cell biology.

De Duve married the former Janine Herman in 1943, and together they had four children: Thierry, Anne, Françoise, and Alain. He became professor emeritus at the University of Louvain in 1985 and at Rockefeller Institute in 1988. His presidency of the ICP ended in 1991, though he continued to give his time. In 1997 the name was changed to *Christian de Duve Institute of Cellular Pathology* to pay tribute to its founder.

In 1976 de Duve was invited to give the Rockefeller Christmas Lectures (now called the Alfred E. Mirsky Christmas Lectures), a series of four lectures given to a selected audience of 550 high school students from the New York area. In 1984 de Duve published a book, *A Guided Tour of the Living Cell,* originally based on the series of lectures. The book took several years to write, as he was busy with laboratory research at the same time, but it was wildly successful, in part because of its numerous illustrations. After retiring from laboratory research, he had more time on his hands. De Duve decided to write a second, more condensed version about all the structures within a living cell. As he wrote this book, *Blueprint for a Cell* (1991), he felt the last chapter should address the origin of the cell. By the time he finished composing it, the content of what was to be the final chapter not only took up half of his book, but led to a new interest—the origin, evolution, and meaning of life. He has since authored several books exploring this subject, including *Singularities: Landmarks on the Pathways of Life* (2005), *Life Evolving: Molecules, Mind, and Meaning* (2002), and *Vital Dust: Life as a Cosmic Imperative* (1995).

The course of de Duve's career demonstrates the close relationship between biochemistry and modern cell biology. During his biochemical analysis of cell fractions, his laboratory did not even have a microscope, the primary tool for cell biologists at the time. While attempting to correlate certain enzymatic activities with different cell fractions, de Duve discovered a new membrane-bound organelle, the lysosome. The enzymes within the lysosome digested cellular components when released from the confines of the protective membrane. De Duve also discovered the peroxisome, a membrane-bound organelle that contains enzymes that act in oxidative reactions such as the breakdown of hydrogen peroxide. His discovery of these two cellular structures that play such an important role in cell physiology, along with the methods he developed to study cellular components, helped establish the field of modern cell biology. Before de Duve's contributions, cell biology was mostly a descriptive field, but he demonstrated how scientists could bridge the gap of knowledge between the observations concerning the structure of the cell parts and the biochemical functions occurring within them.

See also CELL BIOLOGY; CENTRIFUGATION; EUKARYOTIC CELLS.

FURTHER READING
Duve, Christian de. *A Guided Tour of the Living Cell.* New York: Scientific American Books, 1984.

———. *Singularities: Landmarks on the Pathways of Life.* New York: Cambridge University Press, 2005.

Duve, Christian R. de, Berton C. Pressman, Robert Gianetto, Robert Wattiaux, and Françoise Appelmans. "Tissue Fractionation Studies. 6, Intracellular Distribution Patterns of Enzymes in Rat-Liver Tissue." *Biochemical Journal* 60 (1955): 604–617.

The Nobel Foundation. "The Nobel Prize in Physiology or Medicine 1974." Available online. URL: http://nobelprize.org/nobel_prizes/medicine/laureates/1974/. Accessed January 23, 2008.

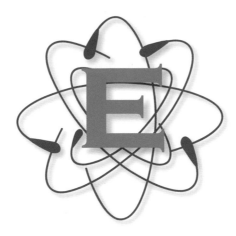

Earle, Sylvia (1935–) American *Marine Biologist* The pioneering American oceanographer Sylvia Earle was one of the first marine scientists to use scuba as an integral part of her research program and has identified many new marine species. Her area of expertise is phycology, or the study of algae, but she is just as comfortable swimming with 40-ton whales. Having logged 7,000 hours underwater, Earle has been aptly named "Her Deepness."

CHILDHOOD AND EDUCATION

Sylvia Alice Reade Earle was born on August 30, 1935, in Gibbstown, New Jersey, to Lewis Reade Earle, an electrical contractor, and Alice Freas Richie Earle, who both raised her to appreciate the natural world. As the middle child of three siblings, Sylvia enjoyed exploring the pond, creek, and orchard around her family's home, a farm outside Paulsboro, New Jersey. She claims she was called to the sea at age three, when she was knocked over by a wave while wading on the shore. Instead of crying, she laughed and got back on her feet, ready to face more. She has harbored this same fearless attitude throughout her career, while setting deep diving records and advocating public education about the oceans.

When she was 12 years old, the family moved to Dunedin, Florida, into a house with the Gulf of Mexico in the backyard. As a birthday present that year, Sylvia received a gift of swim goggles. She enjoyed reading books by her favorite author, the naturalist William Beebe, and yearned to see the creatures he described. Her first diving experience was with a friend and her friend's father in the Weekiwatchee River. The copper helmet she wore was attached to an air compressor on shore. After 20 minutes the air pump stopped working properly and she had to

be rescued, but she had been so enthralled watching a school of fish that she did not let this experience deter her from going under again.

Sylvia enrolled in a summer marine biology course taught by Dr. Harold Humm at Florida State University (FSU) when she was 17 years old. She learned to use scuba gear and loved gliding underwater, following the fishes. After earning her bachelor's degree from FSU in 1955, she applied to and was accepted in several prestigious graduate programs, and she selected Duke University in Durham, North Carolina, where she was awarded a full scholarship and where Dr. Humm was working. As a master's degree candidate, she majored in botany but focused on algae; she became the first person to study the algae in the Gulf of Mexico systematically, amassing a collection with over 20,000 specimens. She received her master's degree in botany in 1956, at only 20 years of age.

That same year she married the zoologist John (Jack) Taylor. They moved to Dunedin, next door to Earle's parents, and set up a makeshift lab complete with specimen cabinets and microscopes in their garage. Their daughter, Elizabeth, was born in 1960 and their son, John (called Richie), in 1962.

In 1964 Earle toured the Indian Ocean on the *Anton Bruun,* for an expedition sponsored by the National Science Foundation. The marine botanist who was scheduled to go had cancelled, and Humm recommended Earle as a replacement on the six-week trip. Though some thought it was bad luck to have a woman on the ship, the 70 members of the all-male crew were impressed with her hard work. She spent as much time as possible exploring underwater and discovered a new species of bright pink algae that reminded her of something Dr. Seuss would invent.

She named it *Hummbrella hydra* in honor of her mentor. Over the next two years, she went on four more expeditions on the *Anton Bruun* and became acquainted with the famous ichthyologist Eugenie Clark.

Researching the role of algae in marine food chains for 10 years familiarized Earle with the far-reaching effects of pollution on aquatic plant life. In 1966 she completed her dissertation, "Phaeophyta of the Eastern Gulf of Mexico." With a Ph.D. from Duke, she assumed a temporary position as resident director of the Cape Haze Marine Laboratories (now the Mote Marine Laboratory), founded by Clark, in Sarasota. The following year Earle accepted simultaneous positions as a research scholar at the Radcliffe Institute and a research fellow at the Farlow Herbarium at Harvard University, which houses algae, fungi, and bryophyte specimens. (Harvard University promoted her to researcher in 1975.) She was particularly interested in ocean ecology, the relationships of plants and animals with each other and with their environment, and studied these relationships by making numerous dives. On one such dive in 1968, Earle dove to 100 feet (30 m) below the surface in the submersible *Deep Diver*; she was four months pregnant with her daughter, Gale, from her second marriage at the time. (She had divorced Taylor and married an ichthyologist from Harvard named Giles Mead.)

TEKTITE II

After seeing an announcement on the bulletin board at Harvard, Earle submitted a proposal for underwater research for Tektite II. Tektite I was a government experiment that put four scientists 50 feet (15 m) underwater in an enclosed habitat near the Virgin Islands for 60 days in 1969. The goals of the Tektite II project were to determine the limitations and practicality of saturation diving and to study undersea habitats, but the National Aeronautics and Space Administration (NASA) was also interested in how people handled living together in tight quarters in an unusual environment. Earle's qualifications and proposal for studying the influence of herbivorous fish on marine plants were impressive. At 1,000 hours, she had more hours of diving experience than all the other applicants, but the navy was not prepared to employ a woman in the project. They decided to hire Earle to lead an all-female team of oceanographers to live in an underwater chamber for two weeks in 1970.

The underwater habitat had carpet, television, bunk beds, showers, and a stove for cooking frozen meals. The upstairs work area had microscopes and a communications panel. A ladder led down to a chamber through which the divers entered the water. Air

Sylvia Earle, shown here holding a crab while scuba diving off the Florida Keys, has dedicated her life to exploring life under water and educating the public about the oceans. *(Connie Bransilver/Photo Researchers, Inc.)*

pressure prevented water from flowing up into the living quarters. NASA psychologists monitored the five scientists constantly and recorded their activities every six minutes. Earle's favorite time to explore was in the predawn darkness, and she especially enjoyed observing the behavior of individual fishes. As part of their experiments, Earle and another team member tested rebreathers instead of scuba tanks. Rebreathers were more expensive and complicated to prepare, but they allowed the divers to stay underwater for four hours rather than one and were quieter, so they could hear the fish grunt and chew on coral.

At the end of two weeks, the five scientists had to spend 19 hours in a decompression chamber to allow their bodies to readjust to normal atmospheric pressure. Earle had documented 154 species of plants, including 26 species never before seen in the Virgin Islands. In addition to confirming that herbivorous fish greatly affected marine plant populations, she learned the sleeping habits of different types of fish and realized that individual fish had food preferences just as people do. The mission stimulated a lot of publicity to which Earle had difficulty adjusting at first, but she decided to use it to her advantage, in order to reach more people with a message about the dangers of ocean pollution. She used her newfound fame to begin writing for *National Geographic* magazine and to produce films that spurred public interest in marine biology. She hoped that greater understanding of the oceans would lead to more positive action toward her goal of protecting them.

STUDIES ON WHALES

In 1976 Earle became a research biologist and curator at the California Academy of Sciences and a fellow in botany at the Natural History Museum of the University of California at Berkeley. Her marriage to Mead had ended, and she had moved to California. She began a project studying humpback whales during winter 1977 in collaboration with Roger Payne and Katy Payne, who were experts on these whales that migrate from Hawaii, where they mate and give birth, to Alaska, where they feed each summer. At the time, most of what was known about whales had been learned from examining carcasses, but Earle knew she could learn much more by studying them in their natural environment. In the 1960s, Roger Payne had set up a microphone in the sea in hopes of recording sounds the whales made and captured scores of songs composed of creaks, grunts, and moans. Biologists have since learned that only males produce the distinct melodies which last up to 20 minutes and can be heard at a distance of 20 miles (32 km). When in the water, the vibrations from the whales' songs made Earle's body vibrate. By following the whales, Earle learned to recognize individual whales by distinctive markings on their faces, flippers, tails, and undersides. Earle collaborated with Payne and the filmmaker Al Giddings to produce a documentary film about the humpbacks, *Gentle Giants of the Pacific,* describing the role the whales play in the oceanic ecosystem.

FREE DIVING AND DEEP OCEAN EXPLORATION

The next adventure involved a Jim suit, which resembled space attire and allowed a person to walk on the ocean floor. Jim suits were typically used for people making repairs to underwater machines or oil rigs; a scientist had never used one for research purposes. Giddings thought it would be remarkable to film Earle outfitted in one strolling along the ocean bottom at 1,000 feet (305 m) for pictures in a National Geographic Society book, *Exploring the Deep Frontier* (1998), and for an upcoming television special. The garments were made of magnesium, a malleable metal that could withstand the intense water pressure at great depths, and had steel pincers as hands. A cable connected the suit's wearer with a submersible that followed behind.

Earle welcomed the challenge, and on October 19, 1979, she set a record for free diving at 1,250 feet (381 m). In the Jim suit, she walked stiffly along the ocean floor and jotted down observations she made while watching the giant seven-foot (2.1-m) rays and long-legged crabs with whom she shared her pedway. She also admired pink sea fans, jellyfish, a cat shark, lantern fish, hatchetfish, and bamboo coral that rippled with gleaming blue rings when nudged. Her two-and-one-half-hour time limit passed quickly, but before returning to the submersible platform for ascension, she planted two flags to mark the historic moment in the ocean floor, a U.S. flag and a National Geographic flag. Since then, more advanced submersibles have replaced the Jim suits for many purposes.

Though Earle was thrilled to have set a new free-diving record, she yearned to go deeper, since the average depth of the ocean is 13,000 feet (3,962 m). In a joint venture in 1982, Earle and the British-born engineer Graham Hawkes (to whom she was married in 1986–89) founded a company called Deep Ocean Technology (later called Deep Ocean Engineering, Inc.) to build state-of-the-art, deep-diving, one-person submersibles. One major obstacle they had to overcome was choosing a material that could withstand the pressure at great depths but was transparent to allow observation. Customers were hard to find, so Hawkes built a large remotely operated vehicle (ROV) that could be used to inspect undersea equipment. After selling one to Shell Oil Company, they started receiving more orders. In 1984 they created *Deep Rover,* a spherical submersible with mechanical arms that held one diver and operated at record-breaking depths of 3,000 feet (914 m). Earle went down in *Deep Rover* at night, an experience she compared to falling into a fireworks display because of the luminescent creatures floating around her. Her mesmerizing experience observing jellies, shrimp, interesting fish, and an octopus was marred by the sight of a soda can on the ocean floor. Today Deep Ocean Engineering continues to design and manufacture ROVs that are sold internationally.

NOAA AND LATER CAREER

President George H. W. Bush appointed Earle chief scientist at the National Oceanic and Atmospheric Administration (NOAA) in 1990. NOAA is the U.S. governmental organization whose mission is to describe and predict changes in the Earth's environment and to conserve and manage the nation's coastal and marine resources. Earle was initially worried about not being able to voice her opinions in public as an upper-level government representative, but she accepted the opportunity in hopes of changing the stubborn governmental mindset that maintained marine research as a low funding priority. She had become increasingly aggravated at the government's apathy about undersea exploration and the unwillingness to invest money for undersea research.

In 1991, as chief scientist of an underwater research team, Earle investigated the effects and aftermath of 500 million gallons (more than 2 billion liters) of oil that Iraq deliberately dumped into the Persian Gulf during the Gulf War. Her task was to figure out how long it would take the ocean and its inhabitants to recover from the devastating effects. The landscape was completely blackened for miles, and the waters were oily and brown. Seeing the marine organisms covered in black slime reaffirmed her commitment to spreading the message of ocean conservation. She also investigated the effects on the ecosystems in Prince William Sound, Alaska, of the oil spills from the ship the *Exxon Valdez*. She resigned from NOAA after 18 months.

In 1992 Earle founded Deep Ocean Exploration and Research (DOER) with her daughter, Elizabeth Taylor. Located in Alameda, California, DOER has the stated mission designing and developing practical and effective technologies to provide working access to the deep ocean and other challenging environments.

Earle accompanied Japanese scientists on an expedition in 1991, when she descended in a three-person submersible named *Shinkai 6500* to 13,000 feet (almost 4 km), deeper than she had ever been before. The Japanese government asked Earle to lend her expertise for building a remote, then manned, submersible that could dive to 36,000 feet (11 km) in 1993.

In 1995 Earle used her reputation as an expert on the deep ocean to champion conservation by publishing a book, *Sea Change: A Message of the Oceans*, which celebrates the variety and abundance of life below the sea's surface. Along with her fascination, she soberly shared her worries about the future of the oceans in the hands of uninformed humans.

National Geographic named Earle the Explorer in Residence in 1998. From 1998 to 2002 she served as project director for the Sustainable Seas Expeditions, supported by the National Geographic Society, the NOAA, and the Goldman Foundation, to explore and document the marine life and conditions of the 12 U.S. marine sanctuaries. A comprehensive survey of these underwater national parks will allow marine scientists to detect changes to the ecosystems and to make recommendations for the preservation of their

health. During this time she also published *Wild Ocean* (1999) and *Atlas of the Ocean* (2001).

In 2006 President Bush established the United States's first National Marine Monument, located in the northwestern Hawaiian Islands, in part because of Earle's persuasiveness. Currently, Earle serves as president of Deep Search International and chair of the Advisory Council for the Harte Research Institute for Gulf of Mexico Studies.

Earle has led more than 60 expeditions and logged more than 7,000 hours underwater. She has earned numerous honors and awards and received 15 honorary degrees from institutions including Florida International University, the Monterey Institute, Duke University, the University of Connecticut, and the University of Rhode Island. She is a member of the American Association of the Academy of Sciences, the California Academy of Sciences, the Marine Technology Society, and the World Academy of Arts and Sciences. She serves on boards and committees for Woods Hole Oceanographic Institute, Mote Marine Laboratory, the World Wildlife Fund, and many more. In 2000 she was inducted into the National Women's Hall of Fame and the U.S. Library of Congress named her a Living Legend, an award presented to people who have made significant contributions to America's diverse cultural, scientific, and social heritage. The sea urchin *Diadema sylvie* and the red alga *Pilina earli* were named in her honor.

Time magazine aptly named Earle the first "hero for the planet" in 1998. Her authorship of more than 150 publications about marine science earned her the right to speak on behalf of the oceans and to educate the public about the oceans and marine life. As an ambassador for the sea, she has strongly and repeatedly expressed the opinion that if people knew about marine life, they would care more about protecting it. Living up to this responsibility, she has made more than 100 television appearances in interviews and in special programs.

See also BEEBE, WILLIAM; CLARK, EUGENIE; MARINE BIOLOGY.

FURTHER READING

Academy of Achievement. "Sylvia Earle, Ph.D." Available online. URL: http://www.achievement.org/autodoc/page/ear0bio-1. Updated January 17, 2008.

Baker, Beth. *Sylvia Earle: Guardian of the Sea.* Minneapolis, Minn.: Lerner Publications, 2001.

Earle, Sylvia A. *Sea Change: A Message of the Oceans.* New York: Ballantine Books, 1995.

Earle, Sylvia A., and Henry Wolcott. *Wild Ocean: America's Parks under the Sea.* Washington, D.C.: National Geographic Society, 1999.

Stanley, Phyllis M. *American Environmental Heroes.* Springfield, N.J.: Enslow, 1996.

eating disorders Eating disorders such as anorexia, bulimia, and binge-eating disorder affect males and females of all ages from all socioeconomic backgrounds, though young females are at the highest risk. Eating disorders are real, treatable medical conditions and are not due to lack of willpower. They are often associated with other illnesses, including depression, substance abuse, or anxiety disorders. The effects of eating disorders can be life-threatening or cause serious damage to one's health. Anorexia nervosa is characterized by extreme weight loss due to self-imposed starvation. People suffering from anorexia have an unrealistic fear of being fat and often see themselves as fat even when dangerously underweight. Bulimia nervosa is characterized by episodes of overeating, followed by an activity to counter the binge, such as vomiting, high doses of laxatives, periods of fasting, or excessive exercising. Binge-eating disorder, also called compulsive eating, is characterized by episodes of uncontrolled, impulsive eating, well beyond the point of becoming full. In contrast to bulimia, the binge is not followed by purging, but after a binge, the person may feel intense shame or feelings of self-loathing. Many people who suffer from eating disorders experience aspects of two or all three of the conditions.

While the symptoms of eating disorders involve obsessions with food, the causes are believed to be a combination of emotional, social, and psychological factors. Food is used to accomplish something other than meeting nutritional needs, such as a means to calm emotions, to cover up one's lack of self-esteem or loneliness, or to symbolize something the person can control. Though in many cases the causes fall within the realm of psychology, in some cases, biochemical or physiological abnormalities also play a role. Hormonal signaling and neural signaling among body tissues affects one's appetite and energy expenditure; thus endocrine or nervous system malfunctions may lead to improper regulation of appetite, which may trigger an eating disorder. Twin studies suggest that anorexia and bulimia have a genetic component, and scientists believe that several genes may interact with the environment to make someone susceptible to eating disorders. Because females are more susceptible than males, and eating disorders most often emerge shortly after the onset of puberty, many scientists suspect that gonadal steroids play a role in the development of these disorders.

In the case of anorexia nervosa, the body responds to starvation by slowing its metabolism to conserve energy. This can lead to a heart rate that is too slow or low blood pressure, and feelings of severe fatigue, weakness, and fainting. The body will eventually break down muscle tissue to utilize the protein stores for energy. Bones will become brittle, skin and hair will become dry, and the hair may start to fall out. A layer of fine hair may develop over the entire body in an effort to preserve body heat, a task that normal metabolism accomplishes in a healthy individual. Dehydration can cause kidney failure.

The health consequences of bulimia differ. One major problem is electrolyte imbalance. As food travels through the digestive tract, the body reabsorbs necessary electrolytes, such as sodium, potassium, and chloride. Induced vomiting prevents the food from ever reaching the intestines, and laxatives move the food through the digestive tract too rapidly for sufficient reabsorption to occur; fluids follow and are lost as well. Electrolyte imbalances can cause heart failure. Over long periods, induced vomiting can cause serious tooth decay from the acidity of the gastric juices, which can also inflame the esophagus. Frequent vomiting can also cause gastric or esophageal rupture. Overuse of laxatives diminishes the digestive tract's ability to maintain regular bowel movements, leading to a dependency on the laxatives to produce bowel movements. The belief that laxatives are effective in controlling weight is erroneous; most of the weight lost by stimulating food to rush through the intestines is water and indigestible fiber that would make its way out on its own eventually. As soon as the person drinks water, the "weight loss" is reversed. If the person does not rehydrate, the effects of dehydration include weakness, blurry vision, fainting, kidney damage, and even death.

The physiological effects of binge eating mirror those of obesity, whether the person suffering from compulsive eating is overweight or not. These risks include high blood pressure, high cholesterol level, high triglyceride levels, Type II diabetes, and disease of the gallbladder.

Females between the ages of 15 and 19 are at the highest risk for developing eating disorders, though every age group of both males and females has seen an increase with each decade since the 1930s. According to the National Institute of Mental Health of the National Institutes of Health,

- Between 0.5 and 3.7 percent of females suffer from anorexia nervosa at some time during their life.
- During their lifetime 1.1 to 4.2 percent suffer from bulimia nervosa.
- Between 2 percent and 5 percent of all Americans suffer from binge-eating disorder within a six-month period.

Experts believe many more cases of eating disorders exist than are reported because of the shame associated with the disorders.

One strategy for prevention and for treatment of eating disorders is to focus on nutritional eating, eating to maintain health. The obsession of dieting is difficult to counter. Instilling healthy eating attitudes and habits is the best means for preventing or treating an eating disorder and for achieving and maintaining a healthy body weight.

See also NUTRITION.

FURTHER READING

National Institute of Mental Health, National Institutes of Health. "Eating Disorders: Facts about Eating Disorders and the Search for Solutions." Available online. URL: http://www.nimh.nih.gov/publicat/eatingdisorders.cfm. Accessed August 2, 2008.

ecology Ecology is the study of interactions and relationships between organisms and their environment. The environment comprises the abiotic, or nonliving, components of the surroundings as well as all the other populations that live in the same area. Abiotic components include factors such as local weather and climate, nearby water bodies, geological structures or features, and the chemical composition of the soil or atmosphere. Learning about these relationships discovers information about the beneficial services ecosystems provide for people and other living organisms. Ecosystem services include vital processes such as purifying water, maintaining ideal atmospheric gas compositions, and pollinating crops. Knowledge gained from ecological studies also reveals how people can utilize the Earth's natural resources more wisely. For example, ecological research led to the discovery that phosphates from laundry detergents and nitrogen from chemical fertilizers cause photosynthetic organisms in lakes and ponds to overgrow, choking waterways and killing fish. The inorganic nutrients also lead to algal blooms in lakes. As the algae and photosynthetic bacteria die, detritivores (organisms that feed on dead and decomposing organic matter) thrive from consuming the organic remains and deplete the oxygen from the waters. This knowledge allowed society to take actions to prevent this negative outcome.

Ecology and evolution are closely related, as the conditions of the environment in which an organism resides select for adaptations that render the organism more suitable for living within that environment. For example, aquatic organisms that live in marine habitats have anatomical and physiological mechanisms that prevent them from losing fluids despite the high solute concentration of the oceans. Bony marine fishes take in water and salt ions through the mouth, excrete excess salt across their gills, and excrete salt ions and very little water from their kidneys. In con-

trast, freshwater fishes take in water and some salt ions through their mouth and gills, and they excrete large quantities of dilute urine from their kidneys to get rid of excess water. Over time, the different environments have favored selection of the anatomical and physiological adaptations that allowed the organisms to survive in either saltwater or freshwater. This is an example of an animal's interacting with an abiotic environmental factor. Ecologists have also discovered numerous examples of interactions between two or more species that have affected the evolutionary histories of all the involved species. Coevolution occurs when two or more interacting species develop reciprocal adaptations. For example, the pollination of plants provides an exemplary model of the mutual evolutionary influence of plants and pollinators. The flowers of yucca plants are a unique shape and require very tiny pollinators. Yucca moths can enter the flower and lay eggs in it, and then the larvae feed off the seeds in the ovary. The unique shape of the flower limits the evolutionary development of the moth, and because the plant clearly benefits from the moth's ability to enter, its shape is influenced by changes in the moth. Other insects or birds have tongues or beaks that have coevolved with the shape of a flower. The flower shape influences the length of the pollinator's tongue or beak, and the need for pollination by the insect or bird limits the amount of change the plant can undergo.

HISTORY OF ECOLOGY

Ecology as a field of life science initially developed as botanists noticed that the environmental conditions affected the distribution of plant species. The German naturalist Alexander von Humboldt was the first to look specifically for relationships between organisms and their environments. While hiking in Colombia in 1801, he noticed that the types of vegetation changed as the altitude increased. His research led to the emergence of the field of plant geography. In 1866 the German biologist Ernst Haeckel coined the term *ecology,* defined as the comprehensive science of the relationship of an organism to the environment. In 1877 the concept of biological or ecological communities emerged as biologists such as Charles Darwin, Alfred Russel Wallace, and Karl Möbius recognized that organisms depend on and interact with one another. Advancements in chemistry led to the exploration of biogeochemical cycles, the combination of biological, geological, and chemical processes that act to convert elements (e.g., carbon, nitrogen, oxygen, and phosphorus) between inorganic and organic and unavailable and available forms. The Austrian geologist Eduard Suess coined the term *biosphere* in 1875, and the Russian geologist Vladimir Vernadsky elaborated on this concept during the 1920s.

The Danish botanist Eugenius Warming authored the first ecology textbook, *Plantesamfund,* in 1895. This book was translated and published in English in 1909 as *The Oecology of Plants: An Introduction to the Study of Plant Communities.* Warming also taught the first university ecology course.

During the 20th century, ecology gave rise to the environmental and conservation movements that aim to protect nature and its life-forms. The American author Rachel Carson changed the way people viewed nature by demonstrating the potential negative impact that humans have on the ecology of planet Earth. Increased education of the public about human-induced destruction of ecosystems and loss of biodiversity has led to a greater push for ecological research and a better understanding of human ecology.

WHAT ECOLOGISTS DO AND WHY

Ecologists ask questions such as, Where do organisms live? What does the particular environment provide for the organism in terms of food, shelter, and climate? How does the organism affect its surroundings by living there? What unique adaptations does an organism have that allow it to survive and reproduce in the specific conditions of that habitat? To approach these sorts of questions, an ecologist must have a good understanding of many branches of life science, such as anatomy, physiology, evolution, and ethology, as ecology is a multidisciplinary field. Ecologists must also be aware of the general characteristics of different life-forms and appreciate their similarities and differences in order to predict possible interactions among species. A zoologist (one who studies animals), a botanist (one who studies plants), a microbiologist (one who studies microorganisms), or a biologist who specializes in any other life-form may all consider themselves to be ecologists.

An ecologist can specialize in a particular type of ecosystem. For example, a marine ecologist focuses on oceanic ecosystems, such as a kelp forest ecosystem. Tropical ecology, desert ecology, Arctic ecology, forest ecology, and stream ecology are other examples of subfields that focus on a particular type of biome or habitat. Another way to divide ecology into subfields is based on the type of organism studied. A microbial ecologist might research the interactions of different bacterial and protozoan species living in an animal's gut. An insect ecologist may investigate factors contributing to colony collapse disorder (CCD), commonly known as "the disappearing honeybees." A plant ecologist may examine the effect of the introduction of an invasive woody vine into a new geographic region on the native vegetation. Ecologists may also specialize in a particular

ecological issue such as maintaining or increasing biodiversity, examining the effects of or ways to prevent acid rain, developing sustainable ecological systems, or restoring ecosystems.

Ecologists can study interactions at numerous levels of biological organization from the individual level up to the biosphere. The subdiscipline concerned with the individual level and how organisms adapt to a particular environment is called behavioral ecology. A behavioral ecologist may examine how an alligator maintains its body temperature within a certain range despite external fluctuations. A population ecologist might explore the factors leading to the steady decline in the numbers of a species of fish. Or at a slightly higher organizational level, the researcher might investigate how the decreasing numbers of fish have affected the populations of their predators, how the numbers relate to the decline in coral reefs in the area, or how the biodiversity of the entire marine community is affected. At the ecosystem level, an ecologist might explore how an increase in atmospheric carbon dioxide levels has affected the flow of energy through trophic levels of the Amazon rain forest. Ecological researchers may also examine how the change in landscaping or geology of a region throughout history has affected the distribution of different life-forms in it, such as in the Arctic. Many ecologists are actively at work trying to figure out the effects of global climate change on the biosphere.

The career opportunities for a trained ecologist vary widely, depending on the specialty. Federal, state, and local governments hire ecologists as consultants for advice on how to reduce negative environmental impacts when planning new developments or industrial projects. Policymakers may consult ecologists for advice on improving conservation efforts or resolving environmental problems. Parks and private nature organizations hire ecologists as resource managers or as developers of educational programs for people of all ages. Academic research institutions hire ecologists to conduct field research or laboratory research and to teach ecology and advise college undergraduate and graduate students. To gain employment as an ecologist one must have a strong background in the life sciences, including zoology, botany, and microbiology, but also in physical and earth science. Chemistry is important for understanding geochemical cycling and how an organism's metabolism affects the surrounding conditions of the soil, the atmosphere, and the hydrosphere. Geology is also relevant since landscaping and geological features, such as altitude, streams or nearby water bodies, and frequency of events such as volcanic eruptions, all affect the type of organisms that live in a certain area. Statistics is useful for determining how closely a sample represents a population and for determining whether observed trends or patterns are significant. Competence in computer software is also necessary, as are written and oral communication skills for conveying results.

Understanding the relationships between different organisms and the environment is important because a change that might seem minor could have potentially drastic consequences. Ecosystems are composed of a community of interacting populations of living organisms and the physical environment of a certain area. The complex ecological interactions form a web of connections that represent beneficial and seemingly harmful exchanges that together maintain the health of an ecosystem. For example, in 2003 scientists reported a domino effect in North Pacific wildlife resulting from the mass slaughter of whales by commercial whaling. As whales disappeared, killer whales, who feed on other whales, started to feed on other prey. They first turned to harbor seals, then when the harbor seal numbers decreased, the killer whales fed off fur seals, then off sea lions and sea otters. The boom in commercial whaling occurred between 1949 and 1969, when more than 500,000 whales were killed in the Pacific. But the decline in other marine mammals has been more recent, since the turn of the millennium. Sea otters normally prey on sea urchins, keeping their numbers in check. The decrease in the sea otter population caused an explosion in the sea urchin populations. Sea urchins feed on kelp, and as their numbers increased, they destroyed the kelp forest ecosystem in southwestern Alaska. In summary, whaling that took place 50 years ago is responsible for the loss of the kelp forest. An interesting lesson from this incident is that food web interconnectivity is too complex to predict all the consequences of one action—the effects can reach farther and last longer than one might expect. Preserving ecosystems and protecting biodiversity benefit all life-forms on Earth, including people. Better working knowledge of the relationships among different species of organisms and with their environment will help people learn how to protect nature.

See also BIOGEOCHEMICAL CYCLES; BIOGEOGRAPHY; BIOMES, AQUATIC; BIOMES, TERRESTRIAL; BIOSPHERE; CARSON, RACHEL; COMMUNITY ECOLOGY; ECOSYSTEMS; ENVIRONMENTAL SCIENCE; HAECKEL, ERNST; POPULATION ECOLOGY.

FURTHER READING

Ecological Society of America home page. Available online. URL: http://www.esa.org/. Accessed January 25, 2008.

Molles, Manuel C. *Ecology: Concepts and Applications.* 4th ed. New York: McGraw Hill, 2008.

ecosystems An ecosystem consists of all the organisms within a biological community and the physical environment with which they interact. The size of an ecosystem may be as small as one rotting log on a forest floor or as large as the floor of the Atlantic Ocean. On a larger scale, ecosystems function to provide many services such as biogeochemical cycling of nutrients, moving water through the water cycle, maintaining atmospheric conditions, and more. Two critical phenomena that characterize ecosystems are energy flow and nutrient cycling.

PRIMARY PRODUCTION AND ENERGY FLOW

The ultimate source of energy for all ecosystems is the Sun. The energy then flows through the trophic levels of the ecosystem. Living systems must follow natural laws, including the law of conservation of energy, which states that energy cannot be created or destroyed. The energy does, however, become transformed as it moves through the ecosystem.

Photosynthetic organisms such as plants absorb the energy in the form of light, or radiant energy. They convert some of it into chemical energy, but energy transformation is an inefficient process and much of the energy is lost as heat. The primary producers, or autotrophs, of an ecosystem are the organisms that support all the other organisms in an ecosystem by providing energy in a usable form. They constitute in which the first level of an ecosystem's trophic structure: the processes energy and nutrients move through the organisms in the ecosystem. Examples of primary producers include plants in terrestrial ecosystems and phytoplankton in aquatic ecosystems. After using sunlight to generate energy-rich organic compounds, organisms belonging to other trophic levels obtain their energy and nutrients by ingesting the autotrophs, or products synthesized by the autotrophs. These other organisms are called heterotrophs because they cannot synthesize their own organic molecules from inorganic carbon sources. Organisms that consume autotrophs directly are called primary consumers, or herbivores. Secondary consumers eat the herbivores and are called carnivores. If they eat the primary producers (plants) in addition to herbivores (other animals), they are called omnivores. Tertiary consumers eat secondary consumers. Decomposers, or detritivores, ingest the decaying organic material, returning the inorganic components to the ecosystem.

As energy flows through a system of trophic levels, much is lost at each step. The gross primary production (GPP) of an ecosystem is the total primary production. The net primary production (NPP) is the GPP minus the amount of energy that the primary producers use as part of their own metabolism. NPP is therefore an indicator of how much energy is available to other organisms in an ecosystem. Two ways to express NPP are as new biomass, or dry weight of vegetation, added to the ecosystem per unit area per unit time, or as energy per area per unit time.

Different factors contribute to an ecosystem's primary productivity. Because sunlight provides the ultimate source of energy, its availability affects the productivity of an ecosystem. Geographical regions surrounding the equator are thus generally more productive than other terrestrial ecosystems. Tropical rain forests rank among the highest. Light cannot sufficiently penetrate deeper than several feet (about a meter) of water. Thus the NPP of oceans is rather low; however, they still contribute significantly to Earth's total productivity simply because they cover so much area.

The availability of nutrients also affects primary production. Even if energy is abundant, without nutrients, organisms cannot convert the radiant energy into chemical energy. Limiting nutrients are nutrients that, when added, allow for continued production. Carbon, oxygen, hydrogen, nitrogen, and phosphorus make up 93 to 97 percent of living biomass. Other essential nutrients include potassium, calcium, sulfur, magnesium, and sodium. Photosynthetic organisms and other autotrophs synthesize organic compounds, such as sugars and amino acids, from inorganic sources. Consumers in an ecosystem obtain their nutrients by ingesting the primary producers and other organisms.

When the organisms die, microorganisms decompose the organic matter. As do all living organisms, deritivores require not only the energy easily obtained from digesting the dead and decaying organic material, but also nutrients. Carbon is plentiful, but nitrogen is scarce in dead plant material, thus is limiting to detritivores. Phosphorus is also limiting in most terrestrial and wetland ecosystems; thus agriculture depends on the use of fertilizers containing the correct balance of both nitrogen and phosphates. During the natural process of decomposition, many nutrients return to the environment, where primary producers can once again fix them into organic compounds. Biochemical processes in living organisms also return nutrients to the environment. For example, carbon dioxide is released as a product of cellular respiration, in which organic molecules such as carbohydrates are broken down to release the energy for the organism to use in order to carry out other life processes. Geological processes also act to cycle nutrients through ecosystems. Erosion and weathering of rocks release phosphorus-rich minerals into the soil, where acids can dissolve them, and make the phosphorus available for plants to uptake them in the form of phosphates for incorporation into biomolecules such as nucleic acids. Upwelling carries nutrient-rich waters to the ocean surface, where primary producers can incorporate them in organic material. Thus, biological, chemical, physical, and geological processes all play a role in cycling nutrients through an ecosystem.

Other physical factors that influence the efficiency of primary production in terrestrial ecosystems are temperature and moisture. Warm and moist environments, such as tropical forests, are more productive than drier and colder ecosystems, such as the arctic tundra or vast deserts.

ECOLOGICAL PYRAMIDS

Primary production is the amount of light energy converted into chemical energy by autotrophs in an ecosystem. Secondary production is the amount of chemical energy in a consumer's food that its metabolism converts into new biomass. As primary production can, secondary production can be examined within a defined geographical area and within a given period. As the energy moves from a lower to a higher trophic level, the majority is lost. Between one-fifth and one-sixth of the energy at one level can be transferred to the next. The rest of the biomass or energy in an ecosystem is either not consumed by organisms of the next trophic level, it is not digested, or it is released as heat, which is unusable by other consumers.

Production efficiency measures the amount of energy stored in food that is not used in cellular respiration. Some is used in growth and reproduction to perpetuate biomass, which may be consumed later by another organism. Production efficiency equals the ratio of the net secondary production to the total amount of energy used in growth, reproduction, and cellular respiration, in other words, the assimilation of primary production.

$$\frac{\text{production}}{\text{efficiency}} = \frac{\text{net secondary production}}{\text{assimilation of primary production}}$$

Of all animals, endotherms, which maintain their body temperature through metabolism, have the lowest production efficiencies (between 1 and 3 percent), as they purposefully burn energy just to keep warm. Insects have a relatively high production efficiency, averaging about 40 percent.

In the early 1940s Raymond Lindeman first proposed the existence of trophic levels as groups of organisms that share a similar role in energy transfer within an ecosystem. He emphasized the complex relationships between organisms in a community and their environment and proposed the concepts of primary producer, primary consumer, secondary consumer, and so on. During his studies on energy flow through ecosystems, he recognized that as energy traveled between trophic levels, much was lost, result-

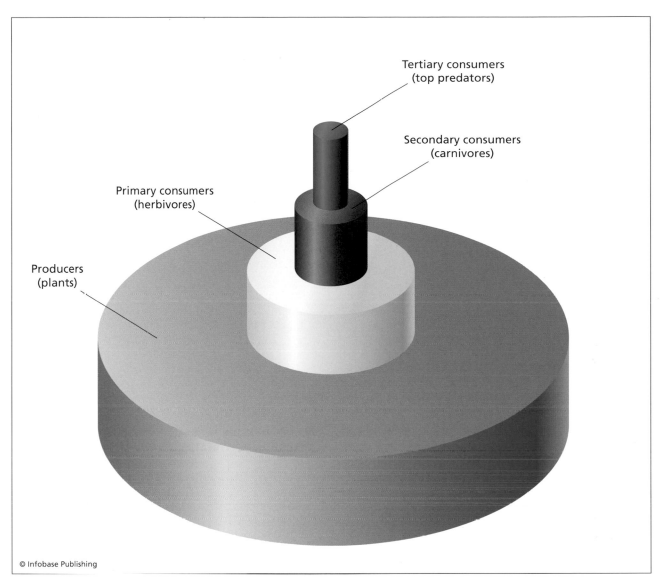

Ecological pyramids illustrate the concept of progressive loss as one moves up through the trophic levels of an ecosystem. The phenomenon of progressive loss holds true for net production, biomass, available energy, and numbers of individuals.

ing in a pyramid-shaped distribution of energy among trophic levels. Another ecologist, Charles Elton, had previously proposed the pyramid concept, but Lindeman substantiated Elton's claim with data from two lake systems in Wisconsin. As a result of Lindeman's work, ecologists began to study energy patterns and influences on energy flow through ecosystems.

Studies have shown that trophic efficiencies range from 5 to 20 percent, because of energy losses through respiration, energy excreted as unused biomass in feces, and the fact that consumers do not intake all the biomass from a lower trophic level. To illustrate, consider a representative ecosystem in which primary producers only capture 1 percent of the solar energy input. The primary consumers incor-

porate approximately 4 to 5 percent of the energy they obtain from the primary producers in their own biomass, making it available for organisms at the next trophic level, the secondary consumers. If only 10 percent of that is made available to the next level, pretty soon, very little energy is available for the next level. The major losses that occur as energy moves between trophic levels limits the total number of trophic levels that most ecosystems possess to four or five.

The loss of energy available to consumers at successive trophic levels can be represented by a pyramid of net production. A decrease in the amount of biomass at each trophic level accompanies the loss of available energy. An exception to this generalization is found in some aquatic ecosystems in which the turnover of phytoplankton is so high that the amount of biomass can never accumulate sufficiently. Predators, tertiary consumers such as hawks or sharks, typically have smaller populations, reflected in a typical pyramid of numbers, because only a fraction of 1 percent of energy flows to this trophic level in an ecosystem.

One may wonder why so much of the biomass and chemical energy made available by primary producers is unused. Scientists have proposed several hypotheses to explain this: 1) Plants have defense mechanisms such as poisonous chemicals to deter herbivores. 2) Nutrients play a larger role in limiting herbivore populations than does energy input. 3) Physical and geological factors help control herbivore populations. 4) Competition keeps the levels of herbivores low. 5) Community dynamics such as predation and parasitic diseases help maintain populations.

See also BIOGEOCHEMICAL CYCLES; COMMUNITY ECOLOGY; POPULATION ECOLOGY.

FURTHER READING

Molles, Manuel C., Jr. *Ecology: Concepts and Applications.* 4th ed. New York: McGraw Hill, 2008.

electrophoresis Electrophoresis is a technique that separates biomolecules according to properties such as size, shape, or charge. Life scientists use electrophoresis both to analyze samples and to purify biomolecules such as proteins and nucleic acids. Samples are placed in a semisolid medium called a gel that is saturated with an ionic buffer solution. An electric current drives the molecules through the medium, which acts as a porous sieve, toward the anode at different rates dependent on size and shape. After running the gel, different mechanisms including staining or radiography allow visualization of the biomolecules. Their position in the gel reveals information about the characteristics of the molecules. If the goal of the electrophoresis is to prepare a substance or reagent for further use, the desired molecules can be extracted from small pieces cut from the gel and subject to purification.

DNA ELECTROPHORESIS

Deoxyribonucleic acid (DNA) or ribonucleic acid (RNA) fragments of different sizes can be separated using either agarose gel electrophoresis or polyacrylamide gel electrophoresis, depending on the approximate molecular weight of the fragment(s) of interest. Agarose is a neutral polysaccharide obtained from the cell walls of algae that is used in a powdered form. When boiled, agarose dissolves in water, and, when cooled, it solidifies into a semisolid gelatinous matrix. Higher concentrations of agarose give firmer gels and are better for separating fragments as small as 100 nucleotide base pairs long, compared to several thousand base pairs for gels made from lower concentrations of agarose. The molten agarose is poured into a rectangular plastic tray with combs positioned to leave indentations called wells at one end of the gel after the gel cools and the comb is removed. Gel trays have a range of sizes and are typically about a few inches wide and slightly longer. Before or sometimes after loading the DNA samples, the agarose gel is submerged in a running buffer that contains a chemical buffer to maintain an optimal pH. DNA electrophoresis requires slightly basic conditions to keep the DNA deprotonated, in other words, to maintain its negative charge.

The wells hold only a small amount of sample; typical samples measure between 10 and 25 microliters (one microliter = 10^{-6} liter). Before placing the DNA samples in the wells, loading dye is added. The loading dye usually contains a dense but water soluble component such as glycerol or sucrose to encourage the sample to sink to the bottom of the well. Otherwise, the loaded sample may float freely and mix with the running buffer. The loading dye also contains tracking dyes, such as bromophenol blue and xylene cyanol, that travel at rates comparable to DNA fragments of a specific size; thus one can follow the DNA as it runs through the gel by watching the dye fragments separate.

After the samples are in the wells, a power source supplies the voltage, establishing the flow of an electric current in the tank containing the gel and running buffer. Because the phosphates in the DNA backbone confer a negative charge on it, the DNA will migrate from the wells located near the cathode end toward the positive electrode, the anode. Within a few minutes the dye will appear slightly in front of the wells, indicating movement of the samples into the gel matrix. The complete run is usually shorter than a few hours, depending on the gel size, concentration, and the molecular weights of the fragments being separated. Smaller fragments will travel through a gel at a faster rate than larger fragments. This is because the larger fragments experience more friction as they migrate through the agarose matrix.

Large DNA molecules, those that are between hundreds of thousands and millions of base pairs long, cannot be resolved using a constant current through the gel. Instead, researchers use a technique called pulsed-field gel electrophoresis (PFGE), in which relatively long pulses of current in the forward direction are interrupted by shorter pulses in the opposite direction. This tightens the bands so they can be viewed as separate entities on the gel rather than appear as smeared streaks.

In order to visualize the DNA, the gel must be stained with ethidium bromide, a chemical that intercalates or inserts itself between the nitrogenous bases of the nucleotides. Exposure to ultraviolet light causes the ethidium bromide to fluoresce a bright orange color. The positions of the DNA fragments are typically recorded by taking a photograph or digital image of the illuminated gel. Running a separate lane with a molecular weight marker that contains fragments of known length parallel with the sample fragments allows one to estimate the size of any fragments present in the sample lanes if they fall within the range of the molecular weight standards. By plotting a curve of the distance migrated versus the log of the molecular weight of the standards, one can read the approximate weight from the curve for any band by measuring the distance it traveled.

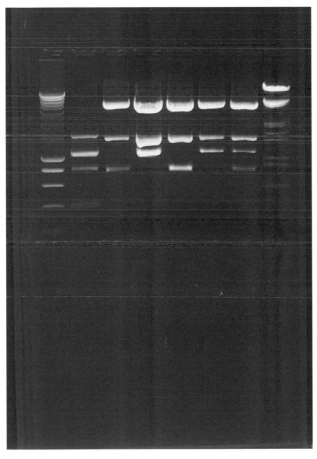

DNA fragments fluoresce when stained with ethidium bromide and viewed under ultraviolet light. After electrophoresis, the larger fragments remain closer to the top, while the smaller fragments travel faster and are farther away from the wells. *(Pascal Goetgheluck/ Photo Researchers, Inc.)*

If the purpose of the gel was preparative, the regions of the gel containing the DNA of interest can be excised. Special techniques allow for the extraction and purification of DNA from the fragment. Biotechnology companies sell kits that facilitate this process. Alternatively, one can use agarose that melts at a low temperature, extract the DNA from the molten solution using organic chemicals, and precipitate the DNA from the aqueous solution using salts and alcohol.

The DNA can also be transferred from the gel to a nitrocellulose or nylon membrane in a procedure called Southern blotting that permits the identification of specific DNA fragments. First the DNA is denatured by treatment with a basic solution, so it is single-stranded. During transfer of the DNA from the gel to the membrane, the relative positions of the DNA fragments are preserved. The membrane can then be treated with different radioactive probes, pieces of DNA that will hybridize by forming hydrogen bonds with complementary base pairs if they are present on the membrane. Autoradiography with X-ray film or phosphorimaging reveals the positions of the bands that contain DNA that hybridized with the probe. This characteristic allows a researcher to look for specific DNA sequences within a fragment or sample of DNA.

Agarose gels are quick, cheap, and easy to prepare and run, but the best resolution is limited to fragments about 100 base pairs or longer. Smaller pieces of DNA appear smeared and fuzzy on agarose gels. Polyacrylamide gel electrophoresis (PAGE) is more useful for separating fragments that are as small as a few nucleotides long. Polyacrylamide is a polymer of acrylamide monomers (amides of acrylic acid). Preparation for PAGE is more complicated, as the acrylamide monomers are neurotoxic, and the apparatus is vertical rather than horizontal as in agarose gel electrophoresis. Dyes are still added to the DNA samples before running the gel so the researcher has an idea of how far certain sizes of fragments have run over a given period. Most often, PAGE results are viewed by autoradiography. If the DNA has been radiolabeled, then laying a piece of X-ray film over the gel will expose regions of the film that correspond to the position of the DNA fragments on the gel. Unless the purpose of the PAGE is preparative, the gel is usually dried before exposing it to X-ray film. After the exposed film is developed, dark bands will appear, indicating the location of the DNA fragments on the gel.

PROTEIN ELECTROPHORESIS

Polyacrylamide gel electrophoresis is also useful for analyzing protein samples. As does DNA, most proteins carry a negative charge in a slightly basic

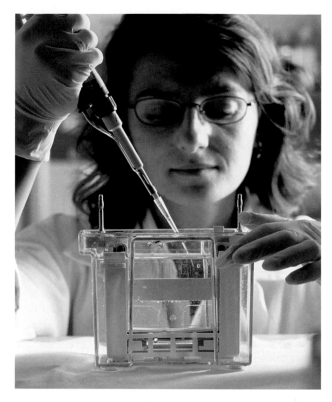

Polyacrylamide gel electrophoresis utilizes vertical apparatuses, and buffer tanks have contact only with the top and the bottom portion of the gel. *(Mauro Fermariello/Photo Researchers, Inc.)*

solution; thus they will migrate through a gel with speed characteristic of their size. Because proteins have complex three-dimensional folded structures held together by hydrophobic interactions, ionic interactions, and hydrogen bonds, it is necessary to add a mild anionic detergent (usually sodium dodecyl sulfate, or simply SDS) to denature the proteins by interfering with those interactions. The SDS also coats the entire polypeptide with negative charges, masking any other charges. Often, a reducing agent is added to disrupt covalent disulfide linkages of different polypeptides of a multisubunit protein. These actions ensure that the rate at which a polypeptide moves through a gel depends only on its size and not its shape or intrinsic charge. In a discontinuous system, a short stacking gel with a very large pore size is layered on top of the resolving gel and contains a different buffer than the resolving gel. This improves the final resolution by compressing the sample before it enters the resolving gel, which then separates the polypeptides on the basis of size. After running a protein gel, staining with a dye such as Coomassie blue or with silver colors the polypeptide bands for direct visualization.

When the mixture of polypeptides is complex, adding a second dimension to the separation proce-

dure provides more resolution. After the sample has been separated by electrophoresis once, the samples are subject to another run at a 90 degree angle to the first, either through a different concentration of polyacrylamide or by use of a buffering system with a different pH. Nondenaturing gels must be used for this, as SDS will mask the native charges, but this means that individual polypeptides cannot be analyzed. In a slightly more complicated version of two-dimensional electrophoresis, the first step separates the molecules according to their native charge. The gel is cast in a tube that has a pH gradient. Application of an electrical field moves the molecules toward the anode until they reach a point in the gradient at which their pH is neutral, then they stop. Their final position along the gradient depends on the number of acidic and basic side chains present in the polypeptide. After this first step, which is called isoelectric focusing, the tube gel is laid on top of a typical slab gel and subjected to typical SDS-PAGE, which then separates the proteins by size.

As DNA can, proteins can be transferred from a gel onto a membrane. In western blot analysis, the proteins are transferred to a nitrocellulose membrane. After being treated with special buffers, the membrane is placed beside the gel, and an electric current forces the polypeptides off the gel onto the membrane.

After the proteins are immobilized onto a membrane, the membrane can be probed with antibodies, proteins naturally made by the immune system

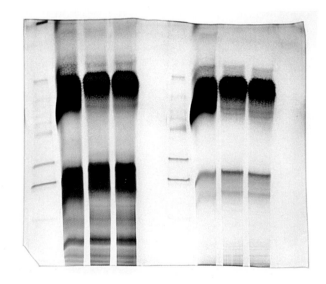

SDS-PAGE separates polypeptides on the basis of their size, with the smaller molecules traveling faster, thus appearing toward the bottom of the gel, while the larger molecules migrate more slowly and stay near the top of the gel. *(R. A. Longuehaye/Photo Researchers, Inc.)*

that specifically recognize and bind to other protein molecules. If the antibodies are radiolabeled, the pattern of the proteins of interest can be visualized by exposure to X-ray film. Otherwise the antibodies can be treated with a series of chemical reagents and fluorescent dyes designed to allow visualization without the use of radioactivity.

One method for detecting protein-DNA interactions is the gel mobility shift assay (GMSA) or the band shift assay. With this technique, one can examine the ability of proteins to bind to segments of DNA containing specific sequences. If a segment of DNA is bound to a protein, its movement through a gel will be much slower than movement by the same segment in the absence of protein. The result is a shift in the position of the band when it is bound to protein; the band will remain closer to the wells of the gel.

See also BIOMOLECULES; RECOMBINANT DNA TECHNOLOGY.

FURTHER READING

Westermeier, Reiner. *Electrophoresis in Practice: A Guide to Methods and Applications of DNA and Protein Separations.* Weinheim, Germany: Wiley-VCH, 2005.

embryology and early animal development

Embryology is the study of embryos and their development. The term *embryo* once encompassed the stages of vertebrate development from the time of conception until birth or hatching, but now it refers only to the earliest stages of animal development. After fertilization, a zygote undergoes a series of rapid divisions, produces the foundation of fundamental tissues, and initiates the formation of organs and organ systems. After the basic body structure is established, the embryo is considered a fetus. In humans, the transition from an embryo to a fetus occurs approximately two months into gestation. The processes involved in early animal development are similar in all species, beginning with cleavage and growth and resulting in the formation of differentiated tissues and organs that perform tasks that contribute to whole-body function.

FERTILIZATION AND CLEAVAGE

Fertilization occurs when two haploid gametes unite to restore the diploid condition and initiate the development of a new individual. Whether this takes place externally (as in marine invertebrates) or internally (most terrestrial animals including mammals), the subsequent events are similar. When a sperm cell from an adult male has contact with an egg cell from an adult female, a sac located at the tip of the sperm head releases digestive enzymes that aid in penetrating the zona pellucida, a matrix surrounding the egg. The plasma membranes of the sperm cell and the egg cell fuse, and all of the sperm except the tail enters into the cytoplasm of the egg. Though many sperm can pass through the corona radiata, the layer external to the zona pellucida that contains cells that nourish the egg, and several sperm can attempt to penetrate the zona pellucida, biochemical mechanisms prevent more than one sperm from passing through. The nucleus of the sperm fuses with the nucleus of the egg, the chromosomes duplicate, and the cell prepares for the first cell division. One variation between species is the stage of meiosis that the egg has reached prior to fertilization. If meiosis is not yet complete, as in humans, fertilization stimulates the completion of meiotic division.

Embryonic development can be divided into three stages: cleavage, gastrulation, and organogenesis. Fertilization initiates the process of cleavage, repeated mitotic divisions in which the cytoplasmic contents of the egg divide into numerous progressively smaller cells called blastomeres, all containing a genetically identical nucleus with one set of chromosomes from each parent. The zygote first divides into two cells, then four cells, then eight, and so on. This can occur very rapidly; for example, a frog zygote produces 37,000 cells in 43 hours, and a fruit fly zygote produces 50,000 cells in 12 hours. The ball of cells resulting from the first five to seven divisions is called a morula. A blastocoel, an intercellular cavity filled

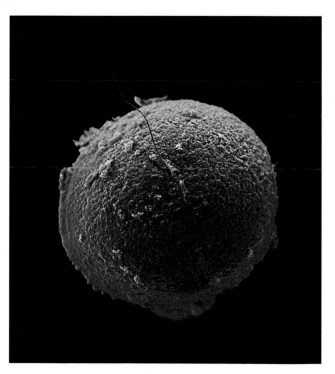

Fertilization occurs when an egg and a sperm unite and their nuclei fuse. *(Eye of Science/Photo Researchers, Inc.)*

with fluid, begins to form, transforming the morula into a blastula, a hollow ball of cells. The process of blastulation usually occurs simultaneously with cleavage, but some embryologists technically consider it part of gastrulation.

The eggs of many animals, not including mammals, show polarity either before or immediately after fertilization. Having polarity means a decided difference exists between opposite ends of the cell or the embryo; for example, the cytoplasmic contents might be distributed unequally. The polarity influences cleavage patterns and persists into the morula and further developmental stages. The sizes of the cells differ, and in different animals the position of the blastocoel also differs.

Some vertebrates and insects have a pronounced yolk, a mass of stored food within the cytoplasm of an egg, positioned at one pole of the egg. The yolk can be larger than the rest of the cytoplasm and displaces other cytoplasmic contents, leading to unequal cell divisions during cleavage, different-sized cells within the morula, and off-center positioning of the blastocoel. At the end of cleavage, the bastula consists of blastomeres with genetically identical nuclei but different cytoplasmic materials. These subtle differences help direct the next stage of embryonic development, gastrulation.

GASTRULATION

Gastrulation involves the movement and rearrangement of the cells in the blastula; different portions of the blastula have different developmental fates. When complete, the result is a cup-shaped embryo with three distinct layers and a primitive gut. The process differs slightly among animal types but generally involves the same mechanisms and results in the foundation of the basic body plan for that species. The three layers are called germ layers and include ectoderm, mesoderm, and endoderm. The ectoderm is the outermost layer, the endoderm lines the digestive tract, and the mesoderm is between the other two. These germ layers develop into the various tissues and organs of the mature animal.

Different mechanisms of cell movement participate in the formation of the primitive gut in different species. In animals such as sea urchins, invagination, the infolding or the simple pushing in of some cells from the surface of the blastula, forms a blind pouch called the archenteron, the primitive gut consisting of endoderm cells. The blastopore, the opening of the archenteron, will become the anus (as occurs in all deuterostomes). Migration of cells called mesenchyme cells accompanies invagination. The mesenchyme cells that will become mesoderm position themselves between the developing archenteron and the outer layer of cells, the ectoderm, of the gastrula. The archenteron eventually reaches ectoderm across

the blastocoel and fuses with it, forming the second opening of a digestive tube. The gastrula becomes a ciliated larval form that ultimately develops into an adult sea urchin.

In other animals, such as frogs, a dorsal lip forms where invagination occurs, then continued invagination extends all the way around the blastula, following the inner surface of the ectoderm until the blastopore forms a complete circle. Accompanying invagination is involution, whereby cells roll inward from the surface of the blastula to form endoderm and mesoderm. When gastrulation is complete, the lip of the blastopore surrounds a structure called the yolk plug.

In chicks, blastulation results in a blastoderm rather than a blastula. The blastoderm consists of two layers, the epiblast and hypoblast, covering a large yolk mass. Only the epiblast contributes cells to the developing embryo. Cells move from the epiblast to the middle of the blastoderm, forming a primitive streak that defines the anterior-posterior axis. Some of the migrating cells form endoderm; others move into the blastocoel, located between the epiblast and the hypoblast, and become mesoderm. The remaining epiblast cells form ectoderm.

ORGANOGENESIS

By the end of gastrulation, cells have become associated with one of the three primary germ layers, and their position within the layers determines their fate. During organogenesis, cells of the germ layers undergo numerous morphological and functional changes to develop into various body tissues and organs. Induction directs many of these changes. The process of induction involves chemical agents secreted by some cells that act on their neighboring cells; thus the position of a cell affects the chemical signals it encounters. The chemicals usually alter the expression of specific genes, resulting in the formation of specific tissues, which then organize and cooperate to form organs and organ systems.

In chordates, the first structures to form are the neural tube and the notochord. Dorsal mesoderm condenses to form a notochord, which then signals the ectoderm to form the neural plate. The neural plate rolls up into a tube, the neural tube, which develops into the central nervous system. In vertebrate embryos, cells released from a band of cells called the neural crest travel to different locations and form peripheral nerves, skull bones, and teeth. Dense clusters of mesodermal cells called somites positioned along the notochord give rise to a variety of structures including the ribs, the muscles of the rib cage and back, and the dermis of the dorsal skin. Throughout organogenesis, the structuring of organs and the shaping of the body continue, as does the differentiation of the tissues.

Among vertebrates, the three primary germ layers develop into the same body parts. In general, the ectoderm gives rise to the epidermis, sweat glands, hair follicles, and epithelial lining of the mouth and nasal passages; sense receptors in the epidermis, cornea, and lens of the eye; and the nervous system,

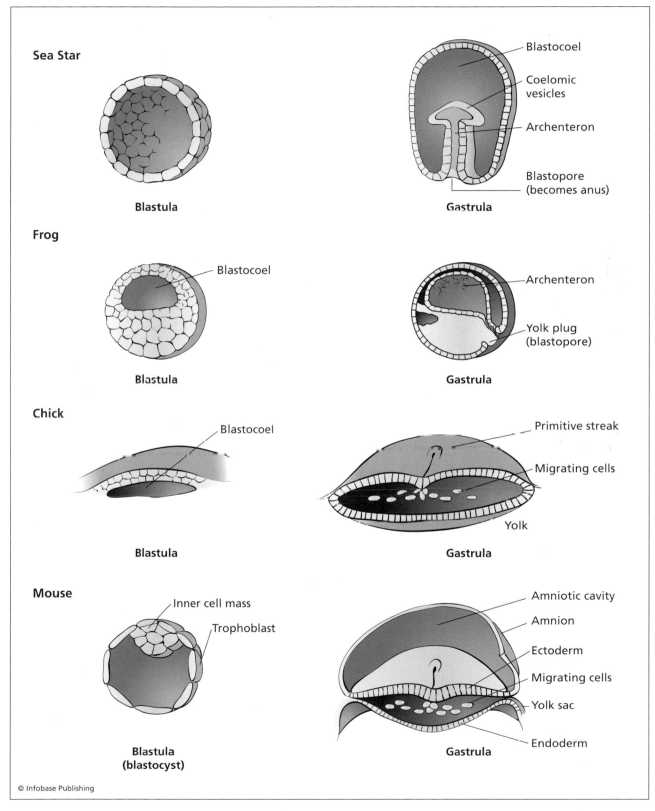

In most animals, gastrulation involves an extensive rearrangement of the cells, converting a roughly spherical blastula into a complex arrangement consisting of three germ layers.

adrenal medulla, tooth enamel, and pituitary glands. The mesoderm develops into the notochord, skeletal system, muscular system, muscular layers of the stomach and intestine, excretory system, circulatory system, lymphatic system, reproductive system, dermis of the skin, lining of the body cavity, and adrenal cortex. The endoderm gives rise to the epithelial linings of the digestive tract and the respiratory system, the linings of the urethra and bladder, the lining of the reproductive system, the liver, pancreas, thymus, thyroid gland, and parathyroid gland. During early cleavage, cells are totipotent, meaning they have the ability to develop into any cell type. The stage at which cells develop unequal potentials varies among species. For example, in amphibians, the developmental potential is restricted after the first cleavage, at the two-cell stage. In contrast, in mammals, totipotency persists until the 16-cell stage. This means that at the eight-cell stage, the blastomeres can be separated and potentially develop into eight equal individuals. After tissue type is determined, cells differentiate further, until their functional potential becomes restricted to one specialized type. In other words, a liver cell cannot change into a thyroid cell.

AMNIOTES

The early development of vertebrates demands an aqueous environment. Animals that live in aquatic environments can lay their eggs directly into their surroundings. Terrestrial vertebrates including reptiles, birds, and mammals require unique adaptations to overcome this obstacle. A fluid-filled sac surrounds the terrestrial vertebrate embryo as it develops either within a shell or in the uterus. Because the membrane that encloses the fluid-filled sac is called the amnion, these animals are called amniotes. Birds and other reptiles have four extraembryonic membranes protecting the embryo inside a shell: the amnion, the yolk sac, the allantois, and the chorion. The amnion encloses the embryo, which is suspended with a cavity filled with fluid that cushions against harsh movements and protects against dehydration. The yolk sac stores the nutrients from the egg. The allantois accepts metabolic waste discharged from the embryo and functions in cooperation with the chorion as a respiratory organ. Gases diffuse from the external environment, through the shell, through the chorion, and between the embryo and allantois.

Mammals are unique because their eggs are much smaller than those of reptiles and birds; they store only minimal nutrients, just enough until the embryo develops a functional placenta, an organ through which nutrients, gases, and metabolic wastes are exchanged between the embryo and the pregnant mother. In humans, cleavage results in a blastocyst, homologous to a blastula, which consists of more than 100 cells seven days after fertilization. An inner cell mass that forms on one end of the blastocyst develops into the embryo, whereas the trophoblast, the cells forming the outer layer of the blastocyst, is responsible for implantation and invasion of the uterus. As implantation ends, gastrulation begins, and the formation of the three primary germ layers and the four extraembryonic membranes is initiated. As in other vertebrates, the chorion functions in gas exchange, and the amnion encloses a fluid-filled cavity. The yolk sac contains no yolk in mammals but acts as the site for blood cell formation. The allantois becomes part of the umbilical cord that transports blood to the placenta, where embryonic (and later, fetal) blood and maternal blood can exchange nutrients and wastes.

See also ANIMAL FORM; CELL COMMUNICATION; CELLULAR REPRODUCTION; HUMAN REPRODUCTION; REPRODUCTION.

FURTHER READING

Bonner, John Tyler. *First Signals: The Evolution of Multicellular Development.* Princeton, N.J.: Princeton University Press, 2000.

Gilbert, Scott F. *Developmental Biology.* 8th ed. Sunderland, Mass.: Sinauer Associates, 2006.

Silverthorn, Dee Unglaub. *Human Physiology: An Integrated Approach.* 4th ed. San Francisco: Benjamin Cummings, 2006.

Wilt, Fred H., and Sarah Hake. *Principles of Developmental Biology.* New York: W. W. Norton, 2003.

endangered species Endangered species are populations of organisms that are at risk of becoming extinct throughout all or a significant portion of their range. Threatened species are those that are likely to become endangered in the near future. Many nations have laws to protect endangered and threatened species and to restore their population size. In the United States, the Endangered Species Act (ESA) of 1973 formally defined the terms *endangered* and *threatened,* combined native and foreign species lists, required federal agencies to implement conservation programs, and outlined procedures for the goal of protecting species that fell into these categories. Several amendments to this act (in 1978, 1982, and 1988) have strengthened it to fulfill its main purpose—to protect and prevent extinction of imperiled species by providing programs to conserve them and the ecosystems that support them. The Interior Department's Fish and Wildlife Service (FWS) and the Commerce Department's National Oceanic and Atmospheric Administration (NOAA) Fisheries administer the law, with FWS managing land and freshwater species, and the NOAA Fisheries managing marine and anadromous species (those that live in the ocean but spawn in freshwater). According to the FWS, as of January 2008, the approximate totals of U.S. and foreign listed endangered and threatened species includes 1,175 animal and 747 plant populations, though some species are counted more than once when more than one distinct population is endangered.

ESA is a U.S. law; thus other nations do not need to follow its regulations. Every country is responsible for enacting its own legislation concerning the protection of endangered and threatened species, killing them, and selling their parts. Protecting endangered species and their habitats is an issue that affects the entire planet, however; thus in order to have a positive impact conservation efforts require global cooperation. In the 1960s growing concern led to worldwide discussions that resulted in the Convention on International Trade in Endangered Species of Fauna and Flora (CITES), an international agreement that entered into force in 1975. CITES regulates the international trade of living or dead endangered and threatened animals and plants and their parts. While the enforcement of such trade restrictions is difficult, the cooperative attitude demonstrated by the 173 participating countries (as of August 2008) is a step in the right direction.

CAUSES OF EXTINCTION

Natural factors that lead to extinction include disease epidemics, lack of genetic variation in a population faced with new environmental challenges, and limited distribution of a species to one or a few geographical areas. Human-induced threats to endangered species include the destruction of the habitats essential to the species' survival, the overexploitation of wildlife for commercial purposes, and the introduction of nonnative species to an environment. Mounting evidence also shows that human activities are affecting the global climate. Estimates of the number of species that could become extinct as a result of global warming by the year 2050 approach 1.25 million.

The destruction of a species's habitat is the leading cause of extinction. A habitat provides the resources for nourishment and energy as well as the optimal physical conditions for an organism to grow and reproduce. Without someplace to live that offers the same, a species will die out. Habitat loss can occur as a result of pollution (such as from chemicals leaching into the soil), a disruption to the balance of an ecosystem that changes the physical environment of the habitat (such as damming a river), or straightforward elimination of the habitat (such as razing for construction and urban development). One serious issue is deforestation, the clearing

of forests, which is occurring at a rate of millions of acres per year. Tropical rain forests contain an estimated half of all extant species, but these species are rapidly losing their habitats through clearing for commercial logging or for agricultural purposes, such as for cattle pastures or planting crops. Draining wetlands, areas where water covers the soil, for development purposes or even flooding them for recreational uses is another example of a human activity that contributes to habitat loss. Wetlands are among the most productive ecosystems, serving as home for a variety of microbial, plant, and animal species that perform important ecosystem services, such as cycling nutrients and capturing and storing carbon, an act that moderates climate. An estimated one-third of endangered species live in wetlands, and many other animals depend on wetlands for at least part of their life cycle or for food. Wetlands are also crucial in watersheds, areas that drain to a common waterway, and in controlling erosion of shorelines. Global warming has also caused major changes to habitats, affecting the species that live there. Glaciers are melting and causing sea levels to rise, tundras are shrinking, deserts are growing, and forests and grasslands are gradually moving toward more appropriate climates. Small geographical regions that serve as home to species with precise temperature and humidity requirements are disappearing, as are coral reef systems and tropical cloud forests.

The ESA protects not only organisms at risk for becoming extinct, but also the habitats that support their survival. Legislation regarding the use of protected areas that are inhabited by endangered species and setting aside land and water bodies as nature reserves is one means of preventing further human-induced habitat destruction. The establishment of wildlife corridors, protected patches that physically connect different habitats, also helps by allowing for wildlife movement between habitats so animals can interbreed and maintain genetic variation, helping to maintain functional ecosystems. Actions aimed at reducing human activities associated with increased global warming, such as the burning of fossil fuels, are also an important factor in protecting many threatened and endangered species in the coming decades.

According to the FWS, the introduction of nonnative species to a new environment is the second greatest threat to native species. This may occur either intentionally or unintentionally. Humans first took domesticated cats to Australia in the 17th century. Because the continent's ecosystem had not evolved mechanisms to limit the population, cats rapidly multiplied and now the carnivorous hunters are blamed for the extinction and endangerment of several native small mammals and birds. In another example, ships from Europe unintentionally carried zebra mussels in their ballast water to the Great Lakes region of North America, where millions of dollars has since been spent to control their population and prevent the decline in native mussel populations. Nonnative species (also called exotic, invasive, or nonindigenous species) may outcompete native species for space or nutrients, affect interspecies dependencies, or change the physical conditions of the ecosystem through their metabolism or behavior. For example, zebra mussels compete with zooplankton for food, disrupting the natural food webs. As filter feeders, the zebra mussels clarify the water, creating advantageous conditions for some photosynthetic organisms, while harming fish such as walleye that prefer turbid water. The large populations of the zebra mussels settle on, suffocate, and starve the native mussel populations.

The overexploitation of animals and plants for commercial purposes also leads to threatening or endangering species. When individuals of a species disappear at a faster rate than they reproduce, the species is at risk of becoming extinct. For example, in the 20th century, whaling was not regulated and many species dropped to endangered status as whalers killed them for their oil and meat. After recognizing this, several nations voluntarily entered into a moratorium on whaling. Some species, such as the gray whale, have recovered, while others remain endangered or threatened as a result. Rhinoceroses are another animal facing possible extinction, not due to habitat loss, but due to poaching. Their horns are used to construct dagger handles that symbolize wealth and status in some Middle Eastern countries, making them profitable enough to worth risking the legal ramifications of selling them on the black market. Plants are often overexploited for their medicinal properties. The first sources of medicines were plants, and today hundreds or even thousands of plants are harvested for their natural products that have therapeutic and medicinal value. Actions to prevent overexploitation of species indigenous to other countries include trade regulations preventing the sale or import of products from endangered species. The perennial herb ginseng is one example of a plant regulated by CITES. Though ginseng is not currently listed as endangered, CITES regulates its trade in order to prevent the overexploitation of this popular herb. Its root has been used for centuries by practitioners of Eastern medicine as a universal medication for fatigue, attrition due to illness, and physical effects of stress. During the past 50 years, Western medicine's use of ginseng root as a mild stimulant and for the relief of indigestion has grown. Its medicinal effectiveness has been disputed, but its popularity led to a demand high enough to warrant oversight by CITES.

WHY SAVE ENDANGERED SPECIES?

Since life first appeared, new species have emerged and others have become extinct as a result of natural geological and biological changes. The five previous mass extinctions on Earth resulted from natural cataclysmic physical events, such as global climate changes, massive lava floods, and dramatic changes in sea level due to glacier formation and melting. Biologists believe the current rate of extinction, which is far higher than any previous mass extinction rates, is the result of activities by the expanding human population. For many, ethical reasons alone are sufficient to justify protection of endangered species. Given that most species evolved after millions of years of natural processes, many question the rights of humans to eliminate them. From another perspective, to what degree is society responsible for protecting and restoring endangered populations? In the text of the ESA, the U.S. Congress cited numerous reasons why people should attempt to restore populations of threatened and endangered species, stating that wildlife and plants are of aesthetic, ecological, educational, historical, recreational, and scientific value.

One scientific reason to protect endangered species is to preserve ecosystems. In addition to the immediate resources that an ecosystem provides, such as timber for construction or fish for food, ecosystems provide services that are not marketed or sold, but whose loss has devastating consequences. Some estimates claim ecosystems provide trillions of dollars worth of services each year. Ecosystems cycle nutrients and water, remove toxins from the soil, filter or purify water, decompose wastes, regulate climate, keep soil fertile, and maintain biodiversity, just to name a few. These fundamental ecological processes support all life-forms on the planet. Because all of the community members in an ecosystem fill unique niches, each species performs an important function. Significant alterations in the size of a particular population can affect the function of the whole ecosystem. This is particularly true for keystone species, species of plants or animals who have a major impact on the structure and therefore the function of an ecosystem. The story of the gray wolf (*Canis lupus*) of Yellowstone National Park demonstrates the far-reaching effects that the decline of one species can have on an entire ecosystem. Believing the gray wolf to be a nuisance that killed livestock and pets and ruined crops, people who lived nearby had hunted and killed off the population by 1926. In 1995 the FWS captured and reintroduced a small population of 14 gray wolves from Canada to the park ecosystem, and 17 more the following year; that population has since grown to more than 1,000 animals, with close to 200 residing in the park itself. Soon after reintroduction of wolves, the elk population began to decline, as

The reintroduction of the gray wolf, *Canis lupus,* to the Yellowstone National Park ecosystem had a tremendous ecological impact. *(Gary Kramer/U.S. Fish and Wildlife Service)*

born to endangered animal species and to observe the vast array of biological diversity on Earth.

Preventing the loss of species is also important for the undiscovered benefits they offer. The genetic resources of millions of species have not yet been explored, and once those species are extinct, they are lost forever. Scientists have discovered that organisms produce many substances that are useful and valuable to humans. The fungus *Penicillium notatum* produces penicillin, an antimicrobial chemical that ushered in the age of antibiotics. The soil bacterium *Bacillus thuringiensis* produces a toxin that kills crop-destroying insects. The bark of the Pacific yew tree contains taxol, a drug used to treat ovarian cancer. Perhaps scientists will uncover a solution to global warming or find a cure for Alzheimer's disease in the genes of a seemingly unremarkable organism.

See also BIODIVERSITY; CONSERVATION BIOLOGY; ECOSYSTEMS; ENVIRONMENTAL CONCERNS, HUMAN-INDUCED.

FURTHER READING

Evans, Kim Masters. *Endangered Species: Protecting Biodiversity.* Farmington Hills, Mich.: Thomson Gale, 2006.

NOAA Fisheries Office of Protected Resources. "Endangered Species Act (ESA)." Available online. URL http://www.nmfs.noaa.gov/pr/laws/esa/. Accessed January 19, 2008.

U.S. Fish and Wildlife Service. The Endangered Species Program home page. Available online. URL: http://www.fws.gov/endangered/. Updated January 17, 2008.

wolves naturally prey on elk. The reduction in the elk population allowed willow trees that elk fed on to grow taller, particularly those near streams, since the wolves scared the elk away from the streams. More vegetation increased available habitat space for species such as birds, and food for other species, such as beavers. More beavers led to more dams, and therefore more ponds and impoundments, which allowed more shrubs to grow and serve as protection for migratory birds' nests. More shade on the streams from taller trees cooled the water temperatures and provided more areas for trout to brood. Many other species, such as large birds, bears, and other scavenger populations, benefited from eating the leftovers of wolf kills, increasing their populations. Thus, the reintroduction of one predatory keystone species had a tremendous ripple effect.

Many regions depend on nature-based tourist attractions that rely on endangered or threatened species, as do many recreational activities that involve seeking, observing, and photographing wildlife. The National Wildlife Federation estimates that Americans spend $59 billion annually while engaging in such nonconsumptive wildlife recreation. Bird watching is an extremely popular hobby, and regions that boast rare birds are tourist hotspots. People also visit zoos and aquariums specifically to see exhibits with new babies

endocrine system   The purpose of the endocrine system is to coordinate and regulate the functions of body cells, tissues, and organs. The endocrine system regulates major life processes, such as metabolism, immune functions, the production of red blood cells, reproduction, childbirth, and lactation, and controls internal body conditions such as osmolarity, water balance, heart rate, blood pressure, and blood sugar levels. Endocrine glands produce and secrete chemicals called hormones that circulate in the blood and induce functional changes in certain cells. Hormones, also called ligands because they bind to other macromolecules, travel throughout the entire body via circulation, but only target cells, the types of cells that have receptors that specifically recognize the ligand, respond to its presence. While both the nervous system and the endocrine system facilitate communication among different body parts, the nervous system normally responds to stimuli that require an immediate response, such as removing one's hand from a hot stove. The endocrine system generally manages slower or sustained responses that occur over minutes, hours, or longer such as adjusting the degree of water retention during the production of urine, stimulating cell growth, or initiating sexual maturation during puberty. The intensity of response to a hormone depends on its concentration, in contrast with the all-or-none action potentials that guide the transmission of nervous impulses. Both systems cooperate, however, to coordinate body functions, respond to external and internal stimuli, and maintain homeostasis. For example, neurons that have hormone receptors respond to hormonal signals. Some neurons secrete hormones directly into the circulatory system, and a number of nerves directly innervate certain endocrine glands.

In addition to endocrine signaling, other types of chemical signals communicate information among cells. Neurotransmitters carry signals across a synapse, a junction between a neuron and another neuron or a neuron and an effector cell. Autocrine signals exert local effects on the same type of cells that release them. Paracrine signals affect cells in the immediate vicinity; they are not transported in the blood. Exocrine glands produce and secrete substances such as mucus, sweat, and digestive enzymes, but the secretions travel through ducts to the outside of the body or to cavities that are continuous with the outside of the body. Special chemicals called pheromones leave the body of one individual and influence the behavior of other organisms. Not all chemical signals fall into a single category. For example, some endocrine glands secrete hormones that act locally and at a distance or that function as both neurotransmitters and hormones.

HORMONE SYNTHESIS, TRANSPORT, AND ACTION

Hormones, chemical messengers that generally act over long distances, can be either proteins or lipids. Both are secreted by endocrine glands and travel in blood circulation, but the mechanisms by which they stimulate a response differ. Protein hormones include polypeptide chains and glycoproteins (polypeptides with attached carbohydrate moieties). Growth hormone, insulin, glucagon, and oxytocin are examples of protein hormones. Amines, amino acid derivatives such as epinephrine, melatonin, and thyroid hormones, also function as hormones. Lipid hormones include steroids such as estrogen and testosterone and fatty acid derivatives such as prostaglandins.

Protein hormones are hydrophilic (water-loving) and therefore are soluble in water. Ribosomes docked on the endoplasmic reticulum (ER) assemble polypeptide chains according to information in the gene that encodes the hormone. During synthesis, the polypeptide is extruded into the lumen (cavity) of the ER and then sent to the Golgi apparatus for processing. As for all secreted proteins, the Golgi packages the protein hormones into vesicles for storage

until the cell receives a signal triggering secretion. At that time, the vesicles merge with the cell membrane and release the contents into the extracellular fluids of the interstitium, the space between cells in a tissue. As are protein hormones, amines are stored intracellularly after synthesis until time for secretion.

From the interstitial spaces, protein hormones move into blood circulation by leaky capillary walls. Because they are water-soluble, unique transport methods are not necessary. The hormones enter the capillary beds of various tissues, and when they encounter cells that express receptors that specifically recognize the hormone, they bind to it. Protein hormones and amines cannot diffuse through the phospholipid bilayer of cell membranes. They bind to the receptors located on the exterior surface of the cell membrane and exert an effect without ever entering the target cell. Binding of a ligand to its receptor stimulates a specific physiological response depending on the hormone and the type of target cell. Most protein hormones function by activating second messenger systems that change the activity of proteins present in the cell. Other surface receptor–binding hormones alter the permeability of the cell membranes by either opening or closing specific membrane channels. Because the proteins or membrane channels are already present and waiting to be activated, the effects of protein hormone action occur within minutes.

Second messenger systems mediate signal transduction pathways, mechanisms that carry information from a cell's exterior into the cytoplasm, causing a cellular response. Signal transduction pathways involve a series of changes to intracellular proteins, resulting in a cascade of cellular events. The response varies, depending on the type of target cell activated by the protein hormone. Possible final effects include the induction of the expression of specific genes or the secretion of substances stored in vesicles.

Steroid hormones are lipids and therefore are not soluble in water. Cholesterol is the precursor for all the steroid hormones. Low-density lipoproteins (LDLs) carry cholesterol in circulation to the body cells, including the ones that synthesize steroid hormones, called steroidogenic cells. Specific enzymes catalyze the stepwise conversion of cholesterol into the various steroid hormones including progesterone, aldosterone, cortisol, dehydroepiandrosterone, testosterone, estrone, estradiol, and estriol. The steroidogenic cells do not store steroid hormones; instead the hormones simply diffuse through the cell membrane into the extracellular fluids. The body regulates the levels of circulating steroid hormones by controlling the rate of synthesis rather than controlling the secretion, as in protein hormones.

Because steroid hormones are poorly soluble in water, they circulate in the aqueous plasma bound to

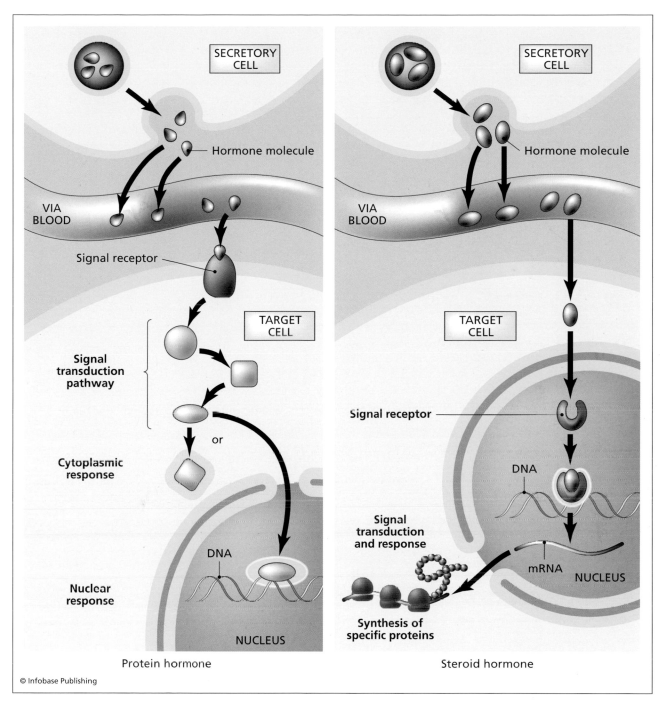

Protein hormones act via surface receptors, whereas steroid hormones bind intracellular receptors.

transport proteins. When bound, the hormone is not active, but the bound and unbound forms are in equilibrium, meaning a constant supply of unbound active hormone is readily available. In the unbound form, the steroid hormone can diffuse through capillary walls into tissue beds, and even into the individual cells directly through the cell membranes. This is in contrast to protein and amine hormones, which bind to receptors located on the outer surface of their target cells. Steroid hormone receptors exist inside target cells. If the cells possess the specific receptors that recognize a particular hormone, the receptor and hormone (ligand) bind to form an intracellular complex.

Steroid hormones act by stimulating the expression of certain genes. The protein products of the targeted genes bring about a cellular or physiological response dependent on the function of that protein. Because new protein synthesis is time-consuming, a delay of several hours precedes observable evidence of the steroid hormone's effect.

Steroid hormone receptors are usually in the nucleus but sometimes are found in the cytoplasm. The steroid hormone diffuses from the bloodstream, through the cell membrane, and into the nucleus, where it finds and binds its receptor, forming a complex. If the receptor is in the cytoplasm, the complex can diffuse into the nucleus from the cytoplasm. Most steroid receptors are transcription factors, regulatory proteins that bind to deoxyribonucleic acid (DNA) and stimulate gene expression. Only certain genes possess the specific DNA sequences, called hormone response elements (HREs), to which the complex binds and activates gene expression. Acti-

vation leads to the synthesis of the protein encoded by that gene. The same hormone can stimulate the expression of different genes in different cell types depending on the presence of other tissue-specific transcription factors.

VERTEBRATE ENDOCRINE SYSTEM

Endocrine glands are organs located throughout the body that produce and secrete hormones into the extracellular fluid surrounding cells. Hormone secretion is the main function of some glands, while other organs contain cells that synthesize and secrete hormones but also have other important functions. For

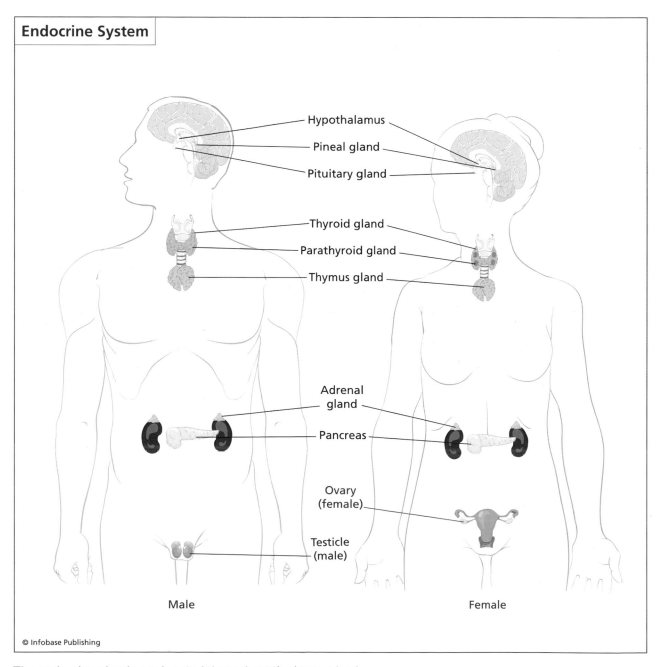

Endocrine System

Hypothalamus

Pineal gland

Pituitary gland

Thyroid gland

Parathyroid gland

Thymus gland

Adrenal gland

Pancreas

Ovary (female)

Testicle (male)

Male

Female

© Infobase Publishing

The endocrine glands are located throughout the human body.

example, the main function of the stomach is to churn food and begin the process of chemical digestion, but certain cells in the stomach also secrete gastrin, a protein hormone that helps control appetite. The testes produce the male gametes, spermatozoa; the Leydig cells inside the seminiferous tubules of the testes produce the steroid hormone testosterone.

Located in the ventral portion of the forebrain, the hypothalamus plays a major role in the coordination of many endocrine and nervous system activities. The pituitary is a lima bean–shaped gland that is located at the base of the hypothalamus and consists of two regions: the anterior and the posterior pituitary. The main functions of the hypothalamus include maintaining homeostasis, controlling secretion of hormones from the posterior pituitary, and releasing chemical factors that regulate the anterior pituitary. The brain relays information about both internal and external environmental conditions to the hypothalamus, and the hypothalamus reacts by giving chemical commands to the pituitary gland. Specialized neurosecretory cells of the hypothala-

mus synthesize two hormones, antidiuretic hormone (ADH) and oxytocin, which are stored in the posterior pituitary, an extension of the hypothalamus consisting of neural tissue, also called the neurohypophysis. The anterior pituitary, also called the adenohypophysis, consists of endocrine cells rather than neural tissue and functions in the synthesis of numerous hormones. Four of the anterior pituitary hormones are tropic hormones, meaning they stimulate activity in other endocrine glands: folliclestimulating hormone (FSH), luteinizing hormone (LH), thyroid-stimulating hormone (TSH), and adrenocorticotropic hormone (ACTH). The anterior pituitary synthesizes and secretes the tropic hormones in response to hypothalamic secretions. The tropic hormones stimulate another endocrine gland to produce yet another hormone that initiates an appropriate physiological change in response to the information originally received by the hypothalamus. The table below, "Human Endocrine Glands and Their Hormones," summarizes the role of several important human endocrine glands.

HUMAN ENDOCRINE GLANDS AND THEIR HORMONES

Gland	Hormone	Main Action
hypothalamus	releasing hormones for growth hormone, thyroid hormone, adrenocorticotropic hormone, the gonadotropins (FSH and LH), and prolactin	stimulates the secretion of the named hormones
	inhibiting hormones for growth hormone and prolactin	inhibits the secretion of growth hormone and prolactin
anterior pituitary gland	adrenocorticotropic hormone (ACTH)	increases glucocorticoid secretion from adrenal cortex
	follicle-stimulating hormone (FSH)	stimulates follicle development in the ovaries of females, stimulates sperm cell production in the testes of males
	luteinizing hormone (LH)	stimulates ovulation and estrogen and progesterone production by the ovaries in females, supports sperm cell production in the testes of males
	prolactin	stimulates milk production in mammary glands of females, increases sensitivity of follicles to FSH and LH
	growth hormone (GH), also called somatotropin	stimulates tissue growth and the necessary metabolic processes for growth
	thyroid-stimulating hormone (TSH)	increases thyroid hormone secretion by the thyroid gland
	melanocyte-stimulating hormone (MSH)	stimulates the melanocytes in the skin to produce melanin pigments, making skin darker

(continues)

(continued)

Gland	Hormone	Main Action
posterior pituitary gland	antidiuretic hormone (ADH)*	promotes water retention in kidney tubules, decreases urine volume
	oxytocin*	increases uterine contractions during childbirth, stimulates milk release in mammary glands
thyroid	triiodothyronine (T_4) and thyroxine (T_3)	increases metabolism, necessary for normal growth and maturation
	calcitonin	inhibits bone breakdown, decreases blood calcium levels
parathyroid glands	parathyroid hormone (PTH)	increases bone breakdown by osteoclasts; stimulates bone breakdown; stimulates calcium reabsorption in kidneys and small intestine, leading to an increase in blood calcium levels; increases vitamin D synthesis
pancreas	glucagon	stimulates liver cells to break down glycogen stores and release glucose into blood
	insulin	stimulates uptake of glucose from blood
	somatostatin	inhibits insulin and glucagon secretion
adrenal glands (medulla)	epinephrine and norepinephrine	increases heart rate and blood pressure, increases blood flow to skeletal muscles and to heart, reduces blood flow to visceral organs and skin, increases blood glucose levels
adrenal glands (cortex)	glucocorticoids (cortisol)	stimulates protein and fat catabolism, increases glucose production, decreases inflammatory response, inhibits immune response
	mineralocorticoids (aldosterone)	increases sodium and potassium reabsorption, increases water reabsorption, increases hydrogen ion excretion
	androgens and estrogens	stimulates the development of some secondary sexual characteristics, but effect is mostly masked by ovarian and testicular steroid hormones
testes	testosterone	aids in sperm production, stimulates development and maintenance of male secondary sexual characteristics, maintains testicular function
	inhibin	inhibits FHS secretion from anterior pituitary gland
ovaries	estrogens	promotes the development and maintenance of female secondary sexual characteristics including the menstrual cycle
	progesterone	stimulates development of the uterine and mammary glands, regulates menstrual cycle, maintains pregnancy
	inhibin	inhibits FSH secretion from the anterior pituitary gland
	relaxin	loosens the connective tissue in the pelvis in preparation for childbirth
pineal gland	melatonin	inhibits gonadotropin-releasing hormone activity, possibly regulates sleep cycles

*These hormones are produced by neurosecretory cells of the hypothalamus but are stored in and released from axon terminals in the posterior pituitary.

See also ANATOMY; ANIMAL FORM; BIOLOGICAL MEMBRANES; BIOMOLECULES; CELL COMMUNICATION; CIRCULATORY SYSTEM; EUKARYOTIC CELLS; GENE EXPRESSION; HUMAN REPRODUCTION; INVERTEBRATES; NERVOUS SYSTEM; PHYSIOLOGY; VERTEBRATES.

FURTHER READING
The Facts On File Encyclopedia of Health and Medicine. Vol. 3. New York: Facts On File, 2006.

Larsen, P. Reed, Henry M. Kronenberg, Shlomo Melmed, and Kenneth S. Polonsky, *Williams Textbook of Endocrinology.* 10th ed. Philadelphia: W. B. Saunders, 2002.

Nagel, Rob. *Body by Design: From the Digestive System to the Skeleton.* Vol. 1. Detroit: U*X*L, 2000.

Norris, David. *Vertebrate Endocrinology.* 4th ed. San Diego: Academic Press, 2006.

Petit, William A., Jr., and Christine Adamec. *The Encyclopedia of Endocrine Diseases and Disorders.* New York: Facts On File, 2005.

Pfaff, Donald W., Ian Phillips, and Robert T. Rubin. *Principles of Hormone/Behavior Relations.* San Diego: Academic Press, 2004.

Seeley, Rod R., Trent D. Stephens, and Philip Tate. *Anatomy and Physiology.* 7th ed. Dubuque, Iowa: McGraw Hill, 2006.

Silverthorn, Dee Unglaub. *Human Physiology: An Integrated Approach.* 4th ed. San Francisco: Benjamin Cummings, 2006.

environmental concerns, human-induced   Human beings are part of the global biological community. As all other life-forms do on the planet, humans interact with their environment. Characteristics of the biosphere influence the way humans live, and conversely, humans affect the environment in which they live. The Earth supplies materials, water, food, and energy humans need not only for basic survival but also for the sake of convenience and entertainment. While the degree to which humans are responsible for leaving the planet in the same condition as they found it is a matter of ethics, the effect that human activities have on the Earth's ecosystems is a matter of science.

Change is natural. Life-forms evolve in response to specific environmental factors, but the environment also changes. At this moment geological processes are eroding mountain ranges and laying new crust on the ocean floor. The chemical composition of the atmosphere and the physical conditions on Earth's surface are vastly different from those when the planet formed 4.5 billion years ago. As an integral part of the planet's ecosystems, natural biological phenomena, such as metabolic processes that participate in the biogeochemical cycling of nutrients or photosynthesis that generates and releases oxygen into the atmosphere, contribute to the chemical and physical conditions of the biosphere. Many of the recent changes to the Earth's ecosystems, however, are the result of artificial human-induced processes. Whether through negligence, ignorance, or apathy, human activities have challenged the Earth's ecosystems beyond their ability to recover without a concerted cooperative effort from society.

A few serious current environmental concerns include the degradation of land, a diminishing clean water supply, depletion of nonrenewable energy sources, and increased pollution. Many environmental issues are interconnected; for example, in order to plant more crops farmers may clear forests, destroying habitats and leading to decreased biodiversity and removing vegetation that helps remove carbon dioxide, which is considered a form of air pollution, from the atmosphere. If the farmer uses artificial fertilizer, excess may run off and pollute the water supply. The severity of all human-induced environmental concerns is compounded by the continually expanding world population and the accompanying increased demand for food, space, and resources. The fields of ecology, environmental science, and conservation biology all aim to understand the causes and effects of the changes occurring in the biosphere so the knowledge can be applied to reduce the negative impact human activities are having on the biological health of the planet and to develop strategies for achieving a sustainable society.

LAND DEGRADATION
Land degradation is a decrease in a land's ability to absorb, store, and recycle water, energy, and nutrients. Natural phenomena such as hurricanes and earthquakes can destroy regions of land, but human activities can have a greater effect. The forest and grassland biomes provide society with food and with land for growing crops, grazing cattle and sheep, providing fiber for clothing and wood that is used for numerous purposes including construction, fuel, and papermaking. In addition to these and other material resources, forests and grasslands perform numerous ecosystem services, such as fixing carbon and producing oxygen via photosynthesis, recycling chemical elements, preventing soil erosion, regulating climate, and providing shade and shelter for thousands of species of animals, plants, and microorganisms. The biological health of these lands is deteriorating. The Millenium Ecosystem Assessment (coordinated by the United Nations Environment Programme) reports

(continues on page 298)

THE EFFECTS OF GLOBAL CHANGE ON TROPICAL FORESTS

by Kenneth J. Feeley, Ph.D.
Center for Tropical Forest Science,
Harvard University Arnold Arboretum

The Earth's environment has changed a lot over the past several decades. Some of these changes are very clear and striking. For example, the concentration of carbon dioxide (CO_2) in the atmosphere has increased by more than 30 percent over the past 150 years and continues to increase steadily by almost 0.5 percent each year. However, most changes are more complicated. Human activities have altered temperature, rainfall, cloud cover, and many other environmental factors, but the rate and even the direction of change can vary widely from region to region. In other words, some parts of Earth are getting hotter while some are getting colder, some parts of Earth are becoming drier while others are becoming wetter, and still other parts of Earth do not seem to be affected much at all. Despite this confusing complexity, scientists now have a fairly detailed understanding of how environments are changing worldwide and which factors are likely driving these changes. They are even able to make some fairly accurate predictions about how the environment is likely to change in the near future, but much more remains unknown. Perhaps most importantly, little is known about the various effects that these changes to Earth's environment are going to have on its living creatures, including human beings.

The limits of ecological knowledge are perhaps best exemplified in tropical rain forests. Tropical rain forests are amazingly rich and complex ecosystems. Within a one-hectare patch of rain forest (approximately 2.5 acres, or a little less than twice the size of an American football field) one may find close to 1,000 different species of trees—more species of trees than in all of the United States and Canada combined. This bewildering diversity is what makes rain forests so

beautiful and so much fun to visit, but it is also one of the primary reasons rain forests are so difficult to study and even more difficult to comprehend. Understanding the complex interactions among the multitudes of organisms inhabiting tropical forests and their rapidly changing environment is absolutely critical. Not only do tropical rain forests support most of the world's plant and animal species, but also they provide many important services to people everywhere. For example, tropical plants supply many types of food and provide much of the wood used in constructing houses and furniture. In addition, the forests play a vital role in regulating Earth's atmosphere and climate. Therefore, disturbances in tropical forests can have implications worldwide.

Ecologists recognize the importance of tropical forests and are actively working to determine how they are responding to changes in CO_2, temperature, rainfall, and other environmental factors. One hypothesis is that trees will benefit from several aspects of global change. Through photosynthesis, plants use CO_2 to synthesize carbohydrates that are then used in the production of wood, leaves, roots, fruits, and seeds. Therefore, as the concentration of atmospheric CO_2 increases, the rate of photosynthesis in plants may also potentially increase, allowing them to grow faster and larger. This would have important implications for the global environment since bigger, faster-growing trees will uptake and sequester lots of carbon from the atmosphere and potentially help to offset some of the carbon emitted from cars, air conditioners, and power plants.

The idea that plants will respond to higher concentrations of CO_2 with faster growth does not apply exclusively to tropical rain forests. In the United States, ecologists have conducted some impressive studies in which they experimentally increased the concentration of atmospheric CO_2 within patches of pine forest and then examined how fast the trees

grew compared to control patches of forest where CO_2 was kept at natural ambient levels. In these studies, the pine trees and other plants responded to increased concentrations of CO_2 by growing faster and also by producing more seeds earlier in life. However, many of the effects were only temporary. The amount of nutrients in the soils could not sustain the accelerated growth, and after several years the trees returned to approximately their normal growth rates. Likewise, in a similar experiment conducted in Switzerland, researchers did not find any long-lasting effects of elevated CO_2 on the growth rates of mature deciduous trees.

Unfortunately, conducting these sorts of large-scale experiments in tropical rain forests is generally not feasible—not only because the forests are so incredibly diverse, but also because they are often too remote and the conditions too rough to allow scientists to set up and maintain all of the necessary equipment. Instead, tropical ecologists investigating the effects of global change have relied primarily on long-term records from tree inventory plots, patches of forest where researchers measure the diameter of every tree at periodic intervals (for example, every year or every five years). The investigators can then calculate tree growth rates by comparing the sizes of individual trees over time. Given enough measurements, they can then determine whether growth rates have changed over time and examine how these changes correspond to any concurrent changes in environmental conditions.

Using an extensive network of tree plots situated throughout the Amazonian rain forests of Brazil, Peru, and Ecuador, ecologists found that some very significant changes have occurred in these forests over the past several decades. Specifically, Amazonian trees appear to be growing significantly faster now than when the studies were initiated 20–30 years ago. Likewise, the total amount of wood, or aboveground biomass, in these forests has increased dramatically over time. These findings are consistent with the predictions of the hypothesized increase in plant productivity due to elevated CO_2.

Studies from tropical rain forests in other parts of the world have produced conflicting results. In Costa Rica, researchers working with a few select species of trees found that contrary to the predictions of so-called carbon fertilization, growth rates actually decreased significantly through time. Likewise, in collaboration with scientists from the Center for Tropical Forest Science and the Smithsonian Tropical Research Institute, this author conducted an extensive set of studies looking at the growth rates of almost a million individual trees representing more than 1,000 different species growing in the rain forests of Panama and Malaysia. This study demonstrated that growth rates have decreased rapidly over the past two decades across the vast majority of species.

Why are trees in some parts of the world growing slower? Why is the widespread increase in CO_2 levels not stimulating tree growth in Central America and Southeast Asia as it is in the Amazon? One possible explanation for the unexpected slowdown is temperature. In Costa Rica, Panama, and Malaysia, the declines in tree growth rates were strongly associated with higher temperatures. Ask most gardeners and they will probably claim that plants benefit from higher temperatures. This is certainly true, but only to a point. In some parts of the Tropics, temperatures are already so high that further increases can cause net productivity to diminish and growth rates to decline. In the Amazon, temperatures have been relatively more stable over the past several decades, so for trees in these forests the benefits of additional carbon may still be outweighing the negative effects of elevated temperatures.

While elevated temperatures can explain the observed patterns, scientists must always be willing to consider alternative explanations and weigh them against the available evidence. For example, changes in cloud cover may also be responsible for the observed results. The number of cloudy days did increase in both Panama and Malaysia, and therefore the decrease in growth may be due to less sunlight. Or maybe the amount of rain has changed? Or perhaps trees in these forests are utilizing the extra carbon to produce more flowers and seeds rather than to grow bigger? The truth is that ecologists will never know the answer with 100 percent certainty. Ecology is a tremendously complex science and involves very few certainties, especially in the extraordinarily diverse Tropics. However, rather than allowing uncertainty to discourage further research, scientists should use it as motivation for their ongoing persistent efforts to gain a better understanding of the relationships within ecosystems.

Tropical ecologists continue to examine the response of rain forests to global change. Using new and improved measuring techniques, they are investigating the patterns of growth and productivity in different types of forests in distant parts of the world. They are looking at patterns of fruit and seed production, how the composition of species within forests is changing through time, and what is happening under the ground with the trees' roots and surrounding soils.

While much additional work is clearly necessary, scientists have already learned a great deal. For example, they now know that tropical forests are not all growing faster in response to increasing CO_2, meaning that society cannot count on them to buffer against ever-increasing carbon emissions. Scientists have learned, in fact, that some forests are actually growing slower, potentially in response to elevated temperatures.

Slower tree growth in tropical rain forests will have very important implications for both the global environment and the global economy. Tropical forests support the majority of terrestrial animal species, all of which depend either directly or indirectly on plant productivity as a source of energy. Consequently, decreased growth will reduce the amount of energy available, thus potentially reducing the number of animal species that these ecosystems can support and thereby diminishing global biodiversity. Tropical forests are also an extremely valuable source of timber. Slower growth may result in decreased standing stocks of timber available for logging. In addition, the rate of forest recovery following logging may decrease, thus lengthening the time that loggers must wait before reharvesting patches of forest. In order for loggers to maintain current yields, they will have to increase either the intensity of the logging or the area of forest that they log. More intensive and more widespread logging does not bode well for tropical forests since logging may negatively impact diversity either directly or indirectly through fragmentation, edge creation, and/or other synergistic effects such as the increased risk of accidental fires. Likewise, an overall decrease in productivity could necessitate an increase in the intensity or extent of extraction for nontimber products. In some cases (such as brazil nuts), harvests are already at or exceeding sustainable levels, and any increase in harvesting is likely to have negative impacts.

To make matters worse, the potential exists for slowing tropical tree growth rates to create a dangerous series of feedbacks. For example, reductions in tree growth may result in reduced rates of carbon uptake from the atmosphere, which, coupled with the extra emissions of CO_2 from the associated increases in logging and deforestation, could accelerate the increase of atmospheric CO_2 and global warming, causing even further reductions in tree growth, and so on and so on.

The world is changing. Most of these changes are extremely complex and can vary greatly from place to place. In some places, the changes may appear positive in the short term, for example, resulting in warmer winters and earlier-blooming flowers. But one must always remember that the living world comprises a delicate web of interactions and that disturbances even in places as seemingly remote as Southeast Asia, the Amazon, or the Congo

(continues)

(continued)

can have important effects for all living creatures, including people. Scientists may never fully understand all the implications of changes in Earth's environment, but they must continue striving to learn as much as possible while trying to minimize human-induced contributions to global change.

FURTHER READING

Clark, D. A. "Detecting Tropical Forests' Responses to Global Climatic and Atmospheric Change: Current Challenges and a Way Forward." *Biotropica* 39 (2007): 4–19.

Clark, D. A., S. C. Piper, C. D. Keeling, and D. B. Clark. "Tropical Rain Forest Tree Growth and Atmospheric Carbon Dynamics Linked to Interannual Temperature Variation During 1984–2000." *Proceedings of the National Academy of Sciences USA* 100 (2003): 5,852–5,857.

Feeley, K. J., S. J. Wright, M. N. Nur Supardi, A. R. Kassim, and S. J. Davies. "Decelerating Growth in Tropical Forest Trees." *Ecology Letters* 10 (2007): 461–469.

Körner, C., R. Asshoff, O. Bignucolo, S. Hattenschwiler, S. G. Keel, S. Pelaez-Riedl, S. Pepin, R. T. W. Siegwolf, and G. Zotz. "Carbon Flux and Growth in Mature Deciduous Forest Trees Exposed to Elevated CO_2." *Science* 309 (2005): 1,360–1,362.

Laurance, W. F., A. A. Oliveira, S. G. Laurance, R. Condit, H. E. M. Nascimento, A. C. Sanchez-Thorin, T. E. Lovejoy, A. Andrade, S. D'Angelo, J. E. Ribeiro, and C. W. Dick. "Pervasive Alteration of Tree Communities in Undisturbed Amazonian Forests." *Nature* 428 (2004): 171–175.

Lewis, S. L., O. L. Phillips, T. R. Baker, J. Lloyd, Y. Malhi, S. Almeida, N. Higuchi, W. F. Laurance, D. A. Neill, J. N. M. Silva, J. Terborgh, A. T. Lezama, R. V. Martinez, S. Brown, J. Chave, C. Kuebler, P. N. Vargas, and B. Vinceti. "Concerted Changes in Tropical Forest Structure and Dynamics: Evidence from 50 South American Long-Term Plots." *Philosophical Transactions of the Royal Society of London Series B-Biological Sciences* 359 (2004): 421–436.

Oren, R., D. S. Ellsworth, K. H. Johnsen, N. Phillips, B. E. Ewers, C. Maier, K. V. R. Schafer, H. McCarthy, G. Hendrey, S. G. McNulty, and G. G. Katul. "Soil Fertility Limits Carbon Sequestration by Forest Ecosystems in a CO_2-Enriched Atmosphere." *Nature* 411 (2001): 469–472.

(continued from page 295)

that dry lands make up 41 percent of the planet's land surfaces, and up to 20 percent of that has become unusable as a result of desertification, the conversion of land into desert. The soil of degraded land is barren, and its conversion into a functional desert biome may take millions of years. Desertification can result from changes in climate, but the increased desertification initiated during the last few hundred years is due to improper land management practices—mainly overgrazing and poor soil practices. In the absence of vegetation that once prevented the soil from eroding, the wind and water wash it away, leaving behind barren, infertile land that cannot be used for growing crops or for growing grass to feed herds. The dust clouds and sandstorms affect neighboring areas as well and are a particular problem in China, Japan, South Korea, and Mongolia.

Archaeological evidence from the Fertile Crescent suggests that even thousands of years ago, people cut down forests to graze cattle on common grasslands. Though grasses can recover from grazing, when too many cattle graze on common grounds, they eat up the metabolic reserves in addition to the top half of the blades of grass, leaving behind insufficient photosynthetic tissue to support the root system. As a result the grasses die, and the previously rich grasslands and forests transform into barren desert.

The trend continues today; farmers have replaced lively ecosystems consisting of grasses, shrubs, and trees with crops of corn, soybeans, wheat, oats, and alfalfa. As of 2005 191 governments worldwide had joined the United Nations Convention to Combat Desertification (UNCCD) with the goal of combating desertification and protecting drylands. Affected countries are developing national, regional, and subregional programs to reverse land degradation. Countries such as those in Africa are the most affected, since two-thirds of the continent is desert or drylands, but the UNCCD reports that even in the United States, desertification affects more than 30 percent of the land, a problem exacerbated by recent severe droughts. In the United States, the Bureau of Land Management, part of the U.S. Department of the Interior, manages 258 million surface acres of public land and 700 million acres of subsurface mineral estate. The bureau tries to balance the use and conservation efforts on these lands by overseeing mineral mining operations, managing commercial forests and woodlands for timber, administering grazing permits, controlling recreational use, protecting wild animal herds, and overseeing other activities and uses.

Tropical rain forests contain an estimated 50 to 80 percent of the world's terrestrial species of plants and animals, though they only cover 6 percent of the land surface. Despite its ecological importance, this biome is rapidly shrinking. Human activities such as harvesting the wood for firewood, construction timber, paper, and plywood and clearing the land for agricultural purposes are degrading the Amazon, which contains more than half of the world's remaining rain forest. In 2006 Britaldo Soares-Filho at the Universidade Federal de Minas Gerais in Brazil and others published the results of a computer simulation that predicted that 40 percent of the Amazon forests will be gone by 2050 if current practices continue (Soares-Filho et al., 2006). Degradation of a forest's canopy reduces precipitation and can lead to the eventual conversion of forest into grassland. Efforts to reduce deforestation are under way. Logging proponents suggest selective logging as one solution, but environmentalists believe that this would lead to a cycle of degradation and land clearing. In addition to logging, people are destroying tropical rain forests to use the land for growing crops. Because of the rapid nutrient recycling within the rain forest ecosystem, and because of the frequent rains, the soil is nutrient-poor. Only the upper superficial layer contains nutrients to support crop growth, so though farming on land that was previously rain forest terrain may initially be successful, the soil soon becomes depleted, and with the vegetation gone, the wind and rains erode the soil, making it unusable.

Another endeavor that essentially vandalizes functional land is oil prospecting. In addition to the Arctic National Wildlife Refuge (ANWR), the Alaskan tundra contains the largest portion (23 million acres) of unprotected wildlife in the nation, part of the National Petroleum Reserve-Alaska (NPRA). Biologists claim that the lush NPRA is critical to the ecological health of the area, which serves as home to caribou, geese, grizzly bears, and bowhead and beluga whales, but the richness in hydrocarbons make the land valuable to the oil industry. Drill pads and pipelines drive away the animals, which must look elsewhere for food. Not only is the disappearance of undisturbed lands an issue, but hundreds of oil spills occur there every year and the amount of pollutants pumped into the air exceeds that of major urban areas.

WATER RESOURCES

All life-forms require water. The availability of a clean water supply determines the distribution of different types of living organisms, migratory patterns of animals, and the locations of cities and human populations. Physical and geological factors drive water through the hydrological cycle, in which water evaporates from water bodies, soil, and plant leaves and rises into the atmosphere, where it affects weather patterns and forms clouds, which the winds move around until they deposit their moisture over the Earth in the form of precipitation. No matter

This aerial photo depicts the stark reality of deforestation by soybean farmers in Terra do Meio, Para, Brazil, in 2004. *(AP Images)*

where the precipitation falls, the water eventually returns to the oceans. The oceans comprise 97 percent of the Earth's water, and of the remaining 3 percent freshwater supply, most exists in the form of glaciers or is inaccessible below the Earth's surface. An estimated 0.003 percent is available for use by humans, and the Food and Agriculture Organization of the United Nations estimates that, globally, 70 percent of that is used for agricultural purposes (for crop and livestock production), 20 percent for industries, and 10 percent for domestic use. Though the quantity of freshwater that falls as rain should be sufficient to meet the world's needs, the distribution of water resources is unequal. *Water for Life Decade (2005–2015),* a booklet published by the United Nations Department of Public Information, states that Asia holds more than 60 percent of the world's population but only 36 percent of the world's river runoff. In contrast, South America has 26 percent of the runoff but only 6 percent of the world's population. Continued population growth in addition to economic growth will worsen the situation. The World Health Organization estimates that 1 billion of the 6 billion-plus population currently do not have access to freshwater, and the United Nations predicts that by 2025, 5 billion of the world's projected 7.9 billion people will not have access to clean water. Inefficient use of water, contamination, and destruction of wetlands, estuaries, and coastlines threaten the already limited supply.

Global warming also contributes to the problem, as many regions depend on snowmelts for freshwater, and the freshwater stored in glaciers melts into the sea. While one might think that warming snow and glaciers more quickly would relieve the problem of water shortage, many glaciers will recede or disappear to the point where certain areas that rely on them will have no source. Also, the shift in the timing of the melting to an earlier point in the spring creates problems later in the summer when the rivers run low.

The saltwater content of the oceans prohibits its use for agricultural or domestic uses. Technological advances have improved desalination processes, the removal of salts from seawater or brackish groundwater. Large desalination plants can process between 25 and 50 million gallons per day; in comparison, in the United States the per capita water withdrawals averaged 1,430 gallons (5,413 L) per day in 2000 (the most recent year for which the United States Geological Survey has published data on estimated water use). Desalination plants utilize a lot of energy, the uptake of seawater kills local marine life, and the resulting hypersaline wastewater presents another problem. Rerouting water supplies may relieve a shortage in one area but then cause a shortage else-where. Even if the area from which water is rerouted is not populated, rerouting the supply can cause ecological destabilization.

In addition to excessive water withdrawal, the introduction of chemical pollution and the construction of dams negatively impact the surface waters. Not only do lakes, ponds, rivers, and bays serve as sources of water, but the habitats they provide are crucial to organisms that play an important role in the food chain of many ecosystems and in the food supply for humans. Wetland ecosystems, both inland and along coastlines, are home to numerous species, some of which are endangered. They also help regulate stream flow, control flooding, collect sediment to prevent it from entering streams, and remove fertilizer chemicals from surface runoff before they enter river systems. Despite this, the area covered by wetlands globally has declined as they are being filled in for urban growth and farming. The worst losses have been in the state of California, Australia, and New Zealand, which have all lost approximately 90 percent of their wetlands. Estuaries, the zones where freshwater rivers flow into the sea, are also at risk. Pollution from sewage treatment plants and industries, oil spills, and erosion sediment chokes out the marine life, and excessive inland water withdrawal has reduced the flow to some estuaries, causing them to dry up completely.

NONRENEWABLE ENERGY RESOURCES

Society's dependence on nonrenewable fuels has led to a multitude of problems. Depletion of the resources is in itself an issue, since most of them require millions of years to produce them, but the manner in which they are harvested from the Earth, are transported, and are used has contributed to air pollution that has led to acid rain and an increase in greenhouse gases, chemical pollution of both lands and waters, and the destruction of habitats and ecosystems. People making decisions related to the use of energy resources must consider not only their renewability or nonrenewability but also their availability and their environmental impact. (Economics, politics, and ethics also play a major role.) Industrial nations fulfill about 90 percent of their energy demands using nonrenewable resources, and less developed countries about 59 percent. According to information posted in September 2007 by the Energy Information Administration, which compiles official energy statistics for the U.S. government, the United States is a major offender, making up less than 5 percent of the world's population, but using more than one-fifth of its primary energy. In 2005 the United States used 22 percent of the world's primary energy. Canada, which has about 0.6 percent of the world's population, uses about 3 percent of its energy. The per capita use is double that of other developed countries including Japan and those in Western Europe.

Renewable forms of energy include hydroelectric, geothermal, solar, and wind. These natural resources are virtually unlimited; they are present in abundance. Society simply needs to commit to utilizing them. With respect to energy, biomass is defined as anything made by a living organism that can be used for energy, such as wood, cow manure, seaweed, crops, or other plant material. Though indiscriminate use of biomass can cause other environmental concerns, biomass is also considered a renewable energy resource. For example, wood can be regenerated by growing new trees, but destroying forests to obtain wood for burning to heat houses leads to other serious problems, such as land degradation and habitat destruction, that contribute to decreased biodiversity.

Nonrenewable energy resources include oil, coal, natural gas, and nuclear power. These resources cannot be regenerated once they are gone, or, if they can, the process would take hundreds of millions of years. Although natural biological and geological processes produced these resources, and although the same processes may still be occurring today, the processes occur at such a slow rate that these resources cannot be replenished for sustained use. *Fossil fuel* is the name given to any fuel created in the Earth from plant or animal remains. Oil, natural gas, and coal are all fossil fuels.

Crude oil, also called petroleum, is an oily liquid made of hydrocarbons. More than 300 million years ago, phytoplankton called diatoms that lived in the sea died, coated the seafloor, became buried, and were converted to oil by the intense pressure and heat. Gasoline and diesel fuel used mainly for transportation purposes are made from crude oil. Combustion of these products during use emits large quantities of pollutants including the greenhouse gas carbon dioxide and sulfur dioxide and nitrogen dioxide, both of which convert to acidic compounds once in the atmosphere. These pollutants lead to numerous other environmental problems, discussed later. Commonly occurring offshore leaks and accidental spills pollute the water, kill marine and coastal lifeforms, and destroy habitats. At the current rate of usage, the world's supply of crude oil will be depleted within 50 years.

Natural gas consists mostly of the flammable gas methane (CH_4), but also of other small hydrocarbons, and it is used to heat homes, heat water, and cook food. The advantage of natural gas is that is burns cleanly and contains very few contaminants, but if it leaks, methane is a powerful greenhouse gas. Gas companies pump natural gas from underground, often near petroleum sources. Though it has no odor and is invisible, gas companies often add a chemical that gives natural gas a bad odor so people know when a leak has occurred. Explosions sometimes occur during extraction or transportation through pipelines.

The solid, black, combustible, rocklike organic substance called coal is the most abundant fossil fuel, and therefore the cheapest. During the Carboniferous, the geological period that occurred between 354 and 290 million years ago, many large leafy plants covered the Earth, and algae filled the seas. After they died, their remains sank to the bottom of swamps and the ocean, where they became peat. Sediment buried the peat and transformed into heavy layers of rock. The pressure forced all the water out of the organic remains and, under conditions of high temperatures, converted them into coal over millions of years. Coal can be mined from the Earth's surface or from underground. In surface mining, bulldozers scrape off the terrain that lies over the coal seam, the horizontal layer containing the coal, so it can be extracted. This type of mining destroys the landscape. Exposed areas erode, and rain washes the sediment away, filling streams, with potential for flooding. Underground mining is hazardous; fires, explosions, and cave-ins are not uncommon and can harm or trap workers, and breathing the coal dust can damage the lungs. Environmentally, cave-ins can lead to cracking or sinking of the land surface, making land unusable for farming. Even when abandoned, coal mines can leak sulfuric acid, a by-product of bacterial metabolism of iron pyrite released from rocks. The sulfuric acid then leaches heavy metals from the rocks, and the acid mine drainage kills plants and animals and corrodes structures. Once extracted from the Earth, the majority of coal is burned to generate electricity or heat. The combustion of coal creates tons of hazardous solid waste in the form of fly ash and bottom ash and releases tons of particulate matter, sulfur oxides, nitrogen oxides, carbon monoxide, and carbon dioxide into the atmosphere, leading to acid rain and contributing to global warming.

To access fossil fuels, geologists must first locate them—not an easy task since they exist trapped underneath rock formations. After extraction, the fossil fuel must be transported by pipelines, over water bodies, or across land to processing plants for cleaning or refinement. The products are then distributed by pipelines, tankers, barges, trucks, or trains to storage depots located both aboveground and underground. Electric, gas, or petroleum industries either transmit the product to customers, use it to generate electricity, or sell it to other companies or to consumers. The numerous steps necessary for the production and use of energy resources themselves require energy, use land and other natural resources, generate pollution, and can cause damage to habitats.

Nuclear energy is another less common energy source. The main fuel for nuclear power plants is the naturally occurring radioactive isotope uranium 235. Radioactive isotopes are atoms that are unstable because they have an excess of neutrons in the atomic nucleus, causing them to emit radiation. Different forms of radiation differ in their ability to penetrate substances, but all of the types of radiation emitted from radioactive isotopes can damage biomolecules, such as deoxyribonucleic acid (DNA) and proteins. Because of this, exposure to radiation can kill cells or cause genetic mutations that can lead to cancer. Fission is the splitting of atoms, accompanied by the release of enormous amounts of energy. When uranium 235 atoms split, they release neutrons that bombard other uranium atoms, splitting them and causing a chain reaction. When these chain reactions occur under controlled conditions in a nuclear reactor, the energy released can be captured and used to perform work. The advantage of nuclear energy is that it does not produce air pollution, and since a small amount of uranium creates such large quantities of energy, less disruption to the environment is necessary for mining. Two major drawbacks are reactor safety and waste disposal. The sources must be contained by special mechanisms. Spills, accidents, and even slow leaks of radioactive material have serious long-term effects on all living organisms and on the surrounding environment. Sources of radioactivity lose their power naturally by spontaneous decay, but this can take hundreds or tens of thousands of years before the source decays to nonhazardous levels.

Researchers are actively seeking alternate energy sources and better methods for providing energy. For example, biofuels are derived from biomass, which contains stored energy. Some crops, such as corn or flaxseed, are grown specifically for conversion into biofuels such as ethanol or anaerobic digestion into biogases. Some automobiles in production today can run on ethanol. Such solutions are not simple, however, since growing crops to use as a substrate for fermentation in alcohol production requires irrigation, contributing to the water shortage crisis; uses land resources; and also requires processing (which also requires large quantities of water, possibly beyond that which water plants can possibly handle), transported, and distributed. Scientists must continue to investigate creative ways to meet society's future energy needs.

POLLUTION

Pollution is the biological, chemical, or physical alteration of the water, land, or air that is harmful to living organisms. Ocean waters and surface waters, such as lakes, rivers, streams, and underground aquifers, can be affected by pollution. Water pollution in different countries takes different forms. Common pollutants found in surface waters of developing countries include human and animal wastes, pathogenic microorganisms, pesticides, and sediment. Water pollution in industrial nations includes all of these and toxic metals (such as mercury), organic chemicals, pharmaceuticals, and acids. Human activities that contribute to water pollution are agriculture, mining, construction, drilling, waste disposal, dam construction, salting of roads and driveways to melt ice, and fertilizing of lawns. Even hot water is considered a pollutant. Thermal pollution most often results when an electric power plant dumps hot waters that have been used to cool liquids or machinery into lakes, rivers, or oceans. In addition to harming aquatic life that cannot tolerate the warmer water, the warmer temperatures lower the amount of dissolved oxygen. The effects of water pollution follow the path of the waters as they trickle through soil, filter through rock beds, and flow down streams and river systems, and into oceans.

When excess inorganic nutrients such as nitrates and phosphates or raw sewage pollute lakes or ponds, algae and aquatic plants overgrow. As they die and fall to the bottom of the water body, bacteria and other microorganisms decompose the remains. As they do so, they use up all the dissolved oxygen in the water, suffocating the other aquatic life in the lake or pond. The anaerobic bacteria thrive, and the by-products of their metabolic processes typically include methane and the foul-smelling gas hydrogen sulfide. This phenomenon is called eutrophication and can happen naturally, but human activities can accelerate this process.

Though contaminants from land activities pollute the ocean, the oceans also suffer from other types of pollution such as oil spills. Almost half of the more than 3.5 million tons (3.2 million metric tons) of oil that enter the oceans every year is from natural offshore deposits. More than half of the rest is from oil dumped into inland water systems, and oil spills or breakages in pipelines or wells account for the rest, though this source receives the most media attention. A significant portion (about 25 percent) of spilled oil, that containing the most toxic components, evaporates and becomes air pollution within a few months. Bacteria naturally metabolize approximately 60 percent of the oil, and much of the rest sinks to the ocean floor. The thick oil or the emulsions it forms when mixed with water physically smothers some marine life near the surface. Marine animals have contact with the surface, where the oil floats, are most affected: marine mammals, reptiles, birds, and animals that live along the shoreline. In estuaries, beaches, and coastlines, the oil kills marine invertebrates and birds that inhale and ingest it. The oil that sinks to the bottom kills benthic life or accumulates in the tissues of filter feeders such as clams and mussels. Sometimes the harsh cleanup techniques cause at least as much environmental damage as the oil itself. The extent of the biological damage caused by an oil spill depends on the particular habitat, the physical and chemical conditions before the spill, and the season of year.

Hazardous wastes need to be disposed of in a manner that will not pollute the groundwater supply. This is often accomplished by storing the waste in underground steel tanks, but over time the tanks corrode and begin to leak into the surrounding earth. Radioactive waste generated by nuclear power plants and nuclear weapons facilities has a long life. Finding perpetual waste storage sites for hazardous and radioactive wastes is especially difficult because people do not want them close to their own cities or communities. Care must be taken to ensure the site is not located near places at risk of earthquakes or volcanoes. A solution that works for some wastes is detoxification, converting the toxic waste into a nontoxic form that allows for disposal in a landfill. Certain plants will uptake and store toxic products; this process is called phytoremediation. The plants, which are often associated with bacteria or fungi, act as biological sponges, absorbing chemicals or toxic elements from the soil or from water bodies.

Air consists of approximately 78 percent nitrogen, 21 percent oxygen, 0.04 percent carbon dioxide, and other trace components present in even smaller amounts. Many air pollutants have natural sources, such as sulfur oxides from volcanic activity, dust from wind storms, methane and hydrogen sulfide from decaying plants, or carbon monoxide, carbon dioxide, nitrogen oxide, and particulates from forest fires. Anthropogenic pollutants are those derived from human sources or activities. Though natural sources contribute more pollution to the environment than anthropogenic pollutants, the latter cause more significant, lasting damage. The U.S. Environmental Protection Agency identifies six major anthropogenic air pollutants as criteria air pollutants, those for which they have set national air quality standards: nitrogen dioxide, ozone, sulfur dioxide, particulate matter (PM), carbon monoxide, and lead. Except ozone and PM, all of these are directly emitted into the environment from three main sources: transportation, fuel combustion, and industrial processes. Ozone forms when nitrogen oxides react with volatile organic compounds (VOCs), and PM can be either directly emitted or formed during chemical reactions in the atmosphere. The effects of exposure to these pollutants on human health include headaches, dizziness, cardiovascular problems, respiratory illnesses,

and eye irritation. Fewer studies on the effects of these pollutants on nonhuman species have been performed, but some are known to be hazardous to plants. Lichens, symbiotic associations of fungi and photosynthetic organisms, are especially sensitive to air pollution, and scientists use them as an indicator of the air quality in a particular area. Contaminants in the air also cause severe damage to metals, buildings, and other structures.

In addition to having local effects on the environment and on human health, air pollutants harm global health. Ozone gas naturally present in the stratosphere, part of the upper atmosphere located more than six miles (10 km) above Earth's surface, screens out much of the Sun's dangerous ultraviolet radiation. The ozone molecules absorb the ultraviolet radiation, temporarily split, and then spontaneously reform ozone. Life on Earth would not be possible without this protection from the ozone layer, and even diminished quantities of ozone cause eye cataracts and skin cancer. In the 1970s scientists discovered that some synthetic chemicals destroy the ozone layer and have been researching and monitoring ozone levels ever since. According to the *Scientific Assessment of Ozone Depletion: 2002,* the most recent assessment carried out by the United Nations Environmental Programme and the World Meteorological Organization, the global average for the period 1997–2001 averaged about 3 percent lower than pre-1980 values, with most thinning occurring in the midlatitude and polar regions. A so-called ozone hole, a large region with localized thinning, exists over the Antarctic and experiences up to a 70 percent decrease in ozone each October, though the levels generally restore themselves by natural processes during the Antarctic winter.

Chlorofluorocarbons (CFCs), once thought to be inert, are the main anthropogenic threat to the ozone layer. Spray cans contain CFCs such as Freon as a propellant, and refrigerators, freezers, and air conditioners use Freon to chill materials. Other CFCs are used in the production of Styrofoam or for cleaning electronic or other types of equipment. Scientists discovered that ultraviolet radiation in the stratosphere causes dissociation of the CFCs, forming a very reactive chlorine free radical. The chlorine free radical reacts with ozone to form chlorine oxide and molecular oxygen. Chlorine oxide also reacts with free oxygen atoms released when ultraviolet radiation breaks down ozone; this results in the formation of more chlorine free radicals and molecular oxygen. One single chlorine free radical can break down 100,000 molecules of ozone.

Jets that fly in the stratosphere also contribute to ozone depletion, though not to the same degree as CFCs. They release nitric oxide gas, which reacts

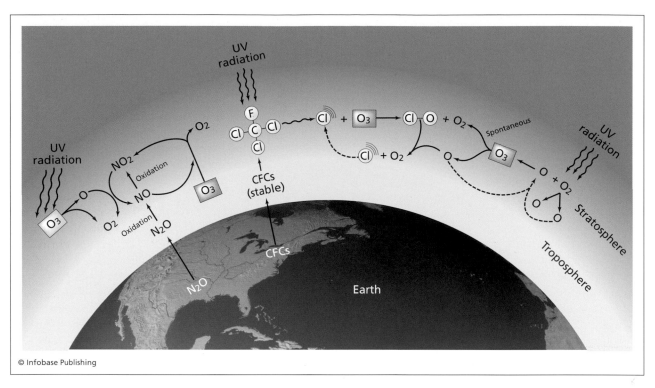

Certain chemical compounds released into the atmosphere cause the breakdown of ozone molecules.

with ozone to form nitrogen dioxide and molecular oxygen.

The depletion of ozone results in a diminished capacity for the stratosphere to block ultraviolet radiation from reaching the Earth's surface. In addition to the hazards this causes to human health, ecosystems could be destroyed as ultraviolet radiation can harm or kill algae, phytoplankton, and plants, the producers within biological communities. In 1987 many countries participated in the development of the United Nations Montreal Protocol, with the goal of addressing this global concern by controlling the levels of ozone-depleting chemicals manufactured and released into the environment. In 1992 nations around the world gathered again and formed the Copenhagen Amendments, with the aim of accelerating the phasing out of CFCs and other ozone-depleting chemicals. Today most CFCs have been replaced with hydrochlorofluorocarbons (HCFCs), which cause less damage, and researchers continue to look for other chemicals that can perform the same functions as CFCs without causing any harm to the ozone layer.

Some pollutants cause acids to be deposited in rain and snow or as dry, dustlike particles or as gases that settle on surfaces and turn into acids when combined with water. Even in the absence of pollution, rainwater is slightly acidic from atmospheric carbon dioxide, but atmospheric contaminants such as sulfur oxides and nitrogen oxides make precipitation even more acidic. While some natural events release these compounds into the atmosphere, the major anthropogenic source is the combustion of fossil fuels for transportation or electric power plants. Acid deposition causes acidification of lakes, which kills aquatic life not only through the decreased pH levels, but also through the increased amounts of metals dissolved from rocks and soils. Not only is aquatic life harmed; life-forms that feed off the aquatic organisms accumulate concentrated levels of the metals in their bodies. Acid rain also causes the destruction of forests and the communities they support by directly damaging the leaves, decreasing photosynthetic capabilities, dissolving nutrients in the soil so they wash away, and dissolving toxic substances that may damage structures involved in water transport.

GLOBAL CLIMATE CHANGE

According to an article, "Global Temperature Change," authored by James Hansen and others and published in *Proceedings of the National Academy of Sciences* in September 2007, the average Earth surface temperature has increased 1.4°F (0.8°C) since the late 19th century, and 1.1°F (0.6°C) in just the past three decades. On the basis of computer modeling, the Intergovernmental Panel on Climate Change (IPCC) predicts the average global surface temperature will continue to rise 2.0°F–11.5°F (1.1°C–6.4°C) during the 21st century. This worldwide phenomenon

Emissions from manufacturing plants and oil refineries can lead to acid rain, which can have a devastating effect on trees and other life-forms. *(Simon Fraser/Photo Researchers, Inc.)*

is called global warming, and scientists attribute it to anthropogenic air pollution. Certain chemical compounds present in the atmosphere are referred to as greenhouse gases because they trap heat in the lower atmosphere. The land, water, air, and life on Earth absorb radiation from the Sun and radiate it back as heat. Instead of moving back into outer space, however, the greenhouse gases reflect the heat back toward the Earth, increasing the surface temperature. Without this natural greenhouse effect, the temperature of the Earth would be too low for many current life-forms to survive—0°F (-18°C) compared to its actual temperature of about 57°F (14°C). Naturally occurring greenhouse gases include water vapor, carbon dioxide, methane, and nitrous oxide. Human activities have increased the levels of these compounds, enhancing the greenhouse effect, in addition to adding anthropogenic compounds including CFCs, HCFCs, and others, compounding the problem. Burning fossil fuels has disrupted the normal carbon cycle by releasing carbon that was stored underneath the Earth's surface back into the atmosphere. Coal mining and

combustion and the use of natural gas and chemical fertilizers have increased the levels of sulfur oxides, nitrous oxide, carbon dioxide, carbon monoxide, and methane released into the atmosphere. Though CFC levels are declining because of the Montreal Protocol and the Copenhagen Amendments, some will remain for at least the next 100 years. The increase in greenhouse gases has also affected rainfall patterns and the severity of storms.

Climate changes will have a tremendous ecological impact. Ecosystems function in a manner dependent on the presence and interactions of life-forms that have evolved through adaptations that are uniquely suited to particular conditions. The rapid changes in temperature, precipitation, and habitats could devastate ecosystems. In Earth's history, mass extinctions of 50–90 percent of species have accompanied global temperature changes of 5°C. Some effects are already apparent, such as the bleaching of coral reefs. Photosynthetic algae called zooxanthellae that live in symbiotic associations within the corals provide organic nutrition for the corals, and

One adverse effect of increased surface water temperatures is the destruction of zooxanthellae that inhabit coral tissues, which leads to the bleaching of coral reefs, as demonstrated by this photo taken in the Florida Keys. *(National Oceanic & Atmospheric Administration/Department of Commerce)*

the corals provide a protected habitat and a supply of carbon dioxide for the autotrophic zooxanthellae. The increased temperatures and violent storms can kill the zooxanthellae of coral reefs, and since these organisms confer the variety of bright colors on the corals, their death causes bleaching of the corals. If the loss of zooxanthellae persists, or if another population of zooxanthellae does not take over, the coral also will die. Coral bleaching can also result from other anthropogenic effects, such as overexploitation, increased sedimentation, chemical pollution in the water, and solar irradiation of shallower coral colonies.

Another significant concern related to the increased surface temperatures is the associated rise in sea levels due to melting glaciers and ice in the Antarctic. Over the past 50 years, the sea level has risen by four to six inches (10 to 12.5 cm), and the IPCC predicts that the sea level will rise by an additional 20 inches (50 cm) by the year 2100. This would not only affect coastal cities, but flood wetlands, draw in salt water that could destroy crops and ruin fertile soils, and lead to worse damage from storms and hurricanes.

SUSTAINABILITY

The best way to solve environmental problems is to prevent them. For all the concerns mentioned and for concerns not addressed here, education is the first step. People need to know the far-reaching impact of their actions. The collective action of many individuals can reverse dangerous trends, for example, reusing and recycling materials, making efficient decisions about energy use, using less water, and buying environmentally friendly products. Society needs to work together to find and implement sustainable solutions, ways to meet needs without harming the environment. Achieving this will require a change in attitude. Many systems society uses today—such as agriculture, energy production, transportation systems—worked well when the world population was small but now have environmental consequences that will affect the Earth for generations.

Society needs to use its resources more efficiently and lessen its dependence on fossil fuels. In the United States, recently ranked at the bottom of the list of industrialized nations for environmental performance (Esty et al. 2008), the EPA has enacted legislation regarding air and water pollution, the use and sale of toxic chemicals released into the environment, hazardous waste, oil spills, and more. Laws will stimulate positive change, but ethical and economic factors will affect attitudes and motivation, which will result in actions that protect the environment longer term.

See also BIODIVERSITY; CONSERVATION BIOLOGY; ECOLOGY; ENVIRONMENTAL SCIENCE.

FURTHER READING

Bourne, Joel K., Jr. "Fall of the Wild." *National Geographic* 209, no. 5 (May 2006): 42–77.

Energy Information Administration. U.S. Department of Energy. "International Total Primary Energy Consumption and Energy Intensity." Available online. URL: http://www.eia.doe.gov/emeu/international/energyconsumption.html. Accessed January 21, 2008.

Esty, Daniel C., Marc A. Levy, Christine H. Kim, Alex de Sherbinin, Tanja Srebotnjak, and Valentina Mara. *2008 Environmental Performance Index*. New Haven, Conn.: Yale Center for Environmental Law and Policy, 2008. Available online. URL: http://epi.yale.edu/Home. Accessed January 30, 2008.

Hansen, James, Makiko Sato, Reto Ruedy, Ken Lo, David W. Lea, and Martin Medina-Elizade. "Global Temperature Change." *Proceedings of the National Academy of Sciences* 103, no. 39 (September 26, 2007): 14,288–14,293.

Intergovernmental Panel on Climate Change. *IPCC Fourth Assessment Report. Climate Change 2007: Synthesis Report*. Released on November 17, 2007. Available online. URL: http://www.ipcc.ch/ipccreports/ar4-syr.htm. Accessed January 21, 2008.

National Oceanic and Atmospheric Administration. "Global Warming Frequently Asked Questions." Available online. URL: http://www.ncdc.noaa.gov/oa/climate/globalwarming.html. Updated March 29, 2007.

Soares-Filho, Britaldo Silveira, Daniel Curtis Nepstad, Lisa M. Curran, Gustavo Coutinho Cerqueira, Ricardo Alexandrino Garcia, Claudia Azevedo Ramos, Eliane Voll, Alice McDonald, Paul Lefebvre, and Peter Schlesinger. "Modelling Conservation in the Amazon Basin." *Nature* 440, no. 7083 (March 23, 2006): 520–523.

United Nations Department of Public Information. *Water for Life Decade (2005–2015)*. Available online. URL: http://www.un.org/waterforlifedecade. Accessed August 18, 2008.

U.S. Environmental Protection Agency. "The Science of Ozone Depletion." Available online. URL: http://www.epa.gov/ozone/science/index.html. Updated August 20, 2008.

Wallace, Scott. "Last of the Amazon." *National Geographic* 211, no. 1 (January 2007): 40–71.

Environmental Protection Agency The U.S. Environmental Protection Agency (EPA) is an independent regulatory federal agency whose mission is to protect human health and the environment. The president of the United States appoints an administrator to lead the agency, which employs more than 18,000 people at the headquarters in Washington, D.C.; 10 regional offices; and one dozen labs across the country. The EPA is involved in all aspects of environmental science research, education, and

MAJOR ENVIRONMENTAL LAWS

Law	Summary
Clean Air Act (CAA), 1970	regulates air emissions from area, stationary, and mobile sources
Clean Water Act (CWA), as amended 1977	regulates discharges of pollutants into U.S. waters
Emergency Planning and Community Right to Know Act (EPCRA), 1986	helps local communities protect public health, safety, and the environment from chemical hazards
Endangered Species Act, 1973	provides a program for the conservation of threatened and endangered plants and animals and the habitats in which they are found
Federal Insecticide, Fungicide, and Rodenticide Act (FIFRA), 1996	provides federal control of pesticide distribution, sale, and use
Freedom of Information Act (FOIA), 1966	allows any person to make requests for government information without requiring identification or explanation
National Environmental Policy Act, 1969	assures that all branches of the government consider the environment prior to undertaking any major federal action that significantly affects it
Occupational Safety and Health Act (OSHA), 1970	ensures that employers provide a place of employment free of recognized hazards to safety and health
Oil Pollution Act (OPA), 1990	strengthens EPA's ability to prevent and respond to catastrophic oil spills
Pollution Prevention Act (PPA), 1990	focuses industry, government, and public attention on reducing the amount of pollution through cost-effective changes in production, operation, and raw materials use
Resource Conservation and Recovery Act (RCRA), 1976	gives the EPA authority to control hazardous waste generation, transportation, treatment, storage, and disposal
Safe Drinking Water Act (SDWA), 1974	protects the quality of drinking water in the United States
Comprehensive Environmental Response, Compensation, and Liability Act (CERCLA or Superfund), 1980	provides a federal "Superfund" to clean up uncontrolled or abandoned hazardous waste sites as well as accidents, spills, and other emergency releases of pollutants and contaminants into the environment
Superfund Amendments and Reauthorization Act, 1986	reauthorizes CERCLA to continue cleanup activities around the country
Toxic Substances Control Act (TSCA), 1976	gives the EPA the ability to track the 75,000 industrial chemicals currently produced or imported into the United States, screen the chemicals, require reporting or testing of certain chemicals, and ban the manufacture and import of chemicals that pose an unreasonable risk

assessment. They develop and enforce regulations, provide financial support for state environmental programs, perform environmental research, sponsor voluntary partnerships and programs, and educate the public to inspire personal responsibility for the environment. Examples of topics that the EPA investigates include acid rain, hazardous waste, asbestos, mold in homes, ozone, radon gas, recycling, wetlands, and many more.

In response to the public demand for a healthier and cleaner environment, stimulated in part by Rachel Carson's writing of *Silent Spring* (1962), which educated people about the potential dangers of indiscriminate pesticide use, President Richard M. Nixon proposed and Congress approved the formation of the EPA in 1970. In order to regulate environmental issues more efficiently, the EPA consolidated several preexisting programs coordinated by the Department of the Interior; the Department of Health, Education, and Welfare; the Department of Agriculture; the Federal Radiation Council; and the President's Council on Environmental Quality.

At least 15 major environmental laws form the legal basis for many of the EPA's programs. These are summarized in the table on page 307.

See also CARSON, RACHEL; ENVIRONMENTAL CONCERNS, HUMAN-INDUCED; ENVIRONMENTAL SCIENCE.

FURTHER READING

U.S. Environmental Protection Agency home page. Available online. URL: http://www.epa.gov. Updated January 18, 2008.

environmental science Environmental science is an interdisciplinary field concerned with the interaction of the natural components of the environment, including the living and nonliving matter and the forces that act on them. In particular, environmental science focuses on the interconnected issues of the human population, natural resources, and pollution. The long-term goal is to achieve sustainability—using and managing the Earth's resources efficiently without depleting or irreparably damaging them. Understanding the environment will help society to use its resources better, minimize or prevent any detrimental impact, and solve environmental problems. Current environmental problems include degradation of the land, overuse of natural resources such as water, air and water pollution, acid rain, depletion of ozone in the stratosphere, production of greenhouse gases in the atmosphere, generation of nuclear waste, and contamination of the seas by oil spills. Many governments have recognized these alarming negative environmental trends and taken steps toward reversing them by developing programs to monitor such trends and to encourage sustainable economic development and sustainable societies. Sustainable development focuses on meeting current needs without compromising the ability of future generations to meet theirs. Measures to improve sustainability include recycling, using renewable energy resources, restoring that which has been used or damaged, and managing population growth.

Almost every other natural science is related to environmental science in some way, and one can approach studies of the environment from any of these perspectives. From the perspective of life science, preserving the atmosphere, the chemistry, and the geology of the environment is necessary in order to sustain healthy living conditions for all of the planet's life-forms. Thus ecology, the study of the relationships and interactions between organisms and their environment, is an important subdiscipline of environmental science. Other subdisciplines include atmospheric science, environmental chemistry, and geoscience. Atmospheric science is the study of the atmosphere. Atmospheric scientists, also called meteorologists, study the physical and chemical properties of the atmosphere. Emissions from automobiles and from industries pollute the air and contribute to the greenhouse gases in the atmosphere, a growing environmental concern. Environmental chemistry is the study of chemicals and chemical reactions that occur in the environment, both naturally, as the way excess rainfall can leach positively charged ions such as calcium and magnesium from the soil, making it more acidic and affecting the biological availability of nutrients, and as a result of human activities, such as the disposal of hazardous wastes. Environmental science can also be studied from a geoscience perspective, as the natural geological processes such as weathering, erosion, climatic phenomena, earthquakes, and volcanic activities all affect landscape structures and the chemical and physical conditions of an ecosystem.

HUMANS AND THE ENVIRONMENT

People are simply part of nature, despite the fact that they exhibit higher intelligence, communicate using language, live in artificially constructed buildings, and depend on numerous technological advancements and inventions to live every day. Many people would have a tough time surviving isolated from the conveniences of modern society—without clothing, tools, electronics, housing, running water, or readily available food. One might think it would be near impossible to depend on nature so completely to provide sustenance and fulfill every other need. In reality, all of society already completely depends on nature. Everything a person sees, touches, makes, and uses is derived from components the environment provides. Though a colorful house with vinyl siding, glass windows, and a concrete driveway might look artificial if placed in the middle of an undisturbed forest or desert, all the building materials, the tools and machinery used to clear the land and construct the house, and the energy that was supplied to perform the work of assembling it all originated in the earth. Not only do food, fibers for clothing, and wood for lumber originate in the environment, but people also depend on natural resources for everything from computer chips to vaccinations whether they recognize it or not. Beyond material needs, Earth's ecosystems provide many services that maintain the conditions necessary to support life on this planet. For example, ecosystems remove carbon from the atmosphere and incorporate it in organic molecules used as food, they produce oxygen necessary for respiration through photosynthesis, they decompose organic matter and return the chemical nutrients to the environment, and they move water through the hydrologic cycle. The inhibition of any of these functions alone would result in the extinction of most of the Earth's life-forms.

Because people cannot be separated from nature, their actions affect nature. When people take in and use natural resources, such as air, water, minerals, plants, fuels, and animals, to meet their needs, they return them to the environment in a different form or alter the environment in the process. This holds true even for natural processes such as breathing: the air people exhale contains more carbon dioxide and less oxygen than the air they inhale. Burning fossil fuels for transportation or for generating of electricity releases particulates, sulfur oxides, nitrogen oxides, carbon monoxide, and carbon dioxide into the environment. In order to grow sufficient quantities of food, farmers must plant increasing acreage of crops for which they clear and cultivate grasslands and forests. The soil becomes depleted of its nutrients, the addition of fertilizer becomes necessary, and the fertilizer runs off, leaches into the soil, and is carried away into the water supply. The ecosystems are destroyed, the biomes altered, and resources ruined.

The goal of environmental science is sustainable development—to management and use of resources in ways that preserve the natural environment. People cannot withdraw from the biological community, but knowledge gained from environmental science can be used to protect the physical environment while meeting people's needs. When environmental issues first emerged into the limelight in the 1960s and 1970s, the initial approach was to treat symptoms. For example, when the dangers of the pesticide DDT became apparent, the public demanded the prohibition of its use. While admirable, this approach will not solve environmental issues in the long term. Without working to develop alternative strategies for pest control, will replacement chemicals be just as dangerous? Will farmers simply use more land to grow more crops to make up for those destroyed by pests? Modern environmental science aims for longer-term solutions by taking a more comprehensive approach that focuses on the root causes of the problem. This is analogous to an individual's adopting a healthier lifestyle that involves nutritious food choices, regular exercise, and weight control rather than treating the symptoms of heart disease or diabetes, diseases associated with obesity. Instead of alleviating the symptoms, or even curing the disease after ill effects occur, the goal is to prevent the disease. For example, consider the effect of an increasing human population on many environmental issues. Decreasing the population growth rate would translate into a smaller population that would require less food and therefore less land to grow the food. Reduced quantities of nonrenewable energy resources would be needed to transport the food and for transportation in general.

In addition to stabilizing the population, other objectives that society must meet in order to achieve sustainability include planning the use of landscapes for purposes based on ecologically sound principles, such as using nutrient-rich soil for agricultural purposes; using resources wisely—for example, not wasting water by watering the lawn in the heat of the day when most of it will quickly evaporate; using renewable clean energy sources such as solar power or wind power; recycling goods and using recycled goods in manufacturing; restoring habitats and ecosystems that human activities have destroyed; and preventing any further damage to the environment. In order to meet these objectives, one must also explore the biological, cultural, psychological, and economic influences on society's current attitudes and behavior. Thus the current need for environmental scientists with a broad-based background is as great as for those trained for more specialized research.

CAREERS IN ENVIRONMENTAL SCIENCE

The training necessary to pursue a career in environmental science depends on the type of career one desires. Knowledge of how to utilize the scientific method is crucial for identifying problems, collecting information about a subject, objectively analyzing data, and designing strategies for preventing and solving environmental problems. A student should consider the aspect of environmental science that he or she finds most interesting—for example, the Earth's resources and the natural biogeochemical means for recycling them, the effect of human activities on biodiversity, the development of new technologies that reduce waste, the mechanisms by which living systems break down toxic chemicals, or other aspects. A future environmental scientist must obtain a strong background in the key principles of environmental science: how organisms interact with biotic and abiotic factors in their environment, the functions of an ecosystem, the unique qualities of different biomes, human impact on the environment, and how living organisms respond to changes in environmental conditions. One might choose to specialize in environmental aspects of a specific subject, such as biology, chemistry, physics, or Earth science, or focus on a specific issue, such as surface water pollution, chemical pesticides, deforestation, biofuels, sustainable transportation systems, and so on. Postgraduate studies or additional training prepares a scientist to become a principal investigator at a research facility, to obtain an academic appointment with a research institution, or to attain a higher-level position with more responsibilities.

Environmental scientists work in a variety of jobs in a variety of atmospheres—at a desk in an office, in a laboratory, in a greenhouse, or outdoors in fields or streams. Governmental agencies might hire an environmental scientist to monitor a specific trend, such as the annual water consumption in a particular area, or to serve as chief naturalist at a national park. A water treatment facility might need an environmental scientist to find ways to eliminate sewage overflows and reduce discharges. An environmental scientist might work at an academic research institution and study the effect of decreasing wetlands on the population of different amphibian species and how that might affect the food web of that community. The agricultural industry employs environmental scientists to develop new crop strains that naturally resist pests in order to decrease the need for chemical pesticides that can adversely affect other species or add to the problem of chemical pollution. An environmental scientist working for a commercial industry might assay gaseous emissions and recommend methods for decreasing the levels released into the atmosphere. A conservation organization might hire an environmental scientist as an expert consultant to testify concerning the likelihood that chemicals released into the soil are harming a neighboring prairie that serves as a habitat for an endangered wildflower species in a trial against a local manufacturing plant.

Knowledge gained through studies in the environmental sciences is applied to enact legislation aimed at preserving the long-term well-being of the planet. Because some of the root causes of current environmental issues are relevant to politics, economics, religious beliefs, ethics, and personal attitudes, environmental scientists often work closely with social scientists to develop and implement sustainable solutions.

See also ECOLOGY; ENVIRONMENTAL CONCERNS, HUMAN-INDUCED.

FURTHER READING
Chiras, Daniel D. *Environmental Science*. 7th ed. Sudbury, Mass.: Jones and Bartlett, 2006.

enzymes Biological catalysts called enzymes speed up the rate of chemical reactions that take place in living cells. Most enzymes are proteins, but some ribonucleic acids (RNAs) have catalytic activity. Enzymes are not consumed or altered during a chemical reaction, and they are specific for the substrates upon which they act. Chemical reactions will only proceed if they are energetically favorable: in other words, an overall decrease in free energy must accompany the reaction. Enzymes cannot force a reaction to proceed if it is energetically unfavorable, but they do lower the activation energy of a reaction, that is, the energy needed to raise a molecule to its transition state so as to undergo a particular reaction. Without a high energy barrier, the velocity of the reaction increases. One molecule of enzyme may catalyze thousands or tens of thousands of biochemical reactions per second.

Enzymes are often named for the type of biochemical reaction they catalyze, and their names often end in *-ase*. For example, oxidoreductases catalyze oxidation-reduction reactions, and transferases catalyze the transfer of certain groups of atoms (such as methyl groups) from one molecule to another. Hydrolases are enzymes that cleave water molecules, often in conjunction with the breakdown of another molecule. Isomerases facilitate the conversion of isomers between forms. Lyases cleave covalent linkages between carbon and another atom, and ligases assist in the formation of bonds between carbon and another atom.

ENZYME FUNCTION

During a chemical reaction, a substrate is converted to a product. Chemists sometimes add a catalyst to speed up the reaction. Catalysts increase the reaction rate but are unchanged during the reaction, meaning after the reaction has occurred, the same amount is present as before the reaction took place. Because of this, only small amounts are necessary. Common catalysts include pure elements such as platinum, compounds such as manganese dioxide, and ions such as copper ions. Heat can also increase the velocity of a reaction. The conditions of living systems with respect to temperature, ionic strength, and concentration of trace elements are very particular, however, and thus common chemical catalysts are not appropriate, effective, or safe. Though all reactions, including biochemical reactions that occur within living organisms, must be spontaneous, they may proceed at rates so slow that no product would be synthesized during the organism's lifetime. Biological catalysts called enzymes facilitate the progress of biochemical reactions, effectively increasing the rate by 1,000 to 100 million times the rate without a catalyst present. The enzyme does not alter the equilibrium of the reaction, just the rate at which the reaction proceeds.

Enzymes work by lowering the activation energy of a biochemical reaction. The activation energy is the energy required to move a substrate (S) from its original form to a higher-energy transition state (S*). The transition form is unstable and soon converts to product (P), which has a lower energy state than both the original substrate and the transition state in a spontaneous reaction. The activation energy serves as a barrier that must be overcome in order for the reaction to proceed.

$$S \rightleftharpoons S^* \rightarrow P$$

Enzymes (E) work by first binding their substrates to form an enzyme-substrate (ES) complex, providing a surface upon which the reaction can take place; then the substrate converts to product. Formation of ES facilitates the transformation of substrate to its higher-energy transition state, so product forms more quickly.

Enzymes are highly specific for the substrate(s) they bind. Substrates interact with one particular region of the enzyme called the active site. Multiple weak bonds such as hydrogen bonds, ionic bonds, and hydrophobic interactions confer specificity to the interaction between the active site of the enzyme and the substrate. The simplest model for the interaction of an enzyme with its substrate is the lock-and-key model, in which the lock represents the enzyme and the key represents the substrate. A more accurate representation is the induced fit model, in which the enzyme and the substrate change conformation slightly as they approach one another. In the more flexible induced fit model, mild distortion of the enzyme and substrate allows the substrate to begin its transformation to the transition state more quickly, further accelerating the progress of the reaction.

Depending on the type of chemical reaction, the formation of the enzyme-substrate complex either puts strain on covalent linkages that must break or holds two substrates in the proper position and orientation to facilitate the formation of the transition state. Once the unstable transition state forms, either the bonds break or new covalent linkages form, creating product.

Some enzymes require additional nonprotein components called cofactors. For example, the enzyme that hydrolyzes the last amino acid in a polypeptide chain, carboxypeptidase A, requires a zinc ion at the active site. When the cofactor is an organic molecule, it is referred to as a coenzyme and is recycled between enzymatic reactions. Many coenzymes are derivatives of vitamins; for example, nicotinamide adenine dinucleotide (NAD$^+$) is derived from niacin, tetrahydrofolate is derived from folate, thiamine pyrophosphate is derived from thiamine, and flavin adenine dinucleotide is derived from riboflavin.

ENZYME KINETICS

Enzyme kinetics refers to studies examining the rate (V) at which an enzyme catalyzes change from substrate to product. In 1913 the German biochemist Leonor Michaelis and the Canadian medical scientist Maud Menten proposed a model explaining the kinetic behavior of many enzymes. Most enzymes (E) follow the Michaelis-Menten model, in which the substrate concentration plotted against the initial reaction velocity assumes a hyperbolic curve. The reaction rate, or the number of substrate molecules converted to product per unit of time, increases as substrate concentration increases until the maximal

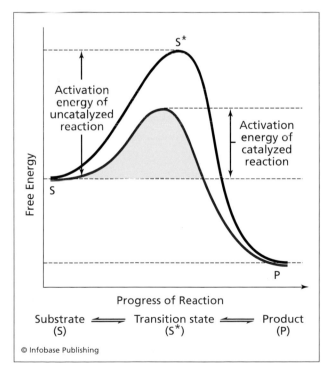

Enzymes work by lowering the activation energy required for the substrate to achieve the transition state.

Lock-and-Key Model

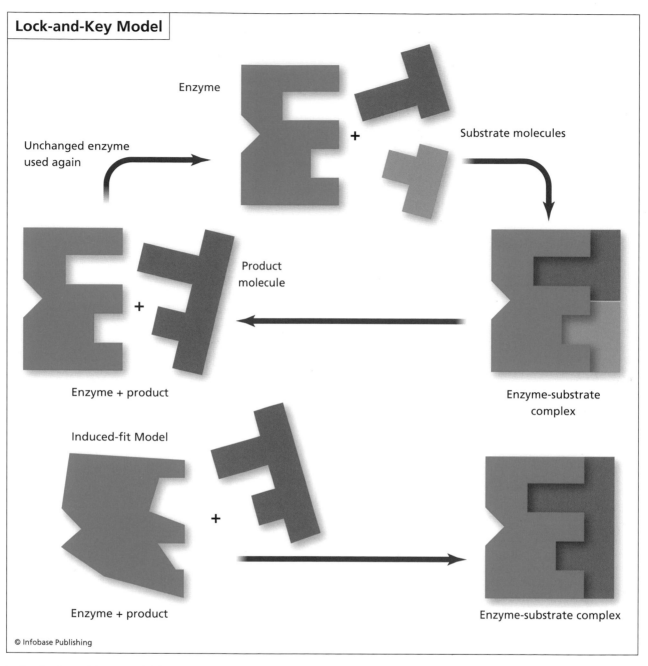

In the lock-and-key model for the interaction of enzyme with substrate, the active site of the unbound enzyme is complementary to the shape of the substrate, whereas in the induced fit model, the enzyme changes shape as the substrate binds.

velocity (V_{max}) is reached. Michaelis and Menten suggested that formation of the enzyme-substrate (ES) complex was reversible or that it could lead to product formation, as depicted by the following equation:

$$E + S \rightleftharpoons ES \rightarrow E + P$$

If k_1 is the rate constant for the formation of ES, k_{-1} is the rate constant for the reverse reaction, and k_2 is the rate constant for the breakdown of ES into product, then K_m is the Michaelis constant, given by

$$K_m = \frac{(k_{-1} + k_2)}{k_1}$$

One can determine K_m and V_{max} by measuring the rate of catalysis at different substrate concentrations ([S]). The reaction velocity (V) is represented by

$$V = \frac{v_{max}[S]}{K_m + [S]}$$

when three assumptions are made: 1) the concentration of substrate is much greater than the concentration of enzyme ([S] > [E]); thus the reaction rate will directly depend on the enzyme concentration; 2) the concentration of the enzyme-substrate complex remains constant; and 3) V is measured immediately after the enzyme and substrate are combined so that the amount of product present is negligible.

K_m is an important aspect of the Michaelis-Menten model. As the substrate concentration at which half of the active sites are filled, K_m serves as a measure of the characteristic affinity of an enzyme for its substrate, the relative attractive force between the two. If K_m is large, then the enzyme has a low affinity for its substrate; if K_m is small, then the enzyme has a high affinity for its substrate. If the reaction velocity is plotted as a function of substrate concentration, then one can determine K_m by taking the substrate concentration that correlates with half of V_{max}. When V_{max} is difficult to determine because of a gradual slope of the curve at high substrate concentrations, a Lineweaver-Burke plot (named after its discoverers, Hans Lineweaver and Dean Burke)

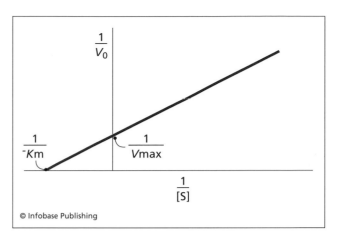

A double reciprocal plot, also called a Lineweaver-Burke diagram, is useful for determining kinetic parameters such as V_{max} and K_m.

can help to determine K_m. A double reciprocal of the Michaelis-Menten equation forms the equation for a Lineweaver-Burke plot.

$$\frac{1}{V} = \frac{K_m}{V_{max}[S]} + \frac{1}{V_{max}}$$

On a Lineweaver-Burke plot, the intercept of the x axis equals $-1/K_m$, and the intercept of the y axis equals $1/V_{max}$.

Allosteric enzymes do not follow Michaelis-Menten kinetics. When the reaction velocity is plotted as a function of the substrate concentration, the result is a sigmoidal (S-shaped) curve rather than a hyperbolic curve. In these cases, the enzyme has more than one active site. After substrate binds to one active site, the affinity of the enzyme for additional substrates increases. The first binding event causes a slight conformational change that facilitates binding at the other sites.

In addition to substrate concentration, other factors also affect the velocity of a reaction. Enzymes function most efficiently within a specific range of temperatures. As the temperature increases up to the optimal temperature, the velocity of the reaction will slowly increase. At temperatures higher than the optimal temperature, the velocity will drop off, first gradually, then more sharply as the enzyme becomes denatured. Enzymes also typically function within an optimal pH range. Ionic interactions often play a role at the active site, and, depending on the pH, ionizable side chains may exist in their protonated forms or deprotonated forms, which affect their ability to participate in ionic bonds. Deviation from the optimal pH is also important because in highly acidic or highly basic conditions, the enzyme itself might denature and therefore lose functionality.

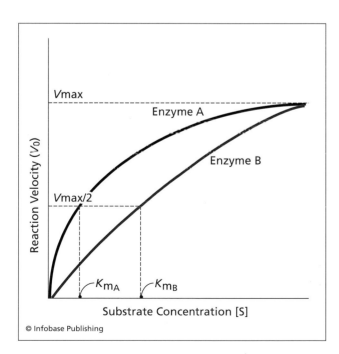

Enzymes that follow Michaelis-Menten kinetics exhibit a hyperbolic curve when the reaction velocity is plotted as a function of the substrate concentration. In this diagram, which compares two enzymes with a similar V_{max}, enzyme A has a higher affinity for its substrate, as reflected by its lower K_m, than enzyme B.

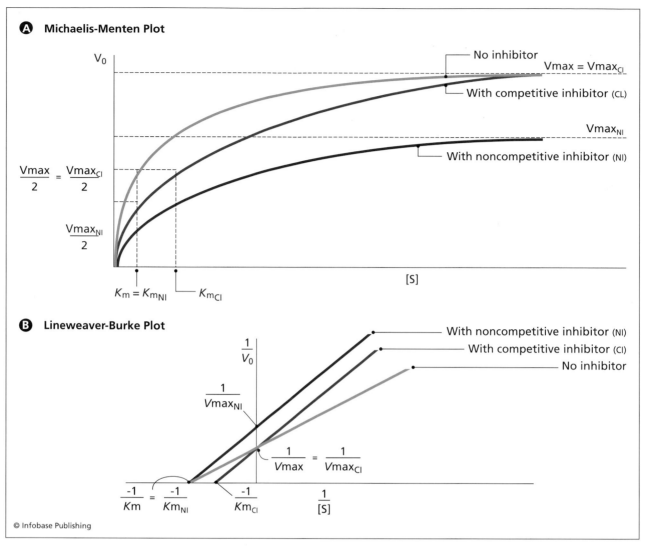

As seen in the Michaelis-Menten and Lineweaver-Burke plots of inhibition, competitive inhibitors decrease K_m but not V_{max}, and noncompetitive inhibitors decrease V_{max} but not K_m.

REGULATION OF ENZYMES

Cells must be able to control the chemical reactions occurring within them in order to respond appropriately to changed conditions, such as different stages during development or in a life cycle, in the presence or absence of various nutritional sources, at different temperatures or altered water availability. In multicellular organisms, cells are organized into tissue types and organs that carry out unique functions, and the proteins each cell type expresses give the cell its specificity. Evolution has selected for cells to be energy efficient, and thus cells typically only make the proteins that they might actually use. Thus one method of controlling which chemical reactions a cell carries out is controlling the expression of an enzyme. This includes whether or not the enzyme is synthesized at all as well as how much is made. If more enzyme is present, more substrate will be converted into product.

Making a protein takes a lot of time, so when the cell needs to be able to start converting substrate to product without delay, a better method of control is to regulate the activity of the enzyme. In this case, the enzyme is already present in the cell in sufficient quantities, but something else needs to happen before it gains function. One common modification used by cells to regulate enzyme activity is phosphorylation. The addition or removal of a phosphoryl (PO_3) group to one or more amino acid side chains can either activate or deactivate an enzyme. Some enzymes are secreted as zymogens, inactive precursors that become functional after the cleavage of one or more peptide bonds.

Allosteric regulation is a type of regulation in which small molecules bind to the enzyme at a site other than the active site, causing the enzyme to change its conformation slightly and affecting its activity. Inhibitors are molecules that reduce the activity of an enzyme after binding. Feedback inhibition is a common regulatory mechanism for stopping the synthesis of a biomolecule that is already present at sufficient levels. In feedback inhibition, a product of a metabolic pathway binds to and allosterically inhibits the activity of an enzyme involved in its own synthesis. In this manner, the molecule self-regulates its production. As the metabolic pathway proceeds, the product accumulates. When the concentration reaches sufficient levels, the product "feeds back" to bind an enzyme that participates in an early step of the metabolic pathway, halting its further production until the concentration of the product decreases to the point where quantities are not great enough to bind and inhibit the enzyme, lifting the inhibition.

Inhibitors can be classified as either competitive or noncompetitive. A competitive inhibitor is a molecule that binds at the enzyme's active site. The inhibitor and the substrate compete for binding to the active site; if one occupies the active site, then the other cannot bind there. If both are present, the relative affinities of the enzyme for the two molecules and the concentrations of each will determine which "wins" the competition. One hallmark of competitive inhibition is the ability to overcome it by the addition of more substrate. In noncompetitive inhibition, the inhibitor binds to a region of the enzyme other than the active site. Upon binding of the inhibitor at a regulatory site, the enzyme shifts its conformation, adopting a slightly different structure that interferes with its ability to function. Because the inhibitor and the substrate do not compete for the same site on the enzyme, increasing the substrate concentration cannot overcome noncompetitive inhibition.

A Lineweaver-Burke plot helps to distinguish between competitive and noncompetitive inhibition. In competitive inhibition, the y intercept ($1/V$ at a substrate concentration of 0) is the same whether the inhibitor is present or not; this is because V_{max} remains unchanged. The slope of the line (K_m/V_{max}), however, increases, because the K_m increases in the presence of a competitive inhibitor. Remember that K_m is the substrate concentration at which half of the maximal velocity is obtained. The increase in K_m reflects the fact that the inhibitor and the substrate compete for binding to the same site and that the addition of more substrate can overcome the inhibition to achieve the maximal velocity. V_{max} does not change; it just requires more substrate to attain it when inhibitor is present. A plot of data measured in the presence of a noncompetitive inhibitor will have a higher y intercept, reflecting a decrease in the value of V_{max}. The slope also increases, but this change is due to a decrease in V_{max}, rather than an increase in K_m, as in competitive inhibition. K_m remains unchanged because the enzyme's ability to bind the substrate does not change in this situation. No matter how much additional substrate is added, the inhibitor can still bind to the regulatory site and reduce the enzyme's activity; thus V_{max} is lower in the presence of noncompetitive inhibitor.

See also BIOCHEMICAL REACTIONS; BIOCHEMISTRY; BIOENERGETICS; BIOMOLECULES; CHEMICAL BASIS OF LIFE.

FURTHER READING

Berg, Jeremy M., John L. Tymoczko, and Lubert Stryer. *Biochemistry.* 6th ed. New York: W. H. Freeman, 2006.

Champe, Pamela C., Richard A. Harvey, and Denise R. Ferrier. *Lippincott's Illustrated Reviews: Biochemistry.* Philadelphia: Lippincott, Williams & Wilkins, 2005.

Copeland, Robert A. *Enzymes: A Practical Introduction to Structure, Mechanism, and Data Analysis.* 2nd ed. New York: John Wiley & Sons, 2000.

Purich, Daniel L. *The Enzyme Reference: A Comprehensive Guidebook to Enzyme Nomenclature, Reactions, and Methods.* San Diego: Academic Press, 2002.

epilepsy Epilepsy is a chronic, noncontagious neurological condition characterized by recurrent seizures, disruptive electrical discharges in the brain. Seizures occur when nerve cells in the brain send out abnormal bursts of electrochemical impulses, causing a range of symptoms from barely noticeable moments of unresponsiveness to violent convulsions. Approximately 0.5–2 percent of all people will experience epilepsy during their lifetime, but it occurs most commonly in young children or the elderly. More than 2 million Americans have epilepsy, and up to 200,000 new cases are diagnosed in the United States each year. Medication successfully treats the majority of cases, allowing people who have epilepsy to live relatively normal lives.

In ancient Greece and Rome, many people believed that seizures were caused by the possession of evil spirits and that the spirits could move from one person to another by touch. Until the 15th century, religious leaders vainly performed spiritual or magical rituals to heal the affected. Society shunned people who had epilepsy and isolated them in institutions. During the 1800s, the English neurologist John Hughlings Jackson proposed that seizures resulted from abnormal discharges in the brain, and although society's acceptance of the illness improved, misunderstandings persisted. In the early 1900s, out of fear that the disease was contagious or was inherited, people with epilepsy were prohibited from some public places or from marrying. A few very rare forms of epilepsy do have a genetic component, but epilepsy is never contagious; nor does it cause mental retardation or mental illness.

BIOLOGY OF A SEIZURE

The brain, one of the most complex and least understood organs of the human body, is responsible for movement, sensory perception, language, physiological processes, emotions, memory, and learning. As the control center of the nervous system the brain has the major function of receiving information from inside and outside the body, processing the input, and coordinating body movements and functions. Cells called neurons form networks that carry signals throughout the body. Specialized cells have the ability to sense environmental stimuli, such as a musical melody or the prick of a rose thorn, and to transform the input into a nervous impulse that a neuron transmits to the brain, which directs the appropriate response. For example, sound waves from a ringing telephone will cause the eardrum to vibrate, an action that disturbs fluid in the inner ear. The motion of the fluid jiggles tiny hairs, a mechanical motion that results in the formation of electrical impulses within auditory neurons (neurons involved in hearing). The impulses travel along the length of the neurons until they reach the portion of the brain involved in auditory responses. The brain processes the information and responds by sending out new impulses to motor neurons (neurons involved in muscle movement). The motor neurons carry the signals to the muscles needed to walk to the ringing phone and pick it up.

A seizure results when nerve cells in the brain fire a surge of uncontrolled electrical impulses. The highly coordinated, precise communications system among neurons becomes disordered commotion. The effect of the misfiring depends on the size and function of the affected area of the brain, and the episodes are characterized by unique feelings, sounds, smells, tastes, and sights. If the confused electrochemical activity occurs within a restricted portion of the brain responsible for sight, then the person might experience temporary blurred vision. If the seizure involves a large portion of the brain, the person might lose consciousness and suffer violent convulsions. Other epileptics (people who have epilepsy) report feeling a tingling sensation or appear to stare into space for a brief period, conscious but unaware of their surroundings. Seizure duration usually varies from a few seconds to several minutes. Though an isolated seizure does not damage the brain, seizures lasting longer than five minutes or immediate repeated seizures can be dangerous.

Neurologists categorize seizures into two broad categories, partial and generalized seizures, based on the degree to which the brain is affected. A person remains conscious during a partial seizure, the more common type, which affects only part of the brain (the focus) and can be simple or complex. The symptoms of a partial seizure depend on the specific part of the brain that is affected: auditory, visual, voluntary muscular movement, and so on. During a complex partial seizure, the person remains conscious but loses the ability to interact with other people. He or she might be aware of the surroundings but is unable to respond to them. The person might mumble, twitch, stare, wander around, perform repetitive or other seemingly deliberate movements, exhibit other unusual behavior, or simply seem confused, but these actions are beyond his or her control. These seizures are the most common type in teenagers and adults, and they last only a minute or two. Some epileptics have only one type of seizure; others have more than one type.

Generalized seizures affect the entire brain and include absence, tonic clonic, atonic, and myoclonic seizures. Absence seizures, also called petit mal seizures, are most common in children and cause someone to have a blank expression and be unresponsive for a few seconds. People may mistakenly assume the child is just not paying attention. Many children outgrow these seizures before reaching adulthood. The most dramatic type of seizure, tonic clonic seizures, also called grand mal seizures, are characterized by a brief initial tonic phase when the person's body extends and stiffens, followed by a clonic phase during which the person's muscles repeatedly contract for a period of one or two minutes. The person may also cry out or foam at the mouth. Atonic seizures cause a person's muscles suddenly to relax, causing the head to droop and the body to drop to the ground if unsupported. Myoclonic seizures are characterized by jerking muscles. Someone might be sitting at the table reading the paper when suddenly his or her arms flail up. They last only a second or two and frequently occur upon waking.

CAUSES OF EPILEPSY

Though epilepsy is more common in people younger than 20 years old or older than 65 years old, people of any age, gender, or race can develop epilepsy, which encompasses a variety of conditions that result in recurrent, unprovoked seizures. More than 70 percent of all epilepsy cases are idiopathic, meaning no obvious structural or functional abnormality can be found. Several known factors that provoke seizures include head injury, tumors, strokes, infections such as encephalitis or meningitis, high fever, toxic chemicals, drug overdose, low blood sugar level, and genetic factors. Approximately 4 percent of children experience febrile seizures, induced by high fevers, but they usually outgrow them. A few women have seizures related to hormone level fluctuations during their menstrual cycles. Seizure frequency tends to increase during pregnancy of women who have epilepsy, probably the result of a reduction in dosage or stoppage of their medication or their body's absorbing or metabolizing the drug at a different rate.

When a person has a seizure, that does not mean he or she has epilepsy. Epilepsy is a condition in which a person has a tendency for recurrent, unprovoked seizures. In the presence of a medical explanation, such as a traumatic head injury sustained during a high-speed automobile accident, the epilepsy is said to be symptomatic. The reason for the brain's susceptibility to seizures is explained. Oxygen deprivation, such as might be sustained by a baby during a difficult childbirth, can disturb the intricate electrical system in the brain, as can developmental defects, neurodegenerative diseases such as Alzheimer's disease, and strokes.

DIAGNOSIS AND TREATMENT

Most people who have seizures are otherwise healthy, and because seizures occur at random, the physician usually cannot get observe them, making diagnosis difficult. A doctor will consider the events and symptoms preceding, during, and following a seizure as well as the patient's medical history to determine whether that person has epilepsy. The brains of patients with epilepsy often exhibit distinctive activity patterns, as shown by an electroencephalograph (EEG), a machine capable of detecting, amplifying, and recording brain electrical patterns (brain waves). The test is painless and only lasts about 30 minutes. During the procedure, electrodes connected to the EEG are pasted to the patient's forehead, temples, and scalp. Sometimes the technician will ask the patient to perform a task to stimulate brain activity, such as rapid breathing, or the technician may provide a stimulus, such as flashing lights, which trigger seizures in some patients. The tracings from an EEG showing normal brain activity are smooth and the peaks are rounded, whereas sharper peaks indicate erratic electrical activity typical of a seizure. The pattern of spikes and waves in the resulting electroencephalogram helps the physician determine the type of seizure and the area of the brain that is affected. Multiple EEGs might be necessary if the patient's brain exhibits normal brain waves between seizures. Because fatigue increases the likelihood of seizures, sometimes the doctor will request the patient stay up all night before the EEG to increase the chance of being able to record the erratic brain activity.

If the EEG shows abnormal activity, further testing may be performed to gain information about the

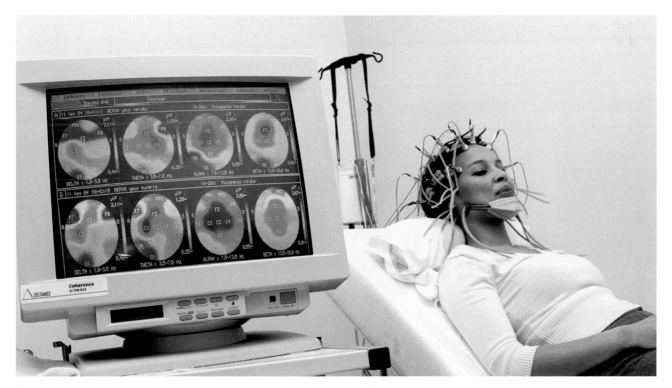

During an electroencephalogram, electrodes placed on the patient's scalp detect electrical activity in the brain. A normal diagram is shown in the screen. *(AJPhoto/Photo Researchers, Inc.)*

anatomy of the brain and to determine the appropriate treatment. Computed tomography (CT), magnetic resonance imaging (MRI), and positron emission topography (PET) are imaging methods that allow a physician to search for possible injuries, tumors, pooled blood, or scars that could be affecting electrochemical activity in the brain and that might be indicative of other dangerous conditions. The quickest method, a CT scan, uses X-rays to produce three-dimensional images of the body's internal structures, such as the brain. MRIs use magnets and radio waves to achieve a more detailed, multicolored, digital image on a computer screen. A more recent advancement in medical technology, the PET scan, allows for observation of metabolic activity in the brain in real time. The prior administration of radioactively labeled glucose, a type of sugar molecule that the brain breaks down for energy, reveals which areas of the brain are utilizing the glucose and are therefore active during monitoring. Areas that are damaged will not utilize the sugar effectively and will appear a different color on the PET computer screen.

Each individual case of epilepsy is unique, and patients and doctors implement a variety of treatment strategies. The goal of any treatment plan is to help the patient reduce the number of seizures while limiting any negative side effects. Although there is no cure, medications successfully control seizures in the majority of epilepsy cases. Having regular sleep, reducing stress, avoiding alcohol, and eating special diets can also reduce the quantity of seizures. Surgery or implantation of an electrical device is helpful in other cases. One rare seizure disorder results from a deficiency of vitamin B_6, and taking vitamin supplements successfully treats this form of epilepsy.

The U.S. Food and Drug Administration (FDA) has approved about two dozen different medications for treating epilepsy, most of which are taken orally and work by providing some control over communication among neurons. Two common antiepileptic drugs (AEDs) are phenobarbital and phenytoin. Because many AEDs are sedatives, drugs that decrease the activity of the central nervous system, common side effects of AEDs include drowsiness, lack of energy, clumsiness, and difficulty concentrating. Sometimes a patient must try several drugs before finding the best one. Half of the patients who do not have any seizures for several years while on medication can stop taking their medication and remain seizure-free. Under a doctor's supervision, the patient takes decreasing dosages over time.

Approximately 20 percent of patients continue to have seizures while on medications. If their quality of life is significantly decreased, brain surgery is another possible form of treatment. The surgeon either removes brain tissue (resection) or, if the focus is in

a region that would prevent it from being removed safely, the surgeon may functionally isolate it by cutting the nerve fibers that connect it to other parts of the brain (disconnection) in order to prevent seizures from spreading. The most common resection surgery, a temporal lobectomy, in which a surgeon removes the damaged portion of the brain where the seizures originate, eliminates partial seizures in more than 60 percent of patients and reduces the number significantly in even more. Two less common operations for epilepsy are a corpus callosotomy and a hemispherectomy. In a corpus callosotomy, performed to treat severe generalized seizures, the surgeon severs the corpus collosum, the bundle of nerve fibers that connect the two sides of the brain. This procedure does not stop seizures but restricts them to one side of the brain. In a hemispherectomy, one-half of the brain is removed. Because this procedure can lead to partial paralysis or loss of some motor function, this rare surgery is only performed for extreme cases in which the brain is already severely damaged. Young children recover best from a hemispherectomy; thus the surgery is rarely performed in children older than 13 years old. Patients usually are discharged from the hospital one week after brain surgery and can return to normal activities within a few months. Most still require AEDs after surgery.

A strict ketogenic diet that is high in fats (as in cream and vegetable oil) and low in carbohydrates helps some people control their seizures. This controversial diet is so named because it causes the body to produce ketones, products that result from utilizing fats rather than carbohydrates for energy. This type of diet is dangerous, must only be initiated under a doctor's supervision, and requires a lot of effort to maintain. Side effects include nausea and vomiting, and the patient must take nutritional supplements to prevent deficiencies.

Since FDA approval in 1997, the implantation of devices that periodically shock the vagus nerve, a nerve believed to be connected to the part of the brain where seizures originate, has decreased the incidence of seizures in a number of patients. A battery-operated pulse generator is placed inside the chest and connected by electrodes threaded underneath the skin to the vagus nerve, located in the neck. Timed periodic electrical impulses reduce the number of seizures by about 20–40 percent, but the mechanism is not understood.

In some cases, epileptics can predict an oncoming seizure. The warning, or aura, resembles a tickling of one's senses and can be in the form of a specific smell, sound, or tingling feeling. The aura is caused by a small partial seizure and is often followed by a larger seizure. Further stimulation of the senses may ward off the oncoming seizure. If the person has a vagus nerve stimulator, holding a special magnet over the implanted device to stimulate it can sometimes halt the oncoming seizure or lessen its severity. Being able to predict a seizure is useful even if the seizure cannot be prevented. The person can get to a safe environment, set aside items that might cause harm, or notify someone for help.

LIVING WITH EPILEPSY

The impact of epilepsy on someone's daily life varies greatly among individuals. Some people with epilepsy require treatment for the rest of their lives, but others completely stop having seizures over time. The majority of patients diagnosed during childhood outgrow their seizures naturally before reaching adulthood. A large number of people who have epilepsy can still attend school, have fulfilling careers, play musical instruments, participate in sports, have families, and drive cars (if they have been seizure-free for a period, from three to 12 months, defined by each state's department of motor vehicles), though changes to one's lifestyle are often necessary. For safety reasons, if seizures become uncontrolled, people who have epilepsy should not work at jobs involving dangerous machinery, jobs that one performs at great heights, or jobs that require driving.

Because no one can predict when a seizure may occur, certain situations could be particularly dangerous for someone who has a seizure disorder. People who experience seizures that cause them to fall to the ground must wear helmets to prevent traumatic head injury. Activities that require complete constant attention, such as mountain climbing or scuba diving, can be fatal if an untimely seizure occurs. All 50 states have laws restricting people with uncontrolled seizures from driving, but most permit a patient with epilepsy to drive if he or she has been seizure-free for a set period—commonly six months, with many requiring three or 12 months depending on the state. Other countries, including Canada and the United Kingdom, permit driving only after a 12-month seizure-free interval. Some countries, such as Australia and Ireland, require two years with no seizures before licensing.

Certain adaptations might be necessary for a person who has seizures and lives alone. The goal is to prevent possible harm if a seizure occurs during certain activities. For example, carpeting hardwood floors will make falls less injurious, and using plastic instead of breakable dishes can prevent accidental cuts if the person is holding them when a seizure begins. Other safety precautions may include abstaining from drugs and alcohol, not swimming alone, wearing a helmet when swimming or playing contact sports, and informing associates and peers what to do or not to do in case of a seizure.

Fallacies prevalent before the disease was better understood have been disproved, but because behavioral symptoms often accompany the physical symptoms, some people wrongly assume that people who have epilepsy are mentally ill or dangerous. Educating others about the illness can help remove the stigma that results from misconceptions and that can be just as traumatic as the seizures themselves. Awareness programs strive to educate the public to generate support for research that could lead to improved methods for diagnosis and strategies for treating for epilepsy. Neuroscientists are constantly seeking more effective AEDs that cause fewer side effects. One possible technological advance includes a device that could be implanted into the brain of someone who has epilepsy to detect an oncoming seizure and send electrical impulses to the brain to avert the seizure. A better understanding of normal brain function and neuron interactions will help advance neurological research toward a cure for epilepsy.

See also NERVOUS SYSTEM; SENSATION.

FURTHER READING

American Epilepsy Society home page. Available online: URL: http://www.aesnet.org. Accessed January 14, 2008.

Epilepsy Foundation home page. Available online. URL: http://www.epilepsyfoundation.org. Accessed January 14, 2008.

National Institute of Neurological Disorders and Stroke. "NINDS Epilepsy Information Page." Available online. URL: http://www.ninds.nih.gov/disorders/epilepsy/epilepsy.htm. Updated October 6, 2008.

Wilner, Andrew N. *Epilepsy: 199 Answers: A Doctor Responds to His Patients' Questions.* New York: Demos, 2003.

ethology Ethology is the scientific study of animal behavior, particularly in natural conditions. The goal of this comparative branch of zoology is to understand how and why animals do what they do. Ethologists must have a solid grasp of anatomy, physiology, and evolution to carry out their research. Because many behaviors affect an animal's ability to survive to reproductive age and pass on its genes to the next generation, natural selection acts on behaviors. The fields of ethology, animal behavior, and behavioral ecology are all closely related. An ethologist generally focuses on innate behaviors, those that have strong genetic components, whereas the focus of someone traditionally referred to as an animal behaviorist resembles the modern field of comparative psychology, which focuses more on learned behaviors or cognitive abilities of animals. The field of behavioral ecology emerged from ethology and concentrates on the role of the environment in shaping modifications to behavioral adaptations. Studies in ethology and behavioral ecology have advanced the field of comparative animal psychology.

Many human behaviors are cultural, meaning they are learned from previous generations; honoring certain religious beliefs and having respect for elders are examples of cultural behaviors. Ethology is mainly concerned with behaviors that have biological components, such as mechanisms animals use to seek and obtain food, communication with other members of the species, migration patterns, and choice of mates. Some behaviors are learned, such as forming a cognitive map of one's immediate surrounding environment, and others are innate, such as a spider's knowing how to build a web characteristic of its species despite having never seen one.

Research performed by the German zoologist Karl von Frisch, the Austrian naturalist Konrad Lorenz, and the Dutch-British biologist Nikolaas Tinbergen formed the basis of the field of modern ethology. In the early 20th century, animal behaviors were explained either as controlled by inexplicable extranatural instincts outside the scope of zoology or as learned; both explanations fell outside the scope of zoology. Frisch, Lorenz, and Tinbergen demonstrated that certain behavioral characteristics contributed to the survival of animals in the same way physiological and anatomical modifications did and therefore resulted from natural selection. While their research specifically addressed behaviors in bees, fish, and birds, the basic principles apply to all animals, including humans. Their efforts birthed the field of ethology and earned Frisch, Lorenz, and Tinbergen the 1973 Nobel Prize in physiology or medicine "for their discoveries concerning organization and elicitation of individual social behavior patterns."

Ethologists combine two different approaches to the study of animal behavior to achieve a more comprehensive understanding: they ask both proximate and ultimate questions. Proximate questions address the mechanisms of different behaviors—determining the trigger of a response, defining the genetic components, and describing the anatomical and physiological mechanisms for carrying out the behavior. In other words, how does it work? Ultimate questions address why animals exhibit the behavior or how it took its current form. What is the evolutionary significance and how did the behavior develop in a species over time? Tinbergen suggested ethology address four central issues, comprising both proximate and ultimate questions, in order to understand a behavior: function, causation, development, and evolutionary history. The first two are proximate questions and the latter two are ultimate questions. The function of a behavior refers to the way the behavior

contributes to an animal's fitness (success in passing on one's genes to the next generation). Causation includes examination of the mechanisms: genetic components, the stimuli that elicit the response, the morphological features that enable the response, and the physiological mechanisms. Developmental issues encompass how the development and maturation of an animal affect the behavior, and which early experiences are required for the animal to exhibit the behavior. Understanding the evolutionary history involves a phylogenetic comparison (a comparison of similar behaviors in related species) in order to outline the origin and modification of the behavioral adaptation over time.

Classical ethology focused on topics such as communication among animals and imprinting. Animals communicate by a variety of mechanisms including visual, chemical, mechanical, and auditory. Von Frisch studied the complicated dance performed by worker bees to communicate the distance, direction, and quality of a feeding source. Other bees observing this dance can interpret its meaning to locate the food. When animals mark their territory by urinating, they leave behind a chemical signal that communicates the boundaries of their claimed territory to other members of their species. Male spiders, who are much smaller than their female counterparts, must approach a female in order to mate, but because of the male's size, the female often mistakes a potential mate for food. To prevent this, the male might vibrate or twitch the web in a particular pattern, a form of mechanical communication to let the female know that he is approaching and that he is a mate rather than a food source (though sometimes he serves as both—a phenomenon known as sexual cannibalism). Auditory communication is also really a form of mechanical communication because information is transmitted through vibrations in the air, detected by special sense organs in the recipient. An example of auditory communication is exhibited by chickadees, whose alarm call communicates information about the size of a potential predator and how rapidly it is moving by the pitch and length of vocalizations they emit. Imprinting occurs when, during the early development of a social animal, a specific event takes place and establishes a behavior pattern. The most famous example of imprinting is the recognition of the first moving object a duckling sees as its mother. In the natural world, the object usually is the mother, but ethologists have demonstrated imprinting on cardboard boxes and on human beings as well.

Other important topics that ethologists research include fixed action patterns, food and feeding behavior, defensive behaviors, movement and migration patterns, courtship and mating behaviors, and parental care. Modern ethology has branched into topics such as playful behavior, honesty and dishonesty in signaling systems, and emotions in animals. Juveniles, and adults to a lesser degree, both exhibit playful behavior, such as chasing and sparring, actions that ethologists believe promote social bonding, reinforce social hierarchies, help improve survival skills by improving locomotor functions, and prepare young animals for sexual, predatory, and defensive behaviors. Deciphering the neuroendocrine basis for behaviors is a popular modern approach to ethology, one that molecular biological techniques have advanced.

The specialty of human ethology, or human behavior, originated with the British naturalist Charles Darwin's text *The Expression of the Emotions in Man and Animals* (1872), in which he examined how human beings and other animals showed emotions such as fear, anger, and pleasure, emphasizing the shared evolution of humans and other animals. One approach to studying human ethology is comparative: examining the behaviors of humans alongside those of chimpanzees and other related species; comparing typical behavior with the behavior of individuals with congenital defects, particularly those with sensory deficits; and comparing human behaviors across cultures, since behaviors exhibited by individuals in a wide variety of cultures are probably biological rather than cultural in nature. The overlapping subfield of evolutionary psychology specifically addresses the evolutionary significance of psychological characteristics such as memory, perception, and language, mostly in humans, though other animals are not excluded. Whereas ethology is concerned with the study of all animal behaviors, sociobiology focuses on the evolution of social behaviors in animals.

Attaining a better understanding of normal or appropriate animal behavior will help zoologists to identify when behavior exhibited by an animal is a result of injury or fear and to decipher the body language animals display to communicate how they feel. Knowing which behaviors are instinctual, which behaviors are learned, and which behaviors are crucial for survival will help in conservation efforts. People whose occupations involve direct interaction with animals—veterinarians, ecologists, zookeepers, forest rangers, wildlife biologists, and conservation biologists—benefit from studying animal behavior but so does the rest of society. Many people keep animals in their homes, voting members of communities affect legislation passed to protect animals and their environments, and as animals themselves, humans can develop a better appreciation of their own behaviors and actions and recognize their relationship to the rest of the living world.

See also ANIMAL BEHAVIOR; ANIMAL COGNITION AND LEARNING; ECOLOGY; FRISCH, KARL VON; LORENZ, KONRAD; SOCIAL BEHAVIOR OF ANIMALS; SOCIOBIOLOGY; TINBERGEN, NIKOLAAS; ZOOLOGY.

FURTHER READING
Barnett, S. A. *Modern Ethology: The Science of Animal Behavior.* New York: Oxford University Press, 1981.

Dugatkin, Lee Alan. *Principles of Animal Behavior.* New York: W. W. Norton, 2003.

Hinde, Robert A. *Ethology: Its Nature and Relations with Other Sciences.* New York: Oxford University Press, 1982.

Lorenz, Konrad Z. *The Foundations of Ethology: The Principle Ideas and Discoveries in Animal Behavior.* Translated by Konrad Z. Lorenz and Robert Warren Kickert. New York: Simon & Schuster, 1981

McFarland, David. *Animal Behavior: Psychobiology, Ethology, and Evolution.* 3rd ed. Harlow, England: Longman, 1998.

Eukarya According to modern biological classification schemes, all living organisms belong to one of three major domains of life: Archaea, Bacteria, or Eukarya. Microfossil evidence suggests that life first appeared about 3.6 billion years ago. The first life-forms were anaerobic and prokaryotic, then oxygenic photosynthesis emerged about 2.5 billion years ago. Phylogenies based on molecular analyses indicate that the universal common ancestor diverged into two lineages. The Bacteria lineage diverged first, and the other subsequently split into two lineages, Archaea and Eukarya. Fossil evidence suggests that eukaryotic cells first emerged approximately 2.1–1.6 billion years ago. The emergence of eukaryotic cells was a significant event, as the complexity of their structure allowed for the evolution of sexual reproduction and multicellularity, two milestones in the history of life leading to the diversity of modern day life-forms.

EUKARYA AS A DOMAIN

Organisms belonging to the domain Eukarya have in common cell structures that are absent in prokaryotes, most notably, a nucleus and a cytoskeleton. Biologists believe that the membrane-bound nuclear compartment resulted from an infolding of the plasma membrane that surrounded the genetic material contained in a region called the nucleoid in prokaryotic cells. The nuclear envelope encloses the deoxyribonucleic acid (DNA), protecting it from damage and allowing for another step in the regulation of gene expression, which conserves energy. Additional invaginations of the plasma membrane led to the membrane-bound endoplasmic reticulum, the site for the synthesis of lipids and some proteins, and the Golgi apparatus, where protein modification, sorting, and packaging occur. The endosymbiotic theory, popularized by Lynn Margulis, a professor of geosciences at the University of Massachusetts, purports that subcellular structures characteristic of eukaryotic cells evolved as a consequence of mutually symbiotic relationships of prokaryotic cells. Smaller cells lived inside other cells and took on specialized functions, and eventually the organisms depended on each other. As the smaller cell lost its ability to carry on independent life functions, it evolved into an organelle within the larger cell. Mitochondria, organelles responsible for cellular respiration, and plastids, organelles that perform photosynthesis, are believed to have developed in this manner.

In 1969 Robert Whittaker proposed a five-kingdom classification system. One kingdom, Monera, consisted of prokaryotic organisms, and the remaining four—Protista, Fungi, Plantae, Animalia—were composed of eukaryotic organisms, characterized by nucleated cells. In 1990, after spending more than two decades conducting labor-intensive research, the American microbiologist Carl Woese proposed a taxonomic category above kingdom. He argued that phylogenies based on molecular evidence supported three different major lineages. He found that the distinction between members of Bacteria and members of his newly proposed group of prokaryotic organisms, Archaebacteria (now known as Archaea), was more significant than the differences between eukaryotic kingdoms. Thus, he placed all four of the eukaryotic kingdoms within a single domain, Eukarya.

Classification schemes continue to evolve as biologists learn more about the phylogenies of different life-forms, including eukaryotic categories. Molecular evidence is revealing that relationships defined on the basis of gross structural appearances differ from those based on DNA sequence comparisons. Molecular data suggest that some organisms once thought to be closely related on the basis of outward appearance are more distantly related, and organisms that look different may have a more recent common ancestor than once assumed. These new data have led to the proposal of several new kingdoms, consisting of organisms once grouped within the traditional four eukaryotic kingdoms. Many biologists continue to employ Whittaker's five-kingdom system, and this encyclopedia often also uses these traditional descriptions because taxonomists are still analyzing data and working to reach a consensus on an improved phylogenetic framework. One must remember that these groupings are used for the sake of convenience and familiarity rather than a demonstrated evolutionary history.

Traditionally, the domain Eukarya includes the four kingdoms Protista, Fungi, Plantae, and Animalia. Protista represents the most ancient eukaryotic kingdom, containing unicellular, colonial, and multicellular organisms that did not fit into Fungi, Plantae, or Animalia for various reasons. The plantlike, autotrophic protists termed *algae* include species as diverse as unicellular diatoms and multicellular kelp. Protozoa are unicellular, heterotrophic animallike protists. Protists have undergone the most drastic changes in classification. Once considered a single kingdom, these organisms are being redistributed into as many as 18 new kingdoms, with some being placed in one of the other three traditional eukaryotic kingdoms. While Protista is no longer considered a valid taxon, the term *protist* is still useful for referring to eukaryotic organisms consisting of a single cell or a few cells.

KINGDOMS FORMERLY INCLUDED IN PROTISTA

The protists consist of approximately 60 different lineages, and comparison of the sequences for the small ribosomal ribonucleic acid (RNA) subunit reveals patterns concerning the evolution of these members of Eukarya. The earliest lineages include the present-day parasites diplomonads (such as *Giardia lamblia*), microsporidia (which causes the silkworm disease pébrine), and parabasalids (represented by *Trichomonas vaginalis*, which lives in the female vagina). These three proposed kingdoms all lack the membranous structures making up the Golgi apparatus, peroxisomes (organelles that contain enzymes that catalyze oxidative reactions), and developed mitochondria. They are adapted to anaerobic environments, and their mitochondria do not have their own DNA, electron transport chains, or enzymes for the citric acid cycle. They do have simple cytoskeletal components and a few membrane-bound vesicles.

An increase in the complexity of membranous intracellular structures, such as those making up the Golgi apparatus in addition to mitochondria and chloroplasts, led to the emergence of new lineages. The myxomycotes are the plasmodial slime molds, which were formerly considered a type of fungus. These protists exist in three different forms during their life cycle. The plasmodial form resembles a patch of wet slime and consists of one enormous fused cell containing thousands of diploid nuclei. In dry conditions, a second form develops—a tall, stalk-like structure containing haploid spores that germinate into the third amoebalike form. The haploid amoebalike cells can fuse with other cells in sexual reproduction and then multiply to form the feeding plasmodial form again.

Euglenozoans are flagellated protozoans that comprise two main groups: the kinetoplastids and the euglenoids. Kinetoplastids are characterized by a kinetoplast, a granule containing a mass of DNA, and include free-living and parasitic species, such as *Trypanosoma,* a blood parasite that causes sleeping sickness. Euglenids, such as *Euglena,* have chloroplasts and are photoysnthetic.

Naegleria make up a separate kingdom. These organisms can exist as an amoebalike form or have a pair of flagella depending on the environmental conditions.

Amoebozoans are characterized by lobe-shaped pseudopodia. Most are free-living, but some, such as *Entamoeba histolytica,* which causes dysentery, are parasitic.

Cellular slime molds, belonging to the tentative kingdom Acrasiomycota, have multicellular fruiting bodies made of separate cells. They exist as unicellular amoebalike cells when organic material is abundant but can form a multicellular sluglike aggregate that migrates as a unit. When it settles, the cells form a stalk with an asexual fruiting body that contains haploid spores that can develop into new amoebalike cells.

Red seaweeds, or rhodophytes, were the first multicellular, photosynthetic organisms. The 4,000-plus species are all marine, reproduce sexually, and possess the photosynthetic pigments chlorophyll a and phycobilins. They are structurally diverse, and some become calcified, like corals do.

The next three types of protists—dinoflagellates, ciliates, and apicomplexans—all belong to a taxon named alveolates, but their exact relationships remain unresolved. They have in common sacs called alveoli, whose function is unclear, under the plasma membrane. The dinoflagellates are mostly marine, planktonic organisms, covered by a hard shell called a test, and motile by a single flagellum. Some are photosynthetic, having chlorophyll a, variants of chlorophyll c, carotenes, and xanthins. Dinoflagellates are responsible for red tide, some produce toxins poisonous to invertebrates and fishes (and humans if they ingest contaminated seafood), and some are bioluminescent. The ciliates are characterized by cilia and two nuclei: one large nucleus for gene expression and a smaller nucleus that functions in sexual reproduction. One example of a ciliate is *Paramecium.* Apicomplexans are sporozoite-forming parasites that have complex life cycles requiring more than one host species. An example is *Plasmodium,* the causative agent for malaria.

The stramenopiles is a clade consisting of six closely related tentative kingdoms. They have in common tiny, hollow hairs that often occur on a flagellum or are derived from organisms that had such hairs. The slime nets (Labyrinthulids) are mostly marine and form colonies on marine plants and seaweed.

TENTATIVE KINGDOMS FORMERLY BELONGING TO PROTISTA

Tentative Kingdom	Examples
Diplomonads	*Giardia lamblia, Giardia intestinalis*
Microsporida	*Nosema*
Parabasalids	*Trichomonas vaginalis, Trichomonas foetus*
Myxomycota	*Echinostelium, Physarum polycephalum*
Euglenozoa	*Euglena viridis, Peranema, Trypanosoma brucei, Trypanosoma cruzi*
Naegleria	*Naegleria*
Amoebozoa	*Entamoeba histolytica*
Acrasiomycota	*Dictyostelium discoideum*
Rhodophyta	*Polysiphonia, Porphyra, Palmaria palmata*
Ciliates	*Paramecium, Stentor*
Dinoflagellates	*Gonyaulax, Pfiesteria shumwayae*
Apicomplexa	*Plasmodium, Toxoplasma, Coccidia*
Labyrinthulids	*Labyrinthula*
Oomycota	*Phytophthora infestans*
Xanthophyta	*Ophiocylium*
Chrysophyta	*Ochromonas, Dinobryon*
Phaeophyta	*Fucus, Sargassum, Postelsia, Macrocystis, Laminaria*
Diatoms	*Thalassiosira*
Foraminiferans	*Globigerina*
Radiolarians	*Actinomma*

They are unique in that they form trails of a mucuslike polysaccharide that their cells move along to find food. The oomycotes, commonly known as water molds, resemble fungi in that they consist of threadlike hyphae and secrete enzymes to digest organic substances then take in the nutrients. Oomycotes were once grouped with fungi, but their similarities are now known to result from convergent evolution.

Their cell walls contain cellulose. Xanthophytes are freshwater photosynthetic protists having various xanthins and chlorophylls a, c, and e. Chrysophytes, commonly known as golden algae, may be unicellular or live in colonies. Some form elaborate skeletons made of silica. The brown seaweeds, kingdom Phaeophyta, include about 1,500 mostly marine species. Some of these grow to be more than 300 feet (about 100 m) long and are called kelps. Though they do not contain true tissues, many brown seaweeds (also called brown algae) have thalli and holdfasts. A thallus is the plantlike body of a seaweed. Leaflike blades extend from a stemlike stipe, and the rootlike holdfast is for gripping and anchoring the alga. Brown seaweeds contain chlorophylls a and c and the pigment xanthophyll. Some multicellular algae undergo an alternation of generations similar to that in plants. Diatoms have shells called tests consisting of two siliceous halves; these shells form diatomaceous earth, which is mined to serve as a filter medium. The diverse diatoms make up a major portion of phytoplankton. Most are photosynthetic and contain chlorophylls a and c, carotenes, and xanthins.

The phylogenies of foraminiferans and radiolarians are not as well understood. Both are protists that have threadlike pseudopods. Foraminiferans have porous calcified tests, or shells. The pseudopods extend through the pores to function in motility and feeding. Radiolarians are marine protists with fused siliceous tests. Their threadlike pseudopods are called axopodia, and they are surrounded by microtubules. When smaller microorganisms land on the axopodia, the radiolarian phagocytoses it, then carries it into the main part of the cell by cytoplasmic streaming down the axopodia.

KINGDOM FUNGI

The kingdom Fungi includes eukaryotic, heterotrophic multicellular organisms that have cell walls. They mostly feed on dead or decaying organic matter by secreting digestive enzymes and then absorbing the smaller organic compounds. Biologists used to think fungi were plants because they were immobile, had structures that resembled roots, and had cell walls. Even after biologists recognized Fungi as a distinct kingdom, they included plasmodial slime molds, cellular slime molds, slime nets, and oomycotes as members. More recent molecular evidence suggests that each of these groups also warrants its own kingdom. Distinguishing features of true fungi are cell walls composed of chitin, a nitrogen-containing polysaccharide, and a unique biochemical pathway for the metabolism of lysine, one of the 20 amino acids that cells use to build proteins. One surprising finding from comparisons of small ribosomal RNA between fungi and other kingdoms of Eukarya is that fungi may be more closely related to animals than plants.

Fungi are traditionally categorized in terms of the anatomy of their sexual reproductive structures. The major fungal phyla include Chytridiomycota, Zygomycota, Ascomycota, and Basidiomycota. The chytrids, believed to be the most ancestral form, are aquatic, do not have septa dividing the filamentous hyphae, and include both saprobes that feed on decaying matter and parasites that feed off living hosts. They produce flagellated gametes that fuse during sexual reproduction to form a zygote that can remain inactive in unfavorable conditions. The polyphyletic phylum Zygomycota includes nonseptate fungi that sexually reproduce by forming zygospores. When hyphae of opposite mating types meet and fuse, a tough zygosporangium forms. Inside, the nuclei fuse and meiosis occurs. Germination leads to the formation of a sporangium that releases genetically variable haploid spores that grow into new organisms. Ascomycotes, also known as sac fungi, reproduce sexually by growing specialized hyphae that have spores called conidia on the tips. Conidia can fuse with hyphae of the opposite mating type, leading to the formation of an ascus, inside of which the parental nuclei fuse. Meiosis follows, forming four genetically variable nuclei. One round of mitosis then creates eight ascospores that disperse and germinate. Basidiomycotes include mushrooms and bracket or shelf fungi. When two hyphae of opposite mating types fuse, dikaryotic mycelia form and grow. (*Dikaryotic* means that the two haploid nuclei remain separate inside each cell of the mycelia.) Under certain environmental conditions, basidiocarps develop, and their gills contain dikaryotic cells called basidia. The nuclei fuse and undergo meiosis, forming four haploid nuclei. Four basidiospores develop, each containing a single nucleus. When mature, basidiospores disperse and germinate to form haploid mycelia. A fifth group, called Deuteromycota, or more appropriately the "imperfect fungi," is not a true clade, but rather a group of fungi whose sexual state is not yet known or may not reproduce sexually at all.

KINGDOM PLANTAE

The remaining two kingdoms of the domain Eukarya—Plantae and Animalia—once were thought to encompass all life but are now known to describe a very small portion of it. The kingdom Plantae includes immobile, multicellular, photosynthetic eukaryotic organisms that have cell walls made primarily of cellulose. Botanists agree that plants evolved from a green algal ancestor, but they have not reached a general consensus of whether or not green algae should indeed be placed in the kingdom Plantae, or whether

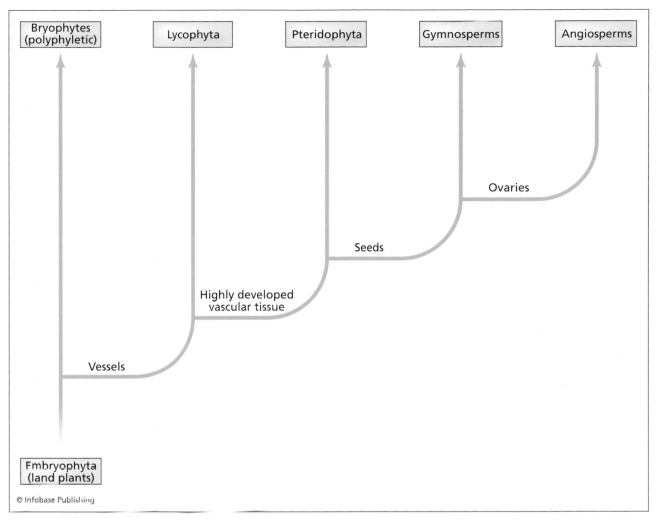

Major structural innovations in the history of land plants include vessels for transporting water and nutrients, seeds, and ovaries to enclose the seeds.

they warrant their own kingdom. Characteristics shared by green algae and land plants include the same type of chlorophyll, the presence of carotenoids as accessory pigments, and starch storage inside the chloroplasts. Many green algae also have cellulose as the main component of their cell walls.

The first major transition in the evolutionary journey from green algae to flowering plants is multicellularity, which developed independently in different lineages. Green algae include unicellular, colonial, and multicellular forms, but land plants are derived from a different photosynthetic protist ancestor. The next major transition, occurring approximately 460 million years ago, was the colonization of land, a move that required numerous adaptations. The common terrestrial ancestor was probably a form of green alga that formed a symbiosis with fungi. This evolved into the Embryophyta lineage, the land plants. The term *bryophyte* informally refers to the nonvascular lineages that include the earliest land

plants, but Bryophyta (with a capital *B*) is a clade of bryophytes consisting of mosses. Vascular plants have organized systems for transporting water and nutrients throughout the plant. The living vascular plants evolved from an ancestor in common with the mosses. This branch diverged into Lycophyta and Tracheophyta. Lycophyta consists of the club mosses and their relatives. The tracheophyte lineage diverged into groups. One group led to the evolution of Pteridophyta, the ferns and horsetails, which as do all other plants up to this point, reproduce by spores. The other branch developed an innovative adaptation approximately 360 million years ago that led to widespread success, the seed. Seed-bearing plants surround their embryos with a supply of nutrients inside a tough, protective covering, and thus they do not depend on water to reproduce. Two major divisions of seed plants are gymnosperms, whose seeds are not enclosed by chambers, and angiosperms, whose seeds develop inside structures called ovaries.

KINGDOM ANIMALIA

The kingdom Animalia includes multicellular, heterotrophic eukaryotic organisms that have no cell walls, are capable of motility during at least one stage of their life cycle, and digest their food internally. Animal life cycles are characterized by haploid gametes that fuse during sexual reproduction to form a diploid zygote that undergoes numerous rounds of mitotic division and a complicated differentiation and development stage to form the complete, mature multicellular adult. More than 35 phyla of animals with similar evolutionary traits exist, with the majority of animals belonging to 11: Porifera, Cnidaria, Ctenophora, Platyhelminthes, Nematoda, Rotifera, Mollusca, Annelida, Arthropoda, Echinodermata, and Chordata.

Animals developed from a common ancestral protist that diverged into the modern-day choanoflagellates and the lineage that evolved into all animals. The monophyletic clade including extant and extinct animal lineages is called Metazoa, animals with bodies composed of differentiated tissues and a digestive cavity lined with specialized cells. Whether or not choanoflagellates should be included in the animal kingdom is debatable. Choanoflagellates are sometimes unicellular and sometimes colonial. Each cell has a collar of microvilli at one end, with a single, protruding cilium that beats to move water toward the circle of microvilli. The collar catches microscopic pieces of food for ingestion by the cell. Sponges, organisms belonging to the phylum Porifera, contain very similarly structured cells called choanocytes. Porifera are said to be parazoan, meaning they do have specialized cells, but they are not organized into true tissues separated from other tissues by membranes—the cells retain some of their independence. The remaining metazoans are eumetazoans, meaning their cells are organized into true tissues that cooperate with other body tissues and combine with other tissues to form organs.

The common ancestral eumetazoan diverged into three lineages, the phyla Cnidaria and Ctenophora and the Bilateria. Cnidaria includes sea anemones, jellyfish, and corals, and Ctenophora includes the comb jellies. Members of both phyla exhibit radial symmetry and are diploblastic, meaning they have only two germ layers, the endoderm and the ectoderm. During the blastula stage of early animal development, the embryo resembles a hollow ball of cells. The next stage, gastrulation, involves the invagination of one side to form a gastrula defined by two cell layers. The inner layer becomes the endoderm, and the outer layer becomes the ectoderm. The third branch, Bilateria, consists of animals that develop from three cell layers; they are said to be triploblastic. The third layer is called the mesoderm, and it offers a new degree of complexity.

The bilaterians branch into two clades that have different patterns of early development: the protostomes and the deuterostomes. After fertilization, the zygote undergoes a series of mitotic cell divisions called cleavage, resulting in a multicelled embryo. Protostomes undergo a spiral cleavage, in which the layers of cells look twisted, whereas deuterostomes undergo radial cleavage, resulting in cells neatly stacked one over another. In addition, the cleavage in protostomes is determinate, meaning the fate of specific cells in the early embryo is already determined. If a cell is removed, it cannot develop into a complete organism. In deuterostomes, cleavage is indeterminate, so each cell taken from an early embryo, say, at the four-cell stage, can develops into a complete individual. During gastrulation, when one side invaginates, the indentation forms the blastopore, a structure involved in formation of the gut. In protostomes, the blastopore extends to the other end of the embryo, the original site of blastopore formation develops into the mouth, and the anus forms at the other end of the embryo. In deuterostomes, however, the blastopore becomes the anus, and the mouth develops from another opening. The larvae that form from protostomes and deuterostomes also differ, especially in the arrangement of the cilia. Protostome larvae are called trochophores, and deuterostome larvae are called tornarias. Last, the arrangement of the nervous system differs between the two. While zoologists are still working on the phylogenetic details within the two major branches of protostomes and deuterostomes, they agree which phyla belong to which group. The protostomes include molluscs, annelids, rotifers, platyhelminthes, arthropods, and nematodes. Echinoderms and chordates are deuterostomes.

Recent molecular evidence suggests that the protostome ancestor diverged into two major sister branches, the Ecdysozoa and the Lophotrochozoa. The molecular studies group phyla differently than traditional placement based on morphologies does. The ecdysozoans secrete exoskeletons, tough outer coverings that the animal outgrows and sheds in a process called ecdysis. Though some animals in other phyla do molt, sequence comparisons of ribosomal RNA genes and the *Hox* genes (which play a role in animal development) link the nematodes and arthropods by a common ecdysozoan ancestor and group the rotifers, platyhelminthes, mollusks, and annelids together in the lophotrochozoan group. Traditionally, the arthropods have been most closely linked with the annelids because their body plans share similar characteristics, particularly with respect to segmentation and appendages. If the molecular evidence accurately reflects the evolutionary history, then segmentation arose separately in these taxa. Zoologists

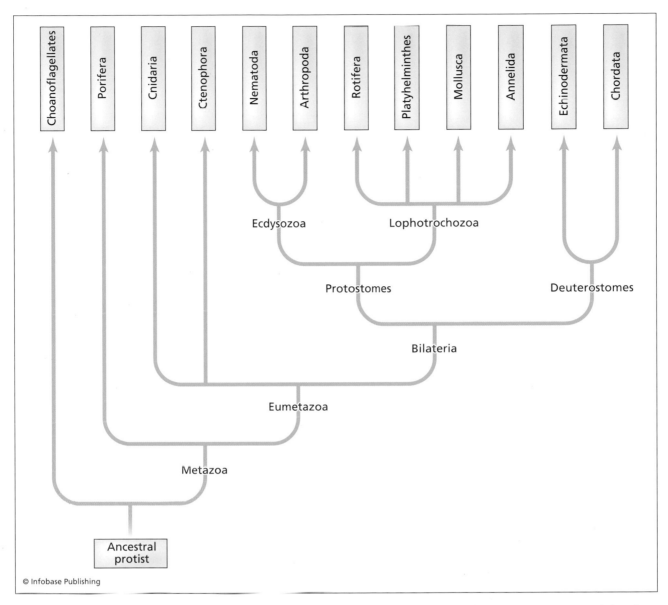

Major developments leading to the current diversity of members of the kingdom Animalia include specialized tissues, bilateral symmetry, and unique developmental processes.

need to examine molecular data from more species from different phyla before drawing reliable conclusions regarding this classification scheme.

Beyond the divisions created by the milestones mentioned so far—multicellularity, true tissues, three rather than two germ layers, and radial versus bilateral symmetry—other innovations upon which classification is based include cephalization, presence of a coelom, and segmentation. Cephalization is the concentration of nerves and sensory organs in an anterior head. A coelom is a space between the body wall and the gut that results when the mesoderm forms a layer of tissue that wraps around the internal body organs. Animals with a coelom (annelids, mollusks, arthropods, echinoderms, and chordates) are called coelomates, and animals without one (poriferans, cnidarians, ctenophorans, and platyhelminthes) are called aceolomates. Pseudocoelomates are animals with a body cavity that is only partially lined with tissue derived from mesoderm (nematodes and rotifers). The coelom allows for freer movement, more complex structure, and larger body size. Segmentation is a body plan divided into repeated segments, a system of organization that underlies the body plans of all coelomates except mollusks. In animals such as earthworms, the segments are identical, with all containing the same structures, or have different structures and perform specialized functions, as in the head, thorax, and abdomen of insects. The presence or absence of these different body plan innovations does not necessarily imply common ancestry, as several features, such as coeloms and segmentation, have evolved independently.

Informally, animals are largely divided into two main groups, the invertebrates and the vertebrates, based on the absence or presence of a backbone. More detailed discussion of the organisms in the animal kingdom in this encyclopedia follows this format. Though this popular manner of dividing the animal kingdom seems to imply equal degrees of significance of the two groups, one must remember that vertebrates are only a subphylum of Chordata, one of the 35 or so recognized phyla within the kingdom Animalia.

See also ALGAE; ARCHAEA; BACTERIA (EUBACTERIA); BIOLOGICAL CLASSIFICATION; EMBRYOLOGY AND EARLY ANIMAL DEVELOPMENT; EUKARYOTIC CELLS; FUNGI; HISTORY OF LIFE; INVERTEBRATES; PLANT DIVERSITY; PROTOZOA; SLIME MOLDS; VERTEBRATES.

FURTHER READING
Adl, Sina M. et al. "The New Higher Level Classification of Eukaryotes with Emphasis on the Taxonomy of Protists." *The Journal of Eukaryotic Microbiology* 52, no. 5 (October 2005): 399–451.

Maddison, D. R., and K. S. Schulz, eds. *The Tree of Life Web Project*. Available online. URL: www.tolweb.org/tree/. Accessed January 25, 2008.

Margulis, Lynn, and Karlene V. Schwartz. *Five Kingdoms: An Illustrated Guide to the Phyla of Life on Earth.* 3rd ed. New York: Henry Holt, 1998.

Tudge, Colin. *The Variety of Life: A Survey and a Celebration of All the Creatures That Have Ever Lived.* New York: Oxford University Press, 2000.

eukaryotic cells Organisms and the cells they comprise can be broadly classified as either one of two major types based on structural complexity: prokaryotic or eukaryotic. Prokaryotic organisms include the unicellular archaea and bacteria, neither of which contains nuclei or other membrane-bound organelles. Eukaryotic organisms can be either unicellular, as in the case of protozoa or yeasts, or multicellular, as in the case of plants and animals. The cells that make up eukaryotic organisms share many common structures, most notably, a nucleus to house the deoxyribonucleic acid and other organelles as part of their endomembrane system.

Eukaryotic cells are generally larger than prokaryotic cells, but certain constraints limit their maximal size. Smaller cells have a larger surface area to volume ratio, and therefore can more efficiently exchange nutrients and waste products across the cell membrane. As cell size increases, this ratio decreases to the point of no longer being able to meet the cell's needs. Eukaryotic cells typically range from 0.0004 to 0.004 inch (10 to 100 µm).

CELL WALL AND CELL MEMBRANE
Whether eukaryotic or prokaryotic, all cells are enclosed by a cell membrane, also called a plasma membrane, that acts as a barrier between the cell and the external environment. The basic structure of the cell membrane is a phospholipid bilayer containing numerous embedded proteins. The main function of the cell membrane is to keep the cellular contents enclosed while preventing unwanted materials from entering and allowing waste materials to exit. Because the interior of the bilayer is hydrophobic, polar or charged substances cannot easily cross. Specialized channels in the membrane allow materials to move in and out in a controlled manner. Receptor molecules located on the exterior surface receive signals via the binding of specific ligands in a process called signal transduction, by which cells communicate with the environment and other cells within a multicellular organism. (The structure and function of cell membranes are covered more fully in the entry on biological membranes.) The characteristics of the cytoplasm, the entire contents bounded by the cell membrane excluding the nucleus and other membrane-bound organelles, can differ greatly from the external environment as a result of the selectivity of the cell membrane for the substances it allows to pass through. Because of this, eukaryotic cells can live in a wide variety of environments.

Plant cells, fungal cells, and some protist cells also have a cell wall external to the cell membrane. Whereas the main function of the cell membrane is to allow or prevent selected substances from passing into or out of the cell, the major functions of the cell wall are protection and structural support. The composition varies, depending on the organism. Plant cell walls are constructed from cellulose, fungal cell walls are made from chitin or cellulose and mixed glycans, and algal cell walls contain polysaccharides such as cellulose, pectin, and mannans and minerals such as silicon dioxide and calcium carbonate.

NUCLEUS
The presence of a nucleus distinguishes eukaryotic cells from prokaryotic cells. Bound by a double membrane called the nuclear envelope, the nucleus holds the cell's chromosomes. About 7.87×10^{-7} to 1.58×10^{-6} inch (20 to 40 nm) separate the two membranes, and pores of approximately 3.97×10^{-6} inch (100 nm) in diameter span the envelope, connecting the interior of the nucleus with the cytoplasm. A nuclear lamina made of protein filaments lines the envelope and gives it support. Because an organism's

genetic information is contained within the chromosomes, the nucleus is referred to as the control center of the cell. The directions for all cellular activities ultimately originate in the nucleus. DNA replication occurs here, as does transcription, the first stage of gene expression and protein synthesis. The nucleolus is a specialized region of the nucleus where ribosomal ribonucleic acid is synthesized.

ENDOMEMBRANE SYSTEM

Eukaryotic cells have an endomembrane system, an elaborate arrangement of internal membranes that compartmentalize cellular functions and play a role in protein processing. These membranes are all related either because they are physically connected or because they transfer tiny vesicles to one another. The plasma membrane and the nuclear envelope are considered part of the endomembrane system.

The endoplasmic reticulum (ER) is continuous with the nuclear envelope and extends outward from it into the cytoplasm. Consisting of approximately half of the cell's total membranes, the ER is organized into connecting tubules and folded stacks called cisternae. The ER serves as the site for both protein and lipid synthesis, though regions that accomplish the two tasks are separate. Ribosomes perform the job of protein synthesis, and, depending on the type of protein being made, a ribosome can do this while floating freely in the cytoplasm or while bound to the cytoplasmic surface of the ER. Generally, proteins that are destined to remain in the cytoplasm are synthesized by free ribosomes, whereas proteins that will be embedded in a membrane, secreted out of the cell, or compartmentalized within the cell are synthesized by ribosomes sitting on the ER. As the ribosomes make the proteins, the polypeptide chains cross the membrane into the ER. Small chains of carbohydrates called oligosaccharides are attached to secretory proteins. Portions of the ER that are covered in ribosomes are termed rough ER (RER) because of the grainy appearance they give the membranes. Other parts of the ER are called smooth (SER) because they are not covered in ribosomes and appear smooth. SER is the site for lipid synthesis, for the detoxification and breakdown of toxic substances, and for carbohydrate metabolism in some tissues.

After making secretory or membrane proteins, parts of the ER containing them pinch off, forming transport vesicles that travel to the Golgi apparatus, another network of flattened, membrane-bound sacs located throughout the cell. The membranes of the transport vesicles fuse with the membrane of the Golgi apparatus, releasing the contents into the Golgi's interior. Enzymes inside the Golgi modify the oligosaccharide chains that were added to the proteins in the ER by removing some of the carbohydrates and replacing others. Different cisterna of the Golgi apparatus perform different tasks and the polypeptides bud off from one cisterna and then fuse with the next, where additional modifications such as the addition of phosphate groups occurs. The Golgi then sorts and delivers the processed proteins to their final destination. Again, vesicles bud off and deliver proteins that are to be secreted or that will become part of the cell membrane to the boundary of the cytoplasm. Other proteins, such as the cell's digestive enzymes, stay packaged in vesicles that remain in the cytoplasm.

Lysosomes are spherical sacs filled with digestive enzymes for breaking down materials taken into the cell by phagocytosis, for recycling a cell's own organic matter, or for carrying out programmed cell death during certain stages of normal development. The pH within lysosomes is slightly acidic, and the enzymes they carry function optimally at a pH lower than in the cytoplasm of the cell.

As are lysosomes, vacuoles are spherically shaped membrane-bound sacs within the cytoplasm. Vacuoles are larger, however, and are used for storage of different substances. Food vacuoles are formed when a cell takes in food particles by phagocytosis. After fusion with a lysosome, enzymes digest the food into smaller molecules that can be used for the cell's metabolic or energy needs. Contractile vacuoles allow certain types of protists to live in freshwater environments by collecting and pumping out excess water that has entered the cell by osmosis. Plant cells have central vacuoles that make up about 80 percent of the cell and perform functions including storage, waste removal, protection, and growth.

OTHER ORGANELLES

Most eukaryotic cells contain numerous mitochondria, organelles involved in metabolism. Bound by a double membrane, mitochondria contain extensive infoldings of the inner membrane called cristae that increase the surface area on which processes of cellular respiration occur. The inner membrane and the intermembrane space between the two membranes contain many enzymes and molecules that play an active role in the harvest of energy from organic compounds. The matrix, the material located within the innermost compartment of mitochondria, contains its own ribosomes and DNA that encodes several mitochondrial specific proteins. Other proteins are imported from the cytoplasm.

Chloroplasts also contain their own ribosomes and DNA. Related to amyloplasts, which store starch in plant cells, and chromoplasts, which give fruits and flowers color, chloroplasts contain chlorophyll, the green pigment involved in photosynthesis. As are mitochondria, chloroplasts are enclosed by a double membrane and contain an internal membranous network in order to increase the productivity of the organelle, whose function is to carry out photosynthesis, the conversion of light energy into organic compounds. Resembling stacks of pancakes, structures called grana consist of individual flattened disks of membranes called thylakoids. Chlorophyll, the pigment that absorbs wavelengths of electromagnetic radiation in the blue and red ranges, is embedded in the thylakoid membranes. (Green light is not absorbed—it is reflected, giving plants a green appearance.) The reactions of photosynthesis that involve the conversion of light energy into adenosine triphosphate (ATP) and nicotinamide adenine dinucleotide phosphate (NADPH) and that produce molecular oxygen, called the light reactions, occur in and on the thylakoids. The dark reactions, which use the energy stored in the form of ATP and NADPH, function to produce organic compounds from carbon dioxide and occur in the stroma, the region of the chloroplast bound by the inner membrane.

Another membrane-bound organelle found in eukaryotic cells is the peroxisome. Roughly spherical in shape, peroxisomes contain enzymes that add hydrogen atoms to oxygen atoms to produce hydrogen peroxide (H_2O_2), which is then broken down into water and molecular oxygen. Peroxisomes function in the metabolism of fatty acids and in liver cell, to detoxify alcohol or other poisons, by removing the hydrogen atoms and transferring them to oxygen.

Ribosomes are the site of protein synthesis. Not bound by a membrane, these structures give a dotted appearance to the cytoplasm of cells and when attached to the endoplasmic reticulum make it appear rough. In eukaryotic cells, the ribosomes comprise two subunits, a 60S subunit and a 40S subunit, that combine to form an 80S complex. Each subunit contains numerous polypeptides and ribonucleic acid.

CYTOSKELETON

Eukaryotic cells have an internal framework that provides cellular shape and support, acts as an anchor for organelles and a track for their movement, and assists in cellular motility. The cytoskeleton is made of three main components: microtubules, microfilaments, and intermediate filaments. Dimers of the proteins α-tubulin and β-tubulin assemble into long, hollow microtubules that grow out of a centrosome, a region near the nucleus. A pair of centrioles made from nine triplets of microtubules sits in the centrosome and functions in cell division. The major function of microtubules is to maintain cell shape. Microtubules also act to pull apart the chromosomes during mitosis so that each daughter cell receives one complete set of genetic information. Organelles and molecules inside a cell use microtubules and micro-

filaments as tracks along which they can move to different locations within a cell. For example, a vesicle containing secretory proteins can be transported along a microtubule from the Golgi apparatus to the cell membrane for fusion and release.

Microfilaments consist of two chains made of the globular protein actin, and they line the inner side of the plasma membrane. Bundles of microfilaments form microvilli, microscopic protrusions from cells whose job is transporting substances across membranes. The microvilli effectively increase the surface area across which transport occurs. Microfilaments also play a major role in muscle contraction by sliding along myosin filaments. Actin-myosin bundles also form cleavage burrows that divide a mitotic cell into two daughters, form pseudopodia that drive amoeboid movement, and function in cytoplasmic streaming, a process that acts to distribute materials within plant cells. Both microtubules and microfilaments constantly disassemble and reassemble within a cell.

The more stable intermediate filaments are constructed from different types of keratins and have different functions. Some form a network that anchors the position of cellular organelles. The nuclear lamina lines the nuclear envelope and is made of another type of intermediate filament. The many different types of intermediate filaments form a framework that reinforces the shape of the cell.

EXTERNAL STRUCTURES

Appendages involved in locomotion include cilia and flagella, both of which are made from microtubules. Cilia are smaller bristlelike structures that cover a cell and can beat in unison to propel unicellular organisms through water. They are approximately 9.84×10^{-6} inch (0.25 μm) in diameter and range from 7.87×10^{-5} to 7.87×10^{-4} inch (2 to 20 μm) in length. When they occur on the surface of stationary cells, such as on the lining of the respiratory tract, their beating moves fluids such as mucus along the surface of the tissue. Flagella are much longer than cilia, measuring between 3.94×10^{-4} and 7.87×10^{-3} inch (10 and 200 μm) in length. Usually only one or a few flagella are found on a single cell. Cilia move using a back-and-forth motion, while flagella undulate in a wavelike motion. Anchored to the cell by a structure similar to a centriole called a basal body, both cilia and flagella possess a common core of microtubules arranged in a "9 + 2" arrangement. When viewed in cross section, nine microtubule doublets form a circle surrounding a pair of singlets.

Most eukaryotic cells have a glycocalyx, a carbohydrate-rich coating that covers the outer surface of the cell. The glycocalyx can form a capsule or a slime layer and functions in protection, adherence

Animal Cell Model

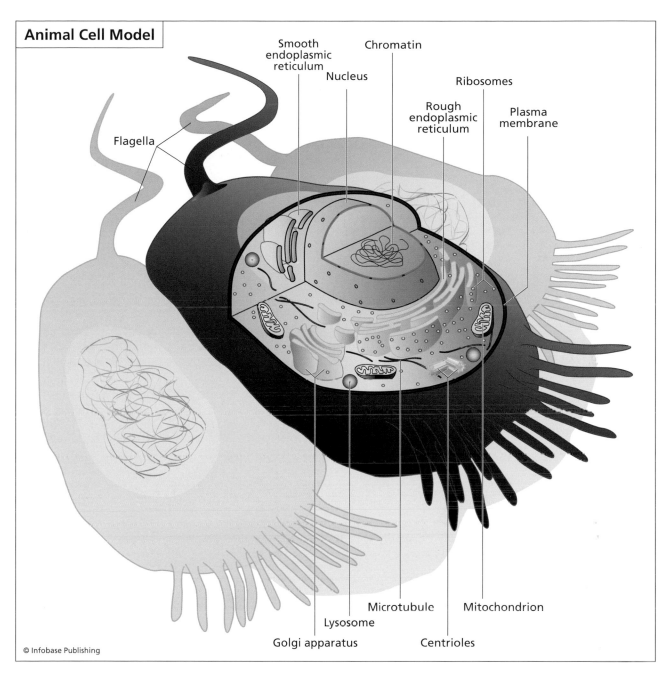

Smooth endoplasmic reticulum

Chromatin

Nucleus

Ribosomes

Rough endoplasmic reticulum

Plasma membrane

Flagella

Microtubule

Mitochondrion

Lysosome

Golgi apparatus

Centrioles

© Infobase Publishing

This diagram of a typical animal cell illustrates the most common eukaryotic cellular structures.

of the cell to other surfaces, and reception of signals from the environment.

ENDOSYMBIOTIC THEORY

Endosymbiosis is an association in which one organism dwells within the body of another. The endosymbiotic theory explains the origin of certain organelles of eukaryotic cells by proposing that smaller prokaryotic cells that had been engulfed by larger prokaryotic cells became permanent intracellular structures, eventually lost their ability to survive independently, and developed into an integral part of the larger

cell. The smaller cells not only survived after being engulfed by the larger cell but became established in the host cytoplasm and carried out essential metabolic functions. In 1966 Lynn Margulis, the main proponent of the endosymbiotic theory, published a landmark paper titled "The Origin of Mitosing Eukaryotic Cells" that drew on 19th-century ideas and on microbial observations and brought this now largely accepted theory to the limelight.

A proposed scheme for the evolution of eukaryotic cells begins with a portion of a prokaryotic cell membrane folding inward to surround the nucleoid region, resulting in a nucleus bound by an envelope consisting of a double membrane enclosing the genetic material. The cell engulfs a smaller bacterium that respires aerobically, and the smaller cell becomes established as an endosymbiont, living in a mutualistic relationship with its larger partner. The smaller cell multiplies within the host, creating several intracellular "machines" that evolve to perform aerobic respiration quite efficiently in an environment where other survival needs are met by the host cell. The host cell becomes dependent upon the early mitochondria to supply its energy needs. Meanwhile, the cell membrane forms numerous infoldings that increase the ancestral eukaryotic cell's ability to synthesize and process proteins and ultimately develop into endoplasmic reticulum. Algal and plant cells developed by the engulfment of photosynthetic bacteria that resembled cyanobacteria. They survived inside the larger cell, harvested energy from light, and converted it onto chemical energy stored in organic compounds, eventually developing into chloroplasts. Flagella and cilia possibly developed from symbiotic relationships of ancestral eukaryotic cells with spiral bacteria. The existence of circular chromosomes and 70S ribosomes inside mitochondria and chloroplasts in addition to other overwhelming structural similarities between prokaryotes and these organelles support the endosymbiotic theory.

See also ARCHAEA; BACTERIA (EUBACTERIA); BIOLOGICAL MEMBRANES; BIOMOLECULES; CELL BIOLOGY; CELLULAR METABOLISM; CELLULAR REPRODUCTION; CHROMOSOMES; EUKARYA; GENE EXPRESSION; MARGULIS, LYNN; MOLECULAR BIOLOGY; PHOTOSYNTHESIS; PROKARYOTIC CELLS.

FURTHER READING

Alberts, Bruce, Dennis Bray, Karen Hopkin, Alexander Johnson, Julian Lewis, Martin Raff, Keith Roberts, and Peter Walter. *Essential Cell Biology.* 2nd ed. New York: Garland Science, 2004.

Panno, Joseph. *The Cell: Evolution of the First Organism.* New York: Facts On File, 2004.

Sullivan, James A. "Cells Alive!" Available online. URL: http://www.cellsalive.com/. Accessed January 14, 2008.

evolution, theory of The theory of evolution is a cornerstone of modern life science. Biological or organic evolution is the change over time within a population in the proportions of individuals with genetically different characteristics. Evolutionary theory purports that currently existing organisms originated from preexisting forms that have been modified in successive generations by the combined action of several processes: genetic variation, the phenotypic expression of genetic characteristics, natural selection, gene flow, genetic drift, and inheritance. Evolutionary changes can be examined at different levels and in different time frames. Microevolution encompasses the changes in the gene frequencies within a limited population or between populations of the same species, often over short periods. Macroevolution refers to major evolutionary changes that occur above the species level, usually over periods of millions of years. Taxonomic groupings are based on macroevolutionary developments.

GENETIC VARIATION AND POPULATIONS

A population is a group of individuals belonging to the same species. Though members of the same species share the same number and types of genes, each individual has a unique set of forms of those genes. The genotype, or genetic composition of an individual, determines the characteristics that individual will have, both those common to all members of the species and those unique to the individual. Environmental factors can also influence the expression of certain traits to varying degrees. The differences among members of the same species are called variations. At the molecular level, genetic variation of individuals includes the existence of different alleles (forms of a gene) occurring at a specific gene locus and different combinations of genes. Parents pass on their genes to offspring through reproduction.

In asexual reproduction, the parent passes on the entire genome to the next generation. Mutations to the genes are the only source of genetic variation. In sexual reproduction, each parent contributes half of his or her genome to the offspring; thus the offspring contain a mixture of genetic material. During the production of gametes (eggs and sperm), meiosis, a specialized form of cell division that halves the genetic material in the sex cells, randomly divides the genetic material into equal quantities. Because the genetic material is partitioned randomly, as described by the law of independent assortment, the resulting cells contain numerous different combinations of alleles. This is similar to shuffling and dealing two hands of 26 cards each from a single deck. Crossing over, or recombination, when pairs of homologous chromosomes exchange segments with one another during meiosis, contributes additional genetic variation to the gametes by altering the associations of alleles located on a single chromosome. Though meiosis maintains genetic variation among individual members of a sexually reproducing population and provides a foundation for evolutionary change, it does not, by itself, change the gene frequencies from generation to generation within the population. Other processes must also take place in order to drive evolution: natural selection, mutation, nonrandom mating, gene flow, or genetic drift.

Because an individual's genotype does not change during one's lifetime, an individual cannot evolve, in the biological sense. Individuals must pass on their genes to create new individuals through reproduction. Populations, groups of members of the same species that live in the same geographic area and interbreed, are the smallest units of evolutionary change. Some populations are well defined by distinct geographic boundaries, such as mountain ranges or islands, and only breed within the population. Other populations are not completely isolated. The gene pool of a population comprises the entire collection of genes present in a given population at one time, including all the alleles at all the gene loci of all the individuals in the population. Some alleles are fixed, meaning they are present in every individual at a given gene locus, whereas others vary. Allele frequency is a measure of the relative proportion of a particular allele in the population. (Fixed alleles have frequencies of 100 percent.) Stable gene pools do not change over time; while individuals exhibit genetic variation due to meiosis, the allele frequencies are stably maintained from generation to generation. No mutation, natural selection, genetic drift, or gene flow into or out of the population occurs, and mating is completely random. When these conditions are met, the gene pool will stabilize and the population is said to be in Hardy-Weinberg equilibrium. In nature the five conditions are rarely met, and as a result, populations evolve.

MECHANISMS OF EVOLUTION

Mutations to genes can occur spontaneously because of errors during replication of the deoxyribonucleic acid (DNA). Exposure to certain chemicals and environmental conditions can also cause genetic mutations. As mentioned previously, sexual reproduction increases genetic variability by altering the associations of certain alleles with one another through independent assortment, recombination, and random fertilization. Sexual reproduction alone does not change allele frequencies in a population; it simply alters the combinations that occur in individuals. New genetic mutations can alter allele frequencies, but the change from one generation to the next is minimal. When natural selection acts on mutations, however, evolution can occur.

Genetic variation can lead to phenotypic differences in individuals. If the phenotype associated with a specific genotype increases an individual's fitness, or reproductive success, then individuals with the fitter phenotype will produce more offspring than individuals with alternate phenotypes. Since offspring receive genes from their parents, the next generation of the population will contain a higher frequency of the alleles responsible for the favorable phenotype, since those individuals produced more offspring. This process, called natural selection, depends on the environmental conditions. Phenotypes that are advantageous in one environment might not be in a different environment. To illustrate, consider a few of the many modifications to desert plants that help them withstand long periods of drought. For example, the roots of desert plants do not penetrate deep into the soil but remain close to the surface, where they can absorb water before it evaporates. Cacti have no leaves to prevent water loss through transpiration; instead the stems of cacti carry out photosynthesis. These same characteristics would not confer an advantage if the plants were grown in a moist environment.

Adaptations are traits that increase fitness, the driving force for natural selection. The level of fitness associated with an allele can only be ascertained by comparison with alternate alleles. Traits that increase the survival rate of a species contribute to an animal's fitness, but selection will only favor such traits insofar as survival improves the reproductive success of the organism. Consider that some species commit suicide in order to reproduce. The male Australian red-back spider allows a female to cannibalize him during the mating process. Natural selection has favored this adaptive behavioral trait because it increases the male individual's reproductive success even though it kills him. Individual fitness, the number of offspring produced per individual over a lifetime that survive and go on to reproduce, is typically stronger than selection at the group or species level. Another factor affecting reproductive success is sexual selection, variation in the ability to obtain a mate for reproduction. The existence of preferences for certain mates leads to nonrandom mating. Certain characteristics may increase an organism's reproductive success by increasing its ability to attract a mate, sometimes even at the expense of a decreased survival rate. A male peacock's feathers increase his attractiveness to females and therefore increase his reproductive success, even though the same feathers hinder his movement, making it more difficult to escape from possible predators. Some social animals exhibit altruistic behavior, in which they appear to sacrifice their own reproductive success for the success of a relative. For example, many bird species help rear young born to others. One might think that natural selection would disfavor genetic variation that leads to such altruistic behavior, but the phenomenon of kin selection helps explain this. The young are often related and thus share the same genes. By helping to defend and feed the young, the nonparental caretaking birds are ensuring the reproductive success of others sharing the same genes: in other words, the allele for this apparently altruistic behavior gains higher fitness by being passed on through relatives.

(continues on page 335)

ADAPTATION OR NOT: HOW TO TELL THE DIFFERENCE

by James Wagner, Ph.D.
Transylvania University

In 1835 Charles Darwin was only 26 years old, and he was already on the fourth year of his five-year voyage around world on the H.M.S. *Beagle.* At this point in the voyage the *Beagle* had sailed into a small chain of equatorial islands about 500 miles (805 km) off the west coast of South America. These islands, known as the Galápagos Islands, were inhabited by a variety of unusual animals, such as the large Galápagos tortoises and the seaweed-eating marine iguanas. However, there were also a number of familiar animals, most of which were various varieties of common birds and plants.

Contrary to popular belief, while visiting the Galápagos Islands Darwin did not have a "Eureka!" moment when he envisaged his concept of evolution. (It was actually not until eight years later, comfortable at his English home in the country, that Darwin concluded, quite reluctantly, that species evolve.) In fact, the 26-year-old Darwin found the small chain of islands to be fascinating but no more so than the jungles and pampas of South America where he had spent the past four years exploring. In his book *Voyage of the Beagle,* Darwin's first published work, he recounts an unusual observation he made while on the Galápagos Islands.

In Charles Island, which had then been colonized about six years, I saw a boy sitting by a well with a switch in his hand, with which he killed the doves and finches as they came to drink. He had already procured a little heap of them for his dinner; and he said that he had constantly been in the habit of waiting by this well for the same purpose.

Darwin then remarks that the birds of the islands had not yet "learnt that man is a more dangerous animal than the tortoise or the *Amblyrhynchus* (marine iguana), disregard him in the same manner as in England shy birds, such as magpies, disregard the cows and horses grazing in our fields." An avid bird hunter, Darwin showed great restraint when he recounts a time he pushed a hawk off a branch using the muzzle of his gun.

When Darwin later developed his idea of "evolution by natural selection" he noted that those aspects of the organism that influence its survival and reproduction will become prevalent in the population because those individuals who lack these traits will disappear from the population. In terms of wild birds, Darwin noticed that it was not that birds were learning to be afraid of humans as much as that humans were killing those birds that were not already afraid of humans. Over time only those birds that had an innate fear of humans survived to reproduce, a trait they passed on to their descendants, leading to the cautious nature of most birds today.

Darwin called those traits that influence an organism's survival and reproduction adaptations. Darwin recognized that adaptations could be physical traits like sharp teeth or poisonous flesh or physiological traits like the ability to digest wood or an immune system that fights infections. Adaptations could also be behavioral traits such as exhibiting caution around humans, a trait that many of the Charles Island birds originally lacked.

Early evolutionary biologists (and many nature shows on television today) viewed organisms as an assemblage of adaptive traits. If the organism has a trait, that trait must have a function and therefore it must be an adaptation. In 1979 two famous evolutionary biologists, Stephen J. Gould and Richard C. Lewontin, challenged this view, which they called the adaptive program, in their paper "The Spandrels of San Marco and the Panglossian Paradigm." Gould and Lewontin argued that biologists frequently explained the existence of all traits of an organism as if they were adaptations optimized by evolution. In contrast, Gould and Lewontin proposed that some traits are just by-products of building an organism and are not adaptive per se. Or if they are adaptive, they may not be perfectly adaptive because not all traits can be optimally created since trade-offs must occur. Although the paper had some critics, it was influential in inducing biologists to broaden their view of the structure and form of organisms and allow for non-adaptive explanations to be considered as part of the process of evolution.

One can apply Gould and Lewontin's idea of trade-offs to the bird example. In the cautious bird scenario, imagine that those birds that were absolutely terrified of humans and always flew away whenever they saw a person would have a low probability of ever being shot and therefore enjoy a high survival rate. But although their mortality rate from hunting would be low, the birds' constant avoidance of people could prevent them from finding food and acceptable nest sites, which may be near people. For birds in people's backyards, a trade-off exists between being cautious of the boy with a stick and tolerating him to a degree in order to land at the bird feeder. Gould and Lewontin argued that for most adaptations perfection is not to be expected because of these types of trade-offs.

The other major contribution made by Gould and Lewontin was provoking scientists to question the basic assumption that all traits must be adaptations. For example, in humans on the underside of the wrist are veins that reside just below a thin layer of skin. These are the blue median antebrachial veins that return the deoxygenated blood from the hand and lower forearm back toward the heart and lungs to be reoxygenated. Having a circulatory system consisting of veins and arteries is clearly adaptive (imagine trying to survive without one), but is having *blue* veins an adaptation? Human veins appear blue because the shorter, blue wavelengths of light more readily reflect off the shallow veins, thereby giving them a bluish tint. The blue color does not result, as many believe, from the blood's changing color from blue to red when it is oxygenated. Human blood is always red; it just shifts from dark red in the absence of oxygen to bright red when it is oxygenated. So is having blue veins an adaptation per se? And how does one determine whether blue veins are an adaptation?

To approach this question, an evolutionary biologist would posit how the color of veins influences the organism's ability to survive or reproduce. In terms of the blue veins it would seem that having blue veins is a nonadaptive by-product of the interaction among the reflective properties of different wavelengths of light, the depth of the veins in the skin, and the collagen from which the veins are constructed. If the veins more readily reflected yellow light than blue, it is hard to imagine how having yellow veins would reduce your survival.

One could argue, however, that in some cultures the appearance of the blue veins may act as an indication of beauty since their visibility indicates clarity of skin, which itself is an indication of health. Having blue veins is the norm, and though having yellow veins would not negatively affect one's health, it might affect the

ability to find a mate, a circumstance that could reduce one's ability to reproduce. In this imaginary scenario, the blue color of veins was originally a nonadaptive trait that later became adaptive because it could be used to signal some other trait—such as an individual's health.

One of Darwin's brilliant insights in developing his idea of adaptation was that some traits may improve an organism's ability to survive while other traits may improve an organism's ability to reproduce. In both situations, those adaptive traits should increase in the population unless they work in opposition to each other. Darwin recognized that adaptive traits favored in sexual selection may in fact be opposed by natural selection. The classic example of the male peacock's tail is a case in point. The large conspicuous tail of the male peacock reduces the male's ability to avoid capture by tigers, their natural enemies. So in terms of natural selection this trait seems to be maladaptive since the tails can increase male mortality. But research has shown that those males with the largest tail feathers, most brightly colored tails, and tails with the most eyespots have more successful matings than their less endowed male neighbors. Thus, sexual selection favors tail feathers, while natural selection opposes it. This exemplifies one of the challenges in determining whether a trait is really an adaptation since one must consider the trait in the contexts of both natural and sexual selection. In addition,

some traits are only adaptive for one sex of the species but both sexes possess the trait. Nipples in male humans exemplify that phenomenon. For female humans the nipple is part of the mammary structure, a specialized gland system used in the production and delivery of milk to their young. However, both males and females are built from the same body type, which is simply modified by hormones during development. Therefore, males possess nonfunctional nipples, making their nipples a nonadaptive trait.

The task of determining whether a trait in an organism is an adaptation or a nonadaptive trait presents a challenge for biologists. Humans naturally desire to explain the world around them, and it is easy to propose an adaptive explanation for why an organism looks or behaves the way it does, but one must be cautious and ask, What is the adaptive advantage of the trait? And what is the evidence for that advantage?

FURTHER READING

Cronin, Helena, and John Maynard Smith. *The Ant and the Peacock: Altruism and Sexual Selection from Darwin to Today.* New York: Cambridge University Press, 1991.

Darwin, Charles. *Voyage of the Beagle.* New York: P. F. Collier & Son, 1909.

Gould, Stephen J., and Richard C. Lewontin. "The Spandrels of San Marco and the Panglossian Paradigm." *Proceedings of the Royal Society* 205 (1979): 581–598.

(continued from page 333)

Another instigator of evolution is genetic drift, a fluctuation in allele frequency due to random chance. The likelihood of genetic drift increases in smaller populations. The bottleneck effect occurs after much of a population dies as a result of a radical change in the environmental conditions, such as caused by a natural disaster or severe climate change. The gene composition of the few individuals that survive the disaster and reproduce becomes amplified in the next few generations. The founder effect occurs when a few individuals become isolated from a larger population and establish a new population. This situation also leads to the reduction of genetic variability due to genetic drift.

Gene flow is the movement of genes into or out of a population, as when individuals migrate. They leave one population and enter another population, taking their genes with them. This increases the variability of the new population and reduces the degree of genetic differences between populations.

MACROEVOLUTION

The microevolutionary genetic changes caused by mutation, selection, nonrandom mating, genetic drift, and gene flow lead to adaptations within a population or species, but macroevolutionary changes lead to the creation of new species and new lineages. Speciation, the creation of new species, is a cornerstone of evolutionary theory and must not only explain changes to a limited gene pool, but also account for the current and past biological diversity. Biological species are defined as groups of similar organisms that have the ability to interbreed. Other definitions of the species concept exist: morphological species share similar body size, shape, and function; phylogenetic species are clusters of organisms that share parental ancestry and descent; paleontological species are groups of organisms with distinct morphological characteristics as evidenced by the fossil record; and ecological species are groups that share the same niche in a community. Since sexual reproduction plays a key role in variation and adaptation, the concept of biological species is most useful when considering the process of speciation, which can occur by two patterns: anagenesis or cladogenesis. Anagenesis occurs when genetic variation accumulates, resulting in the eventual origin of a new species. Cladogenesis occurs when heritable changes result in the branching off of a new species from a parent lineage. The parent species may or may not also change.

The development of new species depends on reproductive isolation, when barriers prevent two different species from reproducing to create viable, fertile offspring. The preventative mechanisms can be classified as either prezygotic or postzygotic. Prezygotic barriers include factors that hinder one species's sperm from fertilizing another species's eggs. When the habitats are geographically separated, such as when fish occupy two different lakes or a mountain range separates two populations of squirrels, the two populations rarely if ever encounter each other in order to mate. Some species do not mate because of temporal isolation, when the two species mate at different times of the day or year, or behavioral isolation, when differences in courtship or mating behaviors prevent two species from coming together. An attempt at mating by two species does not ensure reproductive success. Structural differences or incompatibility of sperm and egg may prevent successful fertilization. Postzygotic barriers prevent the development of the zygote into a viable, fertile offspring when fertilization occurs. The zygote might not complete development, or, if it does, the offspring or the offspring of the next generation might not be fertile. In these cases, even though new organisms are produced, without the production of viable, fertile offspring, the new phenotypes cannot continue to reproduce; thus the two original species remain reproductively separated.

Speciation can occur after geographic isolation, a process called allopatric speciation, or within the same geographic boundaries, called sympatric speciation. Neither of these processes is directed; speciation occurs as a by-product of the reproductive isolation. Allopatric speciation, which appears to be more common, may result when one population of a species migrates to a different area or when a lake recedes and forms two separate isolated lakes. Whatever the physical means for separating part of a population from the rest of the population, after that separation, the populations are reproductively isolated, and the gene pools can diverge. Then selection, mutations, drift, and other mechanisms can result in the eventual inability of the two species to interbreed, even if placed back together physically. Sympatric speciation can occur when populations geographically overlap by changes in chromosomal makeup, by nonrandom mating preferences, or by a change in the environmental conditions of the habitat. Meiosis occasionally fails to reduce the number of chromosomes during cell division, resulting in the production of a polyploidy individual, one that contains more than the typical number of sets of chromosomes. Many crop plants contain four sets of chromosomes rather than two. A polyploidy plant initially may be reproductively isolated but still able to reproduce asexually, then develop the ability to reproduce sexually in future generations.

Macroevolution results from the accumulation of numerous speciation events. Major alterations in morphology, novel adaptations, and new lineages evolve over time. The tempo by which evolution occurs varies. In gradualism, species diverge slowly through the acquisition of new adaptations that accumulate and that can affect major change over long periods. The noted geologists James Hutton and Charles Lyell greatly influenced Darwin's ideas of biological evolution, which incorporated the notion of gradualism into biological processes. Hutton first proposed that the formation of geological features could be explained by the slow but continuous action of geological processes that acted in the past and still act today. Lyell popularized this idea by generalizing it into his theory of uniformitarianism. The premise of Darwinian evolution was that the gradual accumulation of numerous variations over long periods resulted in the creation of new species.

In 1972 Stephen Jay Gould and Niles Eldredge published their hypothesis of punctuated equilibrium, suggesting that throughout evolutionary history, long periods of stasis (no change) were interrupted by short periods of rapid change. They proposed that small subpopulations on the periphery underwent major transitions, then invaded the habitat of the ancestral population. Their paper sparked considerable controversy, as biologists long held their belief in gradualism. Punctuated equilibrium explains the absence of fossil evidence for transitional forms in some lineages, since transitional fossils would not be present in the region into which new forms migrated to join their ancestral population. Transitional fossils

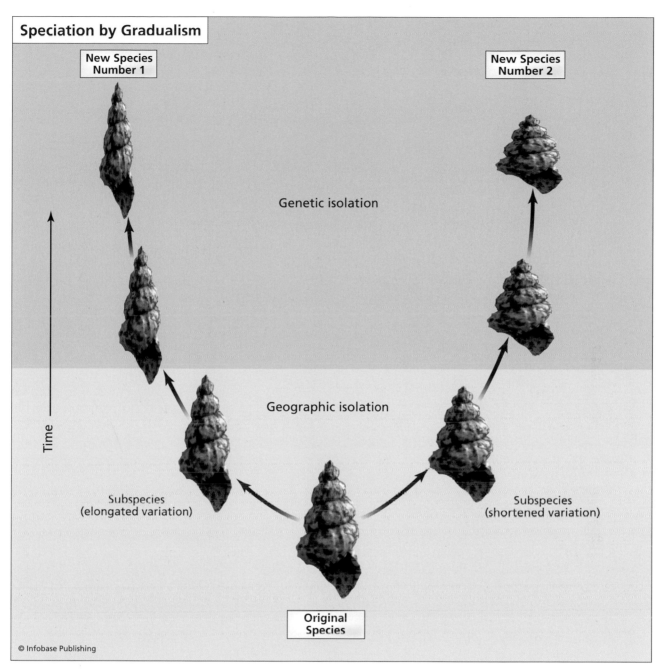

Speciation by Gradualism

New Species Number 1

New Species Number 2

Genetic isolation

Time

Geographic isolation

Subspecies (elongated variation)

Subspecies (shortened variation)

Original Species

© Infobase Publishing

Reproductive isolation may be achieved by geographic separation, resulting in a gradual accumulation of slight variations, eventually leading to the formation of distinct species.

may exist but would be in the peripheral location where speciation occurred. Today biologists consider both gradualism and punctuated equilibrium as valid hypotheses. Periods of stasis may result from stable environmental conditions, during which no selection pressure acts to drive the evolution of adaptations. Intrinsic constraints such as the existence of previous adaptations may also explain the observance of long periods of stasis.

The opposing process of speciation is extinction, resulting in the loss of an entire species and playing a major role in the shaping of evolutionary history.

Extinction occurs when reproductively isolated species fail to produce sufficient numbers to maintain the species. One species in the community may compete for food or resources with another species to the point of exclusion, leading one of the species to extinction. An abrupt change in environmental conditions can kill large numbers of the species, making it difficult for the species to regain its footing. In recent years, human activities have had a profound impact on many species, both directly, by activities such as hunting certain animals for food or for their products, and indirectly, such as by destroying habitats or organisms. Mass extinctions wipe out several groups at one time, leaving many unfilled ecological niches. Surviving species serve as common ancestors that evolve by radiation to fill those niches. Two well-known mass extinctions occurred at the end of the Permian (about 250 million years ago), when plate tectonics moved all the Earth's continents together, leading to the extinction of more than 95 percent of all marine species and 70 percent of all terrestrial species, and at the Cretaceous/Tertiary boundary (about 65 million years ago), when a large asteroid impacted the Earth, leading to the extinction of the dinosaurs.

EVIDENCE FOR EVOLUTION

Biologists can directly test and observe microevolution through the study of population genetics. Macroevolution is inferred from numerous observations and data collected from comparative biochemical and genetic analyses, developmental biology, biogeographical studies, comparative anatomy, and fossil evidence.

Archaeans, bacteria, fungi, protists, plants, and animals carry out many common biochemical processes. The basic cellular process of respiration, by which cells extract energy from molecules to make adenosine triphosphate (ATP), includes the same stages and many similar molecules and enzymes across domains of life. Photosynthesis in plants resembles that of algae and photosynthetic bacteria, though some of the structures involved differ. Molecular processes such as deoxyribonucleic acid (DNA) replication, transcription, and translation occur by very similar mechanisms, with slight adjustments to the enzymes and machinery. These and other biochemical processes support a common ancestor that performed similar tasks.

Comparison of the DNA sequence of the genomes of different organisms provides powerful support for the theory of evolution in addition to a lot of information about lineages and the history of life. Genes that encode proteins contain codons that specify certain amino acids. Natural selection acts on phenotypic characteristics, determined by the function of the protein, which is determined by the sequence of amino acids. Evolutionary theory predicts that the amino acid sequence of proteins that play important basic roles in survival and reproduction would be conserved, that is, exhibit minimal differences between species. Furthermore, closely related species share more sequence similarity than more distantly related species. For example, the sequence similarity between protein coding regions of human and chimpanzee genes approaches 100 percent, of humans and mice 99 percent, of humans and fruit flies 60 percent, and of humans and roundworms 35

percent. Silent mutations are changes to the DNA sequence that result in a different three-nucleotide codon but encode the same amino acid; the sequence has changed, but the protein has not. Because selection acts to preserve function rather than a specific genotype, silent mutations occur more frequently than mutations that alter the amino acid sequence of a protein, which often are deleterious. Other molecular evidence supporting evolution involves introns and pseudogenes. Introns are noncoding regions found within eukaryotic genes. As expected, they contain a higher percentage of differences in nucleotide sequence than the coding regions, as do pseudogenes, genes that that have become inactivated by mutation and no longer encode functional proteins. A phylogenetic tree generated by a computer based on DNA sequence similarities looks practically identical to one generated from fossil evidence and studies of comparative anatomy.

Developmental biology also supports the theory of evolution. Closely related organisms follow similar developmental pathways. Distinguishing related organisms at early developmental stages is difficult, because modifications of the earlier stages of developmental pathways have a greater impact on an organism's overall development. Such changes have a tremendous impact on the formation of the basic body structure and therefore may be more deleterious; thus most modifications occur at later stages of development. Because of this, the developmental pattern of an organism often reflects its evolutionary history, and one can draw inferences about the ancestry of a species by comparing its development with that of other species. The earlier differences in the developmental patterns of two species emerge, the more distantly related the species are. For example, after cleavage, vertebrate embryos consist of a cluster of cells. In amphibians, the cluster of cells forms a roughly spherical ball, whereas in reptiles and mammals, the shape is more disklike. The altered shape is due to an adaptation not present in amphibians but that appeared in the ancestor common to reptiles and mammals, the amniotic egg. Reptiles, however, including the lineage that developed into birds, have heavy yolks, which cause the embryos to spread out over the top of the yolk, being more extended. Mammals and reptiles diverged, and though modern mammals no longer have yolks, the embryo retains a shape more similar to the closely related reptiles than to the more distantly related amphibians.

Comparative anatomy studies of both extant and past life-forms provide additional evidence for evolution. Because natural selection can only act on preexisting traits, one would expect closely related species to share many morphological features. Related species that have major structural differences must have

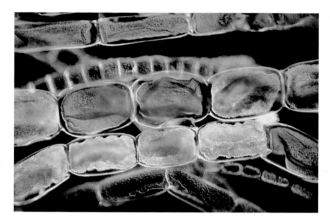

Algae. *(James M. Bell/Photo Researchers, Inc.)*

Gymnosperm *(John J. Mosesso/National Biological Information Infrastructure)*

Moss *(tln, 2007, used under license from Shutterstock, Inc.)*

Ferns *(National Oceanic & Atmospheric Administration/Department of Commerce)*

Angiosperm *(John J. Mosesso/National Biological Information Infrastructure)*

The fact that closely related organisms share more common anatomical features than distantly related organisms supports evolutionary theory. Of the pictured organisms, angiosperms are most closely related to gymnosperms, then ferns, then mosses, then algae.

a more distant common ancestor than species with relatively minor differences in anatomy. To illustrate this, consider some of the structural evidence used to reconstruct the evolution of angiosperms (flowering plants). Biologists believe that modern land plants, including angiosperms, originated from an ancestral form of green alga. As are plants, algae are eukaryotic, can be multicellular, and have cellulose in their cell walls. As land plants evolved, they developed unique structural adaptations that helped them survive on land, such as roots that stabilize the plant and absorb water from the soil. The earliest land plants, called bryophytes, originated about 475 million years ago, and like algae, were nonvascular. About 425 million years ago, plants began to develop specialized systems for transporting water and nutrients throughout the body of the plant, giving rise to vascular plants. Seeds first appeared approximately 360 million years ago. The anatomy of a seed includes a tough protective coating surrounding the plant embryo and nutrition to fuel early development. Gymnosperms (plants that produce naked seeds) and angiosperms diverged from an ancestral vascular seed plant. One unique morphological adaptation in angiosperms is the presence of structures called ovaries, chambers where seeds develop. As evolutionary theory predicts, angiosperm species share most anatomical features with other angiosperm species, then with gymnosperms. Comparing the morphologies of angiosperms and ferns, a type of seedless vascular plant, reveals fewer shared characteristics, and comparison with mosses, a type of bryophyte, even fewer. Though a handful of structural similarities exist between angiosperms and algae, they are mostly limited to cellular anatomy, because they are more distantly related.

The fossil record also supports evolution. Fossils include remnants, impressions, or traces of organisms from past geological ages that have been preserved in the Earth's crust. Recent or young fossils closely resemble extant species, whereas older fossils show less similarity to current life-forms. Examination of fossils sheds light on evolutionary lineages and gives information regarding the time frame for the existence of ancestral forms. Radioactive dating methods reveal when a particular layer of rock formed, and one can infer the age of the fossils found within that rock. Transitional fossils, those that demonstrate transitions between ancestral lineages and modern life-forms, have been particularly useful in constructing phylogenetic trees. A paleobiologist can follow the evolution of certain traits by studying a series of progressively younger fossils within a lineage. Geological studies give information about the conditions on Earth at the time the fossilized organisms lived, and fossils of other organisms that lived in the same region at the same time help evolutionary biologists to understand the ecological features of the organism's habitat and therefore to make sense of the evolutionary processes that led to certain adaptations.

The fact that closely related organisms are often in physical proximity to one another supports evolution, not only extant organisms, but fossil evidence of past life-forms. Biogeography is the study of the geographical distribution patterns of organisms. One would expect knowledge of geographical changes, such as continental drift, rise or fall in sea level, or island formations, to explain the development of unique adaptations of organisms found in those areas, speciation, and extinction events. For example, the isolation of a particular species of bird on an island would select for a beak shape suited to feed off that island's resources. When combined with knowledge of plate tectonics, biogeographical data reveal important patterns regarding the movement or migration between large landmasses at different periods in Earth's history. Conversely, fossil evidence of certain types of fauna or flora in a region gives information about the past climate in that area.

COMMON MISCONCEPTIONS

Many misconceptions concerning evolution are prevalent, not only in general society, but even in biological science literature. One false notion is that the process of evolution is directed toward improvement. This is only the case with respect to a given environment, the conditions of which define "better," and in relation to the other members of the same species living in that environment. An adaptation that once benefited an organism might not in a different environment, and it might constrain future developments or even cause problems under a different set of living conditions. For example, humans have wisdom teeth, a third set of molars that were important in grinding plant tissue that served as the main part of the ancestral human diet. The wisdom teeth allowed for more efficient extraction of the nutrition from the cellulose-laden plant tissue. Today, humans are not as dependent on raw plant material for nutrition and thus do not need the extra set of molars. The jaw size has decreased over time, to the point where the jaw is too small to accommodate the wisdom teeth in some people's jaws. Because of this, impaction often occurs, and extraction is sometimes necessary.

Evolution does not mold "perfect" organisms. Genetic variation results from several random mechanisms: mutations, independent assortment and crossing over during meiosis, random fertilization, and genetic drift. Natural selection acts on these, favoring the best of the available "choices," but adaptations do not appear *because* they will better suit an organism for a particular environment. Selection is only an effect resulting from improved fitness. When variation occurs, it can only modify existing traits. Furthermore, many adaptations are helpful in one aspect, but price is paid elsewhere. For example, endotherms generate body heat through metabolic activity. While this allows one to survive in a wider range of temperatures, endotherms must find and take in more energy in the form of food to fuel their internal heating mechanisms.

Another misconception regarding evolution is summarized by the phrase *survival of the fittest.* Survival is important, but evolution only occurs when successful reproduction accompanies survival. In an evolutionary sense, fitness is a measure of the reproductive success of an organism. Characteristics that are often mistakenly associated with fitness, such as being the strongest, toughest, fastest, or largest, only improve an individual's fitness if they improve that individual's ability to pass on his or her genes to future generations. For example, taller aerial hyphae in fungi might seem advantageous, as increased height would improve dispersal of the spores, but selection limits the height because of the potential for damage to the stalk before the spores are mature and ready for dispersal.

People often associate evolution with the origin of life or, more specifically, with the origin of the human species. The biological process of evolution can only occur in living, reproducing organisms. How life first appeared is a separate issue. Some religions explain the origin of life as the work of a divine being, and some specifically believe that a supernatural being created all life, including humans, in their currently existing forms. Because creationism combines the topics of the origin of life and the creation of individual species, many people assume evolution, the scientific explanation for the existence of the diverse number of species living on the Earth, also encompasses the origin of life. Related to this is the fallacy that humans evolved from apes. Scientific evidence supports that humans and apes share a common ancestor, but extant apes differ from the original common ancestor, as do humans.

See also DARWIN, CHARLES; DOBZHANSKY, THEODOSIUS; EVOLUTIONARY BIOLOGY; GENE EXPRESSION; GOULD, STEPHEN JAY; HISTORY OF LIFE; HUMAN EVOLUTION; ORIGIN OF LIFE; POINT MUTATIONS; REPRODUCTION; VARIATION, GENETIC VARIATION.

FURTHER READING
Futuyma, Douglas J. *Evolution.* Sunderland, Mass.: Sinauer, 2005.

Jastrow, Robert, ed. *The Essential Darwin.* Boston: Little, Brown, 1984.

Rice, Stanley A. *Encyclopedia of Evolution.* New York: Facts On File, 2006.

Ridley, Mark. *Evolution.* 3rd ed. Malden, Mass.: Blackwell, 2003.

Russell, Michael, Andrew Pomiankowski, George Turner, Paul Rainery, and Robin Dunbar. "Evolution: Five Big Questions." *New Scientist* 178 (2003): 32–39.

"The TalkOrigins Archive: Exploring the Creation/Evolution Controversy." Available online. URL: http://www.talkorigins.org/. Accessed January 21, 2008.

University of California Museum of Paleontology. "Understanding Evolution." Available online. URL: http://evolution.berkeley.edu. Accessed January 21, 2008.

evolutionary biology Evolutionary biology is a comprehensive branch of biology that examines the origin of species and how they change over time. The main theme or the central dogma of evolutionary biology is that natural selection shapes the genetic evolution of a population living in a particular environment. This in turn influences the anatomies, physiologies, and behavior of the species over time. The major goals of evolutionary biology are to reveal the history of life on Earth and to understand the fundamental process of evolution. Evolutionary biologists study both the diversity and the similarity among organisms in order to understand better the many different mechanisms by which different species carry out life's processes, the reasons they behave in certain ways, the methods through which they reproduce, and the origin and relatedness of different species. Whereas disciplines such as anatomy, physiology, cell biology, and molecular genetics seek to understand how organisms work, evolutionary biology attempts to explain why. Evolution, or change by descent with modification, unifies all the biological disciplines and gives them deeper significance.

Biologists began recognizing evolutionary biology as its own discipline in the late to mid-20th century as a result of modern evolutionary synthesis. Until the 19th century, society generally accepted that the world existed as it was originally created. Charles Darwin formally described his theory of evolution by natural selection in his seminal book *On the Origin of Species,* published in 1859. He suggested that natural selection caused adaptation, the modification of an organism that renders it more fit for survival and reproduction in a specific environment. Living organisms were dynamic; they changed, as did the Earth they inhabited. After the Newtonian revolution in physics of the 17th century, physical scientists sought mechanistic rather than divine or supernatural causes to explain natural phenomena. As a result of Darwin's work, biologists followed suit. In 1865 the Austrian monk Gregor Mendel published the results of his studies on inheritance in pea plants, laying the foundation for the field of classical genetics. He suggested that individuals possessed pairs of units of inheritance, now called genes. Each parent contributed one gene for each trait, and the combination of genes in the gametes was completely random. Biologists did not recognize the significance of Mendel's work until the early 1900s, when his original paper was rediscovered. Once biologists described the process of random genetic mutation and began to understand how this led to variation among individuals and populations, leaders in the emerging field of evolutionary biology began to formulate the modern evolutionary synthesis.

Also referred to as neo-Darwinism, the modern synthesis combines the concepts of genes as the units of inheritance with evolution by natural selection. Genetic variation resulting from mutation, independent assortment and crossing over during meiosis, random fertilization during sexual reproduction, and other genetic phenomena provides the material for natural selection. Those traits that consistently increase the fitness of individuals are considered adaptive, and selection will maintain them at ascending levels of hierarchal organization: the individual, the family, the population, and the species.

Because evolutionary biology is so comprehensive, scientists approach it from a variety of specialties. Evolutionary biologists often specialize in the study of one particular type of organism such as ornithology, the study of birds, or in different branches of biology such as evolutionary ecology or molecular evolution. The focus may be on mechanisms versus general patterns or over short periods (microevolution) versus long periods (macroevolution). In an attempt to answer both proximate (how) and ultimate (why) questions regarding evolutionary adaptations, evolutionary biologists carry out their investigations both in the laboratory and in the field, and their research often relates to and incorporates knowledge from other fields. For example, population genetics is the study of allele frequencies and distribution within a population. Alleles are alternate forms of a gene. Mutations can result in the creation of new alleles that may or may not persist in the population. Natural selection is one factor that causes allele frequencies to shift when the encoded characteristic affects an organism's fitness. Ecology, the study of the interrelationships of organisms and their environments, is closely related to evolutionary biology, as the interactions with one's environment are what defines whether or not natural selection will favor a particular characteristic. Evolutionary ecology examines how relationships among species affect their evolution, adaptations associated with diet or foraging, the numbers and distributions of organisms, lengths of life spans, ways adaptations affect communities, and other ecological features of evolution. Many evolutionary biologists are paleontologists, scientists who study the history of life on Earth by examining fossil evidence preserved in the Earth's crust. Geological research also provides information about the conditions on Earth, such as climate and atmospheric composition, for evolutionary biologists investigating past life-forms. Computational biology complements lab and field studies by allowing evolutionary biologists to measure changes in the deoxyribonucleic acid (DNA) of a species and to compare DNA among species, facilitating the construction of phylogenies, or evolutionary histories. Advances in bioinformatics have led to the development of genetic algorithms that model evolution. Other subdisciplines of evolutionary biology include behavioral ecology, evolutionary physiology and morphology, human evolution, systematics, and molecular evolution.

One current topic of active research in evolutionary biology is how variations become fixed within a species. In other words, over long periods, what maintains traits that once varied? A better understanding regarding molecular control of development should help biologists solve this problem. A related issue is how the fixed traits constrain the evolution of other variations. Time, another constraint for evolutionary change, raises another puzzle. Laboratory research indicates that genetic changes can occur rapidly under strong selection pressure, but in the natural setting, evolution occurs much more slowly. Another ongoing objective is to reach a better understanding of the concept of species—variations within a species are often continuous, whereas more significant differences exist between species. What factors, other than sexual reproduction, prevent a more continuous pattern of variation between related species?

Evolutionary biology provides a useful perspective for understanding life and relationships among organisms. Past life histories explain why organisms look and function the way they do. In addition to providing a framework upon which all other life sciences can be studied, understanding evolutionary biology is important for many applications. Agriculturists employ artificial selection, the process of choosing parents with a specific characteristic for breeding purposes, to improve crops or domesticated animals. As natural selection does, artificial selection works by the same principles to increase the occurrence of a particular desirable characteristic (or decrease the occurrence of an undesirable one) with the only difference being that humans rather than the environment selects which characteristics are favorable. Also, the problem of resistance to pesticides has increased in recent years, and knowing how to slow this evolutionary process would benefit agriculture. Similarly, bacteria have become increasingly resistant to antibiotics. Viruses mutate rapidly, leading to the development of new strains that can spread rapidly. Thus, knowledge of the evolutionary process benefits the field of medicine. Principles of evolution can also be used in conservation efforts, to try to revive endangered species and preserve biodiversity.

Evolutionary biologists work for many types of companies and institutions. Museums with biological specimen collections, academic institutions, conservation agencies, environmental consulting firms, medical research facilities, the U.S. Department of Agriculture, and resource management agencies in fields such as forestry, fisheries, and wildlife all employ evolutionary biologists.

See also DARWIN, CHARLES; DOBZHANSKY, THEODOSIUS; EVOLUTION, THEORY OF; GOULD, STEPHEN JAY; HISTORY OF LIFE; MENDEL, GREGOR; ORIGIN OF LIFE; VARIATION, GENETIC VARIATION.

FURTHER READING

Futuyma, Douglas J. *Evolution.* Sunderland, Mass.: Sinauer, 2005.

Ridley, Mark. *Evolution.* 3rd ed. Malden, Mass.: Blackwell, 2003.

excretory system   The major functions of the excretory system are the removal of nitrogenous wastes from the body and osmoregulation. Nitrogenous wastes are the waste products from the metabolism of proteins and nucleic acids. The amino groups ($-NH_2$) from catabolized amino acids that are not used to synthesize new molecules form ammonia (NH_3), a highly reactive and toxic substance that dissolves in an aqueous environment to form ammonium ions (NH_4^+) and hydroxide ions (OH^-). Most animals convert the ammonia to either urea or uric acid, both of which are less toxic but still must not be allowed to build up in the body. The excretory system safely eliminates them. Osmoregulation is the regulation of osmotic pressure, the pressure associated with the diffusion of water. Animals that live in marine or freshwater habitats have special adaptations for adjusting to their environments, so their body cells do not dehydrate or uptake too much water and undergo lysis.

DIVERSITY OF EXCRETORY MECHANISMS

Animals use several different mechanisms to rid their bodies of ammonia formed during the metabolism of amino acids and nucleic acids. The availability of water in a particular habitat greatly influences the type of nitrogenous waste an organism excretes.

Animals that live in aquatic environments usually excrete nitrogenous wastes as ammonia. Ammonia is very soluble and can easily diffuse through most membranes into the surrounding water, where it can quickly diffuse to tolerable levels. In aquatic invertebrates, ammonia diffuses directly into the water from all of the body cells. In fish, nitrogenous waste leaves in the form of ammonium ions that exit the body through the gills, and some leaves through the kidneys. In large water bodies, the amount is insignificant, but in a fish tank, the water must be changed to prevent buildup of ammonium ions to lethal levels.

Without access to sufficient quantities of water to dilute ammonia to tolerable levels, terrestrial animals must first convert nitrogenous waste to a less toxic form, then excrete it as a diluted solution. Because marine animals such as sharks, bony fishes, amphibians, and turtles already lose large quantities of water via osmosis, many also do not have sufficient amounts of water to dilute ammonia to tolerable levels and utilize the same strategy. Vertebrate livers convert ammonia to urea by combining it with carbon dioxide. Because urea is 100,000 times less toxic than ammonia, the circulatory system can safely transport and store it prior to elimination without requiring large amounts of water for dilution. The disadvantage of excreting nitrogenous wastes as urea is that the conversion is an energetically expensive process.

The conversion of ammonia to uric acid requires even more energy input; however, animals such as snails, insects, reptiles, and birds that do not have access to sufficient water excrete nitrogenous wastes in this form. Uric acid is insoluble in water, and thus the animals can excrete it as a paste without having to sacrifice much water.

Most systems eliminate nitrogenous waste products as urine, a multistep process involving filtration, reabsorption, secretion, and excretion. The first step is filtration, the passage of fluids through selectively permeable membranes resulting in the formation of a filtrate that consists of water, salts, glucose, amino acids, nitrogenous waste products, and other toxins. The body then selectively reabsorbs valuable nutrients including some of the salts, glucose, and amino acids by active transport. Excess salts and other substances are secreted into the filtrate, the water content of the filtrate is adjusted to maintain the osmotic balance of the bodily fluids, and the urine exits the body.

The mechanisms by which organisms with different anatomies accomplish urine production and excretion differ among animal groups and their various habitats. Members of the phylum Platyhelminthes (flatworms such as planaria) have simple excretory systems composed of protonephridia that function in osmoregulation and the removal of nitrogenous wastes in freshwater species. Metabolic waste products diffuse either out of the body directly or into a gastrovascular cavity and then out of a mouth. Protonephridia consist of networks of branched tubules that terminate in flame bulbs. Vibrating cilia that project from the end of the flame cell into the tubule seem to flicker like candle flames, hence the name. The beating of cilia inside the tube creates a current that draws body fluids into the tubules and sends them out through a nephridiopore in the body wall. The perforations through which fluids enter are too small to permit the passage of large molecules such as proteins. The tubules reabsorb important solutes before excreting the urine into the external environment. Rotifers, some annelids, mollusk larvae, and lancelets also have protonephridia.

Most annelids (segmented worms such as earthworms) have excretory organs called metanephridia. Each body segment contains a pair of metanephridia that are surrounded by capillaries. Cilia near the opening of a metanephridium move fluid from the coelom through a structure called the nephrostome into the tubule. As fluid passes through the metanephridium, solutes are reabsorbed into the blood capillaries. Nitrogenous wastes continue to move through the metanephridium, eventually leaving the body through a nephridiopore. Organisms that live in hypotonic (low-solute) environments excrete very dilute urine to rid the body of excess water taken in by osmosis.

Some mollusks and some arthropods also have metanephridia, but most insects and terrestrial arthropods have Malpighian tubes that function in nitrogenous waste removal and osmoregulation. These thin fingerlike structures extend outward from the digestive tract and are in direct contact with hemolymph, the fluid that circulates in an open circulatory system. Cells of the Malpighian tubes transport substances including nitrogenous waste products, salt, and water from the hemolymph into the lumen of the tubes. The contents of the Malpighian tubes drain into the digestive tract and then the rectum, where the valuable solutes are reabsorbed. Water follows the solutes by osmosis, but the insoluble uric acid is left behind and eliminated with feces. This excretion strategy conserves water very efficiently and was instrumental in the adaptation of insects to land and dry environments.

ANATOMY OF THE MAMMALIAN EXCRETORY SYSTEM

The kidneys are the main excretory organs in vertebrate animals. In addition to excreting wastes, they produce some hormones and help regulate water balance, osmolarity, ion balance, and pH levels. Occurring in pairs and situated in the lower back, the mammalian kidneys contain complex systems of tubules and numerous capillaries. Each human kidney has about 1 million nephrons, the functional units of the kidney. After the kidneys filter the blood to remove wastes and excess water and salts, ureters transport the urine to the bladder for storage. When the bladder is full, the urine passes through the urethra to the exterior of the body. In females the urethra exits near the vaginal opening, and in males the penis contains the urethra.

The mammalian kidneys consist of an outer renal cortex that surrounds the outer and inner renal medulla. The nephron consists of several distinct regions that perform different functions in the production of urine. The Bowman's capsule envelopes the glomerulus, the site of filtration. In mammals and birds, most nephrons (cortical nephrons) exist in the cortex and do not extend into the medulla. In about 20 percent of nephrons (juxtamedullary nephrons), the proximal tubule carries the filtrate from the cortex into the descending limb of the loop of Henle situated in the inner medulla. The ascending limb of the loop of Henle carries the filtrate through the medulla back into the distal tubule, which drains into a collecting duct. Nephrons with this structure allow for the production of concentrated urine to conserve water and are only found in mammals and birds. Cortical nephrons have much shorter loops of Henle, and nephrons of other vertebrates do not have loops of Henle at all. Several nephrons empty into a single collecting duct that leads to the renal pelvis and then the ureter.

Renal arteries supply blood to each kidney, and renal veins carry away the filtered blood. In the cortex, renal arteries diverge into arterioles that branch out and associate closely with each individual nephron. As the blood enters a nephron through an afferent arteriole, it first flows into a ball of capillaries called the glomerulus. As the blood leaves the glomerulus, the capillaries converge into an efferent arteriole, which branches out again into peritubular capillaries that surround the tubules of the nephron and into the vasa recta that extends down into the inner medulla region. The blood vessels are bathed in interstitial fluid, as are the nephron tubules.

NEPHRON FUNCTION

The high pressure of blood in the afferent arterioles forces fluid through the porous capillaries of the glomerulus and across a filtration membrane into the lumen of the Bowman's capsule. The filtrate in the Bowman's capsule contains water, salts, bicarbonate ions (HCO_3^-), hydrogen ions (H^+), urea, glucose, amino acids, and certain drugs. Blood cells and large molecules such as proteins cannot penetrate and remain in the blood vessels. Reabsorption and secretion both occur through a transport epithelium in the proximal tubule via a combination of active transport, cotransport, facilitated diffusion, and simple diffusion. Transport epithelial cells help maintain blood pH levels by secreting excess H^+ into the filtrate in addition to synthesizing ammonia, which passively diffuses into the filtrate, preventing it from becoming

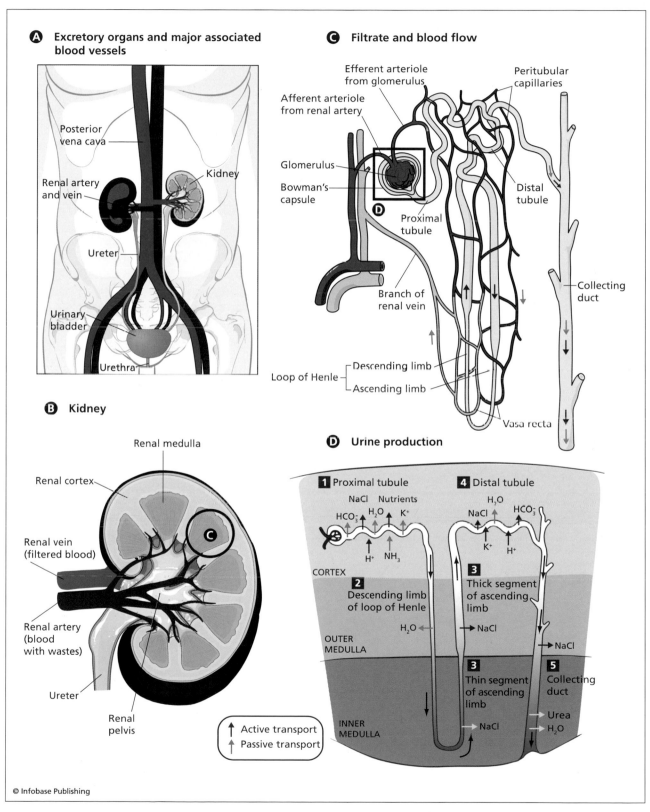

The kidneys are the main excretory organ in humans. The functional unit of the kidney is the nephron, a long tubule that performs the functions of filtration, secretion, and reabsorption in urine production.

too acidic. Sodium ions (Na^+) and chloride ions (Cl^-) from the filtrate diffuse into the transport epithelial cells, which actively transport the Na^+ into the interstitial fluid. Cl^- ions follow to balance the charge gradient, and water follows by osmosis. The salts and water then move back into blood circulation via the peritubular capillaries. The proximal tubule is also the site for reabsoprtion of HCO_3^-, potassium ions (K^+), and nutrients such as sugars and amino acids. Some drugs or other toxins pass from the peritubular capillaries to the interstitial fluid, and the transport epithelium secretes them into the urine.

As the filtrate travels down the descending loop of Henle, reabsoprtion of water continues. The concentration of solutes in the interstitial fluid increases gradually from the cortex to the inner medulla of the kidney. The transport epithelium of the descending loop of Henle is not permeable to salts; thus only water moves out, resulting in significant concentration of the filtrate. In contrast, the ascending loop of Henle is permeable to salt but not water. In the lower portion of the ascending loop, salts diffuse out of the filtrate into the interstitial fluid. This helps maintain the solute concentration gradient between the cortex and the medulla. By the time the filtrate reaches the upper portion of the ascending loop, the transport epithelium must actively transport salts into the interstitial fluid. Because the ascending loop of Henle is not permeable to water, the result is dilution of the filtrate. The distal loop plays an important role in regulating pH levels and salt concentrations. The transport epithelium secretes H^+ and K^+ into the filtrate and reabsorbs HCO_3^-, Na^+, and Cl^-. Water moves out by osmosis as the filtrate travels through the distal tube and down the collecting duct, concentrating the filtrate once again. Under hormonal influences, the transport epithelium of the collecting duct actively reabsorbs varying amounts of Na^+ and Cl^-, depending on how hydrated or dehydrated a person is. As the filtrate approaches the end of the collecting duct, the transport epithelium becomes permeable to urea. Though most urea is destined for excretion, its concentration is so high in the filtrate at this point

that some diffuses out, contributing to the high solute concentration in the medulla. The high levels of solute in the medulla allow mammals to conserve water very efficiently, resulting in the production of hyperosmotic urine, urine that is much more concentrated than body fluids. Urine drains from the collecting duct, into the renal pelvis, and down the ureters to the bladder.

The average human bladder can hold approximately one pint (close to 500 mL) of urine. As the volume increases, the bladder wall stretches, resulting in micturition, or emptying of the bladder. Stretch receptors communicate the information that the bladder is full to the nervous system, which responds by sending signals that relax the skeletal muscles around the urinary sphincter, opening it and allowing the fluid to drain out. Contractions of smooth muscle surrounding the bladder help force the urine out. In infants, micturition is a reflex, but as one grows older, the ability to inhibit micturition voluntarily develops.

See also ANATOMY; ANIMAL FORM; BIOLOGICAL MEMBRANES; CIRCULATORY SYSTEM; HOMEOSTASIS; INVERTEBRATES; PHYSIOLOGY; VERTEBRATES.

FURTHER READING

Alexander, Ivy L., ed. *Urinary Tract and Kidney Diseases and Disorders Sourcebook.* 2nd ed. Detroit: Omnigraphics, 2005.

Field, Michael J., Carol A. Pollock, and David C. Harris. *The Renal System.* London: Churchill Livingstone, 2001.

Nagel, Rob. *Body by Design: From the Digestive System to the Skeleton.* Vol. 2. Detroit: U*X*L, 2000.

Seeley, Rod R., Trent D. Stephens, and Philip Tate. *Anatomy and Physiology.* 7th ed. Dubuque, Iowa: McGraw Hill, 2006.

Sherwood, Lauralee, Hillar Klandorf, and Paul Yancey. *Animal Physiology: From Genes to Organisms.* Belmont, Calif.: Thomson/Brooks/Cole, 2005.

Silverthorn, Dee Unglaub. *Human Physiology: An Integrated Approach.* 4th ed. San Francisco: Benjamin Cummings, 2006.

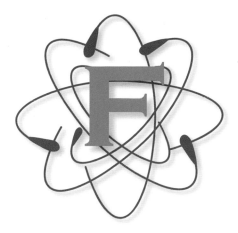

Fleming, Sir Alexander (1881–1955) Scottish *Bacteriologist* Throughout history, sudden outbreaks of disease periodically decimated populations. Though the 19th century introduced the discovery of vaccines to prevent certain illnesses caused by infectious microorganisms, medicine had no way to fight the plague, influenza, smallpox, gonorrhea, tuberculosis, malaria, yellow fever, and other infectious diseases once someone was already sick until a Scottish-born bacteriologist, Sir Alexander Fleming, discovered the antibacterial properties of a substance produced by the mold *Penicillium notatum*. This breakthrough led to a revolution in medicine, stimulating the discovery of several lifesaving antimicrobial compounds that can be used to destroy pathogenic (disease-causing) microbes after they have infected a host.

CHILDHOOD AND MEDICAL TRAINING

Alexander Fleming was born to Hugh Fleming and his second wife, Grace Morton Fleming, on August 6, 1881, in Lochfield, in Ayrshire, Scotland. Nicknamed Alec, he was his father's seventh of eight children and grew up on an 800-acre farm, where he spent his younger days tending the family's sheep, playing in the barns, and fishing in the river. Surrounded by nature, he developed keen observational skills while learning to hunt for peewit eggs and rabbits with his bare hands. He started attending school when he was five, and his father died when he was seven.

At age 13 Alec went to London to live with one of his older brothers, who was an eye doctor. Alec took business classes at the Regent Street Polytechnic Institute for two years and by age 16 had passed all of his exams. Not particularly interested in any specific career, he took a job as a junior clerk in a shipping office, where his duties included hand copying records, bookkeeping, and keeping track of all the cargo and passengers on the ships. In 1900 he joined the London Scottish Regiment, but the Boer War between the United Kingdom and the southern African colonies ended before he was sent overseas. For enjoyment, Alec played on the regiment's water polo team and entered shooting competitions, which he often won. He remained a member of the regiment until 1914. His shipping job bored him, so when his uncle left him an inheritance, he decided to spend it studying medicine, as his brother had.

Fleming was almost 20, older than most embarking on the path to a medical career, but he hired a private tutor and, in less than one year, passed his exams ahead of all the other British candidates. In October 1901 Fleming entered St. Mary's Hospital Medical School on a scholarship. He chose St. Mary's over the other 11 London medical schools because he had once played against them in water polo. He became enthralled by his studies of anatomy and physiology and excelled with minimal effort while also participating in the school's water polo team, drama society, debate team, and rifle club. In 1906 Fleming received his Conjoint Board Diploma, which granted him permission to practice general medicine, but at the suggestion of one of his teammates, he joined the inoculation department as a junior assistant so he would be eligible to participate as a school team member in an upcoming national rifle competition.

The inoculation department was headed by Almroth Wright, a staunch believer in vaccine therapy. Wright had been influenced by the French chemist

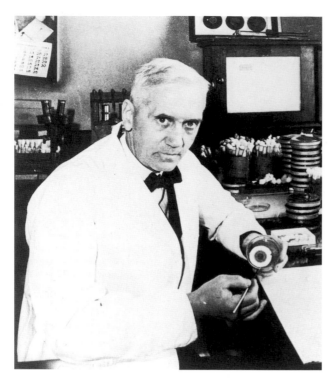

Sir Alexander Fleming is famous for his studies on the antibacterial properties of penicillin and lysozyme. *(Library of Congress)*

Louis Pasteur's work on vaccines. Vaccines stimulate the body's immune system to produce antibodies against disease-causing microbes by the introduction of weakened or killed microbes or parts of the microbes into the body. Antibodies are proteins produced by white blood cells that defend against or help prevent diseases. Some vaccines called toxoids stimulate production of antibodies against a poison produced by a microorganism. Wright was convinced that all infectious diseases could be cured by antibodies either made by the patient or by the introduction of serum from another person. His department extensively examined how vaccinations worked and studied phagocytes. Found in body tissues and fluids, phagocytes are cells that are capable of ingesting and destroying harmful substances or disease-causing microbes. In 1908 Fleming passed his final medical examinations and was awarded the Gold Medal of the University of London. Though Fleming was interested in department's research, he decided to take the exam necessary to specialize in surgery. He passed the surgical examinations in 1909 but continued to work for Wright and developed a good reputation.

One of Fleming's earliest medical accomplishments was the development of a diagnostic test for syphilis, a potentially fatal sexually transmitted disease. The German bacteriologist Paul Ehrlich identified a compound effective in treating syphilis, salvarsan, in 1910. Fleming became an expert in intravenously administering salvarsan to treat syphilis. Intravenous injections were uncommon at the time, and many doctors did not know how to give them.

In 1914 Fleming and several other members of Wright's team joined the Royal Army Medical Corps and established a research center in Boulogne, France. He had to pass through the patient wards on the way to his laboratory, and the surgeons often showed him severe cases of septicemia, tetanus, and other infectious diseases. Fleming was disturbed by the astounding number of infections suffered by wounded soldiers and by the apparent ineffectiveness of the antiseptics used to treat them. He was particularly horrified by the deadly gas gangrene, which caused high fevers, brownish pus at the infection site, and the production of gas below the skin. In these cases, amputation of the infected limbs was necessary in order to save the patient's life. Fleming researched the effect of antiseptics such as carbolic acid, boric acid, and hydrogen peroxide on wounds and found that in the case of deep wounds, they actually did more harm than good because the chemicals killed the white blood cells that naturally fight off infection and did not penetrate deep enough into the wound to be effective. Wright and Fleming encouraged rinsing wounds with only saline solution and letting the body combat bacterial infection naturally, but most ignored their recommendations.

While on leave in 1915, Fleming married an Irish nurse, Sarah Marion McElroy, whom he called Sareen. In 1921 they bought a country house that they called The Dhoon, where they spent weekends together. In 1924 they had one cherished son, Robert, who later became a physician. Sareen died in 1949, and in 1953 Fleming married a Greek bacteriologist, Dr. Amalia Coutsouris-Voureka, who had begun to work at St. Mary's in 1946.

BACTERIOLOGY AND LYSOZYME RESEARCH

In January 1918 Fleming returned to London and to his studies of bacteriology. His experiences treating wounded soldiers with severe infections motivated him to search for an effective antiseptic. The British physician Joseph Lister founded antiseptic surgery, in which the environment, medical instruments, and the surgeon's hands were sterilized before surgery. This practice required instruments to be soaked in carbolic acid to kill any contaminating microorganisms and greatly reduced infection rates after surgeries; however, the acid damaged living tissues. Fleming wanted to find an antiseptic that would not cause harm to the patient's tissues but would kill potential microbial invaders. Meanwhile, Wright appointed Fleming assistant director of the inoculation department, which was renamed the Pathology and Research Department.

Eager to obtain cultures of a wide variety of bacteria, Fleming collected many unusual specimens and grew them in the laboratory in petri dishes containing artificial media. One interesting sample was obtained from his nasal mucus, collected during a recent cold. The plate had many golden yellow colonies of the bacteria that he later named *Micrococcus lysodeikticus*. In 1921, while preparing to dispose of the culture dish, he examined it once more and noticed the bacterial colonies immediately surrounding the mucus itself appeared dissolved. He wondered whether the mucus contained an antibacterial agent.

Further investigation showed that mucus did indeed contain a substance that naturally killed bacteria. He named it *lysozyme,* since it lysed, or broke open, the bacterial cells. Lysozyme acts by punching holes in the cell wall that encircles bacterial cells. With the integrity of the cell disrupted, the cellular contents leaked out and the cells perished. After examining many other bodily fluids, Fleming also found lysozyme in tears (collected by squeezing lemon juice into his own eyes and the eyes of others), saliva, blood serum, pus, and egg whites. Lysozyme obviously was not harmful to the host's living tissues, or to the host's own immune system components, unlike chemical antiseptics. The substance was harmful only to the invading bacteria, acting as a first line of defense to prevent them from colonizing the body. He tested microorganisms that were virulent to different degrees and, not surprisingly, found that the microbes that were most susceptible to lysozyme were the least dangerous. Fleming thought this made sense, since if they were not susceptible to lysozyme, they would be more likely to invade the body and cause infection. Fleming was not successful in preparing concentrated extracts of lysozyme, but later others were able to crystallize the bacteriolytic enzyme, which has become an important tool for microbiologists.

DISCOVERY OF PENICILLIN

Fleming was noted for his procrastination in cleaning up his old culture dishes. This habit resulted in one of the most important medical breakthroughs of the 20th century. A plate with staphylococci had become contaminated with a fuzzy-looking mold later identified as *Penicillium notatum.* Staphylococci are spherical bacteria that grow in clusters that resemble bunches of grapes and may cause infections by entering breaks in the skin, leading to pimples, boils, or a skin disease called impetigo. Though bacteria were found throughout the plate, there were no colonies in the immediate vicinity of the mold; instead, there was a clear halo surrounding the mold growth. Fleming recognized what none of his colleagues noticed when he showed them the petri dish—that the mold must have been secreting an antibacterial substance. He named it *penicillin.*

He systematically investigated his observation, first by culturing the mold, growing it in the lab, and then trying to duplicate the bactericidal action. He inoculated a plate of agar with the mold at the center and then streaked different bacterial cultures like radii of the circular petri dish. After incubation, some bacteria grew near the mold, while others did not. The mold thrived on a broth containing meat extract, which he poured into sterile bottles and then inoculated with tiny pieces of the mold. He filtered some of the broth in which the mold grew and applied it to plates containing healthy staphylococcal cultures. He performed a series of dilutions to determine the strength necessary to destroy the bacteria and looked for negative effects on living tissue. To see whether penicillin was effective in killing other types of bacteria, he added it to plates of several other species. Penicillin proved effective against bacteria that caused pneumonia, syphilis, gonorrhea, diphtheria, and scarlet fever, but not against microorganisms that caused influenza, whooping cough, typhoid, dysentery, or other intestinal infections. Injection of penicillin into mice and rabbits caused no ill effects.

In 1929 Fleming, now a professor of bacteriology at the University of London, reported that penicillin did not harm white blood cells and elicited no negative responses in laboratory animals. In "On the Antibacterial Action of Cultures of a *Penicillium,* with Special Reference to Their Use in the Isolation of *B. influenzae,*" published in the *British Journal of Experimental Pathology,* Fleming described not only the possible use of penicillin as an injectable antiseptic agent, but

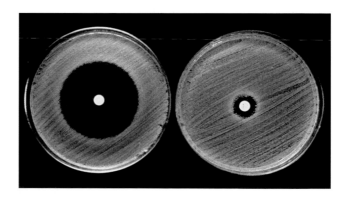

Penicillin diffuses from the center pellet of the *Staphylococcus aureus* cultures. The antibiotic inhibits growth of the culture on the left, while the bacterial strain shown on the right demonstrates resistance to the drug, as evidenced by a smaller zone of inhibition. *(John Durham/Photo Researchers, Inc.)*

also a unique application of penicillin, namely, in the establishment of pure cultures of other bacteria. Because penicillin was very effective against some bacteria and not at all against others, it could be added to the media used to grow the penicillin-resistant cultures to ensure pure cultures, that is, cultures containing only the desired bacteria.

Fleming and his assistants tried to develop a method for the extraction and concentration of penicillin so they could attempt clinical trials; however, it kept losing its potency. By 1932 Fleming stopped actively researching penicillin, but he maintained a culture of *Penicillium* in the laboratory at all times and generously provided specimens to scientists who requested it. In 1935 the German physician Gerhard Domagk announced the identification of the sulfonamide prontosil as a cure for systemic streptococcal infections, stimulating others to search for additional "magic bullets." Fleming switched his focus to the antibacterial properties of sulfonamides, a new class of chemically related drugs found to be effective in preventing the multiplication of some types of bacteria, but he maintained his hope that penicillin would one day be an effective treatment for fatal bacterial infections.

FLOREY AND CHAIN'S WORK

The Australian pathologist Howard Florey and the German biochemist Ernst Chain at Oxford University spent time in the late 1930s characterizing lysozyme. In researching antibacterial substances, they came across Fleming's journal article describing penicillin. By 1940 they successfully developed a procedure involving lyophilization (freeze-drying) and dissolution in methanol for purifying stable penicillin in quantities large enough to test on animals. After a few preliminary investigations, Florey injected 50 white mice with lethal doses of virulent streptococci. He then injected penicillin into 25 of those mice at three-hour intervals for two days and nights; the other 25 were not treated. Within 16 hours, the 25 untreated mice were all dead, but 24 of the treated mice survived. These results were published in "Penicillin as a Chemotherapeutic Agent" in *Lancet* in 1940. Fleming was thrilled when he read the article and soon traveled to Oxford, where the researchers congratulated one another and exchanged information.

The Oxford team performed a series of experiments to determine the best mode of treatment and the optimal dosages. The next step was a human trial, but they needed 3,000 times more penicillin than they used in the mice trial; thus they geared up their production methods. They converted every bit of free space to grow massive amounts of *Penicillium*, creating a makeshift factory in the pathology building and using a variety of everyday objects to extract and concentrate the penicillin juice until they had enough.

Their first patient was Albert Alexander, a policeman who had scratched his face on a rosebush and, as a result, developed severe staphylococcal infections covering his head and potentially fatal blood poisoning. Sulfonamides were not effective, and without treatment he was sure to die. On February 12, 1941, the team began a series of injections, and the patient showed marked improvement. Unfortunately, the bacteria began multiplying again, and this time they did not have any more penicillin to give him, and he died. Slowly, they produced enough penicillin to test on several more humans, and in all cases it proved effective.

In 1942 Harry Lambert, the director of an optical lens–making business owned by Fleming's older brothers, was dying at St. Mary's Hospital of meningitis, an infection of the membranes that surround the brain and spinal cord. Fleming removed a sample of fluid from Lambert's spinal column and examined it under the microscope. He found streptococci. Though doctors had administered sulfonamides to Lambert, he showed no improvement. Fleming wrote to Florey to ask for some penicillin, which Florey provided. Every three hours for one day, Fleming injected penicillin into Lambert, and for the first time in six weeks, Lambert's body temperature returned to normal. When Lambert became feverish again the following week, Fleming took another sample of fluid from his spinal column and once again found streptococci. After consulting Florey, Fleming decided to inject penicillin directly into Lambert's spinal column. After several injections, Lambert miraculously improved and within one month he was completely recovered.

The Oxford team knew they needed help synthesizing the large quantities of penicillin necessary to treat humans but could not find anybody in Britain willing to help them. An agricultural research laboratory in Peoria, Illinois, agreed to grow the mold and extract large quantities of penicillin; however, once the word spread about the miraculous infection-fighting properties the mold juice possessed, all of the product was devoted to treating war casualties. In 1944 production had finally ramped enough for civilians to benefit from treatment with penicillin.

By this time, news of the wonder drug was widespread and Fleming became famous. Several large pharmaceutical firms began researching penicillin production methods. The most respected scientific society in Great Britain, the Royal Society of London, elected Alexander Fleming to membership in 1943, and the following year he was knighted. In 1945 Fleming was awarded the Nobel Prize in physiology or medicine, shared with Chain and Florey, for the discovery of penicillin and its curative effect in various infectious diseases. Other numerous honors and medals were showered upon Fleming such as the Moxon Medal from the Royal College of Physicians (1945), the Honorary Gold Medal of the Royal College of Surgeons (1946), the Gold Medal of the Royal Society of Medicine (1947), and the Medal for Merit from the United States of America (1947). He spent much of his time traveling around the world making appearances, giving speeches, and accepting almost 30 honorary degrees from European and American universities.

FLEMING'S ACCOMPLISHMENTS

Sir Alexander Fleming died after a heart attack on March 11, 1955, in London and was buried in St. Paul's Cathedral. He had never received any money for his hard work and discovery; he donated it all to St. Mary's for research. Today, the Wright-Fleming Institute at the Imperial College London, named in tribute to Fleming and his mentor, houses scientists dedicated to researching human bacterial and viral infections. Often serendipity is credited for the discovery that launched a medical revolution, but one must remember the man who took notice, believed, and pursued what he observed.

Fleming devoted his life to helping humanity by trying to figure out how to fight infections. He did not invent penicillin and, in fact, was not the first to discover it, as others had earlier noticed that mold had some antibacterial properties, but Fleming was the first to recognize its broad significance and to draw attention to it. Since Fleming discovered the antibiotic penicillin, hundreds of other chemicals naturally produced by microorganisms have been found to have antibacterial properties. Interestingly, Fleming prophetically warned against the improper use of antibiotics, predicting it would lead to antibiotic resistance. Insufficient doses, failure to complete the entire course of an antibiotic treatment, and the widespread use of antibiotics to promote growth in cattle and farm animals, as well as the use of antibiotics to treat colds or other nonbacterial maladies, have resulted in a sharp increase in the number of strains of bacteria that are resistant to a wide variety of antibiotics.

See also ANTIMICROBIAL DRUGS; BACTERIA (EUBACTERIA); MICROBIOLOGY.

FURTHER READING

Horvitz, Leslie Alan. *Eureka! Scientific Breakthroughs That Changed the World.* New York: John Wiley & Sons, 2002.

Macfarlane, Gwyn. *Alexander Fleming: The Man and the Myth.* Cambridge, Mass.: Harvard University Press, 1984.

The Nobel Foundation. "The Nobel Prize in Physiology or Medicine 1945." Available online. URL: http://nobelprize.org/nobel_prizes/medicine/laureates/1945/. Accessed January 21, 2008.

forensic biology   Forensic science is the application of the natural sciences and scientific principles to legal matters. Most often this involves the collection and scientific analysis of physical evidence, as from a crime scene. Forensic science encompasses numerous subdisciplines including forensic biology, forensic chemistry, forensic pathology, forensic toxicology, forensic odontology (dentistry), forensic anthropology, and forensic physics. Forensic biology refers to the application of biological knowledge and methods to matters of the law.

Forensic investigators work as part of a team. A forensic biologist handles the biological evidence that might aid in the legal investigation. A forensic chemist analyzes and interprets nonbiological evidence, such as flammable substances in arson cases, soil composition, or flecks of paint. Someone trained in forensic physics can gain useful information from blood splatter patterns and ballistics finding about how a crime was committed. Forensic drug chemists and toxicologists look for signs pointing to the involvement of alcohol or drugs such as marijuana, cocaine, or heroin by examining blood and urine. Team members who specialize in forensic pathology are also trained to recognize evidence of drug use or poisoning as well as other information gathered from body tissues that can help determine whether the death was caused by an accident, suicide, or homicide. A forensic pathologist can establish how much time has passed since death by changes in body temperature, rigor mortis (the stiffening of muscles following death), liver mortis (the pooling of blood in blood vessels), the degree of decomposition, and the degree to which stomach contents have been digested. Forensic anthropology is a branch of physical anthropology, the study of the physical evolution and structure of human beings. A forensic anthropologist inspects and takes measurements of skeletal remains in order to determine the gender, age, race, and stature of the victim. Because experiences or lifestyle habits can also leave characteristic marks or signs on bones or bone fragments, the forensic anthropologist may be able to draw conclusions about whether or not the person had previously broken any bones, was active or sedentary, made repetitive movements that might be associated with a particular occupation, or was malnourished. The anthropologist can make deductions concerning the time of death, the cause of death, and movements that occurred post mortem as well as perform facial reconstructions from a skull that may help others recognize a missing person. Trace specialists analyze pieces of evidence that are by nature very small in size or quantity, such as carpet or clothing fibers, a strand of animal hair, or a sliver of wood. Forensic scientists collaborate with other investigators trained to analyze other forms of evidence including computers, documents, explosives, firearms, indicators of arson, or impressions or markings made by shoes, tires, or tools. Experts in criminal psychology, personality profiling, and psychophysical detection of deception (use of lie detector tests) also help solve crimes.

Forensic biologists work both at the crime scene and in a crime lab. At the crime scene they collect biological evidence and sometimes perform simple tests, such as determining whether a stain contains blood or not, in order to make sure they obtain samples from anything present that might assist in solving the case. They must follow carefully designed procedures to ensure samples of evidence do not become contaminated with other organic substances or chemicals that might damage the biological material. Storage in plastic bags or sterile containers helps protect and preserve the biological samples, which might include insects, plants, bodily fluids, hairs, or clothing that belonged to the victim or the person who committed the crime. After returning to the lab, forensic biologists process and analyze the samples, prepare a written report summarizing the findings, and occasionally appear in court to testify as impartial expert witnesses in criminal or civil matters.

Forensic botany is a specialty of forensic biology in which the scientist uses knowledge of not only plants as the name suggests but also fungi, algae, and other unicellular plantlike microorganisms to aid an investigation. Biological samples retrieved from the victim or the crime scene provide valuable information. Botanical, fungal, and microscopic specimens can indicate a particular geographic location determined through their knowledge of where the species normally grows or the season of death, indicated by the presence or absence of different structures such as spores or flowers. Forensic botanists also assist in cases that involve drugs or poisons, since these substances are plant products.

Another branch of forensic biology is forensic entomology, the application of knowledge regarding insects to legal investigations. In civil cases, expertise in forensic entomology can aid the prosecution of a property management company for an insect infestation or a food distributor for contamination. In criminal cases, entomologists can help establish the time and location of a crime by applying knowledge of insect life cycles and the succession of types of organisms involved in decomposition of a body post mortem. Female flies lay eggs on an exposed corpse within minutes of death. The eggs hatch into maggots that pass through three stages called instars before maturing into adult flies. From specimens taken from the body and knowledge of the physical conditions where the body was found, forensic entomologists can accurately estimate how long the body was in a particular location. As a body decomposes, other insects such as ants, wasps, and beetles feed off it and the maggots. Information concerning the particular species and stages of development of insects present on a body may indicate postmortem movement of the body or reveal information about the conditions (e.g., temperature, moisture) preceding death. In unusual cases, contents of the stomach of a bloodsucking insect or chemicals present in a maggot that has been feeding off a decomposing body can provide biological evidence about the crime.

Scavengers are animals that eat the remains of other organisms. Vertebrate zoologists who study these types of animals and their behavior are valuable to investigations in which the remains are scattered. Familiarity with the relatively consistent patterns of soft tissue modification and disarticulation and the effects of weather on these patterns allows one to estimate the length of time the body has been exposed. Typically, scavengers remove only soft tissue within the first few weeks, followed by destruction of the abdomen, internal organs, and dismemberment of the arms, and then legs. After several months, scavengers will have disarticulated the entire skeleton, saving the vertebral column for last and leaving only the skull behind. Because animals drag bones away from bodies in predictable patterns, knowledge of animal behavior is helpful. Rodents, cats, and dogs leave distinct tooth markings on the surface of the bones. More insects will be present in warmer temperatures and discourage scavengers from feeding off the corpse. In contrast, fewer insects will be present in colder temperatures, and therefore there will be more tissue damage and animals will start removing the bones within a shorter time span. Because flies cannot lay eggs on a submerged body, water slows the process, as do any coverings, clothing, or partial burial. Aquatic creatures such as crabs, sea louses, and some fish feed on decaying corpses.

Different forms of biological specimens (tissue, blood, bone, hair, saliva, semen, or urine) can link a person to another person, to a piece of physical evidence, or to a specific physical location. The investigator must carefully follow detailed protocols for collecting, documenting, storing, and preserving the sample if it is to be used as legal evidence. Microscopic analysis of a single strand of hair can reveal whether or not it was taken from a human or an animal, whether the person had certain diseases, and which part of the body it came from. Historically, the analysis of blood samples involved determination of the blood type based on different proteins located on the surface of the red blood cells, the identification of different variants of isozymes found in the serum, or the presence of other specific antigenic markers. Since the 1980s, however, forensic biologists routinely analyze any deoxyribonucleic acid (DNA) left behind in tissue samples, bodily fluids, or hair roots. Because the nucleus is the cellular compartment that contains the DNA, biological samples that do not contain nucleated cells, such as sweat, tears, or hair shafts without attached root cells, cannot be analyzed by standard methods. DNA profiling (also called DNA fingerprinting or DNA typing) enables the specific identification of individuals because of the uniqueness of an individual's DNA. During sample collection, the investigator must use prescribed precautionary measures to ensure the biological material does not become contaminated, as the techniques used to analyze the DNA are very sensitive. Sometimes factors beyond the investigator's control such as weather conditions, exposure to radiation from the Sun, chemicals, or other organic material that was already present may destroy a sample. These factors can degrade the DNA or chemically alter it, preventing it from being used. DNA fingerprinting is performed when ample material is available and in fair condition. The scientist compares the resulting pattern, or DNA fingerprint, to the DNA fingerprints of any known suspects. If there are no suspects, the sample can be compared to a database of DNA profiles maintained by the Federal Bureau of Investigation. This database, called CODIS (for Combined DNA Index System), contains DNA profiles of numerous convicted felons, DNA profiles from unidentified suspects from other investigations, and other individuals whose DNA has been previously analyzed. When sample quantity or quality limits standard DNA analysis, forensic biologists can use mitochondrial DNA analysis. Mitochondria are cellular structures that contain DNA and are inherited through maternal lineages. One advantage of mitochondrial DNA analysis is that it is particularly sensitive because a single cell contains numerous copies; thus mitochondrial DNA analysis can be performed on old skeletal remains or on hair shafts without roots.

Though forensic investigations have come to rely on DNA analysis to establish the identity of a suspect, investigators still also use traditional methods. Individuals, even identical twins, have unique fingerprints. Forensic investigators can collect fingerprints from all sorts of objects, create digital images from them, and use a computer to compare the prints with those stored in a national database. Bite marks are also unique, and preparing casts or molds from indentations or impressions left on a victim may help establish a suspect's identity.

Training for a career in forensic biology requires a bachelor's degree and a strong background in biology, chemistry, and statistical analysis. Technical training includes microscopy, proficiency in the use of computers, and methods in biotechnology, particularly with respect to DNA manipulation and analysis. Employers often provide additional specialized training for their forensic scientists because technology advances rapidly in this field. Forensic biologists work for law enforcement agencies, for other governmental agencies such as the Postal Inspection Service, and in laboratories of colleges and universities. In order to obtain a position in a reputable crime lab, a forensic investigator must pass a background check and random drug testing. Because of the environment in which they work, investigators must receive training to work safely with hazardous chemicals. Crime lab directors often have doctoral degrees.

See also DNA FINGERPRINTING.

FURTHER READING

American Academy of Forensic Sciences home page. Available online. URL: http://www.aafs.org. Accessed January 25, 2008.

Hunter, William. *DNA Analysis*. Philadelphia: Mason Crest, 2006.

Walker, Maryalice. *Entomology and Palynology: Evidence from the Natural World*. Philadelphia: Mason Crest, 2006.

Franklin, Rosalind (1920–1958) British *Biophysicist* Rosalind Franklin was an expert X-ray crystallographer whose experimental data were instrumental in solving the structure of deoxyribonucleic acid (DNA), though she did not receive proper recognition for her role during her lifetime.

ATTENDS CAMBRIDGE

Rosalind Franklin was born on July 25, 1920, to Jewish socialist parents who lived in London. As a schoolgirl she excelled at arithmetic and science and played hockey, cricket, and tennis. She decided to become a scientist when she was 12 years old, and at age 17 she earned the highest score on the Cambridge entrance examination in chemistry. In October 1938 she entered Newnham College on scholarship. At the time, Cambridge did not accept women as members of the university but did allow them to attend lectures. She took classes in chemistry, physics, mathematics, and mineralogy; joined a mathematics society; and became interested in X-ray crystallography, a method that uses X-rays to reveal the structure of crystals. Crystals form from the regular arrangement of the atoms inside, and when X-rays are directed at the crystal, the atoms cause the X-rays to scatter, creating patterns of spots on X-ray film. Mathematical analysis allows one to determine the position of atoms within a molecule from the spacing and the intensity of the spots on the film.

By her third year at Cambridge, many of the male faculty and students had joined the war effort. Her studies focused on the structure of atoms and molecules. She performed well enough on her exams to earn a college scholarship for another year and, more importantly, a research grant from the Department of Scientific and Industrial Research. She began graduate studies in gas-phase chromatography, but her social life was lacking, and she did not get along with her colleagues or her professor. To escape, in August 1942, she took a job as an assistant research officer for the British Coal Utilisation Research Association studying the physical and chemical properties of coal and charcoal. She submitted a thesis to Cambridge University on this work, and she received a doctorate in physical chemistry in 1945. The following year she moved to Paris to work for the Laboratoire Central des Services Chimiques de l'État, where she continued studying carbon and became proficient in X-ray crystallography. Her reputation as a skilled crystallographer grew as she published several papers with important industrial applications. One significant finding made by Franklin was the discovery of a class of carbons that can never be transformed into graphite by heating.

STUDIES NUCLEIC ACID AT KING'S COLLEGE

In 1951 Franklin joined the Biophysics Unit of the Medical Research Council at King's College of the University of London. The original purpose of her appointment was to study protein structure, but just before she arrived, the direction of the laboratory changed. The head of the physics and biophysics unit at King's College, Professor J. T. Randall, explained that he thought the examination of some nucleic acid fibers obtained from the Swiss scientist Rudolf Signer was of greater immediate importance. This change of plans led to a misunderstanding of Franklin's role at the lab. She thought the X-ray diffraction work on the fibers was to be her own project, whereas Maurice Wilkins, a biophysicist at King's who had been working on nucleic acids for years, thought she was going to assist him.

Wilkins and Franklin initially were cordial as they worked separately studying the DNA fibers, but their quiet bitterness grew into bare tolerance of one another, and then into open hostility. They shared a common research goal of solving the structure of DNA, but they did not share information. Though they might have made a formidable team in the race

to solve the structure of DNA, their bitter relationship hindered any progress. Franklin did work closely with Raymond Gosling, a graduate student who had been assigned to her. One of Franklin's achievements while working at King's was the identification of two different forms of DNA, an A form and a B form. The A form converted into the B form when the humidity was increased, as the phosphate groups absorbed the water molecules, causing the fiber to become longer and thinner. To facilitate peace, in October 1951 Randall suggested that Franklin focus her research efforts on the A form and Wilkins on the B form. As part of this agreement, Franklin also was given the DNA that had been supplied by Signer and the better camera equipment. She used it to determine that the phosphates were positioned on the outside of the molecule and to obtain several more beautiful X-ray photographs of DNA.

When James Watson learned that Linus Pauling was publishing a model for DNA structure in February 1953, even though it was wrong, he rushed to King's College to share the information with Franklin and Wilkins in hopes of collaborating. But Franklin thought their modeling approach was elementary and wanted no part of it. She believed that model building was only appropriate after collecting sufficient hard evidence from X-ray diffraction studies. A little more than a year earlier, Watson and Francis Crick had invited her and Wilkins to see a DNA model they believed to be the answer, but the metal plate and stick arrangement was completely wrong and only confirmed her belief that model building was useless without first having experimental data. According to Watson, Franklin also denied the existence of any evidence that DNA was helical, but her personal laboratory notes divulge she believed otherwise. In fact, it was one of her X-ray photographs that showed the clear central X pattern that suggested a helical form for B DNA. Gosling had shown this photograph to Wilkins, who in turn showed it to Watson after his confrontation with Franklin. Wilkins also divulged figures obtained from the photograph, namely, the repeat length of the helix, 34.4 angstroms.

This information stimulated Watson to resume model building seriously. Within a few weeks, Franklin learned that Watson and Crick planned to publish a model for DNA structure. She and Gosling quickly adapted a paper they had already written but not yet submitted for publication to accompany Watson and Crick's paper. Franklin and Gosling's paper, titled "Molecular Configuration in Sodium Thymonucleate," appeared in the same issue of *Nature* and provided strong experimental evidence for the double-helical model with an exterior phosphate backbone. She was unaware that her data had provided the impetus for the final solution of the structure of DNA by Watson and Crick and certainly had never been properly consulted on its use. Wilkins and his colleagues Alex Stokes and Herbert R. Wilson also published an accompanying paper presenting additional X-ray diffraction evidence for the model proposed by Watson and Crick. In July 1953, Franklin and Gosling published another paper in *Nature* detailing the differences between two different forms of DNA, "Evidence for 2-Chain Helix in Crystalline Structure of Sodium Deoxyribonucleate."

Despite her scientific successes in a field in which she had only been a part for two years, Franklin was not happy at King's. She believed that her coworkers were not serious researchers and was frustrated that she had no academic appointment. After the DNA reports were published, Franklin transferred to John Desmond Bernal's laboratory at Birkbeck College, another college of the University of London, where she remained for five years until her death. By December 1953 she was immersed in studies of tobacco mosaic virus (TMV) structure. She made remarkable advances in the field of viral structure, including the discovery that the ribonucleic acid of TMV was embedded in the inner groove of the helix formed by the protein subunits. She published many papers on the structures of several plant viruses and was admired for her technical competence and expertise. Her reputation was demonstrated when Sir Lawrence Bragg, head of the Cavendish Laboratory at the University of Cambridge, asked Franklin to create virus models for the general public to view at the Brussels World Fair. One of her collaborators and close friends from Birkbeck, Aaron Klug, later won the Nobel Prize in chemistry in 1982 "for his development of crystallographic electron microscopy and his structural elucidation of biologically important nucleic acid-protein complexes." In his Nobel lecture, he acknowledged Franklin, crediting her with introducing him to viral research and setting an example of going after the difficult problems of science.

Despite Watson's negative portrayal of Franklin as stubborn, confrontational, and uncooperative in his personal narrative, *The Double Helix,* which described his perception of the events leading to the discovery of the structure of DNA, she had productive and meaningful collaborative engagements with her scientific teams at Birkbeck College and in Paris. She was not only a serious scientist, but an avid mountain hiker and an expert cook and enjoyed playing with the children of her friends and relatives. Because Franklin tragically died of ovarian cancer on April 16, 1958, she was not eligible for nomination for the 1962 Nobel Prize for her role in the discovery of DNA structure. After her death, Bernal described Franklin's photographs as "among the most amazing beautiful X-ray photographs of any substance ever taken." In March 2002 King's College dedicated the Franklin-Wilkins building in her honor.

See also CRICK, FRANCIS; DEOXYRIBONUCLEIC ACID (DNA); PAULING, LINUS; WATSON, JAMES D.; WILKINS, MAURICE H. F.; X-RAY CRYSTALLOGRAPHY.

FURTHER READING

Franklin, R., and R. Gosling. "Evidence for 2-Chain Helix in Crystalline Structure of Sodium Deoxyribonucleate." *Nature* 172 (1953): 156–157.

———. "Molecular Configuration in Sodium Thymonucleate." *Nature* 171 (1953): 740–741.

Maddox, Brenda. *Rosalind Franklin: The Dark Lady of DNA.* New York: HarperCollins, 2002.

Frisch, Karl von (1886–1982) German *Ethologist* Karl von Frisch is considered one of the founders of the comparative study of animal behavior. During his 60 years as a research scientist, he studied the senses, communication, and social organization of honeybees. He is most famous for demonstrating that bees have color vision and that they communicate information about the distance and location of food sources to their colony mates by genetically programmed characteristic dance movements. His research on the chemical and visual senses of bees stimulated the interest of other zoologists in ethology.

VISION AND SMELL IN HONEYBEES

Karl von Frisch was born on November 20, 1886, in Vienna, Austria, to Anton Ritter von Frisch, a university professor, and Marie Exner. He began university studies in medicine at the University of Vienna but switched to zoology. For his doctoral thesis he studied the color adaptation and light perception of minnows. After receiving his doctorate from the University of Vienna in 1910, he joined Richard von Hertwig at the University of Munich of the Federal Republic of Germany, where he obtained his teaching certificate in zoology and comparative anatomy. In Munich, Frisch began studying the behavior of bees, the major focus of his six-decade-long research career.

Honeybees are social insects that live in hives consisting of approximately 60,000 individuals. A single bee, called the queen bee, lays eggs. Drones are male bees that fertilize the eggs. The infertile female worker bees perform all the work to support the colony: feed larvae, build the honeycomb, and gather nectar and pollen for the colony to eat. Bees obtain food by visiting flowering plants, and in turn they carry pollen from one plant to another, assisting the plants in reproduction. Many biologists believed that the bright colors of many flowers serve to attract insect pollinators. One of Frisch's first studies at the University of Munich was to examine whether or not bees had color sense. Biologists knew that bees could distinguish colors from the following experiment: Experimenters placed honey on a piece of blue cardboard and bees were allowed to feed off it. After a period of time, they removed the cardboard and replaced it with two new pieces, one red and one blue, neither of which contained food. The bees landed on the blue piece, indicating they could distinguish red from blue.

Frisch next wondered whether the bees perceived red and blue, or simply different shades of gray. To examine this, he placed a blue card on a table and surrounded it with cards of several shades of gray. All of the cards contained tiny watch glasses, but only the glass on the blue card contained sugar water. He allowed the bees to visit the cards, rearranging the placement of the blue card containing the food at frequent intervals. Then he replaced all the cards with clean cards and placed empty glass dishes on all the cards. The bees went right to the blue card, indicating they could perceive the color blue. Frisch obtained the same result when using orange, yellow, green, violet, or purple cards. The bees could not, however, distinguish red from black and had some trouble distinguishing blue from violet or purple when used together. They also confused yellow with orange and green, when those colors were tested together. Research performed by others later showed that bees have a sense of four different colors: yellow, blue-green, blue, and ultraviolet. Interestingly, hummingbirds and honey birds typically pollinate red flowers. These studies suggest that insect pollinators affect the adaptation of flower color.

In a related set of experiments, Frisch examined the effect of coloring hive entrances in houses where beekeepers kept many beehives. If a queen bee leaves the hive and returns to the wrong one, the residents will kill her, leading to the demise of her colony. Beekeepers often paint the entrances to prevent this from happening. Frisch tested whether or not the colored sheets surrounding the hive entrances were useful in orienting the bees to the correct hive. By painting the sheets different colors, moving them around, and observing the bees' response, he demonstrated that the colors do help bees distinguish their own hives from others nearby.

In the bees' natural environment, flower color clearly plays a role in the search for food, but flowers also exhibit a variety of shapes. Frisch examined this phenomenon by placing colored cards with cutout patterns over the entrance of boxes containing sugar water. After a short period of training the bees to enter a certain box by recognizing the shape at its entrance, he placed new cards with the cutout patterns in front of the boxes. The bees flew into the boxes with the familiar patterns, even though no sugar water was inside. Frisch found that the bees had trouble distinguishing some shapes from others (e.g., a triangle from a square) but were more successful at recognizing broken patterns such as the shape of the letter X. Following Frisch's example, other animal behaviorists performed more detailed analyses of bees' ability to distinguish certain shapes and patterns between the late 1920s and the 1950s.

CHEMICAL SENSES OF BEES

Frisch realized that the types of flowers pollinated by bees far outnumbered the different colors that bees could distinguish and was curious about which other features helped bees recognize different plants. Bees typically visited the same species of plant. In order to determine whether bees could distinguish the odors of particular flowers as humans can, Frisch set up experiments similar in method to his color experiments. He set up boxes on a table and inside one of them placed a dish of sugar water and either a fragrant flower or a few drops of an essential oil. Every so often he rearranged the position of the boxes on the table to ensure that only scent directed the bees to a particular box. After giving the bees sufficient time to associate the smell with the presence of the sugar water, Frisch replaced all the boxes with clean empty boxes. He placed a flower or oil in one of those boxes. The bees hovered around the entrance to all the boxes but only entered the box with the scent, indicating they could smell the odor that they learned to associate with food.

To see whether the bees could distinguish among numerous scents, he next arranged 24 boxes on the table. All the boxes contained different scents, but only one contained sugar water. After a training period, Frisch and his coworkers placed 24 clean, scented boxes on the table, but none of them contained food. They counted the number of times that bees entered each box during a five-minute period. Many more bees entered the box containing the scent that was associated with food during the training period. A significant number of bees entered three of 46 other boxes (they performed the experiment twice, for a total of 48 boxes, two containing the training scent). These three boxes contained oils that were made from fruits belonging to the same genus as the training scent, and the odor was similar to the training scent, as determined by the human nose.

From these experiments Frisch concluded that bees have a well-developed sense of smell. He found this intriguing, since the olfactory organs of insects and humans are so different. Today, biologists have a better understanding of the cellular and molecular processes involved in olfaction; thus the results are not so surprising. To locate the sense organs in bees, Frisch cut off their antennae after training them in a particular odor. After the surgery, the bees could no longer distinguish the training scent. Cutting off the antennae had no effect on the bees' ability to distinguish colors, as expected. Thus, he concluded that the olfactory organs in bees are located on their antennae.

The sense of taste also depends on detection of chemicals by certain organs. Frisch researched different aspects of this sense. Bees can recognize different degrees of sweetness—different concentrations of sugar in a solution—and even display individual preferences for acceptable concentrations. The threshold of acceptance is the concentration at which a bee refuses to suck up a solution after tasting it. The threshold of perception is the lowest concentration that stimulates the sense of taste. When conditions are poor and the bee struggles to find food, the threshold of acceptance lowers considerably, approaching the threshold of perception.

Frisch tested bees' sensitivity to different concentrations of salt and to different types of sugars. He found that bees do not like salty substances and that they will consume a greater volume of a substance if it is sweeter than another less concentrated sugar solution. Other findings included the fact that bees tolerated bitter substances more than salty substances and that they could also distinguish sourness.

LANGUAGE OF BEES

During the course of his experiments examining the senses in bees, Frisch often found himself waiting for the bees to find his food source. Sometimes it took a few hours, other times it took a few days, but once a single bee found his supplied sugar water, hundreds would soon follow. This made him wonder how the other bees knew that the first bee found a food source and how the first bee communicated this information to the other bees.

The experimental setup required to study this phenomenon was quite complex. Because one cannot observe the events occurring between the honeycombs within a beehive, Frisch built an observation hive in order to learn how the bees shared information regarding a food source with the other members of their colony. The observation hive consisted of one large comb that could be observed through glass windows. He also devised a marking system so he could identify individual bees. He used different combinations of five colors of paint spotted onto different body parts of the bees to indicate the numbers 1–599. For example, bee 431 had a yellow spot on the abdomen to indicate 400, a blue spot on the front

of the thorax to indicate 30, and a white spot on the hind part of the thorax to indicate the digit 1 in the ones place. Frisch set out a glass dish filled with sugar water near the observation hive and proceeded to wait and observe.

After finding the glass dish and sucking up sugar water, a bee flew back to the hive and transferred most of the sugar water to other bees. Then the worker bee performed a round dance, in which she remained in the same general spot and spun in circles to the right then to the left for up to 30 seconds or even longer. She then moved to another spot on the honeycomb and repeated her performance. While she was dancing, the other bees swarmed around her, then they exited the hive and found the food source. When they returned to the hive, they also performed the dance, and more bees found the food source.

To figure out whether and how the dancing conveyed information about direction, Frisch fed several bees from a dish exactly 32.8 feet (10 m) to the west, north, east, and south of the observation hive. Minutes later bees appeared at all the dishes regardless of direction. To see whether the bees could give more definite directions regarding the food source, he put out two large dishes of flowers, cyclamen and phlox. The cyclamen flowers contained a coating of sugar solution. After several bees fed on the cyclamen and returned to the hive, the new bees flew straight to the cyclamen and showed no interest in the phlox. Clearly the bees were giving more specific information than "Go look outside several meters away from the hive," and Frisch suspected scent played a role.

When the phlox contained the sugar solution, the bees sought out the phlox. When Frisch repeated the experiments but used flowers with different fragrances, he learned that after the bee returned to the hive, she carried the fragrance on her body, and other bees detected the particular fragrance. Not only does the upper body of the bee retain the scent for a long period, but she also regurgitates a bit of the nectar gathered from the bottom part of the flower. That nectar gives off the characteristic scent. When new foraging bees went out, they sought that specific scent to find the sugar water. Odorless flowers were untouched. When Frisch set glass dishes containing sugar water on cardboard sheets dabbed with peppermint oil, then bees from the hive would go out and seek sugar water near anything that smelled like peppermint. The new foragers learned the scent from the dancer. The dances were more vigorous when the sugar solution was more concentrated or when the quantity was abundant.

After performing these sorts of experiments for a few decades, Frisch hypothesized that the bees were communicating information about the distance in addition to the scent of the food source. He arranged food sources at 32.8 feet (10 m) and 984 feet (300 m) from the hive. Glass dishes containing sugar water were placed on scented cards. In the first experiment, food was only available at the nearby food source. After bees visited the nearby food source, they returned to the hive and danced for the other worker bees. The new bees went out and fed mostly at the near source (174 bees), although a few (12 bees) visited the distant source. In the second experiment, food was available at the distant food source, but not the nearby one. After bees fed at the distant food source, they returned to the hive and danced, and new bees flew out to seek food from the distant source. Within a one-hour observation period, 61 bees visited the distant source, while only eight visited the nearby source. Thus, the bees somehow communicated information regarding distance during their dance.

The bees that fed from the nearby source performed the same round dance described earlier. Bees that fed from the distant source performed a different style of dance, one described as a tail-wagging or waggle dance. In this dance, the bee moves a short distance forward while moving its abdomen from side to side, then turns 360 degrees to the left, continues straight forward a bit, then turns 360 degrees to the right, and so on. The particular style of dance conveys information regarding the distance of the food source. Frisch found that the waggle dance changed to a round dance between 164 and 328 feet (50 and 100 m). A relationship also exists between the tempo of the dancing and the distance to the food source. The dance tempo (turns per unit of time) is faster when the food source is nearer.

The dance of the bees also conveys information about the direction of the food source, as demonstrated by the following experiment: Frisch fed bees from a scented card 656 feet (200 m) south of the hive, then placed similarly scented cards in all directions at a similar distance from the hive. After observing the dance by the first few bees, worker bees traveled to the food source south of the hive and to nearby cards that carried the same scent but without food on them. When they shifted the food source to another direction, such as east, the bees soon followed. Frisch and his coworkers noticed that when several bees returned from the same food source, they all "waggled" in the same direction between turns. They figured out that the direction of the food source was related to the direction of the short straight path during the waggle dance.

Amazingly, as the day progressed, the direction of the straight path during the waggle dance shifted slightly but continuously even when the location of the food source was kept constant, suggesting that

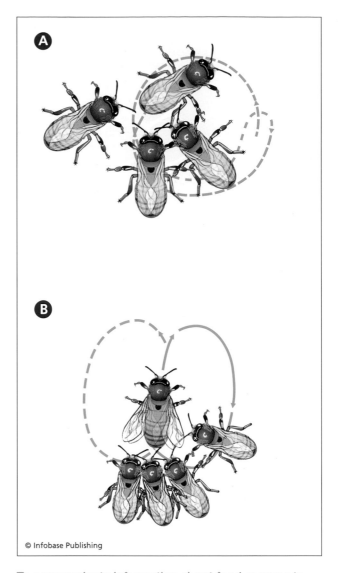

© Infobase Publishing

To communicate information about food sources to other workers, bees perform characteristic dances such as A) the round dance and B) the tail-wagging or waggle dance.

the Sun was involved. Even more surprising was the fact that the bees performed the dance inside a hive near complete darkness; the dancing bees relied on gravity to orient themselves and the observing bees recognized the angle of the dance movement with respect to gravity. The directions seemed to be as follows: a vertical movement during the dance indicated a location toward the Sun: a straight path that points downward indicates away from the Sun, a straight path of the waggle dance that heads 60 degrees to the left of vertical means fly 60 degrees to the left of the Sun, and so on.

Even when the Sun was not out the bees were able to communicate the correct location efficiently. Frisch observed bees when the sky was overcast and found that the bees could still perceive the position of the Sun when it was hidden behind the clouds. When Frisch placed a black plate that allowed the passage of ultraviolet light between the Sun and the beehive, the bees were still able to perform a dance that gave accurate location information. When Frisch obstructed the rays with a glass plate that absorbed ultraviolet radiation, the bees were disoriented. Thus the bees were able to detect the direction of the polarized (traveling in the same direction) ultraviolet light and interpret it to determine the position of the Sun.

This work on communication in bees earned Frisch the most coveted scientific award. In 1973 he received the Nobel Prize in physiology or medicine, shared with two other pioneering ethologists, Konrad Lorenz of Austria and Nikolaas Tinbergen of the United Kingdom, "for their discoveries concerning organization and elicitation of individual and social behavior patterns." Though many obviously considered Frisch's work outstanding, others seriously doubted his conclusions. Many scientists questioned that if the coded dance truly provided all of the specific information that Frisch suggested, then why did it take so long for the newly informed bees to reach the destination of the communicated food source? They proposed that the bees just traced the smell carried by the original dancing bee or that they simply followed the original bee back to the source. In May 2005 scientists published a paper in *Nature* supporting Frisch's conclusions. The group attached radar transponders to the bees and tracked them as they flew to food sources. They found that the bees flew immediately to the vicinity of the source, then flew around a bit in search of the exact location, accounting for the time lag. Their results support Frisch's original conclusions.

In addition to his work on communication in bees, Frisch studied pigments in fish skin, color changes in animals, hearing in fish, sensation of chemicals in fish and in insects, and vision in insects. He also authored numerous books, including a general biology textbook.

In 1921 Frisch briefly joined the faculty at the University of Rostock and served as the director of zoology. After only two years he took a similar position at the University of Breslau, and in 1925 he returned to Munich to assume Hertwig's former position. Frisch oversaw the building of the Zoological Institute at Munich, which was destroyed during World War II. While the institute was being rebuilt in 1946–50, he worked at Graz. After returning to the Zoological Institute, he remained there the rest of his career. He became a professor emeritus in 1958 but continued his studies. Frisch received the Balzan Foundation Award in 1963 and was a member of the

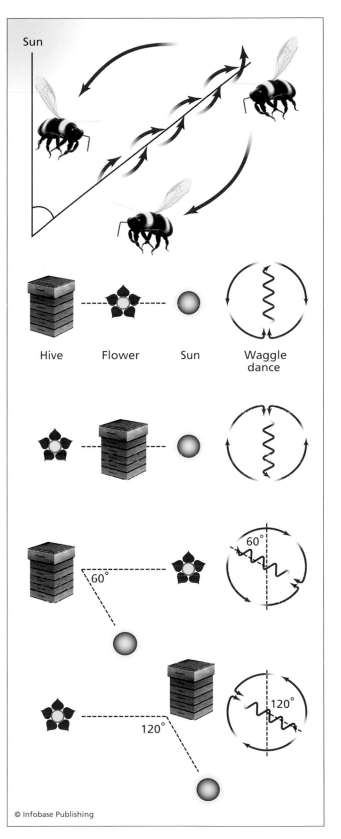

© Infobase Publishing

Bees indicated the direction of the food source by the angle at which they moved in a short, straight path during their waggle dance, using the Sun as a reference.

National Academy of Sciences and the Royal Society of London.

Karl von Frisch died on June 12, 1982, in Munich, Germany. Biologists and psychologists alike remember him for his pioneering work on the sensory capacities and behavioral patterns in bees and other animals. His most significant research demonstrated that bees perform genetically programmed dances to communicate information regarding the distance and direction of a source of nectar to other bees in the colony. This is one of the most complex communication methods to be demonstrated in invertebrates. Not only did Frisch help found the field of ethology, but his discoveries forced zoologists to reconsider the notion of intelligence and the concepts of language and communication.

See also ANIMAL BEHAVIOR; ETHOLOGY; LORENZ, KONRAD; TINBERGEN, NIKOLAAS; TURNER, CHARLES HENRY.

FURTHER READING

Frisch, Karl von. *Bees: Their Vision, Chemical Senses, and Language.* Rev. ed. Ithaca, N.Y.: Cornell University Press, 1971.

———. *A Biologist Remembers.* Oxford: Pergamon Press, 1967.

———. *The Dance Language and Orientation of Bees.* Translated by Leigh E. Chadwick. Cambridge, Mass.: Belknap Press of Harvard University Press, 1967.

The Nobel Foundation. "The Nobel Prize in Physiology or Medicine 1973." Available online. URL: http://nobelprize.org/nobel_prizes/medicine/laureates/1973/index.html. Accessed January 25, 2008.

fungi Members of the eukaryotic kingdom Fungi are unique and diverse. Fungi include molds, mushrooms, yeasts, rusts, smuts, blights, morels, and truffles. Biologists once considered fungi as plants because they were immobile, appeared to have roots, and had cell walls. Plants, however, are photosynthetic, meaning they can harvest light as an energy source and use it to synthesize organic molecules. Fungi are chemoheterotrophs, meaning they cannot synthesize their own organic molecules from inorganic substances and therefore must obtain energy and nourishment from organic substances present in the environment. The bodies of fungi consist of long, slender filaments and are mostly multicellular, though one group, the yeasts, are unicellular. The polysaccharide chitin, the same material found in arthropod exoskeletons, composes fungal cell walls, compared to cellulose in plants. Unlike other eukaryotic organisms, fungi are haploid. They only exist as diploids during a brief phase of sexual reproduction.

Fungi cannot undergo photosynthesis; nor can they engulf food. Instead, they obtain their nutrition by secreting, into the environment, enzymes that digest organic matter. After the enzymes break down the organic matter into smaller components, the fungi absorb the organic molecules into their cells. Many fungi are saprophytic, meaning they obtain nourishment from decaying organic matter, dead organisms, or organic material from other organisms such as animal carcasses, leaf litter, and eliminated waste. Parasitic fungi absorb nutrients from living hosts, including both plants and animals, and can cause infectious diseases. Fungi can also grow on foods such as bread or fruit and even in substances such as house paint. Slightly acidic conditions (pH of about 5) that are unfavorable for most bacteria favor fungal growth. Fungi are also more resistant to low-moisture and high-salt environments.

Molds and fleshy fungi consist of filaments called hyphae, which range in size from microscopic to covering acres of land. Septate hyphae are composed of uninucleate (containing one nucleus) cells separated by septa, or walls, though openings between cells join the cytoplasms of adjacent cells. Coenocytic hyphae do not have septa; they consist of extended cells containing many nuclei. Hyphae grow by elongation at their tips, but if a piece of hyphae breaks, the new fragment can grow into another organism. Specialized types of hyphae carry out the functions of obtaining nutrition and reproduction. Vegetative hyphae must contact the surface on which they grow, so nutrients can be absorbed. When environmental conditions favor growth, the fungus spreads over the surface by branching out to form an intertwining network, or colony, called a mycelium. Reproductive hyphae produce reproductive spores and project outward into the air, so when the spores are released, the air can carry them away.

Yeasts are nonfilamentous, unicellular spherical or oval-shaped fungi. Budding yeasts, such as *Saccharomyces cerevisiae* (baker's yeast), can reproduce asexually by forming a small protrusion from the parent cell. After the parent nucleus divides, the bud receives one-half, continues growing, and eventually pinches off into a smaller but fully functional yeast cell. Structures called pseudohyphae form when the buds fail to separate completely and allow pathogenic yeasts such as *Candida albicans* to penetrate into tissues. Fission yeasts divide into two equally sized daughter cells. Yeasts are facultative anaerobes, meaning they can grow in the presence or absence of oxygen. When grown anaerobically (in the absence of oxygen), they undergo fermentation, a catabolic process that results in the production of ethanol and carbon dioxide. Yeasts are also easy to grow in the lab and are widely used in cell, molecular, and genetic research.

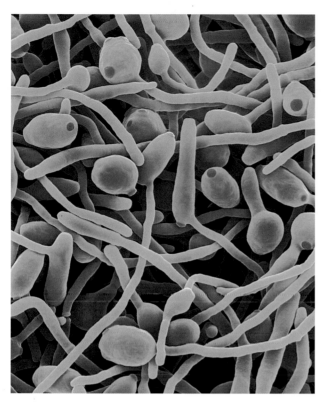

Yeastlike fungi such as *Candida albicans* have dimorphic life cycles and can grow in yeast or hyphal stages. (© *Dennis Kunkel Microscopy, Inc.*)

Some fungi are dimorphic—they can grow like molds with hyphae or like yeasts that bud. Temperature serves as the deciding factor. At 77°F (37°C) they resemble yeasts, but at 98.6°F (25°C) they resemble molds.

FUNGAL REPRODUCTION AND CLASSIFICATION

Fungi reproduce by forming and releasing spores from hyphal tips. Because fungal spores are very small, air currents easily transport them over long distances. Fungal spores differ from the more resistant bacterial endospores in that fungal spores are true reproductive structures. Each spore gives rise to a new organism; thus the production of fungal spores increases the number of organisms. Bacterial endospores allow cells to survive harsh or unfavorable conditions but do not increase the number of bacterial cells.

In asexual reproduction, the offspring are genetically identical to the parent. Filamentous fungi can reproduce asexually by fragmentation, when a separated piece of mycelium develops into an entire new colony. Asexual fungal spores form by mitosis and cell division. The two main subtypes of asexual fungal spores are conidiospores (also called conidia) and the less common sporangiospores. Conidiospores occur in chains at the end of aerial hyphae called

conidiophores and are not enclosed by sacs. They develop by pinching off of the tip or by segmentation of hyphae. The several forms of conidiospores include arthrospores, chlamydospores, blastospores, phialospores, microconidia, macroconidia, and porospores. Sporangiospores develop by successive cleavages inside a sac called a sporangium at the end of aerial hyphae called sporangiophores. When the sporangium breaks open, the spores are released.

In sexual reproduction, two different mating types (a donor cell "+" and a recipient cell "-") of hyphae join together, fuse, and develop into a new organism that forms spores containing genetic information from both parents. The formation of sexual spores occurs in three main stages: plasmogamy, karyogamy, and meiosis. During plasmogamy the haploid nucleus from a "+" mating type invades the cytoplasm of a "-" mating type. The two haploid nuclei fuse in a process called karyogamy, resulting in a diploid nucleus. Meiosis occurs, forming nuclei of sexual spores.

Because sexual spores develop in a variety of diverse ways, one means of classifying fungi is by their mode of sexual reproduction. Three of the most common divisions are Zygomycota, Ascomycota, and Basidiomycota. Members of the phylum

The green mold *Aspergillus flavus* forms asexual spores called conidiospores at the end of aerial hyphae called conidiophores. (© *Dennis Kunkel Microscopy, Inc.*)

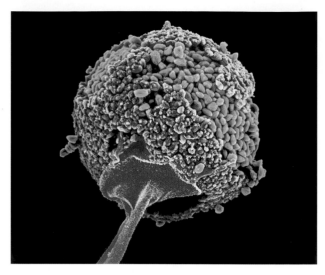

The common bread mold *Rhizopus stolonifer* forms asexual spores called sporangiospores that are enclosed in a sac at the end of an aerial hypha called a sporangiophore. (© *Dennis Kunkel Microscopy, Inc.*)

Zygomycota usually reproduce asexually, but when opposite mating types fuse, sexual spores called zygospores form. The zygospores are large, thick-walled spores that undergo meiosis and germination. The resulting mycelium gives rise to haploid sporangia that resemble the asexual sporangium, except that it contains nuclei with genetic information from both parents. Hyphae of zygomycetes usually have no cell walls. Black bread molds such as *Rhizopus stolonifer* belong to the phylum Zygomycota. Members of the phylum Ascomycota generally produce ascospores inside a tubelike sac called an ascus that forms when two mating types join to form a diploid nucleus. Subsequent meiosis and germination result in the formation of haploid ascospores that disperse when the ascus breaks open. Examples of ascomycetes include *Histoplasma* (the causative organism of Ohio River Valley fever, or histoplasmosis), truffles, and the yeasts. The phylum Basidiomycota includes the mushrooms, rusts, puffballs, and smuts. The fusion of hyphae of different mating types forms a fruiting body (such as a mushroom) that is common to members of this phylum. Club-shaped structures called basidia line the gills underneath the cap of the fruiting body. Fusion of haploid nuclei occurs within a basidium. Meiotic division produces haploid spores that germinate to form haploid hyphae.

Two less well-known phylum are Deuteromycota and Chytridiomycota. Deuteromycota, commonly called the imperfect fungi, includes various fungi that are incapable of sexual reproduction or organisms for which the means of sexual reproduction is not yet known. The organisms that cause athlete's foot and

that give Camembert and Roquefort cheeses their unique flavors belong to this phylum. Members of the phylum Chytridiomycota are usually unicellular and aquatic and produce flagellated, motile gametes and spores. Slime molds and water molds were once classified as fungi because they appeared to share similar life cycles and under certain conditions form structures resembling sporangia. Biologists now believe they are unrelated to fungi.

ECOLOGICAL IMPORTANCE

Fungi closely interact with other organisms in many different ways. A mycorrhiza is a mutualistic symbiotic association formed between a fungus and a plant. The fungus grows inside or wraps around the roots of the plant, and the mycelia branch out to increase the surface area through which absorption of water and nutrients such as phosphorus and minerals occurs. The fungus benefits from carbohydrates synthesized by the plant by photosynthesis. Plants with mycorrhizae can inhabit soil that is less fertile or live in environments that are drier than plants without them.

Lichens are symbiotic associations formed by a fungus with a photosynthetic organism such as a photosynthetic bacterium or an alga. The photosynthetic organism provides a food source for the fungus, while in return the fungus protects its partner from the environment, allowing it to live in habitats that would normally be too harsh, such as on the surface of a rock, in an arid desert, or on a tree trunk. The mycelia absorb moisture and carbon dioxide from the atmosphere for use in photosynthesis. Lichens play an important role in ecological succession, the replacement of one type of community by another at a single location over time, such as after a forest burns down. Lichens help prepare the previously barren terrain for other species by secreting an acid that breaks down rocks into soil and liberates nutrients. Over time, the organic material from dead lichens nourishes future inhabitants. Lichens also serve an important role monitoring the air for pollutants such as sulfur dioxide. Because they are sensitive to such manufactured pollutants, their growth is a good indicator of air quality.

Because most fungi are saprophytes and obtain their nutrition from decomposing organic material of other species, they play a crucial role in the food chain. Plants contain cell walls made of cellulose, a carbohydrate that animals are incapable of digesting. Fungi break down the decaying leaf litter on the forest floors, recycling the nutrients back into the environment in a form that other organisms can utilize. Without fungi, many nutrients would become depleted as they became incorporated into forms unusable to other organisms. Fungi are essen-

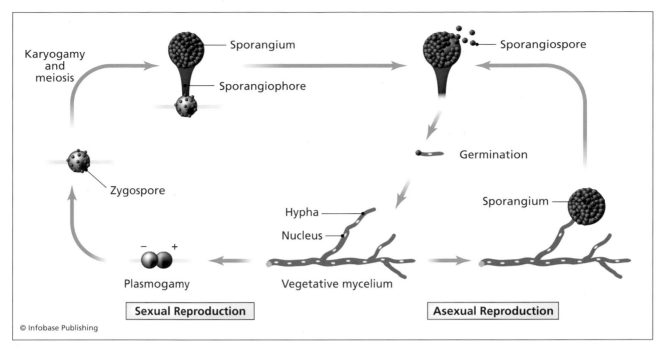

Fungi reproduce both asexually and sexually, as depicted in this generalized life cycle of a zygomycete. In asexual reproduction, spores form by mitosis and cell division. Sexual reproduction involves the fusion of two different mating types and the formation of spores by meiosis.

tial to maintaining conditions that support other life-forms.

IMPACT OF FUNGI

The ecological role played by fungi translates into important economic consequences. The maintenance of soil conditions that support many crops depends on relationships that fungi engage in with plants and other microorganisms. Some fungi inflict devastating damage to crops by causing diseases such as potato blight, black stem rust of wheat, covered smut of barley, powdery mildew, fruit rots, and Dutch elm disease. Other fungi are used in the biological control of costly plant diseases. For example, the yeast *Candia oleophila* protects fruits from harmful molds. The fungus *Trichoderma harzianum* not only protects fruits and vegetables from botrytis, a fungus that causes gray mold, but is also believed to enhance plant growth, degrade pesticides, and prevent the synthesis of toxins produced by other fungi. A mycorrhiza that grows on the roots of trees helps extract calcium from the mineral apatite in forest soils that are calcium poor because of the leaching effects of acid rain. The fungus *Entomophaga maimaiga* attacks the caterpillars of gypsy moths, insects that destroy entire forests by defoliation.

The production of many types of foods depends on fungi. Food and beverage manufacturers exploit the ability of yeasts to produce ethanol and carbon dioxide, two end products of fermentation. The for-

mation of carbon dioxide gas causes bread dough to rise, and ethanol is the form of alcohol found in beer, wine, and liquor. Different fungi impart unique flavors and textures to cheeses, and edible fungi such as mushrooms are a food source themselves. Other fungi cause fruit to rot or bread and cheese to mold. Acidic conditions, as in jams and jellies and inside fruit, inhibit bacterial growth but encourage fungal growth. Fungi can also withstand high solute concentrations used to preserve many foods.

Antibiotics are chemicals produced by one organism that kill or inhibit the growth of another. Many antibiotics are produced by fungi or are derivatives of fungal products. The mold *Penicillium* produces penicillin, the first antibiotic discovered. Other fungi synthesize other commercially available chemical compounds. *Taxomyces andreanne*, a fungus that grows on the bark of Pacific yew trees, secretes taxol, an important anticancer drug. The immunosuppressant drug cyclosporine is derived from the fungus *Tolypocladium inflatum*. *Aspergillus niger* is the source for the flavoring ingredient citric acid and proteases, enzymes that digests proteins.

Less than 1 percent of fungal species are pathogenic to humans and animals. Some species of fungi that are pathogenic to humans cause conditions such as athlete's foot, ringworm, yeast infections, *Pneumocystis* pneumonia (commonly found in acquired immunodeficiency syndrome [AIDS] patients), and histoplasmosis. Diseases caused by fungi are called mycoses and are usually chronic as a result of the slow-growing nature of fungi. Many mycoses are superficial or penetrate to just below the surface of the skin, but some, including histoplasmosis and coccidioidomycosis, are systemic and affect a number of organs.

See also EUKARYA; EUKARYOTIC CELLS; INFECTIOUS DISEASES; MICROBIOLOGY; REPRODUCTION; SLIME MOLDS.

FURTHER READING

Blackwell, Meredith, Rytas Vilgalys, Timothy Y. James, and John W. Taylor. "Fungi: Eumycota: Mushrooms, Sac Fungi, Yeast, Molds, Rusts, Smuts, etc." in *The Tree of Life Web Project*. Available online. URL: http://tolweb.org/Fungi. Content changed February 21, 2008.

Burnett, John H. *Fungal Populations and Species*. Oxford: Oxford University Press, 2003.

Cowan, Marjorie Kelly, and Kathleen Park Talaro. *Microbiology: A Systems Approach*. New York: McGraw Hill, 2006.

Gadd, G. M. *Fungi in Biogeochemical Cycles*. Cambridge: Cambridge University Press, 2006.

Money, Nicholas P. *Mr. Bloomfield's Orchard: The Mysterious World of Mushrooms, Molds, and Mycologists*. Oxford: Oxford University Press, 2002.

Walting, Roy. *Fungi*. Washington, D.C.: Smithsonian Books, in association with the Natural History Museum of London, 2003.

gene expression Gene expression comprises all the events that lead to the synthesis of a functional gene product from a gene on the deoxyribonucleic acid (DNA). Most genes encode for proteins, and the rest for ribonucleic acid (RNA) molecules that play a role in the synthesis of proteins. Genes, the basic units of inheritance, take the form of stretches of DNA on a chromosome. The sequence of the approximately 1,000 or more nucleotides that make up a gene determines the structure of the product it encodes. The central dogma of molecular biology outlines the flow of genetic information as follows:

$$\text{DNA} \xrightarrow{\text{transcription}} \text{RNA} \xrightarrow{\text{translation}} \text{protein}$$

During the process of transcription, an RNA molecule is made from DNA. If the gene encodes a protein, then the RNA is translated into a sequence of amino acids via the process of translation.

TRANSCRIPTION

The location of the DNA in a cell determines where the first step of gene expression, transcription, occurs. In prokaryotes the DNA exists in the nucleiod region of the cytoplasm, whereas in eukaryotes the DNA exists on chromosomes located in the nucleus. During transcription, basically, an enzyme called RNA polymerase scans along the DNA template and constructs a molecule of RNA based on the sequence of DNA it reads.

Transcription results in the synthesis of one of three different types of RNA, all of which participate in the synthesis of proteins from genes. Messenger RNA (mRNA) carries the information necessary to build a protein. Ribosomal RNA (rRNA) is a component of ribosomes, the sites of protein synthesis.

Transfer RNA (tRNA) molecules carry the amino acids to the ribosomes during assembly of the nascent polypeptide.

Though DNA is double-stranded, only one strand within a gene encodes the information necessary to build a protein. The two strands of DNA are complementary, meaning the nitrogenous bases of the deoxyribonucleotides form specific base pairs held together by hydrogen bonds. Adenine always pairs with thymine and guanine always pairs with cytosine. RNA contains uracil instead of thymine and the sugar ribose rather than deoxyribose as part of the nucleotide subunits, but otherwise RNA can participate in complementary base pairing with a single strand of DNA. The major difference is that an adenine on the strand of DNA will pair with a uracil on the strand of RNA. Just as the two strands of DNA run antiparallel to one another (one strand runs 5'→3' and the other runs 3'→5'), a hybrid DNA-RNA molecule will also contain antiparallel strands.

A region called a promoter marks the beginning of a gene. Consisting of a segment approximately 60 base pairs long in prokaryotic organisms and more than 100 base pairs in eukaryotic organisms, the promoter signals to the RNA polymerase where to start transcription and plays an important role in the regulation of transcription. Specific regions of the RNA polymerase recognize and bind to the promoter region of a gene, and the DNA opens up. At the transcriptional start site, the RNA polymerase finds the first ribonucleotide from the pool in the cytoplasm of prokaryotes or the nucleus of eukaryotes. RNA polymerase scans the DNA template in the 3'→5' direction and because it builds an antiparallel complementary strand, it synthesizes RNA in the 5'→3'

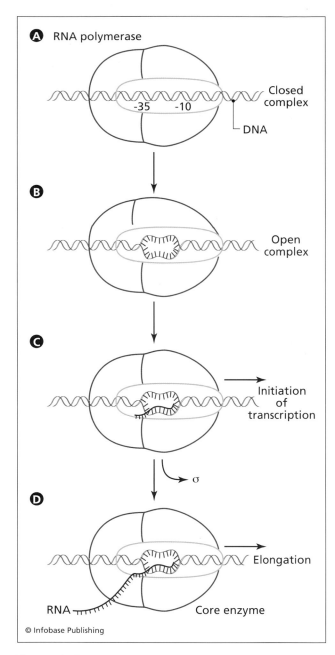

Transcription begins when RNA polymerase recognizes the promoter, opens up the DNA, and begins adding ribonucleotides complementary to the DNA template. Elongation proceeds until RNA polymerase reaches a terminator sequence.

nator sequences usually contain an inverted repeat, a specific sequence followed by four intervening deoxyribonucleotides, then the same specific sequence but in the reverse order on the complementary strand. The result is a segment in the newly transcribed RNA that can form a stem-loop structure by forming base pairs with itself. In some cases, a sequence of thymine-adenine base pairs follows the inverted repeat segment of the terminator. This results in a string of uracils on the transcribed RNA. In some prokaryotic genes and most eukaryotic genes, additional protein molecules assist in the process of termination. The hydrogen bonds between the DNA template and the new RNA transcript denature, the RNA molecule falls free, and the RNA polymerase becomes available to start transcribing another gene.

The next step depends on the type of RNA that the cell has synthesized. If the RNA is rRNA, the molecules combine with other molecules to assemble into ribosomes. Prokaryotic ribosomes contain three segments of rRNA: a 16S (1,540 bases) segment in the smaller 30S subunit and a 5S (120 bases) and a 23S (2,900 bases) segment in the larger 50S subunit. Typically, the smaller subunit also contains 21 polypeptides, and the larger subunit also contains 31 polypeptides. Eukaryotic ribosomes contain four segments of rRNA: an 18S molecule (1,900 bases) in the smaller 40S ribosomal subunit and a 5S (120 bases), 5.8S (160 bases), and 28S (4,800 bases) segment in the larger 60S ribosomal subunit. Except for the 5S segment, the rRNAs all belong to the same gene and are transcribed as one molecule. The smaller subunit also contains approximately 33 polypeptides, and the larger subunit also contains approximately 50 polypeptides. The exact number may differ among organisms.

If the newly transcribed RNA is tRNA, then the molecules must fold into a unique three-dimensional structure to become functional. The length of tRNA molecules ranges between 74 and 93 nucleotides, and at least 20 different types exist, one for each amino acid. The tRNAs carry amino acids to the ribosome during the synthesis of polypeptides, and the presence of a specific anticodon (discussed under the subheading Translation on page 367) ensures that the correct amino acid is added to the growing chain. All tRNAs have the same general shape, with three stem-loop structures formed by base pairing between ribonucleotides within the RNA molecule. After transcription of the tRNA, modifications occur, including the removal of extra sequences at the beginning and the end of the molecule and the chemical conversion of some of the nucleotides to unusual nucleotides.

If the RNA serves as a message encoding for the synthesis of a protein, as is most often the case, then in eukaryotic cells, several posttranscriptional modifi-

direction. As it scans the template, RNA polymerase adds ribonucleotides, following the rules of complementarity, pairing adenines with uracils (or with thymines on the DNA) and guanines with cytosines. The enzyme forms a covalent linkage between each incoming ribonucleotide and the nucleotide previously added to the new strand of RNA.

Transcription halts when the RNA polymerase reaches a terminator sequence on the DNA. Termi-

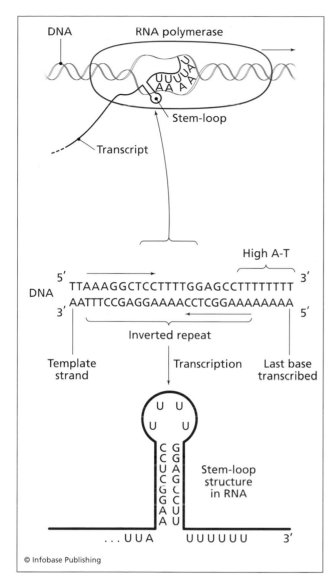

Transcription terminator regions in DNA consist of stem-loop structures formed from inverted repeat sequences.

close proximity. The intervening intron is excised, and the ends of the exons are joined. Numerous factors consisting of both polypeptides and RNA participate in the splicing process. The resulting mature mRNA contains only the coding region that is translated and a leader sequence that precedes the coding region of the mRNA and a trailer sequence, neither of which is translated.

TRANSLATION

During the process of translation, the protein synthesizing machinery translates the information embedded in the sequence of ribonucleotides of the mRNA into a protein. In prokaryotic cells, as soon as RNA polymerase has synthesized a long enough segment of mRNA, a ribosome can attach to the 5' end and begin translation. Then, as soon as that ribosome progresses a certain distance along the length of the mRNA, another ribosome can join and begin translation of a second polypeptide. The result is the formation of a polysome, a cluster of ribosomes simultaneously translating the same mRNA transcript. In eukaryotic cells, however, the processed mRNA must first move to the cytoplasm, where the ribosomes, tRNA molecules, and amino acids are located.

In order for translation to occur, the supply of "charged" tRNA molecules must be sufficient. Enzymes called aminoacyl-tRNA synthetases charge the tRNAs by attaching the appropriate amino acid to the ribose sugar at the 3' end of the tRNA molecule. A different tRNA synthetase specifically recognizes each of the 20 different amino acids and covalently links them to a tRNA molecule that has the corresponding anticodon based on the genetic code. An anticodon is the sequence of three consecutive nucleotides on one loop of a tRNA molecule that forms complementary base pairs with a codon on the mRNA. Each codon contains three ribonucleotides that specify one of the 20 amino acids. Because there are four different ribonucleotides, and they are read in sets of three, there are 64 (4^3) different codons. One of these is the start codon, AUG, and always encodes methionine. Three codons serve as stop signals for translation: UAG, UGA, and UAA. The remaining 60 triplet codons encode for the 19 other amino acids. Because there are many more possible codons than amino acids, several amino acids are represented by more than one codon, a characteristic termed redundancy. When the anticodon of the tRNA binds to the complementary codon on the mRNA, the ribonucleotide at the third position of the codon is often flexible. This reduced constraint at the third position is referred to as wobble. To illustrate these concepts, consider the codons CUU, CUC, CUA, and CUG. The fact that they all encode

cations must precede translation. A methyl-guanosine residue called a cap is added to the 5' end of primary transcripts. This cap aids in ribosomal recognition of the transcript before translation. The enzyme poly-A polymerase adds between 20 and 200 adenine residues to the 3' end of the transcript. The purpose of the polyadenylate tail is to stabilize the transcript and to aid in its transport from the nucleus, where transcription occurs, to the cytoplasm, where translation occurs. Another major posttranscriptional modification is splicing. Eukaryotic genes contain introns, intervening sequences that are not translated into proteins. The regions of the gene that are expressed in proteins are called exons. During splicing, the ends of the exons in the primary transcript are brought into

Codons in mRNA

First base		Second base			Third base
	U	C	A	G	
U	UUU ⎫ UUC ⎬ Phenylalanine UUA ⎫ UUG ⎬ Leucine	UCU ⎫ UCC ⎬ Serine UCA ⎪ UCG ⎭	UAU ⎫ Tyrosine UAC ⎬ UAA ⎫ Stop UAG ⎬	UGU ⎫ Cysteine UGC ⎬ UGA — Stop UGG — Tryptophan	U C A G
C	CUU ⎫ CUC ⎪ Leucine CUA ⎪ CUG ⎭	CCU ⎫ CCC ⎬ Proline CCA ⎪ CCG ⎭	CAU ⎫ Histidine CAC ⎬ CAA ⎫ Glutamine CAG ⎬	CGU ⎫ CGC ⎪ Arginine CGA ⎪ CGG ⎭	U C A G
A	AUU ⎫ AUC ⎬ Isoleucine AUA ⎪ AUG — Start	ACU ⎫ ACC ⎪ Threonine ACA ⎪ ACG ⎭	AAU ⎫ Asparagine AAC ⎬ AAA ⎫ Lysine AAG ⎬	AGU ⎫ Serine AGC ⎬ AGA ⎫ Arginine AGG ⎬	U C A G
G	GUU ⎫ GUC ⎪ Valine GUA ⎪ GUG ⎭	GCU ⎫ GCC ⎬ Alanine GCA ⎪ GCG ⎭	GAU ⎫ Aspartic acid GAC ⎬ GAA ⎫ Glutamic acid GAG ⎬	GGU ⎫ GGC ⎪ Glycine GGA ⎪ GGG ⎭	U C A G

© Infobase Publishing

The genetic code specifies the amino acid represented by each triplet codon.

for the amino acid leucine demonstrates redundancy in the genetic code, and actually, UUA and UUG also encode leucine. The fact that the anticodon of the tRNA that carries leucine binds to any codon that has CU as its first two ribonucleotides no matter which ribonucleotide occurs in the third position demonstrates wobble.

Translation begins when the small subunit of a ribosome attaches to the 5' leader sequence of an mRNA transcript. The 16S rRNA of the 30S subunit in prokaryotes facilitates this interaction by forming temporary base pairs with a complementary sequence in the leader portion of the mRNA, the Shine-Dalgarno sequence, which also serves to position the ribosome at the translation start codon. In eukaryotes, the 5' cap helps the small subunit of the ribosome bind to the mRNA, and then it typically scans the RNA moving in the 3' direction until it reaches the first AUG, where the large subunit assembles to complete formation of the ribosome. An initiator tRNA charged with methionine, which in prokaryotes contains a formyl group attached to it, joins the complex, followed by the large ribosomal subunit. Formation of the initiation complex and ribosome assembly also involve the participation of several initiation factor proteins and require energy.

The ribosome contains three sites: the aminoacyl (A) site, the peptidyl (P) site, and the exit (E) site. The mRNA runs along the bottom of these sites. After the initiator tRNA carries in the first amino acid, methionine, the ribosome is positioned such that the codon that occurs immediately after the start codon is positioned under the A site, and the tRNA with the methioinine is in the P site. Elongation occurs when the tRNA that has an anticodon that pairs with the mRNA codon in the A site takes the next amino acid to the ribosome. Proteins called elongation factors assist in the correct positioning of the ribosome. If the correct tRNA is sitting in the A site, then a peptide bond forms between the amino acid in the P site and the amino acid in the A site. The enzymatic activity that performs this reaction is part of the 50S subunit of the ribosome and contains both protein and RNA components. After the peptide bond forms, the dipeptide is attached to the tRNA in the A site. Next, with the aid of another elongation factor, the mRNA moves relative to the ribosome such that the third codon is now positioned under the A site, the tRNA with the attached dipeptide is in the P site, and the initiator tRNA sits in the E site. The uncharged tRNAs fall from the E site into the cytoplasm, where they become recharged. Elongation proceeds with the next tRNA moving to the A site, a peptide bond forms between the amino acids attached to the tRNAs in the P and A sites, translocation of the mRNA with respect to the ribosomes, and so on.

The energy burning process of elongation continues until a stop codon sits in the A site of the ribosome. No tRNAs exist that have anticodons complementary to the stop codons. Proteins called release factors help free the nascent polypeptide from the final tRNA. The ribosome disassembles and the mRNA dissociates. The polypeptide adopts its three-dimensional conformation. Sometimes, before becoming functional, the polypeptide must combine with other polypeptides to form a complete protein.

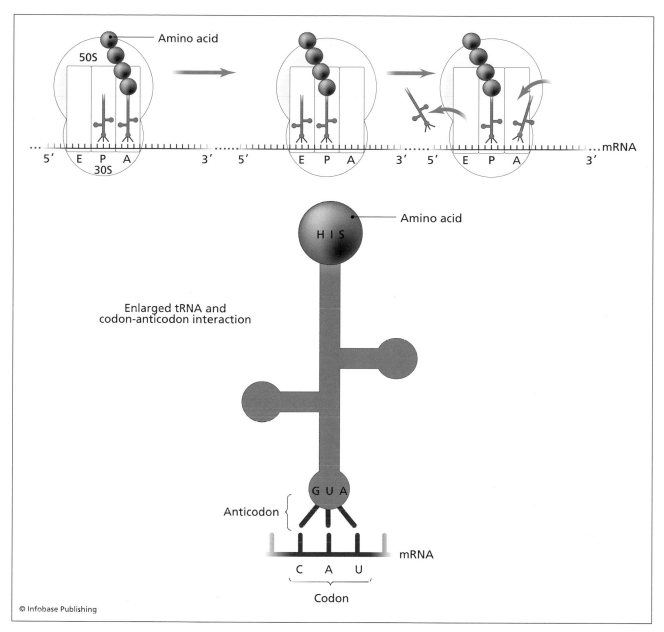

During elongation of translation, tRNA molecules carry the correct amino acids to the growing chain, peptide bonds form, and the mRNA slides over one codon at a time for the addition of the next amino acid.

Modifications such as the addition of sugar moieties or phosphate groups may also be necessary before the protein is fully functional.

Molecular geneticists have encountered several exceptions to the typical processes involved in gene expression. RNA editing is one such exception, in which uracil residues are added to the mRNA or ribonucleotides are chemically modified after transcription but before translation. The discovery of reverse transcription in retroviruses challenged the dogma regarding the flow of genetic information. These viruses, including the human immunodeficiency virus and the Rous sarcoma virus, have RNA as their genetic material but synthesize a complementary DNA molecule using the RNA as a template during the viral replication process. Other viruses, RNA phages that infect bacteria, make an enzyme that enables the single-stranded RNA genome to self-replicate.

DIFFERENCES IN PROKARYOTIC AND EUKARYOTIC GENE EXPRESSION

As briefly mentioned in the preceding discussion, gene expression in prokaryotes and eukaryotes differs in several ways. First of all, the structure of the genes on the DNA varies. In prokaryotes, genes are often polycistronic, meaning they encode for several proteins to be translated from a single product of transcription. In eukaryotes, one gene typically encodes one protein. Another major difference is that in prokaryotic cells, both transcription and translation occur in the cytoplasm and therefore can occur simultaneously. Ribosomes can attach to and begin translation of a molecule of mRNA that is still being transcribed by RNA polymerase. In eukaryotic cells, transcription occurs in the nucleus, where the DNA is housed, and translation occurs in the cytoplasm. Many more factors are involved in the initiation of transcription in eukaryotes than in prokaryotes, and eukaryotic genes often involve regions called enhancers far upstream of the coding region of the gene in addition to the promoter that play a role in regulation of the expression. Prokaryotes have a single RNA polymerase that transcribes the DNA template, whereas eukaryotes have three different RNA polymerases. RNA polymerase I transcribes the nucleolar organizer, the region around which the nucleolus forms and where ribosomal RNA is made. RNA polymerase II transcribes most genes that encode proteins. RNA polymerase III transcribes the 5S and the tRNA genes. After transcription, eukaryotic RNA must undergo several modifications, including the addition of a 5' cap, a poly A tail, and splicing, before it is transported to the cytoplasm. The first amino acid of a polypeptide in prokaryotes is chemically modified by the attachment of a formyl group to its ribose sugar. The ribosomal structure differs between prokaryotes and eukaryotes also, with slightly different components and overall sizes. Despite all of these differences, the general processes of transcription and translation are remarkably similar in prokaryotic and eukaryotic cells.

REGULATION OF GENE EXPRESSION

Cells utilize many strategies to achieve a balance between synthesizing sufficient quantities of gene products when needed and conserving energy. The mechanisms that cooperate to perform this function are collectively termed the regulation of gene expression. Some genes are constitutive, or are always expressed at relatively constant levels in active cells. These are often called housekeeping genes, and they are necessary to maintain the cell's normal activities. For example, rRNA genes are necessary for the construction of ribosomes, which are always needed by active cells to synthesize proteins. Other gene products are only necessary under specific conditions, during certain stages of development or differentiation, or only in specialized cell types.

Regulation of these gene products are made and of the quantity produced can occur at any step during expression. The DNA itself can be chemically altered or packaged into chromatin in a way that either allows or completely prevents the recognition or access of those genes. Transcription initiation is the most commonly regulated step of gene expression. Proteins called transcription factors can bind regions of gene promoters to up-regulate, or increase the expression of certain genes. In some cases, such as genes that are only expressed in specific tissues, certain factors beyond the basal factors are required to initiate transcription. Repressors down-regulate expression by binding to promoters and interfering with RNA polymerase binding or progression through a promoter. Activators are proteins that bind to enhancer regions far away from the promoter but can still interact with proteins bound at the promoter by looping out of the intervening DNA.

Though the mechanisms that regulate the expression of genes are numerous and complex, the end result is that cells express proteins when they need them. For example, steroid hormones such as testosterone, produced only at certain times in mammalian development, bind to tissue-specific protein receptors that are only present in target tissues or organs. Upon recognition of the hormone, the receptor then binds to the DNA and activates the expression of genes that result in the onset of sexual maturation. In another example, to prevent wasting energy, bacterial cells only synthesize the enzymes involved in certain metabolic pathways when the appropriate substrates are available. If the carbohydrate lactose is not present, then a repressor protein binds to the regulatory region of the operon that encodes the genes for lactose utilization and inhibits RNA polymerase from transcribing those genes. (Operons are sequences of adjacent genes that are all under the same regulatory control.) When lactose is present, it binds to the repressor protein and inhibits it from binding to the DNA and interfering with RNA polymerase. The genes are transcribed and translated into the proteins that enable the cell to utilize lactose as an energy source.

See also BIOMOLECULES; CHEMICAL BASIS OF LIFE; DEOXYRIBONUCLEIC ACID (DNA); DNA SEQUENCING; GENETIC DISORDERS; GENETICS; GENOMES.

FURTHER READING

Clark, David P. *Molecular Biology Made Simple and Fun.* 3rd ed. St. Louis: Cache River Press, 2005.

Lewin, Benjamin. *Essential Genes.* Upper Saddle River, N.J.: Prentice Hall, 2005.

Mattick, John S. "The Hidden Genetic Program of Complex Organisms." *Scientific American* 291 (2004): 60–67.

Ptashne, Mark. "Gene Regulation by Proteins Acting Nearby and at a Distance." *Nature* 322 (1986): 697–701.

Raineri, Deanna. *Introduction to Molecular Biology.* Malden, Mass.: Blackwell Science, 2001.

Tjian, Robert. "Molecular Machines That Control Genes." *Scientific American* 272 (1995): 54–61.

Watson, James D., Tania A. Baker, Stephen P. Bell, Alexander Gann, Michael Levine, and Richard Losick. *Molecular Biology of the Gene.* 6th ed. San Francisco: Benjamin Cummings, 2007.

Weaver, Robert F. *Molecular Biology.* 4th ed. New York: McGraw Hill, 2008.

gene therapy Diseases and disorders result from a variety of causes: pathogenic microorganisms cause infectious diseases; degenerative diseases result from either aging, excessive wear and tear, or continuous exposure to harmful substances; several intrinsic and extrinsic factors can lead to developmental disorders; and genetic disorders result from harmful mutations in one's hereditary material, or genome. Genes are the units of heredity that encode all of an organism's characteristics, from skin tone to the ability to tolerate dairy products. Deoxyribonucleic acid (DNA) is the biomolecule that physically carries the genes and consists of four different nucleotides (abbreviated *A, C, G,* and *T*). The specific order of these nucleotides along the length of a DNA molecule encodes the information necessary to build proteins, the molecules that carry out most cellular activities and ultimately determine the characteristics of an individual. Mutations are changes to the sequence of the nucleotides of DNA, and if they occur within a region that encodes a protein, they can affect the protein's structure and possibly destroy the ability of that protein to perform its function. Researchers have identified specific mutations responsible for many disorders including cystic fibrosis, muscular dystrophy, Huntington's disease, hemophilia, severe combined immunodeficiency (SCID), and some types of cancers. People who have genetic disorders such as these may obtain relief someday from gene therapy, as might people who have other chronic diseases that have a genetic component. Disorders that result from multiple mutations or from other extrinsic factors, such as Alzheimer's disease, diabetes, heart disease, and high blood pressure, are not good candidates for gene therapy.

Gene therapy is an experimental treatment that involves the insertion of normal genes into an individual to replace defective genes. Medical researchers are currently investigating gene therapy to treat certain genetic disorders, and a few clinical trials are under way. Unsolved difficulties currently prevent the widespread use of gene therapy, but the technique offers hope. Though researchers have launched hundreds of clinical trials around the world since the first one in 1990, the only gene therapy product currently in use other than in clinical trials is one approved in China for cancer. Safety concerns and ineffectiveness plague human trials, but researchers continue to develop new strategies and techniques.

In order to attempt treating a genetic disorder using gene therapy, the specific mutation that causes it must be identified and the normal gene must be cloned. The normal gene is packaged inside a vector that carries the gene into the targeted cells. Ex vivo methods introduce the DNA to host cells that have been removed from the body, such as blood cells or bone marrow cells. The cells are returned to the body intravenously after exposure to the DNA. In vivo methods involve the transfer of the vector or liposome to cells in the patient's body. Once inside the cells, the gene must be expressed, meaning the cells must produce the protein that the gene encodes. The goal is that once the cells receive a normal copy of the gene, they produce the protein and restore the function that was lacking in the individual with the disorder. Viruses are the common choice for vectors—specifically, viruses that have been genetically engineered so they cannot cause disease. Scientists knock out the genetic information that allows the virus to replicate itself once it has infected a human cell. Theoretically, once the virus transports the recombinant DNA into the human cell, its job is complete, and the virus cannot cause illness. A disabled adenovirus, one of the many types of viruses that cause the common cold, is the most frequently used virus vector. Adeno-associated viruses can insert their DNA directly into one of the human chromosomes rather than simply releasing it into the cytoplasm of a cell and hoping the cell incorporates it on its own. Retroviruses are also useful because they make double-stranded DNA that integrates itself into the host cell chromosomes, becoming a permanent part of the genome for those cells and their progeny. Another approach is to introduce naked DNA directly to the target cells, but this method is inefficient, requires a lot of DNA, and can only be used for certain tissues. A nonviral method for delivering DNA is to package it into tiny spheres surrounded by artificial lipids, called liposomes. The lipid exterior allows the spheres to penetrate the membrane of the host cell and release the DNA upon entry.

Sometimes the therapeutic gene being introduced is not a gene that is mutated in the individual, but rather a gene that encodes a protein that has the potential to overcome or reverse the effects of the individual's symptoms. This strategy may be useful in treating cancer or autoimmune disorders. More than 67 percent of gene therapy trials are for treating cancers. In addition to inserting normal copies of genes that are known to cause cancer when mutated, researchers are examining ways to improve

(continues on page 374)

GENE THERAPY AND CYSTIC FIBROSIS

by Caleb Hodson, Ph.D.
Howard Hughes Medical Institute, Yale University

In an ideal world the air people breathe everyday would be clean. In reality, each breath carries particles of dust as well as other potentially harmful substances such as bacteria into the lungs. To remain healthy, the body must remove or clear these foreign materials from the lungs where they can irritate the airways, reduce lung performance, or cause infection. People with the genetic disease cystic fibrosis, among other symptoms, are unable to clear their airways efficiently. Mutations in their DNA lead to abnormal secretions, which impair an important defense mechanism responsible for maintaining a sterile lung and respiratory environment. As a result, these patients suffer from reduced lung capacity and are more likely to acquire airway infections. Normally, removal of unwanted debris from the airways occurs through a defense process called mucociliary clearance. The two essential elements of mucociliary clearance are found within its name—mucus (*muco-*) and cilia (-*ciliary*).

Mucus, a mixture of water and protein, forms a protective coating over the cells of the respiratory tract. This layer of mucus traps foreign materials such as dust or bacteria before they have a chance to contact and damage the underlying cells. To produce mucus, specialized cells along the airways generate the protein component of mucus, called mucin, which the cells export onto the airway surface. Once at the surface, mucin absorbs water and becomes hydrated into the characteristic gellike consistency of mucus. If not enough water is present to hydrate the mucin properly, the resulting mucus becomes thick and sticky, similar to paste. Mucus must be the correct consistency in order for ciliary action to move it up and out of the airways successfully. Cilia are microscopic hairlike projections that extend out from the surface of airway cells. The millions of cilia present along the respiratory tract beat in a coordinated motion, providing the force necessary to propel mucus up and out of the airways. Watery mucus does not efficiently trap bacteria and particulates, and they remain inside the airways. If mucus is too viscous, then cilia are unable to move it out of the airways.

Individuals who have cystic fibrosis possess a mutation that reduces the amount of water available to hydrate mucin along the airway surface. As a result, the mucus of cystic fibrosis patients is too viscous and is inefficiently cleared from the airways. Over time, mucus accumulates in the lungs, forming large masses or cysts of thick and fibrous mucus, hence the name *cystic fibrosis*. As the mucus accumulates, so do bacteria and other foreign materials that often cause life-threatening infections. Despite treatment with antibiotics and other strategies, half of the people diagnosed with cystic fibrosis die before reaching their upper 30s. Daily life with cystic fibrosis is often difficult and requires strict adherence to medication schedules and participation in physical therapy exercises to increase lung performance.

While the treatment options for patients who have cystic fibrosis have improved over time, at best they provide temporary fixes to a permanent problem, mutations in a gene critical for mucus production. Because the DNA sequence of genes specifies the information needed to construct cellular proteins, changes or mutations within genes can result in changes within the corresponding proteins. Such changes sometimes alter the proper function of proteins as happens with the gene *CFTR*. Specific mutations in the *CFTR* gene sequence are responsible for cystic fibrosis. For example, the most common mutation in the *CFTR* gene that causes cystic fibrosis results in the deletion of a single amino acid from the CFTR protein. However, without this one amino acid, the CFTR protein has very little to no functional activity. Replacement of the defective *CFTR* gene with a normal copy could possibly reverse the ultimate cause of cystic fibrosis. This concept forms the basis for gene therapy and genetic therapeutics. Overcoming the harmful effects of mutated genes by transferring normal copies of the genes into cells holds great promise for the treatment of numerous genetically based diseases. In order for this strategy to work, several obstacles must be overcome. First, the mechanism for transferring the genes must efficiently target the affected cells. Second, the expression of the normal genes must be long-lasting or even persist for the rest of the patient's lifetime. Third, the transfer of genetic material must not cause unmanageable side effects such as inappropriate activation of the body's immune responses. While several clinical trials have been performed to test gene therapy treatments in cystic fibrosis patients, none showed a significant improvement linked to the therapy. One major reason for this lack of success is the incomplete understanding of how to balance each of these requirements for optimal gene therapies.

In order to introduce new DNA into cells, the DNA molecules must be contained within a delivery mechanism to prevent damage and degradation as the DNA is sent to the target cells. In addition, the surface of the target cells must be accessible to the delivery mechanism. The cells lining the airways directly contact the external environment, unlike cells in other organs such as the liver. This exposed location means targeting the affected airway cells in order to treat cystic fibrosis is relatively straightforward and could occur by directly applying gene therapies in an aerosolized form that can be inhaled. Although this characteristic is definitely an advantage over other tissues, there are still significant challenges to targeting airway cells. A substantial hindrance is the protective barrier of mucus within the airways that blocks contact between the DNA targeting mechanism and the cell surface.

Since DNA cannot be directly applied to cells because of damage and degradation, technologies have been developed to assist in the transfer of DNA into cells. The most promising approach for use in gene therapy is to package DNA within virus particles, which are then used to infect the cells of interest. In the simplest terms, viruses are nucleic acid molecules enclosed in a coat of proteins that allow viruses to associate selectively with receptor proteins on the surface of host cells. Viruses attach to the plasma membrane of the target cells and the genetic material is transferred into the cells. Once inside, the viral genes encode the information needed to generate new virus particles that are ultimately released and go on to infect other cells. Through genetic engineering, one can create viruses that contain only a few essential genes plus normal copies of human genes affected in specific disorders such as cystic fibrosis. Additionally, viruses for gene therapy are unable to replicate inside human cells, therefore limiting the possibility of the virus's spreading throughout the body. Ideally, exposure of the airway cells to these engineered viruses should lead to the transfer of the normal gene copies into the airway cells. Subsequently, these genes will produce normal versions of the encoded proteins to compensate for the presence of mutated ones.

While sound in theory, the application of viral technology to gene therapy in the lungs is far from perfect. A well-studied type of virus for gene therapy, adeno-associated virus (AAV), provides high-efficiency transfer of genes to the targeted airway cells; however, its effects are short-lived and production of normal protein diminishes over time. Nonetheless, several clinical trials using AAV-based therapy have been performed. In a 2004 study, patients receiving virus containing the human *CFTR* gene sequence all showed efficient targeting of the gene to airway cells (Moss et al., 2004.) Furthermore, modest improvement in lung function occurred in patients receiving *CFTR*-containing virus compared to patients receiving placebo virus when tests of lung function were made 30 days after the treatment. Unfortunately when the patients were retested at two and three months after viral therapy, the investigators found no difference between patients in the test group and placebo control group patients. The virus can be administered multiple times; however, efficiency of gene transfer is reduced in subse-

quent applications, limiting its usefulness. Importantly, adeno-associated viruses do not seem to activate the immune system in significant ways, and that is a strong advantage over other types of viruses previously considered for gene therapy. While continuing research may provide solutions to the current shortcomings of AAV-based gene therapy, alternative strategies are also being investigated. Another class of viruses known as lentiviruses offers the advantage that genes transferred by this viral class can integrate into the existing genome. If genes integrate, they may be more stable and remain functional for longer periods. This distinction, in contrast to AAV technology, would alleviate the need for repeated administration of virus to the airways, which loses effectiveness over time.

Besides viral technology, several non-viral strategies show promise. DNA can be packaged within synthetic vesicles called liposomes made of compounds that mimic the biochemistry of the cellular plasma membrane. Fusion of liposomes with the plasma membrane introduces the contained gene sequence to the interior of the cell, where the gene has access to the normal cellular machinery. Current clinical trials show mixed results concerning activation of the immune system, and the efficiency with which cells uptake DNA appears to be lower than with viral techniques. Advances in the engineering of the chemical properties of liposomes should increase efficiency, making this approach more appealing for gene therapy. Finally, DNA may be transferred into cells by complexing a gene sequence with nanoparticles. These small molecular compounds are readily taken up by cells, and they rapidly deliver the transported genes to the cell nucleus. A 2004 study using this emerging technology demonstrated measurable CFTR function in patients treated with nanoparticles versus saline solution–treated controls (Konstan et al., 2004). The study showed no adverse side effects to the application; however, the improved CFTR function only persisted a few days, meaning lasting effects require repeated administration of the therapy. More recently, in 2007, a gene therapy strategy using nanoparticle technol-

ogy successfully reduced the growth of cancerous tumors in laboratory animals. (Deng et al., 2007). Further development of nanoparticle applications is hoped to increase their usefulness as treatments for diseases such as cystic fibrosis. Although no gene therapy treatment currently balances all the requirements needed to cure cystic fibrosis, the lessons that have been learned, the continued research, and future technological developments will put society closer to effective gene therapies and truly novel ways to conquer genetic diseases.

FURTHER READING

Deng, Wu-Guo, Hiroyaki Kawashima, Guanglin Wu, Gitanjali Jayachandran, Kai Xu, John D. Minna, Jack Roth, and Lin Ji. "Synergistic Tumor Suppression by Coexpression of FUS1 and p53 Is Associated with Down-Regulation of Murine Double Minute-2 and Activation of the Apoptotic Protease-Activating Factor 1-Dependent Apoptotic Pathway in Human Non-Small Cell Lung Cancer Cells." *Cancer Research* 67 (2007): 709–717.

Konstan, Michael W., Pamela B. Davis, Jeffrey S. Wagener, Kathleen A. Hilliard, Robert C. Stern, Laura J. H. Milgram, Tomasz H. Kowalczyk, Susannah L. Hyatt, Tamara L. Fink, Christopher R. Gedeon, Sharon M. Oette, Jennifer Payne, Osman Muhammad, Assem G. Ziady, Robert C. Moen, and Mark J. Cooper. "Compacted DNA Nanoparticles Administered to the Nasal Mucosa of Cystic Fibrosis Subjects Are Safe and Demonstrate Partial to Complete Cystic Fibrosis Transmembrane Regulator Reconstitution." *Human Gene Therapy* 15, no. 12 (2004): 1,255–1,269.

Moss, Richard B., David Rodman, L. Terry Spencer, Moira L. Aitken, Pamela L. Zeitlin, David Waltz, Carlos Milla, Alan S. Brody, John P. Clancy, Bonnie Ramsey, Nicole Hamblett, and Alison E. Heald. "Repeated Adeno-Associated Virus Serotype 2 Aerosol-Mediated Cystic Fibrosis Transmembrane Regulator Gene Transfer to the Lungs of Patients with Cystic Fibrosis: A Multicenter, Double-Blind, Placebo-Controlled Trial." *Chest* 125 (2004): 509–521. Available online. URL: http://chestjournal.org/cgi/content/full/125/2/509. Accessed August 3, 2008.

(continued from page 371)

the effectiveness of a patient's own immune system to attack cancer cells by introducing genes into white blood cells to help them recognize and attack tumor cells. The insertion of genes encoding for the production of certain cytokines, chemicals produced by the immune system that increase its efficiency, may also help. Two other ideas for treating cancer by gene therapy are inserting genes that cause the target cells to self-destruct or that prevent the growth of blood vessels to feed tumors, depriving them of nutrition. An alternative approach is the introduction of genes to a patient's blood-forming stem cells, with the aim of making the individual more resistant to the side effects of chemotherapy, allowing him or her to withstand traditional treatments better.

PROBLEMS AND PROGRESS

Major problems with current gene therapy protocols include inserting the normal genes into the targeted cells within the body and having them function properly after insertion. The body can naturally suppress their expression, preventing the cells from synthesizing functional protein, even if the gene is successfully inserted. In order for gene therapy to cure a condition, the therapeutic DNA must remain functional, and the target cells must be long-lived and stable. Currently, moderately successful therapies involve multiple rounds of treatment. Another problem is that the host's immune system often attacks the viral vectors. This is particularly troublesome when the host's immune system has previously encountered the virus used as a carrier, because the immune response is quicker and stronger upon the second and subsequent exposures. This characteristic hinders the effectiveness of multiple rounds of treatment.

According to the National Institutes of Health (NIH), hundreds of gene therapy trials have been registered, but the trials have had mixed results. Theoretically and in lab animal experiments, gene therapy

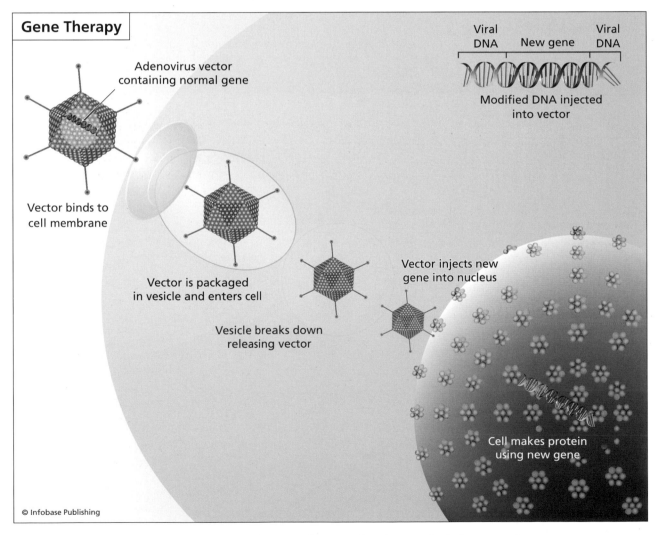

Gene Therapy

Adenovirus vector containing normal gene

Vector binds to cell membrane

Vector is packaged in vesicle and enters cell

Vesicle breaks down releasing vector

Viral DNA   New gene   Viral DNA

Modified DNA injected into vector

Vector injects new gene into nucleus

Cell makes protein using new gene

© Infobase Publishing

Viruses are often used as vectors to carry therapeutic genes into host cells during gene therapy.

looks promising, but when it is tested on humans, success is rare.

The first trial, launched by an NIH research physician named William Anderson in 1990, involved two girls who had inherited a form of SCID that left them unable to fight infections because of the lack of the enzyme adenosine deaminase (ADA). Researchers took some of the girls' cells, cultured them in vitro, inserted a normal copy of the ADA gene, then introduced their own genetically engineered white blood cells back into their bodies. Both girls were able to live normal lives afterward. Critics claim, however, that their apparent success is due to injections of a synthetic ADA enzyme that the patients started receiving shortly after the gene therapy treatments, not to restored immune function.

In one well-publicized case from 1999, a patient died of a reaction to gene therapy treatment for a potentially fatal metabolic disorder. The patient, 18-year-old Jesse Gelsinger, had a metabolic condition caused by the lack of an enzyme, ornithine transcarbamylase, that the liver uses to break down ammonia. The disorder led to frequent hospitalizations during childhood and required him to take numerous daily medications to maintain his health. After treatment, his immune system attacked the adenovirus vector designed to carry the normal gene to his liver cells. He suffered from multiple organ failure and died four days after receiving treatment. Investigations by the U.S. Food and Drug Administration (FDA) revealed questionable practices by the researchers at the University of Pennsylvania, where the trial was conducted.

In January 2003 the FDA placed a hold on all active gene therapy trials that used retroviral vectors to insert genes in blood stem cells after a second child who had been treated for X-linked severe combined immunodeficiency disease (X-SCID) in France developed an illness similar to leukemia. Initial results from the trial, which began in 1999, were promising, and all of the 10 treated children left the hospital with functional immune systems and seemed to be living normal lives, but in 2002 some of the children developed a condition resembling leukemia, in which some of the white blood cells started growing uncontrollably. After discussing appropriate safeguards for using retroviral vectors, in April 2003 the FDA lifted the ban on the use of such vectors in blood stem cells.

After learning that gene therapy researchers were making unsupported claims and not reporting unexpected adverse reactions of gene therapy trials, the public support and willingness to volunteer for trials diminished. In 2004 the NIH and the FDA, the governmental agency that oversees gene transfer research, launched a Genetic Modification Clinical Research Information System (GeMCRIS), a public database containing accurate information related to ongoing research and clinical trials of gene therapy. GeMCRIS helps the NIH and FDA, researchers, physicians, patients, and the general public to monitor progress, share successes, and address concerns related to safety of gene transfer research.

In 2003 researchers at the University of California at Los Angeles reported the successful delivery of genes across the blood-brain barrier, opening the door for gene therapy treatments for disorders such as Parkinson's or Alzheimer's disease. They used liposomes coated with a polymer called polyethylene glycol as the vector. Another avenue of active research involves RNA interference, a new potential type of gene therapy that uses double-stranded RNA to block the production of certain proteins. This form of gene therapy may be useful for disorders such as Huntington's disease, in which the symptoms result when the presence of a mutant protein rather than the simple lack of a protein's ability to function causes disease symptoms. Recent successes in gene therapy include the following:

- In 2005 the delivery of the *Atoh1* gene, which stimulates hair cell growth, into the cochlea of deaf guinea pigs using adenovirus triggered growth of the hair cells and restored 80 percent of the hearing to the animals.
- In 2006 two adults were successfully treated for a myeloid disease with gene therapy.
- Also in 2006 patients with advanced metastatic myeloma were treated with genetically engineered immune cells that were altered to target and attack cancer cells.
- In 2007 the delivery of two tumor suppressing genes via lipid-based nanoparticles reduced the number and size of lung cancer tumors in mice.
- Also in 2007, human gene therapy trials for a type of inherited childhood blindness commenced.

Though gene therapy has suffered many difficulties and setbacks, the potential benefits are worth continued research. Among the attraction is the promise of curing disease by attacking the underlying cause, rather than simply treating the signs and symptoms.

See also CLONING OF DNA; GENE EXPRESSION; GENETIC DISORDERS; POINT MUTATIONS; RECOMBINANT DNA TECHNOLOGY.

FURTHER READING

Center for Biologics Evaluation and Research, U.S. Food and Drug Administration, Department of Health and

Human Services. "Cellular and Gene Therapy." Available online. URL: http://www.fda.gov/cber/gene.htm. Updated July 26, 2007.

Judson, Horace Freeland. "The Glimmering Promise of Gene Therapy." *Technology Review* 109, no. 5 (November 2006): 40–47.

Panno, Joseph. *Gene Therapy: Treating Disease by Repairing Genes.* New York: Facts On File, 2004.

Preuss, Meredith A., and David T. Curiel. "Gene Therapy: Science Fiction or Reality?" *Southern Medical Journal* 100, no. 1 (2007): 101–104.

U.S. Department of Energy Office of Science, Office of Biological and Environmental Research, Human Genome Program. "Human Genome Project Information: Gene Therapy." Available online. URL: http://www.ornl.gov/sci/techresources/Human_Genome/medicine/genetherapy.shtml. Updated September 19, 2008.

genetic disorders The human genome contains more than 3 billion nucleotide base pairs, and each nucleated cell, except gametes, contains 46 total chromosomes. Abnormalities can occur at the molecular level or the chromosomal level, leading to genetic disorders. Some genetic disorders have been passed down numerous generations while others result from errors that occur during gamete formation. Genetic disorders can be classified into four main types: single-gene disorders, multifactorial disorders, chromosomal disorders, and mitochondrial disorders.

SINGLE-GENE DISORDERS

More than 6,000 genetic disorders characterized to date result from the absence of a functional protein product from one gene. A mutation at that locus might result in the production of abnormal protein that cannot perform its cellular function, the gene might be expressed at inappropriate levels (either too much product or too little product), or the gene might be missing altogether. Many single-gene disorders involve a mutant protein product resulting from an altered nucleotide sequence within the coding region of the gene, and therefore an altered amino acid sequence within the protein. These so-called monogenic disorders follow the simple Mendelian patterns of autosomal dominant, autosomal recessive, or X-linked inheritance. Autosomal disorders are encoded for by genes located on any of the 22 chromosomes other than the X or Y chromosome. Genes responsible for X-linked disorders are found on the X chromosome.

In autosomal dominant disorders, a person needs only to receive one mutant gene to inherit the disorder. If one parent has a mutant copy, 50 percent of the gametes from that parent will contain the mutant version of the gene. Thus, there is a 50 percent chance that any child born to the affected parent will have the same disorder. Huntington's disease follows an autosomal dominant pattern of inheritance and affects a person's ability to think, talk, and move by destroying parts of the brain. The responsible gene encodes the protein huntingtin, which helps direct the transport of secretory vesicles. The gene normally contains between 10 and 26 repeats of the triplet CAG. In people who have the disorder, the repeat occurs too many times, more than 40, and affects the protein's function. In the case of Huntington's disease, a person may reach middle age before exhibiting any symptoms and may have children before knowing he or she has a mutant gene. Other examples of autosomal dominant genetic disorders include neurofibromatosis, which causes the growth of noncancerous tumors called neurofibromas; achondroplastic dwarfism, which causes short stature; and polydactyly, which is characterized by the presence of extra digits.

Autosomal recessive conditions only manifest when a person inherits two mutant copies of the gene. If even one copy of the normal gene is present, the person is said to be a carrier and exhibits the normal phenotype (the observable characteristics and properties of an organism as determined by a individual's genes and their interaction with the environment). Someone who is a carrier has a 50 percent chance of contributing the mutant gene to offspring. When both parents are carriers, the probability that a child will receive two mutant alleles and express the mutant phenotype is 25 percent. The probability of the child's being a carrier but not having the disorder is 50 percent, and the probability that the child inherits two normal alleles is 25 percent. Sickle-cell disease, characterized by misshapen red blood cells, is an example of an autosomal recessive disorder. The protein hemoglobin carries oxygen throughout blood circulation within red blood cells, which are normally flexible. They must be able to travel through tiny capillaries that penetrate the body tissues. A single nucleotide replacement causes a different amino acid at the sixth position in one of the hemoglobin polypeptides. This causes a sickling of the red blood cells and hinders their ability to carry oxygen to vital organs efficiently. The red blood cells have a shorter life span; thus people who have sickle-cell disease suffer anemia, and when the blood cells become stuck in blood vessels, episodes of severe pain can occur. Other autosomal recessive conditions include cystic fibrosis, adenosine deaminase deficiency (ADA), phenylketonuria, and galactosemia.

X-linked disorders are encoded for by genes located on the X chromosomes. Normally, people with two X chromosomes are female, and people with one X chromosome and one Y chromosome are male.

A mutation in the hemoglobin protein causes sickle-cell disease, characterized by sickle-shaped red blood cells. (*Eye of Science/Photo Researchers, Inc.*)

The inheritance patterns of X-linked disorders, which are most often recessive, are unique since females have two copies of all the X-linked genes and males only have one. Because of this, X-linked disorders more commonly afflict males, who will have the disease even if they inherit only a single copy of a recessive mutant gene. Mothers contribute only X chromosome to their offspring, whereas fathers contribute either X or Y chromosomes equally. Fathers who are carriers of a mutant allele will pass the allele to half of their daughters but to none of their sons, since they contribute a Y chromosome to their sons. Mothers who are carriers will pass the mutant gene to half of their offspring, whether male or female. Because females will also receive an X chromosome from their father, they usually also get a normal copy of the gene. But sons only receive the one X, so 50 percent of males born to a carrier mother will inherit an X-linked disorder. One example is hemophilia A, a disease in which afflicted individuals have a reduced ability to clot blood because of the lack of a key factor involved in the clotting process. Because clotting is slow, even minor injuries are serious. Duchenne muscular dystrophy is an X-linked neuromuscular condition in which afflicted individuals have a mutant dystrophin gene and suffer progressive muscular weakness and a shortened life span.

MULTIFACTORIAL OR POLYGENIC DISORDERS

Multifactorial genetic disorders, also called polygenic disorders, result from numerous mutations and often also have an environmental component. Breast and ovarian cancer exemplify many aspects of multifactorial disorders. Between 5 and 10 percent of breast and ovarian cancers are believed to be inherited. Two genes associated with these cancers are BRCA1, located on chromosome 17, and BRCA2, located on chromosome 13. The BRCA gene products are involved in DNA and chromosomes repair, so individuals who have mutant versions accumulate mutations and damaged chromosomes, because cancer results from the accumulation of mutations that ultimately compromise the regulation of the cell cycle, leading to uncontrolled cell growth and division and potentially cancer. Other examples of complex multifactorial genetic disorders include heart disease, hypertension, Alzheimer's disease, diabetes, colon cancer, and obesity.

CHROMOSOMAL DISORDERS

In chromosomal disorders, pieces of chromosomes can be duplicated or missing; an incorrect number of genes can be encoded on the segment or whole chromosomes can be present in too many or too few copies as a result. Because many genes exist within a section of or on an entire chromosome, the effects are severe and often lead to spontaneous miscarriages early in a pregnancy. Some chromosomal disorders do result in live births, though they may reduce the life span of afflicted individuals. Aneuploidies are conditions in which the number of chromosomes is abnormal. Human aneuploidies include trisomy 21 (47, XX or XY, +21), trisomy 18 (47, XX or XY, +18), trisomy 13 (47, XX or XY, +13), Turner syndrome (45, X), Klinefelter syndrome (47, XXY), triple-X female (47, XXX), and XYY karyotype (47, XYY). The most common is trisomy 21, commonly known as Down syndrome, which occurs in about one in 700 live births and is characterized by moderate to severe mental retardation, slanted eyes, a broad short skull, and broad hands with short fingers. Aneuploid disorders result from chromosomal nondisjunction during meiosis, the failure of a pair of homologous chromosomes to separate during the first meiotic division. Maternal age is one factor that increases the risk of nondisjunction in gametes.

Other chromosomal abnormalities result when a segment of a chromosome is absent or duplicated. In Williams syndrome, a portion of chromosome number 7 is missing as a result of breakage during gamete formation. Among other genes, the deleted region includes the gene for the protein elastin, which is important in the formation of blood vessels; thus individuals who have Williams syndrome suffer from heart defects and circulatory problems. Other symptoms of this rare disorder include mental retardation and distinguishing facial features. Cri du chat (meaning "cry of the cat" in French) syndrome is a disorder

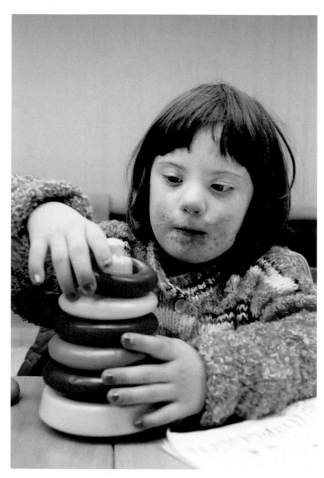

Individuals with Down syndrome have distinguishing facial morphologies including a round head with slanted eyes and a flat nose. *(PhotoCreate, 2007, used under license from Shutterstock, Inc.)*

resulting from a deletion in chromosome number 5. Children born with this disorder have a distinctive cry resulting from abnormal larynx development.

MITOCHONDRIAL DISORDERS

Mitochondrial disorders result from mutations in mitochondrial proteins. The mitochondria are organelles that carry out cellular respiration, the metabolic processes by which cells break down energy-rich organic molecules to extract energy from them. Mitochondria contain their own DNA, a closed, circular form resembling that found in prokaryotic organisms, but this DNA only encodes approximately 13 of the 1,000 proteins found in the mitochondria. Because the egg contributes most of the cytoplasmic contents, including organelles, to an embryo, mitochondrial genes are passed to children through the mother. Symptoms of mitochondrial diseases are similar whether the mutation is on a gene encoded for by the mitochondrial DNA or by nuclear DNA. The range and severity of symptoms vary tremen-

dously as do the genetic causes, complicating diagnosis. Symptoms may include brain dysfunctions (such as seizures, mental retardation, and developmental delays), neurological problems, heart defects, hearing or vision deficits, short stature, diabetes, kidney disease, and metabolic problems.

See also CANCER, THE BIOLOGY OF; DEOXYRIBONUCLEIC ACID (DNA); GENE EXPRESSION; GENETICS; GENOMES; INHERITANCE; POINT MUTATIONS; REPRODUCTION.

FURTHER READING

Centers for Disease Control and Prevention. "Single Gene Disorders and Disability." Available online. URL: http://www.cdc.gov/ncbddd/single_gene/default.htm. Accessed January 21, 2008.

U.S. Department of Energy Office of Science, Office of Biological and Environmental Research, and the Human Genome Program. "Genetic Disease Information—Pronto!" Available online. URL: http://www.ornl.gov/sci/techresources/Human_Genome/medicine/assist.shtml. Updated July 21, 2008.

genetic engineering Genetic engineering is the purposeful manipulation of genes or of an organism's genome. Using recombinant DNA technology, genetic engineers can cut up the deoxyribonucleic acid (DNA) from one organism, isolate genes, study them in vitro, modify them, replace them, or insert them into the genome of another organism. The result is often an organism with a different genotype, or genetic makeup, and sometimes a different phenotype, or observable characteristic. Genetic engineering has revolutionized agriculture, medicine, and biotechnology and significantly impacted other fields as well.

AGRICULTURAL APPLICATIONS

The purpose of agriculture is to produce crops, raise animals, or cultivate and prepare the products for market. For hundreds of years farmers have performed controlled breeding to alter the genetic makeup of plant and livestock populations. The careful selection and breeding of parents with desirable traits increase the probability that their offspring will exhibit the same characteristic or an increased degree of expression of that characteristic. This process of artificial selection and breeding is slow and does not guarantee results. In addition, because of independent assortment of alleles during meiosis and the random nature of fertilization, the individual offspring exhibit variable success. Another problem is that inbreeding, the mating of closely related individuals to preserve their desirable characteristics or to eliminate unfavorable characteristics, can lead to decreased vigor or health of the individuals and populations. Genetic engineering gives farmers more control in the manipulation of an organism's genetic makeup and achieves results more rapidly than traditional breeding.

Genetically modified organisms (GMOs) are organisms whose genomes have been artificially altered by recombinant DNA technology. Most of the genes inserted into crop plants function to increase yield, improve nutritional value, or protect crops against adverse conditions such as drought, frost, salty soil, disease, and pests. Hardier plants that can withstand environmental stresses and healthier plants with higher yields will help meet the demand for staple crops such as rice while reducing the need for additional acreage and freshwater. Genetic engineering may also reduce the need for chemical pesticides, which can leach into groundwater and harm ecosystems. Hawaiian papaya growers credit genetic engineering with saving the industry from the insect-borne papaya ring spot virus in the late 1990s. The soil bacteria *Bacillus thuringiensis* naturally produces a protein toxin that kills numerous lepidopteran insect pests when ingested. Genetic engineers have inserted the gene that encodes this toxin into several crops, including corn, cotton, potatoes, and tomatoes, increasing the yields and reducing the need for chemical pesticides. *Agrobacterium* is a type of bacterium that naturally causes tumor formation in plants but is used as a tool for agricultural biotechnology because it can transfer DNA into plant cells via a plasmid called Ti (for tumor-inducing). For such purposes, the tumor causing genes are removed from the plasmid, and the genes that produce the desired trait are inserted. After reintroduction of the recombinant plasmid into *Agrobacterium*, the plants are purposely infected with the bacteria. When the process is successful, the plant genome takes up the DNA and expresses the desired trait.

One specific recent example of genetic engineering of crops involves trypsin modulating oostatic factor (TMOF), a hormone from mosquitoes that acts as a natural insecticide and has no effect on humans. Scientists have engineered a harmless strain of tobacco mosaic virus to contain this gene. Plants inoculated with this virus start producing TMOF, which prevents insects that ingest it from being able to digest food. As a result, the insects that eat the leaves of the virus-infected plant starve to death. Another potential benefit of genetically engineering crop plants is the elimination of common allergy-inducing substances in foods. For example, many young children and infants are allergic to soy, which is a common ingredient in many processed foods, including baby formula and cereals. Researchers are exploring means of silencing the gene products from soy that commonly trigger allergic reactions.

Crops can also be engineered to make them more nutritious. Vitamin A deficiency, which can lead to blindness, is common in poorer regions of Asia, but the insertion of genes for vitamin A production into rice crops may help to alleviate this problem. The insertion of genes for the production of healthy fatty acids into plants and animals is another potential beneficial application of genetic engineering to agriculture.

Agriculturists have also used genetic engineering to improve the quality of farmed animals. For example, aquaculturists genetically engineered salmon to produce greater than normal levels of growth hormones, so the fish grow large enough to sell faster. Scientists in the United States and Japan claim to have engineered cows so they are resistant to bovine spongiform encephalopathy (BSE). Also known as mad cow disease, BSE causes progressive, fatal nervous system degeneration in cattle, and the ingestion of contaminated meat products from cattle with BSE can lead to Creutzfeldt-Jacob disease, a similar neurodegenerative disease in humans.

Some people worry about the safety and possible negative environmental impact of genetically engineered foods. The U.S. Food and Drug Administration, the Environmental Protection Agency, and the U.S. Department of Agriculture work together to oversee the introduction of genetically modified plants into the environment and the sale of genetically modified foods, which have been declared to be as safe for consumption as their traditionally bred counterparts. The first genetically modified food entered the market in 1994—the Flavr Savr tomato, developed by the biotech company Calgene. The tomato contained a gene that inhibited the softening process that naturally occurs during ripening, so they could theoretically be vine-ripened for better flavor and then shipped without bruising and rotting. (In reality, the original Flavr Savr tomato was more difficult to ship than anticipated, thus was withdrawn from the market a few years later.) In the United States, soybeans, corn, and cotton are the major genetically engineered crops, and their products are found in substances such as cornstarch, soy protein, and canola oil. Estimates of the percentage of processed foods that contain these or other products from genetically engineered plants available in U.S. grocery stores range from 70 to 75 percent.

APPLICATIONS TO MEDICINE

Genetic engineering has already benefited the field of medicine in numerous ways, and the potential for future applications continues to grow. Because of genetic engineering, pharmaceutical companies can produce many drugs more cheaply and in practically unlimited supplies. In 1982 the U.S. Food and Drug Administration approved the first drug produced as a result of genetic engineering—insulin, a hormone necessary for treating many diabetics. Previously, the hormone had to be extracted from the pancreas of slaughtered animals and thus was expensive and available in a limited supply, but now biotech companies manufacture large quantities using genetically altered microorganisms. Microorganisms can be quickly, easily, and cheaply grown in large quantities. Insertion of the gene for insulin production turns the microorganisms into insulin-producing factories. The hormone can then be isolated and purified from the bacterial cultures. Biotechnology companies now manufacture many other therapeutic agents by using genetically engineered microorganisms: human growth hormone for stimulating growth in children who do not adequately produce the hormone on their own, interferon for treating various viral diseases and cancers, the cytokine interleukin-2 for use in treating some cancers, blood clotting factors to treat hemophilia, tissue plasminogen activator to dissolve blood clots, erythropoietin to treat anemia, and tumor necrosis factor to treat tumors.

Pharmaceutical companies also use microorganisms to produce antibiotics, chemical substances that some microorganisms naturally produce to inhibit the growth of other specific microorganisms. Genetic engineering is employed to make recombinant vaccines to stimulate immunity against specific diseases. In contrast to whole-agent vaccines made from weakened or inactivated viruses, recombinant vaccines are made by recombinant DNA technology and only contain a few genes or proteins of the pathogen. Though they are not as effective as whole-agent vaccines, recombinant vaccines are generally safer.

Scientists have engineered some plants to make proteins called antibodies that help the immune system fight illnesses such as cancer and heart disease. Researchers have also inserted genes from viruses that cause measles, hepatitis B, Norwalk virus, cholera, and diarrhea into potatoes and bananas. The hope is that ingestion of these foods will stimulate people to develop a specific immune response against those viruses, just as if they had received a vaccine by injection.

Since scientists completed the sequencing of the human genome, the number of genes identified and associated with certain diseases has increased and will continue to do so. Once a mutation within a particular gene has been identified as the cause of a genetic disorder, researchers can explore treatment through gene therapy. Though medical researchers have not yet demonstrated the successful treatment of a genetic disorder through gene therapy, the potential exists and many clinical investigations are currently under way.

Xenotransplantation is the transplantation of a tissue or organ from one species to another species. Some tissues and organs from animals such as pigs can function effectively in humans, but the recipient's immune system often attacks the donated tissue or organ. Medical researchers are trying to use genetic engineering to create organisms that express molecules that will prevent the recipient's immune system from recognizing the transplanted tissue as being foreign, in hopes of preventing rejection.

In a recent example of genetic engineering with medical applications, scientists at the Roslin Institute in Edinburgh, Scotland, have modified chickens so they produce and secrete certain proteins in their egg whites. Specifically, these engineered chickens make proteins that are used to treat skin cancer and multiple sclerosis

OTHER BIOTECHNOLOGICAL APPLICATIONS

Broadly defined, *biotechnology* refers to any practical application of life science, and especially the application of technology to living organisms or the products of living organisms. Any use of genetic engineering thus falls under the umbrella of biotechnology. Other applications not discussed previously include the manipulation of organisms for the commercial production of various enzymes or chemicals.

The benefits of genetic engineering to basic research are enormous. Cloned genes, meaning genes that have been isolated for further use or study, are typically inserted into plasmids or viral vectors, which are maintained in bacterial strains. They allow molecular biologists to examine the gene structure, to determine its sequence, and to study how it is regulated. The researchers might use the altered bacteria to synthesize the protein for analysis of its structure and function or to observe the effects of various mutations. Some microorganisms naturally digest chemicals, such as spilled oil, that are considered pollutants. Bacteria may be genetically engineered to enhance their ability to degrade such chemicals for bioremediation purposes. Genetic engineering may also aid in the recovery of threatened or endangered species by inserting genes that make them heartier or more resistant to disease.

In January 2008 Craig Venter announced that a team of scientists at the J. Craig Venter Institute in Rockville, Maryland, had successfully created an entire bacterial genome from chemicals in a test tube. They have not yet been able to insert the genome into a living organism, where it would serve as the hereditary information. In 2002 scientists had created and assembled an active poliovirus in vitro, but viruses are not considered living organisms. This achievement of manufacturing an entire bacterial genome puts genetic engineers one step closer to creating synthetic forms of life, a pursuit known as synthetic biology. Applications of synthetic biology include designing organisms to perform specific tasks, such as bioremediation of a certain pollutant, and helping biologists attain a better understanding of the minimal requirements for life.

See also AGRICULTURE; BIOINFORMATICS; BIOTECHNOLOGY; CLONING OF DNA; DEOXYRIBONUCLEIC ACID (DNA); DNA SEQUENCING; MOLECULAR BIOLOGY; RECOMBINANT DNA TECHNOLOGY.

FURTHER READING
Grace, Eric S. *Biotechnology Unzipped: Promises and Realities.* Rev. 2nd ed. Washington, D.C.: Joseph Henry Press, 2006.
Spangenburg, Ray, and Diane Kit Moser. *Genetic Engineering.* New York: Benchmark Books, 2004.
U.S. Food and Drug Administration, Center for Food Safety and Applied Nutrition. "Biotechnology." Available online. URL: http://www.cfsan.fda.gov/~lrd/biotechm.html. Updated November 4, 2008.
Wexler, Barbara. *Genetics and Genetic Engineering.* Farmington Hills, Mich.: Thomson Gale, 2006.

genetics Genetics is the branch of the life sciences concerned with heredity, the transmission of characteristics down generations, and variation, the structural or functional divergence of characteristics among organisms. The smallest unit of heredity is the gene, a segment of deoxyribonucleic acid (DNA) consisting of a specific sequence of nucleotide subunits that encodes information for building biomolecules such as proteins. The types of proteins that a cell expresses determine the functions a cell can perform. Since cells are the basic unit of life, an organism's genes ultimately determine many of the characteristics an organism will possess, including traits inherent to that species, though environment also plays a significant role. Genetic information passes from a parent cell to daughter cells during cellular reproduction and from parents to offspring via DNA, which exists in cells in the form of chromosomes. Genetic variation among individuals results from differences in the alleles, or forms of a gene, which are simply different sequences of nucleotides within a gene. These variations among organisms can lead to adaptations upon which natural selection acts to drive evolution.

HISTORY OF GENETICS
For thousands of years, humans have recognized and exploited concepts of genetics. People observed that offspring often resembled their parents and extended this phenomenon to efforts growing crops and domesticating animals. Society established plant and animal breeding as tools for increasing the quantity and quality of food production long before the term *genetics* was coined. In 1665 the English scientist Robert Hooke discovered and described cells in his book *Micrographia,* but more than 200 years passed before scientists recognized their role in inheritance or variation within or between species. In 1833 the British botanist Robert Brown discovered cell nuclei, spots within plant cells. Today biologists know that nuclei contain and protect DNA, the genetic information. In the late 1930s, the German botanist Hugo von Mohl described mitosis, the duplication of cellular nuclei, and Theodor Schwann and Matthias Schleiden articulated the cell theory, stating that cells are the basic unit of living organisms. In 1858 Rudolf Virchow articulated that cells arise from preexisting cells.

During the 1860s, an Austrian monk named Gregor Mendel, now considered the father of classical genetics, performed numerous studies on inheritance in pea plants. His work led to the description of dominance, the law of segregation, and the law of independent assortment. Though these concepts form the basis for understanding transmission genetics, his work was largely unnoticed until 1900. During this time, Walther Flemming in Germany found a structure that bound to basic dyes and named it chromatin. Flemming also observed threadlike strands splitting and separating during mitosis. These strands were later called chromosomes, and now Flemming is considered the father of cytogenetics. The German biologist Oskar Hertwig first described meiosis in 1876, the Belgian zoologist Edouard Van Beneden described the actions of chromosomes during meiosis in *Ascaris* (a type of roundworm) eggs in 1883, and the German biologist August Weismann noted that two cell divisions were necessary to create haploid cells that would fuse during fertilization. Between 1887 and 1890, Theodor Boveri published findings that supported that chromosomes are individual entities that persist throughout the cell cycle and that sperm and egg contribute equal numbers of chromosomes during fertilization. Thus by 1900 the events of mitosis and meiosis, even with respect to chromosomes, had been delineated, but because nobody was aware of Mendel's work, the connection between these cellular events and inheritance was not made.

In 1900 three biologists (Hugo de Vries, Carl Correns, and Erich von Tschermak) independently rediscovered Mendel's work from 1865. This led to a flurry of activity repeating Mendel's experiments and widespread recognition of their significance. In 1902 the American biologist Walter Sutton figured out that the behavior of chromosomes during meiosis mimicked Mendel's factors of heredity, and thus genes were located on chromosomes. The American geneticist Thomas Hunt Morgan observed crossing over (a process resulting in genetic exchange between chromosomes) during meiosis in fruit flies in 1911 and obtained experimental evidence for phenomena including sex linkage, genetic distance, and nondisjunction. This work clearly established the chromosomal theory of inheritance.

Researchers at the Rockefeller Institute, Oswald Avery, Colin MacLeod, and Maclyn McCarty, provided proof that the DNA component of chromosomes was the molecular carrier for genetic information in 1944, ushering in the era of molecular genetics. James Watson and Francis Crick revealed the structure of DNA in 1953, paving the way to determine how the molecule transmits genetic information between generations and how it encodes for the synthesis of proteins. With the discovery of restriction enzymes and the use of cloning vectors such as plasmids, recombinant DNA technology has taught biologists much about how genes confer characteristics on an organism and the effects of different types of mutations. Through genetic engineering, researchers have cloned genes and introduced them into other species, resulting in the large-scale production of biochemicals, transgenic animals, and genetically modified foods. Most recently, the field of genomics has grown explosively, culminating in the mapping of numerous genomes that will further advance the applications of biological and medical research.

SUBDISCIPLINES OF GENETICS

The broad field of genetics can be divided into a few major subdisciplines, each encompassing several fields of specialty: classical genetics, molecular genetics, and evolutionary genetics. Classical genetics includes topics concerned with the transmission of traits from one generation to the next as described by the chromosomal theory of inheritance. Genes exist as physical entities on chromosomes and segregate with the chromosomes during the first division of meiosis. In sexual reproduction, gametes (eggs and sperm) each contain one-half of the normal number of chromosomes (they are said to be haploid), so when they unite during fertilization, the normal number is restored, and two copies of each type of chromosome exist (a condition called diploid). The members of a pair of chromosomes, called homologues, contain all of the same genes, but they can contain different alleles, or forms of those genes. In order to create haploid cells from diploid cells during gamete formation, the homologous chromosomes pair up, and an event called crossing over occurs; it can result in recombination, the exchange of pieces or

sections of a chromosome with its homologue. In the end, the haploid eggs contain random combinations of the chromosomes from the mother, and the sperm cells contain random combinations of chromosomes from the father. The unique combination of chromosomes an individual receives from his or her parents determines the biological gender and the allele combinations (the genotype) that the offspring will possess. The collective effect of all the specific alleles an individual inherits and the manner in which the gene products interact will affect that individual's characteristics, or phenotype. Classical geneticists study the processes by which all of these events occur.

Cytogenetics is a subfield of classical genetics. Cytogeneticists study genetics at the cellular level, in other words, chromosomes. They study the processes involved in chromosomal duplication and division during mitosis and meiosis, both when the events occur normally and when problems, which may lead to aberrations in chromosomal structure or number, arise. Many abnormalities of chromosomal number, such as too many or too few, do not result in viable offspring. Abnormalities in the number of sex chromosomes are viable more often than abnormalities in autosomal chromosomes (the ones not involved in sex determination). In humans, the sex chromosomes are called the X and the Y chromosomes. Genes located on these are said to be sex-linked and follow unique, gender-specific patterns of inheritance.

Classical genetics also encompasses gene mapping studies. The farther apart two genes are along the length of a chromosome, the more likely it is that recombination will occur between them. Geneticists can use the frequency with which two genes are inherited together to determine the genetic distance between them, which in turn can be used to map the position of various genes linked on the same chromosome.

Another major subdiscipline of genetics is molecular genetics, which focuses on genetics at the molecular level. Molecular geneticists study the structure and function of nucleic acids and the molecular events involved in synthesizing new DNA, called DNA replication, and in the expression of genes, a process involving multiple steps and resulting in the synthesis of new, functional proteins. The first step in the expression of a protein is transcription, the synthesis of a messenger ribonucleic acid (mRNA). As the first step in the process of making proteins, this step is highly regulated. In eukaryotic cells, the mRNA is processed and modified before transportation to the cytoplasm. Next, ribosomes assemble a polymer of amino acids based on the sequence of ribonucleotides in the mRNA transcript. This new molecule, called a polypeptide, undergoes folding and sometimes additional modifications or joins with additional polypeptide chains to become a functional protein. Changes in the DNA sequence, called mutations, may occur spontaneously or as a result of the action of chemical agents or radiation exposure. Cells have mechanisms for repairing molecular mutations to the DNA but are not always 100 percent effective. If not repaired, the mutations may affect the sequence of the final protein and therefore its function. All aspects relating to the regulation and synthesis of functional proteins are of interest to molecular geneticists.

The major subdiscipline of evolutionary genetics encompasses the subfields of quantitative and population genetics, which involve the examination of frequencies of certain alleles in a population and ways they change over time. Changes to allele frequencies within a population can occur over time as a result of biological processes including mutation, migration, and natural selection in addition to circumstances such as population size and randomness of mating. In stable populations, allele frequencies may reach equilibrium, a situation allowing one to make predictions concerning the genotypic and phenotypic frequencies of alleles and traits. Evolutionary geneticists also study the rate of evolutionary change and can use this information to establish the degree of relatedness among different species and to draw conclusions about the history of biological evolution.

APPLICATIONS OF GENETICS

Genetic research complements the research performed by other kinds of biologists. Geneticists observe organisms that have mutations and look for the effects of the improperly functioning genes. This approach allows for the determination of the function of the gene product as well as its importance and relationship to other genes—for example, is the protein crucial for a specific anatomical or physiological characteristic, is the mutation dominant or recessive, or does the gene interact with other genes? This information helps provide a more complete picture of how organisms develop, reproduce, and perform other life functions. The advent of genetic engineering and recombinant DNA technology has allowed researchers to examine these questions more directly. They can isolate specific genes and control the conditions under which the gene product and its action are studied.

Genetic research impacts the field of medicine tremendously. Using knowledge of cytogenetics and molecular genetics, physicians can diagnose genetic diseases caused by chromosomal abnormalities and molecular mutations. Genetic counselors educate couples and assist them in understanding their risks of conceiving a child who has a genetic disorder on the basis of family medical histories or gene testing. Research related to the cell cycle and its regulation has led to a better understanding of diseases such

as cancer, which results from the accumulation of many mutations to DNA. With the human genome mapped, medical researchers will be able to relate many more specific genes and mutations to certain medical disorders or conditions. Genetic research in other organisms has also improved the ability of physicians to treat human disease. For example, knowing the mechanisms by which bacteria develop resistance to certain antibiotics and the relative frequencies of the genes in different types of bacteria that confer resistance helps a physician decide which medications to prescribe.

Agriculture and horticulture have also benefited immensely from an understanding of genetics. Controlled breeding programs have been used for centuries to improve yields, such as more wool from sheep, greater quantities of milk from dairy cows, or more grain per acre of wheat. The tools of molecular genetics allow agriculturists to use molecular markers such as restriction fragment length polymorphisms to follow the transmission of different traits without having to wait a season for a plant to grow and mature to the point where the characteristic can be directly observed, saving time and increasing efficiency. They can also use recombinant DNA techniques to transfer genes that make a plant more resistant to harsh environmental conditions or to pests, increasing vigor and yields.

Forensic analyses often rely on genetic techniques to investigate crimes. For example, detectives can collect blood or other bodily fluids from a crime scene as evidence that a suspect was present because every person has a unique DNA fingerprint, left behind in samples containing nucleated cells. These same fingerprints can be used to determine paternity or the degree of relatedness of individuals.

The most important benefit of genetic research may be attaining a greater understanding of life itself. Many cell and molecular characteristics and events are shared by all organisms, and biologists can use the degree of similarity to construct phylogenies. All living things use DNA to carry genetic information from one generation to the next. The steps involved in protein synthesis (transcription of DNA to make RNA, followed by translation using the genetic code to build a polypeptide) are very similar even across kingdoms. Vertebrate animals, plants, bacteria, fungi, protists, archaeans, and even viruses share a practically universal genetic code, meaning the same nucleotide codons encode for the same amino acids. The field of genetics has granted scientists insight into life itself and the evolutionary processes that led to the development of the species present in the world today.

See also AVERY, OSWALD; CELL BIOLOGY; CHROMOSOMES; CLONING OF DNA; CRICK, FRANCIS;

DEOXYRIBONUCLEIC ACID (DNA); DNA FINGER-PRINTING; DNA SEQUENCING; FRANKLIN, ROSALIND; GENE EXPRESSION; GENE THERAPY; GENETIC DISORDERS; GENETIC ENGINEERING; GENOMES; HOOKE, ROBERT; INHERITANCE; MACLEOD, COLIN MUNRO; MCCARTY, MACLYN; MCCLINTOCK, BARBARA; MENDEL, GREGOR; MOLECULAR BIOLOGY; MORGAN, THOMAS HUNT; POLYMERASE CHAIN REACTION; RECOMBINANT DNA TECHNOLOGY; VARIATION, GENETIC VARIATION; VIRCHOW, RUDOLF; WATSON, JAMES D.; WILKINS, MAURICE H. F.

FURTHER READING

Carlson, Elof Axel. *Mendel's Legacy: The Origin of Classical Genetics.* Cold Spring Harbor, N.Y.: Cold Spring Harbor Laboratory Press, 2004.

genomes A genome is the complete DNA content of an organism. Characterization of an organism's genome includes a description of the number of chromosomes in addition to the size of the genome expressed as number of kilobase (kb) pairs (one Kilobase equals 1,000 base pairs) in a haploid genome. Eukaryotic organisms generally have between two and 10 times as many genes as prokaryotic organisms, and their genomes are often thousands of times larger. The amount of deoxyribonucleic acid (DNA) in a cell, however, is not necessarily related to the evolutionary complexity of the organism.

Genomics refers to the discipline of life science concerned with the characterization of genomes and the methods used to achieve that goal. Scientists are actively working toward obtaining complete genome sequences for many different species. According to Genomes OnLine Database (GOLD), as of January 2008, almost 700 completely sequenced genomes had been published, including numerous viruses, bacteria, and archaeans as well as higher organisms including the yeast *Saccharomyces cerevisiae*, the nematode *Caenorhabditis elegans*, the fruit fly *Drosophila melanogaster*, the mouse *Mus musculis*, and humans. This information will be useful for numerous applications such as identifying genes that cause disorders, examining evolutionary relationships among organisms, determining the function of proteins, and achieving a better understanding of life at the molecular level.

PROKARYOTIC GENOME ORGANIZATION

The majority of DNA in a prokaryotic genome exists in one closed, circular molecule of DNA and includes between 1,000 and 10,000 genes. The size ranges between 580 kb and 1.0×10^4 kb, with most prokaryotic genomes averaging 4–5×10^3 kb, and most single genes averaging 1,000–2,000 base pairs (bp).

The widely studied bacterium *Escherichia coli* has 4.5×10^3 kb. Obligate intracellular parasites have characteristically small genomes since they rely on host organisms to fulfill many metabolic needs and therefore do not need to encode or express the genes necessary for those functions. Exceptions to the norm do exist. The genome of *Bacillus megaterium* is very large, 3×10^4 kb. The bacterium responsible for Lyme disease, *Borrelia burgdorferi*, has linear chromosomes. The genome of the organism that causes cholera, *Vibrio cholerae*, consists of two circular chromosomes, and the archaean *Methanococcus jannaschii* has three. Archaean chromosomes exhibit a range in chromosome size from about 500 kb to 5.8×10^3 kb, similar to that of bacteria. Compared to eukaryotic genomes, prokaryotic genomes are characterized by very little junk DNA, the term for DNA that does not encode for anything. The gene density in prokaryotes is high—approaching one gene per kilobase of DNA in *E. coli*. Introns, noncoding sequences found within the coding region of a gene, are rare in bacteria and only found in transfer RNA genes in archaeans. Topoisomerases, enzymes that wind up and twist the DNA, package the prokaryotic chromosome, which attaches to the inner cell membrane. Few proteins remain associated with the chromosome, a condition described as naked.

Many bacteria also have plasmids, small extra-chromosomal pieces of DNA that can carry additional genes. A few linear plasmids also exist, and these have specialized ends to protect the DNA from degradation. Plasmids replicate independently of the bacterial chromosome and can be present in multiple copies, ranging from one to hundreds of copies per cell. Copy number is a property of the type of plasmid, and genes encoded by the plasmid control the number of copies present in a cell. The size of plasmids averages between 2,000 and 100,000 base pairs, and they often encode for genes that confer antibiotic resistance. The range of potential host bacteria varies among plasmids: some have narrow host ranges, and some can exist and replicate in many different bacterial types. Bacteria can transfer plasmids from one cell to another through bacterial conjugation, a property that the plasmid being transferred encodes. Plasmid DNA can also be introduced into bacteria cells by transformation, a process in which naked molecules of DNA are taken up by competent recipient cells.

Prokaryotic chromosomes contain operons, organized units of several genes that are transcribed as one molecule of messenger ribonucleic acid (mRNA). An operon typically includes regulatory protein binding sites and a few structural genes. The structural genes are usually functionally related; for example, they may all encode proteins necessary for one par-

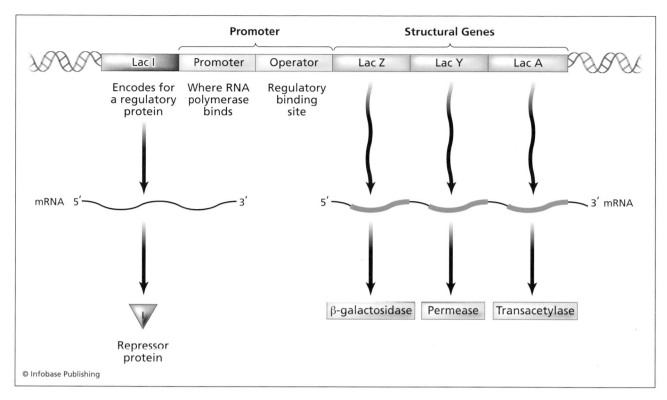

Bacterial genes are often organized into operons, sets of structural genes whose mRNA is synthesized as one molecule. The *lac* operon contains genes necessary for the utilization of lactose.

ticular metabolic pathway to proceed. By transcribing several of the genes onto a single mRNA, they are regulated as a unit by a single regulatory element. In bacteria, the lactose (lac) operon encodes for proteins necessary for the metabolism of lactose and includes the gene that encodes its own regulatory protein, the *lac* repressor.

EUKARYOTIC GENOME ORGANIZATION

Eukaryotic genomes vary in size by up to four orders of magnitude but typically contain between several million and several billion base pairs in a haploid genome. Examples of genome size in a few well-researched eukaryotic organisms are humans, 3.0×10^6 kb; *Drosophila melanogaster* (fruit fly), 1.7×10^5 kb; maize (corn) 2.0×10^6 kb; *Arabidopsis* (a small flowering plant) 7.0×10^4 kb; lily, 1.0×10^8 kb. The wide range in size is due to varied numbers of introns and repetitive sequences. The chromosomes of eukaryotic organisms are linear, occur in pairs, and contain thousands of genes each. The number of chromosomes varies among species and ranges from two to hundreds in polyploid organisms. Humans have 46 chromosomes, fruit flies have eight, and maize has 20. A karyotype reveals the structure of a genome at the chromosomal level, with the chromosomes paired with their homologues and arranged according to shape,

size, and centromere position. Within a species, particular genes always reside at the same gene locus, the position on a specific chromosome where a gene resides. With the exception of the roundworm *Caenorhabditis elegans*, genes are not organized into operons as in prokaryotes. Each mRNA transcript encodes for a single protein.

The cells of eukaryotic organisms contain mitochondria, organelles that function in cellular respiration. Plant and algal cells also contain chloroplasts that function in photosynthesis. Both mitochondria and chloroplasts possess their own DNA, believed to be derived from a prokaryotic endosymbiont. According to the endosymbiotic theory, eukaryotic mitochondria and chloroplasts evolved from smaller prokaryotic organisms that lived in an endosymbiotic relationship with a larger prokaryotic cell. The smaller organisms performed specialized tasks, cellular respiration in the case of mitochondria and photosynthesis in the case of chloroplasts, while living inside the larger cell. Eventually both cell types lost their ability to live independently. The DNA inside mitochondria and chloroplasts resembles a prokaryotic genome, being closed, circular, and naked. Many of the genes on the DNA from these organelles share sequence similarities with prokaryotic organisms, lending further support to the endosymbiotic theory. These genes encode for proteins that perform critical cellular functions, thus must not be omitted from discussions about eukaryotic genomes.

Though eukaryotic genomes are millions or billions of base pairs long, the majority of the DNA is noncoding, meaning it does not encode for the synthesis of a protein or RNA molecule. Humans, for example, have approximately 30,000 genes, with one gene averaging 27 kb in length, though the average-sized mRNA transcript is only 2,000 nucleotides long. Less than 2 percent of the human genome encodes instructions for the synthesis of proteins. In comparison, between 5 and 10 percent of the *Drosophila* genome encodes for proteins, and 85–90 percent of bacterial genomes are coding. Compared to that in prokaryotes, the overall gene density in eukaryotes is low, but eukaryotic genes are organized into clusters on the chromosomes.

REPETITIVE DNA IN EUKARYOTIC GENOMES

The sequence organization of eukaryotic genomes is very complex but can be generally divided into three main classes. The single-copy, functional genes encode for proteins necessary to carry out cellular functions. Spacer DNA, as the name implies, are regions of DNA that are not transcribed and do not have any characteristics of the known types of repetitive elements, described later. The third type of DNA is repetitive DNA—DNA that is present in multiple copies in the genome. Some repetitive DNA sequences are functional, and others have no known function.

Satellite DNA, which consists of highly repetitive DNA sequences, makes up about 5 percent of the human genome and 10 percent of the mouse genome. Density gradient centrifugation of eukaryotic DNA results in two bands, one main band and a second, smaller and less dense band. The lighter band results from satellite DNA, which contains a higher percentage of adenine-thymine base pairs than the rest of the DNA. The separation of the two bands on the gradient represents the difference in nucleotide composition. Prokaryotic genomes do not exhibit this phenomenon. The sequences found in the satellite band are present in heterochromatic regions surrounding the centromeres of chromosomes and in telomeres, the ends of chromosomes. These inactive regions are packed more tightly than regions of the chromosomes that contain actively transcribed genes; thus they appear darker when stained and viewed under a light microscope. Though the role of highly repetitive sequences is not known, the heterochromatic region near the centromeres binds the kinetochore, the structure to which spindle fibers attach in order to separate homologous chromosomes during mitosis and chromatids during meiosis. The sequence that composes the centromere regions has

been defined for several species and contains numerous repeats extending up to 1 million base pairs. In *Drosophila* the repeated sequence is 10 bp long, and in humans it is 170 bp. Telomeres also contain highly repetitive DNA. Telomeric DNA sequences contain short tandem repeats (six nucleotides long in humans) that provide stability to the chromosome. Other telomeric-associated sequences are found near and within the telomere. Their function is unknown.

Another type of repetitive DNA for which there is no known function is middle repetitive DNA, including tandemly repeated sequences and interspersed sequences. Variable number tandem repeats (VNTRs) are 15–100 bp long and repeat up to 1,000 times at locations throughout the genome, forming minisatellites. The human genome has about 30,000 minisatellite DNA loci. Because the number and the size of the repeats vary among individuals, minisatellite DNA is used in the technique of DNA fingerprinting. Microsatellites are also dispersed throughout the genome; they consist of shorter tandem repeats than minisatellites. For example, dinucleotides contain just two nucleotides, and they repeat between five and 50 times. The human genome has approximately 200,000 microsatellite loci.

Transposons are repetitive DNA sequences that can replicate and insert themselves into other locations within the genome. Short interspersed elements (SINEs) are less than 500 bp long and can have as many as 500,000 copies in a genome. One SINE family in humans, called the *Alu* family, makes up more than 5 percent of the genome. Long interspersed elements (LINEs) are similar to SINEs but are longer, between 1,000 and 5,000 kb, and occur in multiples of 20,000 to 40,000 in humans. LINEs are referred to as retrotransposons, because they use the enzyme reverse transcriptase to propagate themselves by first making a strand of DNA using a piece of RNA as the template, a strategy used by retroviruses.

Some middle repetitive DNA is functional, for example, genes that are present in multiple copies to increase their level of expression. Ribosomal RNA (rRNA) is one example of a multiple copy gene that occurs as tandem repeats in the nuclear organizer, the region of the chromosome set that is physically associated with the nucleolus, a dark-staining, spherical region located inside the nucleus. In contrast to the organization of the rRNA genes, some proteins or families of proteins are encoded by copies of genes that are dispersed throughout the genome. For example, actins are encoded by between five and 30 gene copies, keratins by more than 20, and the variable region of immunoglobulins (antibodies) is encoded for by 500 different genes. When members of multiple copy gene families become nonfunctional, they are called pseudogenes. These can also result from duplicated genes that have undergone destructive mutations. The presence and location and the sequence of pseudogenes in a species reveal information about evolutionary relationships.

GENOMICS AND PROTEOMICS

Genomics is the branch of biotechnology concerned with the genetic mapping and DNA sequencing of the genes and genomes of organisms, identifying the genes, developing organized databases of this information, and applying this knowledge. *Bioinformatics* refers to the development of hardware and software and the use of computers to acquire, store, organize, and analyze biological information such as genetic data. Many databases and software tools are freely available online to assist genetic researchers at public and private institutions. Several specialized fields within genomics explore different aspects of genome structure and function. Functional genomics focuses on how the genome determines the characteristics of individuals and species and how the DNA and proteins encoded by the DNA interact with each other within and among organisms and their environment. *Transcriptomics* refers to the large-scale study of mRNAs and the regulation of their synthesis.

A proteome is the entire set of proteins expressed in a cell, tissue, or organism, and proteomics is the study of their expression, regulation, structure, modification, cellular location, function, and interactions and the organization and application of this information using bioinformatics. One tool for exploring protein function is knockout studies, in which genes encoding certain proteins are inactivated in a living organism so the effect can be studied. While knockout studies are limited to model organisms, such as yeast, *Drosophila,* or mice, bioinformatics facilitates the comparison of sequences between humans and these model organisms. Thus information obtained in other species benefits human proteomics as well. Comparative genomics involves close examinations of the DNA sequences and patterns between genes and species. The aim of structural genomics is to construct three-dimensional models of proteins, which will complement functional studies. Traditional biochemistry techniques to separate and characterize proteins such as chromatography and gel electrophoresis are still useful, but the emergence of bioinformatics has greatly advanced proteomics. During the 21st century, the applications of the knowledge gained from genomics studies will not only improve human health care, but provide much deeper insight into many aspects of living systems ranging from how species evolve to how to sustain livable conditions within the Earth's biosphere.

See also CHROMOSOMES; DEOXYRIBONUCLEIC ACID (DNA); GENE EXPRESSION.

FURTHER READING
Campbell, A. Malcolm, and Laurie J. Heyer. *Discovering Genomics, Proteomics, and Bioinformatics.* 2nd ed. San Francisco: Benjamin Cummings, 2006.

Gerstein, Mark, and Deyou Zheng. "The Real Life of Pseudogenes." *Scientific American* 295, no. 2 (August 2006): 48–55.

Jeffreys, A. J., V. Wilson, and S. L. Thein. "Hypervariable Minisatellite Regions in Human DNA." *Nature* 314 (1985): 66–73.

Korenberg, J. R., and M. C. Rykowski. "Human Genome Organization: *Alu,* LINEs, and the Molecular Organization of Metaphase Chromosome Bands." *Cell* 53 (1988): 391–400.

Richards, Julia E., and R. Scott Hawley. *The Human Genome: A User's Guide.* 2nd ed. Burlington, Mass.: Elsevier Academic Press, 2005.

Ridley, Matt. *Genome: The Autobiography of a Species in 23 Chapters.* New York: HarperCollins, 2000.

U.S. Department of Energy Office of Science, Office of Biological and Environmental Research, Human Genome Program. "Facts about Genome Sequencing." Available online. URL: http://www.ornl.gov/sci/techresources/Human_Genome/faq/seqfacts.shtml. Updated September 19, 2008.

Geoffroy Saint-Hilaire, Étienne (1772–1844) French *Naturalist* The French naturalist Étienne Geoffroy Saint-Hilaire was a skilled comparative anatomist who put forth the principle of unity of composition, claiming that all animals shared an archetypical structural design. His anatomical research was highly regarded, but Geoffroy Saint-Hilaire's greatest influence was on the development of the study of pre-Darwinian evolution.

Étienne Geoffroy was born on April 15, 1772, in Étampes, near Paris. His father, Jean-Gérard Geoffroy, was a procurator at the tribunal and later became a judge. Étienne had 13 older siblings. He received the surname *Saint-Hilaire* as a child and later added it to his family name. At age 15 he began preparations to devote his life to the church, but the onset of the French Revolution in 1789 disrupted his plans. At his father's recommendation, he pursued the study of law. After receiving a law degree in 1790, he followed his own interests and studied medicine and science at the Collège du Cardinal Lemoine. During the revolutionary turmoil in 1792 Geoffroy attempted to free an imprisoned priest. Because of his efforts, in 1793, when Geoffroy was only 21, one of the priest's friends arranged for him to be appointed professor of vertebrate zoology at the Jardin des Plantes (the royal botanical gardens), which became the Muséum National d'Histoire Naturelle (National Museum of Natural History) in June 1793. There he became friends with Jean-Baptiste Lamarck, who was the museum's professor of invertebrates.

After a brief correspondence, Geoffroy invited the up-and-coming comparative anatomist Georges Cuvier to Paris, and they collaborated on many projects. Shortly after his arrival in 1795, Cuvier was appointed an assistant professor of animal anatomy at the museum. Together the two men discussed the notion that all animals were derived from a single type of animal, an idea that Cuvier later rejected. In 1796 Geoffroy formally proposed this principle of unity in his *Mémoire sur les rapports naturels des* Makis lémur *L*. (Memoir on the natural relations of the *Makis lemur* L.) He said, "It seems that nature is confined within certain limits and has formed living beings with only one single plan, essentially the same in principle, but that she has produced variation in a thousand ways in all her accessory parts." He initially applied this principle to mammals but then later extended it to vertebrates and eventually to invertebrates as well.

From 1798 to 1801 Geoffroy joined Napoleon's scientific staff in Egypt. He visited archaeological sites and collected mummified animals and humans for scientific research. His analysis of these materials was reported in *Description de l'Égypte par la commission des Sciences* (1808–24). One of his conclusions was that animals from 3,000 years ago were similar to living forms. Scientists who believed in the immutability of species used this finding to defend their viewpoint, though geologically speaking, 3,000 years is very short.

Geoffroy became a member of the French Academy of Sciences in 1807, and in 1809 he joined the zoology faculty at the University of Paris.

During the early 19th century, biologists held divided opinions on the nature of anatomical differences in animals. Formalists believed that form determined function and that all vertebrates developed from a single archetype and thus shared the same underlying organization, from which functions were derived. In contrast, the so-called functionalists thought that form developed as the manifestation of a particular function: in other words, form followed function. The controversy boiled down to whether organization or activity was more important. Cuvier was a functionalist who believed that God had created all organisms with the structures necessary for their survival in their particular habitats and that the form of each species was permanently fixed. Geoffroy was a formalist, as were Georges-Louis Leclerc, comte de Buffon, and Lamarck. He believed that species transformed over time. In Geoffroy's books, the two-volume *Philosophie Anatomique* (Anatomical philosophy, 1818–22) and *Histoire naturelle des mammifères* (Natural history of mammals, 1819), he described three laws that summarized his philosophy of anatomy. The law of connections stated that similar structures retained the same pattern of connections to other structures. The law of permanence said that new organs are not created, but structures are simply modified from preexisting structures. The law of balance required that when one organ developed, another degenerated or became reduced.

Geoffroy claimed that all vertebrate animals shared the same underlying anatomical organization with only slight modifications to accessory parts. He cited vestigial organs, body parts that are degenerate or imperfectly developed in comparison with ancestral forms or related species, and embryonic stages of development as proof that organization dictated function. Because he believed that all of an organism's habits, activities, and functions were the result of the animal's structure, he focused on the overall body plan or major anatomical features of different organisms. Varied structures in different animals were considered similar if they had the same pattern of organization. The modern term for this concept, *homology*, illustrates Geoffroy's and the formalists' viewpoint. Homologous structures of two different organisms are derived from the same structure in a common ancestor. For example, an arm of a human, a front leg of a dog, and a wing of a bat are all homologous. Though the structures appear different in these three animals, they are all organized in a similar manner, with bones in the same positions relative to other bones. Geoffroy outlined a set of rules for determining whether two varied structures were homologous.

In *On the Origin of Species* (1859), the British naturalist Charles Darwin used the concept of homologous structures to illustrate and identify evolutionary relationships. Modern evolutionary biologists do the same. Unfortunately, Geoffroy carried his examples of related connections too far and ended up incorrectly describing some homologous relationships that biologists have since dismissed as false. For example, he suggested the carapace of insects corresponded to the vertebrae.

A colleague named Robert Edmund Grant studied marine invertebrates in Edinburgh, Scotland, in the late 1820s. While examining mollusks, Grant found they had a pancreas and wrote to Geoffroy. This led Geoffroy to extend his unity principle to invertebrates. In 1829 a formalist paper was published. The authors, Laurencet (first name unknown) and Pierre Stanislas Mayranx, made structural comparisons between cephalopods (squids, cuttlefish, and octopuses) and vertebrates and proposed an account for a transition from cephalopods to fish. Cuvier thought this was ridiculous and tried to interfere with the Academy of Science's review of the paper. This led to a series of debates that took place in 1830 between Cuvier and Geoffroy regarding the unity of composition principle. Cuvier took advantage of this opportunity to point out several inaccurate descriptions of uniformities in structure. He declared that similarities in structure resulted from the need to perform similar functions; they were not due to common descent. Geoffroy defended his plan of uniform structure among animals, claiming that philosophically only a single animal existed. In *Principles de philosophie zoologique* (Principles of zoological philosophy), published in 1830, Geoffroy attempted to explain his debate with Cuvier and their conflicting views on the form of species.

Today, aspects of both formalism and functionalism permeate the fields of anatomy and physiology; thus the controversy was never resolved, but rather, the two perspectives were merged or synthesized. An organism's current form restricts or limits the possible modifications that may alter its function over time, but at the same time, the conditions of a particular environment strongly influence the functions or activities an organism exhibits. Similar environmental conditions can lead to the emergence of analogous structures that are not homologous. For example, the wings of a bat and the wings of a mosquito perform similar functions, but they are not derived from a single structure in a common ancestor, and their structural strategies are dissimilar.

Geoffroy is not considered an evolutionary biologist, but rather an accomplished comparative anatomist. He did, however, make comments later in his career that suggested he believed that the environment shaped the evolution of animals, a kind of precursor to Darwin's proposed theory of natural selection. Another claim made by Geoffroy, that morphological changes occurred in quick bursts, rather than slowly, foreshadowed the theory of punctuated equilibrium proposed by the American paleontologists Stephen Jay Gould and Niles Eldredge in 1972. Punctuated equilibrium theory states that evolution occurs in sudden, rapid spurts followed by long periods of no substantial change, and today punctuated equilibrium is considered an important process in large-scale evolutionary change.

Geoffroy became blind in 1840 and later suffered an attack that left him paralyzed. He resigned his chair at the museum in 1841, and his son, Isidore Geoffroy Saint-Hilaire, succeeded him. His health failed, and Étienne Geoffroy Saint-Hilaire died in Paris on June 19, 1844. His prominence and influence on 19th-century science and on other 19th-century scientists, including Charles Darwin, have engraved a permanent place for him in the history of life science. Even his adversaries such as Cuvier could only praise Geoffroy's talents for description and classification.

See also ANATOMY; BUFFON, GEORGES-LOUIS LECLERC, COMTE DE; CUVIER, GEORGES, BARON; DARWIN, CHARLES; EVOLUTION, THEORY OF; GOULD, STEPHEN JAY; LAMARCK, JEAN-BAPTISTE.

FURTHER READING

American Philosophical Society. "Étienne Geoffroy Saint Hilaire Collection (1811–1840)." Available online. URL: http://www.amphilsoc.org/library/mole/g/geoffroy.htm. Accessed January 25, 2008.

Guyader, Hervé le. *Geoffroy Saint-Hilaire: A Visionary Naturalist.* Translated by Marjorie Grene. Chicago: University of Chicago Press, 2004.

Travis, John. "The Ghost of Geoffroy Saint-Hilaire." *Science News* 148, no. 14 (September 30, 1995): 216–217.

germ theory of disease   The germ theory of disease states that microorganisms cause infectious diseases. Specific microorganisms cause specific diseases and can be spread directly from person to person or indirectly through shared contact with other objects. The germ theory of disease is one of the most important historical contributions of science to the welfare of society. Its acceptance opened the door to many other developments that advanced medicine: hygienic practices, aseptic techniques during surgery, development of vaccines and antibiotics. Knowledge and understanding of the causes of infectious diseases have saved millions of lives over the past 120 years.

Until the end of the 19th century, physicians were ignorant as to the cause of illness and disease. From the time of Hippocrates (460–370 B.C.E.), classical medicine taught that the body contained four humors (fluids) that must be kept in balance in order for a person to be healthy. The relative amounts of the four humors—black bile, yellow bile, phlegm, and blood—were also associated with one's personality. Different illnesses resulted from an imbalance in the levels of the fluids, a condition that could result from variations in diet or activity. For example, an excess of yellow bile caused illnesses associated with vomiting, which was considered the body's mechanism for reducing the levels of yellow bile to normal. Physicians attempted to restore the balance of these fluids by administering substances that induced vomiting and diarrhea or by forcing starvation or blood loss through the use of blood-sucking leeches or even intentional cutting of a vessel, procedures now considered primitive and barbaric as well as ineffective. Another hindrance to the advancement of medicine was society's belief that disease resulted from supernatural causes or divine powers. People believed that those who practiced witchcraft could cast a spell to make someone sick, that demons could cause illness by inhabiting or possessing individuals, and that God punished sinful behavior by causing disorders or diseases.

Numerous people contributed to the development of the germ theory of disease. A Dutch draper, Antoni van Leeuwenhoek, first discovered microorganisms in pond water and in 1674 and subsequently found them to be ubiquitous. Many people, scientists included, believed that these simple life-forms arose spontaneously from nonliving matter. This belief, called spontaneous generation, originated thousands of years before and offered an explanation as to why maggots appeared on rotting meat and rats appeared in garbage piles.

Until the 1840s, puerperal fever, also called childbed fever, killed so many women that many preferred to give birth on the streets rather than in a hospital out of fear of contracting this disease. Ignorant of the cause of this condition, physicians routinely delivered babies without first washing their hands, often after performing autopsies on women who had died of the condition. The Hungarian physician Ignaz Semmelweis demonstrated that hand washing significantly reduced the incidence of puerperal fever, a practice that reduced the mortality rate at his hospital from 12.24 to 2.38 percent.

In 1859 the French chemist Louis Pasteur performed the last of a series of experiments that debunked the theory of spontaneous generation. He showed that microorganisms are present in the air, carried by airborne dust particles, but do not arise from the air. Afterward, scientists accepted microorganisms as independent life-forms that reproduced to perpetuate their species. Pasteur also showed that microorganisms caused fermentation and putrefaction, and that microorganisms could be killed by heating or treatment with harsh chemicals.

An English physician named Joseph Lister was aware of both Semmelweis's and Pasteur's research. Lister made a connection between their work and his studies on gangrenous wounds, wounds of soft tissue that died and decayed, believed by most to be caused by the exposure of the tissues to chemicals in the air. Lister believed that microorganisms caused the tissues to rot and applied the knowledge shared by Semmelweis and Pasteur to eliminate the microorganisms from wounded tissues in hopes of decreasing gangrene after surgery. In 1867 Lister published his findings that spraying a solution of carbolic acid onto surgical instruments and onto wounds prior to and during surgery markedly decreased the incidence of gangrene. Today Lister is considered the father of antiseptic surgery.

Also during the late 1860s, Pasteur examined the causes of diseases in silkworms and demonstrated an association between particular microorganisms and specific diseases. He soon turned his efforts toward demonstrating a relationship between microorganisms and diseases in humans and large animals. He developed the first vaccines against cholera, anthrax, and rabies. Vaccines are injections of microorganisms or parts of microorganisms that prevent future infection of a disease by stimulating specific immunity.

The German physician Robert Koch studied many of the same diseases as Pasteur including anthrax and cholera. They each benefited from knowledge of the other's experiments, and competition between the two hastened the advances that led to the establishment of the germ theory of disease. In the 1870s, Koch outlined a set of procedures for ascribing a particular microorganism to a specific disease. First, one must isolate the suspected microorganism from a disease victim. Second, one must grow a pure culture of the isolated organism in the lab. Third, injection of the microorganism into a healthy individual must cause the same specific disease as that in the individual from whom the organism was first isolated. Last, the same organism must be isolated from the injected individual. Meeting these conditions, referred to as Koch's postulates, demonstrates that the isolated organism is the etiological agent for the particular disease. Koch performed experiments that helped to identify the causative organism for anthrax to be *Bacillus anthracis,* a relationship experimentally proved by Pasteur. Koch also determined the causative organism for tuberculosis to be *Mycobacterium tuberculosis* and for cholera to be *Vibrio cholerae.*

History credits both Pasteur and Koch for major contributions to the development and acceptance of the germ theory of disease. Since then, improvements in sanitation, public health mandates, the development of vaccines, the discovery and use of antibiotics, and other improvements in health care and medicine have ended the era in which infectious diseases caused the majority of all deaths. As a result, the average life expectancy has increased from slightly less than 40 years for someone born in 1850 to almost 80 years for someone born today.

See also ANTIMICROBIAL DRUGS; HOST DEFENSES; KOCH, ROBERT; LEEUWENHOEK, ANTONI VAN; MICROBIOLOGY; PASTEUR, LOUIS; SPONTANEOUS GENERATION; VACCINES.

FURTHER READING

Adler, Robert E. *Medical Firsts: From Hippocrates to the Human Genome.* Hoboken, N.J.: John Wiley & Sons, 2004.

Waller, John. *The Discovery of the Germ: Twenty Years That Transformed the Way We Think about Disease.* New York: Columbia University Press, 2002.

Golgi, Camillo (1843–1926) Italian *Cell Biologist*
Camillo Golgi was the first Italian to receive the Nobel Prize in physiology or medicine (1906). He made a true scientific breakthrough when he developed a method for staining neurons. His work made possible the elucidation of the innermost anatomy of the nervous system, which eventually led to a basic understanding of its function.

TRAINING AND CAREER

Camillo Golgi was born on July 7, 1843, at Corteno (now called Corteno Golgi), in the province of Brescia, in northwestern Italy. His father, Alessandro, was a physician, and Camillo also decided to study medicine. He graduated in medicine in 1865 from the University of Pavia with a thesis on hereditary factors in mental illness. After obtaining his medical degree, he worked for seven years at the Hospital of St. Matteo in Pavia, where he studied the cause of mental diseases. Meanwhile, he worked at the Institute of General Pathology with the experimental pathologist and histologist Giulio Bizzozero, who taught him histological techniques. Golgi never practiced medicine but devoted his life to medical research instead. He left Pavia briefly in 1872–76 to serve as chief physician at the Hospital for the Chronically Ill at Abbiategrasso, where he started investigating the nervous system using a microscope set up on the kitchen table in his small living quarters. He returned to Pavia and in 1876 was appointed professor of histology; then in 1881 he succeeded Bizzozero as the chair of general pathology at the University of Pavia, and he remained there for the rest of his career. Many distinguished scientists spent time in Golgi's laboratory at the Institute of General Pathology. Golgi served the university as dean of the faculty of medicine and also as rector of the University of Pavia for several years. He became a senator of the kingdom of Italy in 1900. Later in his life, during World War I, he assumed responsibility for a military hospital in Pavia, and he created a center for the study and treatment of peripheral nervous lesions and for rehabilitation of the wounded. He retired in 1918 but remained as professor emeritus.

THE BLACK REACTION

Golgi's earliest scientific article, suggesting organic lesions in the neural centers caused psychiatric disorders, was published in 1869. Determined to find facts to support his own claims, he initiated histological studies of the nervous system; however, the methods at the time were inadequate for examining the complex nature of nervous tissue. The axons and dendrites of neurons are too thin to be seen using common staining methods. While at Abbiategrasso, he experimented with different staining techniques until in 1872 he finally had success at developing a procedure that left darkened outlines in some neurons. He published a brief note with the translated title "On the structure of grey matter," in the *Gazzetta Medica Italiana* in 1873. This contained the first mention of the black reaction, the fundamental discovery in neural anatomy that now bears his name and that gave Golgi Nobel fame. Golgi described the microanatomy of a nerve cell body including its cellular processes in much better detail than previously possible. A few years later he published the first sketches of neural structures in olfactory bulbs as seen by the Golgi staining method. In 1885 he published a monograph that contained the detailed illustrations of the microanatomy of the central nervous organs. Between 1877 and 1880 he was extremely prolific and published many more observations made possible by using his staining technique.

Histologists today still use Golgi staining or Golgi impregnation. A series of numerous attempts and adjustments led to the following procedure: After hardening, or fixing, a piece of neural tissue with potassium dichromate, he immersed it with silver nitrate. The salt silver nitrate formed, crystallized, and filled in the neuron along its entire length. For an unknown reason, only between 1 and 5 percent of cells become impregnated and stain black using this method. This turns out to be beneficial since nervous tissue is complex and contains many overlapping and intertwined fibers and cells. A comparison demonstrating its utility would be observing a shadow painting of a distant densely populated forest. If every trunk, branch, and leaf were painted black, the entire canvas would appear black. But if the structures of only one or two trees were outlined in black and the rest all remained colorless, the observer would be able to distinguish the detail and structure of the individual trees within the forest. Golgi later developed modifications of his basic silver staining technique by altering the length of time the tissue was immersed in the dichromate solution in order to emphasize the neurons, the glia, the cell process, or all at once.

One major ramification of Golgi's staining method was that it revealed morphological features hidden by previously used histological techniques. For example, in 1837 the Czech anatomist Jan Evangelista Purkinje described unique cells in the cerebellum, the area of the brain that maintains the body's position in space and subconsciously coordinates motor activity. He was only able to visualize the bodies of these cells, now called Purkinje cells, but Golgi staining later revealed that these cells contained more extensive branching than any other type of neuron. They contain intricate, oak-tree-like branching that is integral to their function of carrying all outputted cerebellar information and is necessary for coordination of motor activities. Until the application of Golgi's technique, the branching had been invisible—only the bottommost part of the trunk had been observable. Being able to see the organization of the neuronal extensions of the Purkinje cells allowed neurobiologists to trace the projections and draw conclusions about their function.

The Golgi staining method also impregnated glial cells (also called neuroglia), allowing Golgi to describe their structure. These cells provide support and nutrition to the neurons. Golgi differentiated between two neuron types, Golgi type I and Golgi type II. The former possess long axons that extend long distances from the cell body, whereas axons of the latter cells project locally into the vicinity of the cell body to form local circuits or interneurons.

In 1838–39 Matthias Schleiden and Theodor Schwann had proposed the cell theory, purporting that tissues were composed of cells, but anatomists had not yet applied this concept to nervous tissue. This was mainly due to the fact that its structure and organization were still so mysterious. Until the late 1890s, two competing hypotheses described the structure of nervous tissue: the reticular theory and the neuron theory. According to the reticular theory, nervous tissue consisted of a large syncytial network made of many physically connected fibers. This diffuse arrangement would enable nervous impulses to flow continuously through the system. The neuronal hypothesis proposed that the nervous system was composed of numerous individual cells called neurons. Golgi preferred the reticular theory. Understanding how a nervous impulse could travel between individual separate cells, as the neuron theory demanded, was more difficult to imagine, yet the cellular detail disclosed by Golgi's staining method supported this school of thought. Biologists have since elucidated the method by which nervous impulses travel across synapses, the spaces between neurons. The presynaptic cell secretes chemicals called neurotransmitters that carry the information across the junction to the next cell, where they bind to receptors and initiate an impulse in the postsynaptic cell.

After he had been considered every year since its inception in 1901, on the basis of the sharply perceptive observations of neurons that biologists were able to make using Golgi's staining method, Golgi was awarded the Nobel Prize in physiology or medicine in 1906 for his work on the structure of the nervous system. The Nobel Prize was shared with the Spanish anatomist Santiago Ramón y Cajal, one of the main supporters of the neuron theory, the correct view that the nervous system consists of anatomically and functionally distinct cells.

During the Nobel ceremonies, Golgi delivered a distasteful lecture that cost him the respect of others thereafter. In the first sentence of his lecture, Golgi admitted that he had always been opposed to the neuron theory (for which his corecipient Cajal had gathered much supportive evidence) and then stated that "this doctrine is generally recognized to be going out of favour." From that point on, he made incorrect, combative, and narcissistic statements until nearly all of the audience members and his colleagues were insulted. Those who were there reported that Golgi seemed oblivious to the hostile nature of his comments. After Golgi's lecture and in sharp contrast, Cajal spoke and humbly acknowledged the contributions of others (including Golgi) and described the brilliant research that supported the neuron theory. Scientific observations backed Cajal's claims, and soon after, the neuron theory became widely accepted.

INTERNAL RETICULAR APPARATUS

During the course of his cellular examinations, Golgi observed an intracellular structure that he called an internal reticular apparatus. He published his discovery of this structure in 1898, but for half a century many scientists doubted its existence, believing its appearance was an artifact from the staining procedure. In the 1950s, electron microscopy confirmed its existence, and today cell biologists recognize the importance of this cytoplasmic structure in eukaryotic cellular physiology. Referred to as the Golgi apparatus, the Golgi complex, or simply the Golgi,

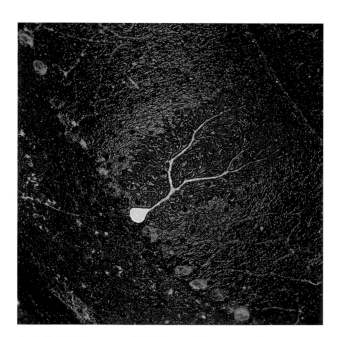

Golgi staining revealed the treelike branching pattern of axons in specialized nerve cells of the cerebellum, called Purkinje cells. *(David Becker/Photo Researchers, Inc.)*

this structure is responsible for modifying, sorting, packaging, and routing substances made in the cell. Portions of the Golgi pinch off to form membrane-bound vesicles that deliver membrane proteins and other components to the cell membrane by fusing with it. Substances produced by secretory cells are delivered to the cell's exterior by a similar mechanism. The Golgi apparatus also packages potentially destructive substances, such as lysosomal enzymes.

OTHER DISCOVERIES

In 1878 Golgi discovered nerve endings embedded among the fibers of a tendon that perceive changes in muscle tension, or stretch. These nerve endings are a type of proprioceptor and are called Golgi's tendon organs. Proprioceptors are specialized sensory organs that help individuals know the position of their body parts in space by providing information about joint angles and muscle tension and length.

Golgi's early research included discrimination of the different species of parasites that cause malaria, a disease characterized by periodic attacks of chills and fever. Between 1882 and 1892 he examined the life cycle of the causative agent, *Plasmodium,* and he developed a method for photographing blood cell preparations to identify different stages. When a female *Anopheles* mosquito bites a human, she releases *Plasmodium* sporozoites into the bloodstream with her saliva. The sporozoites quickly travel to the liver, and over five to seven days they multiply to form thousands of merozoites. The liver cells burst open, and the merozoites invade red blood cells, where they multiply again. After two days the newly infected red blood cells burst open, and the released merozoites invade more red blood cells. Golgi correlated the recurrent chills and fever with the release of the parasites from the blood cells. Some of the merozoites develop into gametocytes within the blood cells, and a mosquito may take these up during a blood meal. The life cycle continues in the mosquito, and if the mosquito bites another human, that person may become infected. Golgi's observations made it possible to diagnose the correct form of malaria afflicting a patient and to treat the disease more effectively.

Golgi's magnum opus was *Opera omnia,* consisting of four volumes that contained most of his publications. He published three in 1903, and his coworkers edited and published the fourth after his death.

In 1877 Golgi had married the former Donna Lina Aletti, Bizzozero's niece. The couple adopted his niece, Carolina. Golgi died on January 21, 1926, in Pavia.

The method of silver staining invented by Camillo Golgi enabled physiologists to study the anatomy of the nervous system at the cellular level and to observe subcellular structures clearly. Golgi's efforts transformed the mass of tangled fibers that made up the central nervous system into a structured network of neurons that exhibited complex anatomies. Investigators gained the ability to dissect individual components of nerve fibers from tissue specimens. Only after becoming able to define the structures could scientists efficiently explore the cellular mechanisms responsible for the functions of the nervous system. His work significantly impacted the debate over the neuron versus reticular theory of nervous tissue structure by providing evidence to confirm the neuron theory, which, ironically, Golgi did not support. In addition to the staining method and cellular structure that bear his name, the Historical Museum at the University of Pavia dedicated a hall to Golgi. In 1994 his portrait appeared on an Italian postage stamp.

See also EUKARYOTIC CELLS; INFECTIOUS DISEASES; NERVOUS SYSTEM.

FURTHER READING

Golgi, Camillo. "Sur la structure des cellules nerveuses." *Archives Italiennes de Biologie* 30 (1898): 60–71.

The Nobel Foundation. "The Nobel Prize in Physiology or Medicine 1906." Available online. URL: http://nobelprize.org/nobel_prizes/medicine/laureates/1906/index.html. Accessed January 25, 2008.

Rapport, Richard. *Nerve Endings: The Discovery of the Synapse.* New York: W. W. Norton, 2005.

Gould, Stephen Jay (1941–2002) *American Evolutionary Biologist, Paleontologist* Trained in paleontology, Stephen Jay Gould was the 20th century's most prominent interpreter of evolutionary thought. In 1972 his fellow American paleontologist Niles Eldredge and Gould proposed punctuated equilibrium as a modification to the model of natural selection originally expounded by Charles Darwin. Though at first controversial, punctuated equilibrium is now considered an important process in large-scale evolutionary change. Gould was a prolific, award-winning author, particularly concerning the origins and diversity of life, who wrote for the general public as well as for a scientific audience.

AN EARLY INTEREST IN PALEONTOLOGY AND EVOLUTION

Stephen Jay Gould was born on September 10, 1941, in New York, New York. Steve was the older of two sons born to Eleanor Rosenberg Gould and Leonard Gould, a court stenographer who enjoyed natural history. At the age of five, Steve resolved to become a paleontologist after seeing the *Tyrannosaurus rex* exhibit at the American Museum of Natural History

Stephen Jay Gould was a prominent 20th-century evolutionary biologist and prolific author. *(Time & Life Pictures/Getty Images)*

in Manhattan. When he was 11 years old, he read *The Meaning of Evolution* (1949), written by the curator of the Department of Geology and Paleontology at the American Museum of Natural History, George Gaylord Simpson, who helped establish the modern synthesis of Darwin's theory of evolution by natural selection. Though he only minimally understood what he read, he was fascinated by it. The high school did not provide adequate instruction on evolution, so Steve began reading Darwin's work independently. He would later unite his two interests of paleontology and evolution.

Steve spent the summer after high school at the University of Colorado and then enrolled at Antioch College in Yellow Springs, Ohio. The intellectual and creative genius of the evolutionary biologist Charles Darwin impressed Gould, though he would later challenge his description of the progression of evolution. He completed his bachelor's degree with a double major in geology and philosophy in 1963. At Columbia University, Gould pursued his doctorate in evolutionary biology and paleontology by researching fossil land snails in Bermuda. To trace the evolutionary history of the snails, he searched strata representing millions of years but found basically no changes. A fellow graduate student and future collaborator, Niles Eldredge, observed a similar phenomenon in trilobites.

In 1965 Gould married an artist named Deborah Lee, with whom he had two sons, and accepted a position as an assistant professor of geology at Antioch College in 1966. The following year he completed his doctorate in paleontology and became an assistant professor of geology and assistant curator of invertebrate paleontology for the Museum of Comparative Zoology at Harvard University, where he continued researching the evolution of snails. He

remained at Harvard his entire life, becoming an associate professor in 1971, and a full professor only two years afterward. In 1982 he was named the Alexander Agassiz Professor of Zoology.

PROPOSES THE THEORY OF PUNCTUATED EQUILIBRIA WITH NILES ELDREDGE

As an undergraduate student of geology, Gould dared to question the constancy inherent in uniformitarianism, the principle that asserts the Earth's physical features result from geological processes that have operated steadily and in the same manner since its formation 4.5 billion years ago. Why assume rates were constant and unchanging? As a student, he wrote a paper titled "Hume and Uniformitarianism" examining the assumption of constancy of natural laws in order to reach scientific conclusions about the past. He published a revised version, "Is Uniformitarianism Necessary?" in the *American Journal of Science* in 1965. Gould continued thinking about uniformitarianism and the other extreme, catastrophism, the belief that Earth's geological formations, such as mountains and lakes, resulted from tremendous catastrophes, such as floods or earthquakes. Years later, he described the concepts of time and direction in geology in the technical book *Time's Arrow, Time's Cycle* (1987).

Gould supported evolutionary theory, but without convincing evidence for slow, gradual transitions between species, he questioned the widely accepted manner by which it occurred. Phyletic gradualism, rooted in Darwinism but slightly modified, maintained that speciation occurred from the slow and steady transformation of entire populations over a large geographic range, but data indicated that evolution occurred in sudden, rapid spurts followed by longer periods with no substantial changes. Eldredge and Gould published "Punctuated Equilibria: An Alternative to Phyletic Gradualism" in *Models in Paleobiology* in 1972. This paper attempted to explain the tempo and pattern of evolution by fossil evidence and helped revitalize paleontology by generating a vast amount of literature in response. Although the German-born evolutionary biologist Ernst Mayr had initially proposed the basis of the concept, Eldredge and Gould's more thorough and extended presentation of the model that they named is considered the foundation of the punctuated equilibrium school of thought. Their paper also contained a warning to scientists about the danger of seeing only what an accepted theory dictated and suggested that paleontological research and comprehension of Earth's true history had been hindered by the assumption of phyletic gradualism.

Gould was disappointed in Darwin for attributing the absence of nongradualistic fossil evidence to imperfections in the fossil record. He also noticed that some variations did not provide adaptive improvements and therefore could not be explained by Darwinian logic. Eldredge and Gould proposed that the abrupt appearance and stasis of species in the stratigraphical record were consistent with allopatric speciation, speciation occurring in small geographically separated subpopulations. Within a large population, new genetic variations became lost within an even larger mix of forms for a specific characteristic, but if an individual that carried the genetic variation were isolated from the rest of the population so that gene flow was reduced, then that variation had a greater chance of becoming established in a separated subpopulation, resulting in lineage splitting. One would expect transitional fossils to be rare if speciation occurred abruptly in small peripheral populations.

Eldredge and Gould believed that the breaks in the fossil record accurately depicted the past. New species did not evolve within the same geographic area, and entire populations did not gradually transform into new species. In contrast, they proposed that speciation was a rapid event that occurred in a small isolated population, followed by a long period of stasis, with no change, a model called punctuated equilibrium (PE).

Reactions to PE varied. Some paleontologists felt the need to defend gradualism and flaunted the few well-documented examples. One classic example in paleontology is the slow, progressive increase in the number of whorls in the Liassic oyster *Gryphaea*. Biologists were surprised at the abundant evidence demonstrating that species remain virtually unchanged for millions of years, even in the face of rapid geological or climatological change. Organisms appeared to migrate rather than adapt to sudden climactic shifts. Critics of PE, those who are known as Darwin fundamentalists, labeled it "evolution by jerks." Gould wittingly responded by calling gradualism "evolution by creeps."

Though the title of the original paper introducing PE called it an alternative to Darwin's gradualism, Gould later explained that the two methods did not operate exclusively. Examinations of the overall fossil fauna overwhelmingly supported PE, whereas support for gradualism usually was found by investigating a specific lineage. PE explained why intermediate fossils connecting related species were absent, although transitional fossils between major lineages did exist.

Gould published several follow-up papers expanding the theory of PE. While some criticized his ideas, saying he made PE out to be more important than it actually was or periods of stasis were simply a weak attempt to explain missing links in the fossil record, others recognized his work as brilliant. The Paleontological Society awarded him the Schuchert Award in 1975 for excellence in paleontological research by a scientist less than 40 years of age.

PROLIFIC AUTHOR

In addition to more than 1,000 scientific papers, Gould authored at least two dozen books, interestingly, all using a manual typewriter. From 1974 to 2001, Gould published a series of 300 consecutive monthly essays for a column titled "This View of Life" in the magazine *Natural History*. His enlightening discourses ranged in scope over topics of science, philosophy, history, art, and literature and were collected and republished in 10 volumes under intriguing titles such as *Hen's Teeth and Horse's Toes* (1983) and *I Have Landed* (2002), explaining complex topics such as evolution and other natural phenomena without oversimplifying them. *The Panda's Thumb* (1980), a book that described a wrist bone that pandas use to help them strip bark from bamboo shoots and allowed them to switch from eating meat to eating plants, received the 1981 American Book Award in science. The following year Gould won the National Book Critics Circle Award

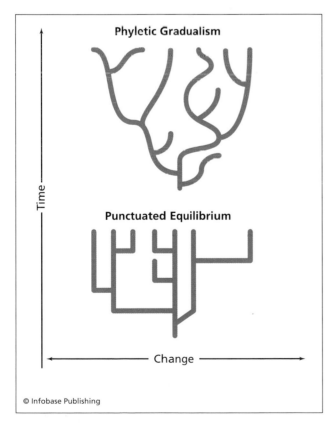

Phyletic Gradualism

Punctuated Equilibrium

Time

Change

© Infobase Publishing

In phyletic gradualism, slight variations accumulate over long periods, whereas in punctuated equilibrium, speciation occurs rapidly and is followed by long periods of stasis.

for his book that attacked the misuse of standardized intelligence tests to discriminate against certain races and religions, *The Mismeasure of a Man* (1981). His *Wonderful Life: The Burgess Shale and the Nature of History* (1989) won the Rhône-Poulenc Prize, a literary award for the best nonfiction science book written for the general reader. The best-seller described a British Columbia limestone quarry that formed 530 million years ago and holds a variety of unusual and complex fossil remains. Gould used this geological structure as an illustration of the characteristic randomness of evolution. He believed that evolution did not purposefully strive toward perfection and encouraged people to wonder "what if" biological history had proceeded down a different path.

Because of his recognized expertise on evolution and his ability to communicate scientific concepts lucidly, Gould served as a witness in an Arkansas state trial challenging the teaching in public schools of so-called creation science alongside evolution. Gould demonstrated that intelligent design, the belief that a higher being created the Earth, had no scientific basis and that, in fact, scientific evidence discredited many biblical stories that creationists interpreted as literal truths. The court ruled in favor of eliminating creationist teaching on the basis that it was religion and did not meet the criteria to qualify as science.

WRITES MAJOR TREATISE ON EVOLUTIONARY THEORY

Gould's last contribution to scientific libraries was a mammoth 1,433-page treatise, *The Structure of Evolutionary Theory,* written over two decades and published in 2002. In the book, he reviewed the undeniable facts of Darwinian evolution: more offspring are produced than can survive given competition for resources within a population, variations occur within individuals, and the variations are passed on to the next generation. Natural selection provided the mechanism by which variants that were better adapted to a particular environment achieved better reproductive success and passed on the favorable characteristics to their offspring. A master of analogy, Gould likened the frame of Darwin's evolutionary theory to a piece of coral with three major limbs branching from a central trunk. The trunk represented the core of Darwinian logic, the theory of natural selection, and the three branches represented a tripod of agency, efficacy, and scope. The central branch represented the agency by which natural selection worked—the claim that natural selection worked on organisms, not genes or clades or any other level in the hierarchical organization of life. The second branch represented the efficacy of natural selection—that natural selection alone was the mechanism for adaptive evolutionary change. The

third branch symbolized the scope of natural selection, the extrapolation that small microevolutionary variations such as those that transformed wolves into dogs explained all taxonomic diversity given the immensity of geological time. Gould explained that severing the central trunk or disproving natural selection as an evolutionary force would destroy the theory (as it would kill the organism). Severing close to the branch points of the three major limbs would significantly compromise the theory, but excision and regrafting other parts would maintain its essential nature.

Gould proceeded to expand, add, and redefine aspects of classical evolution to restructure the symbolic coral, allowing growth of stronger branches to occur. The accumulation of new and different types of data over the 30 years since Eldredge and Gould first proposed PE allowed Gould to revise the structure of evolutionary theory by regrafting upon the original foundation. The recognition of species as Darwinian individuals capable of participating in natural selection led to a generalization of the hierarchical theory and the expansion of the agency branch by permitting selection to act on multiple levels in the hierarchy of life: genes, cells, organisms, demes, species, and clades. Gould cut back the efficacy branch a little, maintaining that creativity was necessary to build "evolutionary novelties" but allowing for a variety of additional mechanisms to guide evolutionary pathways by imposing some constraints (such as structural or developmental). For example, as the diameter of a single-celled organism increases, the ratio of its surface area to its volume exponentially decreases. Physical forces limit the maximal size of a cell since the surface area of the membrane of a very large cell could not support the required amount of material exchange with the environment. Gould did not feel microevolutionary processes were sufficient to explain the extent of diversity of life despite the vastness of geological time, so he modified the scope branch to include the role of broader-scale operations in the establishment of new species. The discovery of a catastrophic mass extinction occurring 65 million years ago supported this modification to evolutionary theory. Scientists found an unusually high level of iridium, an element detected in meteors, comets, and the Earth's mantle, within a layer of sediment deposited around that time, suggesting that an extraterrestrial impact and the resulting environmental disturbances caused the extinction of 85 percent of all species at the end of the Cretaceous period.

Gould exhibited an extraordinary breadth of knowledge, and people often asked his opinions on divisive issues. Though frequently questioned about the possibility of life on other planets, he never responded with a simple yes or no but took care to explain that given the variety of earthly life-forms and the vastness of the universe, it seemed improbable that only Earth provided conditions permissive of the origin and support of life. Gould staunchly opposed biological determinism, the assumption that biology determines individual differences, making them unchangeable. Believing that science could never be detached completely from a personal dimension because scientists are human, Gould often spoke about the cultural embeddedness of science, the fact that society influences scientific endeavors.

In 1981 the MacArthur Foundation awarded Gould a fellowship, nicknamed the "genius grant," awarded to U.S. residents or citizens who show exceptional merit and promise for continued and enhanced creative work. *Discover* magazine named Gould Scientist of the Year for 1981 for developing the theory of punctuated equilibrium. In 2000 the U.S. Library of Congress named him one of 83 living legends who have "advanced and embodied the quintessentially American ideal of individual creativity, conviction, dedication, and exuberance." Gould also received the Medal of Excellence from Columbia University in 1983, the Silver Medal from the Zoological Society of London in 1984, the Gold Medal for Service to Zoology from the Linnean Society of London in 1992, and the Distinguished Scientist Award from the Center for the Study of Evolution and the Origin of Life at the University of California Los Angeles in 1997. He helped found *Paleobiology,* a journal that publishes articles focusing on processes and patterns in biological paleontology. He also received 44 honorary doctoral degrees and belonged to numerous scientific organizations including the National Academy of Sciences, Royal Society of Edinburgh, Paleontological Society, American Society of Naturalists, and American Association for the Advancement of Science, for which he served as president in 2000. In 1996 Gould became the Vincent Astor Visiting Research Professor of Biology at New York University and divided his time between New York and Cambridge.

In 1982 Gould was diagnosed with mesothelioma, a type of abdominal cancer. Though the median survival time was eight months, he lived two more decades and died of an unrelated lung cancer on May 20, 2002. His is survived by his second wife, Rhonda Roland Shearer, and his two sons, Jesse and Ethan, from his first marriage.

Gould was fluent in several languages and could read sources in their original languages. He was also a New York Yankees fanatic and a gifted baritone who sang with the Boston Cecilia Society. Though the public recognized him as a famous paleontologist, he described himself as a historian at heart and never limited his studies to a particular field. Wherever answers could be found, Gould ventured there to find them. His intellect surpassed the ordinary, and associates described him as both brilliant and arrogant. Gould's research and writings on punctuated equilibrium strongly influenced ideas about macroevolution and loosened the restrictions of classical evolutionary theory. He demonstrated that stasis is an important phenomenon worthy of examination and that punctuation is an interesting model of change. Alongside discussions on Darwin and his contributions to modern evolutionary theory, modern textbooks of Earth science, paleontology, biology, and evolution all include descriptions of punctuated equilibrium. Time will reveal how Gould's more recent ideas concerning the restructuring of evolutionary theory will influence future lessons on the history of life.

See also DARWIN, CHARLES; EVOLUTIONARY BIOLOGY; EVOLUTION, THEORY OF.

FURTHER READING

Chavez, Miguel. "The Unofficial Stephen Jay Gould Archive." Available online. URL: http://www.stephen jaygould.org. Accessed January 21, 2008.

Eldredge, Niles, and Stephen Jay Gould. "Punctuated Equilibria: An Alternative to Phyletic Gradualism." In T. J. M. Schopf, ed., *Models in Paleobiology.* San Francisco: Freeman, Cooper, and Company, 1972.

Gould, Stephen Jay. *The Structure of Evolutionary Theory.* Cambridge, Mass.: Belknap Press, 2002.

The Official Stephen Jay Gould Archive. Available online. URL: http://www.sjgarchive.org/. Accessed January 21, 2008.

Gray, Asa (1810–1888) American *Botanist* Asa Gray was an American botanist who updated the classification and naming of the North American flora from the outdated Linnaean system to a more natural system based on biological similarity. He was one of the first paid professional botanists, *the* botanist at Harvard University for three decades, and the taxonomic authority on plants for the 19th century. Known for his staunch support of Charles Darwin, Gray helped prepare American scientists to accept the idea that species were mutable. His series of botany textbooks served as the standard and his *Manual of Botany of the Northern United States* survived well into the 20th century. When he retired in 1873, Harvard University hired four botanists to replace him.

CHILDHOOD AND TRAINING

Asa Gray was born on November 18, 1810, in Sauquoit, New York, to Moses and Roxana Howard Gray. When Asa was still an infant the family moved to Paris Furnace, where Moses set up a successful tannery, allowing him to purchase a farm back in Sauquoit in 1823. Asa was introduced to Greek and Latin at Clinton Grammar School between the years of 1823 and 1825. A student of the nearby Hamilton College ate at the house where Asa lived during the school year, and he introduced Asa to the college's Phoenix Society librarian, who let Asa borrow novels, which he devoured. Asa entered Fairfield Academy in 1825 and started attending lectures at the College of Physicians and Surgeons of the Western District of the State of New York in Fairfield, in 1826. He received a doctor of medicine degree in 1831, at the age of 20.

Dr. James Hadley, the chemistry and materia medica teacher at Fairfield, had introduced Gray to botany and stirred in him an interest in science. Gray began collecting floral specimens. On a trip to New York City to purchase medical books with a letter of introduction in hand from Hadley, Gray visited the home of the foremost American botanist, Dr. John Torrey. Unfortunately, Torrey was out of town, but upon his return he admired the plant specimens that Gray left for him. Gray abandoned practicing medicine after only one year to devote himself to the study of plants. To earn money, Gray taught science part-time and worked as a librarian. He increased his scientific collection by sending dried specimens from America to overseas locations and receiving numerous samples in return. In 1833 he began assisting Torrey in studying the vegetation of the northern region of the United States and in reorganizing his herbarium by the natural method. He even moved in with the Torrey family for a while. For a few years he continued his involvement in other endeavors such as teaching at Hamilton College. Frustrated with the lack of suitable textbooks, he decided to write one based on the natural system of classification, with the goal of providing accurate information for the beginner as well as a useful reference for botanists. He published *Elements of Botany* in 1836, presenting botany as a balanced science. The well-reviewed book increased Gray's reputation and his communication with other natural historians. He later published revised and updated editions named *Botanical Textbook* (1850, 1853) and then *Introduction to Structural and Systematic Botany* (1858). Believing the available print resources sacrificed content and accuracy, Gray also published *First Lessons in Botany and Vegetable Physiology* (1857), a less technical version of his college textbook written for high school students, and *How Plants Grow: A Simple Introduction to Botany* (1858) for schoolchildren. The publisher's reports and sales of these books indicate his attempt at simplifying science while maintaining integrity was successful.

A CAREER BOTANIST

Gray also became a full collaborator with Torrey in researching and writing the two-volume *Flora of North America* (1838–43). Having established himself as a botanist, he was invited to join the U.S. Exploring Expedition (also known as the Wilkes Expedition because the commander was Charles Wilkes), a scientific naval expedition to explore the West Coast of North America, Oceania, and Australia. Two years later, after repeated delays and unending bickering between the scientists and the politicians involved, Gray opted to resign in favor of another opportunity; thus he never sailed with the survey team. The state of Michigan had joined the Union and the enthusiastic young new governor, Stevens T. Mason, planned to found a new university. In 1838 the board of regents of the University of Michigan appointed Gray the first permanent professor of the university. Because the institution only existed on paper at this point, Gray spent the first year of his professorship in Europe collecting books for the library, examining herbaria, and getting to know other botanists.

After his return to the United States, the university was not in immediate need of his services since still no students were enrolled, so he continued working on *Flora of North America* with Torrey while providing advice to the administration from afar. The voyage proved to his advantage for this task as he had many new notes and observations from his experiences and interactions with the scholars of Europe. Their major goal was to complete a survey and classification of the plants of North America. The continent lagged far behind Europe in cataloguing the indigenous flora. For almost a century, natural historians had been following the classification system proposed by the Swedish botanist Carl Linnaeus, which was based on the easily observable but superficial characteristics of number of male and female parts of a flower. European botanists had already switched to a more modern system for classifying plants (called the natural system), influenced by the work of the French botanist Antoine Laurent de Jussieu and and the Swiss botanist Augustin-Pyramus de Candolle. Gray and Torrey attempted to restructure the groups of North American flora on the basis of biological characteristics rather than arbitrarily selected flower parts. Collectors aware of their efforts sent specimens from all over North America and overseas. Gray and Torrey diligently worked, cataloguing specimens, recording observations, and writing summary reports of the samples from the numerous explorations.

The position at the University of Michigan never materialized because of the state's poor financial situation. After being on leave without salary for two years, he formally resigned in 1942 when Harvard University offered Gray the Fisher Professorship of Natural History, a position that he accepted with the condition that he could focus on instruction in botany and the restoration and maintenance of the botanical gardens. At the time, botany was still considered an interest rather than a profession. Typical training consisted of finding a mentor willing to share his knowledge, and botanists struggled financially to support their botanical activities. Gray's arrangement at Harvard made him one of the first paid professional botanists. He remained at Harvard for the rest of his career, teaching introductory botany and a few upper-level courses while enjoying the ability to devote himself to his botanical researches.

Though both Gray and Torrey persisted in their efforts of reporting on specimens sent to them from expeditions, their formal collaborative efforts ended with Gray's move to Boston. The shadow of the incomplete *Flora* hung over Gray, but he set to work collecting seeds and roots to renovate the Harvard Botanic Garden. By the time he passed away, the Botanic Garden had become a hub of botanical science in the United States.

One of Gray's most successful endeavors was the *Manual of Botany of the Northern United States* (first edition published in 1848), which covered a more limited geographical range than the never-finished *Flora of Northern America*. The volume included all the flowering plants as well as some lower plants. The eighth edition, now called *Gray's Manual of Botany*, is still popular for identifying plants in the northeastern United States.

As America's leading botanist, Gray lent his expertise to many others, spreading himself too thin and not having sufficient time to devote to projects that he considered worthy. For the government he analyzed and reported on samples collected during the Pacific Railroad Surveys (1855–57). He corresponded with numerous individual collectors who obtained samples for him from all over the United States, but in return, he had to engage in the time-consuming activities of maintaining the correspondence, advising them, and sending them supplies. By 1948 Wilkes, the commander of the U.S. Exploring Expedition for which Gray worked for two years in the 1830s, practically begged Gray to help organize, analyze, and report on the botanical specimens that the so-called botanist who sailed on the expedition had collected. Many samples had been lost and others were damaged. No American botanist had expertise in tropical plants or other unique plants collected from the vast areas covered by the expedition, but everyone agreed Gray was the best choice for the job. Knowing this, Gray requested that the government support his traveling to Europe, where he could work in a well-equipped herbarium with access to scientific publications to aid him in his analysis. Wilkes agreed, and Gray spent one year overseas working in Sir William Hooker's herbarium, visiting the Royal Botanic Garden at Kew, and seeking expertise from other prominent botanists. Work continued back in Boston for several years. Other teams worked on special groups such as mosses, ferns, algae, and fungi (which are not plants but were included in this project). Petty disagreements over details of format and style plagued the project. Of the 100 official copies of Gray's first volume on flowering plants, 21 burned in a warehouse fire in Philadelphia.

SUPPORTS DARWIN

The British naturalist and evolutionist Charles Darwin began corresponding with Gray in 1855. Both were interested in the geographical distribution of plant species, a field called biogeography. By corresponding with amateur botanists around the country, Gray was able to define the ranges of many species. In 1857 Darwin shared his developing theory concerning the origin of species by natural selection with Gray, the third colleague to share the privilege. (Charles Lyell and Joseph Hooker were the other two.) One of Gray's most influential works was built upon Darwin's ideas. In response to a letter from Darwin, in which he questioned Gray about the geographical distribution of alpine plants in the United States, Gray performed a statistical analysis of the plants described in his *Manual* and published a paper in the *American Journal of Science* in 1856 on the distribution of plants, "Statistics of the Flora of the Northern United States." Three years later, he followed with a paper, "Diagnostic Characters of New Species of Phaenogamous Plants, Collected in Japan by Charles Wright, Botanist of the U.S. North Pacific Exploring Expedition," in the *Memoirs of the American Academy of Arts and Sciences,* proposing an explanation for why plant species in eastern North America resembled species from Japan more than species from western North America. He suggested that they evolved from a common ancestor in the Bering Strait region that migrated southward during glaciation and diverged into two lineages, one in North America and one in Asia. This work provided solid botanical evidence supporting Darwin's notion that the environment played a key role in the evolution of species, spelled out in his famous book, *On the Origin of Species,* published in 1859. As a result of religious and social conventions at the time, society accepted the idea of evolution in plants more readily than in animals or humans. Gray's work opened the minds of American biologists to the notion of evolution in all living organisms. The correspondence between the two men also served as evidence of Darwin's priority in formulating the theory of evolution by natural selection, as the British naturalist Alfred Russel Wallace independently drew the same conclusion as Darwin around the same time. An outline of Darwin's theory was presented at the Linnaean Society meeting in July 1958.

One of Gray's adversaries was the Swiss-born naturalist Louis Agassiz, a colleague from Harvard whom Gray had hosted during a visit in 1846. Agassiz had been invited to the Lowell Institute, an educational foundation in Boston, to give a series of lectures titled "Plan of Creation in the Animal Kingdom." Agassiz believed that species were divinely created and distinct to each geological age. Gray, who was Presbyterian, also believed in creation but disagreed with Agassiz's point that Caucasians were created separately from African Americans and Malays, especially since this was the sort of argument favored by slavery proponents, who claimed the Caucasian race was superior. This point of contention served as a wedge that grew deeper after Agassiz joined the faculty at the Lawrence Scientific School of Harvard University in 1848. Gray grew to believe Agassiz did not follow scientifically sound logic when drawing conclusions and disliked Agassiz's showman style during lectures. Darwin's upcoming publishing of his fleshed out theory of evolution by natural selection and speciation by descent with modification gave Gray a platform on which he could publicly challenge Agassiz. A series of debates ensued, beginning with one at the Cambridge Scientific Club in December 1858, followed by several at the American Academy meetings. Gray outlined his conclusions about the relationship of North American and Asian flora, and Agassiz politely countered, saying his conclusions were drawn from the animal world. The discussions continued for several months, with comments made concerning the effects of climate on the distribution of plants, fossil evidence of flora, and the continuity of species. When pressed for written records of their arguments, Agassiz did not produce them.

After *On the Origin of Species* was published, Gray openly supported its main contention and collaborated with the Scottish geologist Charles Lyell to develop strategies for responding to religious objections to evolution. *Atlantic Monthly* published several essays written by Gray purporting that natural selection did not conflict with Judeo-Christian beliefs. Gray also published numerous anonymous articles attacking religious opponents and attempting to reconcile Darwin's ideas and the belief in God by proposing that God was responsible for evolutionary development; the book *Darwiniana* (1876) contained many of these essays and articles. Thus Gray's reputation and efforts prepared American scientists for accepting Darwinian evolution. Darwin and Gray's friendship persisted, and in 1877 Darwin dedicated a new book about structural differences between plants of the same species to Gray.

IMPACT

During his tenure at Harvard, Gray published numerous botany textbooks for all levels that were popular not only in academics but also with the general public. He trained and assisted countless collectors and amateur hobbyists to obtain and identify botanical specimens, which they in turn shared with him, expanding the database from which he could draw conclusions. He retired from teaching in 1873 but continued to live in the residence of the botanical garden. He spent his later years expanding *Flora* into *Synoptical Flora of North America* (1878). Illness forced him to give up his botanical studies in November 1887. When Gray passed away on January 30, 1888, in Cambridge, Massachusetts, he was survived by his wife, the former Jane Lathrop Loring, whom he had married in 1848.

Gray served as president of the American Academy of Arts and Sciences from 1863 to 1873 and president of the American Association for the Advancement of Science in 1872. He was also a regent of the Smithsonian Institution in 1874–88 and became a foreign member of the Royal Society of London in 1873. The herbarium that Gray established when he went to Harvard in 1842 grew into a world-class center of botanical research, and in 1864 he gave it to Harvard, in conjunction with his book collections. The Gray Herbarium, which specializes in vascular plants, held 1,939,914 specimens as of 2007. The small brick building that Harvard built around his small library has grown into the Library of the Gray Herbarium, holding more than 63,000 volumes. In 1984 the American Society of Plant Taxonomists established the Asa Gray Award to honor botanists who have made significant contributions to plant systematics.

As the undisputed leader in the field of plant taxonomy for the 19th century, Gray made contributions to plant classification and botanical education both within and outside the walls of Harvard University that also made him one of the most influential scientists. Gray left his mark in classrooms, libraries, herbaria, and botanical gardens around the world. He influenced schoolchildren, amateur plant collectors, and scientists alike. He helped turn botany into an established scientific profession by insisting on the use of scientific reasoning for classifying plant species, by authoring numerous textbooks used to train amateur and professional botanists, and by demonstrating that plants, and animals, evolved in the manner proposed by Darwin. Gray also established a world-class center of botanical study at the Harvard Botanical Garden and Gray Herbarium.

See also BIOLOGICAL CLASSIFICATION; BOTANY; DARWIN, CHARLES; HOOKER, SIR JOSEPH DALTON.

FURTHER READING

Dupree, A. Hunter. *Asa Gray: American Botanist, Friend of Darwin.* (Johns Hopkins Paperbacks edition) Baltimore: Johns Hopkins University Press, 1988.

Rodgers, Denny. *American Botany, 1873–1892; Decades of Transition.* Princeton, N.J.: Princeton University Press, 1944.

Sargent, Charles S., ed. *Scientific Papers of Asa Gray.* 2 vols. Boston: Houghton, Mifflin, 1889.

Griffith, Frederick (1879–1941) British *Microbiologist* Frederick Griffith discovered the transformation principle in 1928 while trying to develop a vaccine to prevent pneumonia. This experiment was crucial in revealing deoxyribonucleic acid (DNA) as the genetic material.

Frederick Griffith was probably born in 1879, though sources also state 1877 and 1881, in Hale, in Cheshire, England. After graduating from the University of Liverpool in 1901, he held many positions as a microbiologist before accepting one as a medical officer in the pathology laboratory of the Ministry of Health in London. At the time, pneumonia was a leading cause of death, and much of his scientific effort involved determining the specific strains of bacteria in sputum specimens from patients diagnosed with pneumonia. In his free time, he studied the differences among the numerous strains he accumulated. He observed that many samples contained four or five different pathogenic strains of bacteria. Thinking it was unlikely for one person to be infected simultaneously with so many different types of pneumococcal bacteria, Griffith hypothesized that one strain could convert into another. Bacteriologists already knew that, over time, a cultured pathogenic pneumococcal strain lost its ability to produce a capsule, a thick extracellular polysaccharide layer that gave the colonies a smooth appearance and contributed to their virulence. If cultured in the presence of antisera that contained antibodies against capsular components, the virulent smooth strain (termed S) changed into the harmless, nonencapsulated rough strain (termed R). Occasionally, the R form reverted to the S form. Griffith wondered which structural remnant allowed this reversion to occur.

To examine this process, he heated a culture of S type pneumococci to kill them and then he injected masses of the dead S cells in combination with a small inoculum of living R cells into mice, within a few days all of the mice were dead; however, mice that were injected with only living R cells survived. After spending considerable time and effort definitively proving that all of the heated S organisms were dead, Griffith published his results, "The Significance of Pneumococcal Types" in the *Journal of Hygiene* in 1928.

Though Griffith was primarily interested in the implications of his results for the epidemiology and treatment of pneumonia, across the Atlantic Ocean, Oswald Avery and his coworkers were interested in determining which substance from the S strain transformed the R strain into the encapsulated, virulent form. After confirming Griffith's results and duplicating them in vitro, Avery, Colin MacLeod, and Maclyn McCarty identified DNA as the molecular carrier of genetic information.

Frederick Griffith died during a World War II air raid in London in 1941 before he could realize the significance of his contributions to the field of genetics.

See also AVERY, OSWALD; DEOXYRIBONUCLEIC ACID (DNA); MACLEOD, COLIN MUNRO; MCCARTY, MACLYN.

FURTHER READING

Griffith, Frederick. "The Significance of Pneumococcal Types." *Journal of Hygiene* 27 (1928): 113–159.

INDEX

Note: Page numbers in **boldface** indicate main entries; *italic* page numbers indicate photographs and illustrations.

A

AAV (adeno-associated virus) 373
abiogenesis 698
abiotic components 75, 275
absence seizures 316
absorption
    in digestion 251, 253–255
    by root system 620–621
abstinence 686, 689
abyssal zone 106, *107*
Acarapis woodi (tracheal mite) 15, 17
acetylation 194, 473
acid 737
acidophiles 40–41, 43
acid rain 304, *305*
acoelomates 35, *36*
acquired immunodeficiency syndrome (AIDS) **1–3**, 466, 468
Acquiring Genomes (Margulis) 530
Acrasiomycota 323, *323*
actin cytoskeleton *644, 645*
Actinobacteria 52
Actinopterygii 716–717, *717*
action potential 579–580, *580, 581, 677*
activation energy 311, *311*
activators 370
active transport 96, *96*–97
adaptation
    of bacteria 50
    in biology 104
    of birds 722
    in coevolution 275
    of desert plants 114
    of digestive systems 255
    in evolution 333, 506, 732–734
    identifying 334–335
    of plants 610–612, 629
    for thermoregulation 423–424
    traits as 334–335
addict (term) 4

addiction, biology of **3–8**, *7*
addition reactions 66, *66*
adenine
    Chargaff's research on 181
    in DNA structure 245
    in nucleic acids 122
adeno-associated virus (AAV) 373
adenosine triphosphate (ATP)
    in cellular respiration 171–172
    from chemiosmosis 172
    energy from 73
    from fermentation 172–173
    free radicals from 11
    from glucose 119
    in muscle function 572–573
    in photosynthesis 603, 604
ADH (antidiuretic hormone) 250, 425
ADHD (attention-deficit/hyperactivity disorder) 576–577
ADP ribosylation 473
adrenal glands *294*
adsorption 728
adult stem cells 751
adventitious buds 659
advertising, and obesity 609
aerobic respiration
    historical emergence of *413*, 413–414
aerosols
    for biological weapons delivery 98
affinity chromatography 189
AFM (atomic force microscopy) 555
Africa
    HIV pandemic in 1
agar 22, 168–169
agarose gel electrophoresis 280, 281, 655, *656*
Agassiz, Louis 400
agency
    in evolution 396
aggression 30, 523
aging **8–13**, *9,* 753
agnathans 715
Agouti gene 474

agriculture **13–20,** *19*
    biodiversity and 71–72
    biology in 105
    biotechnology in 131
    cloning applications in 210–212
    DNA cloning in 207
    environmental science in 310
    evolutionary biology and 342–343
    fungi and 363
    genetic engineering in 378–379, 749
    genetics in 383
    history of 14
    honeybees in 18
    pesticides used in 155
Agrobacterium 51
agronomy 14
AIDS (acquired immunodeficiency syndrome) **1–3**, 466, 468
air pollution 303
Akihito (crown prince [later emperor] of Japan) 203
alcohol 4, 6, 7
alcoholic fermentation 173
aldosterone 425
Alexander, Albert 350
Alexander the Great 43, 44, 45
algae *20,* **20–22,** *339*
algin 22
alkalinophiles 40–41, 43
alleles. *See also* chromosomes
    in addiction 4
    frequencies of 333
    in inheritance 190, 472–476, 542–545, *543*
    polymorphic 713–714
allergies 463–464
alligators 721–722
allopatric speciation 336
allosteric enzymes 313
allosteric regulation 314–315
almonds, bee-pollination of 15
alpha-helix 119, *120*
alpha particles 653
Alphaproteobacteria 51
alpine tundra 117